Lecture Notes in Computer Science 2552

Edited by G. Goos, J. Hartmanis, and J. van Leeuwen

Springer
Berlin
Heidelberg
New York
Barcelona
Hong Kong
London
Milan
Paris
Tokyo

Sartaj Sahni Viktor K. Prasanna
Uday Shukla (Eds.)

High Performance Computing – HiPC 2002

9th International Conference
Bangalore, India, December 18-21, 2002
Proceedings

Volume Editors

Sartaj Sahni
University of Florida, CISE Department
Gainesville, FL 32611, USA
E-mail: sahni@cise.ufl.edu

Viktor K. Prasanna
University of Southern California, Department of Electrical Engineering, EEB 200C
3740 McClintok Ave., Los Angeles, CA 90089-2562, USA
E-mail: prasanna@usc.edu

Uday Shukla
IBM Global Services India Exports, India Software Lab
Golden Enclave, TISL Tower, Airport Road, Bangalore 560 017, India
E-mail: ushukla@in.ibm.com

Cataloging-in-Publication Data applied for

A catalog record for this book is available from the Library of Congress

Bibliographic information published by Die Deutsche Bibliothek
Die Deutsche Bibliothek lists this publication in the Deutsche Nationalbibliographie; detailed bibliographic data is available in the Internet at <http://dnb.ddb.de>.

CR Subject Classification (1998): C.1-4, D.1-4, F.1-2, G.1-2

ISSN 0302-9743
ISBN 3-540-00303-7 Springer-Verlag Berlin Heidelberg New York

Springer-Verlag Berlin Heidelberg New York
a member of BertelsmannSpringer Science+Business Media GmbH

http://www.springer.de

Printed in Germany

Typesetting: Camera-ready by author, data conversion by DA-TeX Gerd Blumenstein
Printed on acid-free paper SPIN: 10871720 06/3142 5 4 3 2 1 0

Message from the Program Chair

Welcome to the 9th International Conference on High Performance Computing. The conference features 10 contributed paper sessions with 57 papers, 2 invited sessions with 9 papers, an industrial track session, a poster session, 5 keynote addresses, 10 tutorials, and 5 workshops. The 57 contributed papers are from 12 countries.

The 10 contributed paper sessions were put together by a distinguished and international program committee that comprised 56 committee members and 5 program vice-chairs: Oscar Ibarra (Algorithms), Vipin Kumar (Applications), Michel Cosnard (Architecture), Mani Srivastava (Communication Networks), and Francis Lau (Systems Software). Following a rigorous review process, the program committee selected 57 of the 145 papers that were submitted in response to the call for papers for presentation at the conference and inclusion in the conference proceedings. The 2 invited sessions were organized by Vijay Kumar (Biocomputation) and Viktor Prasanna (Embedded Systems), the industrial track by Sudheendra Hangal, the poster session by Paul Roe and Rajkumar Buyya, the keynote addresses by Cauligi Raghavendra, the tutorials by Srinivas Aluru, and the workshops by C.P. Ravikumar.

I wish to thank the entire program committee and especially the program vice-chairs for the excellent job they did in the review of the 145 submitted papers and subsequent selection of papers. Thanks also go to those who organized the remaining sessions, mentioned above, that make up the conference as well as to those who performed the administrative functions that are essential to the success of this conference. Finally, I thank Viktor Prasanna for giving me the opportunity to serve as program chair of this conference.

December 2002 Sartaj Sahni

Message from the General Co-chairs

It is our pleasure to welcome you to the 9th International Conference on High Performance Computing and to Bangalore, the IT capital of India. This message pays tribute to many volunteers who made this meeting possible.

We are indebted to Prof. Sartaj Sahni for his superb efforts as program chair in organizing an excellent technical program. We would like to thank the vice-chairs, Michel Cosnard, Oscar Ibarra, Vipin Kumar, Francis Lau, and Mani Srivastava for their efforts in putting together a strong program committee, which in turn reviewed the submissions and composed an excellent technical program. We would like to thank Vijay Kumar for his efforts in organizing the invited session on Biocomputation. Cauligi S. Raghavendra assisted us in inviting the keynote speakers.

Many volunteers contributed their efforts in organizing the meeting: Manish Parashar handled publicity, José Nelson Amaral interfaced with the authors and Springer-Verlag, with assistance from Angela French, to bring out these proceedings. David Bader acted as vice general chair, Paul Roe and Raj Buyya handled the poster/presentation session, Srinivas Aluru put together the tutorials, S.V. Sankaran assisted us with accommodation for student scholars, Ajay Gupta did a fine job in handling international financial matters, Dheeraj Sanghi administered scholarships for students from Indian academia, Sudheendra Hangal acted as industry liaison, Vinod Sanjay, though not listed in the announcements, assisted with publicity within India.

Continuing the tradition set at last year's meeting, several workshops were organized by volunteers. We would like to thank Srinivas Aluru, Rajendra Bera, Suthikshan Kumar, G. Manimaran, C.P. Ravikumar, C. Siva Ram Murthy, V. Sridhar, S.H. Srinivasan, and M. Vidyasagar for their efforts in putting together the workshop programs. These workshops were coordinated by C.P. Ravikumar.

We would like to thank all of them for their time and efforts.

Our special thanks go to A.K.P. Nambiar for his continued efforts in handling financial matters as well as coordinating the activities within India.

Major financial support for the meeting was provided by several leading IT companies. We would like to thank the following individuals for their support:

N.R. Narayana Murthy, Infosys,
Karthik Ramarao, Hewlett-Packard,
Kalyan Rao, Satyam,
Shubhra Roy, Intel India, and
M. Vidyasagar, TCS.

We would like to thank Kemal Ebcioglu for his efforts in obtaining sponsorship from the IFIP Working Group on Concurrent Systems.

Continued sponsorship of the meeting by the IEEE Computer Society, ACM, and the European Association for Theoretical Computer Science are much appreciated. Finally, we would like to thank Henryk Chrostek and Sumit Mohanty of USC and the Banquet staff at the Taj Residency, Bangalore for their assistance over the past year and Albee Jhoney, Nirmala Patel, and O. Pramod for their assistance with local arrangements.

December 2002

Viktor K. Prasanna
Uday Shukla

Message from the Vice General Chair

It was a pleasure to welcome attendees to Bangalore and the 9th International Conference on High Performance Computing. It was an honor and a pleasure to be able to serve the international community by bringing together researchers, scientists, and students, from academia and industry, to this meeting in the technology capital of India.

First let me recognize **Manish Parashar** for his help publicizing this conference, and **José Nelson Amaral** for serving as the proceedings chair. **Srinivas Aluru** did an excellent job organizing tutorials presented by leading experts. HiPC 2002 includes 10 tutorials in areas likely to be at the forefront of high-performance computing in the next decade, such as computational biology, wireless networking, quantum computing, and pervasive computing.

I wish to thank all of the conference organizers and volunteers for their contributions to making HiPC 2002 a great success. I would especially like to thank the general co-chairs, **Viktor K. Prasanna** and **Uday Shukla**, for their enormous contributions steering and organizing this meeting. Their leadership and dedication is remarkable. It is to their credit that this meeting has become the premier international conference for high-performance computing. Special thanks are also due to the program chair, **Sartaj Sahni**, for his hard work assembling a high-quality technical program that includes contributed and invited papers, an industrial track, keynote addresses, tutorials, and several workshops.

December 2002 David A. Bader

Conference Organization

General Co-chairs
Viktor K. Prasanna, University of Southern California
Uday Shukla, IBM Global Services India Pvt. Limited

Vice General Chair
David A. Bader, University of New Mexico

Program Chair
Sartaj Sahni, University of Florida

Program Vice-chairs

Algorithms
Oscar Ibarra, University of California, Santa Barbara

Applications
Vipin Kumar, AHPCRC-University of Minnesota

Architecture
Michel Cosnard, INRIA, France

Communication Networks
Mani Srivastava, University of California, Los Angeles

Systems Software
Francis Lau, Hong Kong University

Keynote Chair
Cauligi S. Raghavendra, University of Southern California

Poster / Presentation Co-chairs
Paul Roe, Queensland University of Technology
Rajkumar Buyya, University of Melbourne

Tutorials Chair
Srinivas Aluru, Iowa State University

Workshops Chair
C.P. Ravikumar, Texas Instruments India

Scholarships Chair
Dheeraj Sanghi, Indian Institute of Technology, Kanpur

Awards Chair
Arvind, MIT

Finance Co-chairs
A.K.P. Nambiar, Software Technology Park, Bangalore
Ajay Gupta, Western Michigan University

Publicity Chair
Manish Parashar, Rutgers, State University of New Jersey

Local Arrangements Chair
Rajendra K. Bera, IBM Global Services India Pvt. Limited

Industry Liaison Chair
Sudheendra Hangal, Sun Microsystems

Publications Chair
José Nelson Amaral, University of Alberta

Steering Chair
Viktor K. Prasanna, University of Southern California

Steering Committee
Jose Duato, Universidad Politecnica de Valencia
Viktor K. Prasanna, Chair, University of Southern California
N. Radhakrishnan, US Army Research Lab
Sartaj Sahni, University of Florida
Assaf Schuster, Technion-Israel Institute of Technology

Program Committee

Algorithms
Cevdet Aykanat, Bilkent University
Yookun Cho, Seoul National University
Omer Egecioglu, University of California, Santa Barbara
Tsan-sheng Hsu, Academia Sinica
Myung Kim, Ewha Womans University
Koji Nakano, Japan Advanced Institute of Science and Technology
Sandeep Sen, Indian Institute of Technology, Delhi
Bhabani Sinha, Indian Statistical Institute
Meera Sitharam, University of Florida
Hal Sudborough, University of Texas at Dallas
Albert Zomaya, University of Western Australia

Applications

Rupak Biswas, NASA Ames Research Center
Amitava Datta, University of Western Australia
Timothy Davis, University of Florida
Ratan Ghosh, Indian Institute of Technology, Kanpur
Anshul Gupta, IBM T.J. Watson Research Center
Hillol Kargupta, University of Maryland at Baltimore County
Piyush Mehrotra, NASA Ames Research Center
Raju Namburu, Army Research Laboratory, Maryland
D.K. Panda, Ohio State University
Srinivasan Parthasarathy, Ohio State University
P. Sadayappan, Ohio State University
Vivek Sarin, Texas A&M University
Jon Weissman, University of Minnesota

Architecture

Makoto Amamiya, Kyushu University
Franck Cappello, Paris-Sud XI University
Kemal Ebcioglu, IBM T.J. Watson Research Center
Michael Frank, University of Florida
Tony Hey, Engineering & Physical Sciences Research Council
Nectarios Koziris, National Technical University of Athens
David Nassimi, New Jersey Institute of Technology
Siva Ram Murthy, Indian Institute of Technology, Madras
Mateo Valero, Universidad Politecnica de Catalunya
Xiaodong Zhang, National Science Foundation

Communication Networks

Prathima Agrawal, Telcordia Technologies
B.R. Badrinath, Rutgers University
Sajal Das, University of Texas at Arlington
Ramesh Govindan, ICSI/ACIRI Berkeley
Sumi Helal, University of Florida
Abhay Karandikar, Indian Institute of Technology, Bombay
Steve Olariu, Old Dominion University
Michael Palis, Rutgers University at Camden
Parmesh Ramanathan, University of Wisconsin, Madison
Ramesh Rao, University of California, San Diego
Krishna Sivalingam, Washington State University

Systems Software

Ishfaq Ahmad, Hong Kong University of Science and Technology
Hamid Arabnia, University of Georgia
Jiannong Cao, Hong Kong Polytechnic University
Elizier Dekel, IBM Research Laboratory, Haifa

R. Govindarajan, Indian Institute of Science
Weijia Jia, City University of Hong Kong
Rajib Mall, Indian Institute of Technology, Kharagpur
Kihong Park, Purdue University
Krithi Ramamritham, Indian Institute of Technology, Bombay
Yu-chee Tseng, National Chiao Tung University
Cheng-zhong Xu, Wayne State University

National Advisory Committee

R.K. Bagga, DRDL, Hyderabad
N. Balakrishnan, SERC, Indian Institute of Science
Ashok Desai, Silicon Graphics Systems (India) Pvt. Limited
Kiran Deshpande, Mahindra British Telecom Limited
H.K. Kaura, Bhabha Atomic Research Centre
Hans H. Krafka, Siemens Communication Software Limited
Ashish Mahadwar, PlanetAsia Limited
Susanta Misra, Motorola India Electronics Limited
Som Mittal, Digital Equipment (India) Limited
B.V. Naidu, Software Technology Park, Bangalore
N.R. Narayana Murthy, Infosys Technologies Limited
S.V. Raghavan, Indian Institute of Technology, Chennai
V. Rajaraman, Jawaharlal Nehru Centre for Advanced Scientific Research
S. Ramadorai, Tata Consultancy Services, Mumbai
K. Ramani, Future Software Pvt. Limited
S. Ramani, Hewlett-Packard Labs India
Karthik Ramarao, Hewlett-Packard (India) Pvt. Limited
Kalyan Rao, Satyam Computer Services Limited
S.B. Rao, Indian Statistical Institute
H. Ravindra, Cirrus Logic
Uday S. Shukla, IBM Global Services India Pvt. Limited
U.N. Sinha, National Aerospace Laboratories

Workshop Organizers

Workshop on Bioinformatics and Computational Biology

Co-chairs
Srinivas Aluru, Iowa State University
M. Vidyasagar, Tata Consultancy Services

Workshop on Soft Computing

Chair
Suthikshan Kumar, Larsen and Toubro Infotech Limited

Trusted Internet Workshop

Co-chairs
G. Manimaran, Iowa State University
C. Siva Ram Murthy, Indian Institute of Technology, Madras

Workshop on Cutting-Edge Computing

Co-chairs
Uday S. Shukla, IBM Global Services India Pvt. Limited
Rajendra K. Bera, IBM Global Services India Pvt. Limited

Workshop on Storage Area Networks

Co-chairs
V. Sridhar, Satyam Computer Services Limited
C.P. Ravikumar, Texas Instruments India
S.H. Srinivasan, Satyam Computer Services Limited

Table of Contents

Keynote Address

Session III – Systems Software I

Chair: *Rajib Mall*

Session IV – Networks

Chair: *Abhay Karandikar*

Keynote Address

Session V – Algorithms II

Chair: *Rajendra Bera*

Session VI – Mobile Computing and Databases
Chair: *Nalini Venkatasubramanian*

Session VII – Applications
Chair: *Shahrouz Aliabadi*

Session VIII – Systems Software II
Chair: *P. Sadayappan*

Keynote Address

Invited Session I – Embedded Systems

Chair: *Viktor K. Prasanna*

Keynote Address

Invited Session II – Biocomputation

Chair: *Vijay Kumar*

Info-Bio-Nano Interface: High-Performance Computing & Visualization

Priya Vashishta, Rajiv K. Kalia and Aiichiro Nakano

Collaboratory for Advanced Computing and Simulations (CACS)
Department of Materials Science & Engineering, Department of Computer Science, Department of Physics, and Department of Biomedical Engineering
Vivian Hall of Engineering
University of Southern California, Los Angeles, CA 90089-0241, USA
priyav@usc.edu

Vision

Follow the advances in computing technologies from teraflop to petaflop (hardware, software, and algorithms) to :

- Perform realistic simulations of nanosystems and devices
- Demonstrate the feasibility of simulating systems not yet attempted
- Incorporate simulation and parallel computing & visualization in physical sciences and engineering education

Computing technology will grow by a factor of more than a thousand in the next ten to fifteen years. Our goal is to follow this computing revolution from teraflops (10^{12} flops) to petaflops (10^{15} flops). Using this unprecedented computing power, available for the first time in the history of science and engineering, it will be possible to carry out realistic simulations of complex systems and processes in the areas of materials, nanotechnology, and bioengineered systems. Coupled with immersive and interactive visualization this will offer unprecedented opportunity for research as well as modifying graduate and undergraduate education in science and engineering. We will start by building a 1,024 processor parallel supercomputer for research and follow the trajectory of developments in hardware, algorithms and visualization technologies.

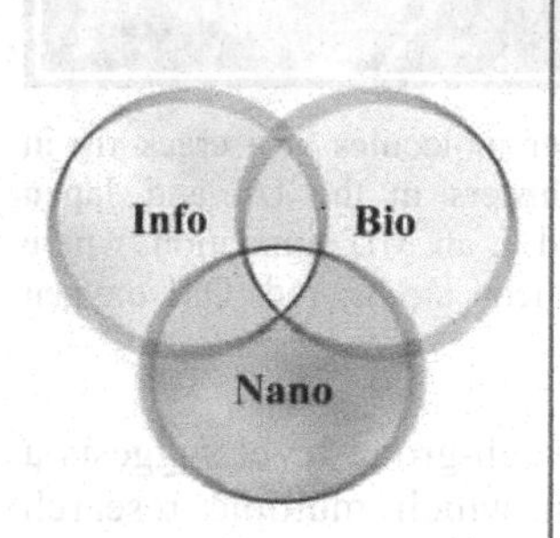

- Bio-inspired paradigms for information processing
- Information processing & nanostructure-inspired applications in life sciences & biotechnologies

Within Reach,

- At the nano-scale (≤100nm)
  – 10 million -10 billion atom nanosystems (inorganic, organic, biochemical) can be simulated & visualized while maintaining their atomistic nature
- At micro-to meso-scales (0.1μm to μms)
  – Seamless transition from discrete to continuum model via connection to finite element approaches
  – Allows examination of systems such as NEMS, MEMS

S. Sahni et al. (Eds.) HiPC 2002, LNCS 2552, pp. 3-5, 2002.

The Tools:

1. New, Efficient Simulation Technologies & Algorithms
   - Scalable
   - Portable
   - Load-balanced
2. 3-D, Immersive & Interactive Visualization
   - Large-dataset walkthrough
   - Parallel & distributed visualization
3. Physical/Chemical/Biochemical phenomena
   - Quantum mechanical (up to ~10^5 atoms)
   - Classical molecular dynamics (up to ~10^{10} atoms)
   - Finite-element approaches (for Materials, Flow dynamics, etc.)

Metacomputing on a Grid of geographically distributed computing platforms connected via high-speed networks could revolutionize computational research, by enabling collaborative, hybrid computations that integrate multiple expertise distributed over wide geographical locations.

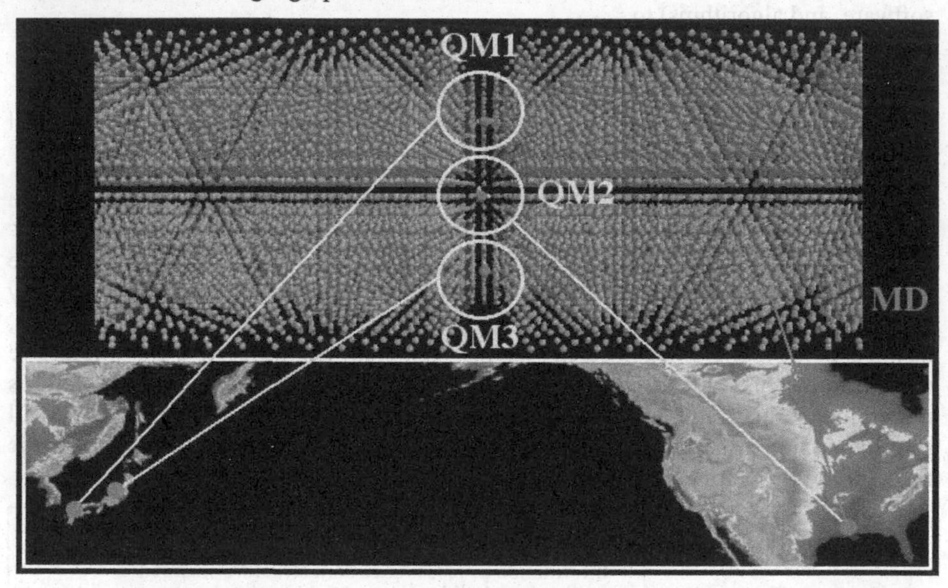

Fig. 1. Multiscale MD/QM simulation of the reaction of water molecules at a crack tip in silicon (top), performed on geographically distributed PC clusters in the US and Japan (bottom). In this example, three QM calculations are embedded in an MD simulation, where green spheres represent QM silicon atoms; blue, handshake silicon atoms; red, QM oxygen atoms; yellow, QM hydrogen atoms; gray, MD silicon atoms

The availability of inexpensive PC clusters at the research-group level suggests a new collaborative mode for computational research, in which multiple research groups of diverse expertise participate in a metacomputing project by providing both expert-maintained application programs and computational resources to run them. Such a multidisciplinary application is emerging at the forefront of computational sciences and engineering. The multiscale simulation embeds accurate quantum mechanical (QM) calculations to handle chemical reactions (below a length scale of 10^{-8} m) within a molecular dynamics (MD) simulation to describe large-scale atomistic processes (up to a length scale of 10^{-6} m), see figure. Modern design of high-performance materials and devices focuses on controlling structures at diverse length scales from atomic to macroscopic, and such multiscale MD/QM simulations are expected to play an important role in scaling down engineering concepts to nanometer scales.

In summary, multiscale atomistic simulation approach implemented with scalable & portable, space-time multiresolution algorithms, and their applications to large scale simulations for the first time provides tools to study challenging scientific and engineering problems at the Info-Bio-Nano interface.

2-D Wavelet Transform Enhancement on General-Purpose Microprocessors: Memory Hierarchy and SIMD Parallelism Exploitation*

Daniel Chaver, Christian Tenllado, Luis Piñuel,
Manuel Prieto, and Francisco Tirado

Departamento de Arquitectura de Computadores y Automatica,
Facultad de Ciencias Fisicas, Universidad Complutense, 28040 Madrid, Spain
{dani02,tenllado,lpinuel,mpmatias,ptirado}@dacya.ucm.es

Abstract. This paper addresses the implementation of a 2-D Discrete Wavelet Transform on general-purpose microprocessors, focusing on both memory hierarchy and SIMD parallelization issues. Both topics are somewhat related, since SIMD extensions are only useful if the memory hierarchy is efficiently exploited. In this work, locality has been significantly improved by means of a novel approach called pipelined computation, which complements previous techniques based on loop tiling and non-linear layouts. As experimental platforms we have employed a Pentium-III (P-III) and a Pentium-4 (P-4) microprocessor. However, our SIMD-oriented tuning has been exclusively performed at source code level. Basically, we have reordered some loops and introduced some modifications that allow automatic vectorization. Taking into account the abstraction level at which the optimizations are carried out, the speedups obtained on the investigated platforms are quite satisfactory, even though further improvement can be obtained by dropping the level of abstraction (compiler intrinsics or assembly code).

1 Introduction

Over the last few years, we have witnessed an important development in applications based on the discrete wavelet transform. The most outstanding success of this technology has been achieved in image and video coding. In fact, state-of-the-art standards such as MPEG-4 or JPEG-2000 are based on the discrete wavelet transform (DWT). It is also a valuable tool for a wide variety of applications in many different fields (image fusion [1], computer graphics [2], etc). This growing importance makes a performance analysis of this kind of transformation of great interest.

Our study focuses on general-purpose microprocessors. In these particular systems, the main aspects to be addressed are the efficient exploitation of the memory

---

* This work has been supported by the Spanish research grants TIC 99-0474 and TIC 2002-750.

S. Sahni et al. (Eds.) HiPC 2002, LNCS 2552, pp. 9-21, 2002.

hierarchy, especially when handling large images, and how to structure the computations to take advantage of the SIMD extensions available on modern microprocessors.

With regard to the memory hierarchy, the main problem of this transform is caused by the discrepancies between the memory access patterns of two principal components of the 2-D wavelet transform: the vertical and the horizontal filtering [2]. This difference causes one of these components to exhibit poor data locality in the straightforward implementations of the algorithm. As a consequence, the performance of this application is highly limited by the memory accesses.

The experimental platforms on which we have chosen to study the benefits of the SIMD extensions are two Intel Pentium-based PCs equipped with a P-III (SSE) and a P-4 (SSE2) processor respectively. Due to portability reasons and in order to prevent long development times, we have avoided coding at the assembly language level. Consequently, all the optimizations have been performed at the source code level. To be specific, in order to allow automatic vectorization, we have introduced some directives, which inform the compiler about pointer disambiguation and data alignment, and we have performed some code modifications such as loop transformations or variable scope changes.

This paper is organized as follows. The investigated wavelet transform and some related work are described in sections 2 and 3 respectively. The experimental environment is covered in section 4. In Section 5 we discuss the memory hierarchy optimizations, then in section 6 our automatic vectorization technique is explained and some results are presented. Finally, the paper ends with some conclusions.

2 2-D Discrete Wavelet Transform

The discrete wavelet transform (DWT) can be efficiently performed using a pyramidal algorithm based on convolutions with Quadrature Mirror Filters (QMF). The wavelet representation of a discrete signal S can be computed by convolving S with the lowpass filter H(z) and highpass filter G(z) and downsampling the output by 2. This process decomposes the original image into two sub-bands, usually denoted as the coarse scale approximation (lower band) and the detail signal (higher band) [2].

This transform can be easily extended to multiple dimensions by using separable filters, i.e. by applying separate 1-D transforms along each dimension. In particular, we have studied the most common approach, commonly known as the square decomposition. This scheme alternates between operations on rows and columns, i.e. one stage of the 1-D DWT is applied first to all the rows of the image and then to the columns. This process is applied recursively to the quadrant containing the coarse scale approximation in both directions. In this way, the data on which computations are performed is reduced to a quarter in each step [2].

From a performance point of view, the main bottleneck of this transformation is caused by the vertical filtering (the processing of image columns) or the horizontal one (the processing of image rows), depending on whether we assume a row-major or a column-major layout for the images. In particular, all the measurements taken in our research have been obtained performing the whole wavelet decomposition using a (9,7) tap biorthogonal filter [2]. Nevertheless, our results are qualitatively almost filter-independent.

3 Related Work

A significant amount of work on the efficient implementation of the 2-D DWT has already been done for all sorts of computer systems. Focusing on the target of this paper, i.e. general-purpose microprocessors, several optimizations aimed at improving the cache performance have been proposed in [3][4][5]. Basically, [4] and [5] investigate the benefits of traditional loop-tiling techniques, while [3] investigates the use of specific array layouts as an additional means of improving data locality.

The thesis of [3] is that row-major or column-major layouts (canonical layouts) are not advisable in many applications, since they favor the processing of data in one direction over the other. As an alternative, they studied the benefits of two non-linear layouts, known in the literature as 4D and Morton [3]. In these layouts the original mxn image is conceptually viewed as an !*m/tr*∀ !*n/tc*∀array of *tr* *tc* tiles. Within each tile, a canonical (row-major or column-major) layout is employed.

The approach investigated in [4] is less aggressive. Nevertheless, they addressed the memory exploitation problem in the context of a whole application, the JPEG2000 image coding, which is more tedious to optimize than a wavelet kernel. In particular, they considered the reference implementations of the standard. By default, both implementations use a five-level wavelet decomposition with (7,9) biorthogonal filters as the intra-component transform of the coding [4]. The solution investigated by these authors, which they dubbed "aggregation", is similar to the classical loop-tiling strategy that we have applied to the vertical filtering in [5] (the one that lacks spatial locality if a row-major layout is employed). In this scheme, instead of processing every image column all the way down in one step, which produces very low data locality (on a row-major layout, rows are aligned along cache lines), the algorithm is improved by applying the vertical filtering row by row so that the spatial locality can be more effectively exploited (see figure 1).

In [6] we have extended these previous studies with a more detailed analysis based on hardware performance counters and a study of the vectorization on an Intel P-III microprocessor.

In the present work we have explored some ideas introduced in the context of special purpose hardware for a wavelet-based image coder [7], but with quite a different goal. Instead of applying them to minimize the memory needs, which is a significant issue in mass-market consumer products due to its cost impact, we have investigated their applicability in the reduction of the computational cost, which is the main concern in general purpose computers.

As explained above, the traditional implementation of the square variant of the 2-D wavelet decomposition consists in applying the horizontal filtering to all the image lines before the column filtering starts. The algorithm proposed in [7], known as on-line computation [8], provides a significant memory saving by starting the vertical filtering as soon as a sufficient number of lines (determined by the filter length) have been horizontally filtered. As opposed to the traditional processing, which requires a memory size of the order of the image, this line-based implementation only needs to store a minimum number of image lines. In our context, this strategy can also provide important benefits since it improves the temporal locality of the algorithm.

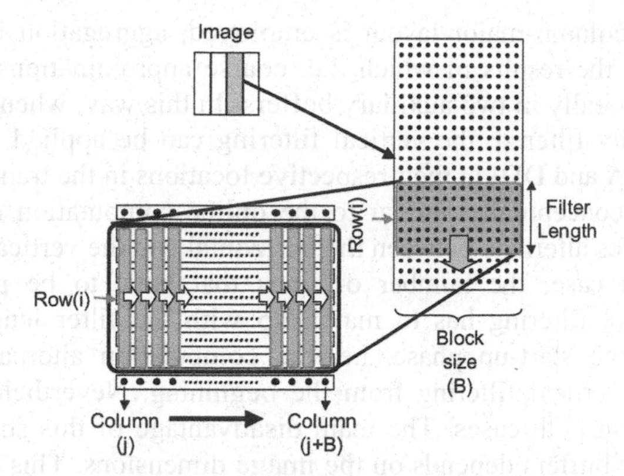

Fig. 1. Aggregation technique in the vertical filtering using a row-major layout

4 Experimental Environment

The performance analysis presented in this paper has been carried out on two Pentium-based PCs, equipped with a P-III 866 MHz and a P-4 1,4GHz, for a more detailed description see [10]. The programs have been compiled using the Intel C/C++ Compiler for Linux (v5.0.1) and the compiler switches "-O3 -tpp6 -xK -restrict" or "-O3 -ttp7 -xW -restrict " depending on the processor (P-III or P-4). In the case of the P-III, our measurements have been made using a high-level application-programming interface, PAPI (v2.0.1) [11]. In the case of the P4 processor, the lack of high level application interfaces to access its performance counters has forced us to carry out the performance analysis using only the OS timing routines.

5 Cache Analysis and Pipelined Computation

As mentioned before, the wavelet transform poses a major hurdle for the memory hierarchy, due to the discrepancies between the memory access patterns of the two main components of the 2-D wavelet transform: the vertical and horizontal filtering [2]. Consequently, the improvement in memory hierarchy use represents the most important challenge in algorithm from a performance perspective. In this work, we have extended two of the approaches studied in [6] with a new technique that we have dubbed pipelined computation, which is conceptually similar to the on-line computation described above.

5.1 Pipelined Computation

The first approach investigated in this work combines aggregation (tiling) and the idea of online computation under a column-major layout. Both optimizations seem very complementary since the former tries to improve the spatial locality of the transform whereas the main goal of the latter is to enhance its temporal locality. In particular, since a column-major layout is employed, aggregation is applied to the horizontal filtering, the results of which, i.e. coarse approximations (A) and details (D), are stored temporally in two auxiliary buffers. In this way, when a whole column has been horizontally filtered, the vertical filtering can be applied, storing now the results (AA, AD, DA and DD) in their respective locations in the transformed image.

This strategy is conceptually similar to the online computation described above, since both approaches alternate between the horizontal and the vertical filtering. In the online computation case, the number of lines that have to be processed before applying the vertical filtering has to match up with the filter length. In this new approach, there is no start-up phase, i.e. the computation alternates between the horizontal and the vertical filtering from the beginning. Nevertheless, the memory behavior is similar in both cases. The main disadvantage of this strategy is that the size of the auxiliary buffers depends on the image dimensions. This means that those buffers become too long for large images, and consequently the temporal locality could be significantly diminished.

5.2 4D Pipelined Computation

The second approach that we have studied combines the 4D layout with the ideas of aggregation and pipelined computation. It divides the processing of every tile column into three different phases (see figure 2).

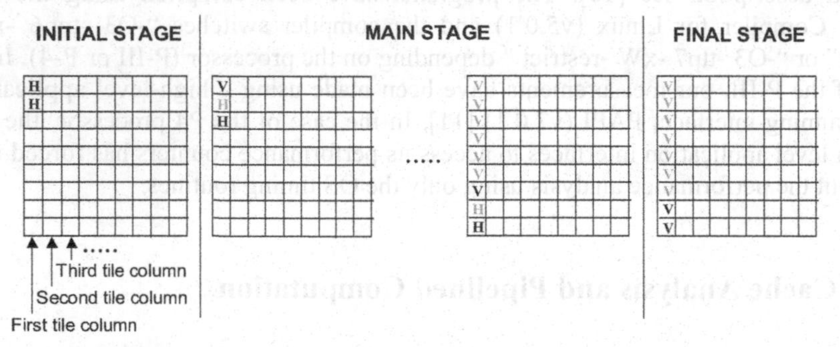

Fig. 2. Pipelined computation combined with a 4D layout

In the initial stage, the first two blocks of the tile column are horizontally filtered (using aggregation). Then, as figures 2 and 3 illustrate, it alternates between the vertical and horizontal processing of the different blocks, concluding with a final stage where the two last blocks are vertically processed. Like the first approach, in order to perform this sequence three auxiliary blocks have to be employed.

In this case, the size of the auxiliary blocks is independent of the image dimensions, and thus they could be significantly smaller than in the previous approach. Besides saving memory, now the temporal locality could be more efficiently exploited since a lower number of elements have to be horizontally filtered before the vertical one can be applied.

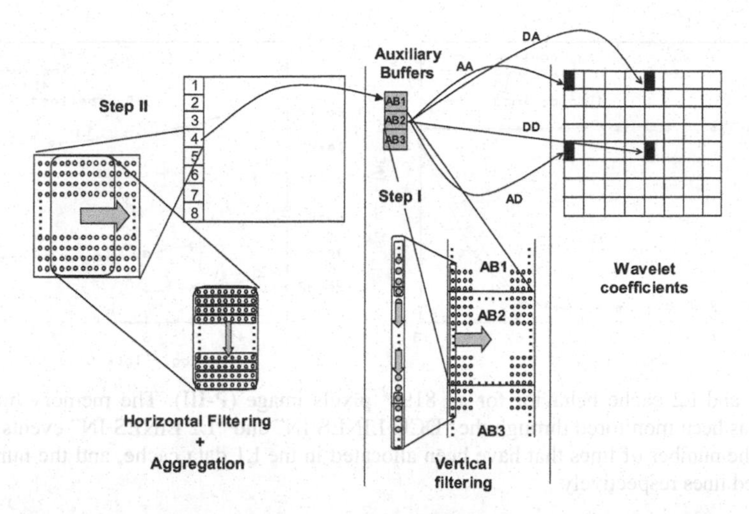

Fig. 3. Pipelined computation combined with a 4D layout (second step of the main stage processing). In the first step the auxiliary buffer AB2 is vertically filtered using neighbor coefficients from the auxiliary blocks AB1 and AB3. The second step describes the horizontal filtering of the next block, which is temporary stored in buffer AB1

5.3 Performance Results

In this section we are interested in assessing the benefits of the proposed new approaches, pipelined computation and 4D pipelined computation. The results reported have been obtained using the experimental framework explained in section 4.

The performance results for the four codes under study, when processing an image of 8192^2 pixels, are displayed in figures 4 and 5. We should firstly remark that the 4D versions present a significantly lower number of misses than the other implementations. This poor behavior of the aggregated and the pipelined versions is mainly caused by the number of conflict misses involved in the computations of the horizontal filters. Since these approaches use a column-major layout, these computations require access to non-continuous memory blocks with a stride that depends on image height. Given that the image sizes considered in this research are to the power of 2, above a certain height these blocks are mapped into the same cache set, provoking an important number of misses (we have employed a 9-tap filter whereas both processors only have 4 lines per set in the L1 data cache and 8 lines per set in the L2). Under a 4D layout the mapping of these blocks does not depend on the image height but on the tile height, which is much lower. Consequently, the number of conflict misses is significantly reduced.

Besides this difference between linear and non-linear layouts, we observe that both versions of the pipelined computation significantly reduce the L2 misses due to their better temporal locality (the pipelined and the 4D pipelined outperform the aggregation and the 4D approaches by about 14% and 42% respectively). This improvement in the L2 exploitation translates into a significant performance gain (as can be seen, the execution time is strongly related to the L2 behavior), especially in the case of the 4D layout, where the pipelined implementation achieves around a 32% speedup on both the P-III and P-4 processors.

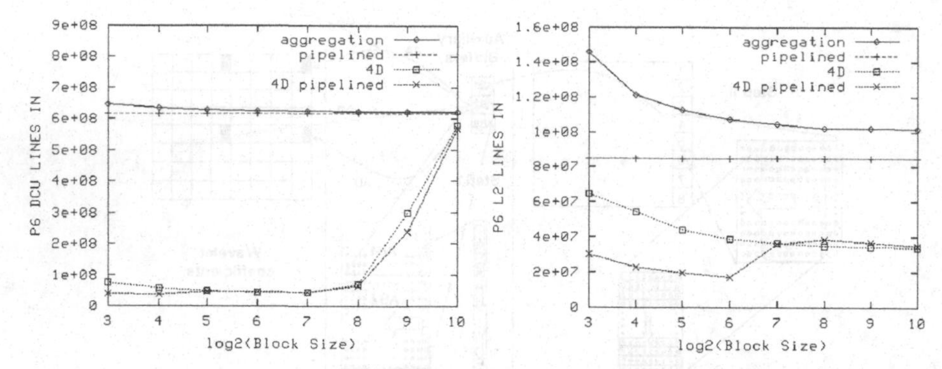

Fig. 4. L1 and L2 cache behavior for an 8192^2 pixels image (P-III). The memory hierarchy behavior has been monitored through the "DCU LINES IN" and "L2 LINES IN" events, which represent the number of lines that have been allocated in the L1 data cache, and the number of L2 allocated lines respectively

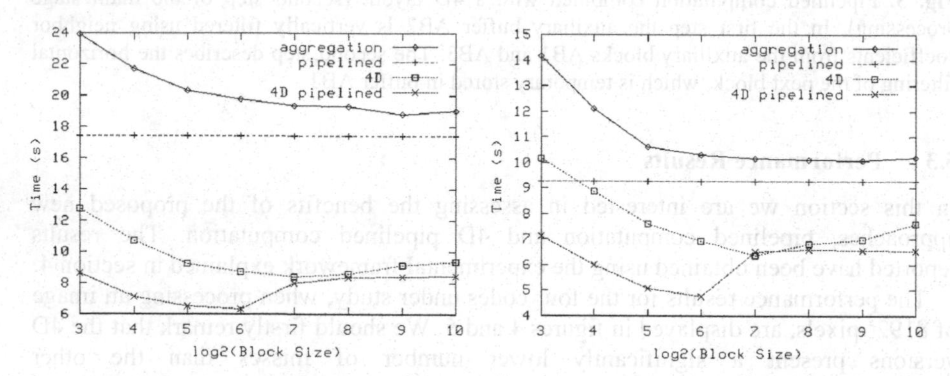

Fig. 5. Execution time for an 8192^2 pixels image on the P-III (left) and the P-4 (right)

Results for a 4096^2 pixels image are shown in figures 6 and 7. Now, we observe a quite different behavior of the L2 cache since, unlike the previous case, the number of conflict misses in the aggregation and pipelined approaches are not so significant. As a consequence, the results of aggregation and 4D are similar. Nevertheless, the benefits of the pipelined implementations are as significant as for the 8192^2 pixels image. In this case, given that the L2 conflict misses problem has disappeared, the pipelined even outperforms the 4D approach.

Table 1 summarizes the results obtained for different image sizes on both platforms. The approach that has always exhibited the more efficient memory access pattern and the higher performance has been the 4D pipelined. The observed speedups over the 4D approach range from a factor of 30% up to 40% on the P-III (independently of the image size), whereas on the P-4 it grows with the image size, from 30% up to 50%. Nevertheless, the simple pipelined version achieves competitive results for small image sizes, since in these cases it also captures the temporal locality.

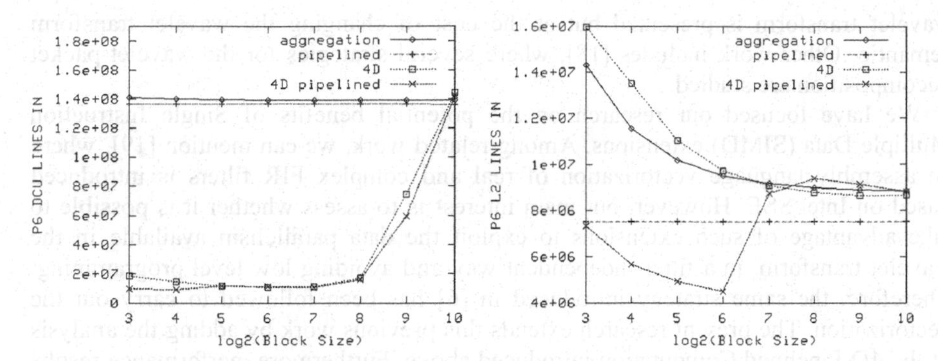

Fig. 6. L1 and L2 cache behavior for a 4096^2 pixels image (P-III)

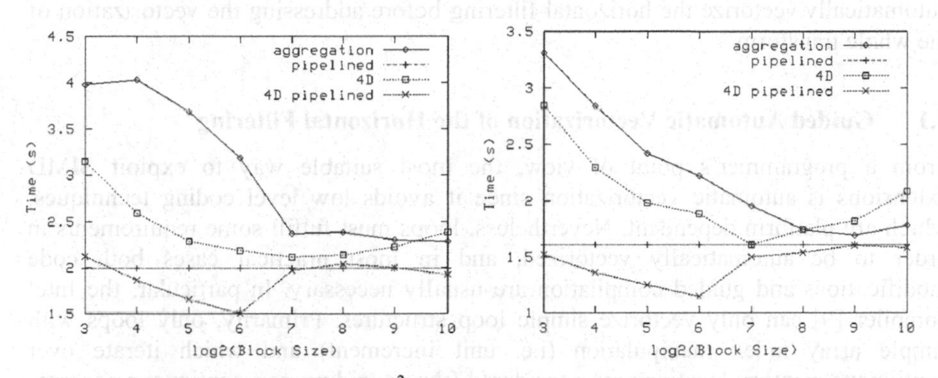

Fig. 7. Execution time for an 4096^2 pixels image on the P-III (left) and the P-4 (right)

Table 1. Percentage of speedup achieved with the investigated optimization

Image Size		512^2	1024^2	2048^2	4096^2	8192^2
P-III	**Pipelined over Aggregation**	17.2	15.2	16.6	14.0	7.9
	4D Pipelined over 4D	30.7	31.7	31.0	40.6	32.5
	4D Pipelined over	11.5	20.2	21.4	32.7	276.5
P-4	**Pipelined over Aggregation**	9.0	8.7	5.4	6.2	9
	4D Pipelined over 4D	47.0	48.5	40.7	35.6	32.6
	4D Pipelined over	29.4	33.8	37.0	39.1	194.7

6 SIMD Optimization

Previous research on the parallel wavelet transform has been concentrated on special purpose hardware and out-of-date SIMD architectures [13][14][15]. Work on general-purpose multiprocessor systems includes [5] and [16], where different parallel strategies for the 2-D wavelet transform were compared on the SGI Origin 2000, the IBM SP2 and the Fujitsu VPP3000 systems respectively. In [17] a highly parallel wavelet transform is presented but at the cost of changing the wavelet transform semantic. Other work includes [18], where several strategies for the wavelet-packet decomposition are studied.

We have focused our research on the potential benefits of Single Instruction Multiple Data (SIMD) extensions. Among related work, we can mention [19], where an assembly language vectorization of real and complex FIR filters is introduced based on Intel SSE. However, our main interest is to assess whether it is possible to take advantage of such extensions to exploit the data parallelism available in the wavelet transform, in a filter-independent way and avoiding low level programming. Therefore, the same strategy introduced in [6] has been followed to carry out the vectorization. The present research extends this previous work by adding the analysis of the 4D Pipelined Computation introduced above. Furthermore, performance results on an Intel P-4 processor are also presented. As in [6], we have first studied how to automatically vectorize the horizontal filtering before addressing the vectorization of the whole transform.

6.1 Guided Automatic Vectorization of the Horizontal Filtering

From a programmer's point of view, the most suitable way to exploit SIMD extensions is automatic vectorization since it avoids low level coding techniques, which are platform dependent. Nevertheless, loops must fulfill some requirements in order to be automatically vectorized, and in most practical cases both code modifications and guided compilation are usually necessary. In particular, the Intel compiler [9] can only vectorize simple loop structures. Primarily, only loops with simple array index manipulation (i.e. unit increment) and which iterate over contiguous memory locations are considered (thus avoiding non-contiguous accesses to vector elements). Obviously, only inner loops can be vectorized. In addition, global variables must be avoided since they inhibit vectorization. Finally, if pointers are employed inside the loop, pointer disambiguation is mandatory (this must be done by hand using compiler directives). Considering these restrictions, under a column-major layout only the "aggregated" horizontal filter can be automatically vectorized. For more details see [10].

6.2 Vectorization of the Whole Transform

The automatic vectorization is relatively straightforward in the horizontal filtering, since image elements are stored contiguously in columns. However, in order to apply the same technique to the vertical filtering, we needed the elements to be stored contiguously in rows. Guided by [19], where an assembly language vectorization of FIR filters is introduced, we have tried to force the compiler to automatically vectorize the filter itself. However, this approach is not possible without dropping the abstraction level, which is not the goal pursued in this work.

The approach investigated in [6] consists in applying a block transposition of the resultant wavelet coefficients from the horizontal filtering, so that vectorization can be performed automatically. In this research, we have also followed this vectorization strategy, which is graphically described in figure 8. As can be seen, the results of the horizontal filtering are temporarily stored in a new auxiliary block before performing the required block transposition. In order to reduce the overheads involved in this operation, an optimized 4x4 matrix transposition based on the Intel compiler predefined _MM_TRANSPOSE4_PS macro [9] has been employed.

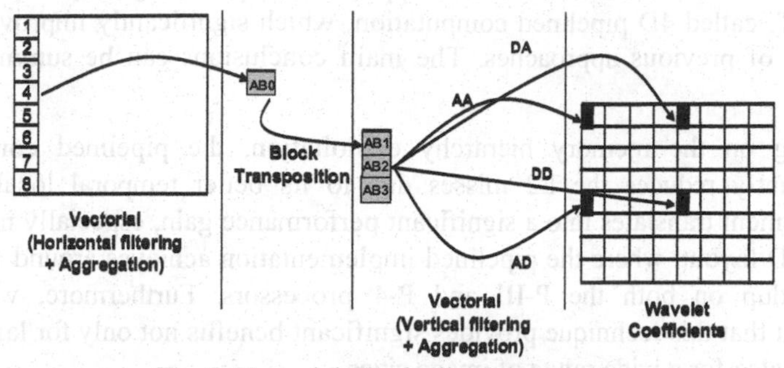

Fig. 8. Vectorial 4D pipelined computation

6.3 Experimental Results

We have restricted this section to the 4D pipelined due to its superior performance. The improvements achieved when vectorization is performed are summarized in table 2. The vectorized horizontal filtering beats the running times of the scalar version for all image sizes. On average, the improvement is better on the P-III (around 35%) than on the P-4 (around 25%). The reason behind this behavior is that for the former processor, the Intel compiler also introduces data-prefetch when vectorization is enabled [6]. Therefore, the 35% achieved on the P-III is due to both data prefetching and SIMD extensions, whereas on the P-4 the 25% is only caused by the SIMD parallelism (the compiler does not introduce data-prefetch instructions in the P-4).

Further improvements are obtained when both filterings are vectorized. On average, the speedup for the P-III only grows to 45% due to the overheads associated with the block transposition. In fact, for a 512^2-image size it is better not to apply the vectorized vertical filtering. However, in the P-4, where we have observed lower transposition overheads (percentagewise), the average speedup extends to 70%. In fact, the improvement grows with the image size, reaching a significant 90% for the 8192^2-image since the impact of the block transposition becomes insignificant.

Table 2. Percetange of speedup achieved by the vectorized versions over the non-vectorized 4D pipelined for the target platforms

Image Sizes		512^2	1024^2	2048^2	4096^2	8192^2
P-III	Horiz.	44.4	35.1	34.8	26.0	31.2
	Horiz+Vert.	36.8	42.5	45.5	45.6	52.9
P-4	Horiz.	30.7	21.4	21.3	27.7	30.8
	Horiz+Vert.	54.5	65.8	71.1	85.4	90.0

7 Conclusions

In this paper we have introduced a novel approach to optimizing the computation of the 2D DWT, called 4D pipelined computation, which significantly improves on the performance of previous approaches. The main conclusions can be summarized as follows:

1. Focusing on the memory hierarchy exploitation, the pipelined computation significantly reduces the L2 misses due to its better temporal locality. This improvement translates into a significant performance gain, especially in the case of the 4D layout, where the pipelined implementation achieves around a 30-50% of speedup on both the P-III and P-4 processors. Furthermore, we should highlight that this technique provides significant benefits not only for large image size but also for a wide range of image sizes.
2. We have introduced a novel approach to structuring the computation of the wavelet coefficients that allows automatic vectorization, which is independent of the filter size and the computing platforms (assuming that similar SIMD extensions are available).
3. In order to apply the vectorization to both filterings (horizontal and vertical) a block transposition is required. However, the performance gain achieved through vectorization by far compensated the transposition overhead, especially on the P-4. The overall speedup obtained with regard to the previous 4D and aggregation approaches is on average around 2 for the P-III and 2.5 for the P-4.

Finally, we should remark that further improvements of the 2-D wavelet transform can be achieved by introducing a lifting-based scheme [2], since this reduces the computational complexity of the algorithm. Preliminary results obtained with this approach are qualitatively comparable to their convolution-based counterpart. However, we should note that in order to apply the investigated optimizations, a similar memory waste is required. In other words, our approach also provides an efficient way to exploit the memory hierarchy and the SIMD parallelism for lifting, but at the cost of ignoring its memory saving characteristics. Taking into account the abstraction level at which the optimizations are carried out, the speedups obtained on the investigated platforms are quite satisfactory, even though further improvement can be obtained by dropping the level of abstraction (compiler intrinsics or assembly code).

References

[1] Z. Zhang and R. S. Blum. A Categorization of Multiscale-Decomposition-Based Image Fusion Schemes with a Performance Study for a Digital Camera Application. Proceeding of the IEEE, Vol. 87(8): 1315-1325, August 1999.

[2] E. J. Stollnitz, T. D. DeRose and D. H. Salesin. Wavelets for Computer Graphics: Theory and Applications. The Morgan Kaufmann Series in Computer Graphics and Geometric Modeling, Morgan Kaufmann Publishers, Inc. San Francisco, CA, 1996.

[3] S. Chatterjee, V. V. Jain, A. R. Lebeck, S. Mundhra and M. Thottethodi. Nonlinear Array Layouts for Hierarchical Memory Systems. Proceedings of 1999 ACM International Conference on Supercomputing, pp. 444-453, Rhodes, Greece, June 1999.

[4] P. Meerwald, R. Norcen, and A. Uhl. Cache issues with JPEG2000 wavelet lifting. In proceedings of 2002 Visual Communications and Image Processing (VCIP'02), volume 4671 of SPIE Proceedings, San Jose, CA, USA, January 2002.

[5] D. Chaver, M. Prieto, L. Piñuel, F. Tirado. Parallel Wavelet Transform for Large Scale Image Processing. Proceedings of the International Parallel and Distributed Processing Symposium (IPDPS'2002). Florida, USA, April 2002.

[6] D. Chaver, C. Tenllado, L. Piñuel, M. Prieto and F. Tirado. Wavelet Transform for Large Scale Image Processing on Modern Microprocessors. To be published in the proceedings of Vecpar 2002, Porto, Portugal, June, 2002.

[7] C. Chrysafis and A. Ortega. Line Based Reduced Memory Wavelet Image Compression. IEEE Trans. on Image Processing, Vol 9, No 3, pp. 378-389, March 2000.

[8] M. Vishwanath, The recursive pyramid algorithm for the discrete wavelet transform. IEEE Trans. Signal Processing, vol. 42, pp. 673-676, March 1994.

[9] Intel Corp. Intel C/C++ Compiler for Linux. Information available at http://www.intel.com/software/products/compilers/c50/linux

[10] D. Chaver, C. Tenllado, L. Piñuel, M. Prieto and F. Tirado. Vectorizing the Wavelet Transform on the Intel Pentium-III and Pentium-4 Microprocessors. Technical Report 02-001. Dept. of Computer Architecture. Complutense University, 2002.

[11] K. London, J. Dongarra, S. Moore, P. Mucci, K. Seymour and T. Spencer. End-user Tools for Application Performance Analysis, Using Hardware Counters. Presented at International Conference on Parallel and Distributed Computing Systems. August 2001.

[13] C. Chakrabarti and C. Mumford. Efficient realizations of encoders and decoders based on the 2-D discrete wavelet transforms. IEEE Trans. VLSI Syst., pp. 289-298, September 1999.

[14] T. Denk and K. Parhi. LSI Architectures for Lattice Structure Based Orthonormal Discrete Wavelet Transforms. IEEE Trans. Circuits and Systems, vol. 44, pp. 129-132, February 1997.

[15] M. Holmström. Parallelizing the fast wavelet transform. Parallel Computing, 11(21): 1837-1848, April 1995.

[16] O.M. Nielsen and M. Hegland. Parallel Performance of Fast Wavelet Transform. International Journal of High Speed Computing, 11 (1): 55-73, June 2000.

[17] L. Yang and M. Misra. Coarse-Grained Parallel Algorithms for Multi-Dimensional Wavelet Transforms. The journal of Supercomputing 11:1-22, 1997.

[18] M. Feil and A. Uhl. Multicomputer algorithms for wavelet packet image decomposition. Proceedings of the International Parallel and Distributed Processing Symposium (IPDPS'2000), pages 793-798, Cancun, Mexico, 2000. IEEE Computer Society.

[19] Intel Corp. Real and Complex FIR Filter Using Streaming SIMD Extensions. Intel Application Note AP-809. Available at http://developer.intel.com.

A General Data Layout for Distributed Consistency in Data Parallel Applications

Roxana Diaconescu

Norwegian University of Science and Technology
Department of Computer and Information Science, 7491 Trondheim, Norway
roxana@idi.ntnu.no

Abstract. This paper presents a general layout for partitioning and mapping data across processors in data parallel applications. Our scheme generalizes the existing schemes (block, cyclic) and enables non-traditional ones (e.g. graph partitioning [7, 17]). A distributed algorithm uses the data layout and the read/write access patterns to ensure consistency for data parallel applications. We show examples of the applicability of our data layout and consistency schemes for different classes of scientific applications. We present experimental results on the effectiveness of our approach for loosely synchronous, data parallel applications.

1 Introduction

This paper investigates the issues of general data layout and consistency for scientific applications in distributed memory systems [3, 4]. Hence, a data layout shows how to partition a large data set into subsets and how to map the subsets onto different address spaces. We address data parallel, SPMD (Single Program Multiple Data) applications, with general access patterns. In this concurrency model, *peer processes*[1] proceed independently and synchronize by exchanging messages. Communication is required to maintain data consistency in the presence of dependences across address spaces.

Previous experience with data parallel applications shows that the synchronization points are loose, they occur at the end of a large computation step, and typically involve data residing on the boundary between partitions. Thus, our data layout takes into consideration the spatial locality of data in order to ensure an efficient consistency scheme. The layout explicitly captures the *relation* between the elements of the set to be partitioned. Furthermore, it uses this extra information to find a general solution to the transformation between local and global data spaces in the absence of a symbolic expression (e.g. as with block and cyclic partitioning). Then, a distributed algorithm uses the mapping between local and global data spaces and the data access patterns to automatically ensure consistency for data parallel applications.

The remainder of this paper is organized as follows: Section 2 introduces the data model we use, and formulates the problems of data partitioning, mapping

[1] Identical processes.

S. Sahni et al. (Eds.) HiPC 2002, LNCS 2552, pp. 22–33, 2002.

and consistency. Section 3 describes a distributed data consistency algorithm based on our layout scheme. Then, in Section 4, we describe concrete applications of our layout and consistency schemes. Section 5 presents results on the efficiency of our approach. Section 6 overviews the related approaches to data layout and consistency. We conclude the paper in Section 7 and indicate future directions for our work.

2 The Data Parallel Model

This section presents three problem formulations for the general data layout and consistency. The purpose of these formulations is to devise a layout that can be applied to both regular and irregular applications such as to reflect data locality.

We thus introduce new formulations for data partitioning and mapping problems that take into account the relations between the elements of the set to be partitioned. Since the key to data consistency is the transformation between the global and local data spaces, we find a solution to the transformation problem and we prove its correctness.

2.1 Data Partitioning

We formulate the problem of data partitioning as follows: Given a data set D of size N, N large, and a symmetric relation R:

$$D = \{d_i \mid i = \overline{1,N},\ R \mid \forall\ d_i\ s.t.\ \exists\ d_j,\ j = \overline{1,N}\ d_j\ =\ R(d_i),\\ \exists R^{-1}\ s.t.\ d_i\ = R^{-1}(d_j)\}.$$

Find a set $S = \{S_p \mid p\ =\ \overline{1,P}\}$ with P bounded and sufficiently small such that the following conditions are satisfied:

$$S_p = \{d_{i_p} | i_p = \overline{0, |S_p|}, R\}\ such\ as \tag{1}$$
$$\forall d_{i_p} \in S_p\ s.t.\ \exists\ d_{j_k} \in S_k,\ S_k \neq S_p,\ R(d_{i_p}) = d_{j_k},\ then\ R(d_{i_p}) = \emptyset$$
$$\forall d_{i_p} \in S_p\ s.t.\ \exists\ d_{j_k} \in S_k,\ S_k \neq S_p,\ R^{-1}(d_{i_p}) = d_{j_k},\ then\ R^{-1}(d_{i_p}) = \emptyset$$
$$\forall p, k = \overline{1,P},\ p \neq k\ S_p \cap S_k = \emptyset \tag{2}$$
$$\forall p, k = \overline{1,P},\ p \neq k,\ abs(|S_p| - |S_k|) \leq \Delta,\\ \Delta \ll |S_p|,\ \Delta \ll |S_k| \tag{3}$$
$$\bigcup_{p=1}^{P} S_p \equiv D. \tag{4}$$

The first condition says that the relations between elements assigned to different subsets (partitions) are set to void. The second condition ensures a non-overlapping [1] data partitioning. The third condition requires that the size of any two different subsets must not differ by more than a bounded quantity which is much smaller than the size of both subsets (e.g. load balance). The last condition is the prerequisite to data consistency: the initial data set and its partitions are interchangeable. That is, the initial data set can be correctly reconstructed from its subsets.

2.2 Data Mapping

We formulate the data mapping problem as follows. Find a bijective partitioning function $\mathcal{P}, \mathcal{P} : D \to S$, such that[2]:

$$\forall\, d_i \in D,\ \mathcal{P}(d_i) = d_{i_p} \in S_p,\ \exists\ \mathcal{P}^{-1} : S_p \to D, \tag{5}$$

$$s.t.\ \mathcal{P}^{-1}(d_{i_p}) = d_i \tag{6}$$

Given a partitioning function that satisfies the above, a solution to the transformation between the initial data set and its partitions is:

$$\forall\, d_{i_r} \in S_p,\ \ R(d_{i_p}) = \emptyset,$$

$$let\ R(d_{i_p}) = (\mathcal{P} \circ R \circ \mathcal{P}^{-1})(d_{i_p}) \tag{7}$$

$$\forall d_{i_p} \in S_p,\ R^{-1}(d_{i_p}) = \emptyset,$$

$$let\ R^{-1}(d_{i_p}) = (\mathcal{P} \circ R^{-1} \circ \mathcal{P}^{-1})(d_{i_p}) \tag{8}$$

Claim: The solution to the transformation problem is correct.
Proof: It is easy to see that $\exists \mathcal{P}^{-1}(d_{i_p}) = d_i \in D$ *and* $\exists R(d_i) = d_j$. Then it $\exists (R \circ \mathcal{P}^{-1})(d_{i_p}) = d_j$. It follows that given it $\exists \mathcal{P}(d_j)$, then $(\mathcal{P} \circ R \circ \mathcal{P}^{-1})(d_{i_p})$ is correctly defined. The proof of correctness for relation (8) is identical.

According to the data partitioning formulation, there cannot exist explicit relations among two different subsets. But since we want to preserve the equivalence and irreversibility of the initial data and its partition into subsets, we can soundly construct a relation through function composition.

2.3 Data Consistency

The data consistency problem requires to ensure that applying the same transformation independently on each of the data subsets leads to the same result as applying the transformation on the initial data set.

Let $T : D \to V$, with T a *a generic transformation* and V a set of values, be a transformation such that:

$$\exists T : S_p, p = \overline{1, P} \to V_p \tag{9}$$

Data consistency problem requires:

$$V = \bigcup_{p=1}^{P} V_p \tag{10}$$

to be satisfied.

[2] Throughout the paper, $\mathcal{P}$ as partitioning function is different from P as the bounded number of partitions.

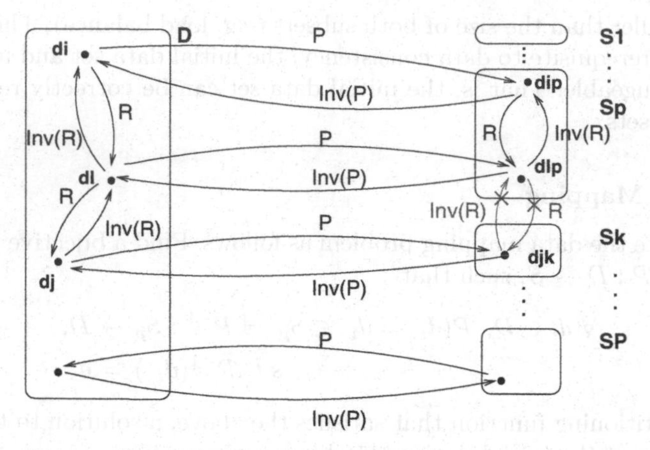

Fig. 1. The partitioning and mapping for a data set

Figure 1 illustrates graphically the problem formulations from above. The data in the global set D is partitioned into disjoint subsets S such as the relation between elements belonging to different subsets is reconstructed to preserve the initial relation governing the global set.

The tuples $\{S_p, R, \mathcal{P}, \mathcal{P}^{-1}, T\}$ contain all the ingredients necessary for being able to ensure data consistency. We call this tuple a subdomain and denote it by $Subd_p$.

3 A Distributed Consistency Algorithm

We present an algorithm for data consistency in data parallel applications with general access patterns. We will show how decomposition, mapping, communication and synchronization can be made implicit by using our algorithm in conjunction with the general data partitioning model. Parallelism remains explicit to the extent that the distributed data type is explicit.

Let D be the distributed data set holding the left values of the distributed data defined by an application. Let $T : D \rightarrow V$ be a composed transformation over the initial set D and producing the final result (values) V. Then we denote a data parallel (sequential) program by the tuple $\{D, T, V\}$.

Let $T_i, i = \overline{1, n}$ be a series of transformations that compose T, i.e. $T = T_1 \circ T_2 \circ T_i \circ ... \circ T_n$. The we can write:

$$\begin{aligned} T(D) &= (T_n \circ ... \circ T_1)(D = D_1) \\ &= (T_n \circ ... \circ T_2)(D_2) \\ &= ... \;=\; T_n(D_n) \;=\; V \\ D_i &= T_{i-1}(D_{i-1}), i = \overline{2, n} \\ V_i &= T_i(D_i), i = \overline{1, n} \end{aligned}$$

There exists a decomposition of T_i into a set of atomic operations such that $T_i = Op_1 \circ ... \circ Op_l$. The atomic operations can be classified into *read* and *write* operations. Let $d_{i_p} \in S_p$ be a reference (address) of a data item in the subset S_p. Then $\forall\, d_{i_p} \in S_p$ *and* $T_i(d_{i_p}) = (Op_1 \circ ... \circ Op_l)(d_ip)$ an operation $Op_l(d_{i_p})$ will be executed according to the algorithm presented in Figure 2.

The algorithm in Figure 2 ensures that all the updates of the remote data that are in relation with the local data are reflected in the local $Subd_p$. For the values that are not directly related to remote data, the computation proceeds independently, in each address space.

The *Dest* pattern keeps track of the local data items that are destinations of a relation from a remote data item. The *Source* pattern keeps track of the local data items that are the origin of a relation with a remote data item. We generate communication on the *Source/Dest* patterns to reflect the changes of a local data item to/from the related remote data item.

This is a relaxed version of the owner-computes rule. That is, since updates to remote related data can affect the locally owned data, consistency is required. With the strict owner-computes rule, the updates can affect only the locally owned data and remote data is read-only. Thus, with the strict owner-computes rule, the consistency problem does not occur. With our approach the *data relation communication patterns* keep track of the related data. An implementation typically replicates these data and uses their values to reflect the changes from remote address spaces locally and vice-versa. With the strict owner-computes rule data cannot be replicated.

$Dest = \emptyset$
$Source = \emptyset$
$\forall\ d_ip\ s.t.\ R(d_ip) \notin S_p$
  Construct $R(d_ip) = (\mathcal{P} \circ R \circ \mathcal{P}^{-1})(d_ip)$
  Add $\{R(d_ip)\}$ *to Dest* pattern
$\forall\ d_ip\ s.t.\ R^{-1}(d_ip) \notin S_p$
  Construct $R^{-1}(d_ip) = (\mathcal{P} \circ R^{-1} \circ \mathcal{P}^{-1})(d_ip)$
  Add $\{R(d_ip)\}$ *to Source* communication pattern
if $Op_l(d_ip)$ *writes* d_ip then
  Gather all the right values corresponding to the addresses (left values) from *Source*
  Generate communication on the *Source* pattern
  Generate communication on the *Dest* pattern
  For all left values (addresses) from the *Dest*
   $d_ik = Address(Dest[k])$
   $v_ik = Value(d_ik)$
   $d_ip = R^{-1}(d_ik)$
   $v_ip = Op_l(v_ik)$

Fig. 2. A distributed algorithm for data consistency

4 Examples

This section presents practical examples of data layouts and consistency schemes for typical computations in numerical applications. Thus, regular and irregular data layouts and computations are presented.

4.1 Regular Applications

We use the term regular to refer to applications that use linear array representations (possibly multidimensional) with affine references and regular distributions (block, cyclic). Let us consider the code excerpt in Figure 3. The data representation is a linear array. The data to be distributed is the array x, i.e. the set $D = \{x+i \mid i = \overline{1,N}, R(i) = i-3, R^{-1}(i) = i+3\}$. Then, assuming a block distribution that assigns b (block size) consecutive entries from x to a subset $p = \overline{1,P}$ (see Figure 4), $b = \frac{N}{P}$[3].

We define $\mathcal{P}, \mathcal{P}^{-1}, R, R^{-1}$ as follows:

$$\mathcal{P} : D \to S_{p=\overline{0,P-1}} = \{i_p = \overline{p \times b+1,\ (p+1) \times b}\},$$

$$\mathcal{P}(i) = (p, i_p), with\ p = \lfloor \frac{i}{b} \rfloor\ and\ i_p = i\ -\ p \times b \tag{11}$$

$$\mathcal{P}^{-1}\ :\ S_{p=\overline{0,P-1}} = \{i_p = \overline{p \times b+1,\ (p+1) \times b}\} \to D,$$

$$\mathcal{P}^{-1}(p, i_p) = p \times b + i_p \tag{12}$$

$$R\ :\ D \to D,\ R(i) = i - 3 \tag{13}$$

$$R^{-1}\ :\ D \to D,\ R^{-1}(i) = i + 3 \tag{14}$$

Let us consider the situation from Figure 4, where the subsets p and $p-1$ are shown. Assume now that Op_l is the assignment operation and we want to update the value $x[1]$ on processor p. Since $R(1) = 1-3 = -2 \notin S_p$, then on processor p, the data consistency algorithm will proceed as follows. Construct $R(i_p)$ becomes:

$$R(i_p = 1) = (\mathcal{P} \circ R \circ \mathcal{P}^{-1})(1)\ =\ (\mathcal{P} \circ R)(p \times b+1),\ (cf.\ 12)$$

$$= \mathcal{P}(p \times b - 2),\ (cf.\ 13)\ =\ (k, i_k),\ (cf.\ 11)$$

$$k = \lfloor \frac{p \times b - 2}{b} \rfloor = \lfloor p - \frac{2}{b} \rfloor = p - 1$$

$$i_k = p \times b - 2 - (p-1) \times b = b - 2$$

Then in the *Dest* pattern the tuple $\{p, i_p, k, i_k\}$ with the values as above will be added. Now, conversely, on processor $p-1$, when the (write) assignment operation Op_l is applied to $x[b-2]$, then $R^{-1}(i_{p-1}) = b-2+3 = b+1 \notin S_{p-1}$ is detected. Construct $R^{-1}(i_{p-1})$ becomes:

[3] For simplicity assume $N = P \cdot b$.

```
for t = 0 to T do
  for i = 1 to N do
    x[i] = x[i-3]
```

Fig. 3. Regular distributed data

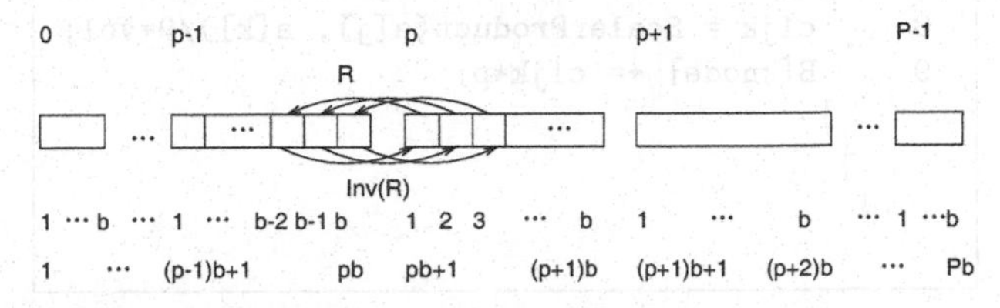

Fig. 4. The partitioning of the x array

$$\begin{aligned} R^{-1}(i_{p-1} = b-2) = (\mathcal{P} \circ R^{-1} \circ \mathcal{P}^{-1})(b-2) \;&=\; (\mathcal{P} \circ R^{-1})(p \times b - 2),\ (cf.\ 12) \\ = \mathcal{P}(p \times b + 1),\ (cf.\ 14) \;&=\; (k, i_k),\ (cf.\ 11) \\ k = \lfloor \frac{p \times b + 1}{b} \rfloor = \lfloor p + \frac{1}{b} \rfloor = p & \\ i_k = p \times b + 1 - p \times b = 1 & \end{aligned}$$

The *Source* set contains the tuple $\{p-1, i_{p-1}, p, 1\}$ with the values as above. The distributed consistency algorithm executes symmetrically, meaning that each processor executes the same sequence of steps. Then, the update on p will complete in the following manner:

$$\begin{aligned} d_{i_{p-1}} &= b - 2 \\ v_{i_{p-1}} &= x_{p-1}[b-2] \\ d_{i_p} &= R^{-1}(b-2) = 1 \\ v_{i_p} &= Op_l(v_{i_{p-1}}) \end{aligned}$$

This would lead to $x_p[1] = x_{p-1}[b-2]$.

4.2 Irregular Applications

We use the term irregular to refer to applications that use non-standard data representations (e.g. general geometries) leading to irregular representations (e.g. graphs or indirection arrays). The code excerpt in Figure 5 shows a typical irregular computation from the FEM (Finite Element Method) iterative Poisson solver for three dimensional tetrahedral meshes. The computation is irregular because of the *indirections* it introduces (such as *knode* and *jnode*).

Figure 6 shows the irregular data layout and partitioning. Let B be a data structure as in Figure 6. Analytically, the value of B at a vertex node adds the

```
1  for (e = 0; e < nelems; e++) {
2   vol = Vol[e];
3   for (j = 0; j < NVE; j++) {
4    jnode = El2MeshNo(e, j, NVE);
5    p = P[jnode];
6    for (k = 0; k < NVE - 1; k++) {
7     knode = El2MeshNo(e, (j+k+1)%NVE, NVE);
8     c1jk = ScalarProduct(a[j], a[k])/9*vol;
9     B[knode] += c1jk*p;
     }
   }
  }
```

Fig. 5. A sequential irregular computation

contributions from all the nodes connected to it within an element. By splitting the elements between subdomains, multiple copies of the same physical vertex reside on multiple data subsets. Therefore, the update operation Op_l, in this case $+=$ in step 9, must be executed according to our distributed consistency algorithm to ensure correct results.

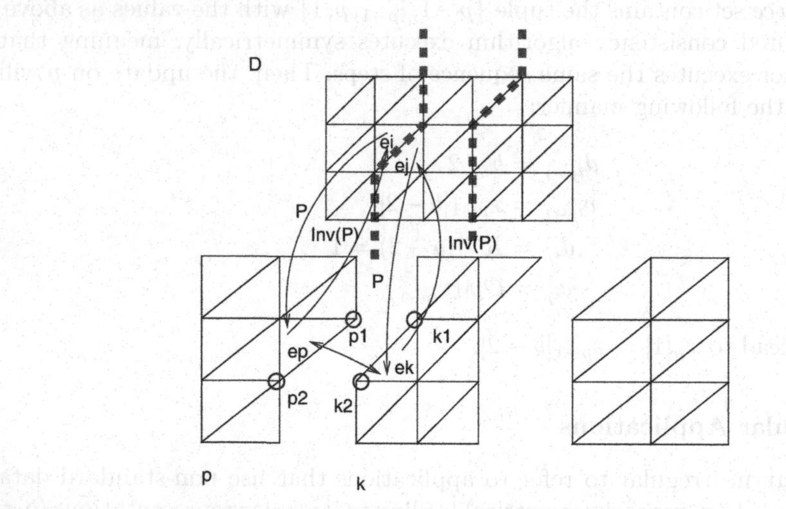

Fig. 6. Irregular data layout and partitioning

In this case the relation between data items connects two vertices of the same element. We have used the following solutions for the data partitioning and mapping:

$$D = \{e_i \mid i = \overline{1,N}, R(e_i) = R^{-1}(e_i) = e_j, j = \overline{1,N}\}$$
$$\mathcal{P} : D \rightarrow S_{p=\overline{0,P}} = \{e_p\}, \mathcal{P}(e_i) = (p, e_p)$$
$$\mathcal{P}^{-1} : S_{p=\overline{0,P}} = \{e_p\}, \rightarrow D, \mathcal{P}(p, e_p) = e_i.$$

Let e_i, e_j be two elements such that $R(e_i) = e_j = R^{-1}(e_i)$ and $R(e_j) = e_i = R^{-1}(e_j)$. This type of relation says that the *edge relation* is not *directed.* Let $\mathcal{P}(e_i) = e_p \in S_p$ and $\mathcal{P}(e_j) = e_k \in S_k$. Let i_{p_1}, i_{k_1} denote the *physically identical* nodes that belong to two different elements(e_p, e_k), that are related (through e_i, e_j). The same for i_{p_2}, i_{k_2}. Then the operation in step 9 of the code excerpt in Figure 5 will be executed as follows, on processor p:

$$i_{p_1}, i_{p_2} \in e_p,\ R(e_p) = e_k \notin S_p$$
$$R(e_p) = (\mathcal{P} \circ R \circ \mathcal{P}^{-1})(e_p) = (\mathcal{P} \circ R)(e_i) = \mathcal{P}(e_j) = (k, e_k)$$
$$\textit{Add } \{(k, i_k)\} \textit{ to the Dest pattern}$$
$$R^{-1}(e_p) = e_k \notin S_p$$
$$R^{-1}(e_p) = (\mathcal{P} \circ R^{-1} \circ \mathcal{P}^{-1})(e_k) = (\mathcal{P} \circ R^{-1})(e_j) = \mathcal{P}(e_i) = (p, e_p)$$
$$\textit{Add } \{(p, e_p)\} \textit{ to the Source pattern}$$
$$Op_l(i_{p_1}, arg) :$$
$$d_{i_{k_1}} = i_{k_1}$$
$$v_{i_{k_1}} = B[i_{k_1}]$$
$$d_{i_{p_1}} = R^{-1}(d_{i_k}) = i_{p_1}$$
$$v_{i_{p_1}} = Op_l(v_{i_{k_1}}, arg) = (+=)(B[i_{k_1}], arg)$$

5 Experimental Results

We have implemented both, a serial and a parallel version to the solution of the Poisson equation using a three dimensional tetrahedra mesh for a FEM discretization. The parallel version was obtained by making few modifications to the serial one and using our prototype system for automatic data layout and consistency. The tests were run on an SGI Origin 3800 supercomputer consisting of 220 R14000s MIPS processors (1000 Mflops per processor).

As input data we have used two meshes, of different sizes. One tetrahedra mesh consists of 2933 vertices, 31818 faces and 15581 elements. The other tetrahedra mesh consists of 1331 vertices, 12600 faces, and 6000 elements. Figure 7 shows the speedup for the parallel Poisson up to 64 processors for the two different data sets.

The shape of the speedup line in Figure 7 shows that there is a threshold number of processor above which the application does not scale well. The

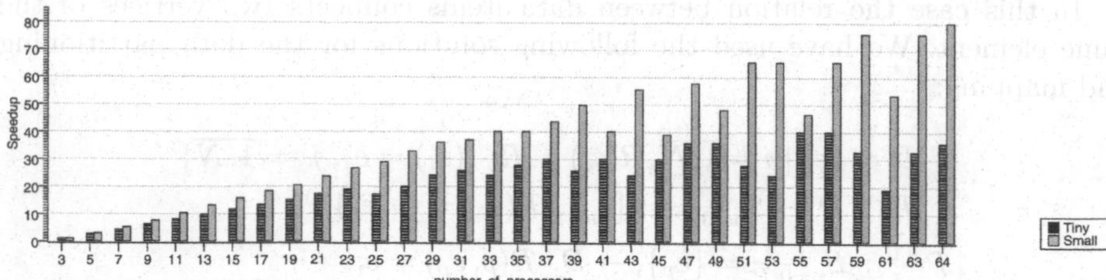

Fig. 7. The speedup of the parallel Poisson problem for different sizes of data

explanation is that above the threshold number of processors the cost of communication becomes significant compared to the amount of computation for small data partitions.

There are several issues with our approach. First of all, the general partitioning scheme requires explicit storage for the information on the transformation between local and global data spaces. With the existing approaches to data partitioning (block, cyclic), a symbolic formula does not incur any extra space overhead. In order to address this efficiency problem, we keep the information distributed over all address spaces. The consistency scheme incurs extra run-time overhead. We try to detect situations where communication can be aggregated and moved outside the loops. Such situations occur often in stencil codes and FEM assembly procedures.

6 Related Work

This section presents other approaches to data layout and consistency. First we review existing approaches to data layout for data parallel applications and point out differences in our approach. Then we review the existing approaches to data consistency and contrast our approach against them.

Data Layout. Several automatic data layout approaches exist for Fortran programs [6, 8, 9, 11, 12, 14, 18, 19, 20]. All these approaches use regular partitioning for multi-dimensional arrays. Moreover, most of the mapping strategies use static program analysis or cost estimations to decide how to place data on different processors. Our partitioning strategy is general and it subsumes regular partitioning. Moreover, the generic relation between the elements of the set to be partitioned is the key to locality constraint considerations.

Data Consistency. The existing techniques for maintaining data consistency across multiple address spaces range from hardware [5, 10, 15, 16], to automated software data consistency, to manually ensured (application level) data consistency [2]. These approaches do not take into consideration the application behavior. Thus, the hardware coherence schemes usually transfer entire blocks of data. Software maintained consistency uses pages as transfer unit. These schemes usually result in fragmentation, false sharing and redundant communication [13]. Few application oriented consistency schemes exist [5, 10]. However, these approaches use replication and transfer entire objects across address spaces. With our approach we ensure consistency only for the related data across address spaces according to the application behavior.

7 Conclusion and Future Enhancements

This paper has explored the issues of automatic data layout and consistency for data parallel applications. We have proposed a general data layout framework that subsumes existing partitioning strategies and applies to non-traditional (irregular) applications as well.

We have presented a distributed consistency algorithm based on the general data layout. The algorithm uses the relations across address spaces to propagate changes to related data in a point-to-point message passing protocol. We have presented results on the effectiveness of our approach to data layout and consistency for a 3D FEM discretization of the Poisson problem for general geometries.

An extension of this work is to evaluate the memory and communication requirements in the context of a concrete programming model. We use objects to encapsulate data partitions and fold synchronization into data accesses. We intend to address efficiency concerns such as space requirements for maintaining the relations across different address spaces, communication generation on read/write data dependences and communication aggregation.

References

[1] J. G. amd Stephane Lanteri. On overlaping partitions. In *Proceedings of the 2000 International Conference on Parallel Processing*, pages 461,468. IEEE, 2000. 23

[2] C. Amza, A. L. Cox, S. Dwarkadas, P. Keleher, H. Lu, R. Rajamony, W. Yu, and W. Zwaenepoel. Treadmarks: Shared Memory Computing on Networks of Workstations. *IEEE Computer*, 29(2):18–28, 1996. 31

[3] G. R. Andrews. Paradigms for process interaction in distributed programs. *ACM Computing Surveys*, 23(1):49,90, March 1991. 22

[4] G. R. Andrews and F. B. Schneider. Concepts and notations for concurrent programming. *ACM Computing Surveys*, 15(1):3,43, March 1983. 22

[5] H. E. Bal, M. F. Kaashoek, A. S. Tanenbaum, and J. Jansen. Replication Techniques for Speeding up Parallel Applications on Distributed Systems. *Concurrency Practice and Experience*, 4(5):337,355, August 1992. 31, 32

[6] V. Balasundaram, G. Fox, K. Kennedy, and U. Kremer. A static performance estimator to guide data partitioning decisions. In *Proceedings of the third ACM SIGPLAN symposium on Principles & practice of parallel programming*, pages 213–223. ACM Press, 1991. 31

[7] J. Chen and V. E. Taylor. Mesh partitioning for efficient use of distributed systems. *IEEE Transactions on Parallel and Distributed Systems*, 13(1):67–79, January 2002. 22

[8] N. Chrisochoides, I. Kodukula, and K. Pingali. Compiler and run-time support for semi-structured applications. In *Proceedings of the 11th international conference on Supercomputing*, pages 229–236. ACM Press, 1997. 31

[9] M. Gupta and P. Banerjee. Paradigm: a compiler for automatic data distribution on multicomputers. In *Proceedings of the 7th international conference on Supercomputing*, pages 87–96. ACM Press, 1993. 31

[10] S. B. Hassen, I. Athanasiu, and H. E. Bal. A flexible operation execution model for shared distributed objects. In *Proceedings of the OOPSLA'96 Conference on Object-oriented Programming Systems, Languages, and Applications*, pages 30–50. ACM, October 1996. 31, 32

[11] S. Hiranandani, K. Kennedy, and C.-W. Tseng. Compiling Fortran D for MIMD distributed-memory machines. *Communications of the ACM*, 35(8):66–80, 1992. 31

[12] D. E. Hudak and S. G. Abraham. Compiler techniques for data partitioning of sequentially iterated parallel loops. In *Proceedings of the 4th international conference on Supercomputing*, pages 187–200. ACM Press, 1990. 31

[13] L. Iftode and J. P. Singh. Shared virtual memory: Progress and challenges. *Proc. of the IEEE, Special Issue on Distributed Shared Memory*, 87(3):498–507, 1999. 32

[14] K. Kennedy and U. Kremer. Automatic data layout for high performance fortran. In *Proceedings of the 1995 conference on Supercomputing (CD-ROM)*, page 76. ACM Press, 1995. 31

[15] D. Lenoski, J. Laudon, K. Gharachorloo, W.-D. Weber, A. Gupta, J. Hennessy, M. Horowitz, and M. S. Lam. The stanford dash multiprocessor. *Computer*, pages 63–79, March 1992. 31

[16] D. J. Scales and M. S. Lam. The design and evaluation of a shared object system for distributed memory machines. In *OSDI94*, pages 101–114, Monterey, CA, November 1994. USENIX Association. 31

[17] K. Schloegel, G. Karypis, and V. Kumar. Graph partitioning for high performance scientific simulations. In J. Dongarra et al., editors, CRPC Parallel Computing Handbook. Morgan Kaufmann, 2000 (in press), 2000. 22

[18] A. Sussman. Model-driven mapping onto distributed memory parallel computers. In *Proceedings of the 1992 conference on Supercomputing '92*, pages 818–829. IEEE Computer Society Press, 1992. 31

[19] S. Wholey. Automatic data mapping for distributed-memory parallel computers. In *Proceedings of the 6th international conference on Supercomputing*, pages 25–34. ACM Press, 1992. 31

[20] A. Zaafrani and M. R. Ito. Partitioning the global space for distributed memory systems. In *Proceedings of the 1993 conference on Supercomputing*, pages 327–336. ACM Press, 1993. 31

A Parallel DFA Minimization Algorithm

Ambuj Tewari, Utkarsh Srivastava, and P. Gupta

Department of Computer Science & Engineering
Indian Institute of Technology Kanpur
Kanpur 208 016, INDIA
pg@iitk.ac.in

Abstract. In this paper, we have considered the state minimization problem for Deterministic Finite Automata (DFA). An efficient parallel algorithm for solving the problem on an arbitrary CRCW PRAM has been proposed. For n number of states and k number of inputs in Σ of the DFA to be minimized, the algorithm runs in $O(kn \log n)$ time and uses $O(\frac{n}{\log n})$ processors.

1 Introduction

The problem of minimizing a given DFA (Deterministic Finite Automata) has a long history dating back to the beginnings of automata theory. Consider a deterministic finite automaton M as a tuple $(Q, \Sigma, q_0, F, \delta)$ where Q, Σ, $q_0 \in Q$, $F \subseteq Q$ and $\delta : Q \times \Sigma \rightarrow Q$ are a finite set of states, a finite input alphabet, the start state, the set of accepting (or final) states and the transition function, respectively. An input string x is a sequence of symbols over Σ. On an input string $x = x_1 x_2 \cdots x_m$, the DFA visits a sequence of states $q_0 q_1 \ldots q_m$ starting with the start state by successively applying the transition function δ. Thus $q_{i+1} = \delta(q_i, x_{i+1})$ for $0 \leq i \leq m-1$. The language $L(M)$ accepted by a DFA is defined as the set of strings x that takes the DFA to an accepting state, i.e. x is in $L(M)$ if and only if $q_m \in F$. Two DFAs are said to be equivalent if they accept the same set of strings.

The problem of DFA minimization is to find a DFA with the minimum number of states which is equivalent to the given DFA. A fundamental result in automata theory states that such a minimal DFA is unique up to renaming of states. The number of states in the minimal DFA is given by the number of equivalence classes in the partition on the set of all strings in Σ^* defined as follows: Two strings x and y are equivalent if and only if for all strings z, $xz \in L(M)$ if and only if $yz \in L(M)$ [6].

Besides being widely studied, DFA has many applications in the diverse fields like pattern matching, optimization of logic programs, protocol verification and specification and modeling of finite state systems [14]. It is known that non-deterministic finite automata (NFA) are equivalent to deterministic ones as far as the languages recognized by them are concerned. Huffman [4] and Moore [10] have presented $O(n^2)$ algorithms for DFA minimization and are sufficiently fast for most of the classical applications. However, there exist numerous algorithms

S. Sahni et al. (Eds.) HiPC 2002, LNCS 2552, pp. 34–40, 2002.

which are variations of the same basic idea. An efficient $O(n \log n)$ algorithm is due to Hopcroft [5]. Blum [2] has also proposed a simpler algorithm with same time complexity.

Traditional applications of the DFA minimization algorithm involve a few thousand states and the sequential algorithms available generally perform well in these settings. But if the number of states in a DFA is of the order of a few millions, then the efficient sequential algorithms may take a significant amount of time and may need much more than the available physical memory. One way to achieve the speed-up is the use of multiple processors.

DFA minimization has been extensively studied on many parallel computation models. Jaja and Kosaraju [7] have presented an efficient algorithm on mesh connected computer for the case when $|\Sigma| = 1$. A simple NC algorithm has also been outlined in their work. Cho and Huynh [3] have proved the problem to be NLOGSPACE-complete. Efficient and close to cost-optimal algorithms are known only for the case of the alphabet consisting of a single input symbol. A very simple algorithm is proposed by Srikant [13] but the best algorithm for this problem is due to Jaja and Ryu [8]. It is a CRCW algorithm with time complexity $O(\log n)$ and cost $O(n \log \log n)$. Unfortunately, no efficient algorithm, which is also economical with respect to cost, is known for the general case of multiple input symbols. The standard NC algorithm for this problem requires $O(n^6)$ processors. This is due the use of transitive closure computation on a cross product graph. In [12], a simple parallel algorithm for DFA minimization along with its implementation is presented.

In this paper we have proposed an efficient parallel algorithm for DFA minimization having $O(kn\log n)$ time complexity on an arbitrary CRCW PRAM with $(\frac{n}{\log n})$ processors. The rest of the paper is organized as follows. Section 2 contains some preliminaries along with the naive sequential algorithm. In Section 3, we discuss a fast parallel algorithm for the problem which uses a large number of processors. An efficient parallel algorithm has been proposed in Section 4. Conclusion is given in the last section.

2 Review

The DFA minimization problem is closely related to the coarsest partitioning problem which can be stated as follows. We are given a set Q and its initial partition into m disjoint sets $\{B_0, \ldots, B_{m-1}\}$, and a collection of functions, $f_i : Q \to Q$. We have to find a coarsest partition of Q, say $\{E_1, \ldots, E_q\}$, such that: (1) each E_i is a subset of some B_j , and (2) the partition respects the given functions, i.e. $\forall j$, if a and b both belong to the same E_i then $f_j(a)$ and $f_j(b)$ also belong to the same E_k for some k.

Assume, Q is the set of states, f_i is the restriction of the transition function δ to the ith input symbol, i.e. $f_i(q) = \delta(q, a_i)$, the initial partition contains two sets, namely F and $Q - F$. The size of the minimal DFA is the number of equivalence classes in the coarsest partition. For the general case of multiple function coarsest partition problem, an $O(n \log n)$ solution is given in [1]. Later, a linear

Sequential Algorithm

1. *for all* final states q_i *do* block_no[q_i] = 1
2. *for all* non-final states q_i *do* block_no[q_i] = 2
3. *do*
4. *for* i= 0 *to* k-1 *do*
5. *for* j = 0 *to* n-1 *do*
6. b_1 = block_no[q_j]
7. b_2 = block_no[$\delta(q_j, x_i)$]
8. label state q_j with (b_1, b_2)
9. *endfor*
10. *Assign* same block number to states having same labels
11. *endfor*
12. *while* <number of blocks is changing>

Algorithm 1: The sequential algorithm for DFA minimization

time solution has been proposed in [11] for the case of the single function coarsest partition problem. The single function version of the problem corresponds to having only a single symbol in the input alphabet Σ.

But the simplest sequential algorithm for solving the problem, given in Algorithm 1, runs in $O(kn^2)$ time, where $|\Sigma| = k$ [1]. It performs as follows. Initially there are only two blocks: one containing all the final states and the other containing all the non-final states. If two states q and q' are found in the same block such that for some input symbol a_i, the states $\delta(q, a_i)$ and $\delta(q', a_i)$ are in different blocks, q and q' are placed in different blocks for the next iteration. The algorithm is iterated for at most n times because, in the worst case, each state will be in a block containing just itself. In each iteration new block numbers are assigned in $O(kn)$ time. Therefore, the total time taken is $O(kn^2)$.

3 A Fast Parallel Algorithm

The fastest known algorithm for the multiple input symbol case is due to Cho and Huynh [3]. The DFA minimization problem is initially translated to an instance of the multiple function coarsest partition problem to yield the set S of states, the initial partition B containing sets of final and non-final states and the functions f_i which are restrictions of the transition function to single input symbols. A graph $G =< V, E >$ may be generated as follows: $V = \{(a, b) | a, b \in S\}$ and $E = \{((a, b), (c, d)) | c = f_i(a), d = f_i(b) \text{ for some } i\}$.

For any pair $x, y \in S$, x and y get different labels in the coarsest partition if and only if there is a path from node (x, y) to some node (a, b) such that a and b have different B-labels. The algorithm is given below. It can be shown that the algorithm can be implemented on a EREW in time $O(\log^2 n)$ with total cost $O(n^6)$. The bound $O(n^6)$ arises because of the transitive closure computation of a graph with n^2 nodes. So far, it has not been possible to find an algorithm with a reasonable cost, say $O(n^2)$, and a small running time.

Parallel Algorithm

1. *Construct* graph G as defined above
2. *Mark* all nodes $(p, q) \ni p$ and q belong to different sets of the B-partition
3. *Uses* transitive closure to mark pairs reachable from marked pairs
4. *comment* Note that all unmarked pairs are equivalent

Algorithm 2: A fast parallel algorithm for DFA minimization

4 An Efficient Algorithm

In this section we have proposed a parallel version of the simple $O(n^2)$ time sequential algorithm outlined in Section 3. We use $O(\frac{n}{\log n})$ processors to achieve an expected running time of $O(\log n)$. Using an arbitrary-CRCW PRAM, we have parallelized the inner for-loop which iterates over the set of states. The the new labels obtained in this for-loop are hashed using parallel hashing algorithm of Matias and Vishkin [9] to get new block numbers for the states. The algorithm is given below.

Lines 1-8 are the same as in the sequential algorithm except that Lines 5-8 of Algorithm 3 are now done in parallel. The n labels obtained in Line 8 are hashed to $[1..O(n)]$ using the hashing technique due to Matias and Vishkin [9].

Theorem 1 (Parallel Hashing Theorem). *Let W be a multiset of n numbers from the range $[1..m]$, where $m + 1 = p$ is a prime. Suppose we have $\frac{n}{\log n}$*

New Parallel Algorithm

1. *Initialize* block_no array
2. *do*
3. *for* $i = 0$ *to* k-1 *do*
4.   *for* $j = 1$ *to* $\frac{n}{\log n}$ *do in parallel*
5.   *for* $m = (j - 1) * \log n$ *to* $j * \log n - 1$ *do*
6. b_1 = block_no[q_m]
7. b_2 = block_no[$\delta(q_m, a_i)$]
8.   label state q_m with (b_1, b_2)
9.   *endfor*
10.   *endfor*
11.   *Use* parallel hashing to map the n labels to $[1..O(n)]$
12.   a number to which a state's label gets mapped to its new block_no
13.   *Reduce* the range of block_no from $O(n)$ to n
14.   *endfor*
15. *while* number of blocks is changing

Algorithm 3: New parallel algorithm for DFA minimization

processors on an arbitrary-CRCW PRAM. A one-to-one function $F : W \longmapsto [1..O(n)]$ can be found in $O(\log n)$ expected time. The evaluation of $F(x)$ for each $x \in W$, takes $O(1)$ arithmetic operations (using numbers from $[1..m]$).

We have to assign new block numbers to states such that states with different labels get different block numbers and states with same label get same block number. Also we do not want the block numbers to become too large. The labels are pairs of the form (b_1, b_2) where $b_1, b_2 \leq n$. Therefore we can map a label (b_1, b_2) to $b_1 * (n+1) + b_2$ and we can treat these labels as numbers in the range $[1..2n^2]$. A number m such that $2n^2 < m \leq 4n^2$ and $m+1$ is a prime can be found in $O(\log n)$ time (see [9]) and this has to be done just once. For instance, it can be done after the initialization phase of step 1.

We want the block numbers to remain in the range $[1..n]$. However, after hashing, we get numbers in the range $[1..Kn]$ some fixed K. This range shrinking can be implemented on an arbitrary-CRCW PRAM in time $O(\log n)$. The procedure is given as Algorithm 4.

First, each processor hashes $\log n$ labels and sets $PRESENT[x]$ to 1 if some label got hashed to the value x, where $PRESENT[1..Kn]$ is an array. Several processors might try to write in the same location in the array but since we have assumed an arbitrary-CRCW PRAM, one of them will succeed arbitrarily. Each of the $\frac{n}{\log n}$ processors now considers a range of $K \log n$ indices of the $PRESENT$ array and computes the number of locations which are set to 1 using $O(\log(\frac{n}{\log n}))$ prefix sum algorithm. New block number for a given location is simply the number of 1's occurring before that location. A processor can easily compute the new block number for a location in its range by adding the number of 1's occurring before that location within the processor's range to the number of 1's occurring in earlier ranges (this has already been computed by prefix-sum).

To find the time complexity, let us consider first Algorithm 4. From the parallel hashing theorem we know that evaluation of the hashing function in line 3 takes $O(1)$ time. Line 4 is an assignment and so the loop in Lines 2-5 takes $O(\log n)$ time. Similarly the loop in Lines 9-11 takes $O(\log n)$. Prefix sum computation in Lines 13-14 also takes $O(\log n)$ time. The loop in Lines 17-21 takes $O(\log n)$ time. Evaluation of the hash function at line 25 takes $O(1)$ time. Therefore, the last loop (Lines 24-27) too takes $O(\log n)$ time.

The outermost for-loop (Lines 3-14) of Algorithm 3 runs exactly k times where k is the size of the input alphabet and the outer do-while-loop (Lines 2-15) of our algorithm can run for at most n times since the minimal DFA does not have more than n states. Therefore, the expected time complexity of our algorithm is $O(kn \log n)$. Since we use $O(\frac{n}{\log n})$ processors, the cost is $O(kn^2)$ which is the cost optimal parallel adaptation of the $O(kn^2)$ sequential method.

5 Conclusion

In this paper, we have considered a well-known problem from classical automata theory and have presented a parallel algorithm for the problem. We have essentially adapted the naive $O(kn^2)$ sequential algorithm and have shown that our algorithm requires $O(n \log n)$ time using $O(\frac{n}{\log n})$ processors. Thus it is a cost optimal parallelization on an arbitrary-CRCW PRAM.

Finally, it will be of immense theoretical and practical importance to come up with impossibility results about the limited parallelizability of the sequential DFA minimization algorithms.

// *Initialize* the $PRESENT[1..Kn]$ array
1. *for* $i = 1$ *to* $\frac{n}{\log n}$ *do in parallel*
2. *for* $j = (i-1) * K \log n + 1$ *to* $i * K \log n$ *do*
3. Let x be the value to which the label of q_j hashes
4. $PRESENT[x] = 1$
5. *endfor*
6. *endfor*

// *Compute* number of 1's in each processor's range
7. *for* $i = 1$ *to* $\frac{n}{\log n}$ *do in parallel*
8. $a_i = 0$
9. *for* j $= (i-1) * K \log n + 1$ *to* $i * K \log n$ *do*
10. $a_i = a_i + PRESENT[j]$
11. *endfor*
12. *endfor*

// *Compute* partial sums
13. $s_0 = 0$
14. *Compute* $s_i = \sum_{k=1}^{i} a_i$ for $1 \leq i \leq \frac{n}{\log n}$ using prefix sum

// *Compute* new block numbers
15. *for* $i = 1$ *to* $\frac{n}{\log n}$ *do in parallel*
16. $a_i = s_{i-1}$
17. *for* $j = (i-1) * K \log n + 1$ *to* $i * K \log n$ *do*
18. $a_i = a_i + PRESENT[j]$
19. *if* $PRESENT[j] = 1$ *then*
20. new_block_no[j] $= a_i$
21. *endfor*
22. *endfor*

// *Update* the block_no array with the new block numbers
23. *for* $i = 1$ *to* $\frac{n}{\log n}$ *do in parallel*
24. *for* $j = (i-1) * \log n$ *to* $i * \log n - 1$ *do*
25. Let x be the value to which the label of q_j hashes
26. block_no$[q_j]$ = new_block_no$[x]$
27. *endfor*
28. *endfor*

Algorithm 4: Reducing the range of block numbers

References

[1] Aho A. V., Hopcroft J. E. and Ullman J. D.: The design and analysis of computer algorithms. Addison-Wesley, Reading, Massachusetts (1974) 35, 36

[2] Blum N.: An $O(n \log n)$ implementation of the standard method of minimizing n-state finite automata. Information Processing Letters **57** (1996) 65-69 35

[3] Cho S. and Huynh D. T.: The parallel complexity of coarsest set partition problems. Information Processing Letters **42** (1992) 89-94 35, 36

[4] Huffman D. A.: The Synthesis of Sequential Switching Circuits. Journal of Franklin Institute **257** (1954) 161-190 34

[5] Hopcroft J. E.: An $n \log n$ algorithm for minimizing states in a finite automata. Theory of Machines and Computation, Academic Press (1971) 189-196 35

[6] Hopcroft J. E. and Ullman J. D.: Introduction to automata theory, languages, and computation. Addison-Wesley, Reading, Massachusetts (1979) 34

[7] Jaja J. and Kosaraju S. R.: Parallel algorithms for planar graph isomorphism and related problems. IEEE Transactions on Circuits and Systems **35** (1988) 304-311 35

[8] Jaja J. and Ryu K. W.: An Efficient Parallel Algorithm for the Single Function Coarsest Partition Problem. Theoretical Computer Science **129** (1994) 293-307 35

[9] Matias Y. and Vishkin U.: On parallel hashing and integer sorting. Journal of Algorithms **4** (1991) 573–606 37, 38

[10] Moore E. F.: Gedanken-experiments on sequential circuits. Automata Studies, Princeton University Press (1956) 129-153 34

[11] Paige R., Tarjan R. E. and Bonic R.: A linear time solution to the single function coarsest partition problem. Theoretical Computer Science **40** (1985) 67-84 36

[12] Ravikumar B. and Xiong X.: A parallel algorithm for minimization of finite automata. Proceedings of the 10th International Parallel Processing Symposium, Honululu, Hawaii (1996) 187-191 35

[13] Srikant Y. N.: A parallel algorithm for the minimization of finite state automata. International Journal Computer Math. **32** (1990) 1-11 35

[14] Vardi M.: Nontraditional applications of automata theory. Lecture Notes in Computer Science, Springer-Verlag **789** (1994) 575-597 34

Accelerating the CKY Parsing Using FPGAs*

Jacir L. Bordim, Yasuaki Ito, and Koji Nakano

School of Information Science
Japan Advanced Institute of Science and Technology
1-1, Asahidai, Tatsunokuchi, Ishikawa 923-1292, Japan
{bordim,yasuaki,knakano}@jaist.ac.jp

Abstract. The main contribution of this paper is to present an FPGA-based implementation of an instance-specific hardware which accelerates the CKY (*Cook-Kasami-Younger*) parsing for context-free grammars. Given a context-free grammar G and a string x, the CKY parsing determines if G derives x. We have developed a hardware generator that creates a Verilog HDL source to perform the CKY parsing for any given context-free grammar G. The created source is embedded in an FPGA using the design software provided by the FPGA vendor. We evaluated the instance-specific hardware, generated by our hardware generator, using a timing analyzer and tested it using the Altera FPGAs. The generated hardware attains a speed-up factor of approximately 750 over the software CKY parsing algorithm. Hence, we believe that our approach is a promising solution for the CKY parsing.

1 Introduction

An FPGA (Field Programmable Gate Array) is a programmable VLSI in which a hardware designed by users can be embedded instantly. Typical FPGAs consist of an array of programmable logic elements, distributed memory blocks, and programmable interconnections between them. Our goal is to use the FPGAs to accelerate useful computations. In particular, the challenge is to develop FPGA-based solutions which are faster and more efficient than traditional software approaches.

Our basic idea for accelerating computations using the FPGAs is inspired by the notion of *partial computation* [10]. Let $f(x, y)$ be a function to be evaluated in order to solve a given problem. Note that such a function might be repeatedly evaluated only for a fixed x. When this is the case, the computation of $f(x, y)$ can be simplified by evaluating an instance-specific function f_x such that $f_x(y) = f(x, y)$. For example, imagine a problem such that an algorithm to solve it evaluates $f(x, y) = x^3 + x^2y + y$ repeatedly. If $f(x, y)$ is evaluated only for $x = 2$, then the formula can be simplified such that $f_2(y) = 8 + 5y$. The optimization of function f_x for a particular x is called a *partial computation.* Usually, a partial computation has been used for optimizing a function f_x in

* Work supported in part by the Ministry of Education, Science, Sports, and Culture, Government of Japan, Grant-in-Aid for Exploratory Research (90113133).

S. Sahni et al. (Eds.) HiPC 2002, LNCS 2552, pp. 41–51, 2002.

the context of software, *i.e.*, sequential programs [8]. Our novel idea is to built a hardware that is optimized to compute $f_x(y)$ for a fixed x and various y. More specifically, our goal is to present an FPGA-based instance-specific solution for problems that involves a function evaluation for $f(x,y)$ satisfying the following properties: (*i*) the value of a fixed instance x depends on the instance of the problem, and (*ii*) the value of $f(x,y)$ is repeatedly evaluated for various y to solve the problem. The FPGA-based instance-specific solution that we propose evaluates $f_x(y)$ $(= f(x,y))$ using a hardware for function f_x. If the problem we need to solve satisfies these properties, it is worth attempting the instance-specific solution.

The main contribution of this paper is to present an instance-specific hardware which accelerates the parsing for context-free grammars [12] using the FPGA-based approach described above. Let $f(G,x)$ be a function such that G is a context grammar, x is a string, and $f(G,x)$ returns a Boolean value such that $f(G,x)$ returns TRUE iff G derives x. It is well-known that the *CKY(Cook-Kasami-Younger) parsing* [1] computes $f(G,x)$ in $O(n^3)$ time, where n is the length of x [1]. The parsing of context-free languages has many application in various areas including natural language processing [5, 14], compiler construction [1], informatics [13], among others.

Several studies have been devoted to accelerate the parsing of context-free languages [4, 9, 11, 14]. It has been shown that the parsing for a string of length n can be done in $O((\log n)^2)$ time using n^6 processors on the PRAM [9]. Also, using the mesh-connected processor arrays, the parsing can be done in $O(n^2)$ time using n processors as well as in $O(n)$ time using n^2 processors [11]. Since these parallel algorithms need at least n processors, they are unrealistic for large n. Ciressan *et al.* [6, 7] have presented a hardware for the CKY parsing for a restricted class of context-free grammar and have tested it using FPGA. However, the hardware design and the control algorithm are essentially the same as those on the mesh-connected processors [11], and they are not instance-specific.

For the purpose of instance-specific solution for parsing context-free languages, we present a hardware generator that produces a Verilog HDL source that performs the CKY parsing for any given context-free grammar G. The key ingredient of the produced design is a hardware component to compute a binary operator $\otimes_G$ such that $2^N \times 2^N \to 2^N$, where N is the set of non-terminal symbols in G. More specifically, let U and V be a set of non-terminals in G that derive strings α and β, respectively. The operator $U \otimes_G V$ returns the set of non-terminals that derive $\alpha\beta$ (i.e. the concatenation of α and β). The CKY parsing algorithm repeats the evaluation of $\otimes_G$ for $O(n^3)$ times. The details of $\otimes_G$ will be explained in Section 2. Our hardware generator provides two types of hardware. The first hardware has one component for computing $\otimes_G$. The second one has two or more components to further accelerate the table algorithm for the CKY parsing.

The generated Verilog HDL source is compiled using the Altera Quartus II design tool [3], and the object file obtained is downloaded into the Altera APEX20K series FPGAs [2]. The programmed FPGA compute $f_G(x)$, i.e. de-

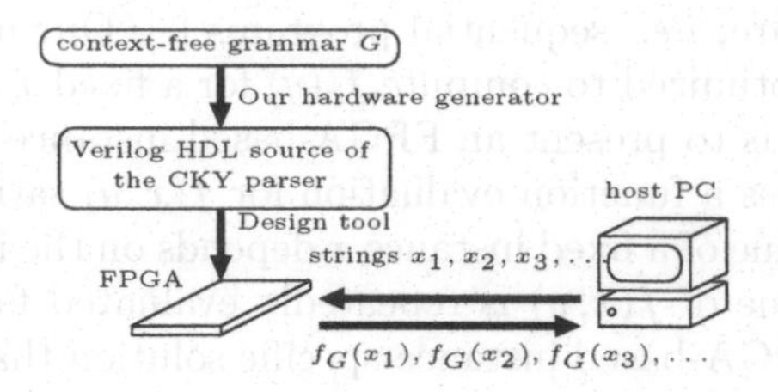

Fig. 1. Our hardware parsing system

termines if G derives x for a given string x. Figure 1 illustrates our hardware CKY parsing system. Given strings $x_1, x_2, x_3, \ldots$ by the host PC, the FPGA computes and returns $f_G(x_1), f_G(x_2), f_G(x_3), \ldots$ to the host.

From the theoretical point of view, our instance-specific solution is much faster than the software solutions. To clarify how our solution accelerates the CKY parsing, we provide the following two software approaches as counterparts:

Naive algorithm: This algorithm computes $\otimes_G$ by checking all p production rules in $O(p)$ time. The CKY parsing using the naive algorithm runs in $O(n^3 p)$ time.

Table algorithm: This algorithm computes $\otimes_G$ by looking up $(\frac{b}{c})^2$ tables of 2^{2c} words with b bits in $O((\frac{b}{c})^2)$ time, where b is the number of non-terminal symbols in G. Although c can take any integer, in practice, c does not exceed 16, possibly $c \leq 8$. The CKY parsing using the table algorithm runs in $O(n^3(\frac{b}{c})^2)$ time.

Our instance-specific solution evaluates $\otimes_G$ in $O(\log b)$ time and the CKY parsing using this approach runs in $O(n^3 \log b)$ time. Since $b \leq p$ always hold, our solution is faster than these software approaches from theoretical point of view.

We have evaluated the performance of our instance-specific solution using the timing analyzer of Quartus II and test it using an APEX20K series FPGA. In order to evaluate the performance of our instance-specific solution, we also have implemented the above software solutions and measured the performance using a Pentium4-based PC.

The timing analysis results show that our instance-specific hardware attains up to 750 speed-up factor over the software solutions. Thus, we strongly believe that our approach for parsing context-free languages is a promising solution.

2 The CKY Parsing and Software Solutions

The main purpose of this section is to briefly describe the CKY parsing and show two software solutions. Let $G = (N, \Sigma, P, S)$ denote a *context-free grammar* such that N is a set of non-terminal symbols, Σ is a set of terminal symbols, P is a set of production rules, and S ($\in N$) is the start symbol. A context-free grammar is said to be a *Chomsky Normal Form* (CNF), if every production rule in P is in either form $A \to BC$ or $A \to a$, where A, B, and C are non-terminal symbols and a is a terminal symbol.

We are interested in the parsing problem for a context-free grammar with CNF. More specifically, for a given CNF context-free grammar G and a string x over Σ, the parsing problem asks to determine if the start symbol S derives x. For example, let $G_{\text{example}} = (N, \Sigma, P, S)$ be a grammar such that $N = \{S, A, B\}$, $\Sigma = \{a, b\}$, and $P = \{S \to AB, S \to BA, S \to SS, A \to AB, B \to BA, A \to a, B \to b\}$. Context grammar G derives *abaab*, because S derives it as follows:

$$S \Rightarrow AB \Rightarrow ABA \Rightarrow ABAA \Rightarrow ABAAB \Rightarrow \cdots \Rightarrow abaab.$$

We are going to explain the *CKY parsing scheme* that determines whether G derives x for a CNF context-free grammar G and a string x. Let $x = x_1 x_2 \cdots x_n$ be a string of length n, where each x_i $(1 \le i \le n)$ is in Σ. Let $N[i,j]$ $(1 \le i \le j \le n)$ denote a subset of N such that every A in $N[i,j]$ derives a substring $x_i x_{i+1} \cdots x_j$. The idea of the CKY parsing is to compute every $N[i,j]$ using the following relations:

$$N[i,i] = \{A \mid (A \to x_i) \in P\}$$

$$N[i,j] = \bigcup_{k=i}^{j-1} \{A \mid (A \to BC) \in P, B \in N[i,k], \text{ and } C \in N[k+1,j]\}$$

A two-dimensional array N is called the *CKY table*. A grammar G generates a string x iff S is in $N[1,n]$. Let $\otimes_G$ denote a binary operator $2^N \times 2^N \to 2^N$ such that $U \otimes_G V = \{A \mid (A \to BC) \in P, B \in U, \text{ and } C \in V\}$. The details of the CKY parsing are spelled out as follows:

CKY parsing

1. $N[i,i] \leftarrow \{A \mid (A \to x_i) \in P\}$ for every i $(1 \le i \le n)$
2. $N[i,j] \leftarrow \emptyset$ for every i and j $(1 \le i < j \le n)$
3. for $j \leftarrow 2$ to n do
4.   for $i \leftarrow j-1$ downto 1 do
5.   for $k \leftarrow i$ to $j-1$ do
6. $N[i,j] \leftarrow N[i,j] \bigcup (N[i,k] \otimes_G N[k+1,j])$

The first two lines initialize the CKY table, and the next four lines compute the CKY table. Figure 2(a) illustrates the CKY table for G_{example} and the string *abaab*. Since $S \in N[1,5]$, one can see that G_{example} derives *abaab*. Clearly, the last four lines are dominant in the CKY parsing. Let T be the computing time necessary to perform an iteration of the line 6. Then, line 6 is executed for

$$\sum_{j=2}^{n-1} \sum_{i=1}^{j-1} \sum_{k=i}^{j-1} T = T \sum_{j=2}^{n-1} \sum_{i=1}^{j-1} (j-i) = \frac{1}{6} T(n^3 - 3n^2 + 2n)$$

times. Let us evaluate the computing time T necessary to perform line 6, i.e., necessary to evaluate a binary operator $\otimes_G$. We will present two approaches that compute $U \otimes_G V$ by sequential (software) algorithms for any given U and V.

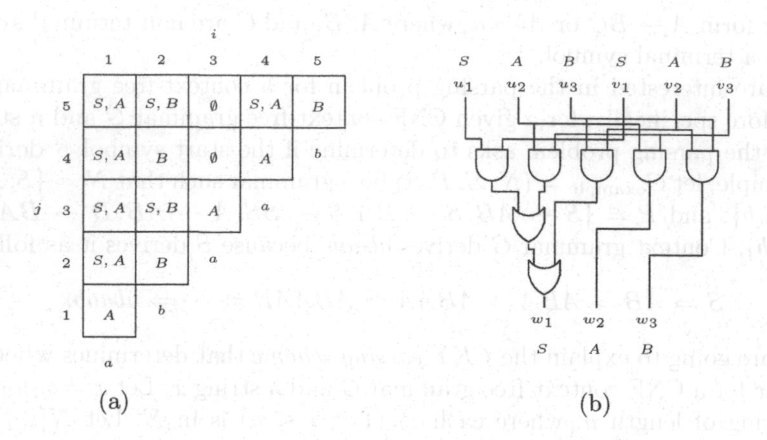

Fig. 2. (a) The CKY table for G_{example} and *abaab* (b) The circuit for computing $\otimes G_{\text{example}}$

In the first approach, named *naive algorithm*, it is checked whether $B \in U$ and $C \in V$ for every production rule $A \to BC$ in P. Clearly, using a reasonable data structure, this can be done in $O(1)$ time. Hence, $U \otimes_G V$ can be evaluated in $O(p)$ time, where p is the number of production rules in P that has the form $A \to BC$. Thus, the first approach enables us to perform the CKY parsing in $O(n^3p)$ time.

Suppose that N has b non-terminal symbols, and let $N = \{N_1, N_2, \ldots, N_b\}$. The second approach that we call *table algorithm* uses a huge look-up table that stores the values of $U \otimes_G V$ for every pair U and V. For a given U $(\in 2^N)$, let $u_1u_2 \cdots u_b$ be the b-bit vector such that $u_i = 1$ iff $N_i \in U$ for every i $(1 \leq i \leq b)$. Similarly, let $v_1v_2 \cdots v_b$ be the b-bit vector for V $(\in 2^N)$. For the purpose of computing $U \otimes_G V$, we use a look-up table of $2^{2b} \times b$ in memory (i.e., the address and the data are $2b$ bits and b bits, respectively). The $u_1u_2 \cdots u_bv_1v_2 \cdots v_b$-th entry of the table stores $w_1w_2 \cdots w_b$, where $w_1w_2 \cdots w_b$ is the b-bit vector representation of $W = U \otimes_G V$. Clearly, if such table is available, $U \otimes_G V$ can be computed in $O(1)$ time. However, the table can be too large even if b is not large. If P has $b = 64$ non-terminal symbols, then the table must have $2^{2\cdot 64} \times 64 = 2^{134} \approx 10^{40}$ bits, which is too large for practical purposes.

We will modify the table algorithm to reduce the table size. Let us partition N into equal-sized subsets such that $N^i = \{N^{c(i-1)+1}, N^{c(i-1)+2}, \ldots, N^{ci}\}$ $(1 \leq i \leq \frac{b}{c})$. We use $(\frac{b}{c})^2$ binary operators $\otimes_G^{i,j}$ $(1 \leq i, j \leq \frac{b}{c})$ such that

- $\otimes_G^{i,j}$ is $2^{N^i} \times 2^{N^j} \to 2^N$, and
- $(U \cap N^i) \otimes_G^{i,j} (V \cap N^j) = \{A \mid (A \to BC) \in P, B \in U \cap N^i, \text{ and } C \in V \cap N^j\}$.

It is easy to see that,

$$U \otimes_G V = \bigcup_{1 \leq i,j \leq (\frac{b}{c})^2} (U \cap N^i) \otimes_G^{i,j} (V \cap N^j).$$

Thus, by evaluating $\otimes_G^{i,j}$ for every pair i and j, we can compute $\otimes_G$. As before, $\otimes_G^{i,j}$ can be computed by looking up a table of size $2^{2c} \times b$. Hence, $\otimes_G$ can be computed in $O((\frac{b}{c})^2)$ time by looking up $(\frac{b}{c})^2$ tables. The total size of the tables is $\frac{b^3}{c^2}2^{2c}$ bits. If $b = 64$ and $c = 8$, then the tables should have $2^{28} = 256M$ bits, which is feasible. However, we need to look up the table for $(\frac{b}{c})^2 = 64$ times. Note that the size of the tables and the number of times needed to be looked up are independent of the number p of production rules. Thus, the second approach is more efficient for large p.

3 Our Instance-Specific Hardware for the CKY Parsing

This section is devoted to show our instance-specific hardware for the CKY parsing. We first accelerate the evaluation of $\otimes_G$ by building a circuit for computing $\otimes_G$ in an FPGA. We then go on to show the hardware details to build this circuit.

Recall that each U and V ($\in 2^N$) are represented by b-bit binary vectors $u_1u_2\cdots u_b$ and $v_1v_2\cdots v_b$, respectively. Our goal is to compute the vector $w_1w_2\cdots w_b$, which represents $W = U \otimes_G V$. For a particular w_k, we are going to show how w_k is computed. Let $N_k \to N_{i_1}N_{j_1}$, $N_k \to N_{i_2}N_{j_2}$, ..., and, $N_k \to N_{i_s}N_{j_s}$ be the production rules in P whose non-terminal in the left-hand side is N_k. Then, w_k is computed by the following formula:

$$w_k = (v_{i_1} \wedge u_{j_1}) \vee (v_{i_2} \wedge u_{j_2}) \vee \cdots \vee (v_{i_s} \wedge u_{j_s}).$$

Thus, w_k can be computed by a combinatorial circuit using s AND-gates and $s-1$ OR-gates with fan-in 2. Furthermore, the depth of the circuit (or the maximum number of gates over all paths in the circuit) is $\lceil\log(s-1)\rceil + 1$. Since we have p production rules of the type $A \to BC$ in P, then $w_1w_2\cdots w_b$ can be computed by a circuit with p AND-gates and $p - b$ OR-gates. Because $s \le b^2$ always hold, the depth of the circuit is no more than $\lceil\log(b^2-1)\rceil + 1 \le 2\log b + 1$. Thus, the CKY parsing can be done in $O(n^3 \log b)$ time using this circuit. Figure 2(b) illustrates a circuit for $\otimes_{G_{\text{example}}}$. Since G_{example} has 5 production rules and 3 non-terminal symbols, the circuit has 5 AND gates and $5-3=2$ OR gates.

The sequential algorithms we have shown in Section 2 take $O(p)$ time or $O((\frac{b}{c})^2)$ time to evaluate $\otimes_G$. On the other hand, our circuit for $\otimes_G$ has the delay time proportional to $O(\log b)$. Since $b \le p \le b^3$ always holds, the circuit for $\otimes_G$ is faster than the sequential algorithms from the theoretical point of view. In what follows, we are going to show the implementation details of our instance-specific hardware. Our first hardware implementation of the CKY parsing uses the following basic components: (*i*) a b-bit n^2-word (dual-port) memory; (*ii*) a b-bit n-word (dual-port) memory; (*iii*) a CKY circuit for $\otimes_G$; (*iv*) an array of b OR gates; and (*v*) a b-bit register.

Figure 3(a) illustrates our first implementation for the CKY parsing. The b-bit n^2-word memory stores the CKY table. The input, $N[1,1]$, $N[2,2]$, ..., $N[n,n]$ is supplied to the b-bit n^2-word memory. The b-bit n-word memory stores a row of the CKY table that is being processed. In other words, it

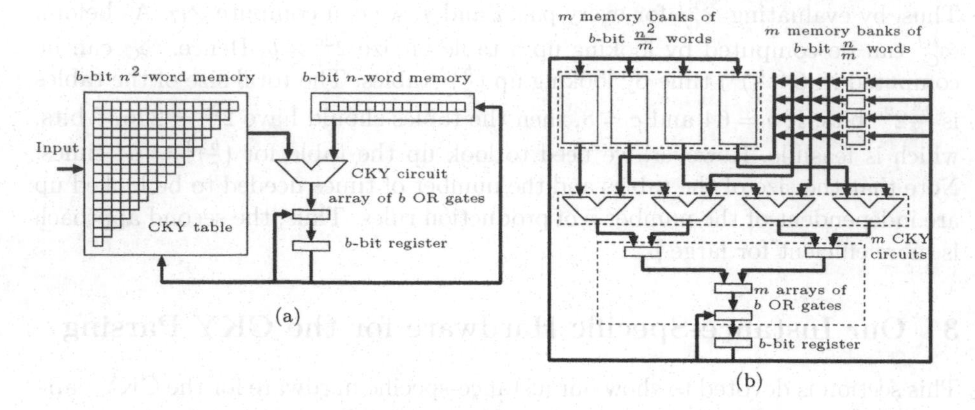

Fig. 3. (a) A hardware implementation for the CKY parsing (b) Our parallel implementation of the CKY parsing for $m = 4$

stores the j-th row $N[1, j], N[2, j], \ldots$ of the CKY table, where j is the variable appearing in line 3 of the CKY parsing. The b-bit register stores the current value of $N[i, j]$, which is computed in line 6 of the CKY parsing. The array of b OR gates is used to compute "$\bigcup$" in line 6. The b-bit n^2-word memory supplies the b-bit vector representing $N[i, k]$ to the CKY circuit. Similarly, the b-bit n-word memory outputs the b-bit vector for $N[k+1, j]$. The CKY circuit receives them and computes the b-bit vector for $N[i, k] \otimes_G N[k+1, j]$. Using this hardware implementation, line 6 of the CKY parsing is computed in a clock cycle. Thus, the CKY parsing can be done in n^3 clock cycles. Furthermore, in a real implementation, a clock cycle is proportional to $O(\log b)$. Thus, the computing time is $O(n^3 \log b)$.

We are going to parallelize the CKY parsing using two or more CKY circuits. For this purpose, we partition the CKY table into m subtables $S(0)$, $S(1)$, $\ldots, S(m-1)$ such that $S(l)$ is storing $N[i, j]$ satisfying $j - i \bmod m = l$. Clearly, for any m consecutive elements $N[i, k], N[i, k+1], \ldots, N[i, k+m-1]$ in a column of the CKY table, these elements are stored in distinct subtables. Thus, the consecutive m elements can be accessed in the same time if each subtable is stored in a memory bank. This fact allows us to parallelize the CKY parsing using m CKY circuits. In order to evaluate the performance of the above approaches, we have implemented the instance-specific hardware CKY parser using a single CKY circuit (*single-circuit*), two CKY circuits (*double-circuit*), and four CKY circuits (*quad-circuit*).

Our parallel implementation of the CKY parsing uses the following basic components: (*i*) m (dual-port) memory banks of b-bit $\frac{n^2}{m}$ words; (*ii*) m (dual-port) memory banks of b-bit $\frac{n}{m}$ words; (*iii*) m CKY circuits for $\otimes_G$; (*iv*) m arrays of b OR gates; and (*v*) a b-bit register. Figure 3(b) illustrates our parallel implementation for the CKY parsing. The m memory banks of b-bit $\frac{n^2}{m}$ words are used to store m subtables, one bank for each subtable. Also, the m

memory banks of b-bit $\frac{n}{m}$ words store a row of the CKY table that is currently being processed. When $N[i,j]$ is computed, these m memory banks are storing the j-th row $N[1,j], N[2,j], \ldots, N[j,j]$ of the CKY table. More precisely, $N[l+1,j], N[l+m+1,j], N[l+2m+1,j], \ldots$ are stored in the l-th bank $(0 \leq l \leq m)$. Thus, m evaluations of $\otimes_G$, say, $N[1,1] \otimes_G N[2,j], N[1,2] \otimes_G N[3,j], \ldots, N[1,m] \otimes_G N[m+1,j]$, can be evaluated in a clock cycle because $N[1,1], N[2,j], N[1,2], N[3,j], \ldots, N[1,m], N[m+1,j]$ are stored in distinct memory banks. This allows us to accelerate the CKY parsing by a factor of m. Thus, the computing time for the CKY parsing is $O(\frac{n^3 \log b}{m})$ for $m \leq n$.

4 Performance Evaluation

We have evaluated the performance our instance-specific solution using the timing analyzer of Quartus II and tested it using an APEX20K series FPGA(EP20K-400EBC652-1X, typical 400K gates with 200Kbits embedded memory and 16K logic elements). In order to evaluate the performance of our instance-specific solution, we have implemented two software solutions and measured the performance on a 1.7GHz Pentium4-PC using Linux OS (Kernel 2.4.9). More specifically, we first evaluate the performance of both software and hardware solutions to compute the function $\otimes_G$. Next, we show the performance evaluation for the CKY parsing algorithm.

Figure 4(a) shows the running time of our hardware and software implementations to compute the function $\otimes_G$. Note that a word of data on a Pentium-based PC is 32-bit. Thus, we have implemented the 32-bit vector using a single word and the 64-bit vector using two words. As a consequence, the two word implementation for the 64-bit vector adds an overhead which makes it slower than the 32-bit vector solution. Recall that the naive algorithm checks whether or not $B \in U$ and $C \in V$ for every production rule $A \rightarrow BC$ in P. Hence, the computing time of the naive algorithm is proportional to the number of production rules.

As for the table algorithm, the computing time obeys a more regular pattern since the running time does not depend on the number of rules but rather it depends on the number of times it has to access the table. Recall that the table algorithm has to perform $(\frac{b}{c})^2$ table look-ups for b non-terminal symbols to compute $\otimes_G$. Thus, by increasing the value of b, the running time of the table algorithm also increases. As expected, for small values of p, the running time of the naive algorithm beats the table algorithm. However, as the number of p increases, the table algorithm is much faster than the naive algorithm.

In computing the function $\otimes_G$, our hardware implementation attained a speed-up of nearly 1000 over the table algorithm, using 64-bit vector approach. A speed-up of nearly 100 is observed using 32-bit vector approach. Comparing the results with the naive algorithm, the gain is even more apparent: for $p = 16384$, our hardware implementation attained a speed-up of nearly 22,000, over the naive algorithm using 64-bit vector approach, and a speed-up of nearly 7,300 using 32-bit vector approach. Since the running time or our hardware im-

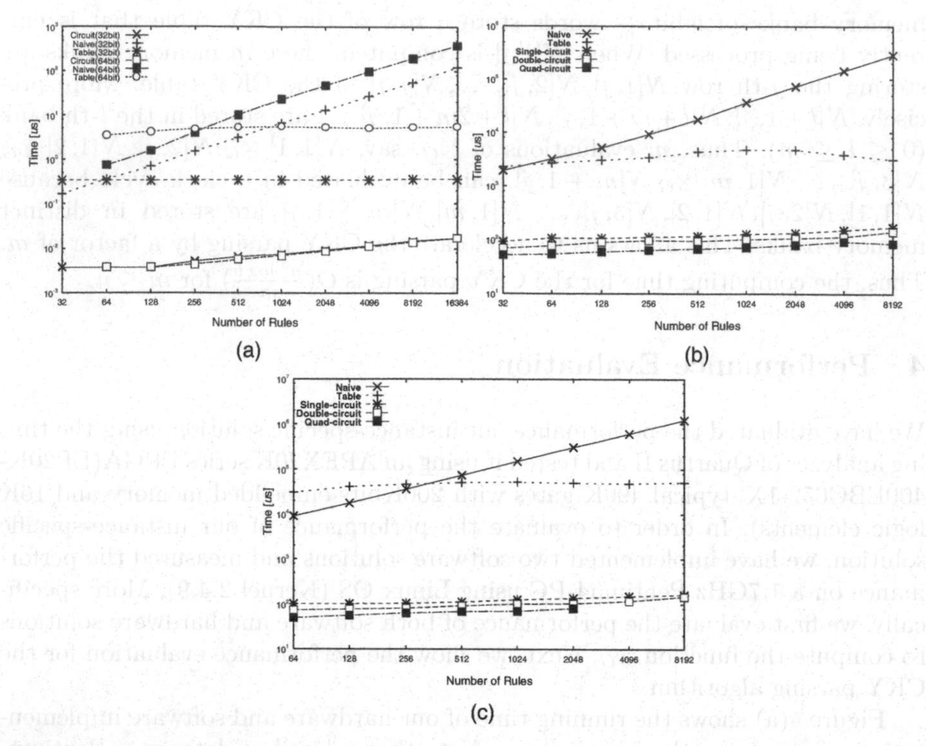

Fig. 4. (a) Computing time to evaluate $\otimes_G$ (b) Computing time of the CKY algorithm with $b = 32$ and $l = 32$ (c) Computing time of the CKY algorithm with $b = 64$ and $l = 32$

plementation is independent of the number of encoding bits, the 32-bit vector and the 64-bit vector approaches have nearly the same running time.

Figure 4(b) shows the computing time of the CKY algorithm for $b = 32$ and $l = 32$ (where l represents the length of the input string). The figure shows the computing time for the sequential algorithms as well as for the hardware implementation. We have implemented the CKY algorithm in hardware using single-, double-, and quad-circuit and plotted the running times of both hardware and software approaches. The software solutions have followed the same pattern observed in Figure 4(a). This is due to the fact that a good portion of the computation time is spent evaluating $\otimes_G$. Our hardware implementation for single-circuit also follows from the previous figure. One can see that the double- and quad-circuit approaches indeed accelerate the CKY parsing as we have predicted in Section 3.

Figure 4(c) shows the computing time of the CKY algorithm for $b = 64$ and $l = 32$. We observed that the running time follows the same pattern of the CKY algorithm for $b = 32$ and $l = 32$. As mentioned before, the 64-bit vector approach adds an extra overhead to the software solutions which does not occur on the hardware implementations. As a result, we observe a degradation on the running time of the software solutions. The number of logic elements necessary to compute $p = 2048$, using a quad-circuit, is nearly 9,600. For $p = 4096$, the number of logic elements necessary to build the quad-circuit surpasses the overall number of logic elements provided by our FPGA. Because of this, we have been able to implement the quad-circuit for p up to 2048.

Table 1 shows the speed-up of the CKY algorithm over the table algorithm (software approach). The computing time of each algorithm can be seen in the previous figures. For $b = 32$ and $l = 32$, our hardware approach achieved speed-up of nearly: 40 using a single-circuit; 50 using a double-circuit; and 70 using a quad-circuit. Our results are even more appealing for $b = 64$ and $l = 32$. In this case, our hardware approach achieved speed-up of nearly: 460 using a single-circuit; 580 using a double-circuit; and 750 using a quad-circuit. Thus, from the above results, we argue that our hardware approach is indeed a promising solution to solve the CKY parsing.

Table 1. Speed-up of the CKY hardware approach over the CKY table algorithm

	$b = 32, l = 32$			$b = 64, l = 32$		
p	Single	Double	Quad	Single	Double	Quad
32	25	33	44	–	–	–
64	25	35	50	304	419	611
128	29	34	51	395	519	731
256	30	35	53	441	577	730
512	37	46	64	454	552	736
1024	38	48	66	413	513	742
2048	36	45	60	362	475	600
4096	26	34	41	326	418	–
8196	18	22	37	314	348	–

References

[1] A. V. Aho and J. D. Ullman. *The Theory of Parsing Translation and Compiling.* Prentice Hall, 1972. 42

[2] Altera Corporation, APEX 20K Devices: System-on-a-Programmable-Chip Solutions, http://www.altera.com/products/devices/apex/apx-index.html. 42

[3] Altera Corporation. Quartus II:system-on-a-programmable chip sowtfware. http://www.altera.com/products/software/quartus2/qts-index.html 42

[4] J. Chang, O. Ibarra, and M. Palis. Parallel parsing on a one-way array of finite-state machines. *IEEE Transactions on Computers*, C-36(1):64–75, 1987. 42

[5] E. Charniak. *Statistical Language Learning.* MIT Press, Cambridge, Massachusetts, 1993. 42

[6] C. Ciressan, E. Sanchez, M. Rajman, and J.-C. Chappelier. An FPGA-based coprocessor for the parsing of context-free grammars. In *Proc. of IEEE Symposium on Field-Programmable Custom Computing Machines*, 2000. 42

[7] C. Ciressan, E. Sanchez, M. Rajman, and J.-C. Chappelier. An FPGA-based syntactic parser for real-life almost unrestricted context-free grammars. In *Proc. of International Conference on Field Programmable Logic and Applications (FPL)*, pages 590–594, 2001. 42

[8] Y. Futamura, K. Nogi, and A. Takano. Essence of generalized partial computation. *Theoretical Computer Science*, 90:61–79, 1991. 42

[9] A. Gibbons and W. Rytter. *Efficient Parallel Algorithms.* Cambridge University Press, 1988. 42

[10] N. D. Jones, C. K. Gomard, and P. Sestoft. *Partial Evaluation and Automatic Program Generation.* Prentice Hall, 1993. 41

[11] S. R. Kosaraju. Speed of recognition of context-free languages by array automata. *SIAM J. on Computers*, 4:331–340, 1975. 42

[12] J. C. Martin. *Introduction to languages and the theory of computation (2nd Edition).* Mac-Graw Hill, 1996. 42

[13] Y. Sakakibara, M. Brown, R. Hughey, I. S. Mian, K. Sjölander, R. C. Underwood, and D. Haussler. Stochastic context-free grammars for tRNA modeling. *Nucleic Acids Research*, 22:5112–5120, 1994. 42

[14] M. P. van Lohuizen. Survey on parallel context-free parsing techniques. Technical Report IMPACT-NLI-1997-1, Delft University of Technology, 1997. 42

Duplication-Based Scheduling Algorithm for Interconnection-Constrained Distributed Memory Machines

Savina Bansal[1], Padam Kumar, and Kuldip Singh

Dept. of Electronics & Computer Engineering, IIT, Roorkee
Roorkee (Uttranchal), INDIA-247 667
{saavidec,padamfec,ksconfec}@iitr.ernet.in

Abstract. Duplication-based scheduling techniques are more appropriate for fine grain task graphs and for networks with high communication latencies. However, most of the algorithms are developed under the assumption of fully connected processor network and with prohibitively high $O(v^4)$ time complexity. An insertion based duplication algorithm is proposed for precedence constrained task graphs, for working with limited interconnection constrained processors. It duplicates only the most important immediate parents of a task, that too if critical. Results are presented for benchmark random task graphs, having widely varying shape and cost parameters for the clique, Hypercube and an extensible and fault tolerant binary de Bruijn (undirected) multiprocessor network. The average performance degradation, due to interconnection constraints, is about 21% in comparison to fully connected processor network. Further, the schedules generated on the fixed degree binary de-Bruijn network are within 5% of the schedules on Hypercube network, whose degree keeps on increasing with size.

1 Introduction

Distributed memory machines (DMMs) are becoming quite popular towards solving the complex scientific problems, which have been shown to possess a vast potential for concurrent computing. Advances in VLSI technology, low latency communication interfaces, networking protocols and routing algorithms have contributed a lot for the same. Scheduling is one of the most challenging problems for such systems. It is a well known NP-complete problem in a broad sense, and polynomial time solutions are known to exist for very limited cases, that too under highly restrictive constraints [4][7]. As a result, many heuristic based algorithms developed over the last four decades tend to sacrifice optimality for the sake of

[1] On QIP study leave from Electronics Engg. Deptt. of GZS College of Engg. & Technology, Bathinda (Punjab –151001, INDIA)

S. Sahni et al. (Eds.) HiPC 2002, LNCS 2552, pp. 52-62, 2002.

efficiency [1][2][3][4][10][13]. Inter Processor Communication (IPC), however, continues to pose the performance limitations on these algorithms.

Scheduling heuristics can broadly be classified as - list or priority based, clustering based, duplication based and guided search based. In the classical list scheduling technique, tasks are assigned static or dynamic priority based on their computation and communication costs [3][4][8][10]. The two frequently used priority terms are *b-level* and *t-level*. The *b-level* of a task in the given task graph, is the sum of communication and computation cost along the longest path from the concerned node to the exit node, while the *t-level* refers to the same value from entry node to this node (excluding its computation cost). The highest priority task among the available tasks (the set of ready to schedule tasks with satisfied precedence constraints) is chosen, followed by the selection of the most suitable processor for assignment to this task.

Clustering-based techniques, on the other hand, generate the schedule in two steps [12]. In the first step, heavily communicating tasks are grouped into a common cluster under the ambience of a clustering heuristic and in the second step, these clusters are mapped onto a set of available processors using communication sensitive or insensitive mapping heuristics. A good comparison work on clustering and mapping heuristics is available in [1] [6].

Duplication-based heuristics are useful for message passing distributed systems, where high communication latency and low bandwidth are the major concern. Here, some/all of the crucial tasks are duplicated on more than one processor with an objective to reduce the starting time of the dependent tasks. Most of the effective duplication algorithms, however, come under the $O(v^4)$ time complexity [2][4][5], and are based on the assumption of fully connected network. The low complexity algorithms, on the other hand, have been suggested with some sort of cost constraints in the application task graph to generate optimal schedules [11]. A comprehensive comparison work on duplication algorithms is available in [5].

Search based techniques, which try to refine an initial schedule, in the available solution space, using some search guiding heuristics have also been suggested at times, though to a limited extent. A recent work by Wu et. al is available in [9].

The objective of the proposed work is to develop a duplication based technique that exploits scheduling holes to generate shorter schedules, for arbitrary cost task graphs with widely varying CCRs (communication to computation cost ratio), and still use limited number of interconnection constrained processors. The paper is organized as follows: Section 2 of the paper outlines the scheduling problem and describes the task and machine model used. Section 3 expounds the proposed technique with Section 4 presenting and comparing the performance results. The section 5 concludes the proposed work.

2 Task and Machine Model

In deterministic scheduling, the problem is assumed to have a natural expression in the form of a graph, which embodies its computational and communication requirements. More formally, the task model can be represented by a quadruple Q = (T, R, $[c_{ij}]$, $[w_i]$), where T = {t_i: i=1, 2..., v} is a set of 'v' tasks, R represents a relation that defines a partial order on the task set T, such that if t_i R t_j then it implies that the task t_i must finish before t_j can start execution. The v x v cost matrix $[c_{ij}]$ gives the amount of dataflow between task t_i and t_j for 1 i, j v and $[w_i]$ is a v vector of computation requirements. A typical task model, as represented by a directed acyclic graph (DAG), is shown in Fig. 1, with vertices representing the tasks and edges the dataflow paths and the corresponding weights on them the computation and communication costs respectively. Tasks without ancestors are called entry tasks and those without descendents are called exit nodes. The term node and task have been used interchangeably in the paper.

The target machine model is represented by a 3-tuple M = (P, $[d_{ij}]$, $[L_{ij}]$) with P $=\{p_i: i=0,1\ldots p\}$ being a set of p processors, $[L_{ij}]$ is a p x p Boolean matrix describing interconnection network topology and $[d_{ij}]$ is a p x p distance matrix giving minimum distance in number of hops between processor p_i and p_j. Communication links are assumed to be contention free to accommodate the deterministic nature of scheduling. Further, IPC is through I/O channels thereby allowing concurrent computation and communication. The communication overhead between two tasks scheduled on the same processor is taken as zero.

The objective function of the task scheduling is to map each task (t_i) in the task graph to a processor (p_j) in the processor network and to assign a starting time such that the overall execution time of the problem i.e. schedule length (SL), is minimized.

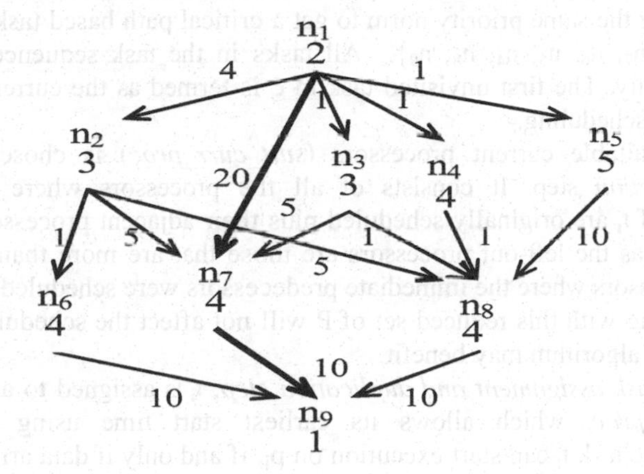

Fig. 1. A simple precedence constrained task graph represented by a DAG[5]

3 Proposed Duplication Technique

Duplication-based task scheduling algorithms are more effective for distributed memory machines where high communication latency is an integral part of the system. Full duplication algorithms [2][5] work by duplicating all possible parents/ancestors of the node under consideration, giving high-quality schedules, though at the cost of prohibitively high time complexity $O(v^4)$ and excessive processor consumption (for a 500 node random graph the average processor consumption is more than 100 [5]). The authors suggested a cost effective duplication scheme in [15] for the fully connected network. In the present work, we are extending the heuristic to take into account the interconnection constraints present in the processor network. In the presence of interconnection constraints, the duplication schemes, however, tend to become more complex.

The algorithm completes in three main steps. In the *priority-assignment* step, an ordered task sequence satisfying the precedence constraints is generated. A critical path (CP), the longest sequential path taking into account computation and communication costs in the task graph, is selected first. The ties are first broken on the basis of higher computation costs, and then on the basis of higher number of immediate predecessors of the critical path nodes i.e. a critical path based on maximum immediate predecessors first (CP/MIPF). It is speculated that such a scheme will benefit the task graphs with multiple independent CPs, as the CP that spans more task graph nodes will be selected and those tasks will be given higher priority for scheduling in the limited resource environment. The task sequence $\{n_1, n_7, n9\}$ constitutes the CP for the task graph of Fig. 1.

To satisfy the precedence constraints, predecessor nodes must be executed before the successor node. So, the predecessors are added in the CP sequence with priority being decided on the basis of higher b-level and the ties being broken on the basis of lower t-level, as in [5]. Nodes that are neither CP nodes nor their parents are appended afterward using the same priority norm to get a critical path based task sequence as, $\xi = \{n_1, n_3, n_2, n_7, n_6, n_5, n_4, n_8, n_9\}$. All tasks in the task sequence ξ are marked unvisited initially. The first unvisited task in ξ is termed as the current task t_i and is considered for scheduling.

A set of suitable current processors (*suit_curr_proc*) is chosen next, in the *processor-selection* step. It consists of all the processors where the immediate predecessors of t_i are originally scheduled plus their adjacent processors. This seems to be rational, as the left-out processors are those that are more than one hop away from the processors where the immediate predecessors were scheduled first of all. It is felt that working with this reduced set of P will not affect the schedules much, while the run-time of algorithm may benefit.

In the last *task assignment and duplication step*, t_i is assigned to a processor p_j in the *suit_curr_proc,* which allows its earliest start time using insertion-based duplication. The task t_i can start execution on p_j, if and only if data arrives from all of its immediate parents (pred (t_i)), so as to satisfy the precedence constraints. The data arrival time for t_i on p_j i.e. DAT (t_i, pj), for a partially connected network is calculated as,

$$DAT(t_i, p_j) \quad max\{\underset{t_y \;\; pred(t_i),\;\; p_k \;\; dupproc(t_y)}{min(FT(t_y, p_k) \quad c_{yi} * d_{kj})}\} \tag{1}$$

The set *dupproc(t_y)* corresponds to a set of processors where task t_y is duplicated. FT(t_y, p_k) is the finish time of task t_y on processor p_k. In case of partially connected networks, unlike clique topology [15], it is desirable to check data arrival times from all the immediate parents and from all the processors where they are duplicated due to unequal distances between the processors. The parent, from which the data arrives last of all on p_j, is termed as the most important immediate parent (MIIP) of t_i for p_j and is given by,

$$t_{MIIP(i,j)} \quad t_y, \ if\ DAT(t_y, p_j) \quad \max\{DAT(t_n, p_j)\} \qquad (2)$$
$$t_n \quad pred(t_i)\ and \quad t_n \quad t_y$$

Earliest start time of t_i on p_j is calculated as,

$$EST(t_i, p_j) \quad max\,\{DAT(t_i, p_j), RT\} \qquad (3)$$

RT corresponds to the start time of the first suitable scheduling hole s_m, if one exists. Otherwise, it relates to the finish time of the last task scheduled on p_j. A scheduling hole is the free time slot left in a processor as a result of non-availability of data earlier. A slot s_m, with time span sufficient to satisfy t_i's computational requirement w_i, is considered suitable, if it meets the following condition.

$SFT\,(s_m) \quad DAT\,(t_i, p_j) \quad w_i$, where $SFT(s_m)$ is the finish time of slot s_m (4)

For the early start of t_i, it is mandatory to reduce the data arrival time of its MIIP. So, we try to bring this MIIP on p_j, if not there already, provided a suitable scheduling hole exists. Start time of t_i is rechecked, for resulting improvements and the process repeated till no more MIIP is there or no suitable slot is there on current processor. Parents or ancestors of an MIIP are not duplicated to reduce the run time of algorithm. Further, if at any stage it is found that the duplication might increase or is not going to improve the start time of t_i, the duplication is not done, as shown in Fig. 2 (step r5).

Reduce_start_time (t_i, p_j)
//input: current task t_i, current processor p_j, Q, M
//output: earliest start time of t_i on p_j

step r1: A *Find the EST (t_i, p_j)*
step r2: t_k *Find the* $t_{MIIP(i,j)}$ *of* t_i *not on* p_j
step r3: If no such t_k *exists, return A*
else find suitable slot for t_k *on* p_j
step r4: If a suitable slot exists find the earliest finish time of t_k *on* p_j *as*
$EFT(t_k, p_j) \quad EST(t_k, p_j) + w_k$,
else return A
step r5: If $EFT(t_k, p_j)$ *is less than A, duplicate it on* p_j *and go to step r1*
else return A

Fig. 2. Pseudo code of Reduce_start_time (t_i, p_j) routine

```
SD algorithm
//Input: Q, M
//Output: Schedule of T on P

Step 1: Construct priority based task sequence ξ, and mark all tasks
as unvisited

0Repeat
    Step 2: Let t_i be the first unvisited task in ξ
    Step 3: Construct a set of suit_curr_proc for t_i
    Step 4: For each p_j in suit_curr_proc, call Reduce_start_time (t_i, p_j)
and record
            EST(t_i, p_j)
    Step 5: Schedule t_i to p_j that gives the minimum EST(t_i, p_j) and undo
duplications
            (if any) done on all other processors in suit_curr_proc
    Step 6: Mark t_i as visited
    Until   all tasks in task sequence ξ are visited

1end
```

Fig. 3. Pseudo code of the proposed SD algorithm for interconnection constrained networks

The pseudo code for the proposed Selective-Duplication based scheduling algorithm (SD) for interconnection constrained networks, is given in Fig. 3. The complexity of the algorithm works out to be $O(pv^2d_{\max})$ for a task graph, with v much greater than the maximum number of immediate predecessors (d_{max})/successors in it.
The running trace of the algorithm, for a Hypercube (HC) network, is illustrated in Table 1 for the task graph of Fig. 1, which was used by Ahmad et. al for comparing various full duplication based algorithm in [5]. The resulting Gantt charts for HC and binary de Bruijn (undirected) (BDM(UD)) [14] networks are shown in Fig. 4.

Table 1. Running trace of SD algorithm on HC network for task graph of Fig. 1

Step	Task sequence	Start times on current processors P_0	P_1	P_2	P_3	MIIP/s duplicated
1	n_1	0	0*	0	0	none
2	n_3	2*	2	NC	2	n_1
3	n_2	5	2*	NC	2	none
4	n_7	8*	8	8	8	n_2
5	n_6	12	5*	NC	6	none
6	n_5	12	9	NC	2*	n_1
7	n_4	12	9	NC	7*	none
8	n_8	17	14	12	11*	none
9	n_9	31	21	22	19*	n_7

* Earliest start time selected NC - not considered

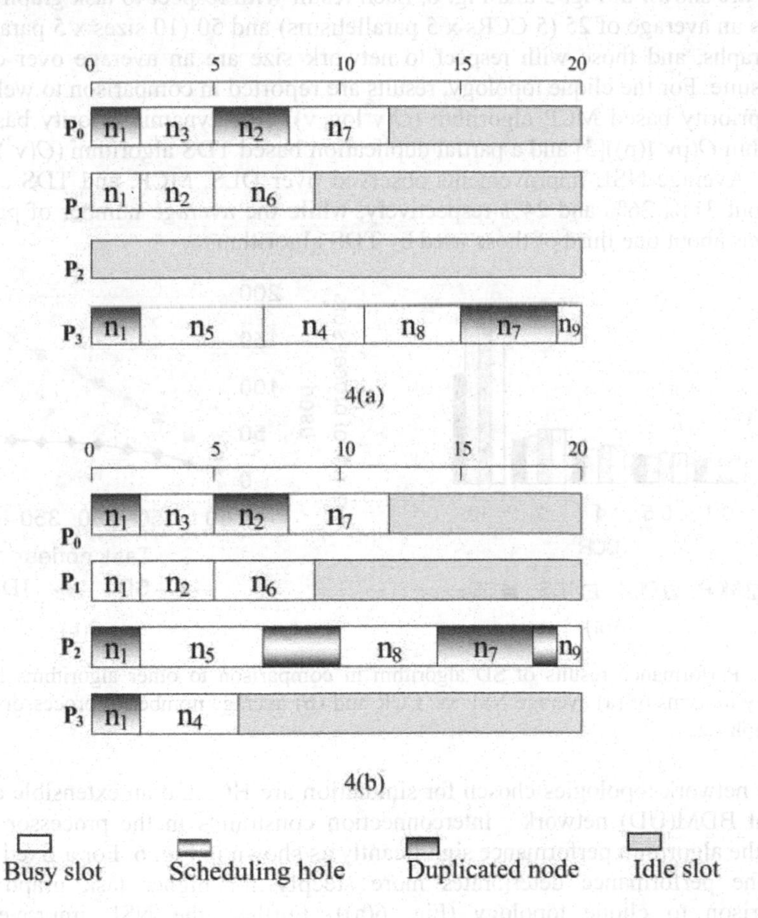

Fig. 4. Gantt charts for the schedule generated by SD algorithm on (a) HC network (SL=20) and (b) BDM(UD) network (SL=20) with diameter=2 for the task graph of Fig. 1

4 Performance Results

The performance of the proposed algorithm has been simulated on the benchmark random graphs with different network topologies. The random test graph suite [13] consists of 250 graphs with ten different sizes, from 50 to 500 with an increment of 50. Corresponding to each size, there are five widely varying CCRs: 0.1, 0.5, 1.0, 2.0, 10.0, and five different average parallelisms as $n\sqrt{v}$ (with n=1, 2, 3, 4 and 5). The performance is measured in terms of normalized schedule length (NSL) and the number of processors used. NSL is defined as,

$$NSL \quad \frac{\text{schedule length obtained by using a particular algorithm}}{\text{maximum computation cost along any path in the task graph}} \tag{5}$$

The denominator represents the lower bound on the schedule length. The performance results are shown in Fig. 5 and Fig. 6. Each result with respect to task graph size and CCR is an average of 25 (5 CCRs x 5 parallelisms) and 50 (10 sizes x 5 parallelisms) task graphs, and those with respect to network size are an average over complete graph suite. For the clique topology, results are reported in comparison to well-known static priority based MCP algorithm ($O(v^2 \log v)$) [8], dynamic priority based DLS algorithm $O(pv^3 f(p))$[3] and a partial duplication based TDS algorithm ($O(v^2)$) [11] in Fig. 5. Average NSL improvements observed over DLS, MCP, and TDS algorithm are about 31%, 26% and 24% respectively, while the average number of processors used was about one third of those used by TDS algorithm.

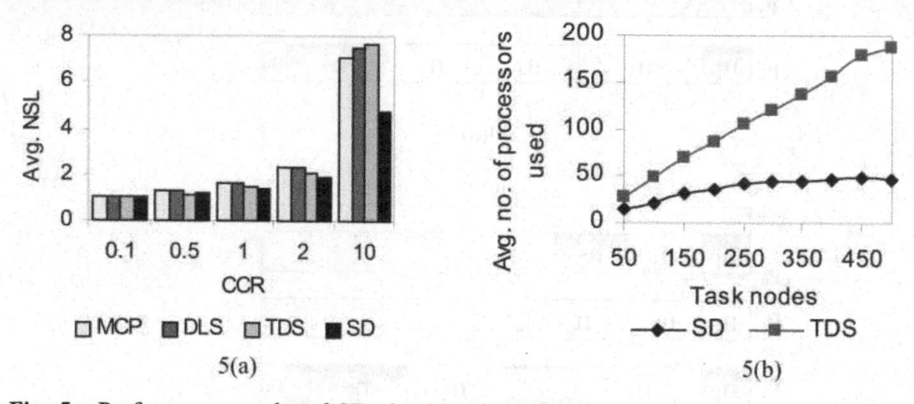

Fig. 5. Performance results of SD algorithm in comparison to other algorithms for clique topology in terms of (a) average NSL vs. CCR and (b) average number of processors used vs. task graph size

The network topologies chosen for simulation are HC, and an extensible and fault tolerant BDM(UD) network. Interconnection constraints in the processor network affect the algorithm performance significantly as shown in Fig. 6. For a fixed network size, the performance deteriorates more steeply for higher task graph size in comparison to clique topology (Fig. 6(a)). Further, the NSL improvement is insignificant beyond certain network size (Fig. 6(c)). This is expected as increase in network size results in increased diameter and hence adds to the communication overheads. The algorithm trades off between available parallelism and IPC, and tends to confine itself to lesser processors. The similar logic applies to performance deterioration with higher task graph size, which may force the heuristic to choose processors that are more than one hop apart. However, an interesting comparison is reflected in Fig. 6(d). The performance of fixed degree BDM(UD) network remains within 5% of that of HC network, whereas the degree of HC increases with the network size ($=2^{degree}$). Results of fully connected network have been supplemented for the sake of network performance comparison.

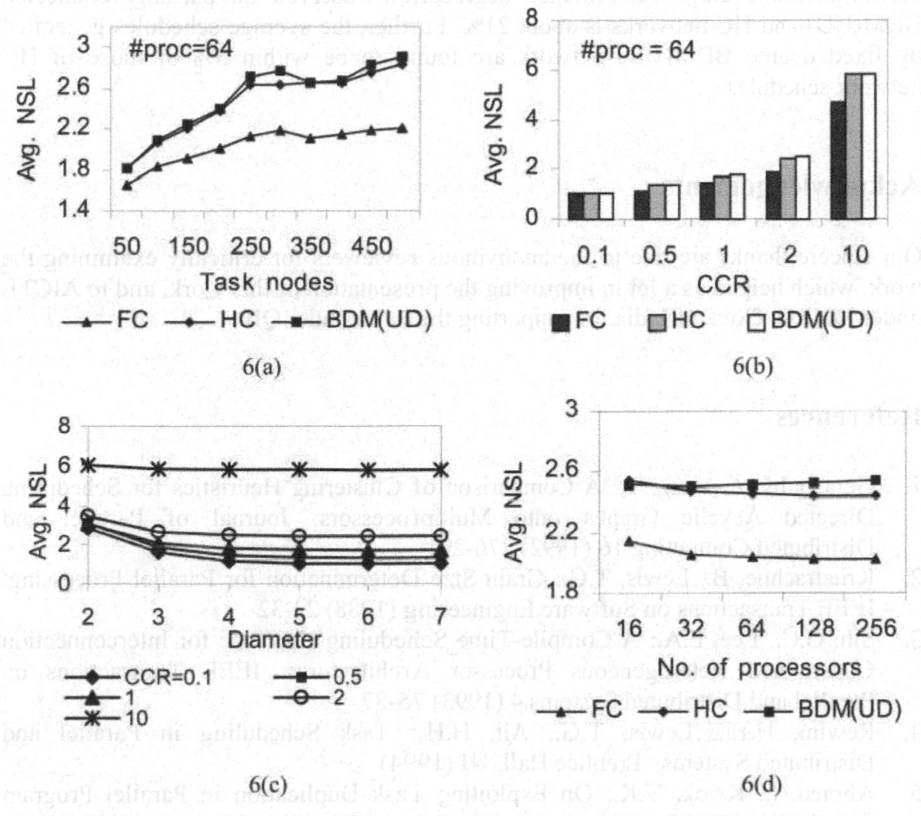

Fig. 6. Performance results of SD algorithm on random graph suite for fully connected (FC), HC and BDM(UD) networks, in terms of average NSL vs. (a) task nodes (b) CCR (c) diameter of HC network for different CCRs and (d) number of processors in the network

Duplication techniques are more effective for fine grain task graphs due to the formation of large sized scheduling holes. Efficient utilization of these holes plays a decisive role in the performance of a heuristic. The better results of the proposed duplication technique are attributed to the amalgamation of best features of the priority selection heuristic in the first step and the judicious exploitation of these scheduling holes in the last step.

5 Conclusions

A Duplication-based scheduling algorithm is developed for interconnection constrained processor networks. The simulation results in terms of average NSL are reported, under the stated assumptions, for benchmark random graphs on the clique, HC and BDM(UD) networks. The average performance results show an impressive schedule length improvement with significantly lesser processor consumption in comparison to the well-known algorithms. In comparison to fully connected networks, the average performance degradation observed on partially connected BDM(UD) and HC networks is about 21%. Further, the average schedules generated by fixed degree BDM(UD) network are found to be within 5% of those of HC network schedules.

Acknowledgements

Our sincere thanks are due to the anonymous reviewers for critically examining the work, which helped us a lot in improving the presentation of this work, and to AICTE under MHRD, Govt. of India for supporting the work under QIP.

References

1. Gerasoulis, A.,Yang, T.: A Comparison of Clustering Heuristics for Scheduling Directed Acyclic Graphs onto Multiprocessors. Journal of Parallel and Distributed Computing 16 (1992) 276-291
2. Kruatrachue, B., Lewis, T.G.: Grain Size Determination for Parallel Processing. IEEE Transactions on Software Engineering (1988) 23-32
3. Sih, G.C., Lee, E.A.: A Compile-Time Scheduling Heuristic for Interconnection Constrained Heterogeneous Processor Architectures. IEEE Transactions on Parallel and Distributed Systems.4 (1993) 75-87
4. Rewini, H.El., Lewis, T.G., Ali, H.H.: Task Scheduling in Parallel and Distributed Systems. Prentice Hall, NJ (1994)
5. Ahmed, I., Kwok, Y.K.: On Exploiting Task Duplication in Parallel Program Scheduling. IEEE Transactions on Parallel and Distributed Systems (1998) 872-892
6. Dikaiakos, M.D., et. al.: A Comparative Study of Heuristics for Mapping Parallel Algorithms to Message Passing Multiprocessors. Tech. Report Princeton University (1994)
7. Garey, M.R., Johnson, D.S.: Computers and Intractability: A Guide to the Theory of NP-Completeness, W.H. Freeman and Co. (1979)
8. Wu, M.Y., Gajski, D.S.: Hypertool: A Programming Aid for Message-Passing Systems. IEEE Transactions on Parallel and Distributed Systems 1(1990) 330-343
9. Wu, M.Y., Shu, W., Gu, J.: Efficient Local Search for DAG Scheduling. IEEE Transactions on Parallel and Distributed Systems 12 (2001) 617-627

10. Sevalkumar, S., Ramamoorthy, C.V.: Scheduling Precedence Constrained Task Graphs with Non-Negligible Inter Task Communication onto Multiprocessors. IEEE Transactions on Parallel and Distributed Systems (1994) 328-336
11. Darbha, S., Agrawal, D.P.: Optimal Scheduling Algorithm for Distributed Memory Machines, IEEE Transactions on Parallel and Distributed systems (1998) 87-95
12. Yang T., Gerasoulis, A.: DSC: Scheduling Parallel Tasks on an Unbounded Number of Processors. IEEE Transactions on Parallel and Distributed Systems 5 (1994) 951-967
13. Kwok, Y.K., Ahmad, I.: Benchmarking and Comparison of the Task Graph Scheduling Algorithms. Journal of Parallel and Distributed Computing 59 (1999) 381-422
14. Samatham, M.R., Pradhan, D.K.: The de Bruijn Multiprocessor Network: A Versatile Parallel Processing and Sorting Network for VLSI. IEEE Transactions on Computers 38 (1989) 567-581
15. Bansal, S., Kumar, P., Singh, K.: A Cost-effective Scheduling Algorithm for Message Passing Multiprocessor Systems. (Accepted for publication in Proc. PDCS-2002, Louisville, Kentucky, USA, Sept. 19-21, 2002)

Evaluating Arithmetic Expressions Using Tree Contraction: A Fast and Scalable Parallel Implementation for Symmetric Multiprocessors (SMPs) (*Extended Abstract*)

David A. Bader*, Sukanya Sreshta, and Nina R. Weisse-Bernstein**

Department of Electrical and Computer Engineering
University of New Mexico, Albuquerque, NM 87131 USA

Abstract. The ability to provide uniform shared-memory access to a significant number of processors in a single SMP node brings us much closer to the ideal PRAM parallel computer. In this paper, we develop new techniques for designing a uniform shared-memory algorithm from a PRAM algorithm and present the results of an extensive experimental study demonstrating that the resulting programs scale nearly linearly across a significant range of processors and across the entire range of instance sizes tested. This linear speedup with the number of processors is one of the first ever attained in practice for intricate combinatorial problems. The example we present in detail here is for evaluating arithmetic expression trees using the algorithmic techniques of list ranking and tree contraction; this problem is not only of interest in its own right, but is representative of a large class of irregular combinatorial problems that have simple and efficient sequential implementations and fast PRAM algorithms, but have no known efficient parallel implementations. Our results thus offer promise for bridging the gap between the theory and practice of shared-memory parallel algorithms.

Keywords: Expression Evaluation, Tree Contraction, Parallel Graph Algorithms, Shared Memory, High-Performance Algorithm Engineering.

1 Introduction

Symmetric multiprocessor (SMP) architectures, in which several processors operate in a true, hardware-based, shared-memory environment and are packaged as a single machine, are becoming commonplace. Indeed, most of the new high-performance computers are clusters of SMPs having from 2 to over 100 processors per node. The ability to provide uniform-memory-access (UMA) shared-memory

* Supported in part by NSF Grants CAREER ACI-00-93039, ITR ACI-00-81404, DEB-99-10123, ITR EIA-01-21377, and Biocomplexity DEB-01-20709.
** Supported by an NSF Research Experience for Undergraduates (REU).

S. Sahni et al. (Eds.) HiPC 2002, LNCS 2552, pp. 63–75, 2002.

for a significant number of processors brings us much closer to the ideal parallel computer envisioned over 20 years ago by theoreticians, the *Parallel Random Access Machine (PRAM)* (see [13, 19]) and thus may enable us at last to take advantage of 20 years of research in PRAM algorithms for various irregular computations. Moreover, as supercomputers increasingly use SMP clusters, SMP computations will play a significant role in supercomputing. For instance, much attention has been devoted lately to OpenMP [17], that provides compiler directives and runtime support to reveal algorithmic concurrency and thus takes advantage of the SMP architecture; and to mixed-mode programming, that combines message-passing style between cluster nodes (using MPI) and shared-memory style within each SMP (using OpenMP or POSIX threads).

While an SMP is a shared-memory architecture, it is by no means the PRAM used in theoretical work—synchronization cannot be taken for granted and the number of processors is far smaller than that assumed in PRAM algorithms. The significant feature of SMPs is that they provide much faster access to their shared-memory than an equivalent message-based architecture. Even the largest SMP to date, the recently delivered 106-processor Sun Fire Enterprise 15000 (E15K), has a worst-case memory access time of 450ns (from any processor to any location within its 576GB memory); in contrast, the latency for access to the memory of another processor in a distributed-memory architecture is measured in tens of μs. In other words, message-based architectures are two orders of magnitude slower than the largest SMPs in terms of their worst-case memory access times.

The largest SMP architecture to date, the Sun E15K [5] (a system three- to five-times faster than its predecessor, the E10K [4]), uses a combination of data crossbar switches, multiple snooping buses, and sophisticated cache handling to achieve UMA across the entire memory. Of course, there remains a large difference between the access time for an element in the local processor cache (around 10ns) and that for an element that must be obtained from memory (at most 450ns)—and that difference increases as the number of processors increases, so that cache-aware implementations are even more important on large SMPs than on single workstations. Fig. 1 illustrates the memory access behavior of the Sun E10K (right) and its smaller sibling, the E4500 (left), using a single processor to visit each 32-bit node in a circular array. We chose patterns of access with a fixed stride, in powers of 2 (labeled `C, stride`), as well as a random access pattern (labeled `R`). The data clearly show the effect of addressing outside the on-chip cache (the first break, at a problem size of size greater than 2^{12} words, or 16KB — the size of L1 cache) and then outside the L2 cache (the second break, at a problem size of greater than 2^{20} words, or 4MB). The uniformity of access times was impressive—standard deviations around our reported means are well below 10 percent. Such architectures make it possible to design algorithms targeted specifically at SMPs.

Arithmetic Expression Evaluation (*AEE*) has important uses in a wide-range of problems ranging from computer algebra and evaluation of logical queries to compiler design. Its classical parallel formulation uses the technique of tree

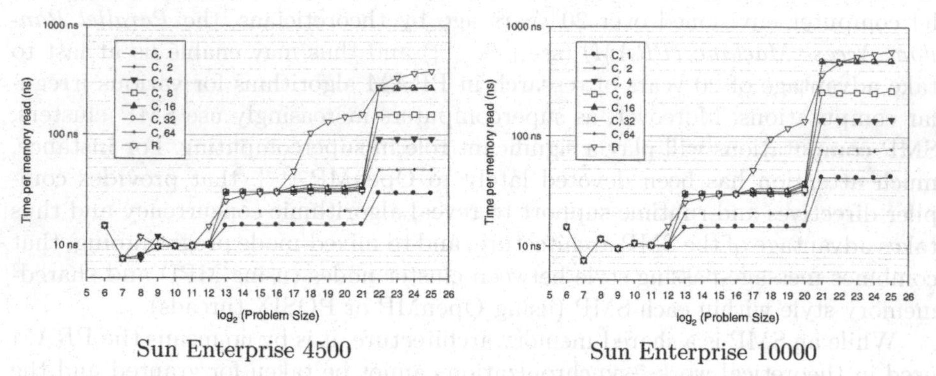

Fig. 1. Memory access (read) time using one 400 MHz UltraSPARC II processor of a Sun E4500 (left) and an E10K (right) as a function of array size for various strides

contraction when the expression is represented as a tree with a constant at each leaf and an operator at each internal vertex. AEE involves computing the value of the expression at the root of the tree. Hence, AEE is a direct application of the well studied tree contraction technique, a systematic way of shrinking a tree into a single vertex.

Miller and Reif [16] designed an exclusive-read exclusive-write (EREW) PRAM algorithm for evaluating any arithmetic expression of size n, which runs in $O(\log n)$ time using $O(n)$ processors (with $O(n \log n)$ work). Subsequently Cole and Vishkin [6] and Gibbons and Rytter [8] independently developed $O(\log n)$-time $O(n/\log n)$-processors (with $O(n)$ work) EREW PRAM algorithms. Kosaraju and Delcher [15] developed a simplified version of the algorithm in [8] which runs in the same time-processor bounds ($O(\log n)$-time $O(n/\log n)$-processors (with $O(n)$ work) on the EREW PRAM). Recently, several researchers in [3, 7] present theoretic observation that this classical PRAM algorithm for tree contraction on a tree T with n vertices can run on the Coarse-Grained Multicomputer (CGM) parallel machine model with p processors in $O(\log p)$ communication rounds with $O\left(\frac{n}{p}\right)$ local computation per round.

2 Related Experimental Work

Several groups have conducted experimental studies of graph algorithms on parallel architectures (for example, [11, 12, 14, 18, 20, 9]). However, none of these related works use test platforms that provide a true, scalable, UMA shared-memory environment and still other studies have relied on *ad hoc* hardware [14]. Thus ours is the first study of speedup for over tens of processors (and promise to scale over a significant range of processors) on a commercially available platform. In a recent work of ours ([2]) we study the problem of decomposing graphs with the ear decomposition using similar shared-memory platforms.

Our work in this paper focuses on a parallel implementation of the tree contraction technique specific to the application of expression evaluation on symmetric multiprocessors (SMPs). The implementation is based on the classic PRAM algorithm (e.g., see [15] and [13]). We begin with the formal description of the parallel algorithm implemented. Next we detail the implementation including the empirical and testing environment and present results of the experiments. The last section provides our conclusions and future work.

3 The Expression Evaluation Problem

AEE is the problem of computing the value of an expression that is represented as a tree with a constant at each leaf and an operator at each internal vertex of the tree. For a parallel formulation, we use the tree contraction technique to shrink the tree to its root. For simplicity, our study is restricted to expressions with binary associative operators. Hence, the trees are binary with each vertex u excluding the root of the tree having only one sibling $sib(u)$. This technique can also be used with expressions having non-binary operators by considering unary operators with their identity elements and converting general trees resulting from ternary operators to binary trees as a preprocessing phase of the algorithm [13]. For the sake of discussion and without loss of generality, let us assume that the internal vertices contain either the addition operator $+$ or the multiplication operator $\times$.

The simple parallel solution of evaluating each subexpression (two sibling leaves and their parent) in parallel and setting the parents of the vertices evaluated equal to the value of the subexpression until the root is reached works well when the tree is well-balanced. In the extreme case of a "caterpillar" tree, when the tree is a long chain with leaves attached to it, this solution requires a linear number of iterations similar to the sequential solution. Hence, an optimal solution should ease the above restriction that each vertex must be fully evaluated before its children can be removed.

The *rake* operation is used to remove the vertices from the tree, thus contracting it. Let T be a rooted binary tree with root r and let $p(v)$, $sib(v)$ and $p(p(v))$ represent the parent, sibling, and grandparent of a vertex v respectively. The rake operation when applied to a leaf v ($p(v) \neq r$) of the tree removes v and $p(v)$ from T and connects $sib(v)$ to $p(p(v))$. The rake operation is illustrated in Fig. 2.

The value of a vertex v, denoted by $val(v)$ is defined as the value of the subexpression (subtree rooted) at v. The value of a leaf is simply the constant value stored in the leaf. To accomplish partial evaluation, each vertex v of the tree T is associated with a label (a_v, b_v) such that the contribution of a vertex to its parent's value is given by the expression $a_v val(v) + b_v$. The label of each vertex is initialized to $(1, 0)$. Let u be an internal vertex of the tree such that u holds the operator $\oplus \in \{+, \times\}$ and has left child v and right child w. The value of vertex u is given by $val(u) = (a_v val(v) + b_v) \oplus_u (a_w val(w) + b_w)$. The value contributed by u to the vertex $p(u)$ is given by $E = a_u val(u) + b_u =$

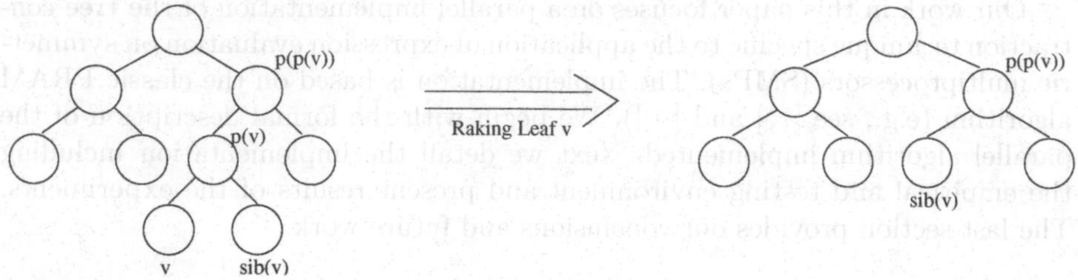

Fig. 2. Rake of leaf v: removes vertices v and $p(v)$ from the tree and makes the sibling $sib(v)$ of vertex v, the child of grandparent $p(p(v))$ of v

$a_u[(a_v val(v) + b_v) \oplus_u (a_w val(w) + b_w)] + b_u$. Say v is the left leaf of u. On raking leaf v, E is simplified to a linear expression in the unknown value of w, namely $val(w)$. The labels of w are then updated to maintain the value contributed to $p(u)$. The augmented rake operation that modifies the sibling labels as described above maintains the value of the arithmetic expression. The original binary tree is contracted to a three-vertex tree with root r and two leaves v_1 and v_2 by applying the augmented rake operation repeatedly and concurrently to the tree. The value of the arithmetic expression is then given by the value of the root which is given by $val(T) = val(r) = (a_{v_1} val(v_1) + b_{v_1}) \oplus_r (a_{v_2} val(v_2) + b_{v_2})$.

The next section formally describes the PRAM algorithm in detail.

4 Review of the PRAM Algorithm for Expression Evaluation

The PRAM algorithm consists of three main steps. The first step identifies and labels the leaves. The second step contracts the input binary tree to a three-vertex tree using the augmented *rake* operation. The resultant tree contains the root and the left- and right-most leaves of the original tree with labels suitably updated to maintain the value of the arithmetic expression. The third step computes the final value of the arithmetic expression from the three-vertex binary tree as described in the previous section. The concurrent rake of two leaves that are siblings or whose parents are adjacent must be avoided as this would lead to an inconsistent resultant structure. Hence, a careful application of the rake operation for tree contraction is required.

Alg. 4 evaluates a given arithmetic expression using the PRAM model (from [13]). After Step (2) of Alg. 4, we have a three-vertex tree T' with a root r holding an operator $\oplus_r$ and two leaves, the left- and right-most leaves, u and v containing the constants c_u and c_v with labels (a_u, b_u) and (a_v, b_v), respectively. These labels hold the relevant information from the vertices that have been raked. Hence, the value of the subexpression at the root r is the value of the given arithmetic expression. This value is given by

$$val(T) = val(T') = val(r) = (a_u c_u + b_u) \oplus_r (a_v c_v + b_v). \quad (1)$$

Data : (1) A rooted binary tree T such that each vertex v has the labels (a_v, b_v) initialized to $(1,0)$, (2) each non-leaf vertex has exactly two children, and (3) for each vertex different from the root, the parent $p(v)$, the sibling $sib(v)$ and an indication if it is a left or right child of its parent.

Result : The value of the arithmetic expression represented by the tree T

begin

1. Label the leaves consecutively in order from left to right, excluding the left- and right-most leaves, and store the labeled leaves in an array A of size n.
2. **for** $\lceil \log(n+1) \rceil$ *iterations* **do**
   - 2.1. Apply the augmented rake operation that modifies the sibling's labels concurrently to all the elements of A_{odd} that are left children.
   - 2.2. Apply the augmented rake operation that modifies the sibling's labels concurrently to the rest of the elements in A_{odd}.
   - 2.3. Set $A := A_{even}$.
3. Compute the value of the arithmetic expression from the three-vertex binary tree.

end

Algorithm 1: PRAM Algorithm for AEE

Alg. 4 ensures that two leaves that are siblings or whose parents are adjacent are never raked concurrently (for proof refer to [15]). Hence the rake operations are applied correctly and the algorithm is correct. Step (1), labeling the leaves of the tree, can be implemented using the Euler tour technique and using the optimal list ranking algorithm to obtain the labels. Hence this step takes $O(\log n)$ time using $O(n)$ operations. Given an array A, A_{odd} and A_{even} (containing the odd and even indexed elements of A respectively) can be obtained in $O(1)$ time using a linear number of operations. The augmented rake operations that modify the sibling labels in Steps (2.1) and (2.2) are done in parallel and hence take $O(1)$ time each. Step (2.3) takes $O(1)$ time. The number of operations required by each iteration is $O(|A|)$, where $|A|$ represents the current size of the array A. Since the size of the array A decreases by half each iteration, the total number of operations required by Step (2) is $O(\sum_i (n/2^i)) = O(n)$. Hence, this step takes $O(\log n)$ time using $O(n)$ operations. Finally, computing the value of the arithmetic expression from the three-vertex binary tree in Step (3) takes $O(1)$ time and $O(1)$ work. Therefore, arithmetic expression evaluation takes $O(\log n)$ time using $O(n)$ operations on an EREW PRAM.

Note that the *rake* operation used for contracting the tree does not create any new leaves. Hence, the data copy in Step (2.3) of Alg. 4 can be completely avoided by replacing Step (2) with Alg. 4. For further details refer to [15]. This modified algorithm has the same complexity bounds as the original algorithm.

5 SMP Algorithm for AEE

This section begins with the description of the programming environment used for the implementation. The methodology for converting a PRAM algorithm to a practical SMP algorithm is then described. The actual implementation details are then given. This section ends with the description of the cost model used for analyzing the SMP algorithm and a thorough analysis of the algorithm.

SMP Libraries Our practical programming environment for SMPs is based upon the SMP Node Library component of SIMPLE [1], that provides a portable framework for describing SMP algorithms using the single-program multiple-data (SPMD) program style. This framework is a software layer built from POSIX threads that allows the user to use either the already developed SMP primitives or the direct thread primitives. We have been continually developing and improving this library over the past several years and have found it to be portable and efficient on a variety of operating systems (e.g., Sun Solaris, Compaq/Digital UNIX, IBM AIX, SGI IRIX, HP-UX, and Linux). The SMP Node Library contains a number of SMP node algorithms for barrier synchronization, broadcasting the location of a shared buffer, replication of a data buffer, reduction, and memory management for shared-buffer allocation and release. In addition to these functions, we have control mechanisms for contextualization (executing a statement on only a subset of processors), and a *parallel do* that schedules n independent work statements implicitly to p processors as evenly as possible.

The PRAM model assumes the availability of as many processors as needed and a synchronous mode of operation. However, SMPs have a limited number of processors and barriers have to be explicitly used to enforce synchronization. Hence the original PRAM algorithm for AEE can be converted to an SMP algorithm by balancing the work between the available processors and synchronizing the processors between the various steps and substeps of the algorithm. These barriers enforce synchronization and hence the algorithm works correctly on a SMP.

Implementation Details Labeling the leaves is done as follows. Each original tree edge is converted into a pair of directed arcs and weighted as follows.

2. **for** *phase i from* 0 *to* $\lceil \log_2(n-2) \rceil$ **do**
   2.1. Apply the augmented rake operation that modifies the sibling's labels concurrently to all the leaves in locations $2^i, 3 \cdot 2^i, 5 \cdot 2^i, 7 \cdot 2^i, \ldots$ of A that are left children.
   2.2. Apply the augmented rake operation that modifies the sibling's labels concurrently to all the leaves in locations $2^i, 3 \cdot 2^i, 5 \cdot 2^i, 7 \cdot 2^i, \ldots$ of A that are right children.

Algorithm 2: Modified Step (2) of PRAM Algorithm for AEE

A directed arc leaving a leaf vertex is assigned a weight of one; all other arcs are given a weight of zero. The Euler tour of the given tree is then constructed. This Euler tour is represented by a successor array with each node storing the index of its successor. This is followed by list ranking to obtain a consecutive labeling of the leaves. Our implementation uses the SMP list ranking algorithm and implementation developed by Helman and JáJá [10] that performs the following main steps:

1. Finding the head h of the list which is given by $h = (n(n-1)/2 - Z)$ where Z is the sum of successor indices of all the nodes in the list.
2. Partitioning the input list into s sublists by randomly choosing one splitter from each memory block of $n/(s-1)$ nodes, where s is $\Omega(p \log n)$, where p is the number of processors. Corresponding to each of these sublists is a record in an array called *Sublists*. (Our implementation uses $s = 8p$.)
3. Traversing each sublist computing the prefix sum of each node within the sublists. Each node records its sublist index. The input value of a node in the *Sublists* array is the sublist prefix sum of the last node in the previous *Sublists*.
4. The prefix sums of the records in the *Sublists* array are then calculated.
5. Each node adds its current prefix sum value (value of a node within a sublist) and the prefix sum of its corresponding *Sublists* record to get its final prefix sums value. This prefix sum value is the required label of the leaves.

For a detailed description of the above steps refer to [10]. Pointers to the leaf vertices are stored in an array A. This completes Step (1) of AEE algorithm.

Given the array A, the rake operation is applied by updating the necessary pointers in the tree's data structure. The concurrent rake operations in Steps (2.1) and (2.2) are handled similarly. Each processor is responsible for an equal share (at most $\lceil R/p \rceil$) of the R leaves in A_{odd}, and rakes the appropriate leaves in each of these two steps, with a barrier synchronization following each step. Step (2.3) copies the remaining leaves (A_{even}) into another array B. Thus iterations of the *for* loop alternate between using the A and B arrays for the rake operation and store the even leaves in the other array at Step (2.3) of the iteration. Finally, the value of the given expression is computed from the resultant three-vertex binary tree. Hence this implementation, while motivated by the PRAM algorithm, differs substantially in its SMP implementation due to the limited number of processors and need for explicit synchronization. In addition, to achieve high-performance, our SMP implementation must make good use of cache.

Alternatively, we could use Alg. 4 for the second step that eliminates the data copy between arrays A and B. While this modification uses less space (only a single array of leaf pointers), it has the disadvantage that due to the increasing strides of reading the leaves from the A array, more cache misses are likely.

SMP Cost Model We use the SMP complexity model proposed by Helman and JáJá [10] to analyze our shared memory algorithms. In the SMP complexity model, we measure the overall complexity of the algorithm by the triplet (M_A, M_E, T_C). The term M_A is simply a measure of the number of non-contiguous main memory accesses, where each such access may involve an arbitrary-sized contiguous blocks of data. M_E is the maximum amount of data exchanged by any processor with main memory. T_C is an upper bound on the local computational complexity of any of the processors and is represented by the customary asymptotic notation. M_A, M_E are represented as approximations of the actual values. In practice, it is often possible to focus on either M_A or M_E when examining the cost of algorithms.

Algorithmic Analysis The first step of the algorithm uses the List Ranking algorithm developed by Helman and JáJá [10]. For $n > p^2 \ln n$, we would expect in practice list ranking to take

$$T(n,p) = (M_A(n,p); T_C(n,p)) = \left(\frac{n}{p}, \mathrm{O}\left(\frac{n}{p}\right)\right). \tag{2}$$

Tree contraction requires $\mathrm{O}(\log n)$ iterations, where iteration i, for $1 \le i \le \log n$ rakes $n/2^i$ leaves concurrently. Since each rake operation requires a constant number of memory accesses of unit size, the tree contraction takes

$$T(n,p) = (M_A(n,p); T_C(n,p)) = \left(\log n + \frac{n}{p}, \mathrm{O}\left(\log n + \frac{n}{p}\right)\right). \tag{3}$$

For $n > p \log n$, this simplifies to the same complexity as in Eq. 2.

Thus, for $n > p^2 \lg n$, evaluating arithmetic expressions with $\mathrm{O}(n)$ vertices on a p processor shared-memory machine takes

$$T(n,p) = (M_A(n,p); T_C(n,p)) = \left(\frac{n}{p}, \mathrm{O}\left(\frac{n}{p}\right)\right). \tag{4}$$

6 Experimental Results

This section summarizes the experimental results of our implementation. We tested our shared-memory implementation on the Sun HPC 10000, a UMA shared memory parallel machine with 64 UltraSPARC II processors and 64 GB of memory. Each processor has 16 Kbytes of direct-mapped data (L1) cache and 4 Mbytes of external (L2) cache. The clock speed of each processor is 400 MHz.

Experimental Data. In order to test the performance of our algorithm, we designed a collection of tree generators that when given an integer d, could generate arbitrarily large regular and irregular rooted, binary trees of $n = 2^d - 1$ vertices. Each generated tree is a strict binary tree, meaning that each vertex has either both subtrees empty or neither subtree empty. Thus, each test tree has n vertices and $(n+1)/2$ leaves. Let the level of a vertex be the number of edges on the path between it and the root. We then generate the following three kinds of strict binary trees:

FULL: These are full, binary trees with 2^l internal vertices at each level l, for $0 \leq l < d-1$, and $(n+1)/2$ leaves at the last level $l = d-1$.

CAT: These "caterpillar" trees are basically long chains of vertices with one leaf attached at each internal vertex, except for two leaves at the bottom (level $(n-1)/2$). Hence, the root is at level 0, and each level l, for $1 \leq l \leq (n-1)/2$, has exactly two vertices.

RAN: Unlike the full and the caterpillar trees, this is an irregular class of binary tree with no specific structure. We randomly create this input using an unrake-like (reverse rake) operation. Each tree is constructed by randomly choosing one of the vertices v in the current tree (except the root) and introducing a new vertex u and a leaf w at the chosen vertex position. More specifically, u becomes the child of $p(v)$ replacing v and w and v become the new children of u. The sub-tree rooted at v remains unaltered. We initially start with a three-vertex tree containing the root and two leaves and build the tree randomly until the tree contains a total of n vertices.

Alg. 4 consists of four main substeps, namely 1) the *Euler tour* computation, 2) *list ranking,* 3) *copying leaves* in the array A, and 4) the *rake* operation that contracts the tree to a three-vertex tree and computes the final value of the arithmetic expression. The running times of our implementation of Alg. 4 for various sizes of **FULL**, **CAT** and **RAN** trees on different numbers of processors are plotted in Fig. 3. The experiments were performed for $d \leq 25$ with the number p of processors equal to 2, 4, 8, 16, and 32, and using 32-bit integers. For $d = 25$ we also give a step-by-step breakdown of the execution time. In all cases, for a particular class of input tree, the implementation performs faster as more processors are employed. The regular trees (**FULL** and **CAT**) represent special cases of expressions and run faster than the irregular input (**RAN**) that represents the most common class of expressions that have an irregular structure. Note that the irregular trees have a longer running time (primarily due to cache misses since the same number of rake operations per iteration are performed), but exhibit a nearly linear relative speedup. This performance matches our analysis (Eq. 4).

Finally, we have implemented a linear sequential time, non-recursive version of expression evaluation to provide sequential timing comparisons. It is important to note that the hidden asymptotic constant associated with the sequential code is very small (close to one), while the parallel code has a much larger hidden constant. Even so, our parallel code is faster than the sequential version when enough processors are used, and in the most general case, the irregular input trees.

In Fig. 4 we compare the performance of the *rake* substep, that is, Step (2) in Algs. 4 and 4 for $d = 25$ on a range of processors. These experimental results confirm that the improved cache hit rate in Alg. 4 overcomes the time spent copying data between the A and B leaf pointer arrays. The improvement, though, is only noticeable when the input trees are regular.

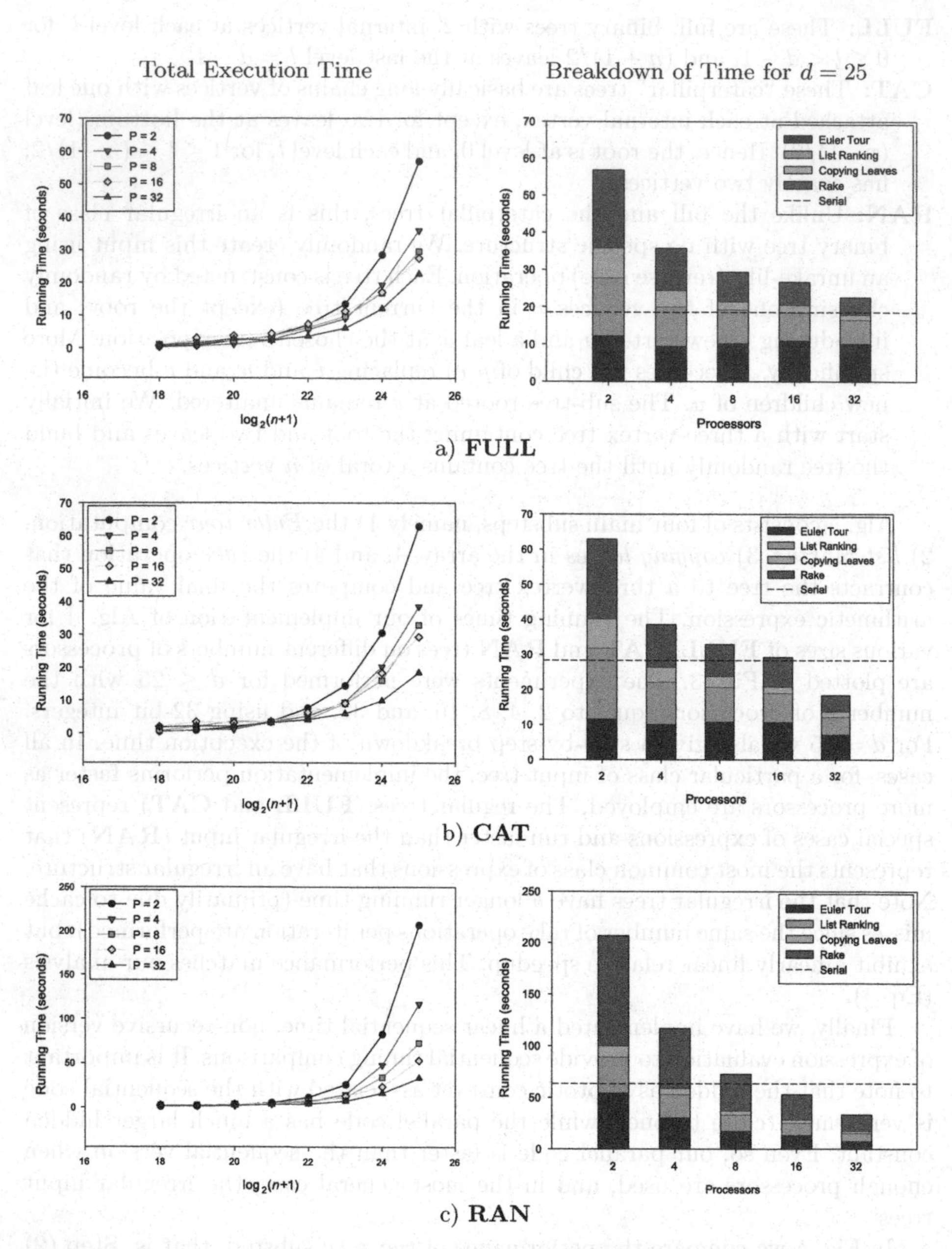

Fig. 3. Expression evaluation execution time for **FULL**, **CAT**, and **RAN** trees. The left-hand graphs plot the total running time taken for varying input sizes and numbers of processors. The corresponding graphs on the right give a step-by-step breakdown of the running time for a fixed input size ($2^{25} - 1$ vertices) and the number p of processors from 2 to 32

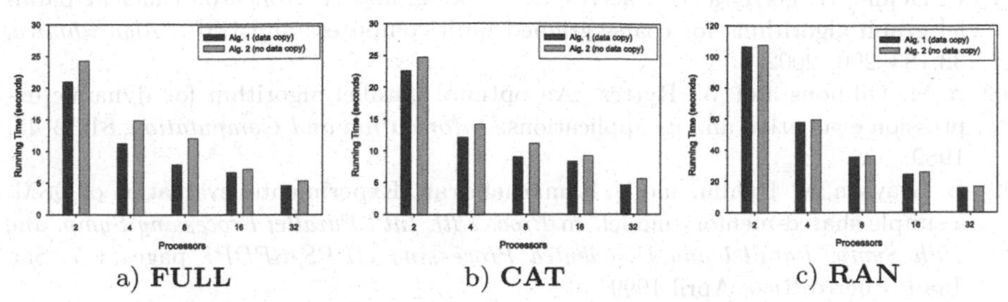

a) **FULL** b) **CAT** c) **RAN**

Fig. 4. Comparison of time for *rake* step for **FULL**, **CAT**, and **RAN** trees for a fixed input size ($2^{25} - 1$ vertices) and the number p of processors from 2 to 32

7 Conclusions

In summary, we present optimistic results that for the first time, show that parallel algorithms for expression evaluation using tree contraction techniques run efficiently on parallel symmetric multiprocessors. Our new implementations scale nearly linearly with the problem size and the number of processors, as predicted by our analysis in Eq. 4. These results provide optimistic evidence that complex graph problems that have efficient PRAM solutions, but no known efficient parallel implementations, may have scalable implementations on SMPs.

References

[1] D. A. Bader and J. JáJá. SIMPLE: A methodology for programming high performance algorithms on clusters of symmetric multiprocessors (SMPs). *J. Parallel & Distributed Comput.*, 58(1):92–108, 1999. 69

[2] D. A. Bader, A. K. Illendula, B. M. E. Moret, and N. Weisse-Bernstein. Using PRAM algorithms on a uniform-memory-access shared-memory architecture. In G. S. Brodal, D. Frigioni, and A. Marchetti-Spaccamela, editors, *Proc. 5th Int'l Workshop on Algorithm Engineering (WAE 2001)*, volume 2141 of *Lecture Notes in Computer Science*, pages 129–144, Århus, Denmark, August 2001. Springer-Verlag. 65

[3] E. Cáceres, F. Dehne, A. Ferreira, P. Flocchini, I. Rieping, A. Roncato, N. Santoro, and S. W. Song. Efficient parallel graph algorithms for coarse grained multicomputers and BSP. In *Proc. 24th Int'l Colloquium on Automata, Languages and Programming (ICALP'97)*, volume 1256 of *Lecture Notes in Computer Science*, pages 390–400, Bologna, Italy, 1997. Springer-Verlag. 65

[4] A. Charlesworth. Starfire: extending the SMP envelope. *IEEE Micro*, 18(1):39–49, 1998. 64

[5] A. Charlesworth. The Sun Fireplane system interconnect. In *Proc. Supercomputing (SC 2001)*, pages 1–14, Denver, CO, November 2001. 64

[6] R. Cole and U. Vishkin. The accelerated centroid decomposition technique for optimal parallel tree evaluation in logarithmic time. *Algorithmica*, 3:329–346, 1988. 65

[7] F. Dehne, A. Ferreira, E. Cáceres, S. W. Song, and A. Roncato. Efficient parallel graph algorithms for coarse-grained multicomputers and BSP. *Algorithmica*, 33:183–200, 2002. 65

[8] A. M. Gibbons and W. Rytter. An optimal parallel algorithm for dynamic expression evaluation and its applications. *Information and Computation*, 81:32–45, 1989. 65

[9] B. Grayson, M. Dahlin, and V. Ramachandran. Experimental evaluation of QSM, a simple shared-memory model. In *Proc. 13th Int'l Parallel Processing Symp. and 10th Symp. Parallel and Distributed Processing (IPPS/SPDP)*, pages 1–7, San Juan, Puerto Rico, April 1999. 65

[10] D. R. Helman and J. JáJá. Designing practical efficient algorithms for symmetric multiprocessors. In *Algorithm Engineering and Experimentation (ALENEX'99)*, volume 1619 of *Lecture Notes in Computer Science*, pages 37–56, Baltimore, MD, January 1999. Springer-Verlag. 70, 71

[11] T.-S. Hsu and V. Ramachandran. Efficient massively parallel implementation of some combinatorial algorithms. *Theoretical Computer Science*, 162(2):297–322, 1996. 65

[12] T.-S. Hsu, V. Ramachandran, and N. Dean. Implementation of parallel graph algorithms on a massively parallel SIMD computer with virtual processing. In *Proc. 9th Int'l Parallel Processing Symp.*, pages 106–112, Santa Barbara, CA, April 1995. 65

[13] J. JáJá. *An Introduction to Parallel Algorithms*. Addison-Wesley Publishing Company, New York, 1992. 64, 66, 67

[14] J. Keller, C. W. Keßler, and J. L. Träff. *Practical PRAM Programming*. John Wiley & Sons, 2001. 65

[15] S. R. Kosaraju and A. L. Delcher. Optimal parallel evaluation of tree-structured computations by raking (extended abstract). Technical report, The Johns Hopkins University, 1987. 65, 66, 68

[16] G. L. Miller and J. H. Reif. Parallel tree contraction and its application. In *Proc. 26th Ann. IEEE Symp. Foundations of Computer Science (FOCS)*, pages 478–489, Portland, OR, October 1985. IEEE Press. 65

[17] OpenMP Architecture Review Board. OpenMP: A proposed industry standard API for shared memory programming. www.openmp.org, October 1997. 64

[18] M. Reid-Miller. List ranking and list scan on the Cray C-90. *J. Comput. Syst. Sci.*, 53(3):344–356, December 1996. 65

[19] J. H. Reif, editor. *Synthesis of Parallel Algorithms*. Morgan Kaufmann Publishers, 1993. 64

[20] J. Sibeyn. Better trade-offs for parallel list ranking. In *Proc. 9th Ann. Symp. Parallel Algorithms and Architectures (SPAA-97)*, pages 221–230, Newport, RI, June 1997. ACM. 65

Dead-Block Elimination in Cache: A Mechanism to Reduce I-cache Power Consumption in High Performance Microprocessors

Mohan G. Kabadi, Natarajan Kannan, Palanidaran Chidambaram, Suriya Narayanan, M. Subramanian, and Ranjani Parthasarathi

School of Computer Science and Engineering
Anna University, Chennai - 600 025, India
{mohan_kabdi,natarajan,palanidaran,mssnlayam,rp}@cs.annauniv.edu

Abstract. Both power and performance are important design parameters of the present day processors. This paper explores an integrated software and circuit level technique to reduce leakage power in L1 instruction caches of high performance microprocessors, by eliminating basic blocks from the cache, as soon as they are dead. The effect of this dead block elimination in cache on both the power consumption of the I-cache and the performance of the processor is studied. Identification of basic blocks is done by the compiler from the control flow graph of the program. This information is conveyed to the processor, by annotating the first instruction of selected basic blocks. During execution, the blocks that are not needed further are traced and invalidated and the lines occupied by them are turned off. This mechanism yields an average of about 5% to 16% reduction, in the energy consumed for different sizes of I-cache, for a set of the SPEC CPU 2000 benchmarks [16], without any performance degradation.

1 Introduction

In the present scenario, silicon area and power have become important constraints on the designers. New technical developments are being implemented to overcome the former constraint. The fabrication technology of VLSI circuit is steadily improving and the chip structures are being scaled down. But, the number of transistors on a chip is increasing at a higher ratio. Also, the drive towards increasing levels of performance has pushed the operating clock frequencies higher and higher, which has resulted in an increased level of power consumption [1].

Power has thus become important, not only for wireless and mobile electronics, but also for high performance microprocessors. It should therefore be considered a "first class" design constraint on par with performance [2].

There has been considerable work on low power processors as evidenced in the literature. Many of these have focused on reducing power/energy in the memory subsystem namely multi-level I-caches [3, 4, 5, 6, 7, 17], d-caches [6, 7, 8, 9, 17]

S. Sahni et al. (Eds.) HiPC 2002, LNCS 2552, pp. 79–88, 2002.

and main memory [10, 11]. Memory subsystems, especially the on-chip caches have caught the attention of designers for the reason that, they are dominant sources of power consumption. Caches often consume 80% of the total transistor count and 50% of the area [12]. Hence cache subsystems have been the primary area of focus for power reduction.

Several techniques have been proposed to reduce the power dissipated by on-chip caches. These techniques can be grouped under (i) architectural alternatives adopting dynamic [5, 6] or static methods [13] and (ii) Software techniques using compiler support [3, 8]. In recent years, software techniques have been receiving more attention in building low power systems. The idea here is that the compiler can assist the microarchitecture in reducing the power consumption of programs either by giving explicit information on program behavior or by optimization. Ideas that have been explored include use of a reduced size cache/buffer to store inner loops [3], and the use of loop transformations to improve the performance of the program, thereby reducing the energy consumed [8]. However, these approaches are effective only for loop intensive programs. This paper presents a more generic, novel, compiler-assisted technique, wherein, selected lines of the cache are turned off under program control, resulting in reduction of power consumption. This approach takes care of both loop-intensive and non loop-intensive programs.

This paper is organized as follows: In section 2, a brief review of the related work relevant to the proposed approach is presented. In section 3, the proposed method, Dead-Block Elimination in Cache (DBEC) is described. In section 4, the hardware modifications required for the proposed method are outlined.

The experimental methodology and the results of the simulation are given in section 5 and section 6 concludes the paper.

2 Related Work

As mentioned in the previous section, compiler supported power minimization is a field of active research. A few of the techniques relevant to the proposed work are presented below. In the approach taken by Nikolaos Bellas et al. [3], a small L-cache similar to the Filter cache [13], is added between the CPU and the I-cache, for buffering instructions (basic blocks) that are nested within loops. However, the basic blocks within loops that contain function calls are not considered for placement in the L-cache. The profile data from previous runs is used to select the best instructions to be cached.

Another approach taken by Hongbo Yang et al. [8], is based on the fact that performance improvement causes a reduction in execution time and hence energy saving can be achieved. Here, the impact of loop optimizations on performance of program vs power are compared. Compiler optimizations namely loop unrolling, loop permutation, loop tiling and loop fusion are shown to improve program performance, which is correlated positively with the energy reduction.

Other techniques that have been explored are hardware-based or architecture-based. Gated-Vdd [4] is a circuit-level technique to gate the supply voltage and reduce leakage in unused SRAM cells. The DRI I-cache [5], dynamically reacts to application demand, and adapts to the required cache size during an application's execution. This work uses Gated-Vdd mechanism at the circuit-level, to turn-off the supply voltage to the cache's unused sections, thus reducing leakage power. The DRI I-cache integrates architectural and circuit level techniques to reduce leakage in an L1 I-cache.

On a similar line as that of DRI I-cache, in the approach used by Kaxiras et al [9, 17], parts of the L1 d-cache are switched off at a much finer granularity (i.e. line granularity) and without resizing. The key idea here is that, if a cache line is not accessed within a predefined fixed interval, the supply voltage to that line is turned off using the Gated-Vdd mechanism. This approach uses a static turn-off interval which is set on an individual application basis, to obtain optimal results.

In the approach followed by Huiyang Zhou et al. [6], a hardware counter called Line Idle Counter (LIC) is used, to keep track of the length of the period a line remains idle, before it gets turned off. The Mode-Control Logic (MCL) compares the LIC value to the turn-off interval stored in a Global Control Register. The miss rate and performance factor are dynamically monitored to set the Global Control Register, periodically, at the end of a statically set predefined interval.

3 Dead-Block Elimination in Cache (DBEC)

The work done in [4, 5, 6] has shown that, dynamically turning off sections of the I-cache and resizing it, results in significant leakage-power savings. Motivated by this approach, the current work explores the "turning-off" of the unused I-cache lines using a software-directed approach. In [5, 6] the time at which a line can be turned off is determined at runtime based on a saturating counter. The choice of the saturation value directly affects performance, as the saturation value is only an estimate of the non-usage of a line. Too large a value would result in reduced power saving, while too small a value would evict the line earlier from the cache, resulting in a miss and thus reducing performance. An optimal value has to be chosen based on the application. Even this application dependent static choice of the saturation value is only an estimate. However, more precise information on when a line is going to be "dead" can be directly obtained from the compiler. The compiler support for pointing out when a block is dead is used in the proposed approach.

The proposed DBEC scheme consists of invalidating and turning-off power to cache lines that are occupied by the "dead" instructions i.e., the instructions that are not "live" at a particular point of program execution. These are the instructions that would not be used again before being replaced in the cache. The information on whether an instruction is "dead" at a particular point of program execution is obtained from the compiler. The granularity at which the dead instructions are handled is a basic block. The compiler identifies basic blocks from the CFG and indicates the beginning and end of the basic blocks in the code. When the program is executed, these indications are used by the

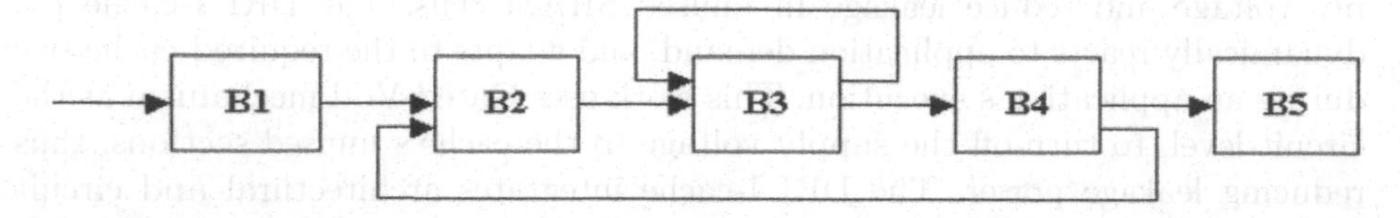

Fig. 1. Segment of a Control Flow Graph

microarchitecture to turn-off dead blocks. The mechanism used, is explained below with an example.

A basic block [14], is a sequence of consecutive statements, in which the flow of control enters at the beginning and leaves at the end, without the possibility of branching, except at the end. Figure 1 shows a segment of a Control Flow Graph. B_is are the basic blocks. B_3 is a loop inside an outer loop containing basic blocks B_2, B_3 and B_4.

The first statement of each basic block is annotated during the compilation stage. The annotations help to keep track of the basic blocks which are executed. B_2 is the block through which control enters the outer loop. The first statement of B_2 is annotated. Whereas B_3 and B_4 are part of the same outer loop and hence, the first statement of these blocks are not annotated. The key idea of DBEC is that the annotated statements help in identifying the blocks executed in the just completed outermost loop. This is done to make sure that B_2 and B_3 are not turned off when B_4 is under execution. When the annotated statement of B_2 is under execution, the block B_1 is completely executed. Hence, the cache lines completely occupied by the instructions of B_1 are turned-off. Similarly, the execution of first statement of B_5 will cause an invalidation of the lines occupied by the previous loop. i.e., as B_2, B_3 and B_4 are turned off. Thus, the instant at which the particular block is going to be "dead" is exactly determined, there is negligible performance degradation in this approach. The earlier approaches [4, 5], [6], have achieved power saving at the cost of performance. Further, this being a static approach supported by the compiler, the runtime architectural overhead of this approach is also negligible.

4 Hardware Modification

The only hardware modification that is required in the DBEC approach is that, with each line of I-cache, one bit called the "turn-off bit" is added to the tag bits. These turn-off bits are initially unset at the start of program execution. The turn-off bits corresponding to lines of cache which contain the code for a loop are set. This is used to turn off these lines, when the execution of the loop is completed. This is achieved using the Gated-Vdd technique. The operation can be explained using the same example given in the previous section. Against execution of the first annotated statement of B_1, the processor sets the turn-off bits of those lines from which the instruction of B_1 are fetched. An instance of the various bytes in each line occupied by the B_1 block is shown in figure 2.

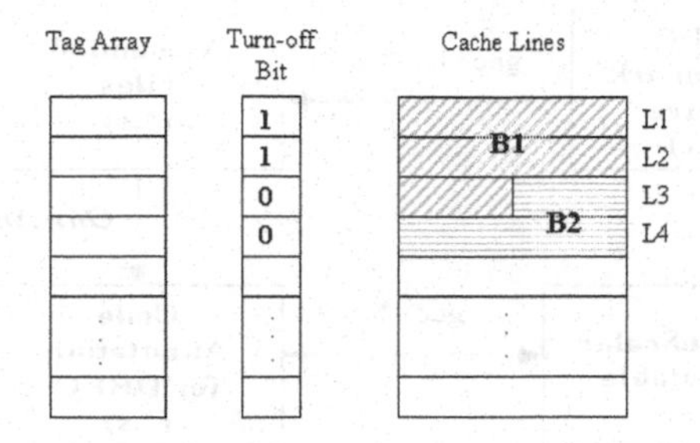

Fig. 2. A segment of the cache with tag bits

Also, the turn-off bits for those lines after the execution of B_1 are shown in the same figure. The turn-off bit of line L3 is not set, indicating that the line is to be retained even after the execution of B_1. This is because line L3 is only partially filled with code from B_1, and contains some code from B_2 also. When the control of execution reaches the beginning of block B_2, the annotated first statement of B_2 will cause the invalidation and turning-off of the lines whose turn-off bits were previously marked for this purpose. Hence, the lines L1 and L2 are now turned-off. The turn-off bit of line L3 is set only when B_2 is taken up for execution. When B_2 completes execution, line L3 is turned off.

5 Experimental Methodology and Results

The SimpleScalar 2.0 toolkit [15] is used to implement the proposed idea. Benchmarks *art, equake, gzip* and *mcf* of the SPEC CPU 2000 benchmark [16] suite have been used to evaluate the performance.

5.1 Experimental Setup

The out-of-order superscalar processor simulator of the SimpleScalar toolkit has been used to simulate switching off of the cache lines when annotated instructions are encountered in the instruction stream. All the benchmarks have been compiled using the C compiler gcc-2.6.3 for the SimpleScalar toolkit which has been modified to annotate instructions that aid switching off of cache lines.

The various phases in the compilation are as shown in figure 3. The source programs of the SPEC benchmark suite are compiled using *gcc* with the *-S* flag to get the assembly files. These assembly files are the input to the *annotator* which annotates instructions as explained below.

The *annotator* annotates instructions in three ways to aid the processor in switching off cache lines. The first type of annotation indicates that the cache lines whose "turn-off" bit has been set are to be switched off. This annotation is normally performed for instructions that begin a basic block. In the case of loops, only the first basic block of the loop is annotated so that the basic blocks within

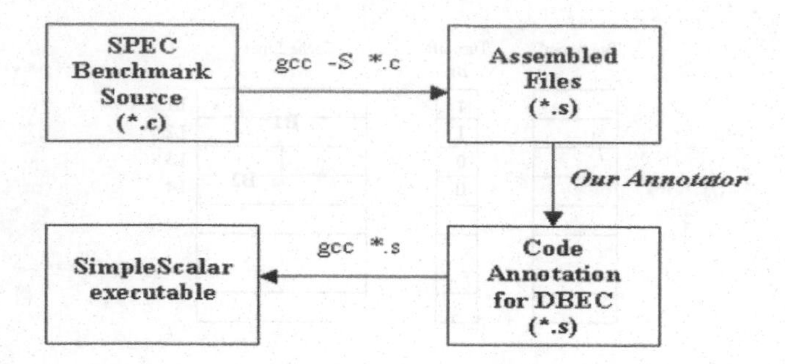

Fig. 3. Phases in the generation of annotated SimpleScalar executable

Table 1. Parameters used in the simulation

Parameter	Value
Fetch width	4 instructions per cycle
Decode width	4 instructions per cycle
Commit width	4 instructions per cycle
L1 I-cache	256, 512, 1024 and 2048 lines
L1 I-cache line size	32 bytes
L1 I-cache associativity	1-way (Direct mapped)
L1 I-cache latency	1 cycle
L1 D-cache	16 K, 4-way, 32 byte blocks
L2 unified cache	256 K, 4-way, 64 byte blocks
L2 unified cache latency	6 cycles

the loop do not switch off other basic blocks of the same loop. This ensures that cache lines containing the loop code are not switched off. The second type of annotation indicates that the next instruction is a "call instruction" within a loop. The third type of annotation indicates that the previous instruction was a function call which was part of a loop. The second and third annotations ensure that the functions which are called from within a loop do not switch off the cache lines that contain the code of that loop. The assembly files thus annotated are converted to SimpleScalar executable using *gcc*.

The implementation of the second and third type of annotation requires simple architectural support in the form of a dedicated counter. The annotation before the call increments the counter and the instruction following it decrements the same. Switching off of the cache lines is performed only when this counter's value is zero indicating that the code under execution is not part of any loop. The instructions to manipulate the dedicated counter, which keeps track of entry and exit of calls, are added. An alternative is to have a dedicated register and use the increment and decrement instructions of the existing ISA before and after the function calls.

Table 2. I-cache miss rates for the base model and the DBEC with 512 lines each

Program	Miss rate for the base model	Miss rate with DBEC
art	0.0001	0.0001
equake	0.0103	0.0103
gzip	0.02	0.02
mcf	0.0045	0.0047

The SimpleScalar simulator is modified to switch off cache lines according to annotations added by the compiler. The simulator is used to collect results regarding power consumption.

5.2 Simulation Results

The main parameters considered in the evaluation of this implementation are the reduction in power consumption and miss rate. The leakage power for the cache is proportional to the total number of cache lines that are switched-on in the cache [9]. Hence the total number of lines switched off is taken to be an estimate of the power savings achieved.

$$\%\ Power\ Saved\ = \left(1 - \frac{\sum(No.\ of\ active\ lines \times Duration\ of\ Activity)}{Total\ number\ of\ lines \times Total\ duration}\right) \times 100$$

Miss rate has been considered to study the impact of this technique on performance. The number of instructions considered while executing the benchmarks is 4 billion. The cache-related parameters used in the simulation model are given in table 1.

Table 2 gives the miss rates of the modified programs and the base programs of the SPEC 2000 benchmark suite. It can be seen that there is no significant difference in the miss rates. This is as expected, because the DBEC will turn-off a line only after an instruction is "dead". Thus the capacity and conflict misses of DBEC model will be same as the base model. Hence, there is no degradation in performance.

Figure 4 shows the power savings obtained for various benchmark programs as the number of cache lines varies from 256 to 2048, with all the cache lines initially on. It can be seen that, while three programs give a modest power saving of about 2-10%, one program *art*, records a maximum power saving of about 50% for 512 lines and about 40% for 1024 lines. This may be explained by the fact that DBEC consumes less power when the program consists of many independent loops, and *art* is one such program. The variation of the power saving for each program, as the number of lines of cache is varied shows an interesting pattern. It increases as the number of cache lines is increased from 256 onwards and then decreases as the number of cache lines is increased beyond 512/1024. One reason for this could be that, the cache lines are assumed to be initially on, and if all the lines of the cache are not used, the power saving decreases.

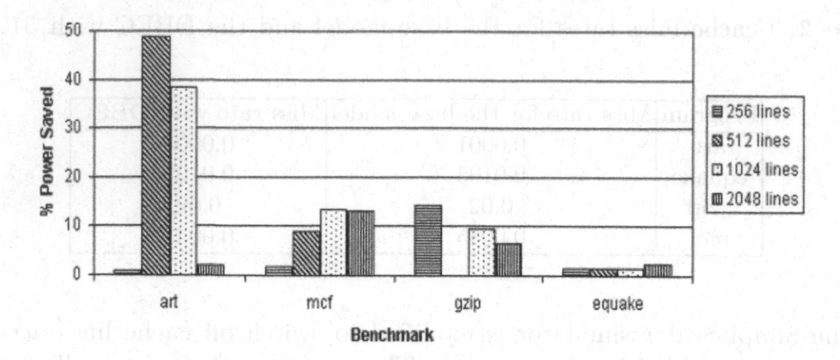

Fig. 4. Variation in I-cache power savings with number of cache lines

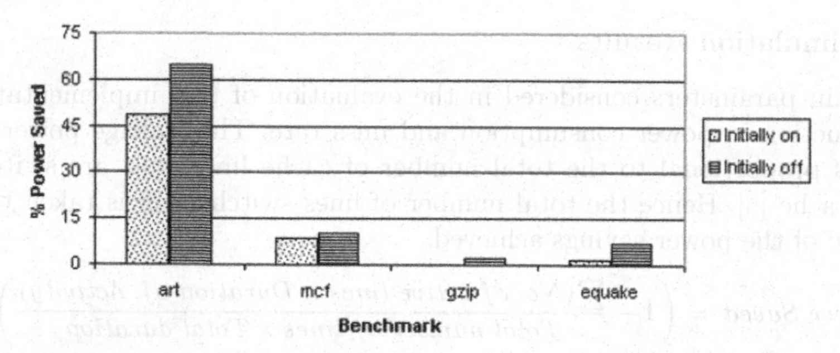

Fig. 5. Power savings for 512 line caches with lines initially switched on Vs off

To study this further, the programs were executed with cache lines initially turned off. One of the programs (*art*), was run for different cache sizes. It was found that, as the number of cache lines increased from 256 to 2048, the power saving consistently increased. Figure 5 shows the comparison of power saved for a cache of 512 lines for the different programs. As expected, the power saving is higher when the cache lines are initially off.

Further, from figure 4 it is observed that the *equake* program shows a power saving of only 1.56% for a cache size of 512 lines. One reason for this could be that with 4 billion instructions *equake* would just be out of its initialisation phase [18]. To explore this further this program was run for 25 billion instructions and the fraction of the power saved was recorded at different instruction counts. The result of this is shown in figure 6. The power saved is not considerable till 7 billion instructions. Beyond 7 billion instructions, power saved gradually increases to reach a maximum of 23% at about 14 billion instructions and starts slowly decreasing. Actually, after 14 billion instructions the number of lines turned-off remain more or less the same but when averaged out over the total number of instructions, the value decreases.

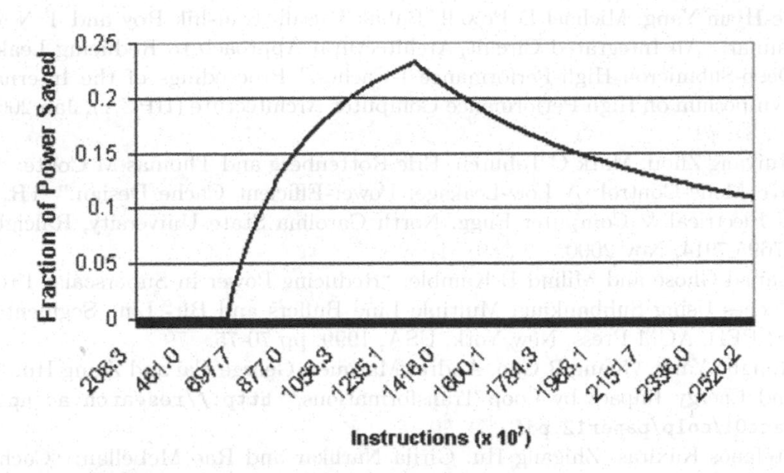

Fig. 6. Fraction of power saved versus instruction count for *equake*

6 Conclusion

The DBEC approach presented here, precisely identifies the blocks which are dead at a particular point of program execution with the help of the compiler. The performance degradation of this approach compared to the base model is almost negligible. Thus, with this approach, it is possible to get an average power saving of about 5% to 16%, and for certain programs a power saving of greater than 40%, with no performance degradation.

Some paths of the programs which are never executed (due to directed branches that are always taken in one direction) will not be turned off by this scheme. Further work has to be done in this direction. Moreover selective annotation can be used to get better saving at the cost of increased miss rate. The DBEC has been implemented as a direct mapped cache. This can be extended for associative caches with a little increase in hardware complexity.

References

[1] Michael K Gowan, Larry L Biro and Daniel B Jackson: "Power Considerations in the Design of the Alpha 21264 Microprocessor," DAC98, San Francisco, CA, 1998, pp 726-731. 79

[2] Trevor Mudge: "Power: A First class Design Constraint for Future Architectures," IEEE Conference, HiPC, India, 2000, pp 215-224. 79

[3] Nikolaos Bellas, Ibrahim Hajj, Constantine Polychronopoulos and George Stamoulis: "Architectural and Compiler support for Energy Reduction in the Memory Hierarchy of High Performance Microprocessors," ISLPED, ACM Press New York, USA, 1998, pp 70-75. 79, 80

[4] Michael D Powell, Se-Hyun Yang Babak Falsafi, Kaushik Roy and T N Vijayakumar: "Gated-Vdd: A circuit Technique to Reduce Leakage in Deep-Submicron Cache Memories," ISLPED, 2000, pp 90-95. 79, 80, 81, 82

[5] Se-Hyun Yang, Michael D Powell, Babak Falsafi, Kaushik Roy and T N Vijayakumar: "An Integrated Circuit/Architectural Approach to Reducing Leakage in Deep-Submicron High-Performance I-Caches," Proceedings of the International Symposium on High Performance Computer Architecture (HPCA), Jan 2001. 79, 80, 81, 82

[6] Huiyang Zhou, Mark C Toburen, Eric Rottenberg and Thomas M Conte: "Adaptive Mode-Control: A Low-Leakage, Power-Efficient Cache Design," TR, Dept. of Electrical & Computer Engg. North Carolina State University, Raleigh, NC, 27695-7914, Nov 2000. 79, 80, 81, 82

[7] Kanad Ghose and Milind B Kamble: "Reducing Power in Superscalar Processor Caches Using Subbanking, Multiple Line Buffers and Bit- Line Segmentation," ISLPED, ACM Press, New York, USA, 1999, pp 70-75. 79

[8] Hongbo Yang, Guang R Gao, Andres Marquez, George Cai and Ziang Hu: "Power and Energy Impact by Loop Transformations," `http://research.ac.upc.es/-pact01/colp/paper12.pdf` 79, 80

[9] Stefanos Kaxiras, Zhigang Hu, Girija Narlikar and Rae McLellan: "Cache-Line Decay: A Mechanism to Reduce Cache Leakage Power," IEEE workshop on Power Aware Computer Systems (PACS), Cambridge, MA, USA, 2000, pp 82-96. 79, 81, 85

[10] Victor Delaluz, Mahmut Kandemir, N Vijayakrishna, Anand Sivasubramanian and Mary Jane Irwin: "Hardware and Software Techniques for Controlling DRAM Power Modes," IEEE Trans. on Computers, Vol.50, No11, Nov 2001, pp 1154-1173. 80

[11] Krishna V Palem, Rodric M Rabbah, Vincent J Mooney III, Pinar Kormatz and Kiran Puttaswamy: "Design Space Optimization of Embedded Memory Systems via Data Remapping," CREST-TR-02-003 GIT-CC-02-011, Feb 2002. 80

[12] John Hennessy: "The Future of Systems Research," IEEE Computers, Aug 1999, pp 27-33. 80

[13] J.Kin, M.Gupta and W.Mangione-Smith: "The Filter Cache: An Energy Efficient Memory Structure," Proc. IEEE Int'l Symp. Microarchitecture, IEEE CS Press, 1997, pp 184-193. 80

[14] Alfred V Aho, Ravi Sethi and Jeffrey D Ulman: "Compilers: Principles, Techniques and Tools," Addison-Wesley, ISBN : 817-808-046-X, Third Indian Reprint 2000. 82

[15] D Burger, Todd M Austin: "The Simplescalar Tool Set, version 2.0 :," CSD Technical Report #1342. University of Wisconsin-Madison, June 1997. 83

[16] "SPEC CPU 2000 benchmark suite," `http://www.spec.org` 79, 83

[17] Stefanos Kaxiras, Zhigang Hu and Margaret Martonosi: "Cache Decay: Exploiting Generational Behavior to Reduce Cache Leakage Power," Proc. of Int'l Symp. Computer Architecture, ISCA, ACM Press, New York, USA, 2001, pp 240-251. 79, 81

[18] Suleyman Sair and Mark Charney: "Memory Behavior of the SPEC2000 Benchmark Suite," Technical Report, IBM, 2000. 86

Exploiting Web Document Structure to Improve Storage Management in Proxy Caches

Abdolreza Abhari[1], Sivarama P. Dandamudi[1], and Shikharesh Majumdar[2]

[1] Center for Parallel and Distributed Computing
School of Computer Science, Carleton University, Ottawa, Canada
{abdy,sivarama}@scs.carleton.ca
[2] Deptartment of Systems & Computer Engineering
Carleton University, Ottawa, Canada
majumdar@sce.carleton.ca

Abstract. Proxy caches are essential to improve the performance of World Wide Web and to enhance user perceived latency. In this paper, we propose a new Web object based policy to manage the storage system of a proxy cache. We propose two techniques to improve the storage system performance. The first technique is concerned with prefetching the related files belonging to a Web object, from the disk to main memory. This prefetching improves performance as most of the files can be provided from the main memory instead of proxy disk. The second technique stores the Web object members in contiguous disk blocks in order to reduce the disk access time. This in turn reduces the disk response time. We have used trace-driven simulations to study the performance improvements one can obtain with these two techniques.

1 Introduction

The growth of the World Wide Web (WWW) has motivated researchers to investigate techniques to improve its performance. Web caching is one of the common techniques used for performance improvement. Several strategies have been proposed for web caching. The most popular ones are proxy caching [5], push caching [14] and Harvest caches [8]. In this paper, we focus on proxy caching in which one or more computers act as a cache for a set of WWW clients.

Proxy caching improves the performance by reducing user latency while obtaining the Web documents. The reason for this improvement is that proxy caches are typically closer to clients than to the web server. This proximity to clients also reduces the network traffic. Furthermore, since the proxy services most of the client requests, web server load decreases substantially, thereby improving its performance.

Although caching is a popular performance improvement technique, the cache hit ratio that can be achieved by most caching algorithms is in the range of 47% to 52% [19]. One way to improve the hit ratio is by prefetching the files into the cache. To implement prefetching, we need to anticipate future requests for files. There is a key difference between caching and prefetching: the purpose

S. Sahni et al. (Eds.) HiPC 2002, LNCS 2552, pp. 89–101, 2002.

of caching is to provide fast access for the second and subsequent accesses by exploiting temporal locality, whereas the goal of prefetching is to provide fast access even for the first time access to the files.

Several prefetching strategies have been proposed for Web caches. One simple technique is based on a geographical push-caching approach [14]. In this method when the load of Web server exceeds a limit, the server replicates (pushes) it's most popular documents to the other cooperating servers. Therefore, clients will be able to make future requests from these servers. The Top-10 prefetching [20] is based on the cooperation between client and server. Server is responsible for preparing the Top-10 list by analyzing the server access log. In the client site, a prefetching agent is responsible for analyzing the client access log to create the prefetching profile for that client. Client profile is the list of servers from which prefetching should be activated. Bestavros [7] proposed a model for speculative service. The notion of speculative service is analogous to prefetching that is used to improve cache performance in distributed/parallel shared memory systems, with the exception that the server, not the clients, controls when and what to prefetch. Mogul [21] proposed a prefetching model in which the server makes predictions and transmits these predictions as a hint to the client. Cohen et al. [10] suggested techniques for grouping resources that are likely to be accessed together into volumes. A volume is included in the server response to the requesting proxy. Hine and Willes [15] investigated the use of knowledge of resource dependencies in improving cache replacement policies. They analyzed the access log for identifying the group of related resources.

We introduced a Web object based prefetching technique for proxy caches [3]. This technique takes the web document structure into account. A *Web object* consists of the main HTML page and all of its constituent embedded files. Our study of popular Web documents to characterize Web objects has shown that most Web objects are small in size and can therefore be cached effectively at proxy or client sites [4]. In the Web object based cache policy, a reference to a member of a Web object is considered as a reference to all the members of the Web object. If the currently referenced file is not in cache, it not only fetches the referenced file but also prefetches all the other members of the Web object into the proxy cache. When a replacement is necessary to accommodate an incoming Web object, all the member files of one or more Web objects are removed. The number of Web objects to be removed depends on the size of the incoming Web object. Selection of a Web object for replacement depends on the basic cache replacement policy. In an LRU based policy (called WOLRU) the least recently used Web objects are removed (more details are available in [3]). Our studies have shown that this prefetching technique significantly improves the hit ratio without generating excessive network traffic [3]. In this paper, we use the Web object concept to devise storage management policies to improve the performance of proxy caches.

In this paper we refer to the delay incurred for serving a hit document (i.e., a document present in the proxy) as cache hit latency. Most proxy caches use main memory along with a disk as the cache. We can reduce the cache hit latency if the proxy provides more documents from its main memory instead of its disk. A recent study by Rousskov and Soloviev [22] reports that, for Squid [23], the disk delay contributes to about 30% of the total hit response time. Note that the disk delay is a part of the hit response time, which is the delay perceived by a client when a proxy serves the hit documents. They added extra code to Squid to enable them measure the disk delay contribution to the hit response time. They concluded that the disk storage subsystem might become a potential bottleneck in proxies. In addition, reducing the disk response time of a proxy reduces the latency of hit documents.

The remainder of the paper is organized as follows. Previous work on proxy cache storage management is briefly reviewed in Section 2. Section 3 presents details on the Web object based storage management policy. Performance of the Web-object based disk placement is discussed in Section 4. The impact on cache hit latency is discussed in Section 5.

2 Related Work on Proxy Cache Storage Management

To improve the performance of the storage system, several methods have been proposed. One common technique is to use a parallel disk system such as RAID in Web/proxy servers. Developing a file system that is specifically designed for the workload of Web/Proxy server is another method. For example, Hummingbird [12], developed at Bell Labs, is a lightweight file system for proxy caches. General distributed file systems such as AFS [16] and DFS [18] are also used to improve storage performance of the Web/proxy servers. Some commercial vendors proposed operating systems with special file systems running on dedicated hardware for proxy caches. For example, CacheFlow [9] provides proxy caches an operating system dedicated for caching, called CacheOS, running on a specific hardware.

Some of the differences between these proposals and our work are summarized below:

- Our strategy does not require a dedicated operating system such as CacheOS.
- Our techniques do not rely on a special file system such as Hummingbird for proxy caches.
- We introduce a new cache management policy that can be used for proxy applications such as Squid or Apache.
- Our proposed caching policy uses the Web object concept to manage the storage. Note that most proxy cache systems manage the storage by file-based policies. This Web object based policy uses the characteristics of the Web objects we introduced in [2, 4].

Based on the results of Web object characterization, we suggested the idea of storing related files in contiguous blocks to reduce disk response time [2]. This idea with some changes is implemented in Hummingbird recently. First, we explain their work and then we discuss how our work differs from theirs.

In a recent study, Gabber et al. [12] report that Hummingbird uses locality hints generated by proxy to pack files into large and fixed-sized clusters. A cluster is the unit of disk access in Hummingbird. Files in a cluster are stored in consecutive blocks for reading and writing in one disk access. Therefore, reading and writing are large operations. Although it is mentioned in [12] that an HTML page and its inline images usually are in the same locality set, it seems that Hummingbird analyzes a Least Recently Used (LRU) list to identify the files that belong to each locality set. Hummingbird stores each locality set into a cluster. From [12] it is not clear how the LRU list is analyzed to form clusters but it is clear that their grouping method is different from ours.

The main difference between our policy and Hummingbird is in the clustering method for finding the group of related files. In our Web object-based policy, the proxy parses each Web page upon receiving it from the server and finds the files that belong to the Web page. The proxy maintains pointers to keep Web object members together when they are in the main memory and when it wants to write them on the disk. Therefore, when our Web object-based policy reads the first member of a Web object from the disk it can prefetch the rest of members of that Web object into the main memory. Thus the subsequent requests for other members of that Web object are served from the proxy main memory. Using the Web object based policy can result in a substantial reduction in cache hit latency even when the Web object members are not stored contiguously (see Section 3 for more details). In our proposed policy, storing the Web object members in consecutive blocks makes the prefetching to be even faster. Hummingbird requires a special file system that uses locality set based clustering. Our technique produces a substantial reduction in cache hit latency by using conventional file systems.

3 Web Object Based Storage Management Policy

A document that is provided by the commonly used proxy applications such as Squid [23] or Apache [6] may exist either in the main memory or the disk. The first step to reduce the latency of hit documents is trying to provide more documents from the main memory rather than the disk of the proxy cache. For that reason we expand the Web object-based caching policy to efficiently manage the documents that exist in the proxy disk. In the proposed policy, we store Web objects instead of files in the proxy disk system. When a document is brought from the disk to the main memory, its related Web object members are also transferred to the main memory. In transferring a document from the main memory to disk we transfer all the members of its related web object. To study the performance of our proposed policy, we conducted several trace-driven simulation experiments.

3.1 Simulation

The programs we developed simulate disk-based proxy caching as follows: When a proxy receives a file or a Web object, it is placed in the main memory of that proxy. If memory is full, Web object based LRU replacement policy is used to create space for the incoming Web object. If we need to free space in the main memory, files/Web objects are transferred from the main memory to the disk if they are not already stored on the disk. Also, if the files/Web objects exist on the disk but are modified (their size has changed since the last access to them) they are transferred (i.e., overwritten) from main memory to disk in case we need free space in the main memory. The modified file is the one whose copy on the disk or main memory of the proxy is stale because it has changed in the server. When a file is changed in a server, proxy fetches it from the server upon the client's request. To find if the requested file is available at the proxy, search for that file is done in both main memory and disk. If the requested file is in the main memory, the number of hits on the main memory is incremented. Otherwise, if the file is on the disk, the number of disk hits is incremented and a Web object containing that file is brought into the main memory. In other situations the program acts according to the Web object based policy as explained in [3]. For example, if a requested file does not exist on the disk and main memory of the proxy, it is considered a miss, and the Web object corresponding to the file is prefetched from the server.

Traces Our trace-driven simulation program uses UC3 and PB traces. These two traces belong to the proxy caches of IRCACHE that is managed by National Laboratory for Applied Network Research (NLANR). UC3 is a log that belongs to a cache at the University of Illinois at Urbana-Champaign (UC). It is an entire access log recorded on March 12, 2000 and includes 453152 request entries. PB cache is located at the University of Pittsburgh (PB) (More details on the logs as well as other caches that are under NLANR administration are available in [17]).

Our PB trace is the first half (the first 12 hours) of the one-day access log that is recorded on December 25, 2001 and is made up of 566539 request entries. We cleaned the UC3 and PB traces to contain only cacheable URLs whose origin server responses contained an appropriate HTTP GET request with status code of 200 that indicates a successful transfer. We developed a program to identify Web objects in the clean UC3 and PB trace (See [1] for details). After running the Web object identification program, we identified unique Web objects for UC3 and PB traces. Note that each Web object is a group that consists of a Web page and all its related embedded objects. The members of Web objects identified by the program are unique and cacheable files that are in the PB and UC3 traces.

Performance metrics We considered two performance metrics to compare the LRU and WOLRU disk-based caching policies: memory hit ratio and disk hit ratio. Memory hit ratios can be computed by dividing the number of memory hits by the total number of hits produced by each policy. Similarly disk hit ratios are computed by dividing the number of disk hits by the total number of hits.

Cache sizes Cache consists of both main memory and disk in our simulation. The experiments are performed with a number of different main memory cache sizes: 16MB, 64MB, 256MB, 1GB, and 4GB. For all these main memory sizes,

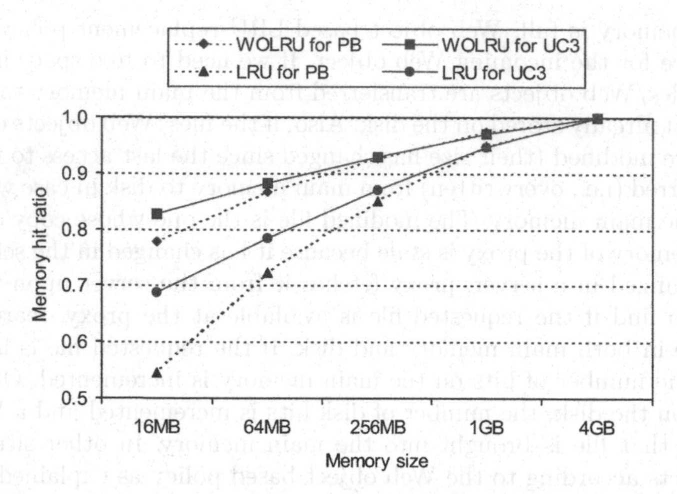

Fig. 1. Memory hit ratios of the WO-LRU and LRU policies for UC3 and PB traces

we assume that the disk size is more than the total size of requested files in each trace. Since the total size of requested files in each trace is less than 4GB this assumption is realistic. Thus, we don't have to evict any files or Web objects from the disk and we can focus on the transfers between main memory and disk.

3.2 Experimental Results

We have conducted several experiments to evaluate the performance of the WOLRU policy. Due to space limitations, we present a representative sample of the results in this section. A complete set of results is available in [1].

Figure 1 shows the memory hit ratios for the UC3 and PB traces. It can be seen from the data presented in this figure that WOLRU offers significant performance gains when the memory size is small. As the memory size increases, the difference decreases. In particular, for memory sizes greater than or equal to 4 GB, the replacement policy does not make any difference as all the files are in the main memory. Cache replacement policies make a difference when there is contention for main memory—thus it is important to replace appropriate files to make room for the new web objects/files. When we use smaller memory sizes, the performance difference between LRU and WOLRU clearly demonstrates that Web object based movement of files between the main memory and disk is desirable. For example, when the memory size is 64MB, the WOLRU policy provides a memory hit ratio improvement of 12.5% for the UC3 trace and nearly 20% for the PB trace. Note that as proxies handle a larger number of files, as they serve more web servers and clients, contention for memory is likely to increase. Memory contention is also expected to increase as the frequency of access for larger pages increases. With more memory contention the memory hit ratios achieved

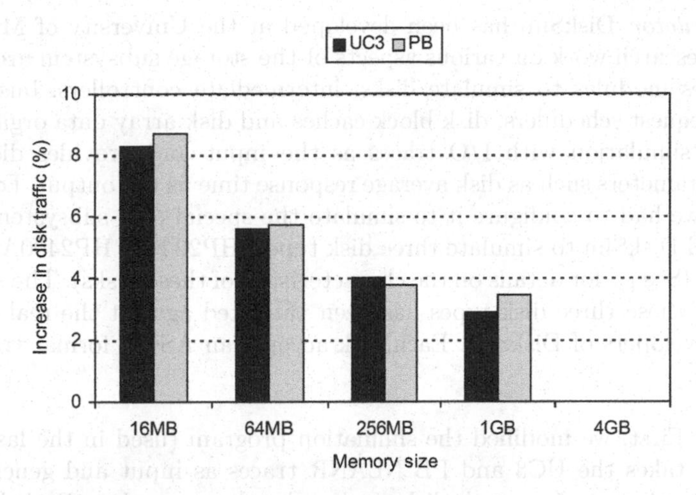

Fig. 2. Disk traffic increase due to WOLRU policy (in comparison to LRU)

with the smaller memory sizes in our simulation are likely to occur with higher memory sizes on such proxy cache systems.

Since WOLRU prefetches members of Web objects from disk to main memory and vice versa, it may transfer a higher number of files between disk and main memory in comparison to LRU. In order to see the increase in the disk traffic, we measured disk traffic as the number of files transferred between disk and memory for the WOLRU and LRU policies. Figure 2 shows the percentage increase in disk traffic due to WOLRU (over LRU). It can be seen from this figure that the increases is limited to less than 6% when the main memory size is 64MB. Even with a 16MB memory size, the increase in disk traffic is less than 9% for both UC3 and PB traces.

4 Web Object Based File Placement Policy

Secondary storage performance is usually measured by the disk drive behavior. Placement of files plays an important role in determining the disk access time. (A detailed discussion on evaluating storage devices is presented in [13, 24].) In this section, we look at how file placement affects the performance of a proxy cache. In particular, we want to know how much additional improvement we can get by using consecutive placement of member files of a Web object over random placement.

4.1 Disk Drive Simulation

To achieve our objectives, we performed another trace driven simulation but this time focusing on the disk performance. We evaluated the disk performance with one of the latest versions of the disk simulator called DiskSim [11]. A brief overview of its characteristics is presented next.

Disk Simulator DiskSim has been developed at the University of Michigan to support research work on various aspects of the storage subsystem architecture. It includes modules to simulate disks, intermediate controllers, buses, device drivers, request schedulers, disk block caches and disk array data organizations. It allows simulation with I/O traces as the input and provides disk performance parameters such as disk average response time as the output. For running DiskSim we had to configure it to simulate the special disk subsystem [11]. We configured DiskSim to simulate three disk types: HP2249A, HP2490A, and Seagate Elite (See [1] for details on the characteristics of these disks). The simulation of each of these three disk types has been validated against the real disk drive by the developers of DiskSim. Each disk accepts an ASCII format trace as the input.

Programs First, we modified the simulation program (used in the last section) so that it takes the UC3 and PB NLANR traces as input and generates high level traces that capture each disk read and write operation. Each line of the resulting traces contains a disk read or write event, time of the event that is read from UC3 and PB traces, the size and the name of the file that is read from or written to the disk and the Web object group information of that file.

We wrote two additional programs, both of which generate the ASCII format input traces. The first program generates a trace in which all the members of a Web Object are located in consecutive blocks. The second program generates a trace in which the members of a Web object are stored with the same device number but in random block positions. For both programs, device numbers are generated randomly and the size of a block is assumed to be 512 bytes.

Simulation Parameters We performed the simulations with the same range of main memory cache sizes mentioned in the last section. We configured DiskSim to simulate the three disks mentioned before. For each disk, we used the default component parameters suggested by DiskSim [11]. We used average disk response time as the performance metric. It is the mean response time for a disk request [11].

4.2 Results

The performance achieved with the placement of the Web object member files on consecutive blocks is compared with that of a random placement. Due to space restrictions, this section presents some representative sample of the results. Figure 3 shows the average disk response times for the WOLRU policy when the Web objects members are stored in consecutive and random blocks for the HP 2490 and Seagate disk systems. (Data for the other disk drives are available in [1].) These results show that the average disk response time can be reduced by approximately 1 to 4 msec for the simulated disks. For example, when memory size is 256MB, we noticed a reduction of about 10.31% in disk response time for the HP disk when consecutive placement strategy is used. The reduction in disk response time for the Seagate disk is about 11%. Note that this is only

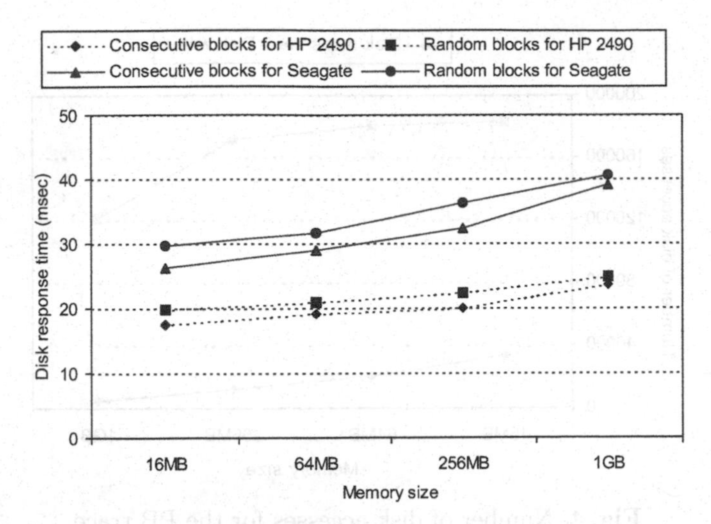

Fig. 3. Impact of consecutive and random block assignment on the disk response time of the WO-LRU policy for the PB trace

a reduction in the disk response time that is a part of the cache hit latency due to consecutive allocation of Web object members. The WOLRU policy has already achieved a major reduction in the hit latency by providing more documents from the main memory. Note that for reducing clutter in the graphs we have presented the data for only two disk types. A similar performance improvement is achieved by consecutive block allocation with the HP 2249 disk.

As expected, the total number of disk reads and writes decreases as memory size increases (see Figure 4). It is interesting to observe that irrespective of the reduction in the number of disk operations, the mean disk response time increases as the memory size increases (see Figures 3 and 4). We observed this phenomenon for all of the simulated disk types. We are currently investigating the reasons for this behavior.

5 Impact on Cache Hit Latency

Since this research focuses on performance improvement when a document is found in a proxy cache, the latency associated with a cache hit is considered. We define the cache hit latency as the delay for serving a hit document by a proxy. Cache hit latency is just a part of the total latency that is perceived by a client when proxy serves a hit document. The hit response time consists of the time for receiving client request after she/he is connected to the proxy, the time for sending the document to the client and the proxy cache hit latency [22]. Since proxy caches use both disk and main memory, cache hit latency can be calculated from the latency of the hit documents that are provided by the main memory and by the proxy disk. We assume that the latency of a document provided from

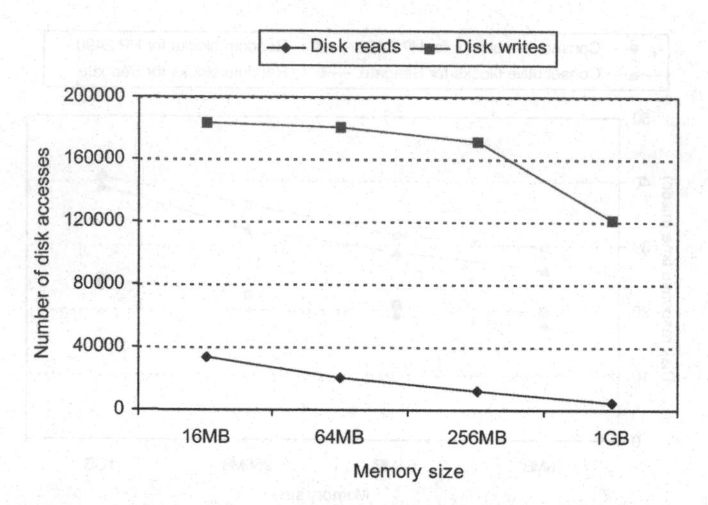

Fig. 4. Number of disk accesses for the PB trace

Table 1. Results for the Seagate disk with 256MB of main memory (UC3 trace)

Policy	Disk hit ratio	Disk response time (msec)
LRU	13%	41.43
WOLRU	7%	38.9
WOLRU (consecutive)	7%	35.05

the main memory is negligible compared to that provided from the disk. Thus, to compute the latency of the documents provided from the proxy disk, we need to know the disk response time of these documents. Under this assumption, the average cache hit latency (HL) for a proxy can be computed by multiplying the disk hit ratio (HD) by the average disk response time (DR).

$$HL = HD \times DR \tag{1}$$

In order to determine the reduction in the cache hit latency, the results for the UC3 trace are presented in Table 1. We found that other results are similar to these results for all traces, memory sizes and simulated disks.

The third column in Table 1 shows the average response time of the disk for the LRU and WOLRU policies. Second column in Table 1 gives the disk hit ratios when using the LRU and WOLRU policies. Therefore, the average latency of the hit documents, given by equation (1), is 5.39 msec and 2.72 msec for LRU and WOLRU, respectively. Thus, WOLRU reduces the latency for hit documents by about 50% for the UC3 NLANR trace.

By using equation (1), we can compute the average cache hit latency for WOLRU as 2.45 msec when the Web object members are stored in consecutive blocks. Thus, by using WOLRU, we reduce the hit latency for the UC3 trace by 55% compared to LRU with consecutive placement. These results suggest that, of the 55% reduction we obtained for the latency for the hit documents, we obtained a reduction of 50% by using WOLRU instead of LRU, and about 5% of the reduction was due to the consecutive placement policy.

6 Conclusion

In this paper, we studied Web object-based techniques for improving the performance of proxy cache systems. Two techniques are proposed for performance improvement. In the first technique a cluster of files that contains the main HTML and the associated embedded files that constitute a page is used as a unit for replacement and prefetching. The second technique attempts to reduce disk response time further by placing the files belonging to the same Web object on consecutive disk blocks.

Using trace-driven simulation we analyzed the performance of these two storage management techniques for proxy caches. Traces collected on real proxy caches are used in our simulation. Results indicate that a large improvement in memory hit ratio, of the order of 50% for example, can be achieved by using the Web object based WOLRU in comparison to the conventional LRU replacement policy. Allocating Web objects on consecutive disk blocks is observed to reduce disk seek time (by 20% for example), in comparison to a random placement of files on disks. The increase in memory hit ratio as well as the decrease in disk seek time, reduce the cache hit latency for the proxy cache.

Acknowledgements

Trace data used in our simulations were collected by the National Laboratory for Applied Network Research under National Science Foundation grants NCR-9616602 and NCR-951745. The disk simulator used to evaluate disk performance was DiskSim. DiskSim has been developed by G. R. Ganger at the University of Michigan to support research work on various aspects of storage subsystem architecture. Financial support for this research was provided by Nortel Networks, Carleton University and Natural Science and Engineering Research Council of Canada (NSERC).

References

[1] A. Abhari. *Web Object Based Policies for Managing Proxy Caches.* PhD thesis, 2002 (expected). 93, 94, 96

[2] A. Abhari, S. P. Dandamudi, and S. Majumdar. Characterization of Web objects in Popular Web Documents. In *ISCA 13th International Conference on Parallel and Distributed Computing Systems*, pages 616–624, Las Vegas, Nevada, U. S. A., Aug. 2000. 91

[3] A. Abhari, S. P. Dandamudi, and S. Majumdar. Using Web Object to Improve the Performance of Proxy Caching. In *Fourth International Workshop on Web Engineering at the World Wide Web WWW10 Conference*, pages 82–92, Hong Kong, May 2001. Available from `http://aeims.uws.edu.au/webe2001/webe-www10-proc.pdf`. 90, 93

[4] A. Abhari, S. P. Dandamudi, and S. Majumdar. Structural Characterization of popular Web Documents. *International Journal of Computers and Their Applications*, 9(1):15–24, Mar. 2002. 90, 91

[5] M. Abrams, C. Standridge, G. Abdulla, S. Williams, and E. Fox. Caching Proxies: Limitations and Potentials. In *Forth World Wide Web Conference '95: The Web Revolution*, Boston, MA, Dec. 1995. 89

[6] Apache Software Foundation. *Apache http server project.* `http://www.apache.org/httpd.html`. 92

[7] A. Bestavros. Using Speculation to Reduce Server Load and Service Time on the WWW. In *Fourth ACM Conference on Information and Knowledge Management*, Baltimore, Maryland, Nov. 1995. 90

[8] P. Bowman, D. Danzing, D. Hardy, U. Manber, M. Schwartz, and D. Wesseles. Harvest: A Scalable, Customizable Discovery and Access System. Technical Report CU-CS-732-95, University of Colorado, Department of Computer Science, Boulder, Colorado, 1995. 89

[9] CacheFlow. *Network cache performance measurement.* CacheFlow White Papers Version 2.1, Sept. 1998. 91

[10] E. Cohen, B. Krishnamurthy, and J. Rexford. Efficient Algorithms for Predicting Requests to Web Servers. In *IEEE INFOCOM*, Geneva, Switzerland, 1999. 90

[11] DISKSim. `http://www.ece.cmu.edu/ ganger/disksim` 95, 96

[12] E. Gabber, L. Huang, E. Shriver, and C. Stein. Storage Management for Web Proxies. In *2001 USENIX Annual Technical Conference*, Boston, MA, U. S. A., June 2001. 91, 92

[13] G. R. Ganger. Generating Representative Synthetic Workloads: An Unsolved Problem. In *Computer Measurement Group (CMG) Conference*, pages 1263–1269, Dec. 1995. 95

[14] J. Gwertzman. Autonomous Replication in Wide-Area Internetworks. B. A. Thesis, Center for Research in Computing Technology, Harvard University, Cambridge MA, Apr. 1995. 89, 90

[15] J. Hine, C. Wills, A. Martel, and J. Sommers. Combining Client Knowledge and Resource Dependencies for Improving World Wide Web Performance. In *INET'98 Conference*, Geneva, Switzerland, 1998. 90

[16] J. H. Howard, M. L. Kazar, S. G. Menees, D. A. Nichols, M. Satyanarayanan, R. N. Sidebotham, and M. J. West. Scale and Performance in a Distributed File System. *ACM Transactions on Computer Systems*, 6(1), Feb. 1998. 91

[17] IRCACHE. For details see `http://www.ircache.net`. 93

[18] M. Kazar, B. Leverrett, O. Anderson, V. Apostolides, B. Bottos, S. Chutani, C. Everhart, W. Mason, S. Tu, and E. Zayas. Decorum File System architecture overview. In *Summer 1990 USENIX Annual Technical Conference*, Anaheim, California, U. S. A., June 1990. 91

[19] T. M. Kroeger, D. D. E. Long, and J. C. Mogul. Exploring the bounds of web latency reduction from caching and prefetching. In *Symposium on Internetworking Systems and Technologies (USENIX)*, pages 13–22, Atlanta, Georgia, U. S. A., Dec. 1997. 89

[20] E. P. Markatos and C. E. Chronaki. A Top-10 Approach to Prefetching the Web. In *ICS4*, Geneva, Switzerland, Jan. 1996. 90

[21] J. C. Mogul. Hinted Caching in the Web. In *1996 SIGOPS European Workshop*, 1996. 90

[22] A. Rousskov and V. Soloviev. A Performance Study of Squid Proxy on HTTP/1.0. *World Wide Web Journal, Special Edition on WWW Characterization and Performance and Evaluation*, 1999. 91, 97

[23] Squid. For details on Squid, see **`http://squid.nlanr.net`**. 91, 92

[24] B. L. Worthington, G. R. Ganger, Y. N. Patt, and J. Wilkes. On-Line Extraction of SCSI Disk Drive Parameters. Technical Report CSE-TR-323-96, University of Michigan, Ann Arbor, Dec. 1996. 95

High Performance Multiprocessor Architecture Design Methodology for Application-Specific Embedded Systems

Syed Saif Abrar

Embedded Processor Dept., Philips Semiconductors, Bangalore, India
saif.abrar@philips.com

Abstract. This paper presents a methodology for the design of application-specific multiprocessor systems. The methodology, named AIM, guides a designer to select the right heterogeneous architecture starting from a set of target applications. The methodology distinguishes between applications and architectures that are modeled as Kahn Process Networks and cycle-accurate bit-true models, respectively. A gradual mapping of applications over architectures is defined to analyze the performance of the resulting system by simulation. As a consequence, functionally complete system is always available with the designer without the overwhelming issues related with the multiprocessor architectures. The described methodology is illustrated through the design of a multiprocessor architecture for an MPEG-1 audio decoder application.

1 Introduction

The increasing digitalization of information in text, speech, video, audio and graphics has resulted in a whole new variety of network and microprocessor technologies. This leads to a demand for new application specific architectures that are increasingly programmable. However, performance requirements, cost constraints and power consumption still require involvement of dedicated hardware blocks. These heterogeneous architectures [1] need a design technology to assist designers in defining such architectures.

This paper presents AIM (Application, Interconnect and Multiprocessor-architecture modeling) methodology for application-specific multiprocessor designs. The term Multiprocessor architecture involves CPUs, DSPs and coprocessors. The computational resources of a multiprocessor architecture can be fully exploited only when the application model has sufficient parallelism. Hence, exploration under AIM methodology starts from executable specifications of a set of target applications. The result is the definition of a multiprocessor architecture capable of executing these applications within predefined constraints with respect to speed, cost, power, etc.

Typically, the design of heterogeneous architectures follows the Y-chart scheme [2]. A limitation of this approach is that the functional resources (CPUs, DSPs) are introduced early in the design stage. The designer faces, all at once, a very wide

S. Sahni et al. (Eds.) HiPC 2002, LNCS 2552, pp. 102-111, 2002.

architecture space to be explored. This may leave out some important questions, like power consumption and minimization, unanswered. This paper introduces the AIM methodology that pushes the involvement of processor(s) or dedicated hardware resources late in the architecture definition. The new methodology is demonstrated through the design of an MPEG 1 Layer 3 audio decoder.

The rest of the paper is organized as follows. Section 2 discusses the related work in the area of performance analyses for applications and architectures. The methodology is discussed in detail in section 3. Section 4 presents the case study and results obtained using AIM. Conclusions are presented in section 5.

2 Related Work

Application modeling has received a lot of interest in order to derive a model of computation. Kahn Process Network [3] is a model of computation that is often used for signal processing applications. Variants like Synchronous Dataflow [4] and Dataflow Process Networks [5] have been studied thoroughly. The Ptolemy system [6] has been designed to support a heterogeneous mix of models of computation for cosimulation. It attempts to combine the semantics of control and data-flow models at their interfaces [7].

Architecture modeling for performance analysis is actively being pursued in different forms. Systolic array [8][9] and wavefront [10] design methodologies transform digital signal processing applications in various steps of refinement into a particular architecture. The hardware/software codesign approach relies on an architecture template to design architectures [11]. However, it lacks the ability to deal effectively with making trade-offs in favor of the set of applications. In order to make domain-specific embedded systems, both programmable [12] and dedicated [13] architectures exist.

Abstract models are also used for quantitative analysis of architectures, as described in [14]. Architecture exploration and performance analysis can be done very fast due to easily construction of abstract models. Cycle-based simulation is also being currently practiced by designers, using frameworks like SystemC and Cynlib. VHDL models have also been used to provide performance statistics. As part of the RASSP [15] project, the ADEPT environment [16] provides constructing a system by interconnecting predefined architecture models. These models contain both functional and timing behavior but do not separate application and architecture.

AIM methodology makes a clear distinction between applications and architectures, and supports gradual mapping of application onto architecture. Immediate mapping of applications onto processor(s) forces a designer to tackle processor-based issues very early in the design-cycle. AIM alleviates this trap by pushing the involvement of processor(s) or dedicated hardware resources late in the architecture definition. As a consequence, a high reusability of both application and processor-independent architecture is obtained. Alternative architectures, with varying system-level parameters, can be efficiently explored. The modeling technique employed by AIM makes it particularly suited for multiprocessor embedded systems.

3 Methodology

3.1 Application Modeling

The objective of application modeling is to expose parallelism while making communication explicit. To model the applications, a process graph is used based on the Kahn Process Networks [3]. Communication between processes, executing in parallel, is via unbounded FIFO channels. These channels are characterized by blocking reads and non-blocking writes. A Kahn process network executes in a deterministic order [17] producing same output data for a given input load irrespective of the schedule. AIM uses the YAPI framework [18] for application modeling which extends the deterministic model of Kahn process networks with non-deterministic events. The process graph under YAPI is modeled as a set of processes and a set of channels/FIFOs. Fig. 1 shows an application model in which processes describe the functionality of the system and communicate via channels.

The processes use READ, WRITE and SELECT primitives for interaction. The READ function consumes data from a process' input port and stores in a local variable of the process. The WRITE function copies the value of a local variable to an output port. The SELECT function selects an input or output port that eventually will produce or consume data, respectively. The application modeling results in the measurement of an application's input-dependent computation and communication workload. The computation workload shows the number of times an instruction is executed by the process. The communication workload is a count of the number of tokens traveling though the channels. Hence, the granularity of parallelism can be modified and studied, irrespective of the architecture.

As shown in Fig. 1, an application model works in a pipeline fashion and can contain bi-directional communication. The designer must aim at balancing the process pipelines, reducing bi-directional communication and identifying synchronization points and mechanisms, if needed.

3.2 Interconnect Modeling

The next step in the AIM methodology is to model the system wide interconnects. With the ongoing advancements in VLSI technology, the performance of an embedded system is determined to a large extent by the communication of data and instructions. Various interconnect topologies are possible for a multiprocessor system: bus, ring, torus, mesh, and others. Interconnect modeling aims at providing quantitative answers to the problems of communication protocol, bandwidth requirements, power dissipation, etc. The AIM proposes the use of TSS [19] to model the communication network. TSS is a cycle-based architecture modeling and simulation framework. This step aims at freeing the designer from the task of application's computation and concentrating on the application's communication. As a result, only the cycle-accurate models of interconnects are deployed and not that of the processes developed during the application modeling phase. High-level abstract models of the processes are used for a functionally correct system. Cycle-accurate interconnect models provide a link to the abstract process models to make the complete system. Fig. 2 shows a shared-memory single-bus system as an example of the interconnect model.

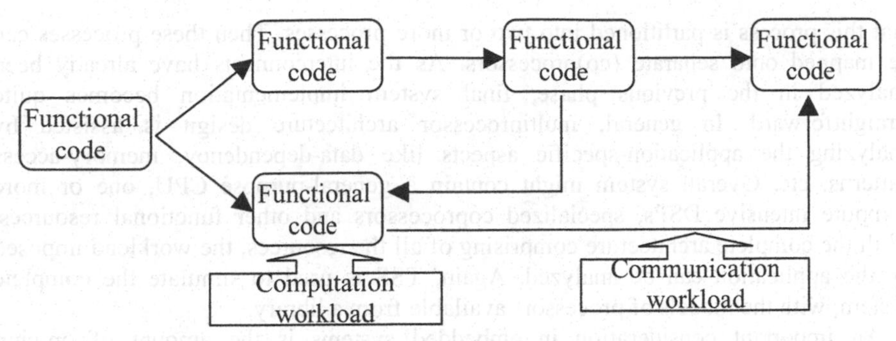

Fig. 1. Application model

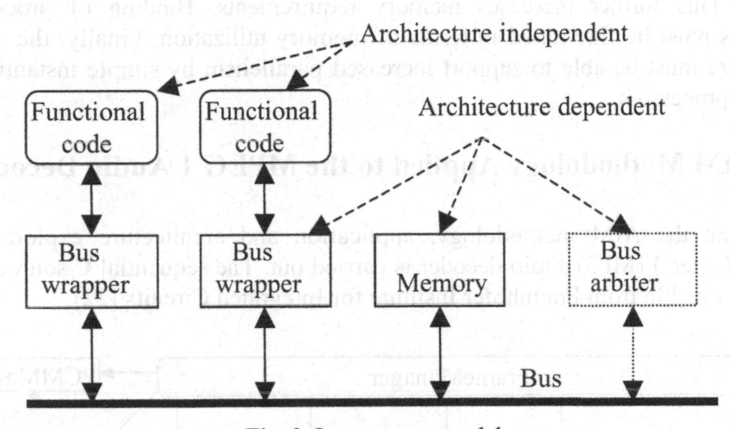

Fig. 2. Interconnect model

The bus wrapper provides a link to the architecture for the processing models and can be changed to study different bus protocols. Other parameters, like bus width, setup delay, transfer delay, that influence a system can also be specified and analyzed.

Application specific mobile embedded systems imposes very high constraints on power consumption and hence, the power minimizing techniques are also to be analyzed at this level. Cycle-true encoder/decoder models for T0 [20] and BI [21] coding techniques have been developed. These are used along with the interconnect models to study reduction in the bit-transitions. At this level, the designer gets an insight into the memory access patterns, system-wide endianness, power metrics, memory ranges, etc. This and the subsequent step form the architecture modeling and mapping steps of the Y-chart, albeit modified.

3.3 Multiprocessor Architecture Modeling

The final task in the AIM methodology is to replace each functional model from the previous step by a hardware resource. This replacement can be one-to-one or many-to-one, depending on the availability and the characteristics of each resource. If it appears that the functionality of a single process needs to be distributed over more than one processing resource, then the designer has to rewrite the application such that this process is partitioned into two or more processes. Then these processes can be mapped onto separate (co)processors. As the interconnects have already been analyzed in the previous phase, final system implementation becomes quite straightforward. In general, multiprocessor architecture design is assisted by analyzing the application-specific aspects like data-dependency, memory-access patterns, etc. Overall system might contain a general-purpose CPU, one or more compute intensive DSPs, specialized coprocessors and other functional resources. With the complete architecture comprising of all the resources, the workload imposed by the application can be analyzed. Again, TSS is used to simulate the complete system, with the models of processors available from a library.

An important consideration in embedded systems is the amount of on-chip memory. In case of multiprocessor architectures, handshaking and context-saving are inherent. This further increases memory requirements. Binding of processes to processors must be fine-tuned in terms of memory utilization. Finally, the resulting architecture must be able to support increased parallelism by simple instantiation of more (co)processors.

4 AIM Methodology Applied to the MPEG 1 Audio Decoder

To validate the AIM methodology, application and architecture exploration for MPEG 1 Layer 3 (MP3) audio decoder is carried out. The sequential C source code is publicly available from Fraunhofer Institute for Integrated Circuits [22].

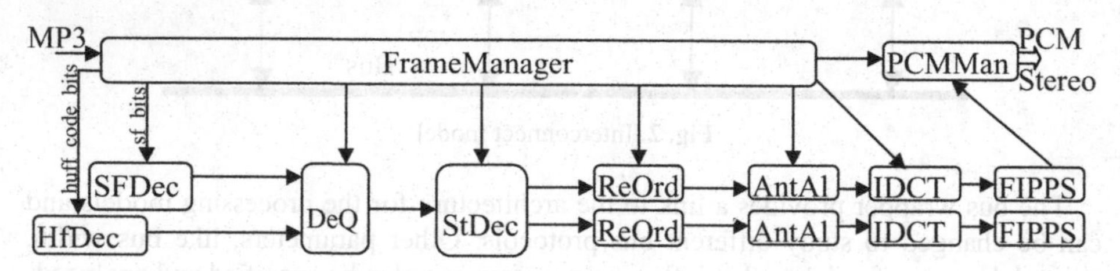

Fig. 3. Process network for the MP3 decoder

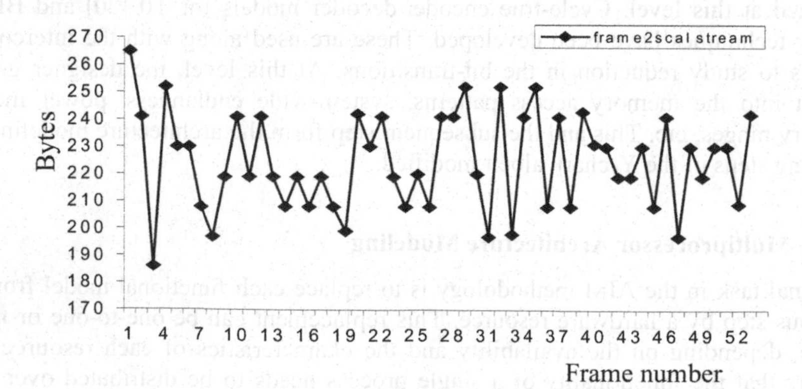

Fig. 4. Communication requirement per frame

As the first step of the AIM methodology, it is partitioned into parallely executing and communicating Kahn Process Network, as shown in Fig. 3. The input to the process network is MP3 stream that is decoded to form the (stereo) PCM output.

Using YAPI to simulate this process graph, statistics for computation and communication workloads are gathered. Fig. 4 shows the variable nature of data stream being passed from the frame-manager (FrameManager) to the scale-factor-decode (SFDec) block.

Workload, in terms of amount of data communicated and number of operations performed, is also analyzed, as shown in Table 1. Number of tokens beings passed through each channel are given in (a) and the call to each procedure being made within a process are shown in (b).

For the next step of Interconnect modeling under the AIM, the process graph is ported onto the cycle-based simulation framework, as shown in the Fig. 2. For simplicity, shared-memory single-bus synchronous design is considered here. The communication part of a process is ported on to a bus model and the computation part is linked to it as a functionally correct C code. Each process from the process-graph shown in Fig. 3 is mapped onto a dedicated resource. In such a setup, it is easy to study bus protocols, to map functional data-types (int/float/byte) onto the supported bus-widths (16/32/64 bits), to define memory ranges for each process, etc. By instantiating alternate models, two busses were analyzed in terms of the total clock cycles required to decode an MP3 stream. Table 2 shows the results for the ARM Ltd. proposed AMBA-AHB [23] and other based on the Philips PRISC [24] architecture.

Table 1. Workload analysis for the MP3 process-graph

(a) Communication workload

Channel	Tokens
Sf_bits	11906
huf_code_bits	148386
...	...

(b) Computation workload

Process	Instruction	Frequency
HfDec	HUFF_DEC	14504
FIPPS	SYNT_POLY	8354304
...	...	...

Table 2. Comparing two bus protocols in terms of simulation cycles

Bus	Clock-cycles
AMBA-AHB	5539238
PRISC Bus	5539229

Table 3. Bus-transition savings through encoding

Coding scheme	FrameManager	IDCT
T0	32.3%	2.2%
Bus-invert	0.29%	26.5%

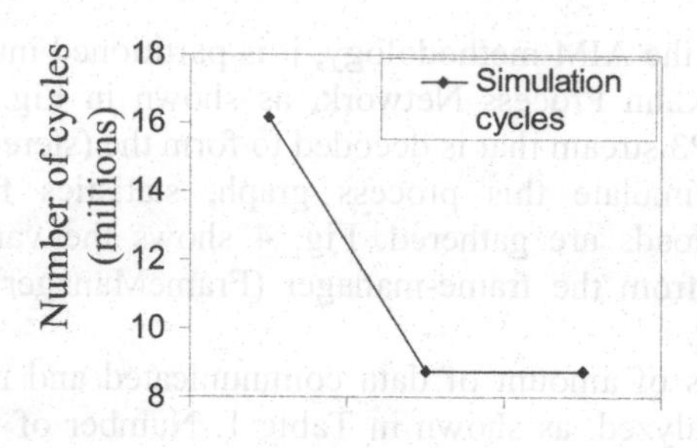

Fig. 5. Simulation cycles taken to decode MP3 stream

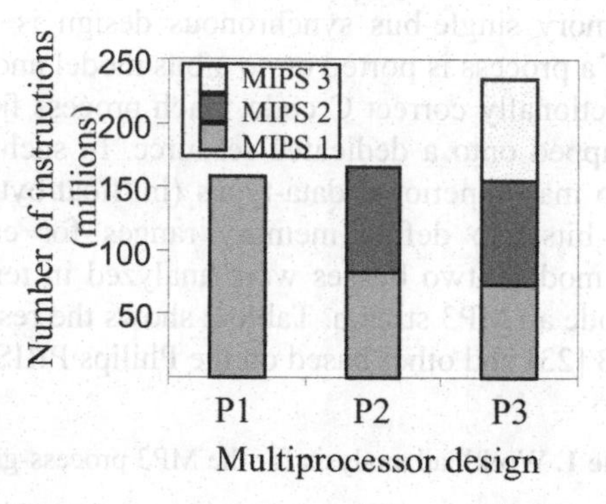

Fig. 6. Number of instructions executed and the contribution of each processor

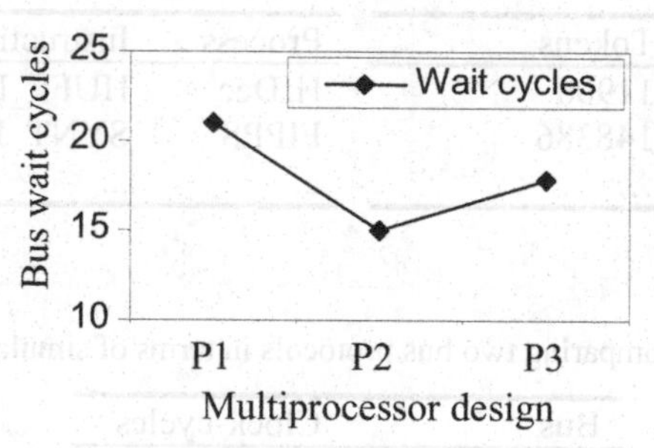

Fig. 7. Number of wait-cycles for each design

It is observed that both the bus protocols consume same clock cycles. It may be due to the fact that both are targeted towards on-chip systems, have been optimized for the same and hence, perform in a similar fashion. Further, bit-toggling minimization techniques, like T0 and BI codings, are studied for the address and data buses. The address bus savings for the frame-manager and the IDCT blocks are shown in Table 3. As the frame-manager takes data sequentially, as opposed to the IDCT, T0 coding offers more savings than the BI coding. The IDCT is at an advantage with the BI coding due to lesser in-sequence addresses as compared to the frame-manager.

Finally, the complete multiprocessor architecture is designed as outlined earlier. As a simplistic decoder architecture, single MIPS processor model is instantiated. All the abstract processes, from step 2 (Application modeling), are executed in software over this MIPS model. This is not a completely parallel realization, as envisioned earlier, but is an important step taken to verify a system's functionality over an embedded processor. Also, speed-up over single processor implementation can be measured after multiprocessor design. After verification with one processor, the design is extended with two and three processors.

Fig. 5 shows the simulator cycles taken to decode an MP3 stream for single (P1), dual (P2) and triple (P3) processor designs. It is observed that there is almost 100% speed improvement for the P2 whereas P3 hardly gives better performance. This implies that incorporating two MIPS processors in the design saturates the performance with respect to speed.

Fig. 6 shows the total number of instructions executed and also compares the contribution of individual processors, in each design. It is evident that each processor is equally involved in the application. This shows that the workload in properly mapped onto MIPS resources. However, with more processors (design P3), the total instruction count increases due to more handshaking and coherence operations.

An important design consideration in shared memory bus-based systems is the time for which the processors are waiting for data. Fig. 7 shows the average number of cycles a processor is blocked due to cache-fill request. P2 has lower waiting time as two processors are keeping the bus busier than was possible with single processor in P1. However, P3 imposes high communication overhead because of three processors and hence has higher waiting time than P2.

From the above results for the three designs (P1, P2 and P3), it is observed that P2 performs better than P1 and P3 in all respects. And must be selected for final decoder implementation.

5 Conclusions

This paper presents a technique for designing multiprocessor architectures for application specific domains. The introduced methodology, AIM, proposes a novel concept of Interconnect modeling. The other techniques used by AIM are Application and Multiprocessor-architecture modeling. Earlier design methodologies concentrated on the functional resources from the start and did not take into considerations the effects of communication structure in detail. This lead to unoptimized system-wide interconnects. The AIM models an application as Kahn process network that provides statistics on the workload. The communication structure between the processes is modeled in cycle-accurate architecture-dependent environment. The processes themselves are executed independent of the architecture. This shows the performance of the interconnection protocol, memory access patterns, power minimization coding effects, etc. Finally, the functionally correct processes are mapped onto the resources like CPUs and DSPs. As the designer moves gradually from the initial application specification to the final multiprocessor implementation, functionally correct system is always available. The methodology is general and hence, broad class of architectures and applications can be modeled. The methodology is particularly suited for signal processing applications executing on multiprocessing architectures.

Design of an MP3 audio decoder using AIM reveals that the considered bus protocols give the same performance whereas applying bus-encoding techniques results in power minimization. An interesting and important result obtained is that the multiprocessor implementation with two MIPS processors performs better than that with three processors.

References

1. Peter, P., Stolberg, H.J.: VLSI implementations of image and video multimedia processing systems. IEEE Trans. on Circuits and Systems for Video technology, Vol. 8. (1998) 878-891
2. Kienhuis, A.C.J.: Design Space Exploration of Streambased Dataflow Architectures- Methods and Tools, Ph.D. thesis, Delft University of Technology (1999)
3. Kahn, G.: The semantics of a simple language for parallel programming, Proc. of the IFIP Congress 74, North Holland Publishing Co. (1974)
4. Lee, E.A., David,G.M.,: Synchronous data flow. Proc. of the IEEE, Vol. 75. (1987) 1235-1245
5. Lee, E.A., Parks, T.M.,: Dataflow process networks. Proc. of the IEEE, Vol. 83. (1995) 773-799
6. Joseph, B., Ha, S., Lee, E.A., David, G.M.: Ptolemy: A framework for simulating and prototyping heterogeneous systems. Int. Journal of Computer Simulation, Special issue on Simulation Software Development. (1992)
7. Chang, W., Ha, S., Lee, E.A.,: Heterogeneous simulation - mixing discrete event models with dataflow. Journal of VLSI Processing, Vol. 15. (1997) 127-144
8. Kung, H.T.: Why systolic architectures?. IEEE Computer, Vol. 15. (1982)
9. Kung, H.T., Leiserson, C.E.: Systolic arrays (for VLSI). Sparse Matrix Symposium, SIAM (1978) 256 –282
10. Kung, S.Y., Arun, K.S., Gal-Ezer, R.J., Rao, D.V. B.: Wavefront array processor: language, architecture, and applications. IEEE Trans. on Computers, Special Issue on Parallel and Distributed Computer, Vol. 31. (1982) 1054 - 1066
11. Michele, G.M., Sami, M.: Hardware/Software Co-Design. Series E: Applied Sciences, NATO ASI Series, Vol. 310. (1996)
12. Chen, D.C., Rabaey, J.M.: Paddi: Programmable arithmetic devices for digital signal processing. Proc. of VLSI Signal Processing IV (1990) 240 - 249
13. Lippens, P.E.R., van Meerbergen, J.L., van der Werf, A., Verhaegh, W.F.J., McSweeney, B.T., Huisken, J.O., McArdle, O.P.: PHIDEO: A silicon compiler for high speed algorithms. Proc. EDAC (1991) 436 - 441
14. Gupta R.K., Liao, S.Y.: Using a programming language for digital system design. IEEE Design & Test of Computers,Vol. 14. (1997) 72 - 80
15. Hein, C., Pridgen, J., Kline, W.: RASSP: virtual prototyping of DSP systems. Proc. DAC (1997)

16. Klenke, R.H., Meyassed, M., Aylor, J.H., Johnson, B.W., Rao, R., Ghosh, A.: An integrated design environment for performance and dependability analysis. Proc. DAC (1997)
17. Kahn, G., MacQueen, D.B.: Coroutines and networks of parallel processes. Proc. of the IFIP Congress 77. North-Holland Publishing Company Co. (1977)
18. de Kock E.A.: YAPI: Application modeling for signal processing systems.Proc.DAC (2000)
19. Kruijtzer, W.: TSS: Tool for System Simulation. ISTNewsletter, Philips Internal Publication (1997) 5 - 7
20. Benini, L., Micheli, G.D., Macii, E., Sciuto, D., Silvano, C.: Asymptotic Zero-Transition Activity Encoding for Address Busses in Low-Power Microprocessor-Based Systems. IEEE 7th Great Lakes Symposium on VLSI. (1997) 77-82
21. Stan, M.R., Burleson, W.P.: Bus-Invert Coding for Low-Power I/O. IEEE Trans. on VLSI Systems, Vol. 3. (1995) 49-58
22. http://www.iis.fhg.de/amm
23. Flynn, D.: AMBA: Enabling Reusable On-Chip Designs", IEEE Micro (1997)
24. Klapproth, P.: PRISC Architecture Framework. Philips Internal Publication (1999)

LLM: A Low Latency Messaging Infrastructure for Linux Clusters*

R. K. Shyamasundar, Basant Rajan, Manish Prasad, and Amit Jain

School Of Technology and Computer Science
Tata Institute of Fundamental Research
Mumbai-400005,India
{shyam,basant,manish,amitjain}@tcs.tifr.res.in

Abstract. In this paper, we develop a messaging infrastructure, called LLM, to arrive at a robust and efficient low latency message passing infrastructure for kernel-to-kernel communication. The main focus is to overcome the high latencies associated with the conventional communication protocol stack management of TCP/IP. The LLM provides a transport protocol that offers high reliability at the fragment level keeping the acknowledgment overhead low given the high reliability levels of the LAN. The system utilizes some of the architectural facilities provided by the Linux kernel specially designed for optimization in the respective areas. Reliability against fragment losses is ensured by using a low overhead negative acknowledgment scheme. The implementation is in the form of loadable modules extending the Linux OS. In a typical implementation on a cluster of two nodes, each of uniprocessor Intel Pentium 400 MHz on a 10/100 Mbps LAN achieved an average round trip latency of .169ms as compared to the .531ms obtained by ICMP (Ping) protocol. A relative comparison of LLM with others is also provided.

1 Introduction

With the advent of high performance network interconnects such as ATM, Fiber Channel, FDDI, Myrinet etc., the possibility that workstation clusters can deliver good performance on a broad range of parallel computations has increased tremendously. However, the most common messaging layers used for clusters such as TCP/IP, PVM (PVM is a higher level library which in turn uses TCP/IP, like most other cluster communication libraries) etc. generally have not delivered a large fraction of the underlying communication hardware performance to the applications. There have been several attempts to improve performance using specialized hardware such as Active Messages [5], Illinois Fast Messages [6], PM [7], U-Net [8] and GAMMA[10]. While these experiments show higher performance,they have the disadvantage of special hardware requirements and hence additional costs. In the design of LLM, we focus on optimizing the software messaging layer at the kernel level to deliver a large fraction of the network's physical

* The work was done under the project *Design and Implementation of Secure Systems for E-Commerce*, supported from MIT, New Delhi.

S. Sahni et al. (Eds.) HiPC 2002, LNCS 2552, pp. 112–123, 2002.

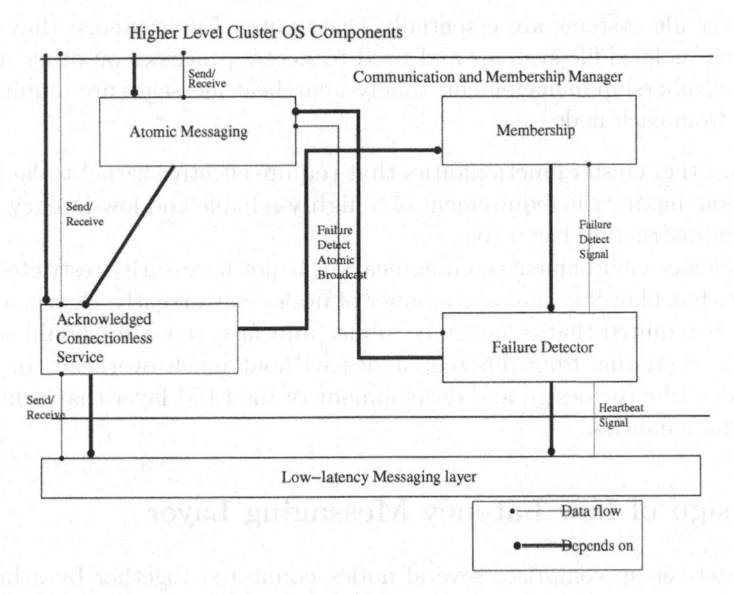

Fig. 1. Cluster messaging architecture

performance, without the use of any specialized hardware. The architecture of LLM and the services it provides shown in Figure 1 consists of:

1. *Low Latency Messaging Layer* (LLM): is responsible to provide the CMM layer complete network transparency by providing an effective and reliable transport of message packets from one cluster node to another.
2. *Communication and Membership Management* (CMM): is responsible for making the higher level cluster OS components transparent to cluster membership changes by employing a failure detection scheme using a two-level heartbeat mechanism, which enhances scalability by reducing the context switches per node. CMM uses the services of LLM for actual transmission of the message and doesn't handle network unreliabilities as it expects that functionality from LLM.

To achieve clustered computing capabilities at the kernel level, a cardinal requirement is a robust and efficient low-latency messaging infrastructure for kernel-to-kernel communication (between two cluster nodes). This infrastructure should make the network errors and cluster membership transitions transparent to higher level kernel components. Some of the requirements expected from this infrastructure are:

1. The distributed lock manager (DLM) is responsible for managing resource allocation and synchronizing concurrent accesses to cluster-wide shared resources and runs on all cluster nodes. Hence, it needs to communicate among peer DLM components on other nodes.

2. Cluster file systems are essentially Distributed File systems; they need to behave as local file systems and need to access processes on other nodes.
3. For membership management, timely heartbeat messages are required to be sent from each node.

There are other cluster functionalities that require effective kernel-to-kernel communication making the requirement of a highly reliable and low-latency message passing infrastructure inevitable.

In a cluster environment, communication is not necessarily restricted to just two nodes but plausible among a number of nodes concurrently. Hence, an infrastructure is required that is not only robust and fast, but also should segregate messages originating from different nodes without much overhead. In this paper, we describe the design and development of the LLM layer that achieves the above functionalities.

2 Design of Low Latency Messaging Layer

Our cluster set-up comprises several nodes connected together by a broadcast LAN and is dynamic where new nodes may join/exit at any time - thus changing the cluster membership. In our LLM design (i) we use a weak asynchronous model in the sense that certain timing considerations are used, and (ii) we assume that interconnecting channels could be lossy and the messages might be reordered/duplicated (although reordering is not possible in a LAN setup). However, we assume that the error rate of the channel under consideration is much lower than a wide area network.

2.1 Design Goals of Low Latency Messaging (LLM) Layer

The LLM is responsible for providing low-latency, reliable transport of message packets from one cluster node to another. The LLM achieves a significant reduction in latency by speeding up memory allocation by object caching [2] using the Linux kernel slab cache [3] and ensures reliability against fragment losses by using a low overhead negative acknowledgment based scheme. The design goals of the LLM layer and guarantees made by it to the layer above are:

i) Reliable transport of message packets from one cluster node to another: This is done using a negative acknowledgment based scheme that requires the destination node to send negative acknowledgments whenever it doesn't receive the expected data fragments.
ii) Multicast filtering based on multicast group identifiers.
iii) Packet fragmentation to provide for the inconsistencies present in allocation of large amount of kernel memory in the Linux OS while copying data from the user to the kernel area.
iv) Packet fragmentation (according to the maximum transfer unit (MTU) of the network) at the sending end and reassembly of fragments back into the original packet in the correct order at the receiving end.

v) Latency reduction in memory allocation by using the Linux kernel slab cache for structured data memory allocations and a simple slab allocator implementation for general purpose memory allocations.

2.2 Interface of LLM with Other Layers

LLM provides its services to the upper layer that is responsible for the communication and membership management of the cluster nodes, and the LLM in turn uses the services provided by the data link layer for the actual sending and receiving of frames. The interface to the actual link layer driver is abstracted by the device independent layer provided by the Linux kernel networking subsystem. The device independent layer provides the higher level protocol layers with a host of generic device-independent functions. The broad ideas used by LLM in gaining advantages over TCP/IP are:
1. *LLM is a minimal TCP and IP rolled into one*: In TCP/IP, TCP is the layer which offers reliability. But this reliability is offered at a packet-level. IP offers a best efforts service, which is unreliable, and it performs fragmentation and reassembly. It also performs routing and a whole lot of other functions; however, these do not come into the picture in a pure LAN environment. It must be noted that in case of a fragment loss, the IP layer (Reliability is not its function) does not retransmit the lost fragment since the only way the peer IP layer at the recipient node would respond to a missing fragment is by aborting the transfer. The onus of assuring reliability is completely on the TCP layer. Hence, it performs packet retransmission on timeout. There are two major disadvantages to this approach are:

a. The amount of duplicate data maintained at any point of time could be enormous in case of big packets, as whole packets instead of fragments are backed up for retransmission.
b. Even the loss of a single fragment in a huge packet results in a retransmission of the entire packet, which means the whole operation of fragmentation and reassembly have to be repeated.

2. *LLM serves as a transport protocol* offering high reliability at the fragment level, which at the same time keeps the acknowledgment overhead low considering the high reliability levels of a LAN, would help eliminate both these disadvantages. The amount of duplicate data maintained at any point of time is low, since backed up fragments which need not be retransmitted are periodically cleared up. Secondly, loss of a single fragment results in the retransmission of just that particular fragment, and not the whole packet.

3 LLM-Implementation

The main purpose of LLM is to reduce the communication time between the nodes of a cluster. For efficient service, it is required that the messaging is fast,reliable and robust. In order to achieve this, the LLM protocol utilizes some

Procedure	Function
*1. sys_clustsend(void *data,unsigned char dst, int exp)*	System call invoked when data at the user area has to be transmitted from one node to another.
*2. build_pkt(struct sk_buff *skb,void *udata,int len, u8 dst)*	This procedre builds a network (skb) packet around the data to be transmitted. The various skb parameters are specified and the buffers are copied from the user to the kernel area depending on the size of the data to be transmitted.
3. sync_send(msg , exp)	If message is to be transmitted synchronously, then this function is invoked. 'exp''indicates wether an acknowledgement is expected for the message.
4.async_send(msg, exp)	This function is invoked for asynchronous messaging.
*5. llm_send(struct sk_buff *msg, int flags)*	This is the main send routine.Here the headers are pushed on the network buffers, data is divided in MTU packets, queued and transmitted.
*6. llm_rcv(struct sk_buff *msg, struct net_device *dev struct packet_typr *pt)*	This routine is invoked at the receiver end whenever the underlying device driver detects an LLM packet. The incoming packet is received in the skb buffer and is passed on to the CMM layer.

Fig. 2. Functional description of procedures

of the Linux kernel architectural features as well as innovative acknowledgment and timing schemes for packet transfer between the nodes. Some of the aspects employed are briefed below:

- Speeding up memory allocation using the Linux kernel slab cache.
- A simple slab allocator for general purpose memory allocation.
- Even though the negative acknowledgment has a some overhead, reliability is assured by requesting acknowledgments at regular intervals.
- By associating each fragment and the whole packet with relative timers, acknowledgment overheads are kept low. Further, backed up fragments that need not be retransmitted are cleared periodically leading to savings in memory space and search for fragments that need to be retransmitted.
- The need to traverse all the fragments to determine whether a particular fragment has arrived, is avoided by using the concept of gap (cf.section 3).
- In case of loss of fragments, only fragments that are lost are retransmitted instead of the entire packet.

In the following, we shall describe the protocol used for communication between the cluster nodes. Consider cluster communication flow from node A to B. Using the operations shown in the table shown in Fig. 2, the protocol designed is depicted in Fig. 3 and Fig. 4 (A-B) which are self explanatory.

The actual data flow through LLM is depicted in Fig. 5. The underlying process is described below (labels correspond to those shown in the figure).

a. If data to be transmitted is greater than 30K (30828 bytes - a multiple of 1468 bytes, the maximum data which can be transmitted in one fragment, excluding the headers), then this data is divided into 30K packets.

b. This packet is then copied from the user area to the kernel area. The entire undivided packet is copied using required number of context switches. A network (`skb`) buffer is built around this packet.

c. At the kernel area, the 30K packets are further divide into MTU packets, each of 1468 bytes and the '`skb`' buffers are built around each packet leaving enough empty space for the headers to be pushed, hence avoiding further copies of the data packets for adding the headers.

d. The headers are pushed into the network buffers and the packet is queued for transmission.

e. At the receiver's end whenever the underlying device driver detects the appropriate packet, LLM receive function is invoked and the incoming packet is received in a similar `skb` buffer.

f. The LLM header is detached and the packet is passed to the CMM layer.

g The CMM header is detached and the fragment is kept in the kernel memory. All incoming packets go through the same procedure and are queued up in the kernel area.

h. After the receipt of the entire packet, the data is copied to the user area.

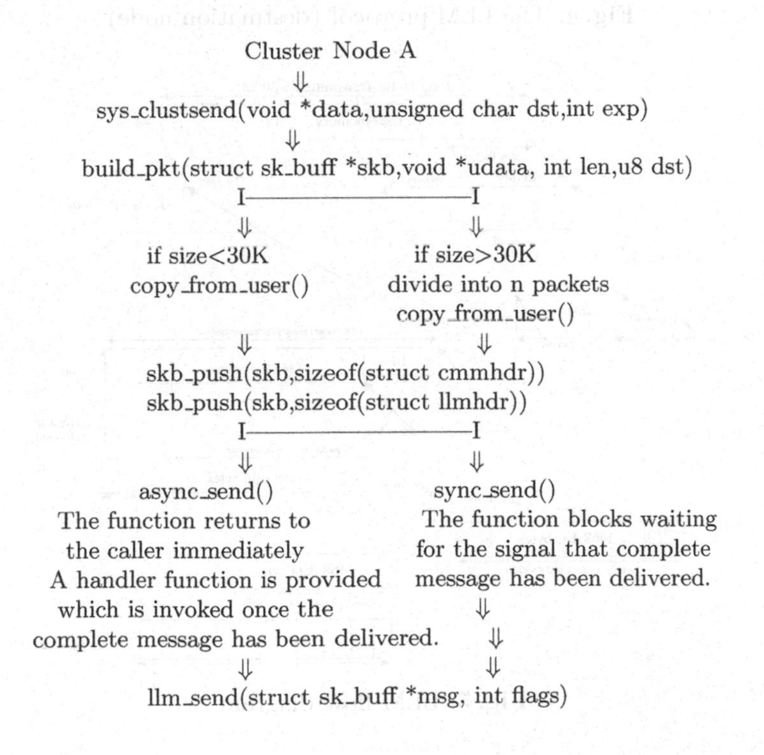

Cluster Node A
sys_clustsend(void *data,unsigned char dst,int exp)
build_pkt(struct sk_buff *skb,void *udata, int len,u8 dst)
if size<30K
copy_from_user()
if size>30K
divide into n packets
copy_from_user()
skb_push(skb,sizeof(struct cmmhdr))
skb_push(skb,sizeof(struct llmhdr))
async_send()
The function returns to
the caller immediately
A handler function is provided
which is invoked once the
complete message has been delivered.
sync_send()
The function blocks waiting
for the signal that complete
message has been delivered.
llm_send(struct sk_buff *msg, int flags)

Fig. 3. The LLM protocol (source node)

{clust_addr src_node
clust_addr dst_node
unsigned char src_hw
[*MAX_ADDR_LEN*]
unsigned char dst_hw
[*MAX_ADDR_LEN*]

unsigned char ack_nak
__u16 pkt_id
__u8 frag_id

unsigned char more_frag
__u8 nfrags }

(A) LLM Header Structure

At Cluster Node B
Incoming Packet
⇓
llm_rcv (struct sk_buff *skb,
struct net_device *dev,struct packet_type *pt)
⇓
llmh = (struct llmhdr *)(skb→nh.raw
if llmh →
acks_exp send appropriate acknowledgement
⇓
skb_pull (skb,sizeof (struct llthdr))
⇓
skb_pull(skb,sizeof (struct cmmhdr)
⇓
sys_clustrcv(void *data)
copy_to_user(data,buffer,size)

(B) Protocol at destination Node B

Fig. 4. The LLM protocol (destination node)

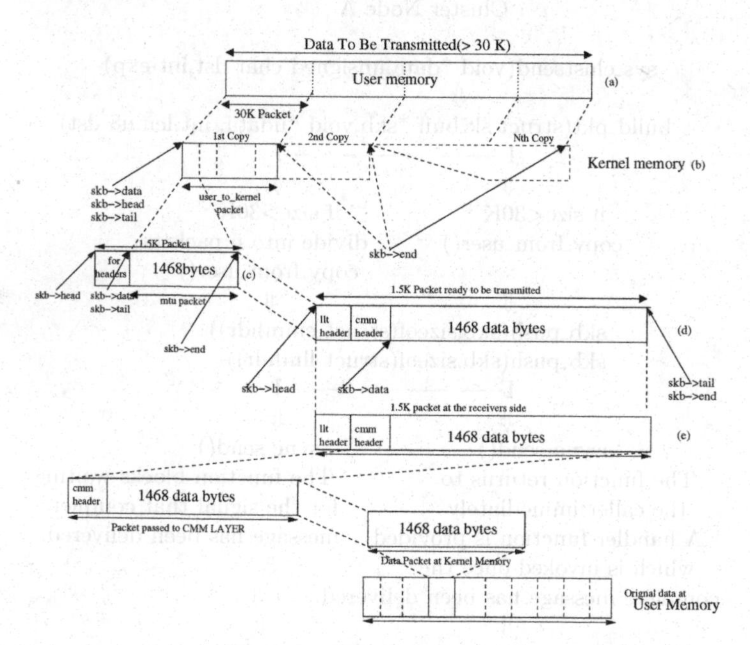

Fig. 5. LLM flow diagram

3.1 Implementation of LLM

The crux of the implementation lies broadly in the following phases: (i) Initialization, (ii) Send and Receive, (iii) Fragmentation and Reassembly, and (iv) Simple slab allocator. These aspects are detailed in the following.
Initialization Phase: The LLM initialization comprises of:

- **Registering the packet type:** LLM has to register its packet type (ETH_P_LLM) with the underlying device driver so that it knows that a packet with protocol id ETH_P_LLM is to be delivered to the LLM driver. This is done by the '`dev_add_pack()`' [4] routine.
- **Cache initialization:** To reduce latency due to memory allocation, the LLM extensively uses the Linux slab cache [2], for creating and storing structured data objects. This significantly reduces object initialization time, which many a times exceeds the cost of allocating and freeing the object. Using the Linux Slab Allocator[3] constructed objects are obtained in which the Linux Network Buffers`skbuff`[4] are already initialized.

Send and Receive (*llm_send* and *llm_rcv*): The LLM provides CMM with interfaces for reliable acknowledged as well as unacknowledged services. These functionalities in the following manner.

- **Reliable acknowledged service:** When a reliable service is expected, LLM sets up a queue data structure at the sending end for the fragments of a packet to be sent. This queue is added on to a hash table which uses the `packet id` as the key. It creates a backup copy of each fragment being transported for retransmission in case a `NAK` (negative acknowledgment) arrives due to a fragment being lost. The backup copy of the fragment is queued up at the end of the appropriate packet queue.
  At the recipient's end, on receiving the first fragment of a packet, if a reliable service is expected, it sets up a packet queue similar to that at the sender's end and adds it on to the hash table, using both the `packet id` and the sender `host id` as the key. Each fragment of a packet that has been received is added to the appropriate queue, taking care to see that the fragments are queued in the correct order.
  Acknowledgment Scheme: The acknowledgment scheme employed by the LLM is a negative acknowledgment based scheme. The high reliability attributed to the communication channel (in a LAN), obviates the need of acknowledging each and every packet sent. Instead if the recipient finds a fragment missing, that is, if it receives an unexpected fragment, it sends a negative acknowledgment (`NAK_FRAG` along with the fragment *id* not arrived to the sending host.
  Timing Considerations: In order to implement the above mentioned negative acknowledgment scheme, the following six timers are employed :
  a. Each fragment in the packet queue has a timer associated with it, on expiry of which (meaning the fragment must have reached the destination properly) the fragment is deleted from the sent packet queue and the memory is freed.

b. Also, every fragment whose number is a multiple of ACK_EXPECTED - a constant value - is an ACK (an acknowledgment has to be sent for this fragment) able fragment. Whenever an ACK able fragment is received, and all other fragments of the packet numbered lower than it have been received, the recipient is required to send an ACK_FRAGS (indicating that all the fragments up to this particular fragment has reached.
c. The last fragment of every packet is also an ACK able fragment. On receiving an ACK_FRAGS, all fragment backups on the sent queue are dequeued and the memory is freed.
d. Once all fragments of a packet have been received, the recipient is required to send an ACK_PKT (indicating all the fragments of this particular packet has reached) to the sending host, indicating the `packet id` for which the acknowledgment is being sent.
e. Every sent packet queue in the hash table has an `ACK_timer` (associated with it which is set whenever an 'ACK' able fragment is sent). On expiry of this ACK_TIMER, a NAK_ACK is sent to the recipient host, indicating the `fragment id` till which an acknowledgment is expected.
f. At the recipient's end each received packet queue has a receive timer associated with it. Every time a fragment of a particular packet arrives, the timer associated with its packet queue is reset. On expiry of this timer, meaning that its been some time (equal to MAX_RCV_LAT) since a fragment of this packet has arrived, the recipient sends a NAK_FRAGS_RANGE(meaning that fragments up to this number have reached and waiting for further fragments) to the sending host specifying the fragment number last received.

This negative acknowledgment has a significantly low overhead as compared to a pure positive acknowledgment scheme, wherein the recipient is required to ACK every fragment it receives. At the same time, reliability is preserved by putting some onus of assuring reliable delivery of the packet on the sender host as well in form of requesting acknowledgments at regular intervals and also for the complete packet. This assures reliable delivery of the packet even in the event when all fragments of a packet get lost or in case of single-fragment packets, which would otherwise fail in a purely negative acknowledgment based scheme (since in such cases the recipient would be unaware of such a packet being sent and hence wouldn't issue a negative acknowledgment).

- **Unacknowledged service:** On the other hand, if reliable service is not expected, there are no such overheads at both the sender and recipient end. It is a best efforts service. If it gets all the fragments right (no losses) in the first attempt it reassembles them into the original packet and forwards it to the CMM. The semantics of the receive timer differ as compared to those in case of a reliable service as in, here on expiry of this timer the earlier fragments are dropped and the communication is aborted.

Fragmentation (*llm_fragment()*) and Reassembly (*reassemble_frags()*): Depending on the size of the packet to be transmitted, LLM fragments the data packet to be transmitted in two different types of fragments.

User_to_kernel packet: While copying the data from the user area to kernel area, LLM fragments the data to be transmitted into packets of 30K each, so that a maximum 32K of memory is allocated on a single call, hence not upsetting the memory management by redefining the buddy system [9] table/list, which may happen if large chunks of memory are alloted in a single call.
MTU packet: If the size of the packet being sent is greater than MTU, then LLM breaks up the packet into fragments of size equal to that of MTU. The LLM header of each fragment sent, bears its fragment *id*, the *id* of the packet of which it is a fragment and indicates whether it is the last fragment of the packet (element `more_frag`); 0 corresponds to the last fragment and 1 corresponds to its negation.

On receiving a fragment, the recipient searches the hash table with the (`pkt_id` and the sender host *id* as the key) for the receive packet queue. If it doesn't exist, i.e., if it is the first fragment of the packet received, a queue structure is created, and if it does exist it adds the new fragment to the queue taking into consideration the correct order of fragments (in this case, the ascending order); the fragment with the lowest *id* is placed at the start of the queue and the one with the largest *id* is placed at the end. On receiving the last fragment (`more_frag` = 0), it checks if there are anymore fragments yet to arrive. If all the fragments of a packet have indeed been received, then the fragments are reassembled into the original packet and forwarded to the CMM.

The time required in deciding if all fragments up to a certain fragment *id* have been received, i.e the time required to traverse all fragments in the queue up to this fragment *id* starting from fragment *id 1* has been avoided by introducing a variable, called `frag_miss`; `frag_miss` is incremented every time a new `gap` is created in the receive packet queue and decremented every time a 'gap' is closed. We say that a new 'gap' is created in the queue, whenever the fragment *id* of the new fragment added differs from that of those placed, both, before and after it, by a value greater than one. A 'gap' in the queue is closed, whenever the fragment *id* of the new fragment added differs from that of those placed, both, before and after it, exactly by one.
Simple Slab Allocator: To expedite general purpose memory allocations, we have implemented a simple 'bucket-with-a-hole' slab allocator. I.e., memory that is freed is returned to the system, and not to the slab allocator's pool. This strategy keeps the task simple, without additional memory management overhead and yet achieves reduction in memory allocation time as compared to kmalloc(). The slab data structure is given below:

```
struct clu_kmem_slab {
        struct clu_kmem_slab  *next;
        struct clu_kmem_slab  *prev;
        size_t                          free;
        /* amount of free space left */
        void                          *next_kmemp;
       /*start of memory to be allocated next*/
        struct timer_list          alloc_timer;
                };
```

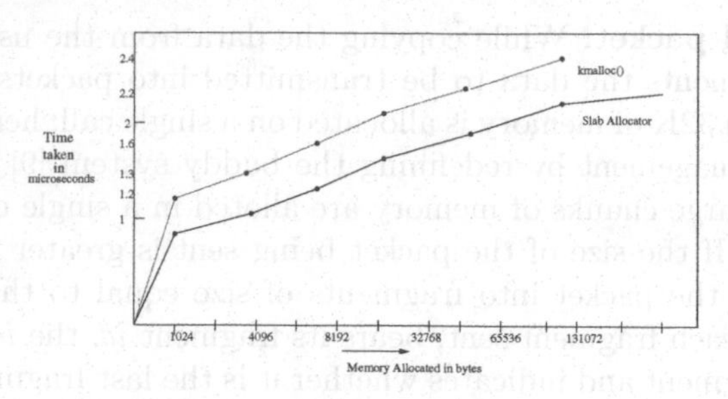

Fig. 6. Slab Allocator vs kmalloc

Table 1. Performance Measurements

Size of data Packet Transmitted (Excluding Protocol Headers)	No. of Packets Sent	Average time taken TCP/IP	LLM
1Kbytes	30	1.02ms	.520ms
2Kbytes	30	2.02ms	1.17ms
4Kbytes	30	3.7ms	2.1ms
16Kbytes	30	15.3ms	7.5ms
32Kbytes	30	31.6ms	14.3ms
64Kbytes	30	60.5ms	27.9ms
128Kbytes	30	133ms	55.5ms

Thus, the slab is primarily a linked list of pointers to free memory locations. The memory allocation interface is given by:

void *clu_kmem_alloc(size_t size, int flags) instead of the **kmalloc()** call. Fig. 6 shows the advantage of this Slab allocator over kmalloc().

4 Performance (Ping-Pong Test)

The prototype was tested on the following configuration: (i) Number of Machines : 2 (Both identical), (ii) Processor Configuration : Uniprocessor Intel Pentium 400 MHz, (iii) Linux Kernel version : 2.4.2, (iv) Broadcast LAN bandwidth : 10 Mbps, and (v) Network Device Controller: Realtek Semiconductor 8139 Ethernet Controller (PCI)

The round trip test was carried out by executing 10,000 round-trips of a 64 byte message. We achieved an average (a round trip time) latency of **.169ms**. The test similar to the ICMP (Ping) protocol yielded a time of **0.531ms**. Performance measurements relative to TCP/IP are shown in Table 1 and Fig. 6.

5 Conclusions

In the paper, we have described the design and implementation of a robust and low-latency message passing infrastructure, LLM, that overcomes some of the overheads of TCP/IP in the context of clusters. It differentiates itself by providing integration within the operating system and thus the capability to interact with the rest of the communication system. In particular, low transportation time has been achieved without the use of any specialized hardware. Our comparison of LLM with TCP/IP and other communication stack management schemes shows a good 40% to 50% reduction in time. Also as the size of the data to be transmitted increases the performance improves further. This can be attributed to the fact that as packet size increases the Linux slab caches come into the picture, and thus, bringing about optimization.

To sum up, LLM provides a fast networking infrastructure without any additional hardware for distributed systems. Further experiments on optimizations and scalability issues are in progress.

References

[1] A. Barak, I. Metrik,*Performance of the communication layers of TCP/IP with the Myrinet Gigabit LAN*, Computer Communications, Vol. 22, No. 11, July 1999, www.mosix.cs.huji.ac.il/ftps/com.ps.gz.

[2] Jeff Bonwick, *The slab allocator: An object caching kernel memory allocator*, USENIX Summer Tech. Conf., Boston, Mass. 1994.

[3] Brad Fitzgibbons, *Linux slab allocator*, www.cc.gatech.edu/people/home/bradf/cs7001/proj2/linux_slab.html.

[4] *Linux kernel sources*: path – /usr/src/linux/net/core/dev.c; /usr/include/linux/skbuff.h, /usr/src/linux/net/core/skbuff.c.

[5] A. Mainwarning, D. E.Culler, *Active Message Application Programming Interface and Communication Subsystem Organization*, TR, Univ. of Calif., Berkeley, 1995.

[6] S. P.V. Karamcheti, A. A.Chien,*Fast Messages: Efficient, Portable Communication for Workstation Clusters and MPP's*, IEEE Concurrency 5(2):60-73, 1997.

[7] H. Tezuka, A.Hori, Y.Ishikawa, M.Sato, *PM: An Operating System Coordinated High Performance Communication Library*, in Proc.Int.Conf. on High-Performance Computing and Networking (HPCN Europe 1997), pp. 708-717, April 1997.

[8] T.von Eicken, A.Basu, W.Vogels, *U-Net: A user level network interface for parallel and distributed computing*, in Proc.15th ACM Symp on Operating Systems Principle, pp. 40 - 53, 1995.

[9] J. Peterson, T. Norman, *Buddy Systems*, CACM, June 1977.

[10] Giuseppe Ciaccio *A Communication system for Efficient Parallel Processing on Clusters of Personal Computers*,PhD Thesis DISI-TH-1999-02, June 1999.

Low-Power High-Performance Adaptive Computing Architectures for Multimedia Processing*

Rama Sangireddy, Huesung Kim, and Arun K. Somani

Dependable Computing & Networking Laboratory, Department of Electrical and Computer Engineering
Iowa State University, Ames, IA 50011, USA
{sangired,huesung,arun}@iastate.edu

Abstract. The demand for higher computing power and thus more on-chip computing resources is ever increasing. The size of on-chip cache memory has also been consistently increasing. To efficiently utilize silicon real-estate on the chip, a part of L1 data cache is designed as a Reconfigurable Functional Cache (RFC), that can be configured to perform a selective core function in the media application whenever higher computing capability is required. The idea of Adaptive Balanced Computing architecture was developed, where the RFC module is used as a coprocessor controlled by main processor. Initial results have proved that ABC architecture provides speedups ranging from 1.04x to 5.0x for various media applications. In this paper, we address the impact of RFC on cache access time and energy dissipation. We show that reduced number of cache accesses and lesser utilization of other on-chip resources will result in energy savings of up to 60% for MPEG decoding, and in the range of 10% to 20% for various other multimedia applications.

1 Introduction

Multimedia and digital signal processing applications demand higher computing power. The widely known 90-10 rule predicates that 90% of the execution time is expended by about 10% of the application code which is computation intensive and that the remaining 10% of execution time is consumed by inner loops in general. Spatial structures excel in the execution of such computation intensive functions as compared to temporal structures [1]. Reconfigurability of such spatial structures provides flexibility to the system for executing a wide range of computation intensive functions. The time for loading configuration of a particular function can be amortized over long execution time and hence can be offset by the speed-up obtained. However, reconfigurable devices get bogged down on large portions of the code that are rarely executed with repetition and

* The research reported in this paper is partially funded by the grants Carver's Trust and Nicholas Professorship from Iowa State University, and grant No. CCR9900601 from NSF.

S. Sahni et al. (Eds.) HiPC 2002, LNCS 2552, pp. 124–134, 2002.

hence loading the configuration for every small segment of computation becomes a bottleneck. A practical compromise is to couple a reconfigurable device with a conventional general purpose processor. Thus, the concept of a general purpose processor with a tightly coupled reconfigurable logic arrays has been widely recognized as the main focus for development of future computing systems.

Current processor designs often devote a largest fraction (up to 80%) of on-chip transistors to caches. However, many workloads, like media processor applications, do not fully utilize the large cache resources due to streaming nature and lack of temporal locality for the data. From these observations, an idea of a different kind of computing machine - Adaptive Balanced Computing (ABC) architecture, has evolved. ABC uses a dynamic configuration of a part of on-chip cache memory to convert it into a specialized computing unit. A reconfigurable functional cache (RFC) operates as a conventional cache memory or a specialized computing unit [2]. The first version of ABC microprocessor [3] has proved the utility of embedding RFCs in superscalar processor, by accelerating the media applications with speedups in the range of 1.04x to 5.0x.

Since cache access is in critical timing path of the processor, it is important to address the impact of making cache reconfigurable on cache access time. In this paper, we show that even with the area overhead, access time of RFC is same as conventional cache at larger cache associativities due to parallel decoding mechanism employed. Further, due to increased concern of power dissipation in the system, current microprocessors are required to deliver higher performance even while keeping the levels of power dissipation at minimum. Area overhead caused due to added routing structure in the cache will result in higher power dissipation in RFC. However, reduced number of cache accesses and lesser utilization of other on-chip resources, due to a significant reduction in execution time of the application, will result in significant savings in energy consumption. In this paper, we demonstrate from the simulation results that overall there is savings in the energy.

The rest of the paper is organized as follows. Section II provides organization of ABC microprocessor and addresses issues related to the impact on cache access time and energy dissipation. Section III presents the performance of ABC processor as compared to a conventional processor for various media benchmarks. In Section IV, we present the analysis for our claim of significant savings in the power consumption, along with the details of power models used. Section V concludes the discussion.

2 ABC Microprocessor

The ABC architecture is built by incorporating a multiple-way set associative data cache memory in a RISC superscalar microprocessor. Some modules in the set associative data cache are built as RFCs. Each RFC module can be configured to a specialized computing function or can be used as a normal data cache memory module. The RFC module is constructed as a two dimensional array of multibit output lookup tables (LUTs).

2.1 RFC Microarchitecture

The components in a conventional cache structure constitute of decoders, data and tag arrays, sense amplifiers (in both data and tag arrays), comparators, multiplexor drivers, and output drivers. It can be noted that, to convert a conventional cache into an RFC, it is necessary to modify only the organization of data array with addition of the routing structure to facilitate the computation of a function. The organization of the rest of the cache components remains unchanged. Hence, in this section we discuss, in detail, only the microarchitecture of the data array in RFC.

The basic cache parameters are C (cache size in bytes), B (block size in bytes), A (associativity), and S (number of sets = C/(B*A)). To alleviate the problem of longer-than-necessary access time, Wada et al. proposed the division of array into subarrays, and presented four parameters, N_{dwl} and N_{dbl} for data array and N_{twl} and N_{tbl} for tag array [4]. The parameter N_{dwl} indicates the number of times data array is split with vertical cut lines, and N_{dbl} indicates the number of times data array is split with horizontal cut lines. The total number of subarrays in the data array of the cache is N_{dwl} x N_{dbl}. Similarly, N_{twl} x N_{tbl} is the number of subarrays in the tag array. Further, Jouppi et al. used two other organization parameters, N_{spd} and N_{tspd}, which indicate the number of sets mapped to a single wordline in data array and tag array, respectively [5, 6]. Thus, for the organization of two-dimensional array of LUTs to build the RFC, parameters N_{dwl} and N_{dbl} are equivalent to the number of rows of LUTs ($= S/N_{lutlines}$) and the number of columns of LUTs ($= 8 * B * A/N_{lutbits}$), respectively. $N_{lutlines}$ is the number of lines in one LUT, 16 for 4-LUT, while $N_{lutbits}$ is the width of one line in an LUT. The organization of the wordlines, bitlines and the parallel decoding structure in RFC module had been extensively discussed in [2]. In this paper, we concentrate more on the impact of such organization on the cache access time and energy dissipation, the two critical parameters in a processor design.

In a 4-way RFC, consider a situation in which one module is active as a computing unit while the other three modules are serving as data cache. During the read and write accesses, the data and tag are read from the sets of all the modules. The bitlines and wordlines are active in the computing module too. However, the set that is selected in the computing module will behave different from the other modules, according to the index decoding performed for set selection. A detailed discussion on the configuration and computation modes of RFC is given in [3].

2.2 Computing Structure in RFC Modules

The routing structure for computation laid out between two rows of LUTs facilitates the computation of a function by allowing flow of execution data from one row of processing elements (PEs) to the other. Discrete Cosine Transform (DCT) is the most efficient technique in image encoding and compression schemes. Convolution, a DSP algorithm, is another common requirement in signal and image processing for pattern recognition, edge detection, etc. Using these two algorithms to map to RFC modules, we can implement the most common multimedia applications like mpegdecode, mpegencode, cjpeg from UCLA mediabench [7] and FIR, IIR from TMS320C6000 benchmarks [8].

The number of pipeline stages for convolution in an RFC depends on the size of the cache. A conventional convolution algorithm (FIR) is shown in Equation 1,

$$y(n) = \sum_{k=0}^{L-1} h(k)x(n-k), \qquad n = 0, 1, 2, ..., \infty \tag{1}$$

where, L is the number of taps and h(k) is the constant coefficient. Similarly, DCT/IDCT function, which is the most effective transform technique for image and video processing is shown in Equations 2.

$$Y = C^T A\, C \qquad and, \qquad A = CYC^T. \tag{2}$$

where, A is a square image (or a square portion of a large image), C is a constant coefficient matrix with same dimensionality as A, and C^T is a transpose of C. Consequently, Y is the transformation matrix.

The one stage of convolution, consisting of a multiplier and an adder, is shown in Figure 1. The first row of LUTs implements an 8-bit constant coefficient multiplier, as a two 4x8 partial products, and the second row implements the addition of the partial products. The next two rows implement a 24-bit adder to accumulate up to 256 taps of FIR filter. In one module (way) of a cache with 256 sets, we can implement four such stages. Similarly, implementation of one PE of DCT/IDCT in two rows of LUTs in an RFC module can be found in shown in [3]. In both the functions, carry select adder scheme is used to implement a faster addition operation, thus minimizing propagation delay for the entire operation. Thus, the routing structure for computation consists of only registers, bus lines and 2-to-1 multiplexers to enable the data flow between rows of LUTs. To address in detail various issues related to the performance of RFC and its impact on the microprocessor in this paper, we avoid the discussion on the details of the LUT-based computational implementation of the above functions. An extensive discussion on the implementation of the functions in LUT based RFC is given in [3].

2.3 Access Time and Energy Dissipation

The propagation delay in various components in a cache contribute towards the cache access time, while the switching activity in them for each cache access contributes towards dynamic power consumption. The organizational parameters discussed in Section 2.1 influence access time and energy dissipation, as variation in each of these parameters results in the requirement of varying number of components. For example, increasing N_{dbl} or N_{spd} increases the number of sense amplifiers required, while increasing N_{dwl} results in the requirement of more wordline drivers. To measure access time and energy dissipation in RFC, we

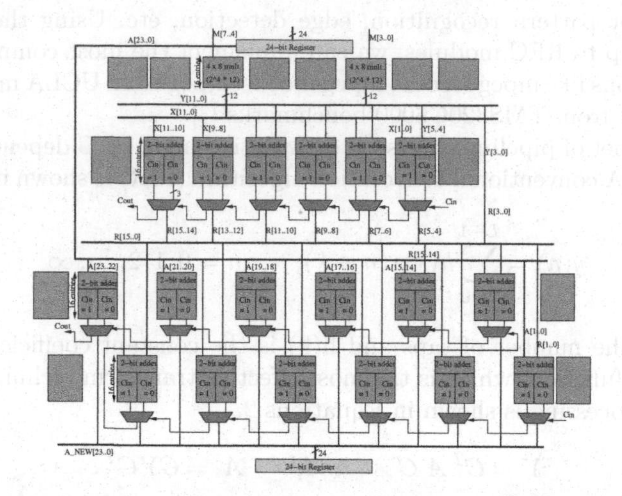

Fig. 1. One tap of FIR filter implemented in four rows of LUTs in an RFC module with 8 columns of LUTs. The blank LUTs do not participate in computation mode, but function normally in memory mode

have used CACTI models [5, 6], with necessary modifications to account for the overhead caused due to the routing structure for computation. The parameters are varied across a spectrum of cache sizes and associativity, there would be a steady increase in the number of decoders in the data array. We incorporated these necessary features in the model for the estimation of access time and energy dissipation in RFC. The values for data array parameters N_{spd}, N_{dwl} and N_{dbl} in the conventional cache are chosen as 1, 2, and 4, respectively. The values for tag array parameters N_{tspd}, N_{twl}, and N_{tbl} are chosen as 2, 1, and 2, respectively. These are the best values obtained from the simulation results in [6] for optimum access time and energy dissipation in the cache. Figure 2(a) shows the estimation of access time, in nano seconds, for base cache and RFC, with varying cache organizations. The estimation of energy dissipation per cache access, in nano Joules, is shown in Figure 2(b). In each of these comparisons, an RFC with implementation of the convolution algorithm, is considered. Similar results are obtained for an RFC with the DCT/IDCT implementation. In a set-associative RFC, when one of the modules is active as a computing unit, the other (A-1) modules will be in memory mode. For this reason, direct-mapped cache is not considered for the implementation of RFC. Hence, the results in Figure 2 are shown for 2- to 32-way set-associative caches with varying sizes.

In RFC, routing structure between the rows of LUTs causes an additional delay, apart from the components mentioned earlier. Since, we have assumed a line size of 32B for all cache sizes, number of sets for caches with lower sizes (8KB, 16KB, 32KB) is smaller and hence number of rows of LUTs is small. Thus the delay caused by routing structure is insignificant. At lower associativities for caches

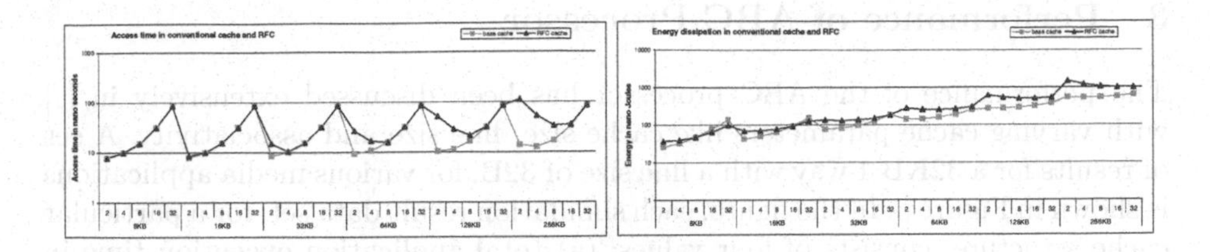

Fig. 2. (a) Access time (b) Energy dissipation, at 0.8μ CMOS technology, in conventional cache and the RFC for Convolution algorithm implementation. x-axis indicates the cache associativity from 2- to 32-way, for each cache size in KB

with larger sizes (64KB, 128KB, 256KB), number of rows of LUTs is higher resulting in a larger amount of routing structure in one module of the cache. Delay caused by the routing structure is larger, and hence larger access time for RFC as compared to conventional cache. As the associativity increases, number of pipeline stages for computation is less and hence smaller delay. Thus at higher associativities, access time in RFC is closer to access time in the conventional cache. In a conventional cache, a major portion of energy dissipated by the set associative configurations is in bitlines and sense amplifiers. As the associativity increases, requirement of the number of sense amps grows and subsequently power consumption also grows. Besides, increase in the number of subarrays at higher values of N_{dwl} and N_{dbl} results in a larger number of decoders and sense amplifiers. This leads to higher amount of energy dissipation in the RFC. Apart from that, the routing structure contributes towards an additional energy dissipation. At smaller cache sizes, difference in energy dissipation in RFC and a conventional cache is smaller due to the small overhead from routing structure and the number of sub arrays in both the models of cache being close. We assume that the size of an LUT remains constant (in all of our implementations we used a 4-LUT as a basic PE). Therefore, as the cache size increases, number of data subarrays increases and hence higher power consumption. However in reality, for higher cache sizes, we do not have to convert complete cache into RFC. Only some modules can be used for computational purposes, and the remaining modules can still be designed as conventional cache modules. The mechanism of computation in RFC and functioning of the ABC microprocessor are discussed in detail in [3]. For the simulation purposes, the ABC processor is implemented with appropriate modifications of the Simplescalar tool set [9]. The source code of simplescalar tool set is modified to incorporate the RFC design and the microarchitecture is modified accordingly to enable the communication between main processor and RFC.

3 Performance of ABC Processor

The performance of the ABC processor has been discussed extensively in [3] with varying cache parameters like cache size, line size and associativity. A set of results for a 32KB 4-way with a line size of 32B, for various media applications is shown in Figure 3. In the figure, each simulation result data set, for a particular cache structure, consists of four values: (a) total application execution time in the general purpose superscalar processor without RFC, (b) portion of time taken for computing the core function in GPP, (c) total application execution time in the ABC processor integrated with RFC, and (d) portion of time taken to execute core function in RFC. Each of these values are normalized to total number of cycles taken for executing the application in conventional superscalar processor. The performance improvement is obtained from the normalized total cycles for the execution of the overall function and also the core function. For example, execution time for MPEG2 decode application is only 25.37% of the time if the application is executed on a conventional superscalar processor. This large speed-up is due to the significant acceleration of the DCT/IDCT function in the RFC.

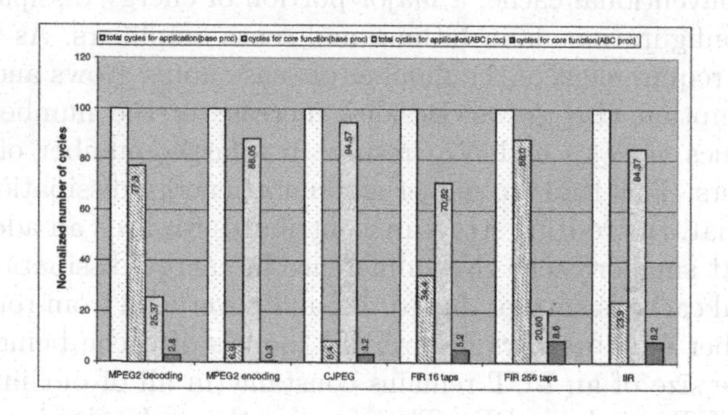

Fig. 3. Performance of ABC processor vs. base processor for a 32KB 4-way cache with a line size of 32B

4 Power Estimation in ABC Processor

In superscalar microprocessors, higher performance is achieved at the cost of higher power consumption due to the complex logic for resolving instruction dependencies. Higher energy dissipation requires more expensive packaging and cooling technology, increases cost, and decreases reliability of products in all segments of computing market from portable systems to high-end servers. Until recently, power efficiency was a concern only in battery powered systems like notebooks and cell phones. Currently, increased microprocessor complexity and frequency have caused power consumption to grow to the level where power has become a first-order issue. Earlier sections discussed the high performance of the ABC processor for various multimedia applications. From simulation results it is observed that when the core function is executed in RFC, total number of instructions executed, number of load and store instructions and number of accesses to L1 data and instruction caches have reduced. In this section, we show that the effect of power dissipation overhead per access in the RFC, is offset by the reduced number of accesses to the cache. We further show that the reduced utilization of other on-chip resources, due to a significant reduction in the execution time of the application, results in savings in energy consumption.

4.1 Power Estimation Models

In CMOS microprocessors, dynamic power consumption is the main source of power consumption and clearly dominates the leakage power dissipation. Therefore, in this paper we take the approach of estimating the dynamic power consumption in the ABC processor as compared to a conventional microprocessor based on the simulation results. Architectural power simulators like Wattch [10] measure the utilization of various processor components and during the simulation, feed these utilization numbers into a high-level power model to estimate the energy behavior of the processor. Using a similar approach, we estimate the component power breakdowns based on the utilization numbers we gather from the simulation of various multimedia benchmarks. For this purpose, we use two models as discussed below. The heuristics developed for estimation of power dissipation in each component are in line with the methodology proposed by Martonosi et al. [10, 11].

Alpha Processor Model and Pentium-Pro Processor Model The reported component breakdown of power consumption in Alpha 21264 processor is shown in Table 1(a). The second column (P_{frac}) indicates power consumed by the component as a fraction of total power in the processor. The third column indicates heuristic for the utilization factor of the component. The reported component breakdown of power consumption in Pentium-Pro processor is shown in Table 1(b). The second column (P_{frac}) indicates power consumed by the component as a fraction of the total power in the processor. The third column indicates heuristic for the utilization factor of the component.

4.2 Power Estimation in RFC

With the heuristics for utilization factors available, as discussed earlier, the component breakdown of total power dissipation for all the components except the L1 data cache can be obtained. The estimation of power dissipation in RFC is more involved since while one module of the RFC is acting as a computing unit, the remaining (A-1) modules are serving the memory accesses. For the estimation of power consumption in RFC under the above mentioned situation, the observations are:

Table 1. Listing of heuristic power estimations for (a) Alpha processor model, (b) Pentium-Pro processor model

Resource	Pfrac	Utilization factor
clock	26.31	total cycles for application
resultbus	2.63	instructions executed
ALU	15.79	instructions executed
L2 cache	2.66	L2 cache accesses
L1 D-cache	12.79	L1 D-cache accesses
L1 I-cache	17.68	L1 I-cache accesses
regfile	2.63	inst. committed - branches comm. + instruction decoded
lsq	6.63	load instrs. + store instrs.
window	8.26	instructions executed
bpred	2.31	branches committed
rename	2.31	instructions decoded

Resource	Pfrac	Utilization factor
Clock	10.5	Total cycles for application
Instruction fetch	18.7	L1 I-cache access
Register Alias Table	4.9	instructions decoded
Reservation Station	8.9	instructions executed
Reorder buffer	11.9	instruction committed
ALU	22.6	instructions executed
L1 Data Cache	11.5	L1 D-cache accesses
L2 Cache	2.5	L2 cache accesses
Memory Order Buffer	4.7	load instructions + store instructions
Branch Target Buffer	3.8	branches committed

- In a 4-way RFC, even when three modules are active as data cache and the fourth module is active for computation, data cache accesses are processed by all the four modules.
- The configuration is loaded into RFC module, treating the instructions to load configuration as normal load instructions that will undergo a miss in the cache.
- The computation in RFC module is LUT based and lookup cost is proportional to the number of data items processed. Power consumption in RFC module that is active in computing is thus proportional to the number of data items loaded into the input buffer. Also, the data is being processed in all the taps of the computing unit simultaneously. Thus, at any instant of time, data being processed simultaneously in all the taps is equivalent to one memory access.
- From the above observations, it can be seen that computation in one module of RFC is performed simultaneously when the RFC is serving the cache memory accesses. However, for a simple and worst-case estimation of the power dissipation in RFC, we assume that the cache accesses and computation in one module of RFC are separated in time.

The heuristic for power estimation in the RFC of the ABC processor is as given below:

$$P_{RFC} = P_{basecache} * P_{factor} * \left(\frac{M2 + conf_{load} + ((1/A) * (1/N_{taps}) * data_{RFC})}{M1}\right)$$

where,

P_{factor} = ratio of per access power consumption in RFC to that in conventional cache. This is obtained as in Figure 2.
M2 = L1 Data cache accesses in ABC processor with RFC
$conf_{load}$ = Number of load instructions to configure one RFC module
N_{taps} = Number of computing stages in one RFC module
$data_{RFC}$ = number of data elements executed in RFC module
M1 = L1 Data cache accesses in base processor without RFC
A = Associativity of the cache.

4.3 Results and Analysis

Figure 4 shows the total power consumed for each of the application under Alpha and Pentium-Pro processor models. The values shown in the figures are normalized to the total power consumed in a conventional superscalar processor for each of the application. The results show that up to 60% reduction in power consumption is achieved for MPEG decoding application, and a reduction in the range of 10% to 20% for various other multimedia applications. The significant savings in the power for the execution of MPEG2 decode application in the ABC processor is due to relatively larger fraction of the application that is mapped to the RFC. The power dissipation is estimated for simulation data obtained for a 32KB 4-way set associative cache with a line size of 32 bytes. It is observed that when the computation intensive core function is mapped and computed in RFC number of instructions executed is reduced. This results in reduced utilization of on-chip resources causing the reduction in overall power consumption. This, along with the reduced number of accesses to data cache, offsets the power dissipation overhead in the RFC. It can be observed that overall power consumption in the processor with RFC is either smaller or almost same as compared to that in the base processor. This is achieved despite a larger amount of power dissipation per access in RFC as compared to base cache. This occurs due to the reduced power dissipation in all the other on-chip components of the processor due to their reduced activity. And, the most important fact is that the reduction in overall power consumption is achieved along with a higher performance in executing the media applications.

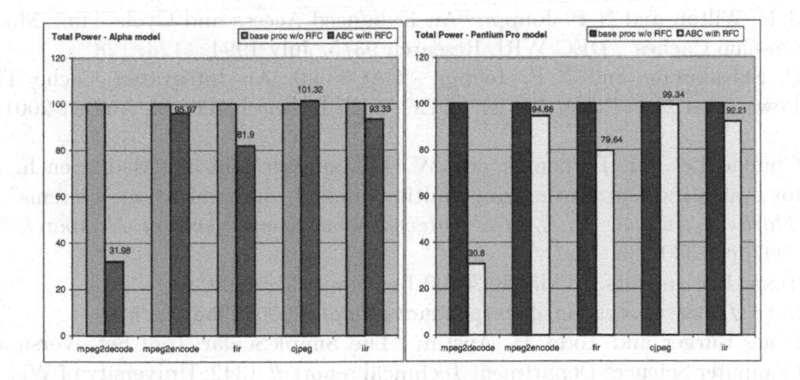

Fig. 4. Total power utilization in ABC processor vs. base processor using (a) Alpha processor model (b) Pentium-Pro model

5 Conclusions

The RFC accelerates computations using a specialized computing unit with minimal modification and overhead in time, power and area domains in the cache and microarchitecture. The paper discusses the high performance of the ABC processor in computing various multimedia applications. Besides, keeping in pace with the current requirement of power-aware architectures, it has been shown that the ABC processor delivers higher computing capacity while providing with significant savings in the energy dissipation in various on-chip components of the processor.

References

[1] A. DeHon, "The Density Advantage of Configurable Computing", *IEEE Computer*, Volume: 33, Issue: 4, April 2000, pp. 41-49 124

[2] H. Kim, A. K. Somani, and A. Tyagi, "A Reconfigurable Multifunction Computing Cache Architecture", *IEEE Transactions on Very Large Scale Integration (VLSI) Systems*, Vol. 9, Issue: 4 , August 2001, pp. 509-523. 125, 126

[3] Huesung Kim, "Towards Adaptive Balanced Computing (ABC) Using Reconfigurable Functional Caches (RFCs)", Ph. D. Dissertation, Dept. of Electrical and Computer Engineering, Iowa State University, July 2001. Available at: http://ecpe.ee.iastate.edu/dcnl/dissertation/HuesungKim.pdf 125, 126, 127, 129, 130

[4] Tomohisa Wada, Suresh Rajan, and Steven A. Przybylski, "An Analytical Access Time Model for On-Chip Cache Memories", *IEEE Journal of Solid-State Circuits*, Vol. 27, No. 8, August 1992, pp. 1147-1156. 126

[5] S. E. Wilton and N. P. Jouppi, "An Enhanced Access and Cycle Time Model for On-chip Caches", DEC WRL Research 93/5, July 1994. 126, 128

[6] P. Shivakumar and N. P. Jouppi, "CACTI3.0: An Integrated Cache Timing, Power, and Area Power Model", DEC WRL Research 2001/2, August 2001. 126, 128

[7] Chunho Lee, M. Potkonjak and W. H. Mangione-Smith, "MediaBench: a tool for evaluating and synthesizing multimedia and communications systems", *Proc. Thirtieth Annual IEEE/ACM International Symposium on Microarchitecture*, 1997, pp. 330-335. 127

[8] Texas Instruments, "TMS320C6000 benchmarks", 2000, `http://www.ti.com/sc/docs/products/dsp/c6000/62bench.htm` 127

[9] Doug Burger and Todd M. Austin, "The SimpleScalar Tool Set, Version 2.0", Computer Sciences Department Technical report # 1342, University of Wisconsin-Madison, June 1997. 129

[10] D. Brooks, V. Tiwari, and M. Martonosi, "Wattch: A Framework for Architectural-Level Power Analysis and Optimizations", *Proc. 27th International Symposium on Computer Architecture*, 2000, pp. 83 -94. 131

[11] R. Joseph and M. Martonosi, "Run-Time Power Estimation in High Performance Microprocessors", *Proc. International Symposium on Low Power Electronics and Design*, 2001, pp. 135-140. 131

Field Programmable Systems

Patrick Lysaght

Xilinx Research Labs
2100 Logic Drive, San Jose, California, CA 95032
http://www.xilinx.com

Abstract. The emergence of Platform FPGAs! marks a significant change in the capabilities of FPGAs. Unlike their predecessors, Platform FPGAs are characterized by heterogeneous architectures. In addition to programmable logic, routing and inputs/outputs, they feature soft and hard microprocessors, bus and memory hierarchies, digital signal processing cores and gigabit communications channels. These devices are targeted at high performance computing applications in embedded systems. The complexity and flexibility of Platform FPGAs is unparalleled. We explore some of the potential implications for current and future design of programmable digital systems.

1 Introduction

The development of FPGAs has reached a point of inflection with the introduction of Platform FPGAs. The key motivation behind these devices is the familiar one of improved performance. However, it is the manner in which much of the performance gain is achieved that characterizes Platform FPGAs. Traditionally, FPGAs were improved by making them bigger and faster and by improving the design of their programmable logic, routing and inputs/outputs.

Platform FPGAs continue to exploit the improvements that have worked so effectively for homogeneous architectures while evolving beyond this approach to become truly heterogeneous architectures. In addition to programmable logic, routing and inputs/outputs, they are characterized by the inclusion of soft and hard microprocessors, bus and memory hierarchies, digital signal processing cores and gigabit communications channels.

1.1 What Is a Platform?

Before proceeding further it is helpful to clarify what we mean by a *platform*. The term is commonly used especially in the context of System on Chip design and its meaning is not always clear. Sangiovanni Vincentelli [1] offers the following general definition: *"... a platform is an abstraction layer in the design flow that facilitates a number of possible refinements into a subsequent abstraction layer (platform) in the design flow."*

S. Sahni et al. (Eds.) HiPC 2002, LNCS 2552, pp. 137-138, 2002.
Springer-Verlag Berlin Heidelberg 2002

The key observation here is that a platform corresponds to the familiar concept of an abstraction layer and that design proceeds by refinement between such layers. Introducing higher levels of design abstraction has long been one of the principal methods of improving design productivity. Automating as much as possible of the refinement process between adjacent layers is the next critical contribution.

2 The Significance of this New Design Abstraction

Platform FPGA is an especially powerful abstraction. It is envisaged that for many embedded applications, the Platform FPGA will be the defining component of the system. In essence it becomes the *field programmable system*, though in reality other system components such as external memories or analogue circuits will be present in almost all cases.

Platform FPGA has many unique characteristics beyond its heterogeneous architecture that suggest that it will be some time before its full potential is fully understood and exploited. For example, because it is designed as a general-purpose platform, it will be used by a large community of designers with varying design capabilities for many different types of application. These designers will have access to unprecedented levels of high performance computing for embedded systems in an entirely configurable and programmable context. How they will interact with and exploit this unique mixture of complexity and programmability is not entirely clear.

It is likely that Platform FPGA will form the basis for future, higher-level abstractions such as a platform for reconfigurable, embedded systems that can be upgraded via dynamic reconfiguration over the Internet.

Going further forward, we may ask what the successor to Platform FPGA will look like five years from now? That it will introduce even greater levels of performance and complexity may be reasonably assumed. The key question is how to make such power accessible to a wide audience in a manner that further enhances their productivity.

References

[1] Sangiovanni Vincentelli, A., Martin, G.: IEEE Design & Test of Computers, Volume: 18 Issue: 6 , (Nov.-Dec. 2001) 23-33

CORBA-as-Needed: A Technique to Construct High Performance CORBA Applications

Hui Dai[1], Shivakant Mishra[1], and Matti A. Hiltunen[2]

[1] Department of Computer Science, University of Colorado
Campus Box 0430, Boulder, CO 80309-0430
[2] AT&T Labs-Research
180 Park Avenue, Florham Park, NJ 07932

Abstract. This paper proposes a new optimization technique called CORBA-as-needed to improve the performance of distributed CORBA applications. This technique is based on the observation that in many cases the client and the server of a distributed application run on compatible computing platforms, and do not need the interoperability functionality of CORBA. CORBA-as-needed dynamically determines if the interoperability functionality is needed for a specific application invocation, and bypasses this functionality if it is not needed. Performance measurements from a prototype implementation in omniORB show that CORBA-as-needed achieves a very significant performance improvement.

1 Introduction

CORBA [3] has been successfully used in the development of a large number of distributed applications. However, it suffers from an important deficiency. There is a significant performance overhead in using CORBA to construct distributed object applications [4, 5]. Therefore, it is difficult to construct high performance distributed object applications using CORBA. The main source of this performance overhead is the large one-way communication delay incurred by CORBA communication methods, compared to the one-way communication delays of transport-level protocols such as UDP or TCP. We propose a technique called CORBA-as-needed that allows distributed applications to bypass CORBA interoperability functionality whenever the client and the server happen to run on compatible platforms. It allows client server applications to first detect if the interoperability functionality of CORBA is indeed needed for a particular invocation, and bypass it if this functionality is not necessary.

In this paper, we present the design, implementation, and performance evaluation of CORBA-as-needed technique in omniORB [1]. We explore four different design alternatives for incorporating CORBA-as-needed in the current CORBA architecture. These design alternatives are called service approach, integration approach, CORBA wrapper approach, and pluggable ORB module approach. These alternatives differ from one another in the exact layer of CORBA architecture where CORBA-as-needed is incorporated. We provide a thorough analysis

S. Sahni et al. (Eds.) HiPC 2002, LNCS 2552, pp. 141–150, 2002.

of the four design alternatives, including their pros and cons. We have implemented a prototype of CORBA-as-needed using the pluggable ORB module approach in omniORB[1]. We describe this implementation and a detailed performance evaluation by comparing the performance of several distributed object applications implemented using our prototype and omniORB [1]. This performance comparison shows that CORBA-as-needed is a very useful technique that can improve the latency and scalability of distributed object applications by as much as 25% when clients and servers run on compatible computing platforms. Also, performance measurements show that this technique has insignificant performance overhead when the clients and servers run on incompatible platforms. CORBA-as-needed is a generic technique that can be used in association with other optimization techniques that researchers have used in the past to improve the performance of CORBA applications.

2 CORBA As Needed

While the ability to operate in a heterogeneous distributed computing environment is an important requirement for modern distributed applications, it is important to note that a majority of the applications do in fact use operating systems, network protocols, and hardware platforms that are compatible with one another. The main idea of this paper is to design a system that dynamically recognizes the situations when the extra support for interoperability provided by CORBA is not needed, and allows applications to simply bypass the interoperability functionality by using a standard TCP/UDP communication mechanism for those situations. In other words, interoperability support of CORBA is used by applications only when it is really needed.

The CORBA-as-needed technique identifies operating conditions when a client and a server are running on compatible computing platforms at the time when the client initiates a connection to the server. If the client and the server are determined to be running on compatible computing platforms, the ORB redirects all future communication requests to the lower transportation layer without passing through the rest of ORB core that implements interoperability functionality. If the client and the server are determined to be running on incompatible platforms, all future communication requests are directed through ORB core in the usual manner.

2.1 Design Alternatives

There are at least four different design alternatives to incorporate CORBA-as-needed technique in the current CORBA architecture. These are (1) service approach, (2) integration approach, (3) CORBA wrapper approach, and (4) pluggable ORB module approach.

In the service approach (Figure 1), CORBA-as-needed is implemented as a new CORBA service called *bypass service*. The bypass service uses a similar

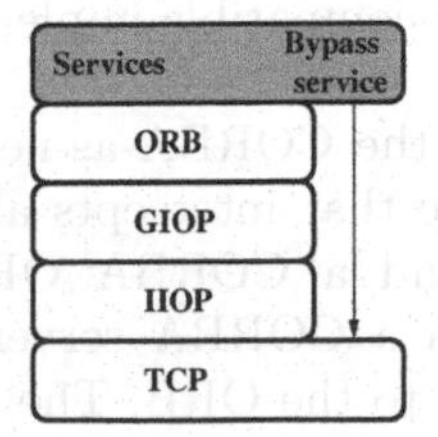

Fig. 1. Service Approach

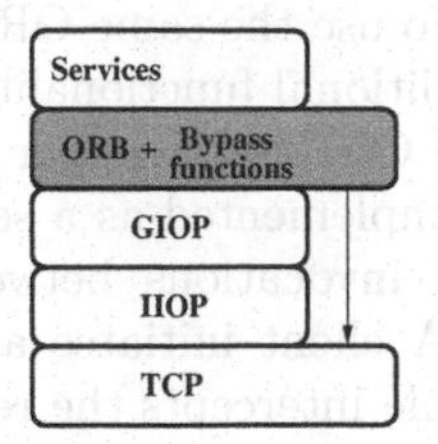

Fig. 2. Integration Approach

interface as other services in CORBA, but uses TCP or UDP as the underlying communication mechanism. The main advantage of this design alternative is that it provides for an easy integration in the CORBA service framework, and can take advantage of other CORBA services such as security. The key problem with this design is that a client needs to have an explicit knowledge of the server's computing environment, so that it can invoke the bypass service whenever appropriate. Thus, this design does not preserve the transparency of the CORBA-as-needed technique from the application. Another problem is the potential performance overhead. In an earlier research project, we observed that providing additional functionality such as group communication as a CORBA service can result in a very significant performance overhead [5].

In the integration approach (Figure 2), an existing ORB is extended to include the CORBA-as-needed functionality. The main advantage of this approach is that it can provide very good performance. Since all state information is available in the ORB, the implementation is also simpler. However, there are several problems associated with this approach. First, this approach violates the basic philosophy behind the design of CORBA: the ORB should provide minimal functionality to allow interoperability between heterogeneous objects. By integrating additional functionality in ORB, the functionality of ORB are overloaded. A more serious problem with this approach is that it may result in the loss of portability and interoperability. Integrating additional functionality in ORB forces applications to be ORB dependent. Application clients and servers are

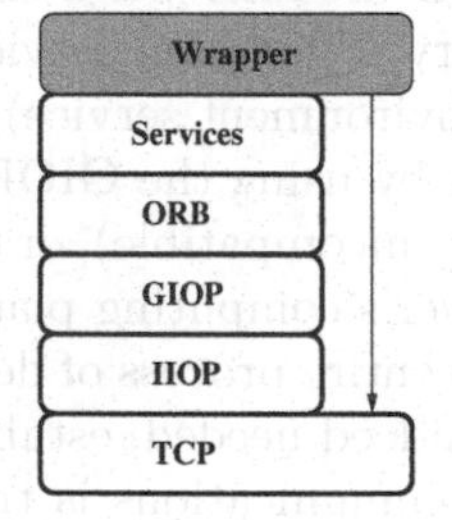

Fig. 3. Wrapper Approach

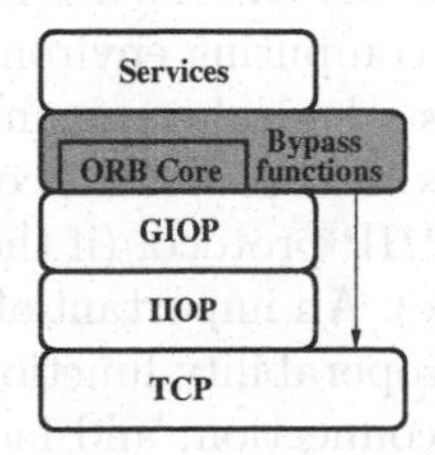

Fig. 4. Pluggable ORB Module Approach

required to use the same ORB implementation (or compatible implementations of the additional functionality).

In the CORBA wrapper approach (Figure 3), the CORBA-as-needed technique is implemented as a separate system module that intercepts all CORBA command invocations between an application and a CORBA ORB. When a CORBA client initiates a new connection with a CORBA server, the system module intercepts the request and forwards it to the ORB. The ORB then calls the named service to get the appropriate information and returns it to the system module. If the system module determines that the client and the server are running on compatible platforms, it redirects all the following communication requests to the transport layer and bypasses the ORB. The main advantage of this approach is its simplicity. It provides a clean separation between an ORB, CORBA services, and the CORBA-as-needed technique. The problem is again the potential for significant performance overhead. Because the system module is a separate module that sits between a CORBA application and an ORB, it may impose significant performance overhead.

Finally, in the pluggable ORB module approach (Figure 4), the CORBA-as-needed is implemented as a separate module that can be plugged into an ORB. The ORB interface to the application is kept untouched. This design approach is based in the idea of pluggable protocols that has been used in many middleware systems, including CORBA [7], to improve a middleware's performance and provide additional functionality. When a client initiates a connection with the server, the ORB activates a component to determine the computing environment of the requested object. If the server's computing environment is determined to be compatible with the client's environment, all future communication are redirected through a TCP/IP connection bypassing the original ORB core. The main advantage of this approach is that it is expected to provide good performance. In addition, because the CORBA-as-needed technique is implemented as a separate module, the existing ORB code is minimally affected. We have used this approach in our prototype implementation.

2.2 Pluggable ORB Module

Figure 5 illustrates the communication model of the ORB in which CORBA-as-needed has been implemented as a pluggable ORB module. On a client's first invocation, the extended ORB first locates the server (location service), determines the computing environment of the server (environment service), and then establishes a logical communication channel either by using the GIOP/IIOP (if the client's and the server's computing platforms are incompatible), or the underlying TCP/IP protocol (if the client's and the server's computing platforms are compatible). An important observation is that this entire process of determining if the interoperability functionality of CORBA is indeed needed, establishing an alternate connection, and facilitating all future communications is transparent to both the client and the server.

Figure 6 illustrates the operational details of CORBA-as-needed implemented as a pluggable ORB module. There are five important components here:

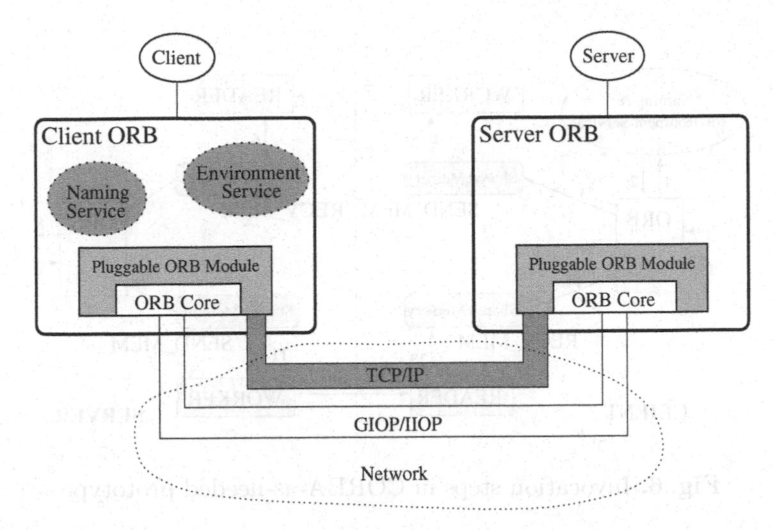

Fig. 5. CORBA-as-Needed as a pluggable ORB Module

(1) shared memories, (2) worker daemons, (3) reader daemons, (4) environment service, and (5) ORB. Shared memory is used on both the client and the server side for interprocess communication on the same machine. There are two types of shared memories. SEND_MEM is used by the ORB to store the request information on the client side and the reply message on the server side. The RECV_MEM is used by the reader daemon to store the information received from the other side (client side for a server, and server side for a client). Local ORB watches RECV_MEM for the incoming messages. The shared memory structure includes a flag to indicate any change in the memory made by the ORB or the reader daemon, the address of the servant, and some other support data structures.

Worker daemons monitor SEND_MEM. When the ORB stores a message in SEND_MEM, the worker daemon retrieves it and sends it to the other side. Reader daemon receives messages coming from the other side and stores them in the RECV_MEM. The environment service fetches the computing environment of the server. Finally, the ORB wrapper provides the CORBA-as-needed functionality as a separate module. The following steps are executed when the ORB receives an initial request from a client for connection with a server:

1. ORB checks with the naming and environment services to get the location of the target object as well as its computing environment information.
2. The naming/environment service responds with the location and environment information of the servant.
3. If the computing environments of the client and the server are compatible, the client ORB puts the request message into the SEND_MEM. This message encapsulates the target class name, method name, arguments to the target object, and server's address. It then sets the flag to indicate that there is new information in SEND_MEM.

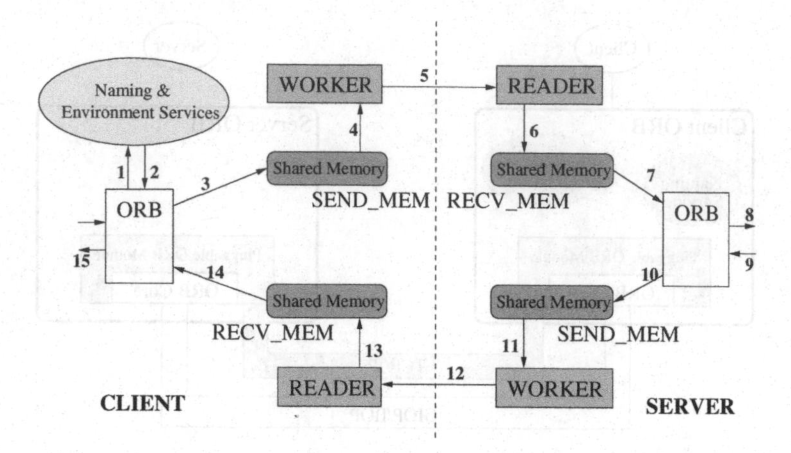

Fig. 6. Invocation steps in CORBA-as-needed prototype

4. The worker daemon notices the change in the SEND_MEM flag. It reads the new message from SEND_MEM.
5. The worker daemon sends the request to the reader daemon on the server side.
6. The server side's reader daemon receives the request message and puts it in the server side's RECV_MEM, and sets the flag.
7. The ORB on the server side notices the change in RECV_MEM's flag, and reads in the request message from RECV_MEM.
8. The ORB calls the dispatch mechanism of the servant.
9. The servant returns the result of the request to the server side's ORB.
10. The server side's ORB puts the reply message in the SEND_MEM. This message encapsulates the result of the request. It then set the flag to indicate that there is new information in SEND_MEM.
11. The server side's worker daemon notices the change in the SEND_MEM's flag, and retrieves the reply message from SEND_MEM.
12. The server side's worker daemon sends the reply message to the reader daemon on the client side.
13. Client side's reader daemon receives the reply message and puts it in the client side's RECV_MEM, and sets the flag.
14. The ORB notices the change in the RECV_MEM's flag, and retrieves the reply message from RECV_MEM.
15. The ORB returns the result to the client.

3 Performance

In order to understand the effect of CORBA-as-needed technique on the performance of a client server application, we have measured the performance of our CORBA-as-needed prototype and compared it with the performance of omniORB. In our performance measurement, we have concentrated on the overall performance of a client server application.

3.1 Computing Environment

The performance measurements were done over two different computing environments: LAN and WAN. In the LAN computing environment, the client and the server were run on two separate HP75 PCs. Both of them have PIII, 450 MHz CPU, with 128 MB RAM. These two PCs are connected by a 10 Mbps Ethernet. Both of these PCs were running Linux. In the WAN computing environment, the client was run on an HP75 PC in the University of Colorado, Boulder, while the server was run on a dual-processor, Linux cluster with PIII, 1 GHz CPUs and 256 MB of RAM in the University of Wyoming, Laramie. The two PCs were connected by the Internet.

We constructed a simple client-server application for all our measurements. Both the client and the server are written in C++. The application is a simple echo string application, where the client invokes a method called `echoString (string str)` on an object `Echo` that resides on the server. When invoked, this method simply returns the string parameter str back to the client.

3.2 Performance Index

We have measured *application latency* to evaluate the effect of CORBA-as-needed on the performance of a CORBA application. Latency is defined as the time interval between the moment a client ORB sends out a request until it gets the reply. Latency consists of two parts: the overhead of the ORB core or pluggable ORB module implementing CORBA-as-needed, and the time spent in the network transmission.

3.3 Performance Results

We measured the latency for five different string sizes: 200 B, 1 KB, 4 KB, 10 KB, and 100 KB. Figure 7 shows the latency measured for these five string sizes. The latency reported in this figure is an average latency over 200 invocations of the method `echoString()`. As we can see from these measurements, CORBA-as-needed results in the improvement of the application latency by as much as 25%. There are two important observations we make from these measurements. First, the percentage improvement in the application latency increases with increase in the size of the string parameter. The main reason for this increase is marshalling and unmarshalling of the string parameter performed by omniORB. Larger the size of the string, longer it takes to do this. As observed in [6], the main source of performance overhead is marshalling and unmarshalling of function parameters. CORBA-as-needed avoids this marshalling and unmarshalling if the client and the server are running on compatible computing platforms.

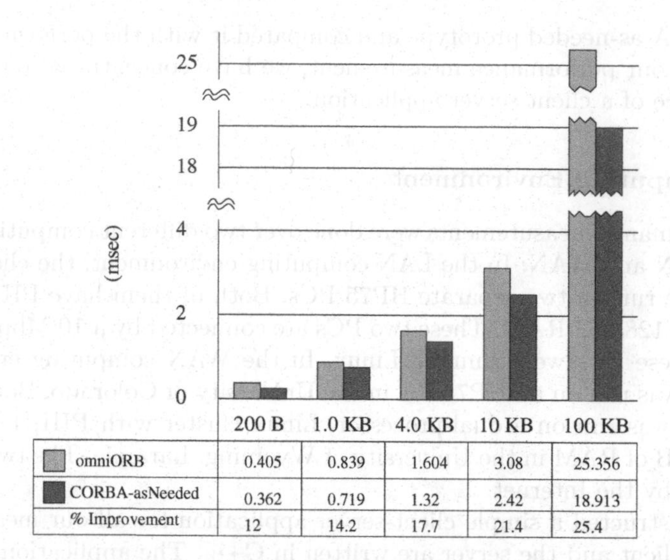

	200 B	1.0 KB	4.0 KB	10 KB	100 KB
omniORB	0.405	0.839	1.604	3.08	25.356
CORBA-asNeeded	0.362	0.719	1.32	2.42	18.913
% Improvement	12	14.2	17.7	21.4	25.4

Fig. 7. Application Latency in LAN

The second observation we make from our performance measurement is that the application latency of CORBA-as-needed was slightly higher for the first invocation compared with the rest of the invocations. Actually, the same was true for omniORB as well. The main reason is that in the first invocation, a connection with the server needs to be established between the client and the server. CORBA-as-needed also interacts with naming and environment services at this time. For all future invocations, both omniORB and CORBA-as-needed reuse the connection established for the first invocation.

Application latency for the WAN environment is shown in Figure 8. Again, the latency reported in this figure is an average latency over 200 invocations of the method `echoString()`. As we can see, the CORBA-as-needed does result in a significant improvement in application latency in WAN computing environment as well. We noticed that the two observations we made from the latency measurement in the LAN environment were also present in the WAN computing environment. These are the increasing improvement in application latency with increase in string size, and a slightly larger latency for the first invocation. The anomaly of lower improvement in application latency when the string size was 10 K, as compared to when the string size was 1 K or 4 K is attributed to the unstable communication characteristics of the Internet, i.e. the communication delays are rather variant in the Internet.

In addition, we noticed that the variation in the application latency between different invocations was much larger than in case of LAN. Again, this is attributed to the unstable communication characteristics of the Internet. Another observation we made was that the improvement in application latency due to CORBA-as-needed was generally lower in WAN than in LAN. This is because the time spent by omniORB in addressing the interoperability issues (marshalling,

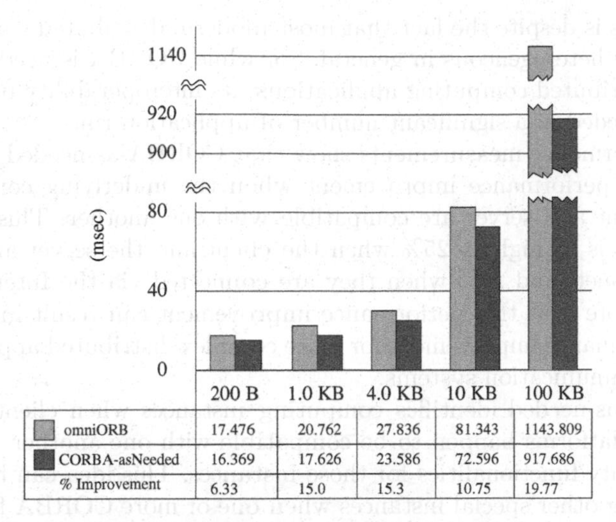

	200 B	1.0 KB	4.0 KB	10 KB	100 KB
omniORB	17.476	20.762	27.836	81.343	1143.809
CORBA-asNeeded	16.369	17.626	23.586	72.596	917.686
% Improvement	6.33	15.0	15.3	10.75	19.77

Fig. 8. Application Latency in WAN

unmarshalling, byte padding, byte alignment, etc.) is constant for a given invocation, and is independent of the network latency. So, as the network latency increases, the percentage of time spent on addressing the interoperability issues, and hence the overhead of omniORB, decreases.

Finally, we also measured the overhead of CORBA-as-needed when the client and the server run on incompatible platforms, i.e. when the CORBA-as-needed technique does not result in improving an application performance. We noticed that the performance overhead was extremely low. It was less than 1% in LAN and less than 0.5% in WAN. As expected, this performance overhead decreased with increase in string size.

4 Discussion

Heterogeneity in modern, distributed computing systems is a fact of life. As a result, the ability to operate in a heterogeneous, distributed computing environment is essential for any current and future distributed computing service. CORBA is a very useful technology, because is enables the construction of distributed applications that can operate in a heterogeneous distributed computing environment. CORBA naturally incurs some performance overhead to provide this support. This extra performance overhead has limited the applicability of CORBA.

In this paper, we have proposed a technique called CORBA-as-needed that can improve the performance of distributed CORBA applications under some common computing scenarios. The main motivation behind this approach is the observation that a majority of the components in modern, distributed computing environments are compatible with one another and can interoperate with one another. This is despite the fact that most modern, distributed computing environments are heterogeneous in general. So, while CORBA is needed to develop modern, distributed computing applications, its interoperability functionality is not really needed in a significant number of application runs.

Our performance measurements show that CORBA-as-needed does result in a significant performance improvement when the underlying computing platforms of client and server are compatible with one another. This performance improvement is as high as 25% when the client and the server are running on the same subnet, and 20% when they are connected via the Internet. It is important to note that this performance improvement can result in even a much higher performance improvement for more complex distributed applications such as group communication systems.

CORBA-as-needed identifies computing instances when client and server's computing platforms happen to be compatible with one another, and bypasses interoperability functionalities for those instances. This idea can be generalized by identifying other special instances when one or more CORBA functionalities are not needed by the application. As an example, another reason for poor performance of distributed, CORBA applications is that CORBA ORBs typically use TCP as a transport-level protocol. The reason for this is that GIOP specifications assume a reliable, stream-oriented protocol as the underlying transport-level protocol. In one of our earlier projects, we observed that this contributes to a significant performance overhead for applications that do not need TCP functionality [5]. Since, in a LAN environment, UDP is mostly reliable and almost always provides a sequenced message delivery, it is worthwhile to experiment with a pluggable ORB module that allows applications to use UDP in a LAN environment. Another possibility is to use an alternate high performance communication protocol, if such a protocol is available.

References

[1] *omniORB: Free High Performance CORBA 2 ORB.* AT&T Laboratories, Cambridge. URL: `http://www.uk.research.att.com/omniOR,B/`. 141, 142

[2] *Visibroker for Java, Programmer's Guide.* Inprise Corporation, 1996.

[3] *The Common Object Request Barker: Architecture and Specification.* Object Management Group, Framingham, MA, 1998. 141

[4] A. Gokhale and D. C. Schmidt. Measuring the performance of communication middleware on high-speed networks. In *Proceedings of SIGCOMM'96,* Aug 1996. 141

[5] S. Mishra, L. Fei, X. Lin, and G. Xing. On group communication support in CORBA. *IEEE Transaction on Parallel and Distributed Systems* 12(2), February 2001. 141, 143, 150

[6] S. Mishra and N. Shi. Improving the performance of distributed CORBA applications. In *Proceedings of the 2002 IEEE International Parallel and Distributed Processing Symposium,* Fort Lauderdale, FL, April 2002. 147

[7] W. Zhao, L. Moser, and P. Melliar-Smith. Design and implementation of a pluggable fault tolerant CORBA infrastructure. In *Proceedings of the 2002 IEEE International Parallel and Distributed Processing Symposium,* Fort Lauderdale, FL, April 2002. 144

Automatic Search for Performance Problems in Parallel and Distributed Programs by Using Multi-experiment Analysis*

Thomas Fahringer and Clovis Seragiotto, Jr.

Institute for Software Science, University of Vienna
{tf,clovis}@par.univie.ac.at

Abstract. We introduce Aksum, a novel system for performance analysis that helps programmers to locate and to understand performance problems in message passing, shared memory and mixed parallel programs. The user must provide the set of problem and machine sizes for which performance analysis should be conducted. The search for performance problems (properties) is user-controllable by restricting the performance analysis to specific code regions, by creating new or customizing existing property specifications and property hierarchies, by indicating the maximum search time and maximum time a single experiment may take, by providing thresholds that define whether or not a property is critical, and by indicating conditions under which the search for properties stops. Aksum automatically selects and instruments code regions for collecting raw performance data based on which performance properties are computed. Heuristics are incorporated to prune the search for performance properties. We have implemented Aksum as a portable Java-based distributed system which displays all properties detected during the search process together with the code regions that cause them. A filtering mechanism allows the examination of properties at various levels of detail. We present an experiment with a financial modeling application to demonstrate the usefulness and effectiveness of our approach.

1 Introduction

The performance of distributed and concurrent applications can be challenging, as an application's performance stems from several factors, including application design, software environment, and underlying hardware. Manual performance analysis is time-consuming, error-prone, and complicated. Many performance tools have been developed, which commonly produce a vast amount of performance data without

---

* This research is partially supported by the Austrian Science Fund as part of Aurora Project under SFBF1104.

S. Sahni et al. (Eds.) HiPC 2002, LNCS 2552, pp. 151-162, 2002.
Springer-Verlag Berlin Heidelberg 2002

interpreting performance results, relating performance information back to the input program, or guiding the user to the essential performance problems.

In previous research we have developed a Java-based specification language, JavaPSL [[5]], for a systematic and portable specification of large classes of experiment-related data (describe performance experiments) and performance properties (characterize a specific negative performance behavior) in distributed and concurrent programs. JavaPSL provides a high-level and portable interface to experiment-related data that enables the programmer to customize existing properties and define new ones without the need to know anything about the underlying implementation details. In addition, we developed SCALEA [[9]], a performance instrumentation and measurement system that computes a variety of metrics (timing and overhead information) for OpenMP, MPI, HPF, and mixed parallel/distributed programs.

In this paper we describe Aksum, a novel multi-experiment tool that tries to automate performance diagnosis of parallel and distributed programs for different problem and machine sizes. Fig. 1 gives an overview of Aksum's architecture. A *user portal* enables the user to input information about the application and the target machine on which the application will be executed. The user can select, modify and create a hierarchy of performance properties to be searched in an application, control the search time, and specify code regions of interest and conditions to stop the search process.

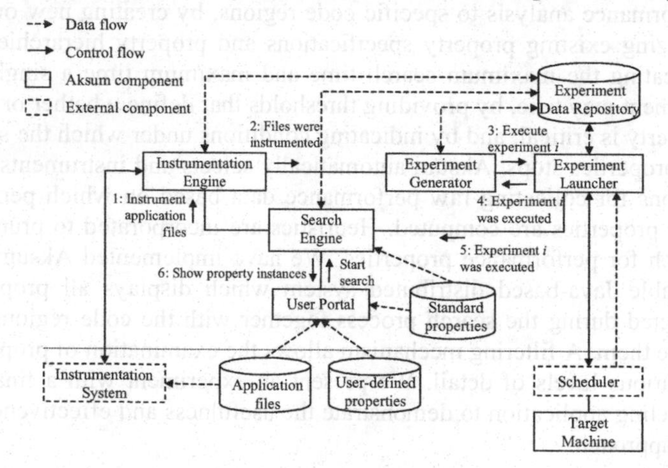

Fig. 1. Aksum's overview

The user-supplied input data is provided to the *search engine*, which is in the center of Aksum and controls the entire search process. By issuing requests to the *instrumentation engine*, the search engine collects the performance information for well-selected code regions for varying user-provided problem and machine sizes. The instrumentation engine of Aksum provides an interface to SCALEA, enabling the search engine to access and traverse application files in a machine-independent way, and to modify makefiles, scripts, and the compilation command line in order to link the instrumented application with the instrumentation library provided by SCALEA. The instrumented code is submitted to the *experiment generator*.

For each instrumented application the experiment generator changes some important execution parameters that control, for instance, the set of input files and the number of threads or processes to be used for the application execution. All files for an experiment are forwarded to the *experiment launcher*, which compiles and executes the application instances on the target machine through an external scheduler. After an experiment has been executed, all performance data are transferred to the experiment data repository. The search engine evaluates the performance data and tries to determine all critical performance properties. A cycle consisting of consecutive phases of application execution and property evaluation is continued until all experiments are done or some system or user-defined condition stops the search process. Every performance property that has been determined to be critical is dynamically displayed (together with the source code) to the user during the search process and stored in the experiment data repository. All components of Aksum can reside on any computing node that is accessible through the Internet. Although the user is given many opportunities to influence the search for performance bottlenecks, Aksum is a fully automatic tool once it is given the application sources as well as problem and machine sizes for which performance analysis should be done.

The remainder of this paper is organized as follows. Aksum's user portal is delineated in Section 2, followed by a description of the underlying search engine in Section 3 and results for a financial modeling application in Section 4. Related work is discussed in Section 5. Finally, we conclude this paper with a summary and outline future work.

2 The User Portal

Aksum's user portal provides the user with a very flexible mechanism to control the search for performance bottlenecks, which is described in this section.

2.1 Property Hierarchy

A performance property (e.g. load imbalance, synchronization overhead) characterizes a specific negative performance behavior of a program. Every property must implement the interface *Property*, thus the term property instance refers to an instance of a class that implements this interface.

The interface *Property* contains three methods:

boolean holds( ): returns true if the property instance holds. In this case the instance is called a true property instance.

float getSeverity( ): returns a value between 0 and 1 indicating how severe the property instance is (the closer to 1, the more severe the property instance is).

float getConfidence( ): returns a value between 0 and 1 that indicates the degree of confidence in the correctness of the value returned by *holds*.

Properties are hierarchically organized into tree structures called *property hierarchies*, which are used to tune and prune the search for performance properties. For example, one may assume that, if an application is efficient, there is no need to compute its load imbalance. This assumption can be encoded in a specific property hierarchy by placing the property LoadImbalance under the property Inefficiency (Fig. 2). Each node in the property hierarchy represents a performance property and is described by three elements:

Performance property name: the name of the performance property associated with this node; the property definition is stored in a property repository (defined by the user or provided by Aksum).
Threshold: a value that is compared against the severity value of instances of the property represented by this node; if the severity value is larger than this value, the property instance is *critical* and will be included in the list of *critical* properties.
Reference code region: the type of the reference code region (currently main program, enclosing outermost loop or subroutine) that will be used to compute the severity of instances of the property represented by this node.

There are three standard property hierarchies provided by Aksum, covering message passing, shared memory and mixed parallel programs. The user can define and store new property hierarchies from scratch or customize predefined hierarchies. The reference code region for every property node in the predefined property hierarchies is per default set to the main program.

2.2 Application Files, Command Lines and Directories

An application consists of various files – denoted *application files* in the remainder of this paper – which are divided into *instrumentable* and *non-instrumentable* files. Instrumentable files are source codes that must be instrumented for performance metrics (overheads and timing information) whereas non-instrumentable files refer to source codes the user does not want to be instrumented by Aksum and any other files necessary to execute an application (e.g., makefiles, scripts, and input files). For instrumentable files, the user can also define which code regions should be instrumented. If not specified, then Aksum assumes that the entire file must be instrumented. The user must also provide the compilation and execution command lines and associated directories for a selected application. This information is later used by the experiment generator and the experiment launcher.

2.3 Application Parameters

An application parameter defines a string that should be replaced in some or all of the application files and in the execution and compilation command lines before the application is compiled and executed. Application parameters are defined by the quintuplet (*name*, *search string, file set, value list, type*), where *name* is a unique name that identifies the parameter, *search string* represents the string to be substituted, *file set* describes the set of application files in which the search string will be searched and replaced, *value list* denotes the list of values with which the search string will be replaced, and *type* indicates if the parameter is machine or problem size related (or neither of them). An application with a set of application files and application parameters is denoted an *application instance* if, for all application parameters, every *search string* has been replaced in the associated *file set* by one of the values in the *value list.* The generation, compilation and execution of an application instance is called an *experiment*. Fig. 3 shows user-provided values for the application parameters under the user portal. Two additional options ("Line by line" versus "Cartesian product") affect the number of experiments generated and the parameter value combinations for each experiment.

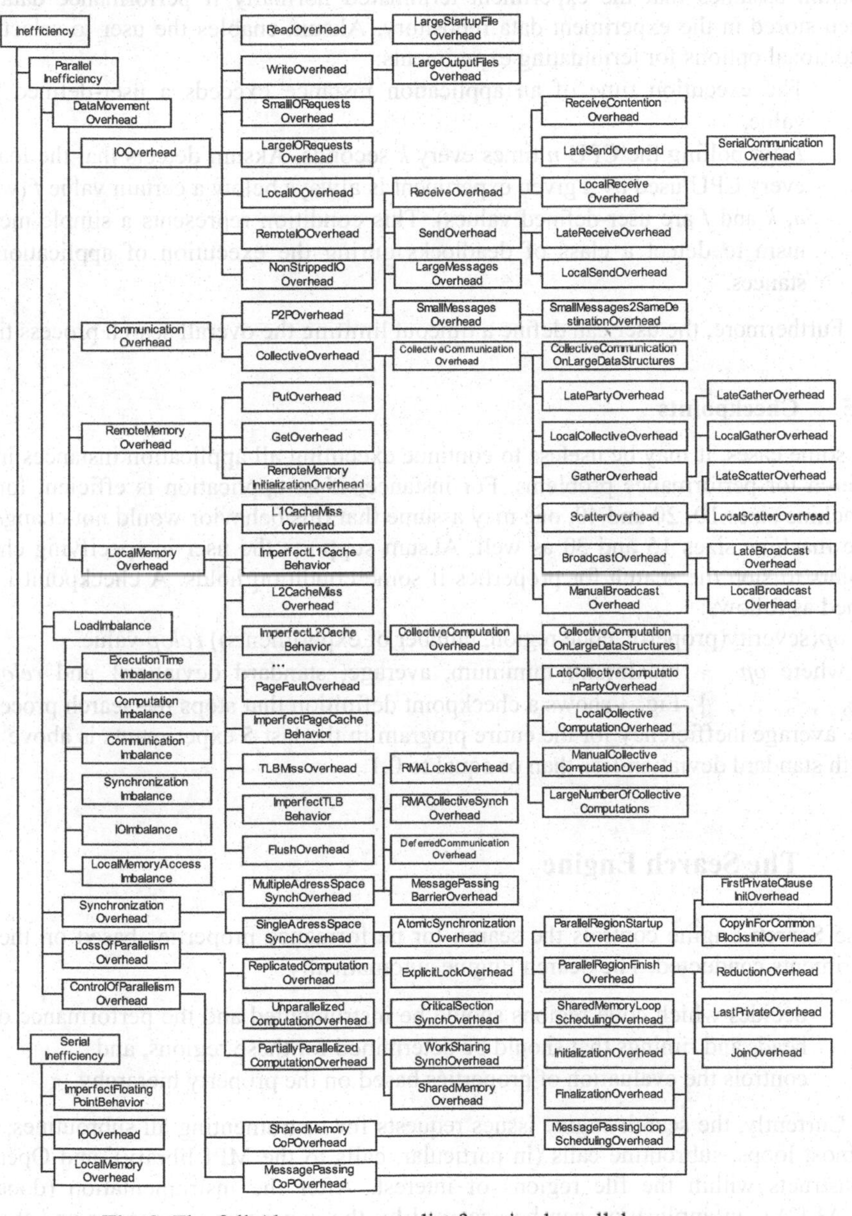

Fig. 2. The full Aksum property list for mixed parallel programs

2.4 Timeouts and Error Recovery

The user can define the maximum number of application instance *compilations* and *executions* that can terminate abnormally (e.g. because of a system error) without canceling the search for properties. As in general it is impossible to determine whether an experiment has successfully finished or not, the current implementation of Aksum assumes that the experiment terminated normally if performance data has been stored in the experiment data repository. Aksum enables the user to select two additional options for terminating experiments:

The execution time of an application instance exceeds a user-defined time value.

After polling the CPU *n* times every *k* seconds, Aksum detects that the load of every CPU used by a given experiment is always below a certain value *f* (where *n*, *k* and *f* are user-defined values). This condition represents a simple mechanism to detect a class of deadlocks during the execution of application instances.

Furthermore, the user can define a timeout limiting the overall search process time.

2.5 Checkpoints

In some cases, it may be useless to continue executing all application instances in the search for performance problems. For instance, if an application is efficient for the machine sizes 10, 20 and 40, one may assume that this behavior would not change for the machine sizes 15 and 30 as well. Aksum supports the user in specifying checkpoints to stop the search for properties if some condition holds. A checkpoint is defined as follows:

op(severity(property, code region, number of experiments)) *relop* value

where *op* {maximum, minimum, average, standard deviation} and *relop* {>, , <, , , }. Fig. 3 shows a checkpoint definition that stops the search process if the average inefficiency for the entire program in the last 5 experiments is above 0.75 with standard deviation less than or equal to 0.1.

3 The Search Engine

The Search Engine controls the search for performance properties based on the experiments conducted. The Search Engine automatically

decides which code regions should be instrumented and the performance overheads and timings that should be determined for these regions, and
controls the evaluation of properties based on the property hierarchy.

Currently, the search engine issues requests for instrumenting all subroutines, outermost loops, subroutine calls (in particular, calls to the MPI library), and OpenMP constructs within the file regions of interest. After the instrumentation (done by SCALEA), an application can be accessed by the experiment generator and the experiment launcher in order to have its parameters replaced (Section 2.3) and to be compiled and executed on the target machine. Every time an experiment terminates normally, the resulting performance data is stored in the experiment data repository and the search engine is notified about that. Thereafter, the search engine evaluates the property hierarchy as follows:

- Every property in the property hierarchy is tried to be instantiated with the data (represented in JavaPSL) stored in the experiment data repository.
- All true property instances are examined against the threshold for the given property in the property hierarchy. If the severity value of the property instance is above the threshold, then the instance is included in the list of *critical property instances,* which is the main output of the search engine.

3.1 Heuristics for Pruning the Evaluation of Property Hierarchies

The search engine traverses the property hierarchy in pre-order; it prunes the evaluation of the property hierarchy and avoids the creation of false and non-critical property instances by using a well-defined heuristic. Assume that an instance of property is true for a code region *q* in an experiment iff an instance of property ’ is also true for *q* in the same experiment. Then, we do not need to test whether holds for a certain code region in an experiment if we already know that ’ does not hold for the same code region in this experiment.

Before we define the pruning heuristics of Aksum we describe some important notation. A *static code region* (or simply code region) is a non-empty set of program statements with multiple control flow entry and exit points. If a code region *q* is executed by a thread *t* in a process *p*, we call a *child* of *q* in that experiment any code region executed by thread *t* in process *p* after the execution of *q* has started but not yet ended. Any code region that has *q* as a child is a *parent* of *q* in that experiment. Moreover, if *q'* is a child of *q* in a certain experiment, and there is no code region *q"* that is a parent of *q'* and child of *q*, then *q'* is an *immediate child* of *q*, and *q* is the *immediate parent* of *q'* in that experiment.

Now, let *q* be a code region, *e* an experiment, and DC(*q*, *e*) = { (*p*, *t*, *q*, *y*) | *t* is a thread in process *p* that executes *q* in experiment *e*, and *y* is the immediate parent of *q* in experiment *e*}. Every element *d* = (*p*, *t*, *q*, *y*) in DC(*q*, *e*) represents a particular execution of a code region and is called a *dynamic code region.* We will denote *y* as *parent(d)*, e as *experiment(d)*, and *q* as *codeRegion*(*d*).

Next, we define a *containment relation* between (dynamic) code regions:

- Given two dynamic code regions, *d* and *d'*, then *d* contains *d'* if *d* = *d'* or there is a sequence of dynamic code regions (d_1,…, d_n) such that $d_1 = d$, $d_n = d'$, and d_i = parent(d_{i+1}) *i*, 1 *i* < n.
- Given a dynamic code region *d* and a static code region *q*, then *d* contains *q* if codeRegion(*d*) = *q* or there is a sequence of dynamic code regions (d_1,…, d_n) such that $d_1 = d$, codeRegion(d_n) = *q*, and d_i = parent(d_{i+1}) *i*, 1 *i* < *n*.
- Given a dynamic code region *d*, a static code region *q*, and an experiment *e*, then *q* contains *d* in experiment *e* if there is a dynamic code region *d'* such that *d'* contains *d*, experiment(*d'*) = *e*, and codeRegion(*d'*) = *q*.

For every node k in the property hierarchy, the search engine loads the definition of the associated property p from a property repository and analyzes its constructor and all property instances that have been created for all properties along the path from the property tree root to k in order to determine the actual arguments that must be used to create instances of p. Based on the definitions above, we can now formally define the heuristics used by the search engine of Aksum:

Let $_k$ and $_k$ respectively denote the property and threshold of a node k in the property tree. For simplicity, we assume that $_k$ has only one constructor. Now, assume that the search engine is just considering k for its search for performance properties and $(a_1, \ldots, a_n)$ is a list of actual arguments compatible with the formal constructor parameters of $_k$. The search engine tries to create an instance of $_k$ using $(a_1, \ldots, a_n)$ as actual arguments if either k has no predecessor node in the property hierarchy, or if there exists a predecessor node g of k and an instance r of $_g$ generated with the actual arguments $(b_1, \ldots, b_m)$ such that *r.holds()* is true, *r.getSeverity()* $\geq$ $_g$, and any of the following conditions is true:

1. i, j, 1 i n, 1 j m: a_i and b_j are dynamic code and b_j contains a_i.

A property can be instantiated with a specific dynamic code region if an instance of a predecessor property in the property hierarchy is critical for either the same dynamic code region or a parent of this dynamic code region.

2. i, j, v, 1 i n, 1 $v < j$ m: a_i is a dynamic code region, b_j is a static code region, b_v = experiment(a_i), and b_j contains a_i in experiment b_v.

A property can be instantiated with a specific dynamic code region a_i if, in the experiment associated to a_i, an instance of a predecessor property in the hierarchy is critical for the static code region that a_i (or a parent of a_i) refers to.

3. i, u, j, 1 $u < i$ n, 1 j m, a_i is a code region, a_u an experiment, b_i a dynamic code region, a_u = experiment(b_j), and b_j contains a_i.

A property can be instantiated for a specific static code region and a specific experiment if an instance of any predecessor property in the hierarchy is critical for a dynamic code region referring to this code region and generated by this experiment.

4. i, u, j, v, d: 1 $u < i$ n, 1 $v < j$ m: a_i and b_j are code regions, a_u and b_v refer to the same experiment, and d is a dynamic code region such that codeRegion(d) = a_i, experiment(d) = a_u, and b_j contains d in a_u.

A property can be instantiated for a specific static code region and a specific experiment if an instance of any predecessor property is critical for the same static code region and the same experiment.

5. i, u, j, d: 1 $u < i$ n, 1 j m: type(a_i) = CodeRegion, type(a_u) = Experiment, type(b_j) = CodeRegion, type(d) = DynamicCodeRegion, codeRegion(d) = a_i, experiment(d) = a_u, a_i contains d for a_u, and v, 1 $v < j$, type(b_v) Experiment.

A property can be instantiated for a specific static code region and a specific experiment if an instance of any predecessor property in the hierarchy is critical for the same static code region independently of the experiment.

4 Experimental Results

In this section, we present a performance analysis for a financial modeling application conducted on an SMP cluster with 16 quad nodes connected through Fast Ethernet. The backward pricing application implements the backward induction algorithm to compute the price of an interest rate dependent product [[2]]. The backward induction algorithm has been implemented as an HPF+ [[1]] code based on which the VFC compiler [[1]] generates a mixed OpenMP/MPI code. Aksum analyses the performance of this code at the OpenMP/MPI program level.

Fig. 3 shows the user portal of Aksum, in which the user pre-selects and modifies the property hierarchy, the application input parameters and their value sets, the execution command line, and the execution directory. For instance, *P* is a machine size parameter that indicates the number of processors (covering 1, 2, 4, 8, and 16 processors) to be considered by Aksum.

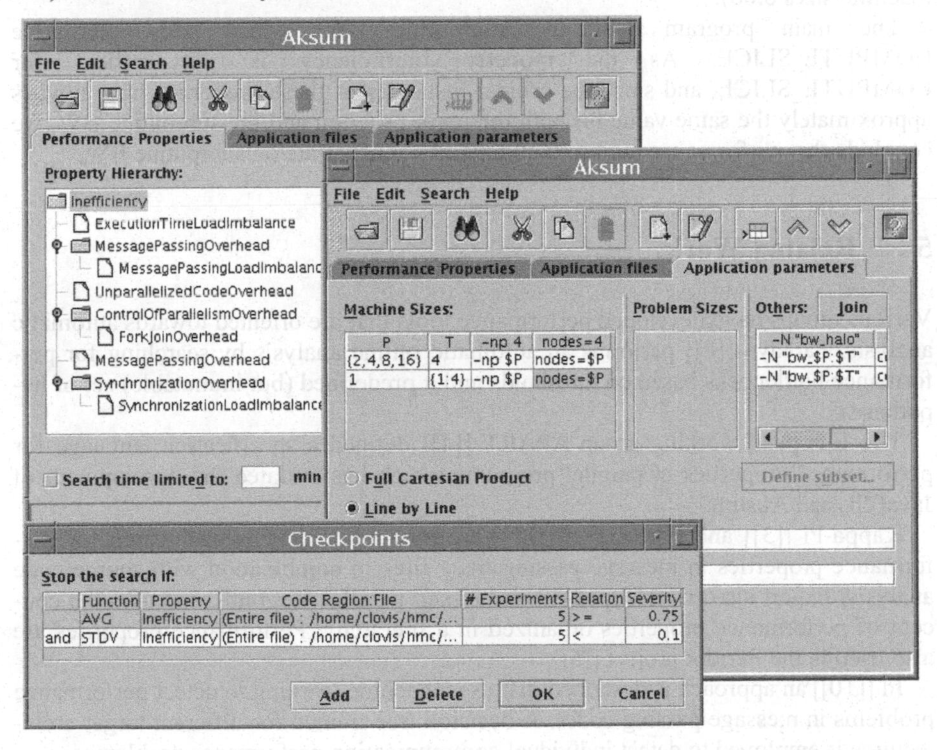

Fig. 3. User-provided input data under Aksum's user portal

Based on the user provided input data, the search engine of Aksum automatically determines that seven performance properties in the property hierarchy are critical for this code (see Fig. 4), where the properties are presented in ascending order of severity. Initially, the user portal displays only the property names for those instances whose severity is above the user-defined threshold. The property instances can be shown by expanding each property name. For every instance the corresponding code region and severity value are indicated. Instrumentation directives automatically inserted by Aksum are also shown. In the backward pricing application, the most serious performance property is ExecutionTimeLoadImbalance, which has an instance that holds for the main (entire) program with severity value 0.80 (see the entry *BW_HALO_3 0.80*). The same property holds for the sub-region of the main program indicated by the entry *BW_HALO_2 0.80*.

The severity of the ExecutionTimeLoadImbalance property instances for the entire application increases with the number of execution threads (not shown in Fig. 4), from 0.01 for 2 CPUs to 0.80 for 64 CPUs. This behavior also explains the increasing severity values for the Inefficiency property (varying from 0.05 for 2 CPUs to 0.79 for 64 CPUs – see the Inefficiency diagram in the upper right window of Fig. 4). All other properties in the property hierarchy have lower severity values (SynchronizationOverhead: 0.01, MessagePassingOverhead: 0.17 with 64 CPUs, for the other machine sizes 0.00).

The main program calls the subroutine BW, which calls subroutine COMPUTE_SLICE. As the property Inefficiency is not critical for COMPUTE_SLICE, and since the critical instances of these properties have always approximately the same value for both the main program and the subroutine BW, we conclude that performance tuning should mainly concentrate on subroutine BW.

5 Related Work

Various groups have developed performance tools that are oriented towards automatic analysis. Paradyn [[8]] performs an automatic online analysis by searching for performance bottlenecks based on thresholds and a predefined (but immutable) set of hypotheses.

The European working group APART [[4]] defined a specification language for performance properties of parallel programs which has initiated the development of JavaPSL and Aksum.

Kappa-Pi [[3]] and Earl/Expert [[11]] are post-mortem tools that search for performance properties in message passing trace files in combination with source code analysis. Expert also covers OpenMP and mixed parallel programs, and uses the concept of performance properties organized in a hierarchy. Performance properties are also used in the Peridot project [[6]].

In [[10]] an approach is described that uses machine learning to detect performance problems in message passing codes. A decision tree trained for different target architectures is employed to detect individual communication performance problems.

All of the previously mentioned tools concentrate performance analysis on single experiments only whereas Aksum examines the performance for multi-experiments based on an arbitrary user-provided set of machine and problem sizes. Aksum allows the user to modify and add new properties and property hierarchies. Aksum also provides a rich set of pre-defined properties and several possibilities to impact the search-for properties. We believe that Aksum is also more flexible to customize the automatic search for parallel and distributed programs compared to other work.

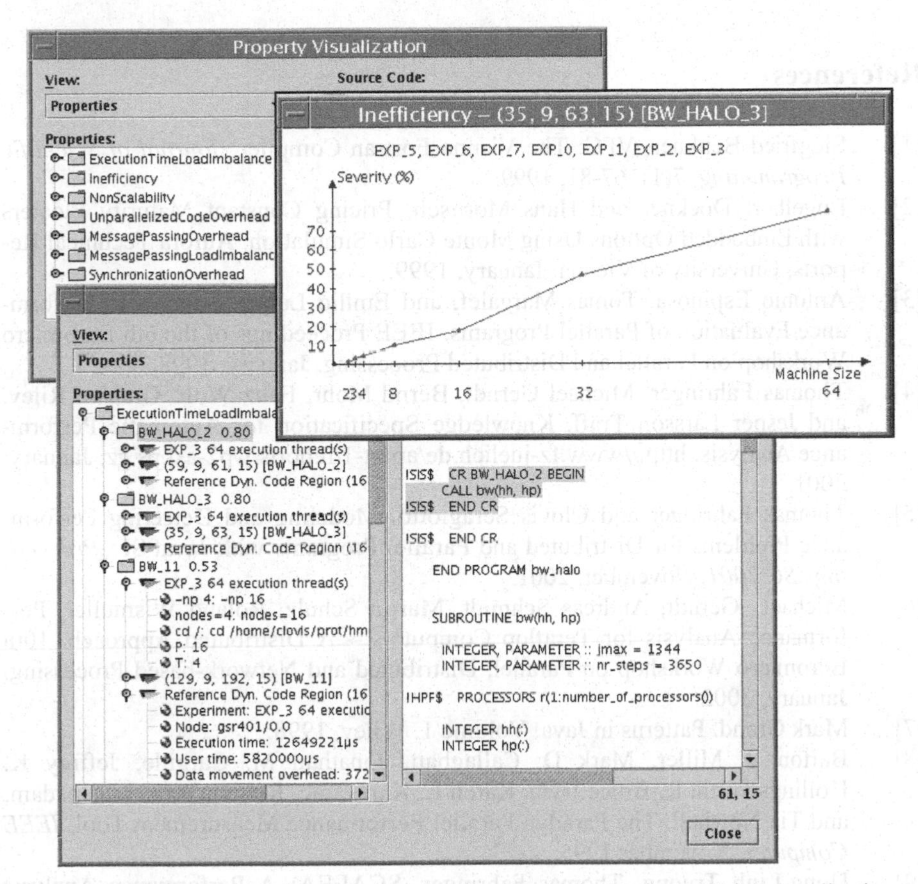

Fig. 4. Snapshots of Aksum's property visualization for the Backward Pricing application

6 Conclusion

We have presented Aksum, which is a highly flexible multi-experiment automated performance analysis tool for distributed and concurrent architectures. Based on a user-provided set of problem and machine size values, Aksum tries to automatically detect all critical performance problems and to guide programmers to performance problems that should be addressed for performance tuning. Aksum provides a very flexible mechanism to create and customize performance properties of interest and to control the search process. Although the user is given many opportunities to influence the search for performance bottlenecks, Aksum is a fully automatic tool once it is given the application sources as well as problem and machine sizes for which performance analysis should be done.

We are currently extending Aksum to enable the specification of search algorithms. Moreover, we plan to extend Aksum by online performance analysis based on dynamic instrumentation.

References

[1] Siegfried Benkner. VFC: The Vienna Fortran Compiler. *Journal of Scientific Programming*, 7(1): 67-81, 1999.

[2] Engelbert Dockner and Hans Moritsch. Pricing Constant Maturity Floaters with Embedded Options Using Monte Carlo Simulation. Aurora Technical Reports, University of Vienna. January, 1999.

[3] Antonio Espinosa, Tomàs Margalef, and Emilio Luque. Automatic Performance Evaluation of Parallel Programs. IEEE Proceedings of the 6th Euromicro Workshop on Parallel and Distributed Processing. January, 1998.

[4] Thomas Fahringer, Michael Gerndt, Bernd Mohr, Feliz Wolf, Graham Riley, and Jesper Larsson Träff. Knowledge Specification for Automatic Performance Analysis. http://www.fz-juelich.de/apart-1/reports/wp2-asl.ps.gz. January, 2001.

[5] Thomas Fahringer and Clovis Seragiotto. Modeling and Detecting Performance Problems for Distributed and Parallel Programs with JavaPSL. *Proceedings SC 2001*, November, 2001.

[6] Michael Gerndt, Andreas Schmidt, Martin Schulz, Roland Wismüller. Performance Analysis for Teraflop Computers – A Distributed Approach. 10th Euromicro Workshop on Parallel, Distributed and Network-based Processing. January, 2002.

[7] Mark Grand. Patterns in Java, Volume 1. Wiley, 1998.

[8] Barton P. Miller, Mark D. Callaghan, Jonathan M. Cargille, Jeffrey K. Hollingsworth, R. Bruce Irvin, Karen L. Karavanic, Krishna Kunchithapadam, and Tia Newhall. The Paradyn Parallel Performance Measurement Tool. *IEEE Computer*, November 1995.

[9] Hong-Linh Truong, Thomas Fahringer. SCALEA: A Performance Analysis Tool for Distributed and Parallel Program. *8th International Europar Conference*, August, 2002.

[10] Jeffrey Vetter. Performance Analysis of Distributed Applications using Automatic Classification of Communication Inefficiencies. *In Proceedings of the 14th International Conference on Supercomputing*, pp. 245-254, Santa Fe, New Mexico, May, 2000.

[11] Felix Wolf, Bernd Mohr. Automatic Performance Analysis of SMP Cluster Applications. Internal Report, Forschungszentrum Jülich GmbH. August, 2001.

An Adaptive Value-Based Scheduler and Its RT-Linux Implementation*

S. Swaminathan and G. Manimaran

Dependable Computing & Networking Laboratory
Department of Electrical and Computer Engineering
Iowa State University, Ames, IA 50011, USA
{swamis,gmani}@iastate.edu

Abstract. In dynamic real-time systems, value-based scheduling aims to achieve graceful degradation during overloads, in addition to maintaining a high schedulability during normal and underloads. The objective of this paper is twofold: (1) to propose an adaptive value-based scheduler for multiprocessor real-time systems aimed at maintaining a high system value with less deadline misses, and (2) to present the implementation of the proposed scheduler in a Linux based real-time operating system, RT-Linux, which in its current form does not employ a notion of task value. We evaluate the performance of the proposed scheduler in terms of two performance metrics, namely, "value ratio" and "success ratio" through both simulation and implementation.

1 Introduction

Multiprocessor and distributed systems are increasingly employed for critical real-time applications. Several heuristic algorithms have been proposed (e.g., [1]) for dynamic scheduling of tasks in multiprocessor real-time systems. These algorithms can maintain a high schedulability during normal loads but do not behave in a predictable manner during overload conditions and can lead to instability due to missing the deadline of critical tasks. Hence, a new scheduling paradigm called *value-based scheduling* [2] is employed, which schedules tasks such that the overall value (i.e., utility) of the system is maximized and maintained above an acceptable level. In such systems, it is assumed that each task offers certain "value" to the system, if the deadline is met; otherwise, a "penalty" is incurred. Thus, value-based scheduling is a decision problem involving the choice of tasks to execute so that the overall system value is maximized [2]. However, like many other practical instances of scheduling, this problem is also intractable and hence several heuristics have been proposed [3][4][5].

Value-based scheduling is of immense use in a flexible real-time system, wherein the system is expected to take decisions at run-time for efficient resource usage and also for system stability, and any design phase decision might

* This research was supported in part by the NSF under grant CCR-0098354.

S. Sahni et al. (Eds.) HiPC 2002, LNCS 2552, pp. 163–173, 2002.

lead to pessimistic usage of resources. An example of a flexible real-time system is an autonomous vehicle controller [2], wherein the system needs to exhibit intelligent and adaptive behavior in order to function correctly in a highly dynamic and non-deterministic environment characterized by unpredictable nature of other vehicles, route information, weather and road conditions. Such a real-time system has two conflicting objectives: (1) guaranteeing mission critical tasks to provide results of acceptable quality and (2) to increase the system utilization (and schedulability) determined by frequency, timeliness and precision, by scheduling as much tasks as possible.

In [6], the tradeoff between *value vs. deadline* scheduling has been studied in detail. The paper concludes that there is a need for different scheduling behaviors under different load scenarios. Consider the following simple example with two tasks in a uniprocessor system, where a task T_i is characterized by $< v_i, c_i, d_i >$ (v_i - value offered by T_i, c_i - computation time of T_i, d_i - deadline of T_i) : T_1 : $< 10, 20, 30 >$, T_2 : $< 100, 50, 80 >$. For scheduling these two tasks in a real-time system at time 0, scheduling based on deadline will meet the deadline of both the tasks (with T_2 scheduled after T_1), if the system is lightly loaded and execute both the tasks before their deadlines. However, if the system is highly loaded, then a deadline scheduling scheme might result in missing the deadline of a higher valued task. Hence during overloads, value-based scheduling scheme is preferred (with T_2 scheduled before T_1). Thus, it can be clearly seen from this simple example that there is a need for different scheduling behavior during different system conditions - such as deadline based scheduling under light loads and value-based scheduling during overloads. However, the value-based scheduler proposed in [3] schedules task using its value-density function irrespective of its workload, hence it performs better in overloaded conditions but cannot maintain a high schedulability during underloads or normal loads. In [7], a robust EDF-based scheduling algorithm is proposed, which is capable of handling overloads. The algorithm requires the system designer to identify the critical tasks apriori and design the system for this, which is quite conservative and can lead to pessimistic usage of resources.

In this paper, we present a scheduler that determines its mode of scheduling by continuous monitoring of its performance (in terms of system value and deadline misses). Based on the monitored values, it switches its scheduling behavior from deadline-based scheduling to value-based scheduling and vice versa, as required. There is no prior work on value-based scheduling for multiprocessor real-time systems, which forms the motivation for our paper.

The rest of the paper is organized as follows: In section 2 we define the system model, present our proposed value-based scheduler and its simulation studies. In section 3, we discuss the implementation of the proposed scheduler in RT-Linux and its performance study. Finally, in section 4, we conclude the paper.

2 Proposed Adaptive Value-Based Scheduler

2.1 System Model

(1) The system consists of m processors. (2) Each task T_i entering the system is characterized by $< a_i, c_i, d_i, v_i >$, where a_i represents the arrival time, c_i represents the computation time, d_i represents the deadline and v_i represents the value offered by T_i to the system. (3) A task T_i is said to be feasible to schedule on a processor P_k, if $EST_{ik} + c_i \leq d_i$, where EST_{ik} is the earliest start time of T_i on P_k, which is max $\{P_k$'s available time, $a_i\}$.

2.2 Proposed Adaptive Value-Based Scheduler

We present an adaptive value-based scheduler for multiprocessor systems, which is a branch and bound search algorithm that can adapt its scheduling behavior based on system load conditions. The objective of proposed scheduler is to search for a feasible schedule for a given set of tasks, such that the overall value of the system is maximized with less deadline misses.

The adaptive scheduler for multiprocessor system is presented in Figure 1. The system consists of a task queue, containing n tasks that are ready to run and ordered based on their deadline (step 1 in Figure 1). The scheduler examines the first K tasks in queue (which we call feasibility check region R) and examines to see if the tasks in R are feasible (step 3). If feasible, then the scheduler applies the branch function (as given in equation (3), which is explained in detail in the next subsection) for each task to find the best task to extend the schedule (search tree). That is, the task with minimum branch function value is selected (step 4 (b)) and the scheduler repeats the same process of task selection until the task queue is empty. If feasibility check fails, then the scheduler backtracks and extends the schedule with the next best task. If the number of backtrack exceeds the maximum backtrack limit then the scheduler drops the task with the least value and begins the search again. The skeleton of the algorithm is similar to that of the myopic scheduler [1] proposed for deadline-based scheduling, except for the heuristic function used in step 4(a). The complexity of the scheduler is $O(n)$.

Value Function The important component of the adaptive value-based scheduler is the branch function (used in step 4(a)), which introduces the adaptive nature in scheduling. The principle idea of this scheduling is to periodically monitor the following performance parameters of the system, for each invocation of the scheduler to schedule a set of tasks (we call this invocation interval as a *scheduling epoch*):

(1) *Value-Indicator (v):* Value of tasks that were rejected as compared to the value of tasks that were accepted, indicated by the value-indicator parameter v. ($v > 1$ implies missing of deadline of higher valued tasks due to scheduling of lower valued tasks.)

Adaptive Value-Based Scheduler()
Input: Task set to be scheduled.
Output: Feasible schedule (maximizing the value).

1. Tasks in the task queue are ordered in non-decreasing order of deadline.
2. Start with an empty schedule (as initial state of search tree).
3. Check if the initial K tasks in the feasibility check region R is feasible.
4. **If feasible**
   (a) Compute the branch function $H(.)$ (as in equation 3) for all tasks in R.
   $H(T_i) = d_i * (1 - F(s)) + (\kappa/v_i) * F(s)$
   (b) Extend the task schedule with lowest branch function value on next available processor.

   else
   (a) Backtrack to the previous schedule (node), if number of backtracks are less than the maximum limit (if so drop task with least value).
   (b) Extend the schedule with next best task.
5. Repeat steps 3-5 until the task queue is empty or all the tasks are either accepted or rejected.

Fig. 1. Adaptive value-based scheduling for multiprocessor systems

$$v = \frac{maximum\ value\ of\ task\ among\ all\ the\ tasks\ rejected}{minimum\ value\ of\ task\ among\ all\ the\ tasks\ accepted} \tag{1}$$

(2) *Utilization (U):* Total system utilization given by $U = \sum_{i=1}^{m} U_i/m$, where U_i gives the utilization of processor P_i.

Then, we define a value weight function $F(s)$ for every scheduling epoch s, which will be used by the scheduler (in step 4(a)) for selecting the best task to schedule, as follows:

$$F(s) = \begin{cases} 1, v \geq 1 \\ v, v < 1\ and\ U \leq 1 \\ F(s-1), otherwise \end{cases} \tag{2}$$

The above defined value weight function, $F(s)$, is calculated for every scheduling epoch and determines the weight that must be given to the *value-based scheduling*, while $1 - F(s)$ determines the weight that must be given to *deadline scheduling.* Hence, this heuristic function is used to select the best task (task with minimum heuristic value) for scheduling. The heuristic function applied for a task T_i is defined as follows:

$$H(T_i) = d_i * (1 - F(s)) + (\kappa/v_i) * F(s) \tag{3}$$

where κ is a constant that normalizes value to the same scale of deadline.

The above heuristic function is dynamic in nature, wherein the first part of the equation represents the deadline scheduling behavior, while the second part determines the value-based scheduling behavior of the scheduler. Thus, change in the $F(s)$ value leads to a change in the scheduling behavior with a value of 1 meaning that the tasks are scheduled just based on value, while a value of 0 meaning that the tasks are scheduled just based on deadline. Since, the value of $F(s)$ changes according to the load, the scheduling behavior also changes accordingly from deadline scheduling to value-based scheduling or vice versa.

Further, it must be noted that the proposed value function can be adopted in any real-time scheduler - dynamic planning-based scheduler (e.g., myopic) or priority driven scheduler (e.g., EDF). The only requirement is to use the proposed heuristic for task selection to introduce adaptive scheduling behavior.

2.3 Simulation Studies

We evaluate the performance of the proposed scheduler through simulation studies for a wide range of workloads by comparing it with the deadline scheduler that uses $H_i = d_i$ as the heuristic function (step 4(a)) in Figure 1). This algorithm is labelled as "D-only" in the performance graphs. The following simulation setup was used for evaluation: (1) Each simulation run generates 10000 tasks. (2) The task inter-arrival time follows exponential distribution with mean θ. (3) A task T_i's execution time (C_i) is chosen at random uniformly in the interval [10,50]. (4) The deadline of a task is assigned to be $\Lambda * C_i$, where Λ is chosen uniformly in the interval [1,4]. (5) The value of a task is chosen at random uniformly between [50,1000]. (6) The system load is characterized by $L = C/\theta$, where C is the average execution time of a task and θ is the arrival rate of tasks in the system. (7) The size of the feasibility check region (K) was chosen to be 6, as the scheduler yielded better performance for this value of K. (8) The value of $F(s)$ was recomputed after successful scheduling of 50 tasks, comprising a scheduling epoch. The following performance metrics were used for evaluation.

- *Success Ratio (SR)*: This is defined as the ratio of the number of tasks that were scheduled to the total number of tasks that arrived in the system.
- *Value Ratio (VR)*: This is defined as the ratio of the value obtained by the scheduler (i.e., the sum of value of scheduled tasks) to the the total value of all the tasks.

The schedulers were evaluated for 2 and 4 (homogeneous) processor systems for loads from 0.5 to 3. The results of the experiments evaluated based on SR and VR are given in Figures 2 and 3. As it can be seen from the figures, the proposed adaptive value-based scheduler performs well in terms of VR during overloads and maintains a high VR than its deadline-based counterpart, which in spite of having a better SR has a poor VR. Further, the performance of the two schedulers during underloads and near full loads are also comparable with respect to SR and both maintain a comparable VR. This is due to the fact that the proposed adaptive scheduler adapts to the workload and switches to deadline-based scheduling during underloads, thereby maintaining a high scheduling success ratio and does value-based scheduling during overloads yielding a high system value.

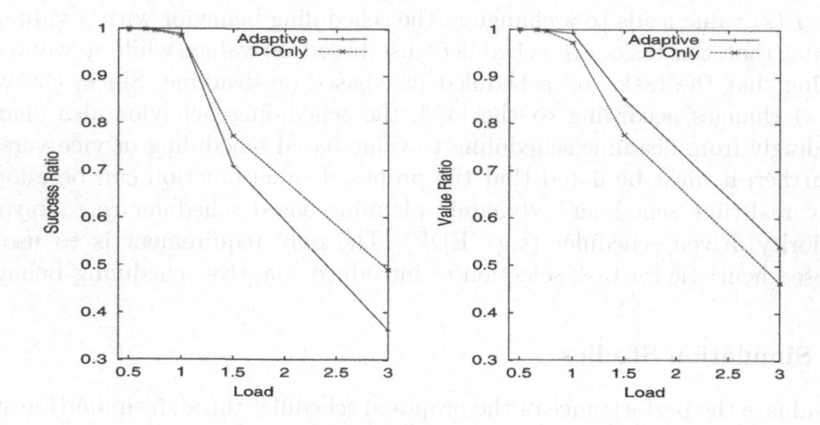

Fig. 2. Performance of the schedulers in a 2-processor system

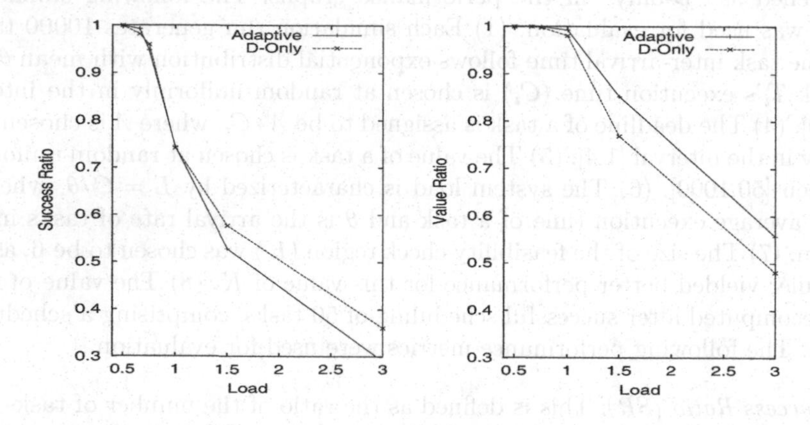

Fig. 3. Performance of the schedulers in a 4-processor system

3 Implementation of Proposed Scheduler

3.1 RT-Linux Architecture

The proposed scheduling scheme was implemented in RT-Linux kernel, one of the most popular open source Linux-based real-time operating systems. RT-Linux is a real-time variant of Linux, which adopts the approach of making Linux run as a low-priority task of a real-time executive. RT-Linux [9] decouples the mechanisms of the real-time kernel from the mechanisms of the general-purpose kernel so that each can be optimized independently and so that the real-time kernel can be kept small and simple.

RT-Linux is module-oriented and relies on the Linux loadable kernel mechanism to install components of the real-time system and to keep the RT-system modular and extensible. The key modules of the system are the scheduler and the one that implements RT-FIFOs. The use of modules ensures the ease of changing policies for scheduling RT tasks, if the deadline requirements are not met. Currently, RT-Linux supports only periodic tasks and has two in-built schedulers - RMS and EDF implemented in it.

3.2 Task Creation and Scheduling in RT-Linux

Real-time tasks are created in RT-Linux using real-time thread library *pthreads*. The pthread's (task's) attributes are stored in a task structure, *rtl_thread_struct*. The important variables of the real-time task structure (rtl_thread_struct) are given as follows:

– *period* - Task Period variable - set during invocation by the task module
– *priority* - Task Priority variable - set during invocation
– *current_deadline* - Task Deadline variable - set at run-time after every period
– *resume_time* - Task Resumption Time variable - set by scheduler at run-time usually as multiples of task period

The kernel maintains a list of tasks that are ready to execute in the system in a global linked list, *task_list* and APIs such as pthread_create(), pthread_delete() and pthread_set_schedparams() are used for task creation, deletion and setting up of task parameters, respectively. Thus, a real-time developer needs to use the above APIs to write a program that creates periodic real-time thread with desired timing characteristics. Scheduler in RT-Linux follows a run-time system model and constructs schedule at run-time without performing a schedulability check. The scheduler can be invoked due to various reasons: such as creation of a new task, completion of currently running task etc. Upon invocation, the scheduler (rtl_schedule()) selects the best task among the ready tasks (task with *resume_time* > *current_time*), based on priority/period/deadline and switches the execution to the selected task. The priority-based scheduler selects the task with highest priority, while RMS and EDF schedulers schedule tasks with lowest period and deadline respectively. In case of overload, the deadlines of few tasks will be missed and it can be observed when $resume\_time < current\_system\_time - period$. Hence the *resume_time* is increased by multiples of *period* until the above condition is invalid. The number of times *period* is added to *resume_time* indicates the number of instances of the task missing the deadline. The complexity of the existing scheduler in RT-Linux is $O(n)$, as the scheduler does a linear search on the task list to select the best task.

3.3 Implementation of Value-Based Schedulers

RT-Linux from the original open sources does not employ a notion of value of a task. Hence for the implementation of value-based scheduling schemes in RT-Linux, the real-time task structure (*rtl_thread_struct*) was changed and a new variable, *value*, was added. Two value-based schedulers, proposed adaptive value-based scheme and HVDF, are implemented in RT-Linux. Further, RT-Linux does not perform schedulability check and hence the proposed scheduler is implemented without schedulability check component. The implementation details of these two schedulers are presented in the next subsections.

Adaptive Value-Based Scheduler For implementation of the proposed adaptive value-based scheduler (Adaptive-Impl), the primary schedule function was changed, wherein task priority is computed with heuristic function as given in equation 3. The task with the minimum heuristic value is selected for scheduling. Thus, the complexity of the scheduler is still linear ($O(n)$). Further, the value of tasks that met and missed their deadlines were profiled through newly added kernel APIs (MM_met_deadline() and MM_miss_deadline()). These APIs update the maximum value of rejected tasks, MM_max_rejval variable, and minimum value of accepted tasks, MM_min_acceptval variable. These variables were used in the calculation of value-ratio for every scheduling epoch determined by MM_invocation_interval variable, which is decremented after every call to rtl_schedule(). Once it reaches zero, FS is recalculated and *MM_invocation_interval* is reset to original feedback value. Thus, recalculation of FS is done every *MM_invocation_interval* times. It must be noted that the *MM_invocation_interval* variable determines the sensitivity of the scheduler to change in workload. The value of v and FS are calculated as shown in equations (1) and (2). (More details can be found in [10]).

HVDF The implementation of this scheduler was done similar to the above scheduler, except that the task with highest value density (v_i/c_i) is selected for scheduling.

Validation of Implementation The correctness of the implementation of the proposed adaptive value-based scheduler was validated by evaluating its performance for various workloads and comparing it with the simulated scheduler for the same set of workloads in terms of success ratio (SR) and value-ratio (VR). The validation method adopted is as follows: Periodic task sets were generated for different loads and the schedulers' (both implementation and simulation) performance were studied. In our studies, we found that the implemented scheduler performs close to the simulated scheduler with a small difference in performance (more details can be found in [10]). The difference in performance observed between these schedulers might be due to extra workload introduced by the Linux OS and its daemons and overheads due to actual scheduler, context switch and interrupt handling, which were not accounted in the simulation.

3.4 Performance Evaluation

The performance of the three schedulers were evaluated by generating random task sets for different loads. Further, it must be noted that RT-Linux supports only periodic threads and in our performance evaluation we studied aperiodic threads by creating periodic threads and running them for a single instance and deleting them upon execution. Further, kernel math libraries are not available in Linux, hence random task sets were generated by user level programs, which were used in the kernel modules.

The performance evaluation was conducted in a Pentium-II 266 MHz machine with RT-Linux running on it. The performance of three schedulers in terms of SR and VR for various workloads and the results are presented in Figure 4. It can be seen from the figures that the proposed scheduler performs better than EDF and HVDF in terms of VR during underloads. Further, it can be seen that during near full loads and full loads, the proposed adaptive scheduler performs better than HVDF in terms of SR as it schedules tasks based on deadline even during near full loads (not based on value as in HVDF). However, during overloads both HVDF and proposed adaptive schemes maintain the same VR as both switch over to value-based scheduling.

The feedback invocation interval (MM_invocation_interval) also affects scheduler's performance as a high feedback rate makes the system more sensitive to transient changes in the system, it also increases the run-time overheads, while a low feedback rate decreases the system's sensitivity to change in workload, yielding a poor performance. The effect of invocation interval on SR and VR was studied for various values of (MM_invocation_interval) for a full (100%) load and the results are given in Figure 5. It can be seen from the figure that the scheduler offers better VR and SR when FS is calculated for scheduling of every 10 tasks in the current experiment setup, which does not introduce too much overhead due to feedback and also does not decrease the system's sensitivity to change in workload.

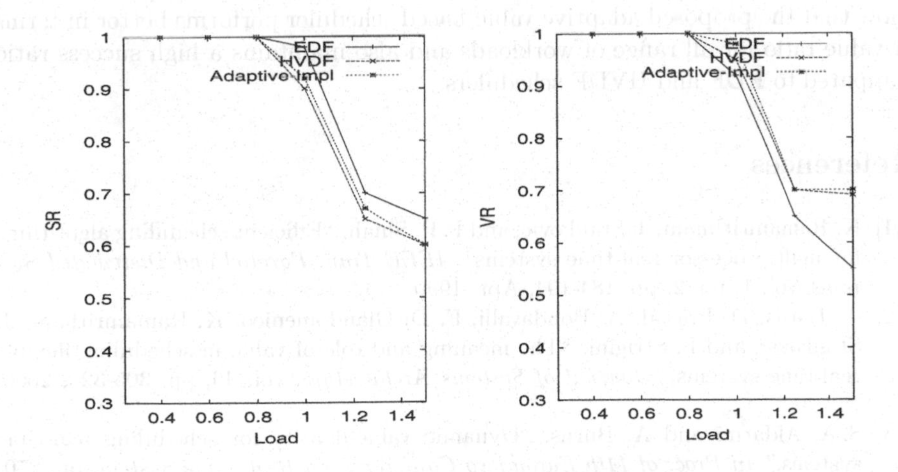

Fig. 4. Performance of the implemented schedulers (uniprocessor system)

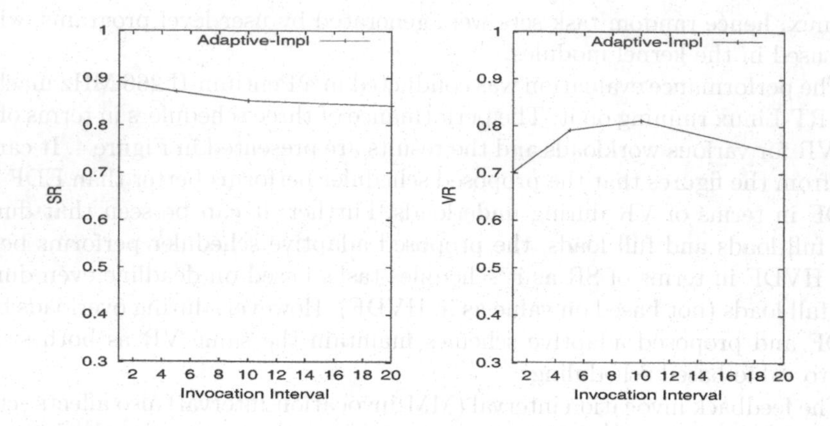

Fig. 5. Effect of scheduler invocation interval (uniprocessor system)

4 Conclusions

In this paper, we have proposed an adaptive value-based scheduler for multiprocessor real-time systems. The proposed adaptive algorithm switches its scheduling behavior from deadline scheduling to value-based scheduling and vice versa based on the system load. We have implemented the proposed scheduling scheme in RT-Linux and evaluated its performance against the EDF and HVDF schedulers. Our performance studies (through simulation and actual implementation) show that the proposed adaptive value-based scheduler performs better in terms of value ratio for all range of workloads and also maintains a high success ratio compared to EDF and HVDF schedulers.

References

[1] K. Ramamrithnam, J. Stankovic, and P. F. Shiah, "Efficient scheduling algorithms for multiprocessor real-time systems", *IEEE Tran. Parallel and Distributed Systems*, vol. 1, no. 2, pp. 184-194, Apr. 1990. 163, 165

[2] A. Burns, D. Prasad, A. Bondavalli, F. Di Giandomenico, K. Ramamritham, J. Stankovic, and L. Strigini, "The meaning and role of value in scheduling flexible real-time systems," *Journal of Systems Architecture*, vol. 46, pp. 305-325, 2000. 163, 164

[3] S. A. Aldarmi and A. Burns, "Dynamic value-density for scheduling real-time systems," *In Proc. of 11th Euromicro Conference on Real-Time Systems*,pp. 270-277, June 1999. 163, 164

[4] S. A. Aldarmi, and A. Burns, "Time-cognizant value functions for scheduling real-time systems," Technical Report #YCS-306, University of York, Oct. 1998. 163

[5] S. Swaminathan and G. Manimaran, "A reliability-aware value-based scheduler for dynamic multiprocessor real-time systems," in Proc. *IEEE Intl. Workshop on Parallel and Distributed Real-Time Systems*, 2002. 163

[6] G. Buttazzo, M. Spuri and F. Sensini, "Value vs. deadline scheduling in overload conditions," in Proc. *IEEE Real-Time Systems Symposium*, pp. 90-99, Dec. 1995. 164

[7] G. Buttazzo, and John A. Stankovic, "RED: Robust Earliest Deadline scheduling," in Proc. *3rd Intl. Workshop on Responsive Computing Systems*, pp. 100-111, Sep. 1993. 164

[8] Real-Time Linux -Available for download at http://www.rtlinux.org.

[9] Michael Barbanov and Victor Yodaiken, "Real-Time Linux ," *Linux Journal*, Mar. 1996. 168

[10] S. Swaminathan, "Value-based scheduling in real-time systems," *M.*S Thesis, Iowa State University, 2002. 170

Effective Selection of Partition Sizes for Moldable Scheduling of Parallel Jobs*

Srividya Srinivasan, Vijay Subramani, Rajkumar Kettimuthu,
Praveen Holenarsipur, and P. Sadayappan

Ohio State University, Columbus, OH, USA
{srinivas,subraman,kettimut,holenars,saday}@cis.ohio-state.edu

Abstract. Although the current practice in parallel job scheduling requires jobs to specify a particular number of requested processors, most parallel jobs are moldable, i.e. the required number of processors is flexible. This paper addresses the issue of effective selection of processor partition size for moldable jobs. The proposed scheduling strategy is shown to provide significant benefits over a rigid scheduling model and is also considerably better than a previously proposed approach to moldable job scheduling.

1 Introduction

The issue of effective scheduling of parallel jobs on space-shared parallel systems has been the subject of several recent research studies [4], [12], [14], [21]. Most of the research to date on this topic has focused on the scheduling of rigid jobs, i.e. jobs for which the number of required processors is fixed. This matches the practice at all supercomputer centers to our knowledge: users specify a specific single value for the number of processors required by a job. However, most parallel applications are moldable [10], i.e. they can be executed on different numbers of processors. If the machine were empty, the fastest turnaround time for a particular job would be obtained by specifying as large a number of processors as possible. But on a machine with heavy load, specifying fewer processors may actually provide a faster turnaround time than specifying the maximum number of processors. Although a run with more processors is likely to take less time than on fewer processors, the waiting time in the queue may be much longer. It would be desirable to have an intelligent scheduler determine the number of processors to allocate to different jobs, without forcing the user to specify a single specific value.

Several studies have considered job scheduling for the case of malleable jobs, where the number of processors for a job can be varied dynamically [1], [2], [3], [15], [16], [17]. In this paper, we consider the scheduling under a moldable job model. Recently Walfredo Cirne [5], [6], [7] evaluated the effectiveness of moldable job scheduling. Using synthetic job traces, he showed improvement in turnaround times of jobs under a moldable scheduling model,

* Supported in part by a grant from Sandia National Laboratories.

S. Sahni et al. (Eds.) HiPC 2002, LNCS 2552, pp. 174–183, 2002.

when compared to a standard conservative backfilling scheme [14], [22]. A greedy strategy was employed at job submission time for selecting the processor partition size for each job - among a set of possible choices identified for each job (using random variables characterizing the job's scalability), the one that gave the earliest estimated completion time was chosen.

Using a subset of a one-year job trace from the Cornell Theory Center [9], we evaluated the effectiveness of a greedy partition selection strategy for moldable job scheduling. Instead of restricting the range of processor choices for each job by use of statistical distributions (as done in Cirne's experiments), we allowed each job a range of choices from one processor to the total number of processors in the system. We found the performance improvement provided by moldable scheduling over standard non-moldable job scheduling to be considerably lower than that reported by Cirne. The results of the experiments highlighted the importance of careful selection of the processor partition size for each job. A greedy selection strategy over a wide range of choices for partition size was problematic - most jobs tended to choose very wide partitions, resulting in deterioration of performance for small jobs.

Using the insights from the initial experiments, we develop a more effective strategy for selection of partition size for moldable job scheduling. We show that the proposed scheme provides considerable overall improvement in job turnaround time of large jobs, without significantly impeding small jobs.

This paper is organized as follows. In Section 2, we provide some background information pertinent to this paper. Section 3 evaluates the previously proposed greedy submit-time moldable scheduling approach. A new approach to selection of processor partition size is presented and evaluated in Section 4. An enhancement to this approach is proposed and evaluated in Section 5 and we provide conclusions in Section 6.

2 Background and Workload Characterization

Scheduling of parallel jobs is usually viewed in terms of a 2D chart with time along one axis and the number of processors along the other axis. Each job can be thought of as a rectangle whose width is the user estimated run time and height is the number of processors requested. The simplest way to schedule jobs is to use the First-Come-First-Served (FCFS) policy. This approach suffers from low system utilization. Backfilling [13], [14], [22] was proposed to improve system utilization and has been implemented in several production schedulers [11]. Backfilling works by identifying "holes" in the 2D chart and moving forward smaller jobs that fit those holes. There are two common variants to backfilling - conservative and aggressive (EASY)[14], [18]. In conservative backfill, every job is given a reservation when it enters the system. A smaller job is moved forward in the queue as long as it does not delay any previously queued job. In aggressive backfilling, only the job at the head of the queue has a reservation. A small job is allowed to leap forward as long as it does not delay the job at the head of the queue.

A job is said to be moldable if it can run on multiple processor request sizes [10]. If the user requests a large number of processors, the execution time of a job may be lower, but it may have to wait for a long time in the queue before all needed processors are available. If the user requests a smaller number of processors, the wait time may be lower but the run time will be higher. There is a need to balance these two factors since the job turnaround time, which is of primary interest to the user is the sum of the job wait time and the job run time. Since the scheduler has the snapshot of the current processor allocation, if the task of deciding the job request size is left to the scheduler, the performance could potentially be better than if the decision is made by the user at submit time.

2.1 Workload Characterization

We use trace based simulation to evaluate the various schemes using the CTC workload log from Feitelson's archive [9]. Any analysis that is based on the aggregate turnaround time of the system as a whole does not provide insights in to the variability within different job categories. Therefore in our discussion we classify the jobs in to various categories based on their weight (i.e processor seconds needed) and analyze the average turnaround time for each category. Since job logs from supercomputer centers only include a single specific number for each job's processor requirement, an important issue in evaluating moldable scheduling approaches is that of job scalability. It is necessary to estimate each job's run-time for different possible processor partition sizes. The user estimated execution time and the actual execution time for the original processor request can be determined from the trace file. We use the Downey model [8] to generate the execution times for new processor request sizes. We assume that the ratio of the user estimated run time with the actual execution time remains the same as the processor request size changes. Cirne [5], [6], [7] used Downey's model of job scalability, with statistical distributions for model parameters.

3 Submit-Time Greedy Selection of Partition Size

We first evaluate the approach to moldable scheduling that has been previously proposed [5]. In this scheme, every job is allowed a range of processor choices, from one processor to the total number of processors in the system. Among these partition sizes, the one that results in the best turnaround time for the job is chosen. The decision of which choice of processor count to allocate is made at the time of job submission - and hence once a choice is made, the job is made rigid and is no longer moldable(Submit-time moldability). Under this scheme, many of the jobs can be expected to choose very wide partition sizes, because of the local greedy nature of the strategy. Consider the arrival of jobs one after the other. For the first arriving job, the widest partition size will give the least turnaround time and hence it chooses a partition size that is equal to the total number of processors in the system. As a result, the next arriving

Table 1. Distribution of jobs based on partition sizes

Partition Size	Non-Moldable	Greedy Moldable
1	2223	1169
2	394	174
3-4	655	212
5-8	494	55
9-16	617	0
17-32	327	0
33-64	172	51
65-128	72	36
129-256	34	70
>256	12	3233

job also chooses a partition size equal to the total number of processors in the system (as it results in the least turnaround time) and so on.

We first evaluated the submit-time greedy strategy under an assumption of perfect scalability of all jobs (i.e. $\sigma = 0$ for the Downey model). Table 1 shows the resulting distribution of partition sizes for the jobs in the system. We observe that a majority of the jobs tend to choose very wide partition sizes. Fig. 1 shows the percentage change in the average turnaround time for the greedy scheme with respect to the standard conservative backfilling scheme.The overall average turnaround time improves in comparison to conservative backfilling. This is because, under the greedy scheme, jobs choose wide partition sizes and hence execute one after the other as opposed to choosing narrow partition sizes and executing in parallel. But the average turnaround time for the small jobs deteriorates to a large extent because these jobs find very few "holes" in the schedule to backfill under the greedy scheme.

As σ is increased (i.e. jobs are less scalable), the performance of the scheme can be expected to deteriorate. This is because there is a resource usage penalty for using wide partition sizes - the total number of processor-seconds needed for a job increases as more processors are used. When job scalability is not perfect, a load-sensitive partition size selection strategy is called for. If there is a single job in the system, it might be appropriate for it to utilize all available processors. However, when there are several queued jobs, wide partition choices for each of the jobs is wasteful of resources; instead narrower choices would be more efficient. Fig. 2 shows performance data for $\sigma = 1$, and performance has indeed deteriorated. Performance deteriorates further as σ is increased, but we omit the data for space reasons.

4 Load-Sensitive Selection of Partition Size

The greedy scheme is not preferable because most of the jobs choose wide partition sizes and as a result, the turnaround times of the small jobs deteriorates to a large extent. Thus no job should be allowed to use up all the processors in the

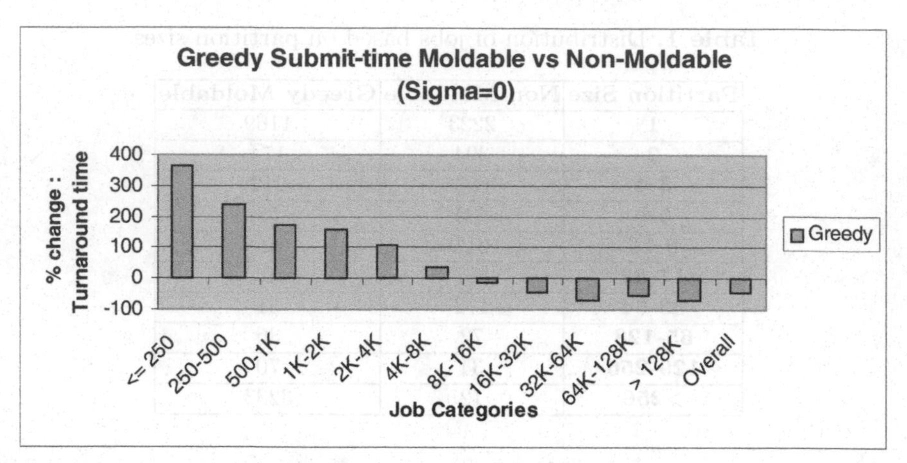

Fig. 1. Category-wise comparison of the performance of the greedy submit-time moldable scheme vs non-moldable scheduling with conservative backfilling. Although the greedy scheme improves the overall turnaround time, it degrades the average turnaround time of the small jobs to a large extent

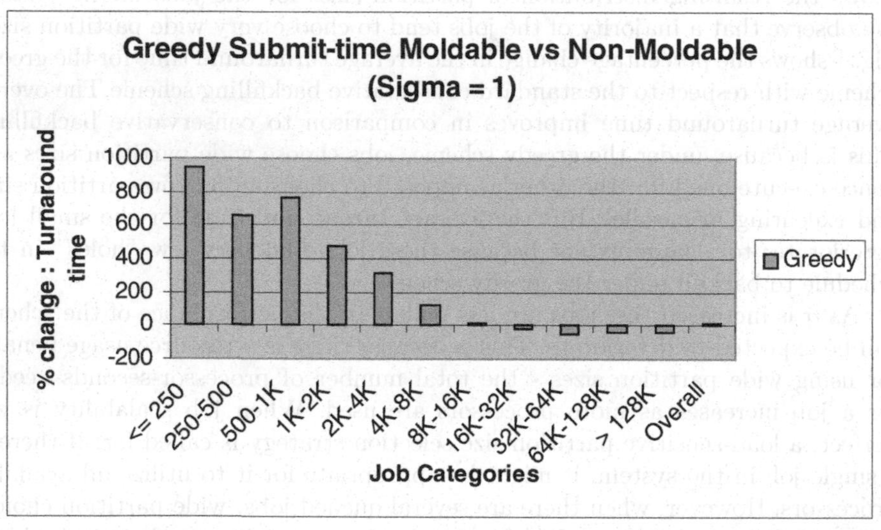

Fig. 2. Performance of the greedy scheme vs non-moldable conservative backfilling. As σ is increased the performance of the greedy scheme further deteriorates

system. Hence we impose a limit of 90% on the maximum number of processors that can be allocated to a single job. This means that no job will be allowed to expand to more than 90% of the total number of processors in the system. Using such a limit alone is not sufficient - although it would improve backfilling opportunities for the small jobs, it would be indiscriminate in allocating the available set of processors to jobs and hence could result in most of the jobs

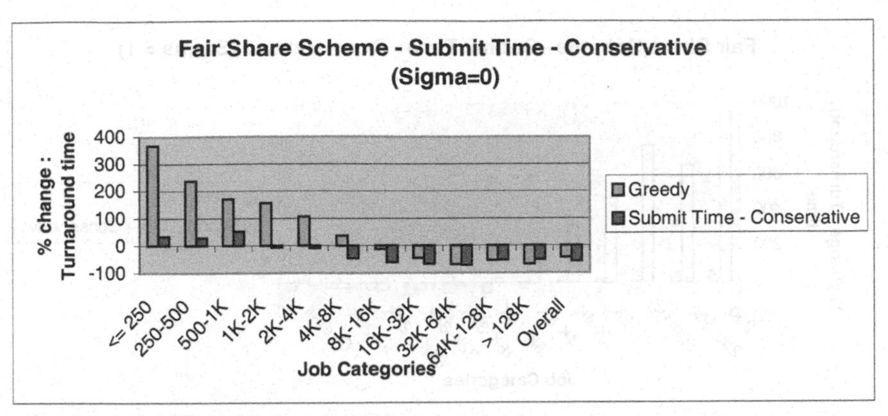

Fig. 3. Performance of the load-sensitive submit-time moldable conservative backfilling scheme. There is a 50% decrease in the overall average turnaround time compared to conservative backfilling. Compared to the greedy scheme the average turnaround time of all categories except the very large jobs improves

expanding to occupy 90% of the processors in the system. Instead, it is desirable that the available set of processors be allocated to the waiting jobs based on load considerations. For example, consider the case where there are four jobs with relative resource requirements of one, four, eight, and twelve. Clearly it would be beneficial to allocate more processors to the heaviest job (i.e. the one with the largest resource requirement) and fewer processors to the lightest job. However, it would not be desirable to allocate most or all processors to the heaviest job and have the others wait. On the other hand, when there is only a single heavy job in the system, we should allow it a wider partition limit. A suitable strategy might be to determine a "fair-share" processor-count for each job, based on the fractional resource requirement of this job compared to the total requirements of all pending jobs. Thus the maximum allowable partition size (called the fair share limit) should be different for different jobs and it should depend upon what fraction of the total weight of the jobs currently in the system (both idle and running) a job constitutes. The fair share limit for a job is defined as follows:

Fair Share Limit of a Job = (Total Processors in the System) * Weight of the Job/Sum of the Weights of all Jobs Currently in the System.

Since the above limit may be overly restrictive, a stretch factor of 2 was used. i.e the maximum allowable processor limit for a job was set to twice its fair share limit. Thus for each job, a range of choices for partition sizes, from one to twice its fair share limit is tested and the one which gives the best expected turnaround time is chosen.

Fig. 3 shows the percentage change in the average turnaround time for the greedy and the fair share scheme with respect to non-moldable conservative backfill scheduling. We observe that the fair share scheme results in a 50% decrease in the overall average turnaround time compared to conservative backfilling. Most

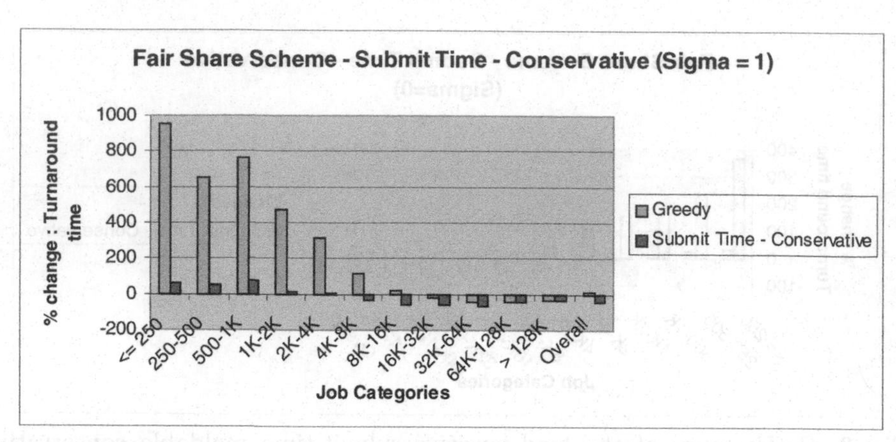

Fig. 4. Performance of the load-sensitive submit-time moldable conservative backfilling scheme. The relative improvement achieved by this scheme compared to the greedy scheme increases with increasing σ

of the job categories improve significantly, with only a slight deterioration for the small jobs. Comparing the greedy scheme with the fair share scheme, the fair share scheme improves the overall average turnaround time and the turnaround time for all the jobs except the very large jobs (whose weight is greater than 128,000 seconds). Fig. 4 shows performance data for σ=1. The relative improvements achieved by the proposed scheme compared to the greedy scheme increases with increasing σ.

5 Schedule-Time Moldability and Aggressive Backfilling

In this section, we attempt to enhance performance by modifying the scheduling strategy in two ways:

- Instead of freezing the partition size choice for each job at job submission time, defer it till actual job start time (Schedule-time moldability).
- Instead of using conservative backfilling, use aggressive backfilling.

There is a fundamental trade-off between conservative and aggressive backfilling. Conservative backfilling provides reservations to all jobs at submission time, while aggressive backfilling has only one reservation at any time. Thus the aggressive scheme has much more backfilling. However, job categories that have difficulty backfilling (such as very wide jobs) suffer from the lack of reservations. Overall, whether conservative or aggressive backfill is better, depends on the mix of the jobs since some some jobs (Short Wide) consistently do better with conservative backfill while others (Long Narrow) do better with aggressive backfill [19], [20].

However, when we consider a moldable job scheduling model, the disadvantage of aggressive backfilling disappears! This is because, there are no longer jobs

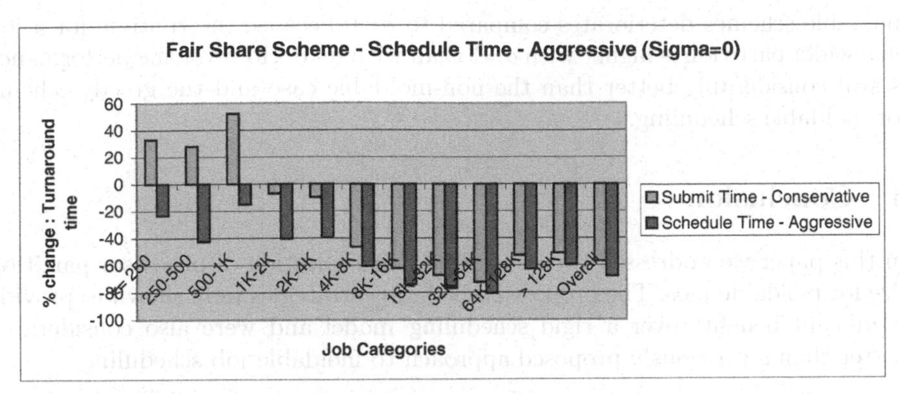

Fig. 5. Performance of the load-sensitive schedule-time-moldable aggressive backfilling scheme. This scheme clearly outperforms standard conservative backfilling and submit-time-moldable conservative backfilling

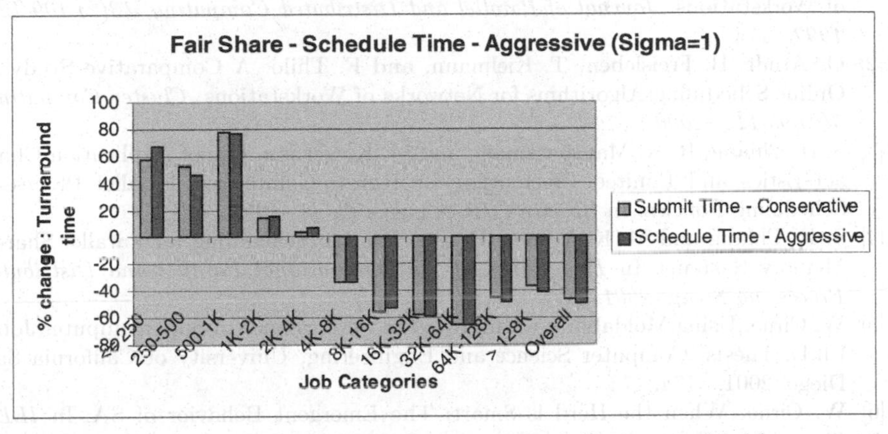

Fig. 6. Performance of the load-sensitive schedule-time-moldable aggressive backfilling scheme when jobs are less scalable

that are forced to be "short and wide" - they can mold themselves to be "long and narrow" instead if that gets them quicker completion. This prompts the development of a schedule-time-moldable aggressive backfilling strategy, that we evaluate next.

Fig. 5 shows the performance of the fair share scheme using aggressive backfill and schedule time moldability. We observe that this scheme clearly outperforms both the non-moldable conservative backfilling scheme and the fair-share scheme using conservative backfill and submit time moldability. The overall average turnaround time improves by almost 70% compared to conservative backfilling. Also, the turnaround times of all job categories improve under the fair share based schedule time moldable aggressive backfilling strategy compared to the other two schemes. Fig. 6 shows performance for σ=1. The performance of the moldable schemes deteriorates compared to σ=0 because the runtime for a job on a wider partition is higher with σ=1 than with σ=0. However the performance is still considerably better than the non-moldable case and the greedy scheme for moldable scheduling.

6 Conclusion

In this paper we addressed the issue of effective selection of processor partition size for moldable jobs. The proposed scheduling strategies were shown to provide significant benefits over a rigid scheduling model and were also considerably better than a previously proposed approach to moldable job scheduling.

References

[1] S. V. Anastasiadis and K. C. Sevcik. Parallel Application Scheduling on Networks of Workstations. *Journal of Parallel and Distributed Computing, 43(2):109-124, 1997.* 174

[2] O. Arndt, B. Freisleben, T. Kielmann, and F. Thilo. A Comparative Study of Online Scheduling Algorithms for Networks of Workstations. *Cluster Computing, 3(2):95-112, 2000.* 174

[3] S. H. Chiang, R. K. Mansharamani, and M. K. Vernon. Use of Application Characteristics and Limited Preemption for Run-to-Completion Parallel Processor Scheduling Policies. In *SIGMETRICS,* pages *33-44, 1994.* 174

[4] S. H. Chiang and M. K. Vernon. Production Job Scheduling for Parallel Shared Memory Systems. In *Proceedings of the International Parallel and Distributed Processing Symp, 2001.* 174

[5] W. Cirne. Using Moldability to Improve the Performance of Supercomputer Jobs. Ph.D. Thesis. Computer Science and Engineering, University of California San Diego, 2001. 174, 176

[6] W. Cirne. When the Herd is Smart: The Emergent Behavior of SA. In *IEEE Trans. Par. Distr. Systems, 2002.* 174, 176

[7] W. Cirne and F. Berman. Adaptive Selection of Partition Size for Supercomputer Requests. In *Workshop on Job Scheduling Strategies for Parallel Processing,* pages 187-208, 2000. 174, 176

[8] A. B. Downey. A Model For Speedup of Parallel Programs. Technical Report CSD-97-933. University of California at Berkeley, 1997. 176

[9] D. G. Feitelson. Logs of real parallel workloads from production systems. `http://www.cs.huji.ac.il/labs/parallel/workload/logs.html`. 175, 176

[10] D. G. Feitelson, L. Rudolph, U. Schwiegelshohn, K. C. Sevcik, and P. Wong. Theory and Practice in Parallel Job Scheduling. In *Workshop on Job Scheduling Strategies for Parallel Processing* , pages 1-34. 174, 176

[11] D. Jackson, Q. Snell, and M. J. Clement. Core Algorithms of the Maui Scheduler. In *Wkshp. on Job Sched. Strategies for Parallel Processing,* pages 87-102, 2001. 175

[12] P. J. Keleher, D. Zotkin, and D. Perkovic. Attacking the Bottlenecks of Backfilling Schedulers. *Cluster Computing,* 3(4):245-254, 2000. 174

[13] D. Lifka. The ANL/IBM SP Scheduling System. In *Workshop on Job Scheduling Strategies for Parallel Processing,* pages 295-303, 1995. 175

[14] A. W. Mu'alem and D. G. Feitelson. Utilization, Predictability, Workloads, and User Runtime Estimates in Scheduling the IBM SP2 with Backfilling. In *IEEE Trans. Par. Distr. Systems,* volume 12, pages 529-543, 2001. 174, 175

[15] E. Rosti, E. Smirni, L. W. Dowdy, G. Serazzi, and B. M. Carlson. Robust Partitioning Policies of Multiprocessor Systems. *Performance Evaluation,* 19(2-3):141-165, 1994. 174

[16] S. Setia and S. Tripathi. A Comparative Analysis of Static Processor Partitioning Policies for Parallel Computers. In *Proc. of the Intl. Wkshp. on Modeling and Simulation of Computer and Telecomm. Syst. (MASCOTS),* pages 283-286, 1993. 174

[17] K. C. Sevcik. Application Scheduling and Processor Allocation in Multiprogrammed Parallel Processing Systems. *Performance Evaluation,* 19(2-3):107-140, 1994. 174

[18] J. Skovira, W. Chan, H. Zhou, and D. Lifka. The EASY - LoadLeveler API Project. In *Wkshp. on Job Sched. Strategies for Parallel Processing,* pages 41-47, 1996. 175

[19] S. Srinivasan, R. Kettimuthu, V. Subramani, and P. Sadayappan. Characterization of Backfilling Strategies for Parallel Job Scheduling. In *Proceedings of the ICPP2002 Workshops,* pages 514-519, 2002. 180

[20] S. Srinivasan, R. Kettimuthu, V. Subramani, and P. Sadayappan. Selective Reservation Strategies for Backfill Job Scheduling. In *Proceedings of the 8th Workshop on Job Scheduling Strategies for Parallel Processing,* 2002. 180

[21] A. Streit. On Job Scheduling for HPC-Clusters and the dynP Scheduler. In *Proc. Intl. Conf. High Perf. Comp.*, pages 58-67, 2001. 174

[22] D. Talby and D. Feitelson. Supporting Priorities and Improving Utilization of the IBM SP Scheduler Using Slack-Based Backfilling. In *Proceedings of the 13th International Parallel Processing Symposium,* 1999. 175

Runtime Support for Multigrain and Multiparadigm Parallelism

Panagiotis E. Hadjidoukas, Eleftherios D. Polychronopoulos, and Theodore S. Papatheodorou

High Performance Computing Information Systems Laboratory, Department of Computer Engineering and Informatics
University of Patras, Rio 26500, Patras, Greece
{peh,edp,tsp}@hpclab.ceid.upatras.gr
http://www.hpclab.ceid.upatras.gr

Abstract. This paper presents a general methodology for implementing on clusters the runtime support for a two-level dependence-driven thread model, initially targeted to shared-memory multiprocessors. The general ideal is to exploit existing programming solutions for these architectures, like Software DSM (SWDSM) and Message Passing Interface. The management of the internal runtime system structures and of the dependence-driven multilevel parallelism is performed with explicit messages, exploiting however the shared-memory hardware of the available SMP nodes whenever this is possible. The underlying programming models and hybrid programming solutions are not excluded, using threads for the intra-node parallelism. The utilization of shared virtual memory for thread stacks and a translator for allocating Fortran77 common blocks in shared memory enable the execution of unmodified OpenMP codes on clusters of SMPs. Initial performance results demonstrate the efficient support for fork-join and multilevel parallelism on top of SWDSM and MPI and confirm the benefits of explicit, though transparent, message passing.

1 Introduction

As clusters of multiprocessor nodes are an attractive platform for high-end scientific computing, the need for two-level thread models was emerged. Currently, message passing, standardized with MPI [8], and Shared Address Space, implemented in software with page-based Shared Virtual Memory (SVM) protocols, are the two leading programming paradigms for these systems. An alternative programming solution is the hybrid-programming model [2], combining MPI for the outer level with multithreading inside each node.

In this paper, we present a general approach for implementing the runtime support of the Nanothreads Programming Model (NPM) [11], a two-level thread model, on clusters of compute nodes. The idea is to exploit existing programming solutions for these architectures and extensively use messaging to minimize the dependence on SVM whenever hardware shared memory is not available. Major architectural features of the runtime system include the adoption of a lazy

S. Sahni et al. (Eds.) HiPC 2002, LNCS 2552, pp. 184–194, 2002.

stack allocation policy, the support of multilevel inter-node parallelism by allocating the user-level stacks in SVM, the use of multithreading for overlapping communication with computation and the exploitation of various communication subsystems. Hybrid programming models are also supported, exploiting only the intra-node parallelism with lightweight threads.

We discuss the implementation of our approach on top of SWDSM and MPI. The result is a programming model with a convenient and powerful API that can efficiently run the same application codes on a very wide range of target platform scales, from SMPs, ccNUMAs to software DSM systems and even across high speed WANs. Moreover, the utilization of SVM stacks, along with a translator for allocating Fortran77 common blocks in shared memory, allows the execution of unmodified OpenMP codes on clusters of SMPs. Initial performance results demonstrate the efficient support for fork-join and multilevel parallelism on top of SWDSM and MPI and confirm the benefits of explicit, though transparent, message passing.

The rest of this paper is organized as follows: Section 2 outlines the Nanothreads Programming Model. Section 3 describes the architecture and the implementation details of our runtime system on top of SWDSM. In Section 4, we discuss the differences when the runtime system is built on top of MPI. An initial performance evaluation of both implementations is reported in Section 5. Finally, in Section 6 we conclude our work and discuss our on-going research.

2 The NanoThreads Programming Model

The primary goal of the Nanothreads Programming Model is the automatic parallelization and efficient execution of standard applications on multiprogrammed, multiprocessing environments. In such environments, the system workload and the resources allocated to a program may change in a dynamic manner. Given a user program, the parallelizing compiler applies control and data dependence analysis and produces an intermediate representation of the program called the Hierarchical Task Graph (HTG). HTG is a data structure that represents the decomposition of a program into parallel tasks at different levels of granularity ranging from the maximum (a single task for the whole program) to the minimum possible, which is specified by the characteristics of the application. Each node at a given level of the HTG is executed as a single lightweight thread and data and control dependence information is used to enforce an execution order for the threads. A runtime library provides the necessary services for thread creation and scheduling. NPM enables the exploitation of multigrain parallelism, which makes it a good candidate for hierarchical architectures, like clusters of SMPs, in which the physical configuration of resources favors the integration of different forms of parallelism in the same program.

The NPM has been implemented on shared-memory multiprocessors in the context of the NANOS ESPRIT project [9]. The NANOS runtime library implements the NANOS API, which is targeted by the NanosCompiler [1]. NanosCompiler is a parallelizing compiler that captures the parallelism expressed by the

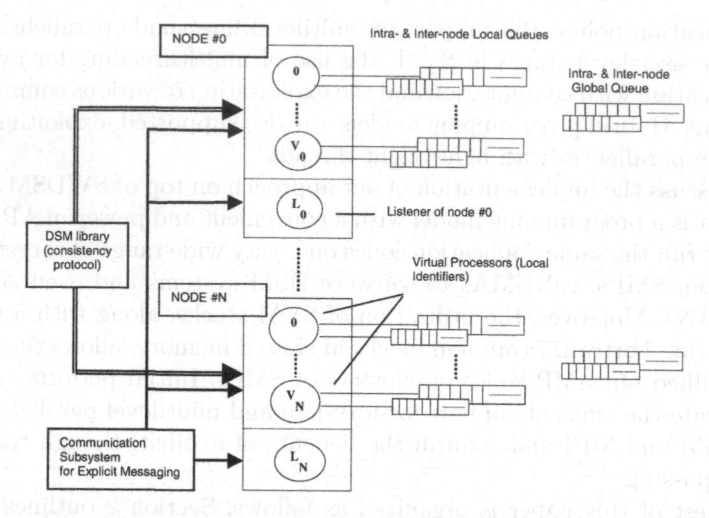

Fig. 1. General Architecture

user through OpenMP directives and the parallelism automatically discovered through a detailed analysis of data and control dependencies. The front-end of the NanosCompiler converts programs written in Fortran77 that use OpenMP directives to equivalent programs with calls to the runtime system. Our runtime system exports the same NANOS API, exploiting thus the existing knowledge and infrastructure.

3 Architecture

Figure 1 illustrates the general architecture of our runtime system on a cluster of multiprocessor nodes. Each DSM process consists of one or more virtual processors, which are DSM kernel threads, and a special server thread, the *Listener*, which is mainly responsible for the dependence and the queue management. The virtual processors execute the application code, which may result in invocations of the DSM protocol. Besides the intra-node queues, there are per-virtual processor inter-node ready queues and per-node global queues. The latter simulate a global distributed queue. Although the queues and the dependencies could be handled through SVM, for performance reasons we use explicit messages and thus the Listener is not required to be a DSM thread. The communication subsystem (used for explicit messaging) of the runtime system can be based on UDP, TCP or even MPI. The latter is possible with the concurrent linking of the DSM and the MPI libraries.

3.1 Work Descriptors

In order to reduce the memory management overhead to a minimum, many user-level thread libraries store the thread descriptor in the stack. In our implementation, we separate the descriptor from its execution vehicle, i.e. an underlying user-level thread, and adopt a lazy stack allocation policy. Actually, the whole runtime system is distinct from the underlying thread package, resulting thus in a modular and portable configuration. All the necessary information (function pointer, arguments, dependencies, successors) is encapsulated in the descriptor and whenever multithreaded execution is necessary, the underlying thread is created right before the first context-switch to it. A recycling mechanism is available for both the descriptors and the underlying threads.

Current SWDSM systems require the arguments of a function that represents inter-node parallelism to be either scalar values or pointers to DSM data. In order to provide true shared-memory functionality on distributed-memory machines, the stacks of the user-level threads must be allocated in SVM. In this case, the lazy allocation policy is clearly beneficial since we avoid unnecessary invocations of the DSM protocol and thread migrations caused by the stealing of a newly created descriptor. This policy also minimizes the peak number of active stacks and leaves larger portions of the shared address space for the application data. Sharing the stacks may introduce excessive complexity regarding the false sharing of data and the interaction with relaxed consistency protocols. We believe that either the programmer should explicitly annotate shared data, or the compiler must discover which variables have to be shared. Moreover, the OpenMP programming model provides good hints concerning the privatization of data.

3.2 Dependence and Queue Management

According to the NPM, When the number of its input dependencies reaches zero, a descriptor is ready for execution. On clusters of multiprocessors, the coherence of this dependence-driven model can be maintained exclusively through shared-memory, by updating the counter of unresolved dependencies through shared memory. However, this requires the allocation of each descriptor in SVM, appropriate protection with locks and invocations of the consistency protocol. A more efficient solution is the combination of hardware shared-memory with explicit messages: each descriptor is associated with an owner node (creator), where a finished descriptor should return in order to update the dependencies of the local successors. If a thread is executed on its owner node, the dependencies of its successors will be updated directly through (hardware) shared memory. Otherwise, its descriptor will return to the owner node and any dependencies will be updated by the Listener. Moreover, explicit messaging allows an implementation of this execution model on top of a pure message passing programming environment.

The runtime system can support hybrid programming solutions which directly reflect the hardware architecture of cluster of multiprocessor nodes (e.g. MPI + OpenMP). This is possible with the adoption of two levels of ready queues, intra-node and inter-node, corresponding to the two possible forms of parallelism. The architecture of queues enables the development of mechanisms that allow good load balancing and exploitation of data locality. On shared-memory machines, the runtime system contains per virtual processor ready queues for the scheduling of parallel loops and a global one for coarse-grain tasks. These queues have been retained for the management of intra-node parallelism and a similar scheme has been adopted for the inter-node parallelism. There are per-virtual processor inter-node queues, while the global queue is distributed among all nodes and consecutive insertions into it result in the cyclic distribution of the descriptors among the available nodes. Every virtual processor has also a "strictly" private ready queue, used internally by the runtime system to allow the execution of code on a specific processor and the switching between the master-slave and the SPMD programming paradigm.

The insertion/stealing of a descriptor in/from a remote queue that resides in the same node is performed through hardware shared memory. Otherwise, the operations are performed with explicit messages to the Listener of the remote node.

3.3 OpenMP Support

As aforementioned, the front-end of the NanosCompiler converts programs written in Fortran77 that use OpenMP directives to equivalent programs with calls to our runtime system. The full support of the NANOS API calls and the use of SVM stacks enable the execution of OpenMP programs on clusters of SMPs. A SWDSM system that maps the common blocks of these codes in SVM, would allows us to run them directly on top of a cluster. Unfortunately, the currently utilized DSM systems require the explicit allocation of the shared data in the shared heap. We have developed an simple translator that takes as input the output of the NanosCompiler and replaces any common blocks with `POINTER` statements, injecting appropriate memory allocation calls (Figure 2). Fortunately, most compilers, with the exception of GCC, support the `POINTER` statement. Moreover, our support of the SPMD model enables the parallel allocation of shared memory even if the underlying DSM library does not provide such functionality.

Our approach, OpenMP support by targeting an appropriate API, differs from the work presented in [5], where a translator converts OpenMP directives to appropriate calls to a multithreaded version of the TreadMarks SWDSM system.

3.4 Implementation Platforms

We have implemented our runtime system on top of three existing software DSM page based systems. Two of them (SVMLib [10] and Millipede [3]) run on Windows NT/2000 platforms, support DSM kernel threads and Sequential Consistency. We support programs written in C and FORTRAN. The development tools were Microsoft Visual C++ 6.0 and Compaq Visual FORTRAN 6.0. The primitives of the underlying non-preemptive user-level threads are based on the POSIX `[sig]setjmp` - `[sig]longjmp` calls. To demonstrate the viability

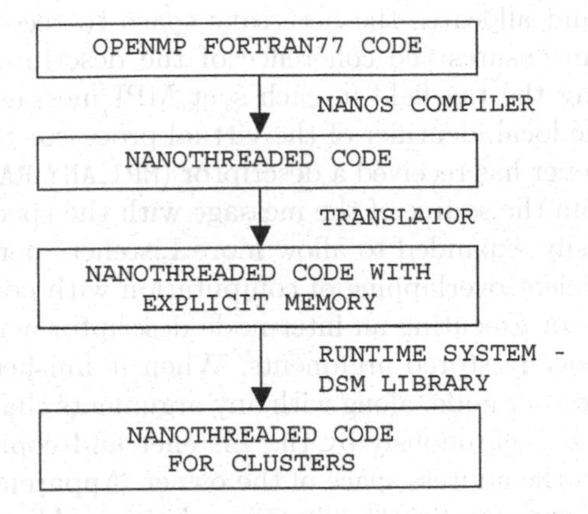

Fig. 2. OpenMP on top of SWDSM

of our work on DSM systems that run on Unix or/and support more relaxed consistency models, we have also used the JIAJIA DSM [6] on Linux 2.4. JIAJIA supports Scope Consistency but unfortunately is not multithreaded. The utilization of SVM stacks requires the execution of the signal handlers on alternative stacks (`sigaltstack`). The nesting of threads and the sharing of stacks may introduce significant complexity when a relaxed consistency model is used. The study of the interaction between the runtime system and the consistency protocol is beyond the scope of this paper.

4 On Top of MPI

The management of the distributed parallelism with explicit messages enables us to apply our approach on top of the message-passing programming paradigm and specifically MPI. We target master-slave MPI programs and the MPI-OpenMP hybrid-programming model. Our initial effort to integrate MPI and NPM is presented in [4].

The general design of the runtime system remains the same and differs in that the communication layer is exclusively MPI and all kernel threads are system scope POSIX threads. There are not shared data or stacks and any data have to be moved explicitly. On master-slave MPI programs, the master distributes the data to the slaves and waits for the processed results to be returned. Based on this observation, we associate each descriptor with the arguments (data) of its corresponding function. This association is similar to Remote Procedure Call (RPC). For each of the function's arguments, the user defines its MPI type and quantity, and the calling convention, which are stored in the descriptor. The only difference in the API is in the creation primitive where the programmer has to provide this description. The insertion of the descriptor in a remote queue corresponds to `MPI_Send` calls for the descriptor and any arguments, based on their description. On the other side, the Listener accepts the descriptors, analyzes their information and allocates the necessary space to receive the arguments. The runtime system ensures the coherence of the descriptor and its data by appropriately setting the tag field in each sent MPI message. Specifically, the tag field denotes the local identifier of the virtual processor that sends the message. Once the Listener has received a descriptor (`MPI_ANY_RANK`), it receives the subsequent data from the source of the message with the specific tag. This multiplexing can be easily expanded to allow more Listeners per node and thus to result in a more efficient overlapping of computation with communication.

A virtual processor executing an inter-node descriptor actually executes the function with the locally stored arguments. When it finishes, it sends the descriptor back to its owner node, along with any arguments that represent results. These are received asynchronously by the Listener and copied on their actual memory locations in the address space of the owner. Apparently, these locations are the function's arguments in the creation primitive. All the aforementioned movement of data is transparent to the user and the only points reminding the underlying MPI programming is the description of the arguments. Currently, the following definitions are used to determine the calling way for an argument:

- `CALL_BY_VALUE`: The argument is passed by value. It is a single value stored in the descriptor.
- `CALL_BY_POINTER`: As above, but the single value has to be copied from the specific address.
- `CALL_BY_REFERENCE`: The argument represents data that are sent with the descriptor and returned as a result in the home node's address space.
- `CALL_BY_RESULT`: No data has to been sent but it will be returned as a result. It is assumed by the target node that receives data initialized to zero.

The standard programming models used (POSIX and MPI) result in an efficient and portable implementation, available and tested on both Windows and Unix platforms. The use of multithreading on top of MPI processes requires a thread-safe MPI library, since each process has one thread (Listener) that executes a blocking receive operation (`MPI_Recv`). As a communication layer, MPI provides portability and it is supported by multiple platforms of operating systems, hardware and interconnection networks.

5 Experimental Evaluation

In this section, we present the behavior of both implementation cases of our runtime system, on top of SWDSM and MPI. Our intention is to demonstrate the functionality of the runtime system rather than to evaluate the underlying DSM/MPI library. Our testbed environment consists of a network of 4 Pentium III 866 MHz dual-processor machines, running Windows 2000 Server SP2. Each machine has 256 MB of main memory and the cluster is interconnected with Fast Ethernet. We use the multithreaded SVMLib and the thread-safe MPI/Pro library (version 1.6.3) for Windows. MPI/Pro implements inter-process communication on a single machine through shared memory queues and between processes on separate machines through sockets.

Table 1. Fork-join overhead of a remotely executed descriptor (μsec)

Comm. Subsystem	One Node	Two nodes
UDP	330	252
TCP	76	73
MPI	45	108

5.1 Applications

Creation & Execution: This benchmark measures the average time a virtual processor needs to create and join a descriptor executed on a remote node. The two DSM processes (virtual processors) may execute on the same or different nodes. The measurements for the three communication subsystems (UDP, TPC, MPI) are depicted in Table 1. We observe that UDP is inferior compared to the connection-oriented and reliable TCP. TCP outperforms MPI on two nodes due to the overhead of the MPI library. On the other hand, when both DSM processes reside on the same node, the MPI communication is performed through shared memory and thus is more efficient. Finally, the two socket solutions perform worse on a single node possibly possible due to inopportune interaction between the two virtual processors (clients) and the two Listeners (servers).

Fibonacci: The multithreaded recursive version of this application is used as a proof of concept of our support for multilevel inter-node parallelism. This application stresses both the DSM protocol and the runtime system. The insertion of the new descriptors in adjacent nodes results in the execution of less parallelism on nodes with larger identifier and thus in load unbalancing. On the other hand, despite the nested parallelism of Fibonacci, no data need to be shared across the nodes. The results are returned back and stored in the local address space of the successor.

Synthetic Matrix Addition (SMA): This application is embarrassingly parallel, follows the master-slave programming paradigm, and adds two matrices of doubles according to the following pseudocode:

```
C = I, A~= 2.0*I, B = 3.0*I (sequential - master)
<T1>
Forall (i=1, N, 1) process(Ai, Bi, Ci)
<T2>
For each i and j do Cij+= 1 (sequential - master)
<T3>
```

The `process` function introduces a fixed computational cost, not enough however to overwhelm the communication overhead and result in linear speedup. The matrices are allocated in shared virtual memory and initialized by the master virtual processor. Since their dimension is 1024x1024 and the additions are performed per-row by individual tasks, we avoid false sharing of the same page. We measure the parallel phase of execution (T2-T1) and the total time until the results have implicitly come back to the master (T3-T1). On top of MPI, the application code differ only at the MPI description of the arguments in the creation primitive. Every descriptor is associated with the corresponding function arguments, i.e. the rows of the three matrices. Since the results are explicitly returned to the node that created the parallelism, the time difference between T2 and T3 is negligible.

NAS - EP: EP [7] is an Embarrassingly Parallel benchmark. It generates pairs of Gaussian random deviates according to a specific scheme. We provide measurements of the native MPI implementation (MPI) and the OpenMP version as generated by the NANOS Compiler (OMP-NANOS). Moreover, we use an SPMD version of EP included in the release of SVMLib, written in C and converted to use kernel threads (SVM-SPMD) and the corresponding version with NANOS API calls (SVM-NANOS).

5.2 Experimental Results

Tables 2 and 3 illustrate the execution times of Fibonacci and SMA on top of SWDSM and MPI. The first column (Procs X Threads) represents the number of DSM/MPI processes and the per-process virtual processors. The second and the third column denote the number of SMP nodes and the total number of virtual processors respectively. The number of processes can differ from the number of nodes because both SVMLib and MPI allow two instances of the same application to be executed on the same node. Such processes are always assigned consecutive identifiers (ranks).

The execution times of Fibonacci (Table 2) demonstrate the excessive overhead introduced by the DSM protocol when the stacks of the user-level threads are allocated in SVM and multiple levels of parallelism are exploited. The last

Table 2. Fibonacci (msec)

Procs x Threads	Nodes	VPs	DSM	MPI
1X1	1	1	46	16
1X2	1	2	47	16
2X1	1	2	1735	15
2X1	2	2	1625	16
2X2	2	4	1329	15
4X1	2	4	2687	78
4X1	4	4	2360	125
4X2	4	8	2594	360
8X1	4	8	3406	1016

Table 3. SMA (sec)

Procs x Threads	Nodes	VPs	DSM (T2-T1)	DSM (T3-T1)	MPI (T3-T1)
1X1	1	1	36.91	36.95	36.84
1X2	1	2	18.50	18.53	18.44
2X1	1	2	19.64	20.13	18.63
2X1	2	2	21.80	23.81	19.33
2X2	2	4	11.80	13.80	10.47
4X1	2	4	11.48	14.14	10.69
4X1	4	4	11.55	14.17	10.53
4X2	4	8	6.19	9.97	6.14
8X1	4	8	6.80	10.20	6.78

Table 4. Execution Times of NAS-EP Class W (sec)

Procs x Threads	Nodes	VPs	MPI	OMP-NANOS	SVM-SPMD	SVM-NANOS
1X1	1	1	33.31	33.06	62.65	62.41
1X2	1	2	-	16.78	31.37	31.34
2X1	1	2	16.96	16.94	31.37	31.39
2X1	2	2	16.59	16.76	31.32	31.34
2X2	2	4	-	8.61	15.71	15.91
4X1	2	4	8.49	8.56	15.73	15.98
4X1	4	4	8.78	8.61	16.03	16.21
4X2	4	8	-	5.03	8.05	8.27
8X1	4	8	4.51	5.26	8.39	8.66

column (MPI) reveals the exact behavior of the Fibonacci application under minimal communication overhead. For up to four virtual processors and two nodes, the execution time remains stable, since the additional processing power is overwhelmed by the algorithm and the increased runtime synchronization overhead. The increase of the execution time on 4 and 8 processes can be attributed and the load unbalance caused by less parallelism being propagated to the nodes with larger rank.

Table 3 presents the time required for the parallel phase (T2-T1) and the effective time (T3-T1) of SMA. Despite the invocation of the DSM protocol for the movement of data, the parallel phase scales quite well for all cases. However, the effective execution time confirms the benefits of multithreading, which provides more processing power with less DSM processes and thus less communication costs. From the last column it is clear that the explicit data movement with MPI calls outperforms the implicit one, based on the paged-based DSM protocol. These performance gains are attained with minimal programming effort while any data movement is performed transparently to the programmer.

Table 4 presents the execution times of the various versions of EP, having applied the most aggressive optimizations provided by the available compilers. We observe that the OMP-NANOS version scales very well even when eight DSM processes are used. The last two columns indicate that the overhead introduced by our runtime system is minimal.

6 Conclusions and Future Work

On shared-memory multiprocessors, the runtime support of NPM takes advantage of the Shared Address Space programming model and provides a framework for application adaptability and load balancing in a multiprogrammed environment. According to our approach, its extension on clusters can be built on top of existing programming models-solutions for these systems. In this paper, we presented the implementation of our approach on top of SWDSM and MPI. The versatility of this work stems from the fact that we work at an intermediate layer; this allows us to focus on the exploitation of the underlying infrastructures rather than on the development of similar solutions. Issues that have to be addressed include load balancing techniques and the interaction with the consistency model. Apart from the study of the supported programming models under a common platform, future work includes the exploitation of the kernel interface of NPM.

Acknowledgments

We would like to thank all our partners within the NANOS Esprit project. Special thanks to Dimitrios Nikolopoulos, Christos Antonopoulos and Constantine Bekas for their valuable help and comments. This work is supported in part by the Hellenic General Secretariat of Research and Technology (G.S.R.T.) research project 99EΔ-566 and the POP IST/FET project (IST-2001-33071).

References

[1] Ayguadé, E., Labarta, J., Martorell, X., Navarro, N., Oliver, J: NanosCompiler: A Research Platform for OpenMP Extensions. In Proceedings of the 1st European Workshop on OpenMP, Lund (Sweden), October 1999. 185

[2] Cappello, F., Richard, O., Etiemble, D.: Investigating the performance of two programming models for clusters of SMP PCs. In Proceedings of the 6th IEEE Symposium On High-Performance Computer Architecture (HPCA-6), Toulouse, France, January 2000. 184

[3] Friedman, R., Goldin, M., Itzkovitz, A., Schuster, A.: Millipede: Easy Parallel Programming in Available Distributed Environments. Journal of Software: Practice and Experience, 27(8): 929–965, August 1997. 188

[4] Hadjidoukas, P. E., Polychronopoulos, E. D., Papatheodorou, T. S.: Integrating MPI and the Nanothreads Programming Model. In Proceedings of the 10th Euromicro Workshop on Parallel, Distributed and Network-Based Processing (PDP 2002), Las Palmas, Spain, January 2002. 189

[5] Hu, C., Lu, H., Cox, A., Zwaenepoel, W.: OpenMP for Networks of SMPs. In Proceedings of the Second Merged Symposium, IPPS/SPDP 99, 1999. 188

[6] Hu, W.W, Shi, W. S., Tang, Z. M.: JIAJIA: An SVM System Based on A New Cache Coherence Protocol. In Proceedings of the High Performance Computing and Networking (HPCN'99), April 1999. 189

[7] Jin, H., Frumkin, M., Yan, J.: The OpenMP Implementation of NAS Parallel Benchmarks and its Performance. Technical Report NAS-99-011, NASA Ames Research Center, October 1999. 192

[8] Message Passing Interface Forum: MPI: A message-passing interface standard. International Journal of Supercomputer Applications and High Performance Computing, Volume 8, Number 3/4, 1994. 184

[9] NANOS ESPRIT Project No. 21097, `http://www.ac.upc.es/nanos`. 185

[10] Paas, S. M., Scholtyssik, K.: Efficient Distributed Synchronization within an all-software DSM system for clustered PCs. In Proceedings of the 1st Workshop on Cluster-Computing, TU Chemnitz-Zwickau, November 1997. 188

[11] Polychronopoulos, C. D.: Nano-Threads: Compiler Driven Multithreading. In Proceedings of the 4th International Workshop on Compilers for Parallel Computing CPC'93, Delft (The Netherlands), December 1993. 184

A Fully Compliant OpenMP Implementation on Software Distributed Shared Memory

Sven Karlsson[1], Sung-Woo Lee[2], and Mats Brorsson[1]

[1] Royal Institute of Technology, KTH, Stockholm, Sweden
{Sven.Karlsson,Mats.Brorsson}@imit.kth.se
[2] Ditto Information Technology Inc., South Korea
swlee@dittotec.com

Abstract. OpenMP is a relatively new industry standard for programming parallel computers with a shared memory programming model. Given that clusters of workstations are a cost-effective solution for building parallel platforms, it would of course be highly interesting if the OpenMP model could be extended to these systems as well as to the standard shared memory architectures for which it was originally intended.
We present in this paper a fully compliant implementation of the OpenMP specification 1.0 for C targeting networks of workstations. We have used an experimental software distributed shared memory system called Coherent Virtual Machine to implement a run-time library which is the target of a source-to-source OpenMP translator also developed in this project.
The system has been evaluated using an OpenMP micro-benchmark suite as to evaluate the effect of some memory coherence protocol improvements. We have also used OpenMP versions of three Splash-2 applications concluding in reasonable speedups on an IBM SP2 machine. This also is the first study to investigate the subtle mechanisms of consistency in OpenMP on software distributed shared memory systems.

1 Introduction

Workstation clusters are cost-effective when it comes to implementing platforms for parallel computation. Initiatives such as the Beowulf project have gained attention to these platforms and there are now several such systems among the top 500 most powerful computers in the world [12, 20]. They are a viable alternative to SMP servers, even for small systems, just because they are cost-effective. There is one major drawback, though. The topology and hardware of a cluster of workstations only support a message-passing programming model. However, it is widely acknowledged that a shared memory model, such as OpenMP [13, 14], is preferred by most programmers when it comes to ease of programming and maintenance of the software base [19]. Software distributed shared memory, software DSM, systems have been developed for quite some time now to provide a shared memory programming model on clusters of workstations [2, 6, 8, 9, 18, 16]. It

S. Sahni et al. (Eds.) HiPC 2002, LNCS 2552, pp. 195–206, 2002.

has been shown that it is indeed possible to achieve reasonable performance for a range of applications using modern software DSM systems even though there is still much work to do as to improve performance [7, 15].

In this paper we present a fully compliant OpenMP implementation of the 1.0 specification for C for a cluster of workstations. It is based on an existing software DSM system called Coherent Virtual Machine [8] and is embodied in the form of a run-time library and an OpenMP compilation system originally developed for an SMP target [5].

The performance of the SMP version of the OpenMP implementation is on par with commercial OpenMP compilers. For the Software DSM version, we have changed the allocation of shared variables since Software DSM systems assume explicit allocation of shared memory and OpenMP programs assumes shared storage by default. The speedup of the same, unmodified, applications on the Software DSM system with eight processors ranges from 4.1 to 6.5 which should be compared to 6.8 to 7.8 achieved on an SGI Origin 3800 which is a distributed shared memory architecture machine. Given the relatively small data set used in these programs, and the difference in inter processor communication time, this difference is not very big. It is important to note here that we are referring to a full implementation of the OpenMP 1.0 specification for C, not a modified version to suit the software DSM system better.

In the next section we briefly introduce the OpenMP specification for C, in section 3 we present the compilation system and the run-time library we developed for shared memory target platforms. This is followed in section 4 with a discussion on software DSM systems. Section 5 outlines the implementation aspects of the run-time library for software DSM followed by a performance evaluation in section 6 before the paper is concluded.

2 OpenMP

OpenMP is a recent effort to provide an informal standard for how shared memory parallel computers should be programmed. It is a thread-level fork-join programming model based on compiler directives. Several types of directives are provided. Parallel directives are used to spawn parallel activity, and there are work sharing directives to make the different threads do different things. Both loop-level parallelism and functional parallelism are supported by the directives. Directives for critical sections and other synchronization primitives are also available as well as a number of directives to control which variables are shared and private to threads. However, in total there are relatively few directives which makes it quite easy to start using OpenMP.

In addition to the directives, OpenMP specifies a number of intrinsic functions that can be used to divide work based on the number of parallel threads and thread identities similar to how the rank is used in MPI programs [11].

The following example shows an excerpt of a C program with a parallel region and one of the more common work-sharing constructs. All examples are in this paper given in C since the software DSM OpenMP compilation system currently only targets C programs. Equivalent structures exist for Fortran programs.

```
1   #pragma omp parallel
2   {
3       foo();
4
5   #pragma omp for
6       for (i = 0; i < 200; i++)
7           a[i] = bar(i);
8   }
```

Line 1 in the example above starts a new parallel region using the parallel directive. OpenMP directives are inserted as pragma compiler directives and typically operate on the lexically next C statement. In this case, a team of threads is created and the compound statement beginning on line 2 is executed in parallel by each of the threads. The number of threads that is created is determined at runtime based on the value of environment variables. The function foo is called by each of the created threads in parallel. The programmer has to make sure that foo does not have any side-effects that can cause race conditions.

Line 5 contains a work-sharing construct, the for-directive. This construct means that the iterations of the for-loop that follows are divided among the threads in the parallel region. Each thread executes its share of the, in this case, 200 iterations. It is the responsibility of the programmer to make sure that the loop iterations are independent.

Notice that without the directives, the program is still a valid program making it easy to maintain a sequential and a parallel version of the code with the same source code files.

We will now have a look at how the SMP-version of the OpenMP translator that we have developed works before we look at what changes we needed to do as to support software distributed shared memory.

3 The OpenMP Compilation System

We have implemented a C compilation system for OpenMP. The basis for this system is a source-to-source code translator which given a source code with OpenMP directives generates a new transformed source code with calls to a runtime library. The library has primitives for spawning and synchronizing threads. The actual translation process is described in detail elsewhere and we only provide a rough sketch here [5]. In essence, each parallel region is transformed into a function called a parallel function. A pointer to one of these parallel functions is passed as an argument to a library function when threads are to be spawned and the spawned threads all execute the parallel function. The translator also inserts various library calls when needed as to implement work sharing and thread synchronization.

There are a number of problems when it comes to implementing OpenMP on a cluster of workstations. First of all we need a shared address space, and secondly there are some particular issues of OpenMP that forces us to modify the OpenMP translator when targeting a software DSM system. We will first discuss the issue of supporting a shared address space at all.

4 Software DSM Systems

A software distributed shared memory, software DSM, system implements a shared memory programming model on top of a machine architecture that does not support shared memory in hardware. The software DSM system is a piece of user-level software that intercepts memory accesses that cannot be satisfied locally, sends messages to the node(s) that has the requested memory contents and resumes execution after that the local node has received what it needs.

We have based our implementation of the run-time library on top of an existing software DSM system, Coherent Virtual Machine, CVM [8]. CVM implements a number of consistency models, but we have focused on using the Home-based Lazy Release Consistency model, HLRC [22]. Like many other software DSM systems [2], HLRC is page based. This means that the system uses the virtual memory page protection mechanism to detect accesses to shared memory and memory contents are replicated on a virtual memory page basis.

In order to simplify the protocol, each page in an HLRC system has a home node which always holds an up-to-date copy of the page. Thus, whenever a processor accesses a page which is not present locally, a copy is retrieved from the home. Several nodes can have write permissions to the same page in which case several data structures are used to locally keep track of changes to the page. These changes are transferred at synchronization points to the home node where they are merged. The synchronization points also forces updated pages to be invalidated. The synchronization points normally occur when synchronization primitives are performed.

We chose to base our OpenMP run-time library on HLRC because of its relative simplicity and because it performs quite robustly. There is, however, no implicit assumption in our system on HLRC so it should be relatively straightforward to port the system also to other types of software DSM protocols, e.g. homeless protocols.

In the next section we describe why we cannot use CVM and HLRC just as they are as target for our OpenMP translator and why we had to augment both the translator and CVM. We also briefly describe some implementation aspects of the translator and the run-time library.

5 OpenMP on a Software DSM System

For obvious reasons we would like to implement the OpenMP run-time library, without changing the interface of the existing SMP implementation. This enables us to keep the exact same OpenMP translator for both the SMP version of the

OpenMP implementation and the software DSM version. Unfortunately, this is not entirely possible since variables declared outside a parallel region, and visible from within that parallel region, are shared among the threads by default. It is possible to explicitly declare variables to be shared, but the OpenMP specification does not require it. This is in conflict to all previous software DSM systems that require shared variables to be explicitly allocated. Therefore, we have augmented the translator to automatically identify all variables that could possibly be accessed from within a parallel region, and thus should be shared, and to allocate them explicitly in the shared address space in the translated program. Automatically allocated variables which usually are stored on a stack can also be shared and to handle this we allocate these variables on a single shared stack. However, automatically allocated variables that are not shared are still put on the respective thread's stack. Naturally, the support for the shared stack and the allocation primitives have been added to the run-time library.

Mapping OpenMP's memory model onto the HLRC model of our software DSM system also caused a few potential performance problems. The only construct that enforces consistency in OpenMP is the flush construct. The flush operation requires that all shared memory variables are written back to the shared memory and it is thus a synchronization point. Most OpenMP constructs that contain some sort of synchronization implies that a flush is performed and there is also the possibility to insert explicit flush statement in an OpenMP program via directives.

A flush with no argument implicitly forces the system to perform consistency operations for the entire shared memory area. The OpenMP flush construct can also contain arguments with named variables in which case the consistency is performed only on those variables. We call the former global-flush and the latter selective-flush. Since OpenMP is intended to be used on a variety of consistency models, including sequential consistency, total store ordering and release consistency [1], the specification requires that implicit global-flush operations are made in connection to many OpenMP constructs, such as entry and exit of parallel regions, barriers, critical sections and work-sharing constructs. On an SMP that implements a sequential consistent memory model, a global-flush is not much more time-consuming than writing all register allocated variables to memory, but for a more relaxed memory consistency model, and in particular on a software DSM system, global-flushes can be very expensive operations. Therefore, it is important that we try to use selective-flush, where only the named variables are affected, as much as possible and that we should reduce the number of global-flushes if possible.

We have implemented the functionality of the consistency operation of global-flush by using a scheme very similar to how locks are handled in a software DSM system. An imaginary lock called the flush-lock is used for this purpose. A flush-operation consists of an acquire of the flush-lock, immediately followed by a release. The acquire and release operations are implemented using message passing making it possible to piggy-back information on the messages sent between the previous flush-lock holder and the acquirer. The information transferred makes it possible to invalidate any updated pages and thus obtaining a coherent memory state on all cluster nodes.

Selective-flush only need to force a coherent memory state for the specified variables and it should thus cause a much lower overhead. This is not considered in current page-based software DSM systems and so we have to augment the coherency protocols. To do this we use the flush-lock again but only the information needed to update the specified variables are sent between the nodes. This greatly reduces the overhead.

It should be noted here that OpenMP also has intrinsic lock functions, however, these locks only perform mutual exclusion and does not enforce consistency.

Work-sharing constructs in OpenMP can have a reduction clause which causes a reduction operation on a variable. The simplest implementation of the reduction operation is to use a critical section where the global reduction variable is updated from the local copies of each process. However, the OpenMP specification states that the result of a reduction is not guaranteed to be visible until a barrier and this made possible for us to merge the reduction operation with the barrier thus eliminating quite a bit of coherency traffic.

The atomic construct in OpenMP has almost the same semantic as a critical section but allows for a more efficient implementation. The code excerpt below is an example of the atomic operation.

```
1 #pragma omp atomic
2  v = v + foo();
```

The update of variable v must be done atomically. The invocation of the function foo, however, is not guaranteed to be atomic and it should therefore not contain any side-effects. The simplest implementation of atomic is to use a critical section with global-flush. However, we have a chance to optimize it using a refined version of our selective-flush approach. We here first execute function foo to get its return value, then acquire the flush-lock asking for updates for variable v only but we will not release it until we have updated v's value exploiting the mutual exclusion properties of the flush-lock.

Before presenting some performance measurements from our prototype we would like to point out that we have through-out the run-time tried to reduce the number of messages sent by merging messages whenever possible.

6 Performance

We have used an IBM SP2 machine with 160 MHz POWER2 uniprocessor nodes as our experimental platform. All the nodes have 256 megabytes of memory and are interconnected with a 110 megabyte/s network. The nodes were running AIX 4.2. A virtual memory page has a size of 8 kilobytes in this version. We have also used a SGI Origin 3800 with 400MHz MIPS R12000 processors for making comparisons.

We evaluated our prototype with three applications from SPLASH-2 [21]. The applications and data sets used were 1D-FFT (2^{20} points), LU (1024×1024

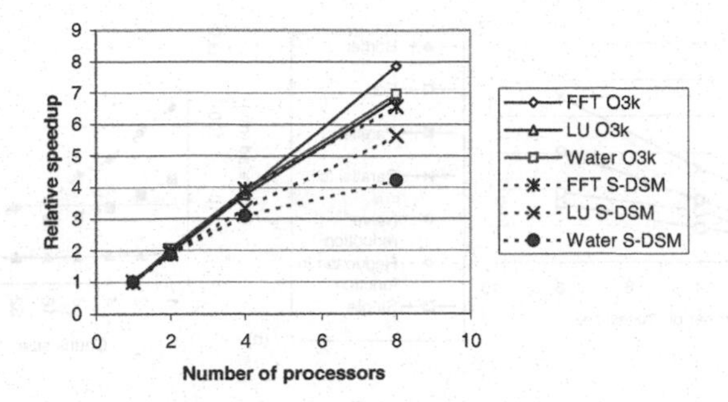

Fig. 1. The relative speedup of the three applications on an SGI Origin 3800 (O3k) and our prototype running on an IBM SP2 (S-DSM)

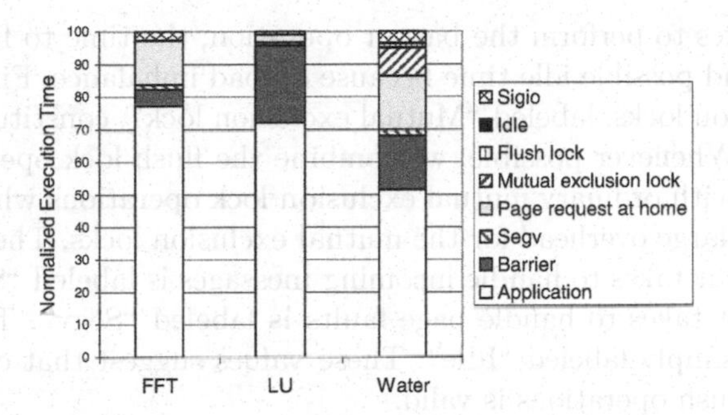

Fig. 2. Execution time breakdown for the three applications

matrix), and Water-spatial (512 molecules, 6 time steps). All applications were run with up to 8 nodes. The EPCC micro-benchmarks were used to measure the overhead of OpenMP constructs and scheduling schemes [3].

Figure 1 presents the relative speedups for the three SPLASH-2 applications. All applications were also run the SGI machine using SGI's own OpenMP compiler. As expected, the speedups are better on the SGI, which is a CC-NUMA and implements shared memory in hardware, but the differences are not extremely large. We believe the good performance of our prototype comes from the coherency protocol enhancements we have made. The OpenMP specification allowed enough freedom to do these optimizations and this indicates that OpenMP is a viable approach for a broad range of parallel computing platforms.

Figure 2 shows the normalized execution time breakdown observed in the software DSM system for the three applications. The fraction labeled "Application" is the application busy time fraction. For 1D-FFT and Water-spatial, the most dominant overhead is page requests which is labeled "Page request at home". Page requests are very common as a page has to be requested from the home node each time a page is accessed after being invalidated. The second largest overhead is the barrier fraction, labeled "Barrier". This time includes

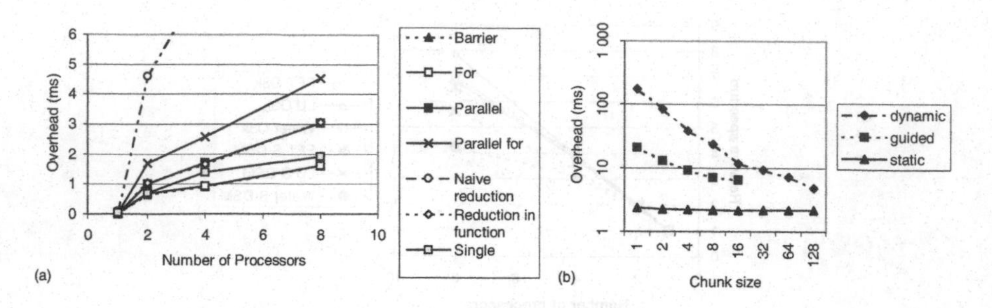

Fig. 3. OpenMP constructs (a) and work sharing scheduling overheads (b) in milliseconds

the time it takes to perform the barrier operation, the time to fully update the home nodes and possible idle time because of load imbalance. Finally, in Water, mutual exclusion locks, labeled "Mutual exclusion lock", constitute a large overhead as well. Whenever possible, we combine the flush-lock operations, labeled "Flush lock", with ordinary mutual exclusion lock operations which is here seen as a relatively large overhead for the mutual exclusion locks. The fraction of the execution time it takes to handle incoming messages is labeled "Sigio" while the time fraction it takes to handle page faults is labeled "Segv". The idle time in the system is simply labeled "Idle". These values suggest that our approach to optimize the flush operations is valid.

Figure 3 (a) shows the measured overheads for a few different OpenMP constructs with implied barriers as reported by the EPCC OpenMP microbenchmarks.

We here clearly see the difference in overhead between using a naïve reduction implementation, using global-flush and locks, and our optimized implementation that delays the reduction operation until the next barrier. The overhead of a parallel construct with a reduction using the naïve implementation, labeled "Naive reduction", is extremely high due to the high cost of flush operations and the serialization caused by the locks. The optimized implementation, labeled "Reduction in function", on the other hand has about the same overhead as a plain parallel construct and we can thus argue that with our implementation the reduction adds no extra overhead.

The parallel for construct is the combination of the parallel directive and a work-sharing construct with static scheduling. It has a much larger overhead than the parallel construct and this is due to an excess barrier introduced by our OpenMP translator. This excess overhead will be removed in the next version of OpenMP translator and was due to the translator treating the parallel for construct as two separated parallel and for constructs and inserted a not needed barrier at the end of the for construct. With these changes we expect the parallel for construct to have essentially the same overhead as a plain parallel construct.

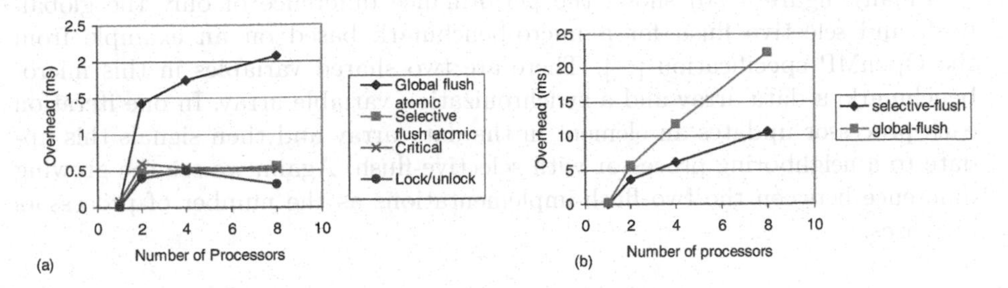

Fig. 4. Mutual exclusion (a) and flush (b) overhead in milliseconds

The barrier construct has approximately the same overhead as a plain work-sharing construct, labeled "For", and this is expected as the static for work-sharing construct is very efficient. We here basically only see the overhead of the barrier at the end of the for construct.

The single construct has a little bit higher overhead than a plain for. Our run-time causes a slight overhead when starting a new work-sharing construct and this leads to the difference in overheads. When measuring the overhead of the for construct the micro-benchmark runs several iterations and so the start-up overhead is not as pronounced in the values of the for construct.

Figure 3 (b) shows the measured overheads of different schedules in work-sharing constructs. The x-axis in the figure shows the chunk size used when distributing iterations of a for-loop on eight processors with 1024 iterations. The static schedule distributes the iterations statically and equally to the different processes and therefore has very little run-time overhead. The dynamic schedule distributes iterations dynamically in chunks and the guided schedule is a dynamic schedule where the chunk size from start is the number of iterations divided by the number of processes. As expected, both the dynamic schedules have a substantial overhead compared to the static scheduling policy. Also, our prototype has the same behavior as a SMP implementation would have although the overheads are higher.

Figure 4 (a) shows the overheads for a few mutual exclusion constructs as measured by the EPCC micro-benchmarks. We see the striking difference in overhead between using global-flush, labeled "Global flush atomic" in the figure, to implement atomic and our optimization using a variant of our selective-flush approach, labeled "Selective atomic flush" in the figure. The overhead of a critical construct is almost as low as the one of just using the intrinsic functions, i.e. lock and unlock. This comes from the fact that the loop that measures the overhead of critical constructs do not touch any shared data and so very little coherency information is sent. The optimized version of atomic has approximately the same overhead as the intrinsic functions. We believe this comes from the fact that the rather large lock latencies in the system can hide the coherency work done in the atomic construct.

Finally, figure 4 (b) shows the performance difference of only the global-flush and selective flush for a micro-benchmark based on an example from the OpenMP specification [14]. There are two shared variables in this micro-benchmark, a data array and a synchronization variable array. In one iteration each processor updates an element in the data array and then signals this update to a neighboring processor with selective-flush. Again, we see the growing difference between the two flush-implementations as the number of processors increases.

7 Related Work

There are three related studies on OpenMP for software DSM systems known to us.

One study was made by H. Lu et al. [10]. They deviate from the OpenMP specification in two ways. Variables in parallel region are treated as private by default and all shared variables must be explicitly declared. Furthermore to alleviate the expensive run-time cost for flush, they introduced condition variables and semaphores, which would replace flushes in pipelined or task-queue-based parallelism.

Y. C. Hu et al. reports on an OpenMP implementation on networks of SMPs [4]. Here no apparent deviations from the OpenMP specification has been made. However, it is unclear how flushes are handled.

M. Sato implemented a OpenMP translation tool which instruments the application's code with communication primitives as to uphold coherency [17]. However, as to work their system required a specialized network interconnect and runtime environment.

In short, our work differs from the related work above in that we fully implement the OpenMP specification in a highly portable way. We have put considerable effort into a correct and efficient implementation of the flush operation. We also evaluate the performance of our system using several benchmarks.

8 Conclusions

Clusters of workstations are becoming more and more sought-after as cost-effective parallel computing platforms. We have in this paper presented a system consisting of an OpenMP translator, a run-time library, and a software distributed shared memory system that together form a fully compliant implementation of the C OpenMP specification version 1.0 for a cluster of workstations. It is one of the first in this kind and we expect to release an open source version of both the compiler and the software DSM system shortly. In contrast to previous work on OpenMP for clusters of workstations/SMPs, we have focused on the compliance aspect of OpenMP and to provide the first published data on the overheads of OpenMP constructs in a software DSM system.

We find the performance of the resulting system quite satisfactory. The speedups are similar to previously reported results and, although the OpenMP construct overheads are high, we find that they exhibit the same general behavior as on SMP systems when the number of processors scale. During the work on the system we have found numerous ways to improve performance. Most notably through a cooperation between the OpenMP translator and the software DSM run-time library. This will be the focus of future research. We will also work on adding support for SMP nodes.

It was surprisingly complicated to get the semantic meaning of some of the more subtle issues of OpenMP correct. Flushes of individual variables turned out to be particularly troublesome to implement correctly.

Acknowledgements

The work in this paper has been partly financed by the European Commission in the Intone project under contract number IST-1999-20252. The OpenMP translator was in part implemented also by Håkan Zeffer, Samer Al-Kassimi and Örjan Friberg. Their contribution is hereby greatly acknowledged. We also gratefully acknowledge the use of computing resources at the Centre for Parallel Computing at the Royal Institute of Technology, Stockholm. Anna Thelin wrote the OpenMP versions of the SPLASH-2 benchmarks.

References

[1] S. V. Adve, K. Gharachorloo, *Shared memory consistency models: a tutorial*, IEEE Computer, Volume: 29 Issue: 12 , Dec. 1996 pp. 66–76. 199

[2] C. Amza, A. L. Cox, S. Dwarkadas, P. Keleher, H. Lu, R. Rajamony, W. Yu and W. Zwaenepoel. *TreadMarks: Shared Memory Computing on Networks of Workstations*, IEEE Computer, Vol. 29, no. 2, pp. 18-28, February 1996. 195, 198

[3] J. M. Bull, Measuring Synchronization and Scheduling Overheads in OpenMP, in *Proceedings of the First European Workshop on OpenMP*, Sept. 1999, pp. 99-105. `http://www.it.lth.se/ewomp99`. 201

[4] Y. C. Hu, H. Lu, A. L. Cox, and W. Zwaenepoel, OpenMP for Networks of SMPs, in *Proceedings of IPPS/SPDP'99*, April 1999, pp. 302-310. 204

[5] S. Karlsson, et al., *A Free OpenMP Compiler and Run-Time Library Infrastructure for Research on Shared Memory Parallel Computing*, Technical Report, Deptartment of Microelectronics and Information Technology, KTH, Royal Institute of Technology, 2002. 196, 197

[6] S. Karlsson, and M. Brorsson, Producer-push-a protocol enhancement to page-based software distributed shared memory systems *Proceedings of 1999 International Conference on Parallel Processing*, September 1999, pp. 291-300. 195

[7] S. Karlsson, and M. Brorsson, A comparative characterization of communication patterns in applications using MPI and shared memory on an IBM SP2, in *Network-Based Parallel Computing. Communication, Architecture, and Applications. Second International Workshop, CANPC '98*, January 1998 pp. 189-201. 196

[8] P. Keleher, *The CVM Manual*, Technical report, Computer Science Deptartment, University of Maryland, May 1995. 195, 196, 198

[9] K. Li, IVY: A shared Virtual Memory System for Parallel Computing. In *Proceedings of 1988 International Conference on Parallel Processing*, 1988, pp. 94-101. 195

[10] H. Lu. Y. C. Hu, and W. Zwaenepoel. OpenMP on Networks of Workstations, in *Proceedings of Supercomputing'98*, Nov. 1998. 204

[11] Message Passing Interface Forum, *MPI: A Message-Passing Interface Standard*, version 1.1, June 12, 1995. 196

[12] H. W. Meuer, E. Strohmaier, J. J. Dongarra, and H. D. Simon, *TOP500 Supercomputer Sites, 18th ed.*, Technical report, Lawrence Berkely National Laboratory LBNL-49122, Nov. 2001. 195

[13] OpenMP consortium, *OpenMP: A Proposed Standard API for Shared Memory Programming*, White paper, `http://www.openmp.org`. 195

[14] OpenMP consortium, *OpenMP C and C++ Application Program Interface*, Version 1.0, October 1998. 195, 204

[15] E. W. Parsons, M. Brorsson and K. C. Sevcik, *Predicting the Performance of Distributed Virtual Shared Memory Applications*, IBM Systems Journal, Volume 36, No. 4, 1997, pp. 527-549. 196

[16] R. Samanta, A. Bilas, L. Iftode, and J. P. Singh, Home-based SVM protocols for SMP clusters: Design and performance, in *Proceedings of the 4th IEEE Symposium on High-Performance Computer Architecture* (HPCA-4), Las Vegas, Nevada, January 1998, pp. 113-124. 195

[17] D. M. Sato, Design of OpenMP Compiler for an SMP Cluster, in *Proceedings of First European Workshop on OpenMP*, Sept. 1999. `http://www.it.lth.se/ewomp99`. 204

[18] D. J. Scales, K. Gharachorloo, and C. A. Thekkath, Shasta: A Low Overhead, Software-Only Approach for Supporting Fine-Grained Shared Memory, in *Proceedings of the Seventh International Conference on Architectural Support for Programming Languages and Operating Systems* (ASPLOS'96), October 1996, pp. 174-185. 195

[19] J. P. Singh, A. Gupta and M. Levoy, *Parallel Visualization Algorithms: Performance and Architectural Implications*, IEEE Computer Magazine, July 1994, pp. 45-55. 195

[20] T. Sterling, D. Becker, D. Savarese, et al., BEOWULF: A Parallel Workstation for Scientific Computation, in *Proceedings of the 1995 International Conference on Parallel Processing* (ICPP), Vol. 1, August 1995, pp. 11-14. 195

[21] S. C. Woo, M. Ohara, E. Torrie, J. P. Singh, and A. Gupta. The SPLASH-2 Programs: Characterization and Methodological Considerations, in *Proceedings of the 22nd International Symposium on Computer Architecture*, Santa Margherita Ligure, Italy, June 1995, pp. 24-36. 200

[22] Y. Zhou, L. Iftode, and K. Li, Performance evaluation of two home-based lazy release consistency protocols for shared virtual memory systems, in *Proceedings of the 2nd Symposium on Operating System Design and Implementation*, October 1996, pp. 75-88. 198

A Fast Connection-Time Redirection Mechanism for Internet Application Scalability*

Michael Haungs[1], Raju Pandey[1], Earl Barr[1], and J. Fritz Barnes[2]

[1] Center for Software Systems Research, Department of Computer Sciences
University of California, Davis
{haungs,pandey,barr}@cs.ucdavis.edu
[2] Vanderbilt University
j.fritz.barnes@vanderbilt.edu

Abstract. Applications that are distributed, fault tolerant, or perform dynamic load balancing rely on redirection techniques, such as network address translation (NAT), DNS request routing, or middleware to handle Internet scale loads. In this paper, we describe a new connection redirection mechanism that allows applications to change end-points of communication channels. The mechanism supports redirections across LANs and WANs and is application-independent. Further, it does not introduce any central bottlenecks. We have implemented the redirection mechanism using a novel end-point control session layer. The performance results show that the overhead of the mechanism is minimal. Further, Internet applications built using this mechanism scale better than those built using HTTP redirection.

1 Introduction

Providing Internet services is increasingly difficult. Popular web sites must effectively handle current user request loads and prepare for increased future loads as their site gains popularity and the Internet audience, in general, grows. Excessive down times or user delay are unacceptable as Internet users will quickly turn elsewhere for their content.

To handle typical web loads, Internet server solutions must provide substantial computational power, I/O throughput and network bandwidth. This is accomplished either by using a high-end server or by clustering commodity workstations. Clustering is by far the most popular solution due to its cost effectiveness and potential for scalability. One of the main challenges with server clusters is distributing load while exporting a single name. Distributing load increases scalability. A single, easily remembered name for the service is necessary for end-user usability. Connection redirection, which maps the exported service name to a single server in the cluster, is the primary way to distribute load while still presenting a single name to the user.

Ideally, a redirection mechanism should be able to redirect requests to servers that are distributed in a wide area network and thereby allow servers to cope

* This research is supported in part by NSF grants CCR-00-82677 and CCR-99-88349.

S. Sahni et al. (Eds.) HiPC 2002, LNCS 2552, pp. 209–218, 2002.

with network congestion and to exploit geographical, temporal and other locality properties for load balancing. Since popular server clusters must scale to millions of simultaneous requests, redirection mechanisms should not introduce any bottlenecks or failure points. A server cluster that provides short, predictable client latencies increases user-perceived quality [4]. Redirection mechanisms, thus, must allow servers to smoothly adapt to dynamic conditions such as flash crowds or machine failures. Finally, redirection mechanisms must be compatible with existing Internet protocols, incrementally deployable as sweeping changes to the Internet's infrastructure are unrealistic, and transparent so that existing applications, such as web and ftp servers, can take immediate advantage of them.

Existing applications use many techniques [3, 6, 8, 12] for dynamically redirecting connections: among these are application-layer protocols, name resolution manipulation, and network packet rewriting. Each solution varies in its client transparency, responsiveness to dynamic conditions, performance, and scalability. DNS redirection is transparent, but only provides coarse load balancing [6] and incurs increased lookup delay [14]. HTTP redirection is scalable, but not application independent. Network packet rewriting is fast and transparent, but limits clustering to servers within a local area network. Also, packet rewriters may become a central point of failure.

In this paper, we present a novel connection-time redirection mechanism, called *Redirectable sockets* (RedSocks). RedSocks adds a redirection operation to the communication channel API that allows a server to redirect channels as needed. During periods of contention, a server sheds load by invoking this operation. RedSocks can redirect communication channel end-points among both LAN and WAN separated servers. This provides servers with a flexible way to respond to network congestion. It is a protocol-based solution that aims at providing a general, application independent redirection facility. Unlike network packet rewriting, RedSocks only requires a single packet manipulation to redirect communication, so it does not introduce a chokepoint. RedSocks is backwardly compatible with client applications making it incrementally deployable.

In Section 2, we describe the `redirect` operation and give examples of its utility. Next, we present our implementation, which is highlighted by our novel session-layer protocol, called *end-point operation protocol* (EOP), that provides communication channel end-point control. In Section 4, we compare RedSocks directly to HTTP redirection and show a significant reduction in server redirection latency. In Section 5, we discuss related redirection mechanisms. Finally, we discuss future research directions.

2 Redirectable Sockets

Redirectable sockets provides a communication abstraction suitable for use in a variety of server architectures. In this section, we discuss the semantics and usage of RedSocks.

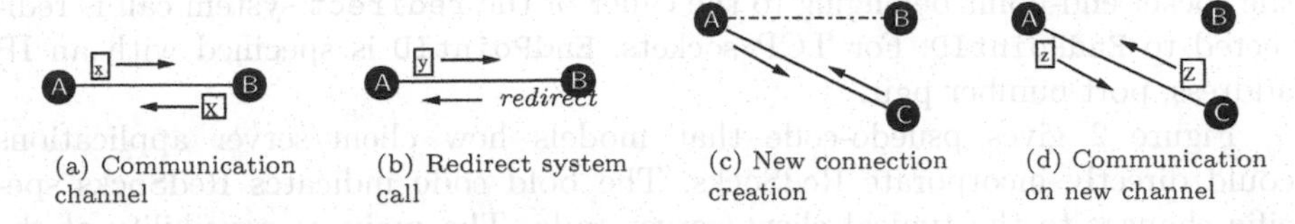

(a) Communication channel (b) Redirect system call (c) New connection creation (d) Communication on new channel

Fig. 1. Redirect end-point B to C

2.1 Definition

Our approach is to provide a primitive, `redirect`, that allows applications to redirect communication channels at connection time. We describe the semantics of a `redirect` system call through a simple example.

As shown in Figure 1(a), a communication channel exists between two nodes, A and B. The application at B then executes `redirect` in order to change the "B" end-point of the channel to C (see Figure 1(b)). C may define an end-point at A, B, or at a different host. The `redirect` system call creates a new channel between A and C, and removes the channel between A and B as shown in Figures 1(c)–(d).

There are two ways to handle the request made by A in Figure 1(b). B has the option to send a response to the request y during the `redirect` system call (See Below). However, if B does not, then A must resend its previous request.

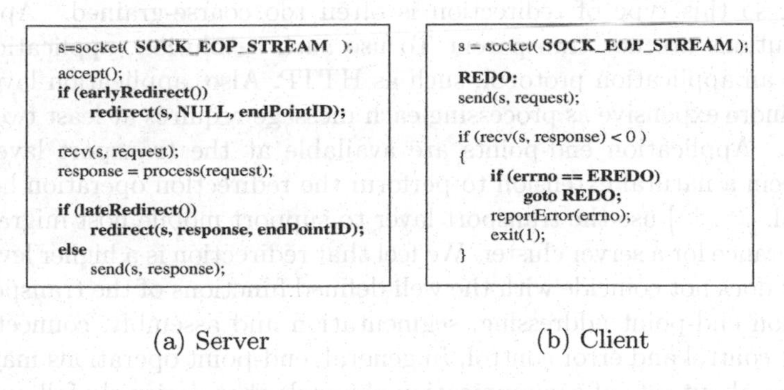

s=socket(SOCK_EOP_STREAM);
accept();
if (earlyRedirect())
redirect(s, NULL, endPointID);
recv(s, request);
response = process(request);
if (lateRedirect())
redirect(s, response, endPointID);
else
send(s, response);
s = socket(SOCK_EOP_STREAM);
REDO:
send(s, request);
if (recv(s, response) < 0)
{
if (errno == EREDO)
goto REDO;
reportError(errno);
exit(1);
}

(a) Server (b) Client

Fig. 2. RedSocks Sample Psuedo-Code

2.2 Usage

An application can use the `redirect` system call to perform the redirect operation. This system call has the following general form: *redirect(data, EndPointID)*. The `redirect` operation first ensures that the data, if given, is delivered. Next, the socket end-point belonging to the caller of the `redirect` system call is redirected to `EndPointID`. For TCP sockets, `EndPointID` is specified with an IP address, port number pair.

Figure 2 gives psuedo-code that models how client/server applications could directly incorporate RedSocks. The bold code indicates RedSocks specific changes to the typical client/server code. The main responsibility of the client is to handle a new error code, EREDO. EREDO is used to distinguish whether or not the server sends a response to the client in the `redirect` system call. If it does not, the client must "resend" its last request. We present a means to relieve the client application of this responsibility in [10].

3 Implementation

We have implemented RedSocks in the Linux 2.4.16 kernel, which implements BSD sockets that conform to version 4.4BSD.

3.1 Design Issues

The first issue in adding operations on communication end-points, such as redirect, is choosing the OSI model layer to enhance. The four choices are the network, application, transport, and session layers. At the network layer, communication end-points are defined by host addresses. Thus, network layer redirection, such as Mobile IP [13], globally applies to all the host's communication channels. Effectively load-balancing Internet services requires redirecting individual requests, so this type of redirection is often too coarse-grained. Application layer solutions are not transparent To use such a solution, applications must agree on an application protocol, such as HTTP. Also, application layer protocols are more expensive as processing each message requires at least two context switches.[1] Application end-points are available at the transport layer and it would seem a natural extension to perform the redirection operation here. Snoeren et al. [16, 15] use the transport layer to support mobile host migration and fault tolerance for a server cluster. We feel that redirection is a higher level operation that does not coincide with the well defined functions of the transport layer: application end-point addressing, segmentation and assembly, connection control, flow control and error control. In general, end-point operations may require knowledge about sets of communication channels that decisively fall outside the scope of the transport layer. At the session layer, the only objects one has to operate on are communication end-points. According to the OSI, the session layer "is the network *dialog controller*" that "establishes, maintains, and synchronizes the interaction between communicating systems." Sychronization is key in redirecting end-points of communication channels. With this in mind, we created a session layer protocol, EOP, to handle dynamic operations on communication end-points.

[1] A context switch occurs at the arrival of a message, at the departure, and additonal context switches can occur during the protocol processing of the message.

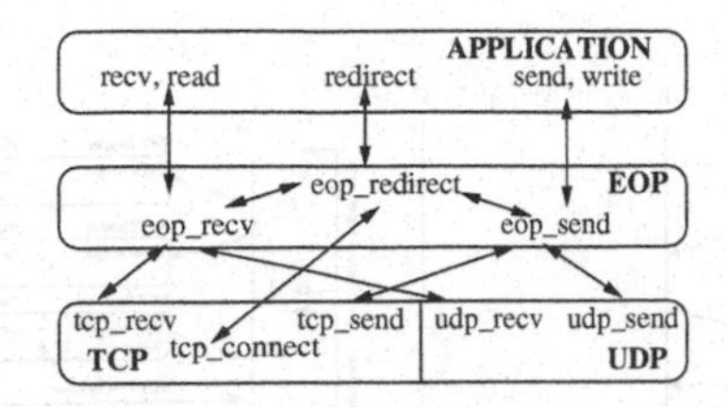

Fig. 3. The architecture for EOP and RedSocks

3.2 Architecture

Figure 3 illustrates our layered architecture. The arrows indicate functional dependencies where data flows through the formal parameters and return values. We defer discussion of the application layer to [10].

End-Point Operation Protocol. EOP is a session layer protocol that wraps transport layer data with an EOP header in order to manage end-point movement for the lifetime of the associated communication channel. When end-point movement is required, EOP invokes several functions in the EOP layer on the client and the server to coordinate and accomplish redirection. The EOP header triggers redirection on the remote end-point, synchronizes activity, and exchanges redirection arguments.

We use a simple 12-byte EOP header (see Figure 4) that contains a 16-bit opcode to specify the current operation. The two operations we use are `normal` and `redirect`. To specify the target of a redirect operation we use two parameters: one 16-bit parameter indicating the port number and the other 32-bit parameter indicating the destination IP address. The 32-bit `slen` field records the size of the user message and is used to delineate messages (described further in the transport layer discussion next).

Figure 5(a) shows the time flow diagrams for connection-time and in-stream redirect operations. After connecting to B, A sends its initial request. B uses the `redirect` system call to redirect requests to C. At A, EOP responds to the redirection request by closing the connection to B, opening a new connection to C, and then returning control to the client. Further requests, possibly including the previous request if B did not send a response with its redirect, are sent directly to C. At this point in time, B is fully divorced from A. RedSocks can also be used as an instream redirection mechanism where requests are idempotent or the servers are sharing application state via an alternative method. Figure 5(b)

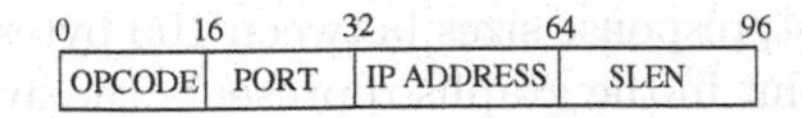

Fig. 4. The EOP Header

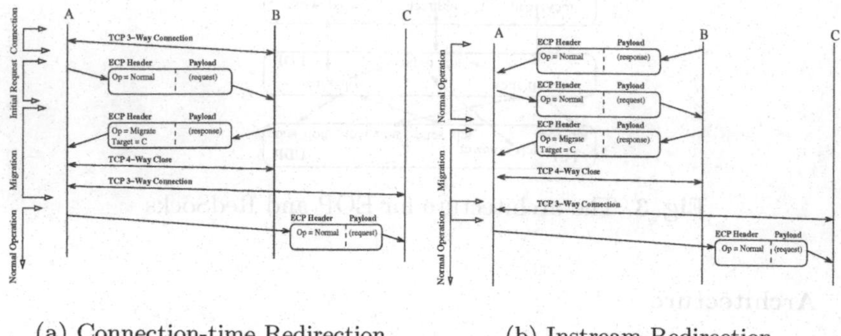

(a) Connection-time Redirection (b) Instream Redirection

Fig. 5. Time line for EOP packet flow

gives the time flow diagram before, during, and after such a redirection. Its flow is very similar to that depicted by Figure 5(a).

Transport Layer. Linux 2.4.15 provides a well-defined interface for the TCP layer. However, we encountered two difficulties in building our session-layer. The first difficulty is due to the way the network subsystem is optimized. At the session layer, Linux assumes the data to be sent lives in user space, but we need to generate our EOP headers in kernel space. Therefore, we had to modify TCP to handle kernel generated data and use a separate `eop_send` call to deliver it.

Protocol headers encapsulate data packets and convey information on how to process the data in the corresponding layers. However, TCP transparently combines data packets and provides a streaming abstraction of the data. Thus, TCP effectively destroys the association between protocol headers and their data for upstream layers. Our second difficulty is that we needed a way to maintain or rebuild this association. We use the `slen` field in Figure 4 for this purpose. The designers of the Stream Control Transmission Protocol (SCTP) describe this difficulty in RFC2960 and propose SCTP as a solution.

4 Experiments

We ran our overhead experiments (see Section 4.1) on four 400Mhz dual-processor Pentium III machines where each has 256MB of RAM, two 18GB hard drives, and a 100 Mbps Ethernet network card. These machines are connected via a 100 Mbps hub. Each machine ran Linux 2.4.16. In these experiments, data was collected for request/response sizes between 100 bytes and 32K, at 100 byte intervals. Each data point in the graphs represents the average over 100 runs.

For our scalability measurements, we used one client machine, located at the University of California at Davis, of the above type and a small server cluster, located at Vanderbilt University in Tennessee. The server cluster consisted of two

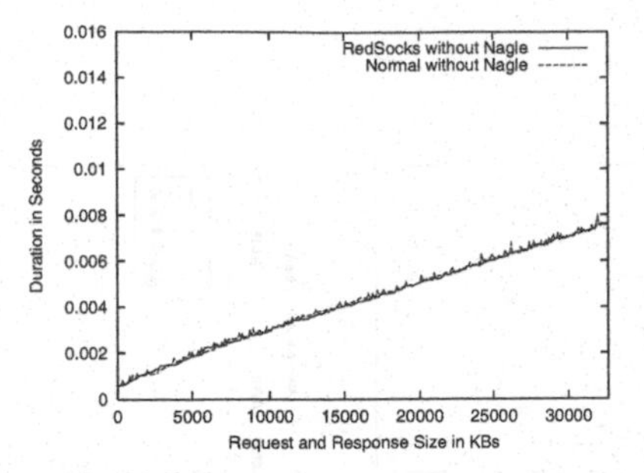

Fig. 6. Overhead of sending and redirection

800Mhz AMD machines with 256MB of RAM and 100 Mbps Ethernet network cards. Both campuses are connected via Internet II.

4.1 Overhead Measurements

To better understand the overhead of using RedSocks when redirection is not performed, we measured request-to-response latency from a single client and single server. This effectively measures pure send and receive overheads. For sending and receiving packets, RedSocks incurs a 6.5% average overhead with the Nagle algorithm enabled and a 1.5% average overhead with it disabled.[2]

The behaviour of RedSocks can be mimicked with traditional socket system calls at the application layer. A client can terminate communication with a host by calling `close` on the socket and then re-initiate communication with a different host via calls to `socket` and `connect`. This solution does not require the use of an EOP header, but fails to provide a number of advantages such as client transparency, server-side load balancing control, and dynamic adaptation. Nevertheless, it provides a good baseline against which to compare RedSocks.

We compare the time it takes to complete two consecutive client requests. With RedSocks, the client sends a request to a host that services the request. In addition, the host redirects its connection end-point to another server that handles the second request. The traditional socket scenario is the same except the client manually reconnects to a host statically specified in the program. We give the transaction time for varying request/response sizes in Figure 6. RedSocks is 0.84% faster when the Nagle algorithm was disabled (see Figure 6).

4.2 Comparing HTTP and RedSocks Redirection Latency

In these experiments, the client generates requests to Server 1. Server 1 redirects the requests to Server 2 which handles them. We modified Lynx 2.8.5 to generate user requests in a manner similar to flood[3]. On the servers, we ran either standard Apache 1.3.22 or our RedSocks-enabled Apache 1.3.22 depending

[2] Interested readers can refer to [10] for more details.

[3] `http://httpd.apache.org/test/flood/`

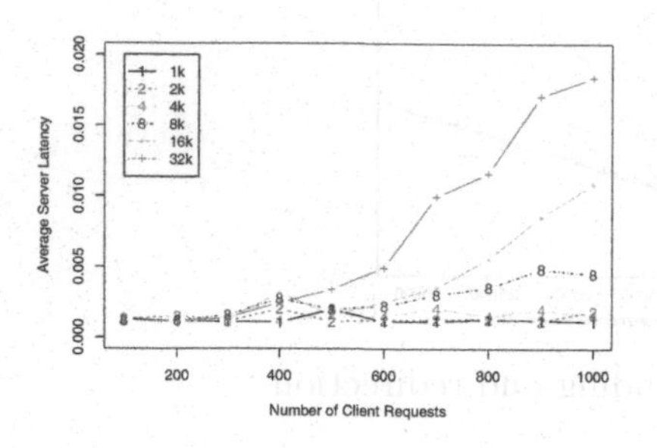

(a) Server Redirection Latency for HTTP redirection for all File Sizes

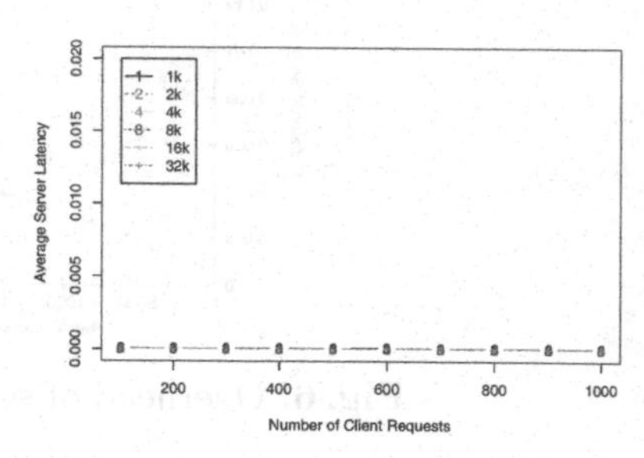

(b) Server Redirection Latency for RedSocks for all File Sizes

Fig. 7. Server Redirection Latency Measurements

on whether we were measuring HTTP redirection or RedSocks. We measured server redirection latency when varying the number of requests and file sizes. Server redirection latency is measured from the time the client connects to the time the `redirect`, or HTTP redirection, operation completes. Figure 7 shows a representative subset of these results.

Figure 7(a) gives the average server redirection latency of the Apache web server using HTTP redirection for differing numbers of requests. The results of this graph are counter-intuitive. One would expect that latency would increase when the number of requests are increased. However, observe the latencies for 1k and 2k files. While they are higher than the latencies seen in the corresponding RedSocks results (see Figure 7(b)), they do not increase with the number of requests. Instead, latency *increases with file size*. In analyzing this phenonemon, we discovered an undesirable trait of HTTP redirection: its performance is correlated to the performance of the clients. Thus, a poorly performing client adversely impacts server scalability as witnessed in Figure 7(a).

The explanation of this lies in the steps taken to perform redirection in both mechanisms. For RedSocks, the steps are to (1) `accept` the connection and (2) `redirect` the request. Both of these steps are efficient and neither the file sizes or numbers of requests for our tests effected server redirection latency (see Figure 7(b)). For HTTP redirection, the steps are to (1) `accept` the connection, (2) `read` the request, (3) create an HTTP redirection response, and (4) `send` the response. The time required to generate the HTTP redirection response adds a constant increase to server redirection latency over that incurred by RedSocks. The key to the non-scalable increase in server redirection latency experienced by HTTP redirection is that it must `read` the request (step 2) and `read` is a blocking system call. Usually, the client is responsive and immediately sends its request resulting in the server having no blocking delay. However, the effects of an unresponsive client can be seen in Figure 7(a) where the client bogs down as file sizes increase. This effect is not specific to HTTP redirection, but occurs in any mechanism that must read a request before redirecting, a common scenario in content-aware redirection.

5 Related Work

DNS can be used to map a single hostname onto multiple hosts [5]. When queried about a name, a modified DNS server returns different IP addresses according to its selection policy. DNS is transparent to both clients and servers making it easy to deploy. It is also convenient because clients must do a DNS lookup to contact Internet services anyway. Unfortunately, DNS-based redirection is coarse-grained, increases load on DNS, and may make poor load-balancing decisions [14]. Several server-based approaches [2, 9] use HTTP redirection, which allows any server to redirect its connections, but is application-specific and has higher overhead.

Dispatcher-based mechanisms, such as Cisco's Local Director, Linux's netfilter, TCP splicing [7] and MagicRouter [1], employ network packet rewriting for both connection-time and in-stream connection redirection. Packet rewriting provides a transparent way to build distributed servers, but at a price: it requires a proxy or router to interpose itself between a connection's end-points and actively manipulate packets. In [11], Hunt et al. introduce TCP handoff (also used in LARD [12]), as part of a content-based load distribution scheme, that avoids bi-directional packet rewriting. However, all incoming packets must still pass through a node, which may become a bottleneck. For a more extensive review of related work, please refer to [10].

6 Conclusion

Building scalable, highly-available web servers requires mechanisms that support multiple machines cooperatively handling requests for a service. RedSocks is a new mechanism introduced to solve some of the issues in building distributed architectures that transparently balance load across servers. With RedSocks, one can manipulate the end-points of the communication. This solution improves upon previous ad hoc mechanisms for balancing load among servers in that it is scalable, fine-grained, and transparent to the client. In future work, we will explore extensions, such as support for fault tolerance and lazy redirection, which allows a server to redirect a connection that was just redirected to it.

References

[1] E. Anderson, D. Patterson, and E. Brewer. The MagicRouter: An application of fast packet interposing. `http://www.cs.berkeley.edu/~eanders/projects/magicrouter/`, October 1996. 217

[2] D. Andresen, T. Yang, O. Ibarra, and O. Egecioglu. Adaptive partitioning and scheduling for enhancing WWW applications performance. *Journal of Parallel and Distributed Computing*, 49(1):57–85, February 1998. 217
[3] M. Aron, D. Sanders, P. Druschel, and W. Zwaenepoel. Scalable content-aware request distribution in cluster-based network servers. In *Proc. USENIX 2000 Annual Technical Conference*, San Diego, CA, USA, 18–23 June 2000. 210
[4] N. Bhatti, A. Bouch, and A. Kuchinsky. Integrating user-perceived quality into web server design. In *Proceedings of the Ninth International World Wide Web Conference*, volume 33(1–6) of *Computer Networks*, pages 1–16, Amsterdam, The Netherlands, 15–19 May 2000. 210
[5] T. Brisco. DNS support for load balancing. RFC 1794, Rutgers University, April 1995. 217
[6] V. Cardellini, M. Colajanni, and P. Yu. Dynamic load balancing on web-server systems. In *IEEE Internet Computing*, pages 28–39. IEEE, May-June 1999. 210
[7] A. Cohen, S. Rangarajan, and H. Slye. On the performance of TCP splicing for URL-aware redirection. In *Proc. 2nd USENIX Symposium on Internet Technologies and Systems*, pages 117–26, Boulder, CO, USA, 11–14 October 1999. 217
[8] O.P. Damani, P.E. Chung, Y. Huang, C. Kintala, and Y. Wang. ONE–IP: Techniques for hosting a service on a cluster of machines. *Computer Networks and ISDN Systems; Sixth International WWW Conference*, 29(8–13):1019–27, 7–11 April 1997. 210
[9] M. Garland, S. Grassia, R. Monroe, and S. Puri. Implementing distributed server groups for the world wide web. Technical report, Carnegie Mellon University, January 1995. 217
[10] Michael Haungs, Raju Pandey, Earl Barr, and J. Fritz Barnes. A fast connection-time redirection mechanism for internet application scalability. Technical Report CSE-2001-10, University of California, Davis, March 2001. 212, 213, 215, 217
[11] G. Hunt, E. Nahum, and J. Tracey. Enabling content-based load distribution for scalable services. Technical report, IBM T.J. Watson Research Center, May 1997. 217
[12] V. Pai, M. Aron, G. Banga, M. Svendsen, P. Druschel, W. Zwaenepoel, and E. Nahum. Locality-aware request distribution in cluster-based network servers. In *Proceedings of the Eighth International Conference on Architectural Support for Programming Languages and Operating Systems (ASPLOS-VIII)*, pages 205–216, San Jose, California, 1998. 210, 217
[13] C. Perkins. IP mobility support. Internet Request for comments (RFC 2002), October 1996. 212
[14] A. Shaikh, R. Tewari, and M. Agrawal. On the effectiveness of DNS-based server selection. In *the Proceedings of IEEE INFOCOM 2001*, pages 1801–10, Anchorage, AK, USA, April 2001. 210, 217
[15] A. Snoeren, D. Andersen, and H. Balakrishnan. Fine-grained failover using connection migration. In *3rd USENIX Symposium on Internet Technologies and Systems (USITS '01)*, pages 221–232, San Francisco, CA, March 2001. 212
[16] A.C. Snoeren and H. Balakrishnan. An end-to-end approach to host mobility. In *Proceedings of the 6th Annual ACM/IEEE International Conference on Mobile Computing and Networking*, pages 155–164, August 2000. 212

Algorithms for Switch-Scheduling in the Multimedia Router for LANs*

Indrani Paul[1]**, Sudhakar Yalamanchili[1], and Jose Duato[2]

[1] Center for Experimental Research In Computer Systems
School of Electrical and Computer Engineering, Georgia Institute of Technology
Atlanta, Georgia 30332
{indrani,sudha}@ece.gatech.edu

[2] Dept. of Computer Engineering (DISCA). Univ. Politecnica de Valencia
46071 Valencia, Spain
jduato@gap.upv.es

Abstract. The primary objective of the Multimedia Router (MMR) [1] project is to design and implement a single chip router targeted for use in cluster and LAN interconnection networks. The goal can be concisely captured in the phrase 'QoS routing at link speeds'. This paper studies a set of algorithms for switch scheduling based on a highly concurrent implementation for capturing output port requests. Two different switch-scheduling algorithms called Row-Column Ordering and Diagonal Ordering are proposed and implemented in a switch-scheduling framework which involves a matrix data structure, and therefore enables concurrent and parallel operations at high-speed. Their performance has been evaluated with Constant Bit Rate (CBR), Variable Bit Rate (VBR), and a mixture of CBR and VBR traffic. At high offered loads both these ordering functions have been shown to deliver superior Quality of Service (QoS) to connections at a high scheduling rate and high utilization.

1 Introduction

Current technology trends indicate a continued increase in network link speeds coupled with an increasing appetite for data as well as timeliness of data. The result has been a demand for high-performance switches that support fast and efficient scheduling of packets to concurrently meet QoS requirements across a diverse set of connections while maximizing the utilization of the link bandwidth.

---

* This research was supported in part by the National Science Foundation under grant CCR-9970720

** Author currently with Dell Computer Corporation, Round Rock, Texas. Email: Indrani_paul@dell.com

S. Sahni et al. (Eds.) HiPC 2002, LNCS 2552, pp. 219-231, 2002.

The Multimedia Router (MMR) project [1][2] is aimed at the design and implementation of a single chip router optimized for multimedia applications in Local Area Network (LAN) environments. The overall goal of the project can be concisely captured in the phrase 'QoS routing at link speeds'. To achieve this goal there is a fundamental trade-off between scheduling rate, scheduling granularity, and Quality of Service (QoS). A very high scheduling rate measured in packets/second can be achieved with very simple scheduling algorithms for small packet sizes such as an Ethernet frame. However, such schemes cannot support distinct QoS requirements across packet streams. Alternatively, implementation of more complex scheduling algorithms will result in a lower scheduling rate typically leading to an increase in the granularity of the units being scheduled. Thus, any scheduling guarantees that can be made are ensured over larger packet arrival times. Therefore, the desired combination of link-speed scheduling, QoS guarantees [11], and high scheduling rate remains a challenging goal.

The MMR scheduling framework separates link scheduling from switch scheduling [2] in order to support fast scheduling decisions. Link scheduling is formulated as the problem of selecting a candidate virtual channel from an input link to transmit a flow-control unit. The switch-scheduling algorithm considers all candidates from all input ports and arbitrates between conflicting requests for output ports. In this paper, we focus on the task of matching the input port requests to output ports by the switch scheduler. We have proposed two switch-scheduling algorithms called *Row-Column Ordering* and *Diagonal Ordering*.

The rest of the paper is organized as follows. Section 2 discusses the state-of-the-art in switch scheduling algorithms and motivates the algorithms presented in this paper. The MMR architecture is briefly reviewed in Section 3. Section 4 discusses the bandwidth allocation scheme and the scheduling framework implemented in MMR. In Section 5, a detailed description of the proposed switch scheduling algorithms, *Row-Column Ordering* and *Diagonal Ordering*, is given. The paper concludes with simulation results and directions for future work.

2 Motivation

The problem of switch scheduling has typically formulated the assignment of input port requests to output ports as a matching problem in bipartite graphs [12][13]. Several algorithms such as maximum-size and maximum weight matching [12] have been proposed. However, from the perspective of implementation in a switch, these matching algorithms lead to complex implementations that reduce scheduling rate and increase scheduling granularity. Similar observations can be made of state-of-the-art scheduling algorithms such as EDF, D-EDD, J-EDD [4], 2-DRR [9], RPA [7], and DWCS [10] that are able to provide features such as QoS, efficiency and fairness for integrated-service networks. Comparatively simpler algorithms (from a hardware implementation perspective) such as RRM, *i*OCF [6], WFA [18], can be readily implemented in hardware but do not provide QoS guarantees. Further, algorithms such as PIM [5][6] result in very high scheduling rates but cannot provide QoS guarantees.

Thus we seek solutions that can provide QoS guarantees at much higher speeds and at a lower cost compared to existing algorithms. We compare our proposed algorithms with several existing ones from the perspective of scheduling rate, QoS guarantees and efficient hardware implementation.

3 The Multimedia Router

The Multimedia Router [1][2] consists of a set of input ports, output ports, an internal crossbar connecting outputs to inputs, routing and arbitration unit, schedulers, and input and output buffers. The architecture of the MMR is shown in Figure 1. The following is a brief description of the various features of MMR.

Switching Technique – The MMR uses a hybrid approach, where the most suitable switching technique is used for each kind of traffic: a connection-oriented scheme [14] for multimedia flows and virtual-cut-through for best-effort messages [17]. Packets are partitioned into flow control digits (flits) that is the unit of scheduling, although equivalently packets can from the unit of scheduling.

Buffer Organization – For each connection, a virtual channel is provided and each virtual channel has its own buffer. This eliminates Head-Of-Line (HOL) Blocking.

Multiplexed Crossbar – The MMR uses a multiplexed crossbar [15] where the internal switch is a crossbar with as many ports as communication links.

Link and Switch Schedulers – Link Scheduler performs arbitration needed at the input side to select one virtual channel from each physical channel. Switch scheduling refers to the problem of matching input ports with requested output ports in a conflict-free manner.

Routing and Arbitration Unit – The routing and arbitration unit implements the routing algorithm and selects the input port of the next router. It also provides some arbitration.

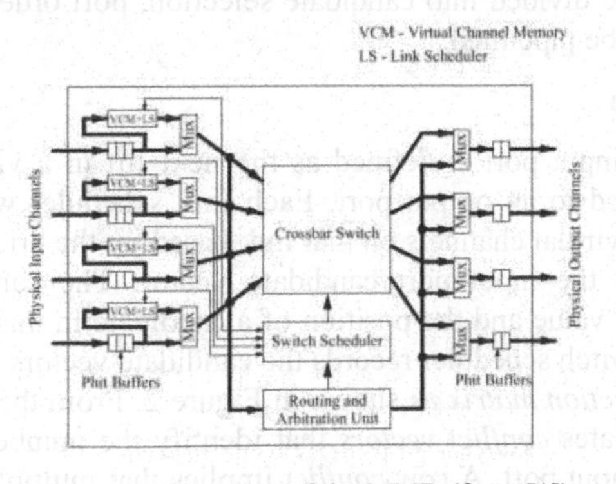

Fig. 1. Architecture of the MMR (from [1])

4 MMR Implementation Aspects

4.1 Dynamic Bandwidth Allocation

When a connection is being established in the MMR the source node generates a routing probe that tries to reserve link bandwidth and buffer space along the path. If resource reservation is successful the connection is established and the request is granted, else the connection fails and all the resources reserved during the construction of the path are released. The MMR supports dynamic allocation of bandwidth to established connections.

4.2 Dynamic Priority Update

The MMR uses a dynamic priority update scheme where the flit priority is updated as a function of the QoS that a flit has experienced up to that point in time relative to the QoS it has requested during connection setup. This is referred to as *priority biasing*. It is very fast and easy to implement in hardware. The biasing function used in this paper is the Inter-Arrival Based Priority (IABP) [2].

Inter-Arrival Based Priority (IABP)

The priority is expressed as a ratio of the queuing delay and the connection's desired inter-arrival time. Delay is computed as the difference between the time a flit arrives at the switch and the time it actually leaves the switch. Specifically:

priority = (queuing delay - ns) /(inter-arrival time - ns), ns: nanoseconds
where, inter-arrival time (ns) = 10^9 (flit size - bits) / (bandwidth – bits/second)

4.3 Scheduling Framework

The MMR scheduling framework is comprised of link and switch scheduling in order to support fast scheduling by pipelining decisions. The basic functions of the link and switch schedulers are divided into candidate selection, port ordering, and arbitration. These functions can be pipelined.

Candidate Selection

A *candidate* on an input port is defined as the next flit in a virtual channel that is ready to be scheduled to an output port. Each link scheduler will produce a list of candidates from the virtual channels on that link based on the priority value of that flit and record them in the input port candidate vector. The candidates are ordered according to priority value and the position of a candidate in this order is the *level* of the candidate The switch scheduler records the candidate vectors forwarded from each input port in the *selection matrix* as shown in Figure 2. From the selection matrix the switch scheduler creates *conflict vectors* that identify the number of conflicts for an output port and an input port. A *row conflict* implies that multiple input ports request the same output port and a *column conflict* implies that the same input port requests multiple output ports.

Port Ordering

Two critical decisions that are taken are the order in which output ports must be examined to resolve conflicts, and the arbitration policy. The former is called the *ordering function.*

Arbitration

Arbitration is a matter of picking the highest priority flit. In this phase the switch scheduler resolves the conflicts strictly based on priority.

5 Row-Column (RC) and Diagonal (Diag) Ordering

The performances of the proposed algorithms are compared to the scheduling algorithm as described in [2], which we call Ord0. Ord0 selects output ports first by level and then in increasing order of row conflicts within a level, with ties broken randomly by selecting one of the ports. The performances of the proposed algorithms are also compared with the *i*SLIP [5] scheduling algorithm. In [5] it has been shown that *i*SLIP performs significantly better than PIM, RRM, *i*OCF [6]. The hardware complexity of *i*SLIP is relatively small, as it requires only round robin arbiters. Moreover, *i*SLIP has a high scheduling rate and high-speed switches [8] have been developed using *i*SLIP.

5.1 Row-Column Ordering

The proposed Row-Column ordering function exploits the spatial property of a matrix where all elements can be manipulated in parallel. This ordering function assigns input port requests to the output ports first in increasing order of conflicts in the columns and then in increasing order of conflicts in the rows. The rationale behind choosing this ordering function is that both input and output ports with the highest number of conflicts should be matched last since those ports have the most options for being matched. Further, row and column conflict vectors can be computed in parallel efficiently and fast. The sums capture global information about input ports and output ports that are used to make fast decisions without having to examine all matrix entries.

Description of Algorithm

The selection matrix is updated with the candidate vectors provided by the link schedulers. The algorithm is described as follows:

1. First check for columns that are conflict-free, i.e. the corresponding entries in the column conflict vector are 1. Assign those requests.
2. Now check for columns in an increasing order of conflicts (i.e. an entry in the column conflict vector is greater than 1). If there are conflicts in a column, i.e. same input port requests multiple output ports, then check the corresponding rows to see if there exists row conflicts i.e. some other input ports have also requested those output ports. Now assign requests in increasing order of row conflicts.

3. If the same number of row conflicts exists for all entries in a particular column, give higher priority to higher-level candidates. In case of all conflicts belonging to the same level, select randomly.
4. After each assignment modify the selection matrix, and row and column conflict vectors accordingly.
5. Repeat the above steps until each input port has been assigned or cannot be assigned.

5.2 Diagonal Ordering

Definition: A ***generalized diagonal*** is a set of N elements in an N x N matrix, such that no two elements are in the same row or column.

The ordering function proposed here results in the sweeping of a generalized diagonal pattern across the selection matrix level by level and selecting the one with the largest sum. It uses N of these generalized diagonals by selecting one basic diagonal and generating the remaining N – 1 ones by shifting the basic diagonal across the matrix (so that each matrix element is covered by one of these diagonals). The rationale behind choosing this function is that diagonals represent conflict-free patterns and the diagonal with the largest sum indicates the one with the maximum number of conflict-free requests. The challenge here is to identify diagonal patterns that can lead to high switch utilizations. Specific patterns capture relationships between input channels and in fact the patterns themselves may request expected or desired relationships between virtual channels.

Description of Algorithm

The selection matrix is updated with the candidate vectors provided by the link schedulers. The algorithm can be described as follows:

1. In the first step compute the diagonal sums of all the generalized diagonals for level 1 candidates. Each diagonal sum represents the number of requests in that particular diagonal.
2. Select the diagonal with the largest sum and grant all requests in that diagonal since they represent conflict-free requests.
3. After each 'grant' modify the selection matrix accordingly and re-compute the diagonals and select them in increasing order of diagonal sums. Grant the requests of all the diagonal elements that have not yet been allocated an output port.
4. Repeat the above step until all the generalized diagonals in the first level of candidates sum to zero, which means there is no other request in the first level which needs to be assigned or can be assigned.
5. Move to the next level and repeat the above steps until each input port that can be assigned is assigned.

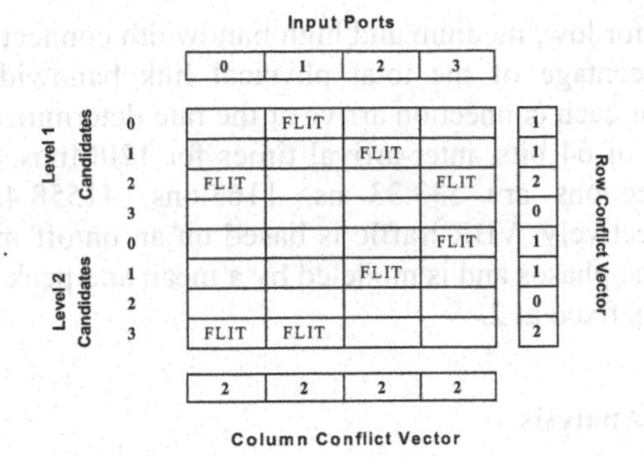

Fig. 2. An Example Selection Matrix and its corresponding Conflict Vector

6 Simulation Results

We have developed a simulation and evaluation environment to implement and evaluate our algorithms. The simulator is a detailed flit-level simulator of the Multimedia Router and is implemented in SystemC [16], a system level C-based simulation environment.

6.1 Experiments

Experiments have been performed with 4x4, 8x8, 16x16 and 32x32 routers. The results in this paper represent 16x16 and 32x32 routers with 16 and 8 virtual channels per physical input link respectively. For results for 4x4 and 8x8 routers, interested readers can refer to [3]. The router implementation is assumed to operate at 100MHz. We have used physical link bandwidth of 1Gbps and flit-size of 64 bits. Link and switch schedules can be computed in a flit cycle. The number of candidate levels considered is 2, as the additional improvement with larger number of candidates has not been found to be substantial. Virtual channel buffer size has been fixed at 1024 flits. All simulations are run for 64000 flit cycles. The simulation clock speed is 10 nanoseconds.

Delay is computed as the average time a flit belonging to a particular connection has spent in the MMR. Typically this includes the receive cycle, virtual queuing time, scheduling time and send cycle. Since the receive cycle, scheduling time and send cycle are same for all the flits of all connections, we measure delay as the virtual queuing time. Jitter is computed as a difference between the successive flit delays. Both delay and jitter are computed in terms of flit cycles.

6.2 Workload Models

The proposed algorithms are tested with constant bit rate (CBR), variable bit rate (VBR) and mixed (CBR and VBR) traffic models. Connections are randomly generated from the set of 120Mbps, 55Mbps, 1.54Mbps, and 64Kbps for each input port for a specified offered load. In this way we can check the behavior of the proposed heuristics for low, medium and high bandwidth connections. Offered load is computed as a percentage of the total physical link bandwidth requested by all connections. Flits for each connection arrive at the rate determined by the inter-arrival time. For a flit size of 64 bits, inter-arrival times for 120Mbps, 55Mbps, 1.54Mbps, and 64Kbps connections are 533.33 ns, 1163.6ns, 41558.4ns, and 1000000ns (1millisecond) respectively. VBR traffic is based on an on/off model, corresponding to burst and non-burst phases and is modeled by a mean and peak bandwidth pair. The peak-to-mean ratio is fixed at 2.

6.3 Results and Analysis

Results for only CBR and mixed workloads for 64Kbps and 120Mbps connections are presented here. For results of VBR and other connections, readers can refer to [3].

6.3.1 Evaluation with CBR Workload

In Figure 3 we illustrate switch utilization as a function of the offered load in a 32x32 switch. We see that the Row-Column ordering function tends to produce utilization approximately 20% higher than that of Ord0 and *i*SLIP, even at high offered loads. This is because of the way RC resolves conflicts and assigns requests. In Ord0 [2] conflicts were resolved by taking only the row conflicts into consideration and using randomization in case of ties. RC improves over that by also considering the column conflicts concurrently, thus eliminating the randomness in Ord0. The Diagonal ordering function produces a systematic search for a match whereas *i*SLIP's behavior is defined as an implementation independent of a specific switch design. It gives better utilization than Ord0 and *i*SLIP in general because it is based on pattern matching and starts with assigning the requests in the diagonal with the maximum number of requests.

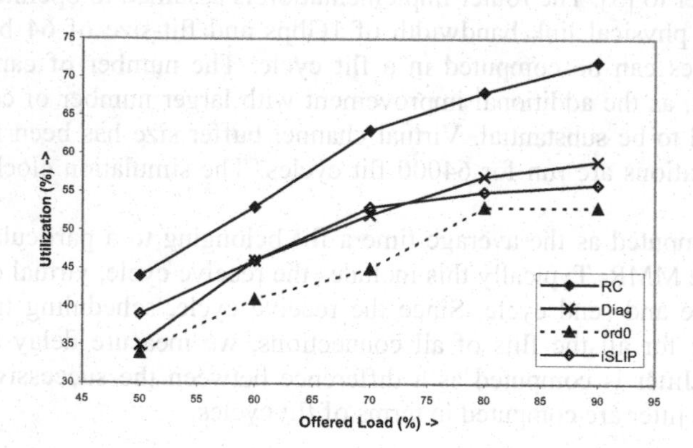

Fig. 3. Utilization vs. Offered Load using CBR workload for 32x32 switch

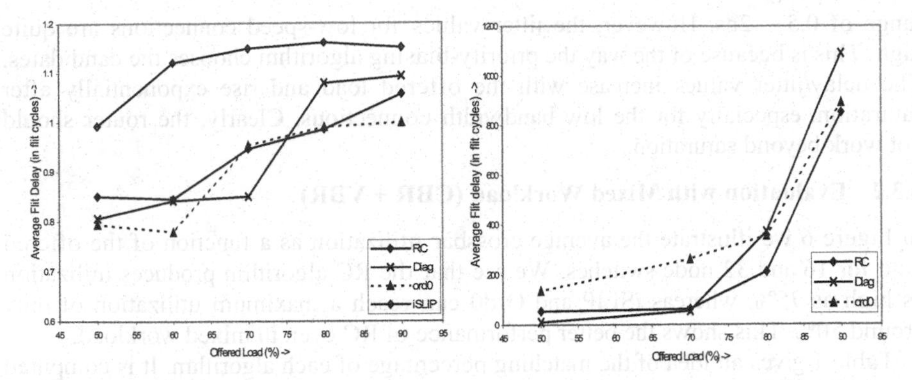

Fig. 4. Delay vs. Offered load for 120Mbps (left) and 64Kbps (right) connections using CBR workload for 32x32 switch

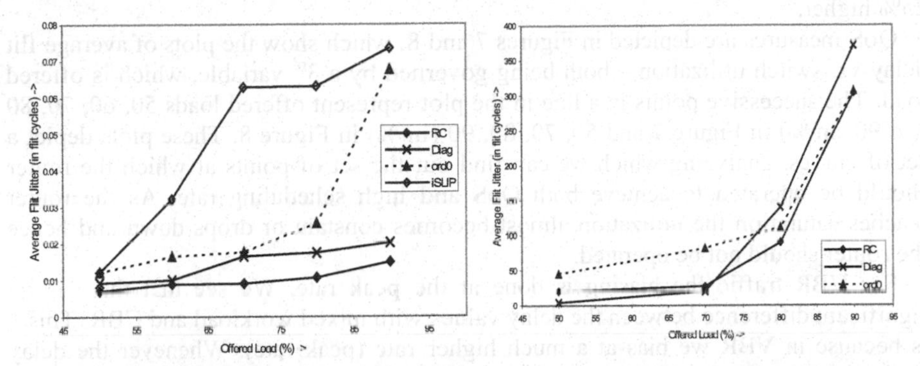

Fig. 5. Jitter vs. Offered Load for 120Mbps (left) and 64Kbps (right) connections using CBR workload for 32x32 switch

The measures of QoS include average flit delay in Figure 4 and average flit jitter in Figure 5. For high bandwidth connections we can see that RC results in better QoS. The Diagonal function also results in good delay and jitter characteristics as compared to Ord0 and *i*SLIP. The average flit delay and jitter using *i*SLIP for high bandwidth connections is almost 2-3 times that of RC at high offered loads. This is because *i*SLIP considers all the requests received for arbitration in a fair way based on the round-robin pointer with a rotating priority. In contrast, IABP-based algorithm residing in the link scheduler biases the priority by the bandwidth of a connection, thus giving higher priority to 120Mbps connections over 64Kbps connections. Hence, our scheme allows a better bandwidth allocation of the output ports, which translates into a better Quality of Service for all the connections.

The average flit delay for a 120Mbps connection lies in the range of 0.2-0.6 microseconds (μs) in case of all algorithms, whereas that of 55Mbps, 1.54Mbps and 64Kbps lies in the range of 0.6 – 2μs, 4-20μs, 50-500μs respectively in case of RC, Diag and Ord0. The delay values of *i*SLIP for 1.54Mbps and 64Kbps connections are much smaller compared to the others and hence are not shown in the plot. In case of RC and Diagonal functions the average flit jitter for 120Mbps connection lies in the range of 0.5 – 2ns. However, the jitter values for low-speed connections are quite high. This is because of the way the priority-biasing algorithm chooses the candidates. The delay/jitter values increase with the offered load and rise exponentially after saturation, especially for the low bandwidth connections. Clearly, the router should not work beyond saturation.

6.3.2 Evaluation with Mixed Workload (CBR + VBR)

In Figure 6 we illustrate the average crossbar utilization as a function of the offered load for 16 and 32 node switches. We see that the RC algorithm produces utilization as high as 75%, whereas *i*SLIP and Ord0 can reach a maximum utilization of only around 50%. This shows the better performance of RC even in mixed workload.

Table 1 gives an idea of the matching percentage of each algorithm. It is computed as a ratio of the number of output ports assigned to the number of requests in the selection matrix, averaged over the total number of simulation cycles. We see that the RC algorithm performs better than Ord0 and *i*SLIP by producing a matching of 20-25% higher.

QoS measures are depicted in Figures 7 and 8, which show the plots of average flit delay vs. switch utilization – both being governed by a 3rd variable, which is offered load. The successive points in a line in the plot represent offered loads 50, 60, 70, 80 and 90 (in %) in Figure 7 and 50, 70, 80, 90 (in %) in Figure 8. These plots depict a set of curves, analyzing which we can find out the set of points at which the router should be operated to achieve both QoS and high scheduling rate. As the router reaches saturation the utilization almost becomes constant or drops down and hence the router should not be operated.

For VBR traffic the biasing is done at the peak rate. We see that there is no significant difference between the delay values with mixed workload and CBR. This is because in VBR we bias at a much higher rate (peak rate). Whenever the delay increases, even by a small amount, the rate at which the priority is increased is higher. With RC the working point of the router can be as high as 85% offered load whereas with *i*SLIP and Ord0 algorithms the router can operate only till approximately 65-68% load and still provide QoS. The average flit delay using *i*SLIP for high bandwidth connections is much higher than that of RC at high offered loads. The jitter characteristics are similar [3].

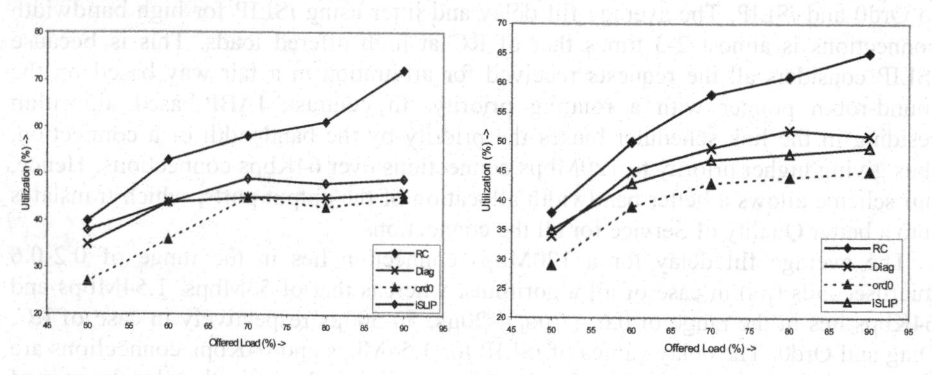

Fig. 6. Utilization vs. Offered Load using mixed workload for 16x16 (left) and 32x32 (right) switches

Table 1. Matching Percentage for different switch sizes using different scheduling algorithms with mixed workload

Switch Size / Algo	RC	Diag	Ord0	*I*SLIP
16x16	77	59	55	56
32x32	75	56	52	52

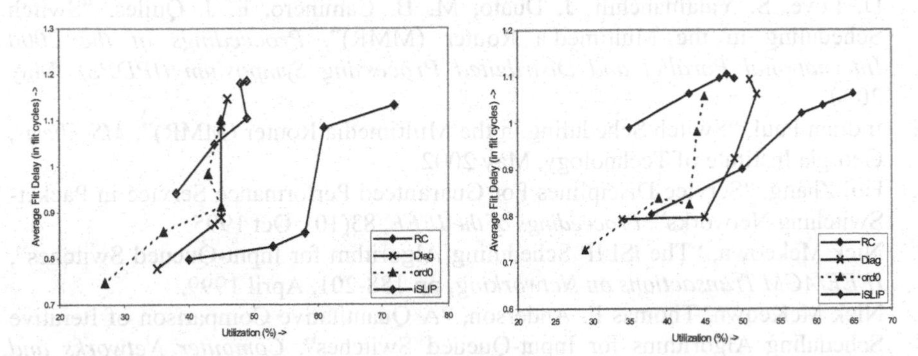

Fig. 7. Delay vs. Utilization for 120Mbps connection using mixed workload for 16x16 (left) and 32x32 (right) switches

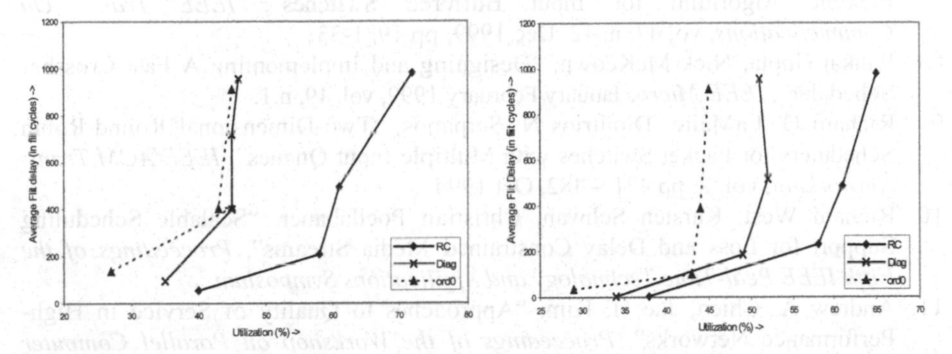

Fig. 8. Delay vs. Utilization for 64Kbps connection using mixed workload for 16x16 (left) and 32x32 (right) switches

7 Conclusions and Future Work

In this paper we have described the design of the Multimedia Router (MMR), which is targeted to provide QoS guarantees to multimedia traffic at link speeds in Local Area Network (LAN) environments. It has been shown that with the Row-Column ordering function a relative improvement in utilization of as high as 20-25% can be achieved at high loads. In terms of implementation complexity they require some matrix manipulations, arithmetic operations such as additions, and parallel binary logic operations that are amenable to concurrent hardware implementation. The next step is to extend this study to a network composed of several MMR's as well as more diverse traffic models including trace driven simulation with traces derived from network applications.

References

1. Jose Duato, Sudhakar Yalamanchili, M. Blanca Caminero, Damon Love, Francisco J. Quiles, "MMR: A High-Performance Multimedia Router – Architecture and Design Trade-Offs", *HPCA – 5*, pp 300-309, January 1999.
2. D. Love, S. Yalamanchili, J. Duato, M. B. Caminero, F. J. Quiles, "Switch Scheduling in the Multimedia Router (MMR)", *Proceedings of the 2000 International Parallel and Distributed Processing Symposium (IPDPS)*, May 2000.
3. Indrani Paul, "Switch Scheduling in the Multimedia Router (MMR)", *MS Thesis*, Georgia Institute of Technology, May 2002.
4. Hui Zhang, "Service Disciplines For Guaranteed Performance Service in Packet-Switching Networks", *Proceedings of the IEEE*, 83(10), Oct 1995.
5. Nick Mckeown, "The *i*SLIP Scheduling Algorithm for Input-Queued Switches", *IEEE/ACM Transactions on Networking*, pp 188-201, April 1999.
6. Nick McKeown, Thomas E. Anderson, "A Quantitative Comparison of Iterative Scheduling Algorithms for Input-Queued Switches", *Computer Networks and ISDN Systems*, vol 30, n. 24, Dec 1998, pp 2309 – 26.
7. Marco Ajmone Marsan, Andrea Bianco, Emilio Leonardi, Luigi Milia, "RPA: A Flexible Algorithm for Input Buffered Switches", *IEEE Trans. On Communications*, vol 47, n. 12, Dec 1999, pp 1921-33.
8. Pankaj Gupta, Nick McKeown, "Designing and Implementing A Fast Crossbar Scheduler", *IEEE Micro*, January/February 1999, vol. 19, n.1.
9. Richard O. LaMaire, Dimitrios N. Serpanos, "Two-Dimensional Round-Robin Schedulers for Packet Switches with Multiple Input Queues", *IEEE/ACM Trans. Networking*, vol. 2, pp 471 – 482, Oct 1994.
10. Richard West, Karsten Schwan, Christian Poellabauer, "Scalable Scheduling Support for Loss and Delay Constrained Media Streams", *Proceedings of the Fifth IEEE Real-Time Technology and Applications Symposium.*
11. Andrew A. Chien, Jae H. Kim, "Approaches to Quality of Service in High-Performance Networks", *Proceedings of the Workshop on Parallel Computer Routing and Communication, Lecture Notes in Computer Science*, Springer Verlag, pp 1-19, June 1997.
12. J. E. Hopcroft, R. M. Karp, "An $O(n^{5/2})$ Algorithm for Maximum Matching in Bipartite Graphs", *Society for Industrial and Applied Mathematics,* 1973, pp 225 – 231.
13. Adwin H. Timmer, Jochen A. G Jess, "Exact Scheduling Strategies based on Bipartite Graph Matching", *Proceedings of the European Design & Test Conference (ED&TC or EDAC–ETC–EuroASIC)*, pp. 42–47, Paris (F), March 6–9, 1995.
14. Patrick T. Gaughan, Sudhakar Yalamanchili, "Pipelined Circuit-Switching: A Fault-Tolerant Variant of Wormhole Routing", *SPDP* 1992, 148-155.
15. W. J. Dally, "Virtual-channel flow control", *IEEE Transactions on Parallel and Distributed Systems*, vol3, no 2, pp. 194-205, March 1992.
16. SystemC – `http://www.systemc.org`

17. P. Kermani, L. Kleinrock, "Virtual Cut-through: A New Computer Communication Switching Technique," *Computer Networks*, Vol.3, 1979.
18. Y. Tamir, H. Chi, "Symmetric Crossbar Arbiters for VLSI Communication Switches", *IEEE Transactions on Parallel and Distributed Systems*, vol 4, No. 1, 1993.

An Efficient Resource Sharing Scheme for Dependable Real-Time Communication in Multihop Networks

Ranjith G[1] and C. Siva Ram Murthy[2*]

[1] Veritas Software India Limited
Pune 411020, India
[2] Department of Computer Science and Engineering, Indian Institute of Technology
Madras 600036, India

Abstract. Many real-time applications require communication services with guaranteed timeliness and fault-tolerance capabilities. Hence, we have the concept of a dependable connection (D-connection), which is a fault-tolerant real-time connection. A popular method used for a D-connection establishment is the primary-backup channel approach, wherein a D-connection is setup by the establishment of a primary and a backup path. But, we note that such a 100% dependable network is very inefficient.
We introduce an efficient scheme called primary-backup multiplexing where D-connections may share their backup paths with other connections. We also describe a method where a D-connection can be given an intermediate fault-tolerance level between 0% dependability (no backup channel) and 100% dependability (dedicated backup). We present an efficient, connection-oriented and fault-tolerant D-connection establishment and resource-sharing scheme for real-time communications. We have noted in the simulation studies, that at light load, our scheme can effect a considerable improvement in the network resource utilization, with a relatively low drop in its fault-tolerance capabilities.

Keywords: Dependable connection, primary channel, backup channel, primary-backup multiplexing, RSVP.

1 Introduction

The advent of high speed networking has introduced opportunities for new computer network applications such as real-time distributed computation, digital continuous media (audio and motion video), and scientific visualization. For such distributed real-time applications, it is of primary importance to guarantee a minimum "quality of service" (QoS) on timeliness, error recovery *etc.*, contracted with the user(s) before setting up the real-time connection. For example, real-time applications have strict requirements on the timeliness of the

* Author for correspondence. This work was supported by the Department of Science and Technology, New Delhi, India.

S. Sahni et al. (Eds.) HiPC 2002, LNCS 2552, pp. 232–241, 2002.

data communicated. Hence, real-time communication schemes which support the above type of applications tend to incorporate the following three properties: QoS-contracted, connection oriented, and reservation based. In this context, a popular scheme for real-time communication in computer networks can be described as follows: When an application has real-time data to transmit, a contract is established between the application and the network. This contract basically limits the input traffic-generation behavior of the application, and further specifies the minimum QoS expected of the network. The network then computes the amount of resources required along any source-destination path so as to satisfy these QoS requirements, and further tries to find such a path. If such a path is available, the necessary resources are reserved along this path. On the other hand, if no such path is available, the network informs the application that its QoS requirements could not be satisfied at the specified input traffic-generation pattern. In the case when the path is available, the network starts accepting data packets from the application, which will be transported only along this selected path. Here, a type of virtual circuit has been established between the source and the destination and this circuit is called a *real-time channel*, a concept elucidated in [1]. Indeed, the concept of this reservation-based real-time channel has gathered approval as opposed to traditional pure packet-switched routing (for example, IP routing).

But one drawback of using such a real-time channel based scheme is that of failure of the channel when any of the components (either links or nodes) along such a real-time channel fails. Any communication network is prone to faults due to hardware failure (router and switch crashes, physical cable cuts *etc.*) or software bugs.

End-to-end detouring [2, 3] is a popular approach that addresses this problem, where a *backup channel* is established between the source and destination of each real-time channel (which is henceforth referred to as the *primary channel*) and is node-disjoint with the primary channel. This channel carries the traffic, in case of a failure in the primary channel. Each backup channel reserves the required amount of resources along the whole path from the source to the destination.

Backup Multiplexing: Assuming the *single link* failure model (channel failure recovery time is much smaller than the network's *mean time to failure* (MTTF)), if (a) primary channels of two connections share no common components and (b) their backup channels with bandwidths $b1$ and $b2$ pass through any link l, it is sufficient to reserve $max(b1, b2)$ for both the backup channels on the link l. This is because of the fact that the backup channels will never be activated simultaneously, under the single-link failure model [2, 3].

Segmented Backups: In [4], the authors propose *segmented backups*, where they find backups for *overlapping segments* of the primary path. We illustrate the terms involved through Figure 1. Links of the primary path are numbered 1 through 9 while those of the backup path are named A through K. Intermediate nodes on the primary path are denoted by $N1$ to $N8$. The primary channel has 3 *primary segments*, each of which has a *backup segment* spanning it. The first primary segment spans links 1 to 3 and its backup segment spans A to C. The

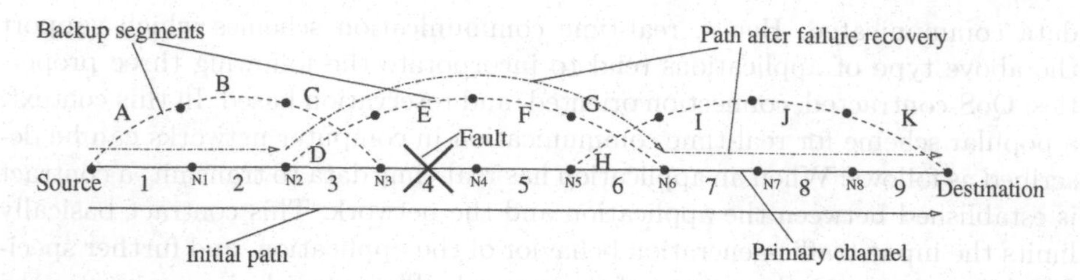

Fig. 1. Illustration of segmented backups

second segment spans links 3 to 6 and its backup segment spans D to G. The third segment spans links 6 to 9 while its backup segment spans links H to K. An extended and improved version of the same concept is given in [5]. In [6] we have outlined the algorithms by which one can establish these segmented backups in a distributed fashion.

The rest of this paper is organized as follows. In Section 2, we explain *primary-backup multiplexing*, which is the basic idea behind our scheme. Details of how different D-connections can achieve various levels of fault-tolerance are given in Section 3. Advantages gained in failure recovery and providing QoS guarantees like end-to-end communication delay are given in Section 4. We briefly discuss the implementation of our scheme in current computer networks in Section 5. In Section 6, we present and discuss the results of the simulation studies conducted on the scheme. Some concluding remarks are made in Section 7.

2 Primary-Backup Multiplexing

In this section, we explain our proposed scheme which reduces the fault-tolerance overhead. Our basic idea is that we allow a *primary* path and one or more *backup* paths to share resources on a link that they both pass through. Then, we say that the primary channel and the backup channel(s) involved is(are) multiplexed in the primary-backup-multiplex mode. A D-connection is termed to be *insecure* when *any link* on its backup path is involved in primary-backup multiplexing. We note that, throughout this work, the word "secure/security" does **not** have any bearing on "network security" issues, but it reflects the fault-tolerance capabilities of connections.

Example: We illustrate the concepts involved below. Consider Figure 2. It shows a multihop network which has a duplex bandwidth of one unit on all the links. At time t=0, there is a dependable connection (D-connection) request between $S1$ and $D1$, and also between $S2$ and $D2$ in the network. The primary path for the request ($S1$,$D1$) is identified to be $p1$, along the nodes 1, 2, 3, 4, 5, and 6. The backup path ($b1$) traverses the nodes 1, 8, 9, 10, 11, and 6. For the connection between $S2$ and $D2$, the primary path is $p2$, along the nodes 13, 14, 15, 16, 17, and 18, and the backup path ($b2$) is along the nodes 13, 8, 9, 10, 11, and 18. Now, we note that the backup paths $b1$ and $b2$ can be multiplexed in the backup-multiplex mode, as $p1$ and $p2$ do not share any common components

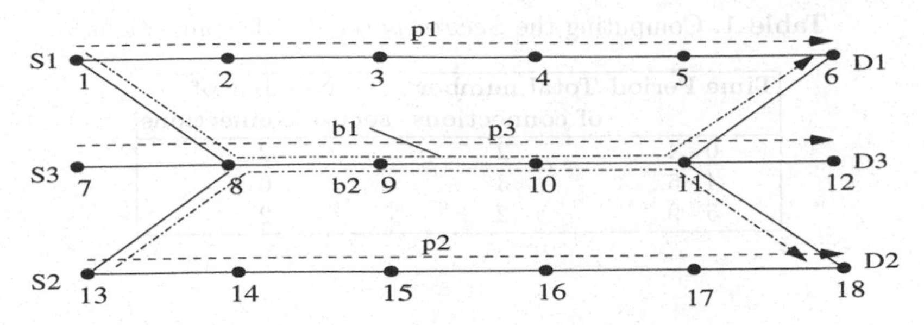

Fig. 2. Example of primary-backup multiplexing

(links). As the connections 1 and 2 are (single-failure) fault-tolerant at this instant of time, they are labeled as *secure* connections. Then, at time t=4, there is a non-dependable real-time connection (needs only a primary path) request from $S3$ to $D3$. In the usual case, this would have been rejected as there is no bandwidth available between $S3$ and $D3$. But, we note that there will be no real-time traffic at all in the links from $S3$ to $D3$ along the nodes 7, 8, 9, 10, 11, and 12, as the resources on these links are reserved in the backup mode. So, in our scheme, we allow the primary $p3$ to be established from $S3$ to $D3$, with the understanding that $p3$ is *primary-backup-multiplexed* with the backup paths $b1$ and $b2$, on the links (8, 9), (9, 10), (10, 11). Thus, connection 3 is established. We also note that both connections 1 and 2 are insecure, *on links* (8, 9), (9, 10), (10, 11) for the time being. Hence, they are labeled as *insecure* at this point of time. Now at time t=5, connection 3 terminates. Then, the resulting release and reallocation of bandwidth causes both connections 1 and 2 become secure on all the links, as they are re-allotted the bandwidth on the links along nodes 8, 9, 10, 11. Then at time t=9, both the connections 1 and 2 terminate.

To characterize the amount of fault-tolerance in a network, we define *Secure_percent* of D-connections from time $t1$ to $t2$ in the following way: $Secure\_percent = \frac{\sum_{t=t1}^{t2} number-of-secure-connections}{\sum_{t=t1}^{t2} total-number-of-connections}$. This effectively represents not only the fraction of connections that are secure, but also the amount of time that they are so, with a higher *Secure_percent* representing a scheme in which *more* number of connections are secure for a *larger* amount of time (the inclusion of the time factor helps to evaluate our dynamic scheme effectively). In fact, *Secure_percent* defined in this way is the probability that a failure-affected D-connection will be recoverable (*i.e.*, it will find a *secure backup path* readily available). So, we evaluate the *Secure_percent* of D-connections in the example as given in Table 1.

Hence, the *Secure_percent* works out to be: $\frac{2*4+0*1+2*4}{2*4+3*1+2*4} = \frac{16}{19} \approx 85\%$. We further note that the ACAR (average call acceptance ratio, *i.e.*, the fraction of the number of calls accepted out of the total number of calls requested) for our scheme is 100%.

Table 1. Computing the *Secure_percent* of D-connections

Time Period	Total number of connections	Number of secure connections
0 - 4	2	2
4 - 5	3	0
5 - 9	2	2

Now consider the case if we had used any 100% secure scheme. *Secure_percent* would have been 100%. But, ACAR would have been only $2/3 \approx 66\%$ as connection 3 would have been rejected. We note that there is a 33% increase in ACAR for a 15% drop in the *Secure_percent.* This is for a situation where 66% of the connections had asked for a backup path. We have observed in our simulation studies that our scheme gives better results when a higher ratio of connection requests are for D-connections, when the demand for network resources are much more dynamic, as we establish two paths for every connection established. We further note that in this example, two backups were *primary-backup-multiplexed* with a primary connection, and that it is possible that *any* number of backups can be multiplexed in this way with a single primary connection.

3 Dependability Classes

In this section, we briefly explain the method by which different connections can get different levels of dependability (fault-tolerance level) as per their demand (which will be dictated by their criticality).

Consider a non-critical application. In our scheme, we establish a backup path even for such an application. But, this backup path can afford to be involved in primary-backup multiplexing to a greater extent, as it is not critical. In this way, more real-time connections can be admitted into the network. Now considering a very critical application, it would be expedient that resources along this backup path is *not* shared with any primary connection.

Security Level Based Classes: In our scheme, we define a number of classes that a D-connection can belong to. Let us assume that they are numbered from 0 to $n-1$. Any D-connection request will have to specify its class as one of these n classes. We shall further define a "*security level*" parameter associated with each connection, and this is i for a connection of class i. An application will have to specify the required security level (henceforth referring to the parameter) of the connection, at the time of requesting a D-connection, which decides the class that the D-connection will belong to. We further define that the class with the highest security level is the class of connections that will be most dependable class. We state that the connections of this class are *guaranteed* to find their backup paths readily available, on occurrence of a single-component failure.

Now, we explain how we provide various levels of security (fault-tolerance) to connections of different classes. This is linked with the idea of primary-backup multiplexing. The basic idea is that backups of D-connections of a (numerically) lower class will be more "willing" to participate in primary-backup multiplexing than backups belonging to D-connections of higher classes. This translates into the fact that a request for a primary-backup multiplex will be honored (at a given link) with more probability if it involves sharing resources with (the backup of) a "lower class" (less critical) D-connection. Moreover, we guarantee that backups of D-connections having the highest security level will not allow any primary to multiplex with them. This makes them 100% secure, at any point of time.

4 Failure Recovery and End-to-End Delay

In this section, we shall examine some advantages that we get in the post-failure scenario, over the current end-to-end backup schemes.

When a fault is detected in the network, the connections passing through the failed component (either link or node) have to be rerouted through their respective backup paths. The process is called *failure recovery*, and is required only when a component in the primary path of a connection fails. Failure recovery is done in three stages: fault detection, failure reporting, and backup activation (or connection rerouting). If the failed component is in the primary segment of a connection that is secure, then the backup segment is readily available. If, on the other hand, if the connection involved is insecure, it means that some parts of its backup path (segment) is being used by some other primary paths, and the whole connection may have to be rerouted.

When a fault is detected, the nodes involved send failure messages towards the source and the destination. The end nodes on the primary segment involved receive these messages and take action (note that faults are managed more locally by the segmented backups scheme). If all the links on the segment are secure (as found out by separate messaging), we send the *BACKUP_ACTIVATE* message. Otherwise, we send the *CONNECTION_FAILURE* message. These can be seen in Figure 3.

The total end-to-end delay along the path of a real-time connection is another important QoS parameter for real-time communication. Now, we shall define the maximum increase in the end-to-end delay that a connection will encounter, as

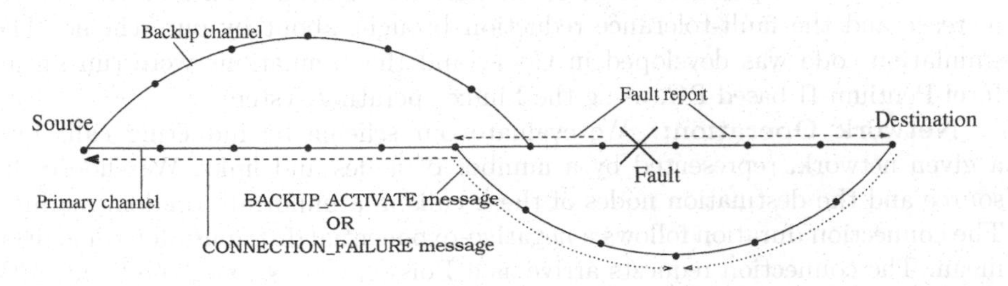

Fig. 3. Propagation of the BACKUP_ACTIVATE or the CONNECTION_FAILURE message

the *failure delay increment* of that connection. The increment will be lesser in our case, as backup (and primary) segments tend to be smaller than a single end-to-end backup path. This aspect will be quantified in our simulation studies, in Section 6.

5 Implementation of Our Scheme

In this section, we first describe the relevant scenarios in the current Internet and other places where our scheme can be used.

Applications of Our Scheme in Current Internet:

1. It can be run at the network level between the backbone routers by Internet backbone providers like UUNET [8] or between the routers in a single autonomous system (AS) by any Internet service provider. It is well known that Internet backbone providers attempt to design their physical networks to ensure that there are disjoint paths between any two routers.
2. It can be run at the application level among a set of routers forming a logical network (overlay) over existing physical inter-network such as in resilient overlay networks (RON) [9].

Applications of Our Scheme in Other Multihop Networks: It is also important to note that the scheme can be implemented in general on any multihop network, in addition to the Internet. This can also include VPN (virtual private networks) based networks that are logically disassociated from the Internet. We also note that as computer communications become more mission critical, QoS-demanding, security conscious, and demand more real-time data support, there is a growing interest in "private networks", *i.e.*, multihop networks that are dedicated to a select set of users. These networks should be free to deploy any relevant protocols. A reservation-based protocol such as RSVP [7], on top of which our schemes can be implemented, is ideal in such a case.

6 Performance Evaluation

In this section, we present results of extensive simulation studies on the ACAR increase and the fault-tolerance reduction brought about by our scheme. The simulation code was developed in C++, and the simulations were run on an Intel Pentium-II based PC, using the Linux operating system.

Network Operation: We evaluate our scheme by inducting calls into a given network, represented by a number of nodes and links. We choose the source and the destination nodes of these calls in a uniformly random manner. The connection duration follows a negative-exponential distribution with a given mean. The connection requests arrive as a Poisson process, and the bandwidth required by these connections also follow a Poisson distribution, both with the mean fixed beforehand. The ratio of (average requested) connection bandwidth to total link bandwidth is fixed at 3%, which can be safely assumed in case of real-time traffic (like multimedia streaming). The connection is setup by routing the primary over the shortest path available, and the backup as described in [6]. The connection duration is an average of 500 units. The connections request for various security levels in such a way that it is a Poisson distribution with the mean as specified. This mean is referred to as the *average security level* (ASL) of the simulation run. We vary the "load" on the network by increasing the inter-arrival rate of the connections (or the number of connections that are inducted during the (fixed) duration of the simulation).

In our simulations, we consider 6 discrete security levels, and correspondingly, there are 6 classes, named class-0, class-1, ...,class-5. The *Secure_percent* that the network is able to provide to these different classes of connections for the considered time duration is also studied, by considering only connections of a certain class and logging their *Secure_percent* parameter. Note that class-5 connections are the most critical. Readings pertaining to ACAR, *Secure_percent* and *average failure delay increment* (Failure Delay Increment, averaged over all connections, see Section 4) are made after the network has been "warmed up" and stabilized. We have simulated the following schemes: 1) the 0% secure scheme, where *no* backup is established for a D-connection, 2) the 100% secure end-to-end backup scheme, 3) the 100% secure segmented backups scheme, and 4) primary-backup multiplexing (with the ASL varied).

Figure 4 shows the ACARs of the various schemes mentioned above, with the ASL of the primary-backup multiplex scheme set as 1. We see when the ACAR of the 100% secure schemes are around 65%, our scheme has augmented the ACAR to nearly 90%. Now, consider Figure 5, where ASL = 3. This means that more D-connections consider themselves critical. We still improve ACAR by 8-10%, on an average. In Figure 6, we study the effect of primary-backup multiplexing on the fault-tolerance levels offered by the network, on an average. In this case, we compute the *Secure_percent* parameter taking into consideration all the D-connections, irrespective of their classes.

In Figures 7 and 8, we plot the *Secure_percent* parameter, for each class (security level based) of D-connection that we have used. Note that different classes have varying levels of willingness to share resources and hence, their fault-tolerance levels will vary. Note from the figures that we are able to give 100% security to class-5 connections. The *Secure_percent* drops on an average by 10-20% from class to class.

In Figure 9, we plot the average failure delay increment, as described before. As expected, this parameter is reduced for both the segmented backups scheme and our scheme, which uses segmented backups. This proves to be an added advantage of our scheme.

7 Conclusions

We have presented in this work the concept of *primary-backup multiplexing*, an efficient resource reservation scheme which augments and effectively harnesses

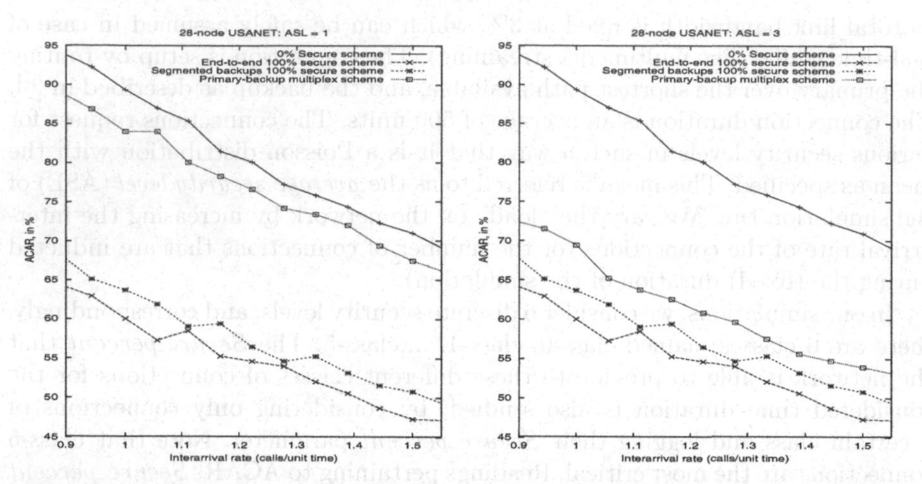

Fig. 4. ACAR for various schemes for ASL = 1

Fig. 5. ACAR for various schemes for ASL = 3

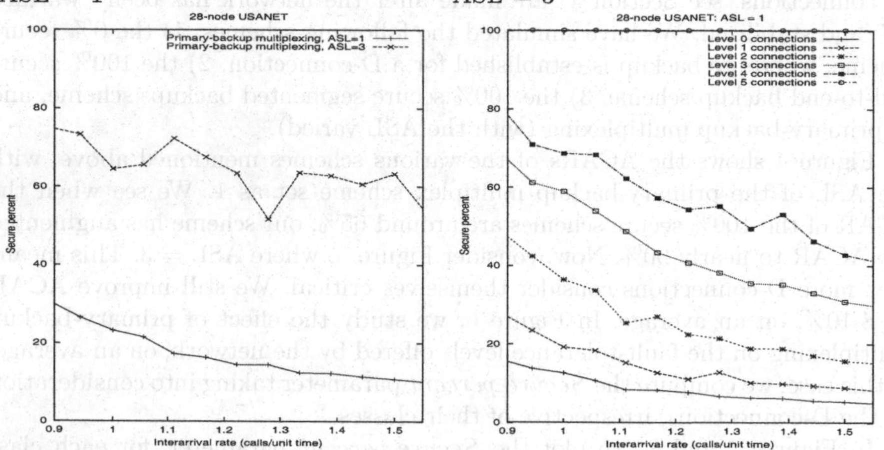

Fig. 6. Secure_percent for ASL = 1,3

Fig. 7. Secure_percent (per connection) for ASL = 1

the power of multiplexing of resources, for establishment of D-connections. Here, applications can balance the trade-off between the fault-tolerance capabilities and admission price, by themselves. We have also shown how D-connections can be grouped into various classes. Our simulation studies bring out the effectiveness of the scheme, compared to previous ones. We further note that our scheme improves the failure recovery characteristics of D-connections.

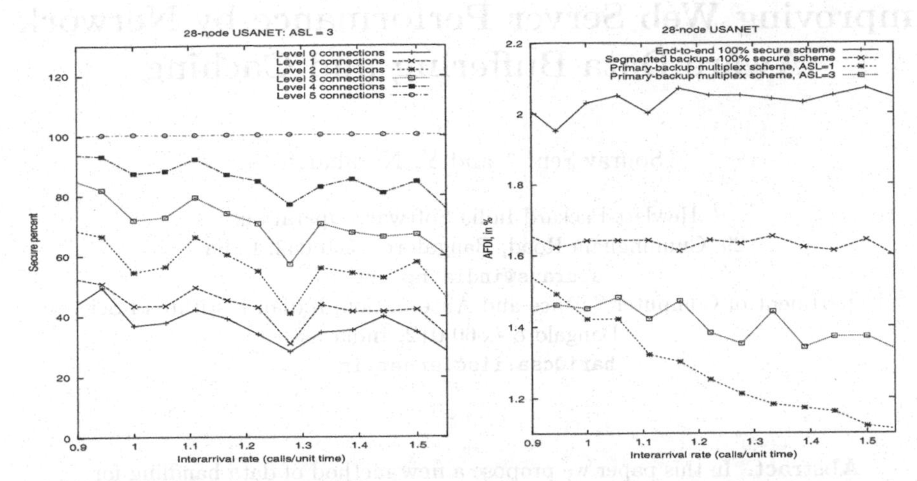

Fig. 8. Secure_percent (per connection) for ASL = 3

Fig. 9. AFDI for various schemes for ASL = 1,3

References

[1] D. Ferrari and D. C. Verma, "A Scheme for Real-Time Channel Establishment in Wide-Area Networks", *IEEE Journal on Selected Areas in Communications*, vol. 8, no. 3, pp. 368-379, April 1990. 233

[2] S. Han and K. G. Shin, "Efficient Spare Resource Allocation for Fast Restoration of Real-Time Channels from Network Component Failures", in *Proc. of IEEE RTSS'97*, pp. 99-108, 1997. 233

[3] S. Han and K. G. Shin, "A Primary-Backup Channel Approach to Dependable Real-Time Communication in Multi-Hop Networks", *IEEE Trans. on Computers*, vol. 47, no. 1, pp. 46-61, January 1998. 233

[4] G. P. Krishna, M. J. Pradeep, and C. Siva Ram Murthy, "A Segmented Backup Scheme for Dependable Real-Time Communication in Multihop Networks", in *Proc. of IEEE WPDRTS 2000*, pp. 678-684, May 2000. 233

[5] G. P. Krishna, M. J. Pradeep, and C. Siva Ram Murthy, "An Efficient Primary-Segmented Backup Scheme for Dependable Real-Time Communication in Multihop Networks", to appear in IEEE/ACM Trans. on Networking. 234

[6] G. Ranjith, G. P. Krishna, and C. Siva Ram Murthy, "A Distributed Primary-Segmented Backup Scheme for Dependable Real-Time Communication in Multihop Networks", in *Proc. of IEEE FTPDS 2002*, Florida, USA, April 15-19, 2002. 234, 239

[7] L. Zhang, S. Deering, D. Estrin, S. Shenker, and D. Zappala, "RSVP: A New Resource Reservation Protocol", *IEEE Network*, vol. 7, pp. 8-18, September 1993. 238

[8] UUNET, "UUNET Technologies," `http://www.uunet.com/network/maps`, October 2001. 238

[9] D. G. Anderson, H. Bala Krishnan, M. F. Kashooek, and R. Morris, "Resilient Overlay Networks," in *Proc. of ACM SOSP*, pp. 131-145, October 2001. 238

Improving Web Server Performance by Network Aware Data Buffering and Caching

Sourav Sen[1] * and Y. Narahari[2]

[1] Hewlett Packard India Software Operations
29, Cunningham Road, Bangalore - 560 052, India
souravs@india.hp.com

[2] Department of Computer Science and Automation, Indian Institute of Science
Bangalore - 560 012, India
hari@csa.iisc.ernet.in

Abstract. In this paper we propose a new method of data handling for web servers. We call this method *Network Aware Buffering and Caching* (NABC for short). NABC facilitates reduction of data copies in web server's data sending path, by doing three things: (1) Layout the data in main memory in a way that protocol processing can be done without data copies (2) Keep a unified cache of data in kernel and ensure safe access to it by various processes and kernel and (3) Pass only the necessary meta data between processes so that bulk data handling time spent during IPC can be reduced. We realize NABC by implementing a set of system calls and an user library. The end product of the implementation is a set of APIs specifically designed for use by the web servers. We port an in house web server called SWEET, to NABC APIs and evaluate performance using a range of workloads both simulated and real. The results show a very impressive gain of 12% to 21% in throughput for static file serving and 1.6 to 4 times gain in throughput for lightweight dynamic content serving for a server using NABC APIs over the one using UNIX APIs.

1 Introduction

Increasing use of the Internet in various forms of business, communication and entertainment has placed an enormous performance demand on its key architectural elements: routers and switches at the middle of the net and web servers and proxy servers at the edge of the net. To get an idea of the magnitude of request burst a web site might be subjected to: it has been reported that the web site of 1998 winter Olympic games in Nagano, Japan has experienced a staggering 110,414 hits in a minute [11]. Similar numbers can be found for other popular web sites as well.

Web servers and proxy servers are extremely I/O intensive applications and it has been reported that a web server spends 70% to 75% of the whole processing

* The author carried out this work while being a full time Master's student of the Department of Computer Science and Automation, IISc.

S. Sahni et al. (Eds.) HiPC 2002, LNCS 2552, pp. 242–251, 2002.

time doing network and file I/O [10]. Therefore, optimizing the I/O processing for a web server is a possible means of scaling its performance.

It has been known that data copy is a bottleneck for high performance I/O [4] [2] [8]. Data copy results because of kernel's intervention in every I/O processing steps. In write/read calls, for example, data gets copied to/from the kernel from/to user applications. Apart from this, the mismatch which exists in the way various kernel subsystems handle data, results in copy while data is transferred from one subsystem to the other. For example, file system keeps data in page cache or in buffer cache while networking subsystem keeps data in fragmented form and needs header information for protocol processing. Therefore, a copy results while data is transferred from file system to the networking subsystem and vice versa.

In this paper we address this problem of data copy solely from the web server's perspective and propose a possible solution. Specifically we ask the question: How far, in a web server's data sending path, can we reduce data copies for both static and dynamic content processing? We address the problem without assuming presence of any special networking hardware.

In achieving the objective we propose a method of handling data which we call *Network Aware Buffering and Caching*, NABC. NABC facilitates layout of data in main memory, at the very first time it is generated, in such a way that protocol processing can be done without data copies (hence the rationale of the phrase *Network Aware Buffering*). NABC also facilitates in-kernel caching of outbound data and safe sharing of that data between processes and kernel. To achieve this, we implement a set of system calls in Linux-2.2.14 kernel and implement a set of C library interfaces for use by the web servers.

1.1 Organization

The rest of the paper is organized as follows. Section 2 discusses NABC in detail and discusses the design rationale behind the APIs. Section 3 outlines kernel and library implementation details. Section 4 presents the performance evaluation of NABC for a range of workloads. In section 5 we discuss related research efforts. Section 6 draws the conclusion.

2 Network Aware Buffering and Caching

2.1 Mechanisms

The main idea in NABC are the following:

- Layout the data in main memory at the very first time it is generated in such a way that protocol processing can be done without touching the data.
- Maintain a unified cache of outbound data in kernel and ensure that related processes and kernel can access it safely.

– When data is generated by one application and is sent out by another, then do not pass the whole data from one application to other by IPC, but pass only the portion needed to be examined and possibly modified (usually the header) and pass a reference to the bulk data body. Device a mechanism by which other application may send it by having a reference to the data body.

The first mechanism tries to reduce the need for data copy in protocol processing and requires that data be kept in fragmented form in memory, the second mechanism reduces the need for duplicate data buffering, and the overhead of data transfer from user application to kernel, and the third mechanism uses the first two and is a domain specific optimization for web servers so that bulk data transfer between processes is reduced.

2.2 The NC_handle ADT

In NABC `NC_handle` ADT effectively hides from the application programmer the intricacies of fragmented memory allocation and data placement. Figure 1 shows essential pointers in NC_handle ADT.

In NABC, a set of related data fragments are called *logical unit* of data. Data transfer from application to kernel takes place by remapping pages from process address space to kernel, this ensures data integrity in the cache. The processes trying to use a particular logical unit of data in the cache, specify that using an unique integer called *cache id*, associated with that unit. An access control mechanism limits access to these data units. NABC facilitates checksum caching on a per fragment basis. Precomputed checksum speeds up protocol processing when data is transferred from the cache.

2.3 NABC - APIs

The APIs can be logically divided into three categories: APIs for memory allocation and data buffering, APIs for cache management, and APIs for network

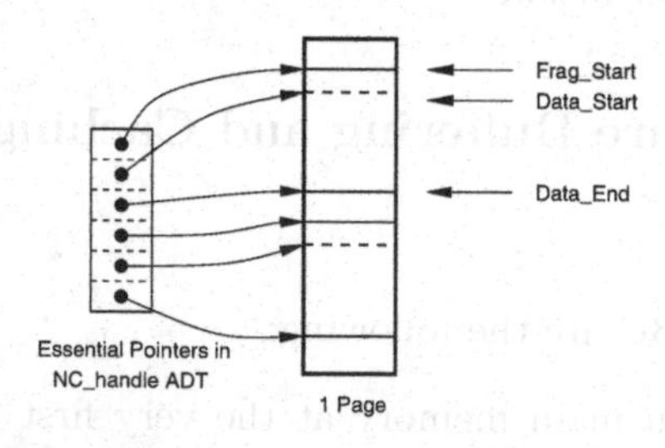

Fig. 1. Essential elements of `NC_handle` ADT. This simplified figure shows a page containing two fragments. There may be multiple pages in one `NC_handle` ADT and many more fragments. The pointers are generated by user library and the pages are mapped by a system call

```
int NC_alloc(NC_handle * handle, int size,
                                    int fraglen);
int NC_read(int fd, NC_handle * handle,
        size_t handle_offset, size_t len);
int NC_insert(NC_handle * handle);
int NC_socksend(int sock_fd,
            size_t cache_id, size_t offset, size_t len);
```

Fig. 2. NC_alloc allocates memory in fragmented form, NC_read reads data to the to the fragments, NC_insert inserts data unit to the cache and NC_socksend sends data from the cache over network

transport initiation. Since the mechanisms proposed are targeted towards a particular kind of applications, the APIs are fairly small in number but rich in semantics.

APIs for memory allocation and data buffering let an application allocate memory in fragments and buffer data onto the fragments. Data may originate from a variety of sources.

The APIs for cache management give applications facilities for creating and destroying data units, collecting the statistics about the cache and controlling access to various data units. A process creating a data unit can give access to it to another process. This facilitates CGI processing where data generated by child process needs to be send out by parent.

Two APIs for sending data over socket are available in NABC. Due to space limitations all the APIs could not be shown. Full details of the APIs can be found in [17].

2.4 NABC and a Web Server

We envision that NABC will be useful for web servers which cache data in main memory for static content serving. In addition to caching static data, efforts are underway where dynamic content is also cached [5]. With dynamic content processing, we envision that the NABC scheme promises performance gain by substantially reducing copy of the data in its sending path.

The most straightforward situation is with static contents. We propose that header and data should be kept separately. That means all the information necessary for HTTP header generation is kept separately in the process address space and the bulk data is kept in the NABC cache and *cache id* is kept alongside the header information. So, when responding to a request, the header information is written separately using conventional `write` call and the data transfer is initiated using `NC_socksend` call.

CGI style dynamic content generation presents some problems for incorporating the scheme, so requires slight change in the way the CGI scripts are to be written. CGI/1.1 specifies meta variables and CGI extension headers. We use meta variables for passing the pid of the main server process to the child and use the extension headers for passing the *cache id* and the data length from the child to the parent. Using these information, the parent can send the dynamic content to the client.

3 NABC – Implementation

NABC implementation was done on Linux-2.2.14 kernel. Apart from changes in kernel we have also implemented a user level C library.

3.1 Implementation Details

Kernel Implementation Details: Kernel implementation consists of implementation of a set of system calls. One set of calls map pinned down pages to process address space and unmap them. One set of calls do cache management and a third set of calls do network transport initiation. The NABC cache interact with the file cache when data is placed directly from file cache to the NABC buffers by copying. For all the data units that are placed in the cache, checksum of data fragments are calculated and cached, this speeds up protocol processing. The NABC cache interact with the networking code when the data is transferred over socket from NABC cache. In the kernel code, all the routines from `tcp_send_skb` and below in the protocol stack remain the same. One asynchronous thread does the job of a garbage collector.

Library Implementation Details: Library implementation is mainly responsible for making fragments out of mapped pages and creating the meta data about the fragments and keeping them in the `NC_handle` structure. When the applications make calls that inserts data into the NABC cache, this meta data is copied to a similar data structure in kernel.

3.2 NABC - Operation

When the modified Linux kernel is booted up, the data structure of NABC is allocated and the garbage collector thread is started as kernel daemon. The applications using the NABC, need to link a static library with their code.

4 Performance

Our experimental setup consisted of one PC acting as server machine with 500 MHz Pentium - III Processor having 320MB of RAM and four 100 Mbps network adaptors running our modified Linux2.2.14 kernel and four PCs acting as client machines each having Pentium - III processor, 64 MB of RAM and one network adaptor running Linux 2.2.18 kernel. All the machines were connected in a switched fast Ethernet.

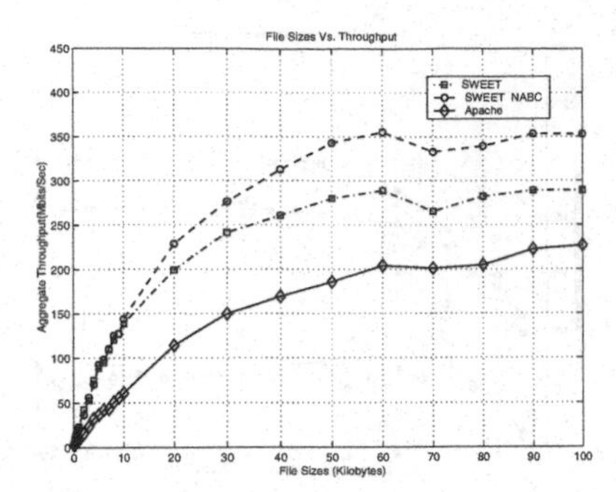

Fig. 3. Aggregate throughput Vs. file sizes when file sizes were varied from 100 Bytes to 100KB. The file sizes are varied in steps of 100 bytes below 1K, 1K between 1K and 10K, 10K between 10K and 100K. Aggregate throughput is the sum of the throughputs observed in individual machines

We have used an in-house web server called SWEET (*Scalable Web SErver using Events and Threads*) as our vehicle for performance evaluation. Very little change was necessary for porting SWEET to SWEET using NABC APIs (SWEET-NABC).

For dynamic data, we have used a scheme similar to fast CGI. But we did not implement fast CGI specification. In our method, a set of pre-forked processes does the dynamic data generation. On receiving the simple request for data, the server chooses a free process to generate the data. In our experiment on dynamic processing, the applications generate data from thin air and no real dynamic processing is involved. Since we wanted to expose the bottleneck the data copy imposes, this suffices our purpose.

As client simulation software, we have used an event driven program similar to *flashtest* available from Rice University [19].

4.1 Fixed Sized File Tests

In fixed sized file tests, a fixed number of files, *all* of same size, were fetched repeatedly in every experimental session. The number of files of a given size was taken as 100, 10 clients per machine was used and 10000 fetches per file size per machine was done. Figure 3 shows the plot of aggregate throughput vs. file sizes for file sizes between 100 Bytes and 100KB. Due to file caching, a file used to get served from the web server's cache except for the first time.

4.2 Modified SPECWeb99 Workload

We have used a modified version of SPECWeb99 guidelines for creating the static workload for evaluating the performance [1]. By doing this, we believe the workload we get is a faithful representative of real life workload.

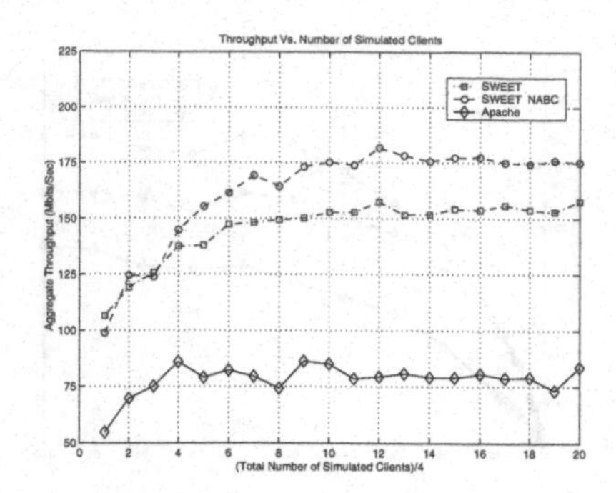

Fig. 4. Throughput Vs. Number of Clients for Modified SPECWeb99 Static Workload

Since we did not implement any cache replacement algorithm, total size of the data set that the server can serve at any experimental session is limited by the cache size that can be accommodated at once in physical memory, so we have kept the number of directories 25 across all experimental run. We have adhered to all the other guidelines of generating load and creating the files. 25 directories occupy around 125 MB of space.

Figure 4 shows change in throughput as the number of clients was varied. The total number of clients were varied from 4 to 80 in steps of 4. 1000 fetches (Different unrelated sequences per machine) per machine were used per experimental run. As can be seen SWEET-NABC almost always outperforms both Apache and SWEET. When the total number of clients are less, then both SWEET and SWEET-NABC operate at under capacity hence the I/O subsystems of OS is not stressed. As the number of simultaneous clients keep increasing, the server is compelled to serve more data. Therefore, data copying and checksum calculation becomes critical bottlenecks and SWEET-NABC with its in-kernel network aware data cache with pre-computed checksum starts to outperform SWEET.

4.3 Experiment With CSA WWW2 Trace

The CSA WWW2 server is used to serve the home page of about 160 students of Computer Science and Automation (CSA) department of IISc. In this experiment we took 1 weeks log of the server and performed the experiments on the sub-traces of that trace based on the prefix cache size. The methodology is as follows: for a prefix cache size of X MB, we fed all the hits which fall within file set which contributed to that prefix cache size. The request was generated by a total of 64 simulated clients for 20 minutes from 4 client machines. we have tried to preserve the original request sequence.

Figure 5 shows the plot of the throughput vs. the prefix data set sizes. At 10MB of prefix size, SWEET-NABC outperforms SWEET by a margin of about 20%. At 50MB the gain is around 17% and at 100 MB it is around 9%. Two reasons contribute towards reduced performance advantage of SWEET-NABC

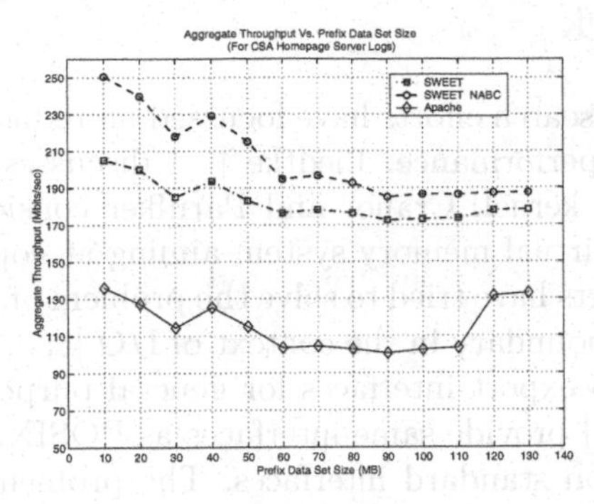

Fig. 5. Throughput Vs. Prefix Data Set Size for CSA WWW2 Server Logs

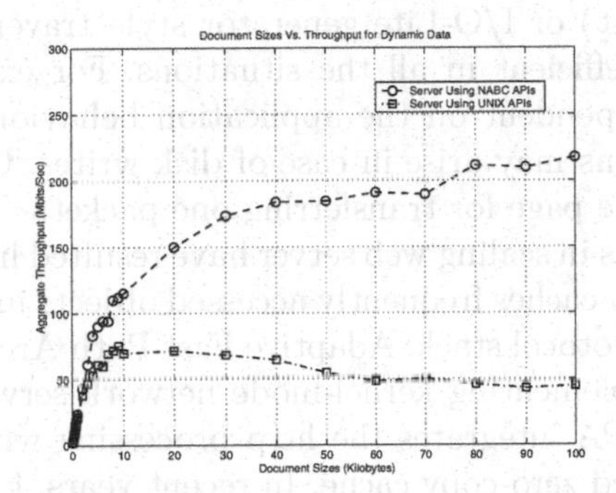

Fig. 6. Throughput Vs. Document Size for Dynamically Generated Documents

over SWEET with increased prefix data set size. Firstly, sweet does not use a very optimized cache lookup algorithm. With increase in prefix data set size, the number of files increase, and cache lookup time increases. Secondly, the file mix, and request mix follow a pattern in which the percentage of small files and request for them both are high [17]. With large number of small files, the per file overhead (connection establishment, teardown etc.) tends to overshadow the performance gain due to copy reduction (This is also evident in Figure 3 where appreciable performance gain is not there for small files).

4.4 Lightweight Dynamic Content Generation

The mechanism of dynamic content generation has been outlined before. We have varied the size of the documents being generated. 10 clients on each of 3 client machines were used to generate load.

As can be seen in Figure 6 that the performance gains in dynamic content processing is even more pronounced. Compared to POSIX APIs, in NABC scheme the number of copies reduced is more, hence we see a throughput gain of almost 4 times for 100KB file and 1.6 times when the document size is 10KB.

5 Related Work

A body of previous research efforts have focussed on reducing data copy for I/O, and enhancing IPC performance. Liedtke [14] discusses improvement of IPC performance in L3 μ kernel. Cranor and Parulker consider design and implementation of a new virtual memory system aiming at copy reduction [7]. Some previous research efforts have tried to solve the problem of copy free data transfer between user/kernel boundary in the context of I/O [4, 2, 3, 6, 18, 16]. All these systems have tried to export interfaces for general purpose applications. Some of the systems e.g., [6] provide same interfaces as POSIX, others like [2, 18, 16] require alternative non standard interfaces. The problems with these systems are that in many cases, the alternative interfaces are complicated to understand and use. For example, there is no parallel in POSIX of the concept of I/O-Lite context (`IOL_Context`) or I/O-Lite generator style traversal. Also, the mechanisms are not very efficient in all the situations. For example, in case of [6], the mechanism is dependent on the application behavior. In case of I/O-Lite, pathological conditions may arise in case of disk writes. Container Shipping [2] may end up using one page for transferring one packet.

More recent efforts in scaling web server have resulted in works like web server accelerator [13] which caches frequently accessed objects in a front node and uses a highly optimized protocol stack. Adaptive Fast Path Architecture (AFPA) [15] is a platform for implementing kernel-mode network servers on production operating systems. AFPA integrates the http processing with TCP/IP stack and uses a kernel managed zero-copy cache. In recent years, kHTTPd and TUX are two implementations of web servers in Linux kernel [12, 9]. Performance benefits notwithstanding, the problem with kernel mode web servers are the lack of fault isolation and lack of development support.

The path we are taking is a mixed one. Our aim is to facilitate the operation of user mode servers. We do this by keeping outbound data in kernel as much as possible so that copy is reduced in protocol processing and provide system call interfaces to the application processes so that the final control over the cached data lies with the application.

6 Conclusion

This paper discusses a new method of data buffering and caching for web servers called NABC. One immediate effort for evaluating NABC could be to evaluate how it performs against the other web server specific optimization efforts discussed in the previous section.

One point where NABC design lacks general appeal is its dependency on the nature of networking subsystem in Linux. But we believe that even if such constraints are removed the principle of NABC can be employed to keep a unified cache of outbound data. Although then file cache can be used as the the cache for static data, for dynamic data we may use the in kernel buffering as is done in NABC to reduce copy and processing time.

References

[1] SPECweb99 Release 1.02.
http://www.spec.org/osg/web99/docs/whitepaper.html. 247

[2] E. Anderson. *Container Shipping: a Uniform Interface for Fast, Efficient, High-Bandwidth I/O.* PhD thesis, Computer Science and Engineering Department, University of California, San Diego, CA, 1995. 243, 250

[3] J.C. Brustoloni. Interoperation of copy avoidance in network and file i/o. In *INFOCOM (2)*, pages 534–542, 1999. 250

[4] J.C. Brustoloni and P. Steenkiste. Effects of Buffering Semantics on I/O Performance. In *2nd Symp. on Operating Systems Design and Implementation (OSDI)*, pages 277 – 291, Seattle, WA, Oct 1996. 243, 250

[5] J. Challenger, A. Iyengar, and P. Dantzig. A scalable system for consistently caching dynamic web data. In *Proceedings of the 18th Annual Joint Conference of the IEEE Computer and Communications Societies*, New York, NY, 1999. 245

[6] H.K. Jerry Chu. Zero-copy TCP in solaris. In *USENIX Annual Technical Conference*, pages 253–264, 1996. 250

[7] C.D. Cranor and G.M. Parulkar. The uvm virtual memory system. In *USENIX Annual Technical Conference*, Monterey, CA, June 1999. 250

[8] P. Druschel. *Operating System Support for High-Speed Networking.* PhD thesis, Department of Computer Science, The University of Arizona, 1994. 243

[9] Answers from planet TUX: Ingo Molnar responds. http://slash-dot.org/articles/00/07/20/1440204.shtml. 250

[10] J.C. Hu, S. Mungee, and D.C. Schmidt. Techniques for Developing and Measuring High-Performance Web Servers over high speed ATM. In *INFOCOM*, San Francisco, CA, 1998. 243

[11] A. Iyengar, J. Challenger, D. Dias, and P. Dantzig. High-Performance Web Site Design Techniques. *IEEE Internet Computing*, 4(2):17 – 26, Mar.-Apr. 2000. 242

[12] kHTTPd Linux http Accelerator. http://www.fenrus.demon.nl. 250

[13] E. Levy-Abegnoli, A. Iyengar, J. Song, and D.M. Dias. Design and performance of a web server accelerator. In *INFOCOM (1)*, pages 135–143, 1999. 250

[14] Jochen Liedtke. Improving ipc by kernel design. In *14th Symposium on Operating Systems Principles (SOSP)*, Asheville, North Carolina, December 1993. 250

[15] R. King P. Joubert, R. Neves, M. Russinovich, and J. Tracey. High-Performance Memory-Based Web Servers: Kernel and User-Space Performance. In *USENIX Annual Technical Conference*, Boston, MA, June 2001. 250

[16] V.S. Pai, P. Druschel, and W. Zwaenepoel. I/O-Lite: A Unified I/O Buffering and Caching System. *ACM Transactions On Computer Systems*, 18(1):37 – 66, 2000. 250

[17] S. Sen. Scaling the performance of web servers using a greedy data buffering and caching strategy. Master's thesis, Department of Computer Science and Automation, Indian Institute of Science, Bangalore, India, January 2002. 245, 249

[18] M. Thadani and Y.A. Khalidi. An efficient zero-copy i/o framework for unix. Technical Report SMLI TR95 -39, Sun Microsystems Lab, Inc., May 1995. 250

[19] Measuring the Capacity of a Web Server. Rice University. Department of Computer Science. http://www.cs.rice.edu/CS/Systems/Web-measurement/. 247

WRAPS Scheduling and Its Efficient Implementation on Network Processors

Xiaotong Zhuang[1] and Jian Liu[2]

[1]Georgia Institute of Technology, College of Computing, 801 Atlantic Drive
Atlanta, GA, 30332-0280
xt2000@cc.gatech.edu
[2]Georgia Institute of Technology, School of ECE, Atlanta, GA, 30332-0280
liu@csc.gatech.edu

Abstract. Network devices in high-speed networks need to support a large number of concurrent streams with different quality of service (QoS) requirements. This paper introduces a new packet scheduling algorithm and its efficient implementation on the novel programmable network processor—Intel's IXP1200. WRAPS algorithm is based on the observation of packet rates within a dynamic window. The window can move continuously or discretely as the system transmits packets. Cumulated packet rates in the window are utilized to predict the next incoming packet of the same flow and reserve resource for the transmission of later packets. The implementation on network processor considers both accuracy and efficiency. To expedite the calculation and avoid the high cost of maintaining an ordered list, we designed a time-slotted circular queue to achieve O(1) insertion and selection time. Our experiments on the real system show good performance in terms of scalability and flow interference avoidance.

1 Introduction

Network technology is being pushed forward by the ever-increasing requirements of future applications. Critical QoS (Quality of Service) properties like packet loss rate, deadline, delay variance, etc. need to be maintained without compromising the processing speed of incoming data streams.

Many papers [2][3][5] have addressed the problem of how to schedule the priority of input packets in a smart way so that the packets from different flows got processed in a proper order and some packets can be dropped to save computational power or bandwidth for more urgent packets.

With the emergence of high-speed programmable network processors, people are looking forward to scheduling streams under severe performance requirements. Network processors are supposed to provide programmable interface for a general network device with relatively cheap prices. It also creates new opportunity to boost the processing speed of network intermediate equipments. Their low price and highly

S. Sahni et al. (Eds.) HiPC 2002, LNCS 2552, pp. 252-263, 2002.

scalable properties make it possible for small business or local service providers to afford and configure them on the fly. Predictably, network processors will become a driving force in network equipment market with their flexibility and high performance/cost ratio.

High processing speed is commonly achieved by exploiting parallelism to hide memory latency. In the Intel IXP1200 network processor, two giga-bit ports are supported by 6 MicroEngines running in parallel. Each MicroEngine contains 4 hardware contexts (threads). Memory operation and on-board computation can be performed at the same time by context switch. However, memory operations become quite time consuming. Roughly, if the network processor is serving two giga-bit ports, one packet should be sent out every 500 – 600 uEngine cycles (uEngine's running at 200MHz on our evaluation board), while each off-chip memory access will take up to 30 cycles to complete.

The price of off-chip memory has dropped tremendously in recent years, although the speed gap remains between memory and cpu instructions. An important consideration in this paper is the trade-off between memory space and memory access frequency. However, many packet scheduling algorithms [5] rely on an ordered queue or list. In these cases, the queue (list) should be updated quickly during the packet arrival and delivery. To keep the queue or list ordered, the number of these operations are always O(n), where n is the number of elements in the queue. It becomes impossible to maintain the ordered queue (list) on network processor without major modifications if we still want to catch up with the packet arrival rate.

In this paper, we present the *time-slotted circular queue* structure to reduce the time of packet ordering and priority adjustments. This structure is shown to be able to shorten the time of both insertion and selection operations of the packets.

This paper is organized as follows. Section 2 describes related works, section 3 presents the WRAPS algorithm , section 4 shows our implementation on the IXP network processor , and section 5 gives results and analysis.

2 Related Work

Recent research has put substantial effort into the design and implementation of efficient scheduling algorithms. Rate-Monotonic scheduler is an optimized scheduler. RM sorts tasks based on the shortest period (service time plus laxity). RM has a system utilization about 69% without missing all the deadlines. EDF (Earliest Deadline first) and MLF (Minimum Laxity First) are two variations of RM scheduler. In common, these scheduling algorithms are borrowed from general system scheduling, where periodic real-time tasks can be scheduled under certain system load.

Scheduling problem was first studied in *General Processor Sharing* [1] system, where a job can be characterized by starting time, finishing time, and deadline. Generalized Processor Sharing (GPS) algorithm forms a basis for most packet schedulers. In GPS, there are a number of periodic tasks. Each task shares a portion of system resources. Weighted Fair Queueing (WFQ)[2] is the packetized version of GPS. WFQ uses the earliest deadline first policy. Other algorithms, such as Virtual

Clock (VC), Weighted Round Robin (WRR), WF^2Q[3], etc, are the variations of WFQ.

R.West, K.Schwan propose *Dynamic Window-Constrained Scheduling* (DWCS)[5]. DWCS is able to share bandwidth among competing streams in strict proportion to their deadline and loss-tolerance. Noticeably, several traditional scheduling algorithms like EDF, static priority, Fair Queueing can be reduced to DWCS.

In this work, we propose a *Window-based Rate Adaptive Packet Scheduling (WRAPS)* algorithm, by which the scheduler can adapt to traffic rates and provide a WFQ-like scheduling to improve bandwidth utilization and fairness among different traffic flows.

3 WRAPS Algorithm

We want to achieve two goals in our packet scheduling algorithm: (1) The scheduler is adaptive to the real-time traffic rate by itself; (2) The scheduler should optimize the utilization of system resources on bandwidth and reduce the overall packet delay as well. The WRAPS algorithm is described below.

3.1 Operation Procedure

Figure 1 shows the internal mechanism of the scheduler. The Measurement and update of traffic rates are based on a window of size W_n . If there is no traffic at a window period, the rate will stay unchanged. Suppose that two flows are fed into the system. Their packet rates, separately R_s and R_p will be monitored by the scheduler. In the simplest case, the average value of the packet rates can be stored as a parameter for the specific flow. A background scheduler is assumed to be running according to the monitored rate for each flow. Noting that the background scheduler works repeatedly according to it's own period without the perception of the arrival times of the packets. It periodically checks the packet queue of a particular flow, sends out one

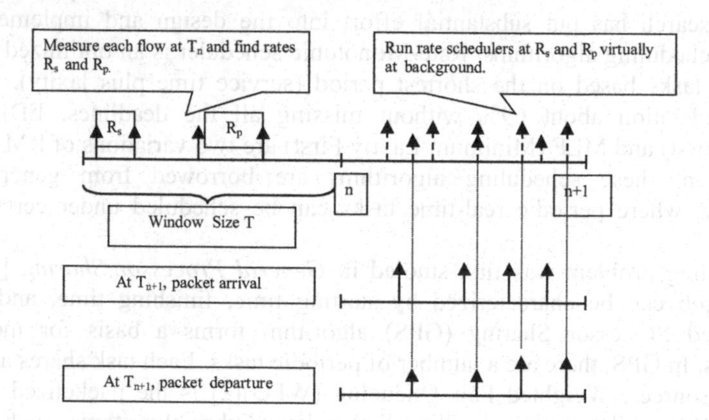

Fig. 1. Window-based Rate Adaptive Packet Scheduling (WRAPS) algorithm

packet (if it's available) or just returns (no packet available for the flow). The period of the background scheduler is adjusted regarding to the monitored rate inside the window. Figure 1 illustrates the arrival and departure pattern of the two flows. The window size can be a fixed number of packets or a fixed time period. Windows move forward when one or more new packets of the same flow arrive so that the packet rate is monitored continuously.

3.2 Formularization of the Algorithm

The rate calculation is simply determined as follows. Suppose r_i is the rate of the ith flow. If the total amount of traffic (number of packets) from the ith flow within time W is P, then $r_i = P/W$. In real systems, we cannot assume there is a background scheduler for each flow, however, the departure time d_i of each packet can be calculated by the formula: $d_i \quad t_{last} \; \frac{\exists a_i \quad t_{last} \forall}{\# \quad r_i \quad !} \; r_i$ where t_{last} is the previous packet schedule time for this flow (when the background scheduler comes), a_i is the packet arrival time. Finally, the packet delay can be written as $d = b \cdot t_{proc}$, where b is the buffer load, and t_{proc} is per packet processing time. The next section will talk about our implementation of the WRAPS algorithm on the network processor.

4 Implementation on the IXP1200 Network Processor

4.1 Platform Description

Our implementation is on the Intel IXP1200 network processor [11][12][13][15]. This product from Intel Corp. aims to provide both high processing speed and flexibility for system developers. Programmers can implement a variety of software features on its six multithreaded micro-engines. Some remarkable characteristics are:

- One strong ARM core and 6 RISC micro-engines. Each micro-engine runs 4 threads.
- 1 KB micro-code space (in on-chip memory) for each of the micro-engine.
- 128 GPRs (general purpose registers), 32 Read/Write transfer registers for each micro-engine. To be fetched in parallel, GPRs are separated into two banks.
- Shared memory resources such as on-chip Scratch pad, hash-engine, IX Bus, etc.
- Supports up to 8MB SRAM (32-bit bus) and 256MB SDRAM (64-bit bus).
- Includes about 50 miscellaneous instructions. All the ALU instruction can be completed in one cycle. But, no division or multiplication instruction is provided.
- The SRAM has lock function to achieve mutual exclusion.
- On-chip RFIFO and TFIFO are buffers connecting to the external IX bus.

4.2 Overall Structure of the Scheduler

The 24 threads are separated into two groups. Each thread of the first 4 micro-engines (16 threads) works on one of the 16 input ports, and the other 2 micro-engines (8 threads) work on the 16 output ports, each serves two ports. More threads are working on the input ports. This is so designed to apply all switching/routing functionalities on the incoming packets with these threads, before putting packets to the queues of the destination ports. The input threads are responsible for the following tasks.

Switching/routing, check the packet header to get the Mac/IP address, and find out the destination port based on the build-in routing/bridge table.

Monitor each flow, identify the flow id of each packet and record the relevant information in the flow information tables and update them after new packets arrive.

Decide the expected rate of current flow and calculate the time it should be scheduled for sending. The packet is then put into the queue.

The output threads are responsible for the following tasks.

Finding out the packets it should send by the time and update the queue.

Put data in transmission buffer and send it via the low-level transmission hardware.

We then use a structure called time-slotted circular queue. There's one time-slotted circular queue for each output port in single queue mode.

4.3 The Time-Slotted Circular Queue

We define a big size circular queue in SRAM. Simply, it is organized as an array in the SRAM memory (Figure 2). Each element in the array (if not NULL) is a pointer to a packet and each position in the array corresponds to a time slot. Two system parameters are associated with the array, i.e. N: the size of the array, S: number of cycles represented by a position in the array. Generally, we take N and S as the power of 2 like $N=2^n$, $S=2^s$.

If a packet with timestamp T comes in, we first omit s least significant bits, then remove the bit over the s+n least significant bits. As shown in Figure 2, we are using the n bits from s to s+n-1 as index to put the packet's pointer into the circular queue. Note that s stands for the granularity of the time slots we consider. The larger the s, the higher the probability two packets may go into the same time slot, i.e. collision happens. With larger s, the flow scheduler will be more inaccurate.

Large s and n are advantageous to represent a broader range of timestamp. We can see, after time 2^{s+n}, the packets are put back to the beginning of the queue, which means a packet scheduled too far away from present may be put in the past. This should be avoided for the correct implementation of the scheduler. n affects the memory space required. With big n, more memory should be allocated to hold the queue, also it will take longer time to search out the next packet to send.

Therefore, big n+s can offer a large time span the queue can represent, whereas n and s should both be kept small to reduce memory access and increase the accuracy. The time-slotted circular queue is a sparse array. Most of the array elements are NULL pointers. Our estimation shows that, on average, only about 5~8% elements are used. However, this brings the complexity of packet insertion to O(1). In most cases, it can be finished with only 2 memory READ/WRITEs. Searching for next packet to send can be completed with only 1 on-chip memory (scratch-pad) operation and 2 off-chip memory (SRAM) operations. And the computational complexity, i.e. number of cycles, is O(n).

In our experiments, we choose n=10. It turns out the needed SRAM space is affordable (64KB for 16 ports), considering there are 8MB SRAM supported by IXP. A more space efficient implementation can further reduce the space consumption by only allocating several bits for each pointer.

To speed up the operations, we have the two-level marking procedure as illustrated in Figure 3. Level 2 marking vectors and level 1 marking vector are added in SRAM and scratchpad.

If n = 10, there are totally 1024 time slots in each queue. Level-two making vectors are 32 long words (1024 bits). Each bit is mapped to a time slot. If the bit is 1, the time slot has a packet, otherwise, no packet pointer (NULL pointer) is in the time slot. Level-two marking vectors are in SRAM, so it still costs certain amount of time to traverse the SRAM. In order to further accelerate the searching of occupied time-slots, level-one marking vector is added to help the searching of level-two marking vectors. The level-one marking vector differs from the level-two marking vectors in that each bit of the level-one marking vector means whether the corresponding long word of level-two marking vector is zero or not. If it's zero, the whole 32 bits of the long word is 0, hence the 32 consecutive time-slots in the queue is unoccupied. For the ease of operation, level-one marking vector is a long word stored in the scratchpad (on-chip memory) for each port. Basically, we start from the last scheduled time-slot, change the time-slot number to the long word number in level-two marking vector, i.e. the bit number in the level-one marking vector. From this bit, find the first bit in the level-one marking vector that is not zero and before the current time slot bit. Then the long word of non-zero level-two marking vector is read out and the least significant non-zero bit is extracted. This bit, along with the long-word position form the index of the time-slot in the queue.

We will describe the whole procedure in later sections.

4.4 Operations on the Queue

Queue operations can be categorized into two parts. After processing the packet, receiving threads put the packet into corresponding slot. There may be conflicts on the slot. Conflicts are resolved in a simplest way. Transmission threads should search from the slot it checked last time to the slot corresponding to the current time until one pending packet is found or it reaches the current time slot. This operation is time-consuming. But with the two-level marking procedure, only one scratch-pad reading and one SRAM reading are needed to locate the next packet to send.

4.4.1 Calculation of the Slot Position and En-Queueing

WRAPS schedule the next packet to send according to their flow properties. Ideally,there should be a scheduler coming at intervals as the rate of the flow. If there is packet available at the point the scheduler comes, the packet is sent at once. In practice, it's not possible to create a scheduler for each flow, because there are too many flows on each port. We have reinterpreted the algorithm for implementation purposes.

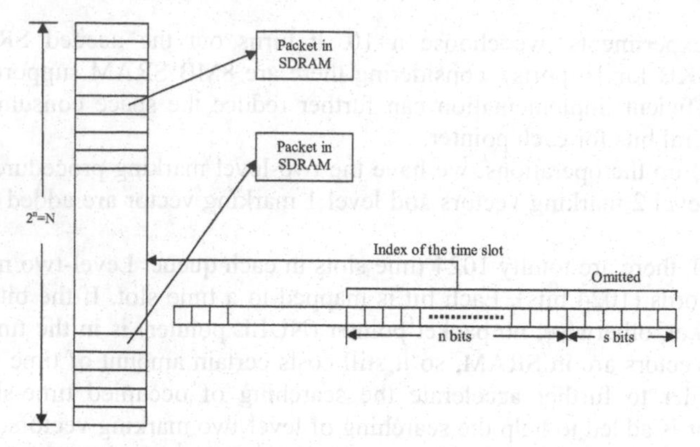

Fig. 2. Structure of the time-slotted circular queue

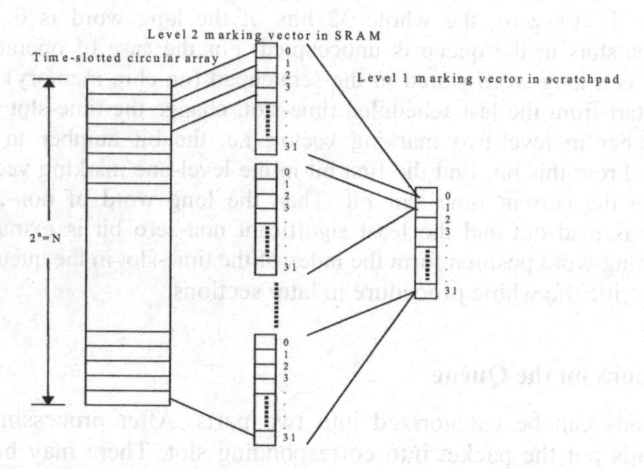

Fig. 3. Data structure of the two level marking vectors

Suppose we know that the last time this flow is scheduled at time T_{last}, and the arrival interval (1/rate) is T_{period}, current time is $T_{current}$. Based on the above information, we can calculate the time when the packet should be sent.

$$T_{sent} \quad \min T : T \quad T_{current}, T \quad T_{last} is.a.multiple.of.T_{period} \quad T_{last} \quad \frac{T_{current} \quad T_{last}}{T_{period}} * T_{period}$$

Knowing that T_{last} is one of the times the scheduler comes, this equation is equivalent to the WRAPS algorithm. Notice that T_{period} can be an arbitrary number. To determine T_{sent}, we need division, which is computationally intensive. In fact, it's not provided by IXP.

In practice, division is done by shift and minus. Surprisingly, it can be done very fast. We can see from the above equation, the division is $(T_{current}-T_{last})/T_{period}$. On a fast processor like IXP, $T_{current}-T_{last}$ is the time interval less than the threads coming back to process the port again. This is actually a small number. In our implementation, it's less than 20 time-slots (or 1280 cycles). After finding the T_{sent}, our next step is to extract the n-bits index from it, as shown in figure 2. With the index handy, we are still not sure whether the slot has been occupied by another packet. Straightforwardly, we read out the level-two vector and check it to see the corresponding bit. If it has been taken, the next bit is checked. Due to the sparsity of the queue, only limited error is introduced by moving the packet to the next available slot. Most probably, a free slot is found at the first check. Our claim is confirmed by the experiment results in the next section. After finding the packet pointer, the bit in the level-two marking vector is set and if a whole long word changes to non-zero, we also set the bit in level-one marking vector.

4.4.2 Finding the Next Occupied Time-Slot and De-Queueing

Transmission threads first read the current time (cycle number), extract the n bit index number. We still take n=10 as example. The 10 bit index number is divided into two parts: the highest 5-bit is the index block number, the lowest 5-bit is the index offset number. In other words, $T=T_{block}*32+T_{offset}$. For level-two marking vectors, T_{block} is the number of long word we should look at and T_{offset} is the bit position inside the long word. For level-one marking vector, we only look at the bit T_{block}.

When a transmission thread begins to find the next occupied time slot, the block number and offset number of last sending (L_{block},L_{offset}) is read from the flow record. The block number and offset of the current cycle (C_{block}, C_{offset}) is got from the index number. The pseudo-codes used to find the next occupied time slot before current cycle are listed below.

This procedure returns either NO_PACKET_TO_SEND or a pair of position in the time-slot queue, where the packet pointer can be found. There is no loop in this code segment. Only two readings from the memory (scratch-pad and SRAM) are invoked. The computational complexity is determined by the sub-function find_least_bit(pos, V), which takes a non-zero long word V and returns the position of the least significant bit that is not zero. After fetching the packet pointer, the bit in the level-two marking vector is cleared and if a whole long word becomes zero, we also clear the corresponding bit in level-one marking vector.

```
Read level-one marking vector V1
Clear all the bits outside the range from Lblock to Cblock
If (V1==0) return NO_PACKET_TO_SEND
Else
        find_least_bit(least_bit_pos,V1);
        Lblock=least_bit_pos

        Read the long word V2 of level-two marking vector at position Lblock;
        if (Lblock == Cblock)
                Clear the bits after Coffset in V2.
                If (V2==0) return NO_PACKET_TO_SEND
                find_least_bit(least_bit_pos, V2)
                return (Lblock, least_bit_pos)
    else
            find_least_bit(least_bit_pos, V2)
                return (Lblock, least_bit_pos)
    endif
endif
```

Fig. 4. Pseudo-code to find the next occupied time slot

Table 1. Some system properties when running at 3 different input speeds

Input traffic rate	50Mb/s	80Mb/s	100Mb/s
Ave. cycle to send a packet	255	189	150
Ave. cycle to find a free slot	4.30	4.55	5.25
Percentage of first slot free	96.7%	94.3%	86.1%
Ave. number of slot to check	1.03	1.05	1.13
Ave. number of element in queue	4.81	5.50	7.62
Ave. cycle to get occupied slot	29.93	26.95	24.22

Table 2. Comparison between the circular queue and link list implementation

CQ—Circular Queue, LL—Link List.

Input traffic rate	50Mb/s		80Mb/s		100Mb/s	
Data Structure	CQ	LL	CQ	LL	CQ	LL
Ave. cycle to send a packet (cycle)	255	445	189	432	150	338
Throughput(Mb/s)	49.36	28.58	78.27	34.60	97.96	37.76

The complexity formula also shows the algorithm can scale very well when n is larger. As n increases, we can first add the number of long words in level-one marking vectors. If the number of level-one marking vectors is more than 32, we need to add another level to solve the problem. The computational complexity will always be O(|n|) and the number of read/write is O(1). By increasing n and decreasing s, we can achieve finer granularity, covering broader range of flow rates and reduce the probability of collision when new packets are put into the queue.

5 Results and Analysis

Our first experiment is based on the model of an eight input/output router. Traffic is fed into the IXP with a constant rate of 50Mb/s, 80Mb/s and 100Mb/s. Inter-packet gap is set to 960 nanoseconds. We implemented a time-slotted circular queue with n=10 and s=6. The test is run for 100000 cycles (Micro-engine is running at 200MHZ).

In Table 1 the first row is a value directly related to the processing speed, i.e. how many cycles are needed before sending out a packet. The second row to the fourth row give the probability and average number to find a free slot by more than one check on the receiving side due to slot conflicts. For the 100Mb/s traffic, about 14% of the first checks fail. Besides, we only observe once that the 3rd check is needed to find a free slot. The fifth row is the average number of elements in the circular queue for each port. Considering the high incoming packet speed, this number is low.

Table 2 shows the comparison between our data structure—the time-slotted circular queue and the naïve link-list. Under light load, CQ can almost keep up with the input traffic rate, while LL degrades significantly—only about 60 percent of the packets can be sent in time. When traffic rate reaches 100Mb/s, LL becomes even worse. We see the packet buffers are filled up quickly. CQ still maintains a good performance to transmit at the full speed.

Next, we experiment on the system scalability with the number of incoming flows. Figure 5 shows the number of flows versus average packet delay. All ports are working on 100Mb/s. Each flow comes from one input port to a designated output port. Flows are distributed evenly among all ports.

We can observe that the delay only increases slightly when flow number is large. Although we have not tested with thousands or millions of flows coming into the system, the trend of packet delay versus number of flows can be reasonably predicted from the figure.

Finally, to test the effectiveness of the WRAP algorithm and the correctness of the implementation, we input two flows with different rate to the same port. By grabbing the input and output time stamp of each packet, the change of behaviors before and after IXP's processing can be read from the figures. To simulate the real traffic flow, we also introduce a certain amount of delay when the flows are fed into the IXP.

Figure 6 and 7 show the shapes of the input traffic and output traffic separately. We calculate the average and standard deviation of the inter-packet gaps for the two flows as listed in Table 3. We can see flow properties are kept and only trivial inter-flow interference exists.

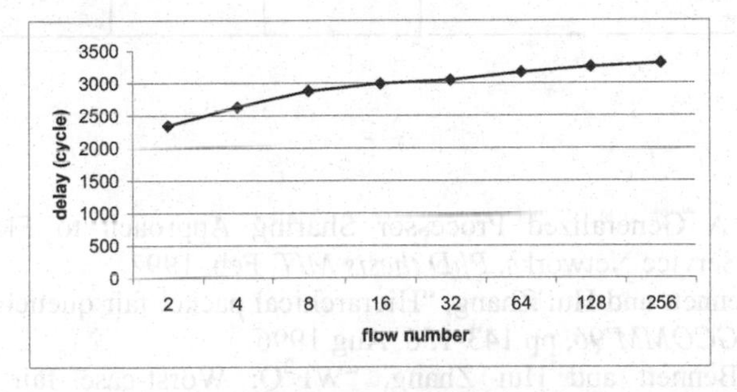

Fig. 5. Flow number versus system delay

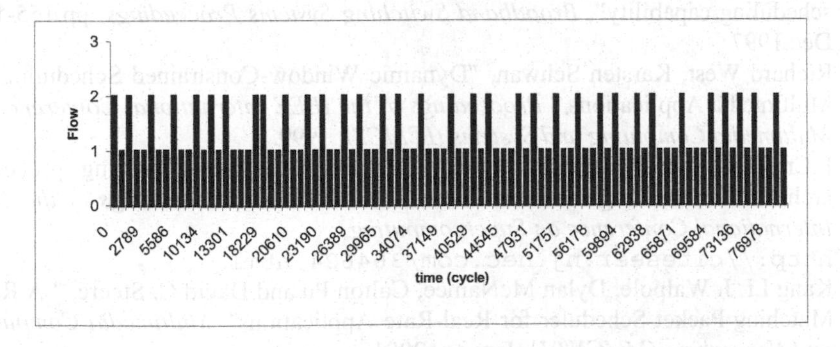

Fig. 6. Input traffic to the IXP with two flows

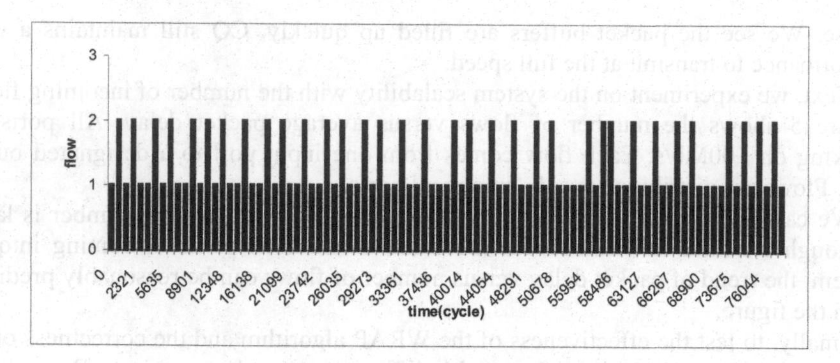

Fig. 7. Output traffic from the IXP with two flows

Table 3. Input/output Flow properties

	Flow 1 input	Flow 1 output	Flow 2 input	Flow 2 output
Avg. inter-packet gap (cycle)	859.48	862.08	1376.86	1385.48
Std. Dev of the inter-packet gap (cycle)	537.65	521.33	515.77	498.30

References

1. A.Parekh, A Generalized Processor Sharing Approach to Flow Control in Integrated Service Networks. *PhD thesis MIT*, Feb, 1992.
2. Jon C.R Bennett and Hui Zhang, “Hierarchical packet fair queueing algorithms”, *In ACM SIGCOMM'96*, pp 143-156, Aug 1996.
3. Jon C.R Bennett and Hui Zhang, “WF^2Q: Worst-case fair weighted fair queueing”, *In IEEE INFOCOMM'96*, pp 120-128, Mar 1996.
4. M.R.Hashemi, A.Leon-Garcia, “A RAM-based generic packet switch with scheduling capability”, *Broadband Switching Systems Proceedings,* pp.155-163, Dec.1997
5. Richard West, Karsten Schwan, "Dynamic Window-Constrained Scheduling for Multimedia Applications," *Proceedings of the IEEE International Conference on Multimedia Computing and Systems (ICMCS)*, 1999.
6. P.Crowley, M.E.Fiuczynski, J.Baer, B.N.Bershad, “Characterizing processor architectures for Programmable Network Interfacess”, *Proceedings of the 2000 International Conference on Supercomputing*, `http://citeseer.nj.nec.com/304624.html`.
7. Kang Li, J. Walpole, Dylan McNamee, Calton Pu and David C. Steere, " A Rate-Matching Packet Scheduler for Real-Rate Applications", *Multimedia Computing and Networking (MMCN'01)*, January 2001
8. Victor Firoiu, Jim Kurose, Don Towsley. “Efficient Admission Control for EDF Schedulers.” *In Proc. of the IEEE INFORCOM'97*

9. L.Y. Zhang, J.W.S. Liu, Z. Deng, I.Philip, " Hierarchical Scheduling of Periodic Messages in an Open System", *Real-Time Systems Symposium, 1999. Proceedings The 20th IEEE*, 1999.
10. Rabindra P. Kar, "Implementing Rhealstone Real-Time Benchmark", *Dr. Dobb's Journal*, April 1990
11. "IXP 1200 Network Processor: Software Reference Manual", Part No. 278306-005. Sep. 2000.
12. "IXP 1200 Network Processor: Programmer's Reference Manual", Part No. 278304-006. September, 2000.
13. "IXP 1200 Network Processor: Development Tools User's Guide", Part No. 278302-005. October, 2000.
14. Richard West and Karsten Schwan, "Dynamic Window-Constrained Scheduling for Multimedia Applications", *IEEE International Conference on Multimedia Computing and Systems, 1999.*
15. Austen McDonald and Weidong Shi, "Intel IXP 1200 Howto", Feb 11, 2001 `http://www3-int.cc.gatech.edu/systems/reading/ixp/austen/`
16. Nicholas Malcolm, Wei Zhao, "Hard Real-Time Communication in Multiple-Access Networks", http://citeseer.nj.nec.com/malcolm95hard.html, 1995.
17. D. Saha, S. Mukherjee, and S. K. Tripathi, "Multirate Scheduling of VBR Video Traffic in ATM Networks", *IEEE J. of Selected Areas in Communications*, Vol.15, No. 6, Aug. 1997

Performance Comparison of Pipelined Hash Joins on Workstation Clusters

Kenji Imasaki, Hong Nguyen, and Sivarama P. Dandamudi

Center for Networked Computing, School of Computer Science
Carleton University, Ottawa, Canada
{kenji,sivarama}@scs.carleton.ca

Abstract. The traditional hash join algorithm uses a single hash table built on one of the relations participating in the join operation. A variation called double hash join was proposed to remedy some of the performance problems with the single join. In this paper, we compare the performance of single- and double-pipelined hash joins in a cluster environment. In this environment, nodes are heterogeneous; furthermore, nodes experience dynamic, non-query local background load that can impact the pipelined query execution performance. Previous studies have shown that double-pipelined hash join performs substantially better than the single-pipelined hash join when dealing with data from remote sources. However, their relative performance has not been studied in cluster environments. Our study indicates that, in the type of cluster environments we consider here, single pipelined hash join performs as well as or better than the double pipelined hash join in most cases. We present experimental results on a Pentium cluster and identify these cases.

1 Introduction

Growth in various application areas has generated demand for high transaction processing rates. This has motivated several researchers to study parallel database query processing, both at the architecture level and at the algorithm level. At the architecture level, several parallel database system architectures including shared-everything, shared-disk, and shared-nothing have been proposed [5]. With the advent of cluster computing environments, cluster-based parallel query processing has been proposed as an alternative to the parallel database systems [3, 6, 8]. Several commercial products including the Oracle Parallel Server [15, 2] and Microsoft SQL Server [7] have been implemented on workstation/PC clusters. Parallel query processing on workstation clusters has been studied in [4, 8, 10, 11, 14, 17, 18, 19, 20].

The use of workstation clusters for database query processing leads a system that exhibits different characteristics than those found in distributed and parallel database systems. One major difference is that the database itself is neither distributed among the workstations (as in a distributed database system) nor relations declustered as in a parallel database system. This is because workstations in a cluster might have individual owners, who can control the behavior

S. Sahni et al. (Eds.) HiPC 2002, LNCS 2552, pp. 264–275, 2002.

of their workstations (e.g., reset the system to remove the "free loaders" on their system). This lack of central control on the system as well as security concerns make it undesirable to distribute relations across the workstations. Thus, the database is kept under the control of a traditional database system, which provides security and maintains the integrity of database. However, we can use the additional processor cycles and memory available on these workstations for query processing on a temporal basis [3]. In particular, we restrict our attention to query processing, which requires read-only access to data.

In such a cluster-based system, everytime a query is executed on a cluster, the workstations involved in the query execution will have to receive the required base relations before query processing can proceed. This overhead is not present in parallel database systems, where the required data usually resides at the processing nodes (for example, in a shared-nothing type architecture).

At the algorithm level, there are three categories of parallelism exploited to speed up query processing: intra-operator, inter-operator, and inter-query parallelism [5]. In intra-operator parallelism, parallelism applies to each operator within a query; several processors execute a single database operation concurrently. In inter-operator parallelism, several operations within the same query are concurrently executed. In inter-query parallelism, several queries are executed simultaneously within a multi-processor system.

In query processing, the join operation has attracted a lot of attention, as it is the most expensive operation. Three basic techniques are used for join processing: nested-loop, sort-merge and hashing. Of the three, the hash join has been found to be the best for equijoin operations [16]. For this reason, we focus here on hash joins on workstations clusters.

This paper focuses on the inter-operator parallelism. For this type of parallelism, hash joins have been used to combine hashing and pipelining techniques. In this paper, we compare the performance of the single- and double-pipelined hash join algorithms on a Pentium-based cluster.

2 Join Algorithms

This section describes the single- and double-pipelined hash joins algorithms used in this paper. The two relations participating in a join operation are traditionally called inner and outer relations. In the traditional, single hash join algorithm, the inner relation is first partitioned into disjoint subsets called buckets, using a hash function applied on the join attribute. This creates the hash table for the inner relation, against which tuples from the outer relation are matched to join the tuples. Thus, hash join algorithms normally involve two phases:

1. *Table-Building Phase:* During this phase, the hash table is built on the inner table. This step involves reading the complete inner relation.
2. *Tuple-Probing Phase:* Once the hash table for the inner relation is available, tuples from the outer relation are used to probe the hash table to see if there is a match on the join attribute. If there is a match, the two tuples are joined to produce the result tuple. Thus, to complete the join operation, a complete scan of the outer relation is required.

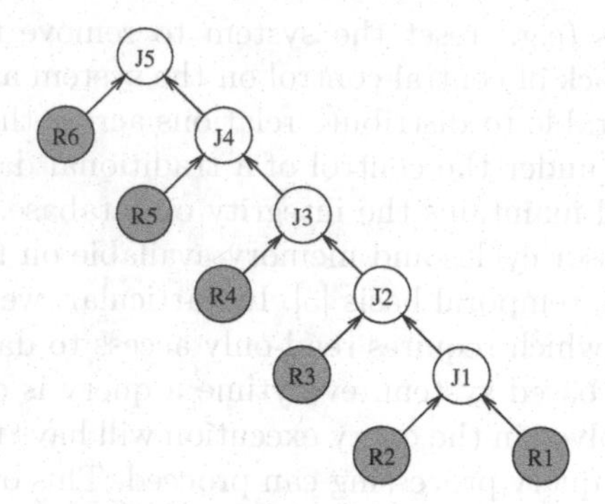

Fig. 1. Right deep query tree

2.1 Pipelined Hash Join

Hash join has been shown to present best opportunity for pipelining [1, 9, 14]. A pipeline of hash joins consists of many stages, each of which is associated with one join operation that can be executed in parallel by several processors, thus substantially speeding up the process. To facilitate pipelined execution, we use a right-deep query tree (see Figure 1). In this tree, the inner relation of a join is shown as the left relation. For example, relations R2, R3, R4, R5, and R6 are inner relations. Relation R1 is the outer relation for join J1. The output produced by J1 will be the outer relation to J2 and so on.

Executing a query by using a pipelined hash join algorithm improves performance, as there is no need to write the intermediate results to disks. These early intermediate results are pipelined to the next join for immediate processing, in parallel. As a result, I/O cost is minimized, and the first results are generated early. However, pipelined hash joins also have setbacks, which include the overhead of setting up hash tables, of filling and depleting the pipeline. Moreover, if one of the stages is stalled for some reason (e.g., due to an increase in the background load on one of the workstations), the whole pipeline becomes slower, affecting the overall query processing performance.

The following notation is used in our description.

- R_i: inner relation,
- R_o: outer relation,
- R_r: result relation.

In this paper, we use the term "segment" to represent part of a relation (often limited by the output buffer size). The following description refers to the software architecture shown in Figure 2. The Database (DBMS) node maintains the actual database. The Master node acts as the coordinator as well as the interface between the slave and DBMS nodes. The slaves actually perform the join operation. More details on the software architecture are given later.

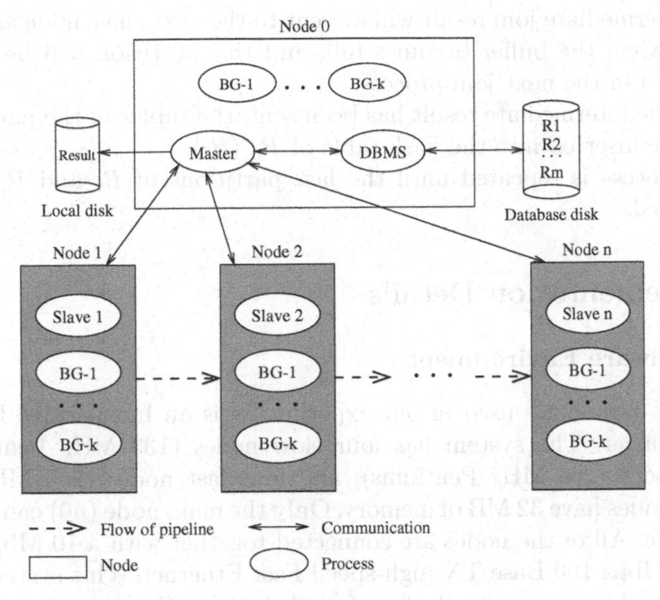

Fig. 2. Software architecture

2.2 Single-Pipelined Hash Join

The single-pipelined hash join works as follows:

1. Each slave node waits for segments from both R_i and R_o. The tuples of each received R_i partition are inserted into the hash table as they arrive. Once the whole R_i has been received and the hash table has been built, the hash join operation begins if a partition of R_o is available at the slave. The result of the join operation R_r is placed in the output buffer.
2. The intermediate join result R_r is shipped to the next slave node in segments when the buffer becomes full. These will act as the partitions of R_o for the next join process.
3. After the intermediate result has been sent to the next slave, the current node waits for another incoming segment of R_o. When this segment arrives, it performs the join as before. This process is repeated until the last partition of R_o has been received (which comes with an end-of-R_o message).

2.3 Double-Pipelined Hash Join

The single-pipelined hash join is asymmetric in that only the inner relation is hashed. In the double-pipelined hash join, both relations are hashed.

1. Each slave node waits for partitions of either R_i or R_o to arrive. If a partition of R_i (respectively R_o) has been received, its tuples are inserted into the hash table of R_i (R_o). If a partition of R_o (R_i) arrives, the join begins between the tuples of R_o (R_i) and the hash table of R_i (R_o).

2. The intermediate join result will be sent to the next slave node as a partition of R_r when the buffer becomes full, and this partition will become an R_o partition in the next join process.
3. After the intermediate result has been sent, the tuples of the partition of R_o (R_i) are inserted into the hash table of R_o (R_i).
4. The process is repeated until the last partitions of R_i and R_o have been processed.

3 Implementation Details

3.1 Hardware Environment

The cluster system we used in our experiments is an inexpensive Linux-based Pentium cluster. The system has four slow nodes (133 MHz Pentiums), two medium nodes (166 MHz Pentiums), and four fast nodes (200 MHz Pentium Pros). All nodes have 32 MB of memory. Only the main node (n0) can be accessed from outside. All of the nodes are connected together with a 10 Mbps Ethernet and a 100 Mbps 100 Base TX high-speed Fast Ethernet. This system was built to simulate a heterogeneous cluster of workstations. Our experiments were run in a dedicated mode with no other users in the system.

In the experiments, node n0 is used to run the Master and DatabaseManager processes. The remaining nodes are used for the slave processes. For the results reported here, we used five slave nodes.

3.2 Software Environment

We have written the programs in GNU C++ using PVM library. The PVM software provides a means to emulate parallel programs in an efficient and straightforward manner. PVM software deals with message routing, data conversion, and task scheduling across a network of heterogeneous computer architectures. Communication between processes is carried out by message passing method, using PVM. We have used `PvmAllowDirect` for the routing method and `PvmDataDefault` for the data packing method.

The relations are stored as Unix files on a local disk. The relations used in our experiments are of sizes 10,000 and 50,000 tuples. A tuple consists of two fields: one is a key attribute, which is a 4-byte integer, and the other is a 10 byte character string.

There are four major software components in the implementation: a Master, a Slave, a DatabaseManager (DBMS), and a Background process. The relationship among these components is shown in Figure 2. The Master process acts as the controller of all processes. When invoked, it spawns the other processes (i.e., Slave processes, Background processes and DBMS process). The Master process monitors the join processes and gathers the final join results, which are then written to a local disk. The execution time is recorded and stored in another file on the local disk.

Slave processes perform the local join operations. The DBMS process is responsible for reading input relations and passing them onto the Master. Background processes control the non-query loads that simulate local user's workload. When configured, Background processes are invoked on the same nodes as the Slave processes and are run during the join operations. A Background process has on-time and off-time parameters, and is activated through a busy loop constant endowed with a sleep function.

In our experiments, the following types of background loads are used to study the impact of background loads on the performance:

- *No background load:* The Slaves perform the join operation without any background processes.
- *Identical background load:* All Salves are assigned the same background load.
- *Proportional background load:* Background load assigned is proportional to the computational capacity of the node.

4 Experimental Results

This section discusses the results of our experiments conducted on a Pentium cluster. The following parameters are used during the experiments. We have used five slave processes and five nodes during the experiments. The number of buckets is kept at 1000. The experiments were conducted with the following buffer sizes: 50, 100, 500, 1000, 1500, 2000, and 2500 tuples.

4.1 Performance with No Background Load

This section establishes the base case in which there is no background load on the slave as well as Database nodes. This represents the case where the cluster is dedicated to database query processing. Performance of the single-pipelined hash join (SPHJ) and double-pipelined hash join (DPHJ) algorithms for 10,000-tuple relations is shown in Figure 3*a*. This data show that SPHJ provides better performance than DPHJ. The performance difference between SPHJ and DPHJ varies from 7.7% to 3.1%. While the performance superiority of SPHJ is marginal, it is important to note that in other environments, DPHJ provided substantial performance improvement over SPHJ. We will discuss the reasons for this later in this section.

When there is no background load to slow down the flow of incoming partitions of the relations, majority of the partitions of the base relations (or inner relations) R_i arrive at the join processes sooner, which are compared to those of the outer relations. As a result, the behavior of the joins performed under DPHJ is similar to that performed under SPHJ. That is, the execution patterns of the two join algorithms are similar, which results in similar execution times. The DPHJ takes more time, as it has to spend additional time to build the second hash table.

This figure also shows that the execution time increases with the buffer size beyond 100 tuples. This is the result of the tradeoff between the communication

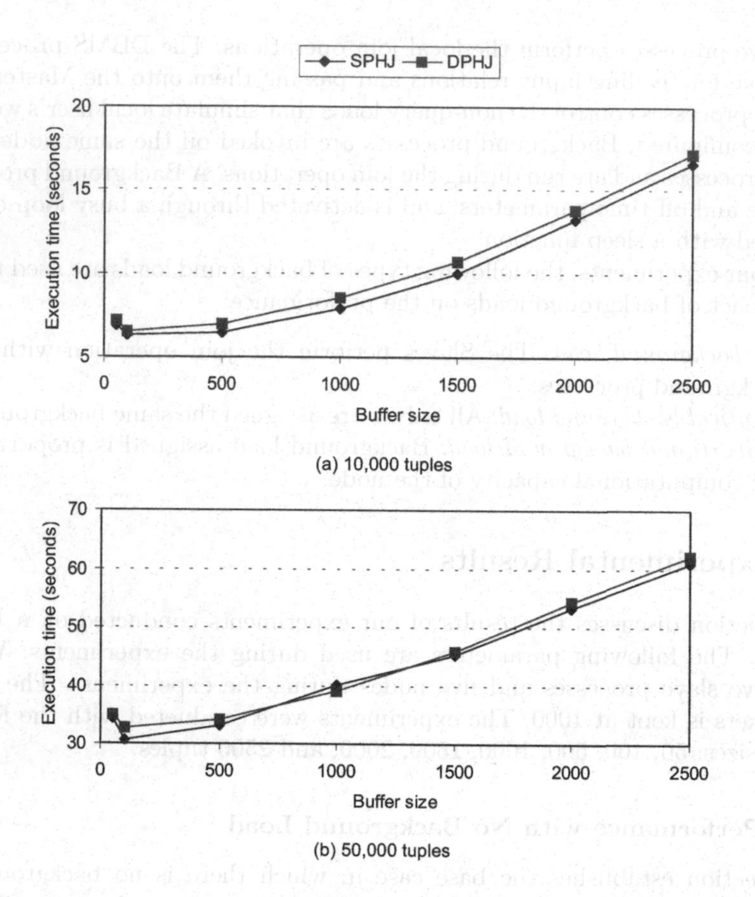

Fig. 3. Performance of the two pipelined hash join algorithms for 10,000- and 50,000-tuple relations

time to send the tuples and the join execution time. We can reduce the communication time if we use larger buffers (thus fewer messages). However, larger buffer also means slaves have to wait for the whole buffer to arrive before working on the joins. Thus, we would like to use smaller buffers to reduce the whole join execution time.

However, as the size of the relations increases, the difference in execution time between SPHJ and DPHJ decreases. Figure 3*b* shows the performance of SPHJ and DPHJ on 50,000-tuple relations. The range of the performance difference between the two join algorithms in this case is from 6.1% to 2.9%. This slight improvement in the performance of DPHJ (compared to the 10,000-tuple relations case) is due to the fact that, when more tuples are involved, there are more segments participating the join processes for the same buffer size. Therefore, SPHJ takes longer to receive the whole inner relation R_i. On the other hand, in DPHJ algorithm, the join process takes place earlier as soon as both segments of R_i and R_o have been received. Thus, the smaller the buffer size, the sooner the join process starts. This leads to better performance with DPHJ as the buffer size decreases. This observation implies that, with very large relations, the bigger the ratio of the relation size to buffer size, the more beneficial DPHJ becomes, compared to SPHJ.

Our results are in contrast to the substantial performance advantages achieved by using DPHJ in wide area network and Internet-based data integration systems [13]. There are several reasons for this performance difference. First, in data integration systems for Internet, data comes from different sources. On the other hand, in a cluster-based system all the data comes form a single source (the database node). Another reason is that Internet-based databases experience long, highly variable delays in delivering the tuples. In this scenario, the double-pipelined hash join performs substantially better by working with the available parts of the relations. The single-pipelined hash join, on the other hand, waits to receive the complete inner relation. Finally, in single pipelined hash join, it is important to identify/predict which of the two relations is the smaller. This is because its performance is sensitive to the selection of the inner relation. Single-pipelined hash join gives better performance if we select the smaller of the two relations as the inner relation. The double-pipelined hash join is symmetric and is not sensitive to the inner and outer relation assignment. In Internet-based databases, it is difficult to predict the size of the relations. Thus, the double-pipelined hash join is preferred for these systems. In contrast, in the cluster-based systems, we can get the relation size information from the system catalog.

Our results suggest that, in the type of cluster-based systems considered in this paper, single-pipelined hash join performs as well as or better than the double pipelined hash join. Implementing single-pipelined hash join is also efficient in the sense that it demands less memory as it builds only a single hash table.

We have conducted a variety of experiments to verify that SPHJ and DPHJ algorithms exhibit similar relative performance. Due to space limitations, the following sections give some sample results. More details are available in [12].

4.2 Impact of Background Load on Database Node

We will now look at the impact of the background load on the performance of these hash join algorithms. In this section, we consider the affect of background load only on the Database node. The next two sections deal with the impact of background load on the slave nodes. For this experiment, we assigned five background processes with an average on-time length of 10 seconds. Note that the background load cycle time is 15 seconds. The results for the 10,000-tuple relations are shown in Figure 4.

From the data presented in this figure, it can be seen that the presence of the background load on the Database node favors DPHJ. As shown in this figure, DPHJ outperforms SPHJ when the buffer size is small. This is due to the fact that the presence of background load on the Database node slows down the flow of R_i segments to the slave nodes, causing the execution to likely follow

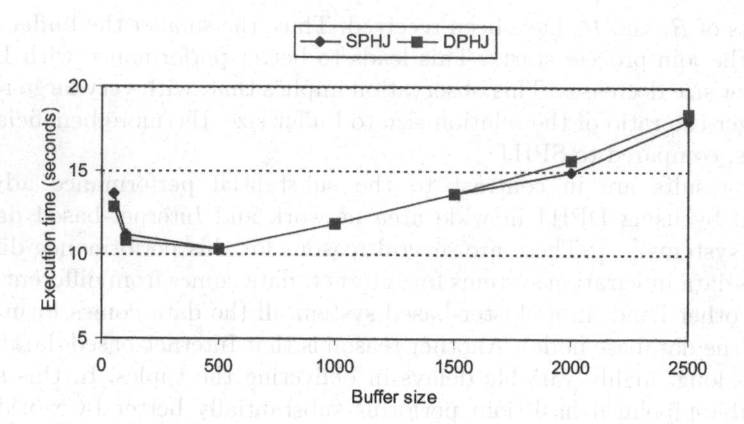

Fig. 4. Performance when there is background load on the Database node (10,000-tuple relations)

the pattern R_i, R_o, R_i, R_o, $\cdots$. This kind of execution pattern improves the performance of DPHJ. The small buffer size also helps boost the performance of DPHJ as smaller buffer sizes means slower SPHJ performance since it takes more messages to transmit the whole inner relation.

4.3 Performance with Identical Background Load

This and the next section look at the impact of background load on slave nodes. In this section we consider identical load on all slaves. This is simulated by assigning one background process with an average on-time length of 10 seconds for each node. The results for 10,000-tuple relations are shown in Figure 5.

As in the case of no background load, the single-pipelined hash join provides better performance than the double-pipelined hash join algorithm. The range of performance difference between SPHJ and DPHJ varies from 11.1% to 1.4%. Since we have already discussed the reasons for the marginal performance superiority of SPHJ in Section 4.1, we will not restate those reasons here.

4.4 Performance with Proportional Background Load

This section presents the results for proportional background load mentioned in Section 3.2. In the proportional background load case, background load assigned is proportional to the computational capacity of the node. This represents the scenario that often occurs in practice (a faster machine often has more load as more users tend to log into the machine). This is simulated by assigning a heavy background load to a fast node, a moderate background load to a medium node and a light background load to a slow node. The weight of the background load is specified by the average on-time period. We simulated proportional background load by assigning a load of 15-second on-time for the fast nodes, 5-second on-time for the medium and 1-second on-time for the slow nodes. The results for

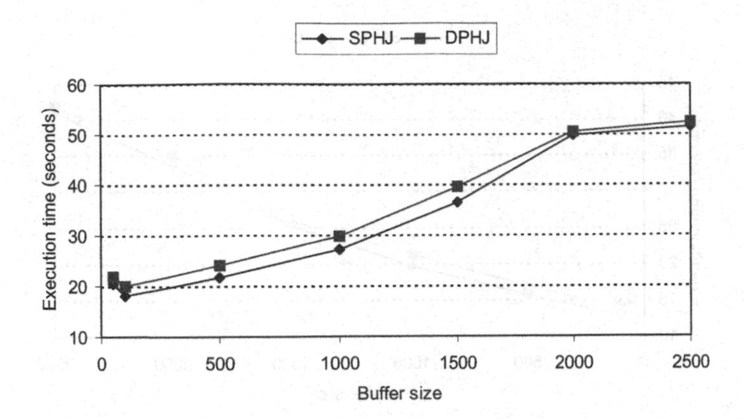

Fig. 5. Impact of identical background load on the slaves (10,000-tuple relations)

10,000-tuple relations are shown in Figure 6. Even in the presence of a proportional background load on slaves, single-pipelined hash join performs marginally better in our cluster environment of the reasons discussed in Section 4.1. The performance difference between the two algorithms is in the range between 8.5% and 1.7%.

We have also conducted experiments with inverse background load in which the fast nodes have light background loads, the medium nodes have moderate background loads and the slow nodes have heavy background loads. The results for the inverse background load are similar to those obtained with the proportional background load. These results can be found in [12].

5 Summary

In this paper, we studied the performance impact of single- and double-pipelined hash joins in a clustered system. Previous studies have shown that the double-pipelined hash join is superior to single-pipelined hash join for systems like the data integration systems for Internet. Our results on a Pentium-based cluster indicate that the single-pipelined hash join performs as well as or marginally better than the double-pipelined hash join. We have identified the reasons for this behavior. Some of the reasons for this difference are due to the system characteristics. For example, in the cluster systems we consider here, all data comes from a single source as the database node supplies the tuples needed for performing join operations of a query. We have presented performance sensitivity to a variety of factors including the background load on the database node and the slave nodes.

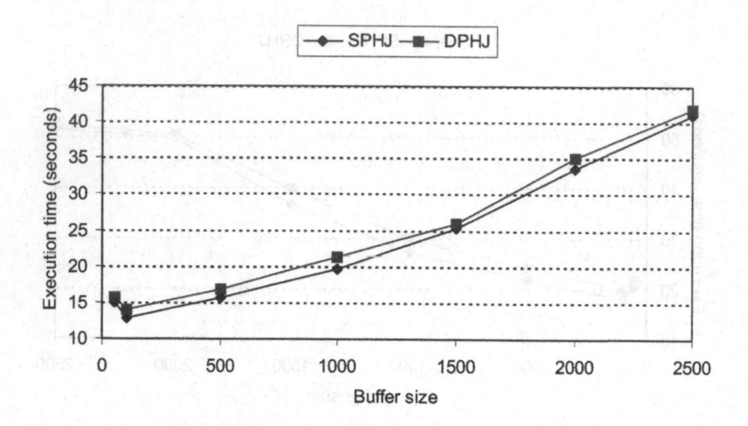

Fig. 6. Performance of the two join algorithms with proportional background load (10,000-tuple relations)

Acknowledgements

We gratefully acknowledge the financial support provided by the Natural Sciences and Engineering Research Council and Carleton University.

References

[1] M.-S. Chen, M. Lo, P. S. Yu, and H. C. Young. Applying Segmented Right-Deep Trees to Pipelining Multiple Hash Joins. *IEEE Transactions on Knowledge and Data Engineering*, 7(4):656–667, Aug. 1995. 266

[2] Compaq Computer Corporation. Parallel Database Clusters for Oracle 8i, 2000. Available at `www.compaq.com`. 264

[3] S. P. Dandamudi. Using Workstations for Database Query Operations. In *International Conference of Computers and Their Applications*, pages 100–105, Tempe, Arizona, Oct. 1997. 264, 265

[4] S. P. Dandamudi and G. Jain. Architectures for Parallel Query Processing on Networks of Workstations. In *International Conference of Parallel and Distributed Computing Systems*, pages 444–451, New Orleans, Louisiana, Oct. 1997. 264

[5] D. J. DeWitt and J. Gray. Parallel Database Systems: The Future of High-Performance Database Systems. *Communications of the ACM*, 35(6):85–98, June 1992. 264, 265

[6] M. Exbrayat and L. Brunie. A PC-NOW Based Parallel Extension for a Sequential DBMS. In *International Parallel and Distributed Processing Symposium Workshops PC-NOW*, pages 91–100, Cancun, Mexico, May 2000. 264

[7] G. Graefe, R. Bunker, and S. Cooper. Hash Joins and Hash Teams in Microsoft SQL Server. In *The 24th International Conference on Very Large Databases (VLDB)*, pages 86–97, Los Altos, CA, 1998.

264
[8] M. A. Haddad and J. Robinson. Using a Network of Workstations to Enhance Database Query Processing Performance. In *Recent Advances in Parallel Virtual Machine and Message Passing Interface, 8th European PVM/MPI Users' Group Meeting*, volume 2131 of *Lecture Notes in Computer Science*, pages 352–359, Santorini, Thera, Greece, Sept. 2001. 264
[9] H.-I. Hsiao, M.-S. Chen, and P. S. Yu. On Parallel Execution of Multiple Pipelined Hash Joins. In *ACM-SIGMOD International Conference on Management of Data*, pages 185–196, Minneapolis, U. S. A., May 1994. 266
[10] K. Imasaki and S. Dandamudi. Performance Evaluation of Nested-loop Join Processing on Networks of Workstations. In *Proceedings of the Seventh International Conference on Parallel and Distrubuted Systems*, pages 537–544, Iwate, Japan, July 2000. 264
[11] K. Imasaki and S. Dandamudi. An Adapive Hash Join Algorithm on a Network of Workstations. In *International Parallel and Distributed Processing Symposium (IPDPS)*, Fort Lauderdale, Florida, Apr. 2002. 264
[12] K. Imasaki, H. Nguyen, and S. P. Dandamudi. Performance Comparison of Piplened Hash Joins on Workstation Clusters. Technical report, Carleton University, School of Computer Science, Ottawa, Canada, July 2002. Avaialble from `www.scs.carleton.ca/ sivarama/publications.html` 271, 273
[13] Z. G. Ives, D. Florescu, M. Friedman, A. Levy, and D. S. Weld. An Adaptive Query Execution System for Data Integration. In A. Delis, C. Faloutsos, and S. Ghandeharizadeh, editors, *The 1999 ACM SIGMOD International Conference on Management of Data: SIGMOD '99*, volume 28(2) of *SIGMOD Record (ACM Special Interest Group on Management of Data)*, pages 299–310, Philadelphia, PA, USA, June 1999. ACM Press. 271
[14] S. Jalali and S. P. Dandamudi. Pipelined Hash Joins using Network of Workstations. In *Parallel and Distributed Computing Systems, 14th International Conference*, pages 422–429, Richardson, Texas, 2001. 264, 266
[15] Oracle Corporation. Oracle9i Real Application Clusters, 2001. Available at `http://otn.oracle.com/products/oracle9i`. 264
[16] D. A. Schneider and D. J. DeWitt. A Performance Evaluation of Four Parallel Join Algorithms in a Shared-nothing Multiprocessor environment. *SIGMOD Record*, 18(2):110–121, June 1989. 265
[17] C. Soleimany and S. P. Dandamudi. Distributed Parallel Query Processing on Networks of Workstation. In *High-Performance Computing and Networking, 8th International Conference (HPCN Europe)*, volume 1823 of *Lecture Notes in Computer Science*, pages 427–436, Amsterdam, The Netherlands, May 2000. Springer-Verlag. 264
[18] T. Tamura, M. Oguchi, and M. Kitsuregawa. High Performance Parallel Query Processing on a 100 Node ATM Connected PC Cluster. *IEICE Transactions on Information and Systems*, 1(1):54–63, Jan. 1999. 264
[19] B. Xie and S. P. Dandamudi. Hierarchical Architecture for Parallel Query Processing on Networks of Workstations. In *The 5th International Conference on High Performance Computing*, Chennai, Madras, India, Dec. 1998. 264
[20] S. Zeng and S. P. Dandamudi. Centralized Architecture for Parallel Query Processing on Networks of Workstations. In *High-Performance Computing and Networking, 7th International Conference (HPCN Europe)*, volume 1593 of *Lecture Notes in Computer Science*, pages 683–692, Amsterdam, The Netherlands, May 1999. Springer-Verlag. 264

Computational Science and Engineering – Past, Present, and Future

N. Radhakrishnan

Computational & Information Sciences Directorate
US Army Research Laboratory

Physics-based computational modeling has become more practical in the past few years due to the availability of high performance computers (HPC), high-speed communications, and more robust visualization and data mining tools. Applications range from understanding fundamental science to solving a diverse set of important practical, real-world problems. The old paradigm of designing, testing, building, and then modeling is being replaced in some cases with a new paradigm of investigating and modeling before designing and building. People are now realizing computational modeling and simulation is an essential ingredient in reducing acquisition time for large systems. Single discipline problems are giving way more and more to the solution of multi-disciplinary problems. HPC, coupled with multi-disciplinary applications, is the driving force behind the challenges we face today in research and engineering.

In today's world, there is an overwhelming need to store, retrieve, fuse, and analyze information in a comprehensive and accurate manner and this requirement has a large impact on HPC technology. This talk will trace the parallel history of computing, as well as computational science, and show the interdependence of these two fields. Discussions will be based on research programs that rely heavily on the use of computational modeling including applications in structural mechanics, fluid dynamics, computational chemistry, and nano-mechanics. The need for better pre- and post-processing tools will be discussed with emphasis placed on advances in scientific visualization and data mining. Then looking towards the future, this talk will touch upon innovative technologies in the areas of computational methods, (i.e. meshless methods), futuristic computers (i.e. quantum computing, biological computing) and natural user interfaces (i.e. voice recognition, gestures).

S. Sahni et al. (Eds.) HiPC 2002, LNCS 2552, p. 279, 2002.

Iterative Algorithms on Heterogeneous Network Computing: Parallel Polynomial Root Extracting

Raphaël Couturier, Philippe Canalda, and François Spies

Laboratoire d'Informatique LIFC, University of Franche-Comté, France

Abstract. This article describes various algorithms on heterogeneous architecture applied to parallel polynomial root extracting. The use of a set of clusters in order to compute great parallel applications is an ongoing challenge. Asynchronous algorithms are well suited for this type of architecture, because no synchronous barriers are needed. The scalability of such architecture is a hard task to achieve. In this paper, we compare local calculus and calculus on the Internet in order to focus on the efficiency of such methods.

1 Introduction

The methods for extracting the roots ω_1, ..., ω_n of a polynomial with complex coefficients:

$$P(z) = \sum_{i=0}^{n} a_i z^{n-i} \; with \; a_0 = 1, \; a_n \neq 0 \; and \; a_i \; in \; \mathbb{C} \tag{1}$$

where n stands for the degree are a well-known subject of research which however is still quite open. There is plenty of literature including theoretical studies and efficient methods [6, 7]. Yet until the past few years, few implementations have been achieved and exploited, none has been in the proportions of algorithms running matrix calculus. There are lots of science fields where these methods are desirable. In fact, it is possible to calculate the eigenvalues of a matrix with high dimensions directly by this means.

Newton's method is the oldest method of root extracting, but it is sequential and cannot therefore be parallelized. At present, there are several simultaneous methods where roots are determined by successive approximations. Some of them use an iterative function such as $Z^{(k)} = H(Z^{(k-1)})$, where H represents a function in $\mathbb{C}^n$ built so that vector $Z^{(k)}$ tends towards the roots, with any initialization of vector $Z^{(0)}$.

Durand-Kerner's method belongs to this category. It uses the Weierstrass operator as an iterative function

$$\forall i \in [1, n] \; ; \; H_i(Z) = z_i - \frac{P(z_i)}{\prod_{\substack{j=1 \\ j \neq i}}^{n} (z_i - z_j)} \tag{2}$$

S. Sahni et al. (Eds.) HiPC 2002, LNCS 2552, pp. 283–291, 2002.

The Weierstrass operator can only work in the case of distinct roots. Another method, not described in this paper, must be used in the case of multiple roots. The vector initialization $Z^{(0)}$ is achieved from Guggenheimer's method.

There are other iterative methods such as Aberth's[1, 2]. Iterative methods raise several problems when implemented e.g. specific sizes of numbers must be used to deal with this difficulty. Moreover, the convergence time of iterative methods drastically increases like the degrees of high polynomials. The parallelization of these algorithms will improve the convergence time. There are several paradigms of parallelization (synchronous or asynchronous calculus, mechanism of shared or distributed memory, data distribution...) which we compare in this article. The study presented here deals more especially with sparse polynomials, i.e. polynomials identified by few monomials which are non nil (fewer than 20 monomials).

Section 2 looks back on the different stages of Durand-Kerner's method. Section 3 describes the adaptations we achieved for our implementation. The last section presents the experimental results we measured by comparing a set of solutions.

2 Looking Back on Durand-Kerner's Method

Durand-Kerner's method is made up of three main stages [9, 3]. The initialization of the polynomial $P(z)$ with complex coefficients is defined in equation 1. **Vector $Z^{(0)}$ Initialization** is important because the vector elements must be distinct from each other. The Guggenheimer's method is used. A ray σ_0 is determined from the polynomial coefficients and the roots are equidistant on the circle. The calculation of σ_0 runs as follows:

$$\sigma_0 = \frac{u+v}{2}; u = \frac{\sum_{i=1}^{n} u_i}{n \cdot \max_{i=1}^{n} u_i}; v = \frac{\sum_{i=0}^{n-1} v_i}{n \cdot \min_{i=0}^{n-1} v_i}; u_i = 2 \cdot |a_i|^{\frac{1}{i}}; v_i = \frac{\left|\frac{a_n}{a_i}\right|^{\frac{1}{n-i}}}{2}$$

Iterative Function H_i corresponds to the equation 2 which will enable the convergence towards polynomial solutions, provided all the roots are distinct. **Convergence Condition** determines the success of the termination. It consists in stopping the iterative function when the whole of the polynomial modules for all roots is inferior to a fixed value ϵ.

3 Adapted Iteration Scheme

Traditionally, comparisons are made between Jacobi and Gauss-Seidel models. For this type of algorithm, the Gauss-Seidel model is more efficient. In our synchronous and asynchronous algorithms, we implement a block Gauss-Seidel scheme. This means that each processor uses a typical Gauss-Seidel scheme into its current block. Thus, each computed block is taken into account for the next computed blocks of the current iteration. Finally, we improved our algorithm convergence by using a Jacobi scheme for the very first iteration. The details of the measurements are in [4].

4 Synchronous Calculation

The difficulty in designing an efficient parallel algorithm for root extracting using Durand-Kerner iteration is due to the fact that at each iteration, the approximation of each root requires approximation of all the other roots. On shared memory architecture the problem can be resolved using shared memory. On distributed architecture the problem of frequent communication arises. Our solution consists in gathering the approximation of some roots and then diffusing it to other processors. This approach requires that each processor knows all root approximations. The frequency of root diffusion among processors is a relevant point in the convergence speed of the Durand-Kerner algorithm. But it is also a limiting factor for the scalability.

This algorithm is designed to be run on heterogeneous architecture or architecture for which load is varying due to use of some computers by other user. A solution to solve this problem is to distribute the number of roots dynamically so that each processor, whatever its speed and its load, can compute its roots in the same time. As we consider that the load of a processor can change, we do not compute the speed of each processor before running our program.

4.1 Description of the Algorithm

After each iteration, a processor determines the time to compute the number of roots it has to approximate. This information is exchanged between all processors so that each processor knows the speed of all processors. In fact we deduce the speed from the number of roots divided by the time required to approximate them in one iteration.

Let N_i^k be the number of roots of processor i at iteration k, N be the total number of roots and T_i be the time required by processor i to compute its roots.

$$\forall k \geq 0, \forall i \in [1 \dots n] \quad N_i^{k+1} = N * \frac{\frac{N_i^k}{T_i}}{\sum_{j=1}^{n} \frac{N_j^k}{T_j}} \tag{3}$$

$\frac{N_i^k}{T_i}$ represents the speed of processor i and $\sum_{j=1}^{n} \frac{N_j^k}{T_j}$ is the sum of the speed of all the processors. Then we can interpret the fraction as the ratio of speed between processor i and all the processors. This ratio, multiplied by the whole number of roots, gives the number of roots that processor i can compute taking into account its speed and the speed of all the processors. To initialize the distribution, each processor receives the same number of roots equal to the whole number of roots divided by the number of processors.

4.2 Remarks on the Algorithm

If we consider that the load of a processor is not modified at each iteration then each processor has the exact number of roots that it can compute with regard to its speed. If during the execution, the load of some processors changes for a long time then these processors may have to compute fewer roots.

However, on the one hand, we can notice that even if the load of some processors varies frequently, the root sharing process will not take it into account as the speed of the previous iteration, which includes the previous load, is considered. On the other hand, a strong variation of the load in one iteration is taken into account in the load balancing process, but only at the next iteration.

The constraint that the load varies few times in one iteration is acceptable in a dedicated computing environment, but if we consider a standard network of workstations, then the load of workstations changes very frequently since users are working on. That is why we have considered a more general environment: a local heterogeneous network of workstations in which we do not make any assumption on the load variation.

5 Asynchronous Calculation

In the previous algorithm all the processors are synchronized several times per iteration in order to exchange the roots they have just computed. If a slow processor is suddenly loaded then all the other processors are delayed until the communication step. As a consequence the efficiency is strongly reduced.

In [3], they show that if the delay between two processors is not too important, then it is possible for each processor to compute its roots in an asynchronous way.

In standard message passing interface, it is often difficult to tackle the asynchronous phase because if a processor is late then all the others have to continue their work and check as frequently as possible if the late processor has sent its results. So, the library's user has to handle this case that makes the program more complex. Furthermore, if we consider a great number of processors n then the number of messages at each communication step is of the order of n^2.

This explains why we prefer to use a server that centralizes all results. This approach has two advantages. The first one is that the asynchronous version is quite easy to implement. The second is that the number of messages decreases when their length increases. As a consequence, the performance over the network can be improved.

5.1 Description of the One-Server Algorithm

We start with the description of the server. Its goal is to schedule the roots among the slave processors. The number of interval roots is fixed at the beginning of the computation and this value is much greater than the number of slaves. The server sends an interval to a slave. Then the slaves compute the interval of roots and send it to the server. Next, the server receives the interval and setup roots. It decides which next interval the slave will compute. Then the server gathers new computed roots that other slave processors have computed since the previous communication and forwards them to the slave. And process goes on.

The time required for a computation of an interval of roots is insignificant since if processors are fast, then they will have more intervals to compute. There is no problem in load balancing because the server manages intervals and distributes them to any idle processor. When an iteration is finished, the server checks if a new iteration is required or if all roots have a good approximation.

6 Synchronous versus Asynchronous

In figure 1 and 2, we represent the load balancing using the synchronous calculation. For that purpose, we use 4 load-free machines. We start the computation, the load balancing is without any effect. At iteration 8 we load processor 3 with a dummy program consuming cpu. At iteration 16 we load the processor 2. At iteration 24 we unload processor 2. Then at iteration 32 we unload processor 3. In figure 1, we can notice that the load variation (at iterations 8, 16, 24 and 32) is clearly visible. Nevertheless, the load balancing is satisfying. In figure 2, the load variation is highlighted by peaks representing an abnormal time consuming iteration.

In figure 3, we display the time required to compute asynchronously one iteration with 4 processors for a server. As previously, we load a processor at iteration 8, a second one at iteration 16. Then we unload one at iteration 24 and another one at iteration 32. The load balancing is done as expected.

Then, we measure the behavior of the synchronous algorithm versus the asynchronous one on a local cluster without any load. We experiment this on a cluster of 15 dual-PIII 1Ghz with a 100Mbps network. We chose a 10,000 degree polynomial and made 10 measurements for each number of processors ranging from 6 to 30.

On figure 4, we report the execution times expressed in seconds for these experiments. This figure shows that the asynchronous calculation is faster than the synchronous one although it has one processor dedicated for the master processor. Therefore, we can conclude that even on a local network without any load, the asynchronous calculation is faster than the synchronous one.

7 Multi-server with Meta-load Balancing

As we show on experiments, there are two factors limiting the scalability. The first one is due to communications latencies if we use remote computers, remote clusters or low bandwidth network. The second limiting factor is due to the time needed for the server to handle all the requests from slaves when the number of slaves increases. That is why we need to develop a multi-master algorithm to deal with these problems. The protocol for server-to-slave communications

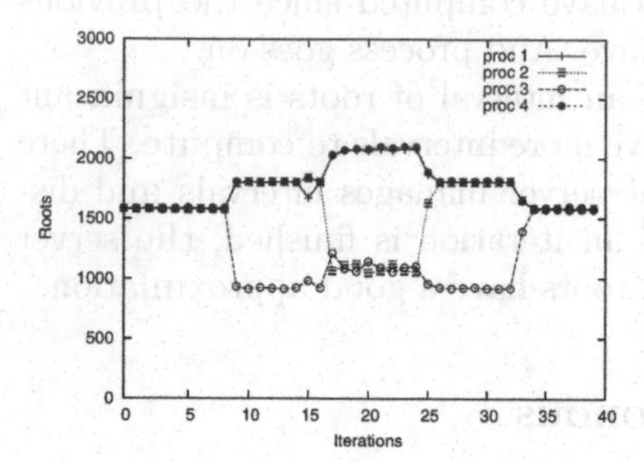

Fig. 1. Load balancing: Roots per iteration (synchronous version)

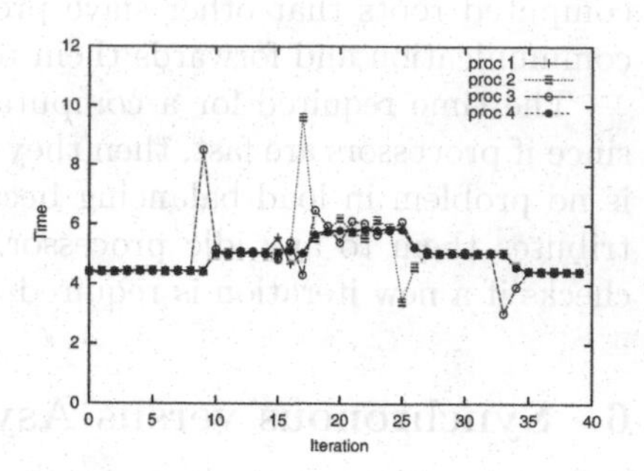

Fig. 2. Load balancing: Time per iteration (synchronous version)

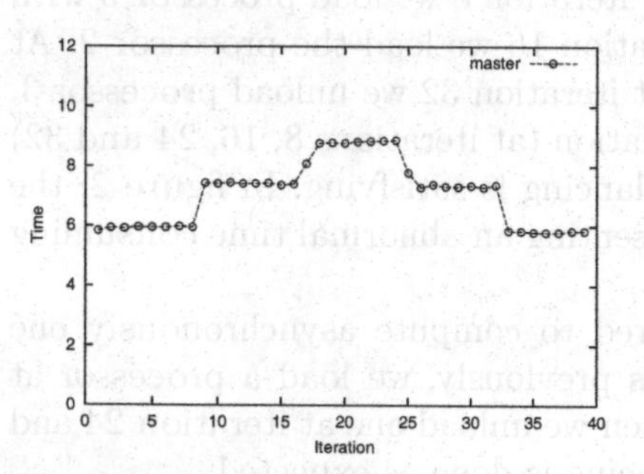

Fig. 3. Load balancing: Time per iteration (asynchronous version)

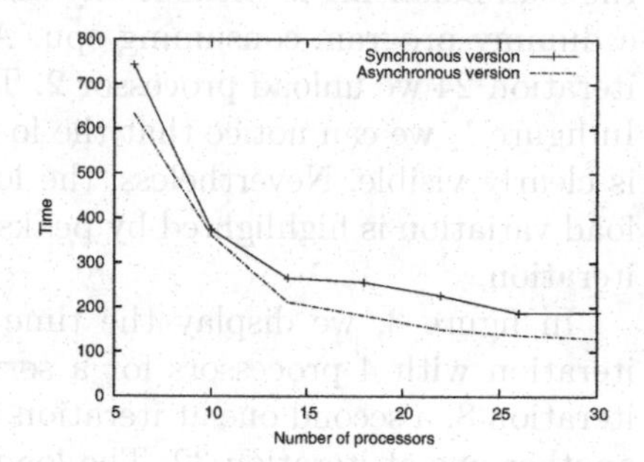

Fig. 4. Execution times of the synchronous and asynchronous algorithms

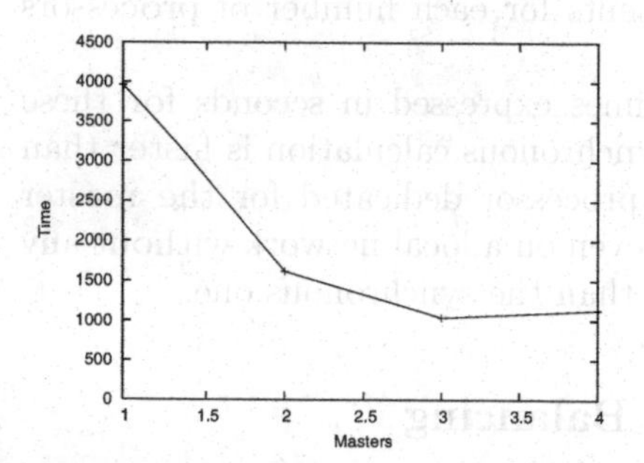

Fig. 5. Execution time of one-master and multi-master algorithms for a 50000 degree polynomial over a homogeneous local cluster

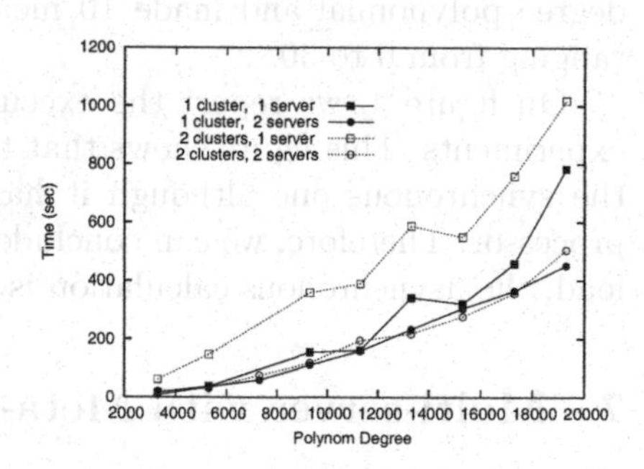

Fig. 6. Comparison times of local and remote clusters with 1 and 2 servers

is the same as the one-server version and we need to define the protocol for server-to-server communications.

We have developed a new version that aims to balance the load between each server in order to keep the number of iterations stable. Indeed, if a master computes roots slower than others then the number of iterations increases and the execution time too.

There are few changes to bring for the algorithm, before it performs a meta-load-balancing functionality. At each end of its local iteration, a master computes the average time required for an iteration. Then it sends this result to the lead master. At a given number of iterations, the lead master balances the load using an equation similar to (3). This equation distributes roots uniformly on each server using the average speed of a server and the total speed of all servers. Having defined the new number of roots that each master should compute, the lead master sends these results.

8 One-Master versus Multi-master Algorithms

8.1 Comparison on a Local Cluster

On a local cluster of homogeneous machines, we have experimented the behavior of the one-master and multi-master algorithm. We made our experimentation on the i-cluster of Grenoble in France. This cluster consists of 216 Pentium III 733Mhz. The network used is 100Mbps from machines to switch and 1Gbps between switches. For this experiment, 90 machines were used and we chose to extract the roots of 50,000 degree polynomial. We measure the behavior of the algorithm using a range from 1 master up to 4 masters. The masters are taken in account into the number of machines. Therefore, with one master, there are 89 slaves and with four masters, there is only 86 slaves. Note that the number of slaves is distributed uniformly for each server. As the algorithm is asynchronous, the number of iterations is not exactly the same from one execution to another. That is why, we have considered the average of 10 executions for each measurement. Figure 5 shows the execution times expressed in seconds for the lead master with the number of masters ranges from 1 to 4. This figure allows us to conclude that it is very important to have more than one master because in this case, the master has a number of requests from slaves that is less important. This figure shows that 3 masters is the best configuration.

8.2 Comparison on 2 Remote Clusters

Experiments have been conducted on two homogeneous clusters. One of them is located in France and the other one is located in the United Kingdom. The communications inside each cluster are achieved on a 100Mbps switch. The link between the two clusters is a 10Mbps on the Internet. This means that the link is not dedicated to this application. Therefore, the communications between the two master processors of each cluster are slow. That implies sometimes an important delay. In order to achieve a significant efficiency, the communications between remote clusters must be asynchronous. Our algorithm uses only "asynchronous" broadcast between each components. When a slave sends its approximated roots to its linked master, this message is relayed to the other masters and then transmitted to all the slaves at the appropriate time. The interesting point of such an algorithm is to improve the scalability. Scalability is limited by the communications. It is crucial to develop a very efficient communication strategy in order to improve performances adding processors. With such an algorithm, the key point is not to transmit a message quickly, but to avoid wasting of time on slaves.

Figure 6 shows that 30 processors locally connected to one master is too much for this type of algorithm. "1 cluster, 1 server" curve has slower performances than the "1 cluster, 2 servers" curve. This indicates that 2 servers reduce network activities and each server has more "ready to respond" time for slaves. Meanwhile, the "2 clusters, 1 server" curve does not deal with the network topology. Half of the slaves have to share the Internet link to be in relation with their linked master. Therefore, the algorithm topology must deal with the network topology. Finally, the "2 clusters, 2 servers" curve has approximately the same results as on one local cluster. These results have been achieved with a low external load on the Internet link. Some experiments with significant external load on the Internet link give us a 20% overtime around. The four curves have been made with the same number of processors.

9 Conclusion

Experiments show encouraging results with asynchronous algorithm on remote clusters. The main factor to improve the global calculation is to reduce waiting times on slaves. The propagation delay is not essential for converging an asynchronous algorithm. Thanks to these results, it will be possible to improve the scalability of this type of algorithm with a multi-server topology.

From now on, we want to extend our results to other types of asynchronous algorithm by developing a dedicated distributed platform. This platform will integrate two types of communications: asynchronous broadcast and asynchronous neighbor exchange.

References

[1] D. A. Bini, *Numerical Computation of Polynomial Zeros by Means of Aberth's Method*, Numerical Algorithms, 13:179-200, 1996. 284

[2] T. C. Chen and W. S. Luk *Alberth's Method for the Parallel Iterative Finding of Polynomials Zeros*, APL94 Conf., Sept, 1994. 284

[3] M. Cosnard and P. Fraignaud, *Analysis of Asynchronous Polynomial Root Finding Methods on a Distributed Memory Multicomputer*, IEEE Trans. on Parallel and Distributed Systems, 5(6):639-648, Jun 94. 284, 286

[4] R. Couturier and F. Spies, *Extraction de racines dans des polynômes creux de degré élevé*, Calculateurs parallèles, Vol. 13(1), pp. 67-81, (2001). 285

[5] M. Lang and B. C. Frenzel, *Polynomial Root Finding*, IEEE SP Letters, 141-143, Oct, 1994.

[6] L. Leger, *Calcul parallèle des racines d'un polynôme à l'aide de la méthode de Weyl, phD Thesis*, University of Rouen, Jun, 1998. 283

[7] W. S. Luk, *Finding Roots of Real Polynomial Simultaneously by means of Bairstow's Method*, BIT, 36(2), 1996. 283

[8] W. Groppn, E. Lusk and A. Skjellum, *Using MPI: portable parallel programming with the message passing interface*, MIT Press, 1994.

[9] K. Rhofir, F. Spies and J. C. Miellou, *Perfectionnements de la méthode asynchrone de Durand-Kerner pour les polynômes complexes*, Calculateurs Parallèles, 10(4):449-458, 1998. 284

Efficient Tree-Based Multicast in Wormhole-Routed Networks

Jianping Song, Zifeng Hou, and Yadong Qu

Wireless Communication Technology Lab, Legend Corporate R&D
P. O. Box 8688, Beijing, 100085, P. R. China
{Songjp,Houzf,Quyd}@Legend.com

Abstract. This paper presents two tree-based multicast algorithms for wormhole-routed *n*D torus and 2D mesh networks, respectively. Both of the two algorithms use multiple spanning trees with an up-down routing strategy to provide deadlock-free routing. In contrast with single spanning tree algorithms that use non-tree links only as shortcuts, the presented algorithms use all links (for *n*D tori, all but *n* links) as tree links and, therefore, better balance network traffic and resource utilization. For the algorithm in 2D mesh networks, shortcuts can still be used to shorten routing path for both unicast and multicast messages. Simulation results show that both of the proposed algorithms outperform the best single spanning tree approach obviously.

1 Introduction

In a multicomputer network, processors often need to communicate with each other for various reasons. The design and resulting performance of communication protocols greatly depend on the underlying switching mechanism used in the multicomputers. Although packet switching and circuit switching have been used widely in the first- and second-generation parallel computers, the more recent parallel computers have adopted wormhole routing [1], which is known to be quite insensitive to routing distance if the network is contention-free.

In this paper, we study the multicast problems in wormhole-routed networks, where each source may send a message to any set of destination nodes at any time. Solutions for multicasting can be classified into three categories: unicast-based, path-based, and tree-based. The unicast-based solutions make use of unicast (one-to-one) communication and add additional software to support multicast. Disadvantages of this approach include necessary involvement of intermediate nodes in message propagation and required latencies to startup messages in source and each intermediate node.

The path-based solutions either construct a single path spanning all nodes of the network or use an underlying deadlock-free unicast routing algorithm to construct multiple paths that combine to reach all destinations. In the former approach, the path lengths can be extremely long, which leads to high latency variation. While in the latter approach, multiple costly message startups are required and this also may cause worm transmissions to be serialized.

S. Sahni et al. (Eds.) HiPC 2002, LNCS 2552, pp. 292-301, 2002.

Tree-based multicast has low latency and small latency variation. Sivaram et al. [2] have shown that tree-based techniques offer a very promising means of achieving extremely efficient multicast routing. However, most existing tree-based algorithms for direct networks avoid deadlock by either requiring complex synchronization mechanisms or by requiring that intermediate routers be able to buffer the message in its entirety, thereby limiting the length of packets to be no longer than the size of these buffers.

In this paper, two tree-based multicast schemes are proposed for *n*D torus networks and 2D mesh networks, respectively. By constructing multiple spanning trees, these two algorithms can balances well the traffic through the network. Simulation results show that both of them significantly outperform the best single tree algorithm.

The remainder of this paper is organized as follows. In Section 2 we review related work on tree-based multicasting. In Section 3 we describe a multicast algorithm in *n*D torus networks, and in Section 4 we describe another multicast algorithm in 2D mesh networks. The simulation results of the two algorithms are given in Section 5 and the conclusion is made in Section 6.

2 Related Work

Tree-based multicast for wormhole routing has been considered difficult in terms of avoiding deadlock. In [3], a deadlock-free tree-based multicast algorithm is presented for short message. The restriction to short messages results because the algorithm sometimes stores entire messages within a router. Hence, this approach is adequate for cut-through switching and does not provide a general solution for multicast in wormhole-routed networks.

A general tree-based multicast scheme can be implemented by modifying the up-down routing algorithm [4]. In up-down routing, a spanning tree rooted at an arbitrary node is constructed. All tree links are classified as 'up' links and 'down' links with respect to the root. A tree link is 'up' if it goes from a node at a lower level of the tree to a node at a higher level. A tree link is 'down' otherwise. Unicast routing is done by traversing zero or more up links until the lowest common ancestor of source and destination is reached, followed by zero or more down links to reach the destination. This algorithm was proved to be deadlock-free. Some non-tree links (known as cross links) can be used as short cuts while still maintaining deadlock-freedom. Up-down routing can be applied to multicast if replication at tree branch points is assumed and the routes are restricted to only tree links. However, to prevent deadlock, this restriction must also be applied to unicast messages. That is, no cross links will be used, and most unicast messages can't route along shortest paths. As a result, links close to the root may get extremely congested and become bottlenecks.

It is desirable to use as many links as possible to better balance traffic. In [5], a new algorithm is presented which allows unicast messages to use cross links but still requires that multicast messages use only tree links. For multicasts, a unicast message is first sent to the root of the tree. The root then sends a multicast message to all destinations. Thus, multicast messages are totally ordered by the root and so deadlock will not occur between two multicast messages. When a multicast message is blocked by a unicast message at a link, it suspends its transmission and yields its path to the unicast message. After the unicast message finishes using the link, the multicast is resumed and some kind of reassembly is needed at the destination for the multicast message. The suspension, resumption and reassembly make this algorithm quite complex to implement in hardware.

Then, a more efficient tree-based multicast algorithm known as SPAM is proposed in [6]. In this algorithm, links are divided into up tree links, up cross links, down tree links and down cross links. A unicast message under SPAM is routed along zero or more up links, followed by zero or more down cross links, followed by zero or more down tree links. A multicast message is first routed to the least common ancestor (LCA) of the set of destinations using the unicast algorithm. Once the message has arrived at LCA, all subsequent routing is restricted to down tree links. A complete proof of deadlock freedom is given in [7]. Although SPAM makes use of cross links, it still places most of the network traffic on the tree links of a single spanning tree. This places a heavy load on tree links, particularly those near the root of the tree.

In order to better balance the traffic, a new algorithm named DSTM is proposed in [8]. In this algorithm, two edge-disjoint spanning trees are constructed and each executes SPAM algorithm respectively. For a unicast message, the tree with the shorter path is picked to transmit it. While for multicast messages, a single tree is picked randomly for each multicast. Since almost all links in the network are used as tree links, the algorithm balances well the traffic through the network. However, to avoid deadlock, for both unicast and multicast messages, only tree links are used, no cross links are allowed. Moreover, the DSTM algorithm can only be applied to 2D torus networks.

In this paper, we proposed two spanning tree-based multicast algorithms, STBM-torus and STBM-mesh, in *n*D torus networks and 2D mesh networks, respectively. In STBM-torus algorithm, *n* edge-disjoint spanning trees are constructed in an *n*D torus network and each executes SPAM algorithm respectively. To avoid deadlock, both unicast and multicast messages can only use tree links. For 2D mesh networks, it is impossible to construct two edge-disjoint spanning trees. However, we will show that the STBM-mesh algorithm still maintains deadlock-freedom although the two spanning trees have common links. What's more, both unicast and multicast messages can use up cross links, and unicast messages can even use the down cross links. As a result, the unicast messages can always route along shortest paths. Simulation results demonstrate that both of STBM-torus and STBM-mesh algorithms significantly outperform the SPAM algorithm for both pure multicast and combined unicast/multicast traffic.

3 Multicast in Torus Networks

Using SPAM algorithm as the underlying tree-based deadlock-free multicast algorithm with wormhole routing, our investigation concentrated on finding the edge-disjoint spanning trees to achieve better performance. A N-node n-dimensional torus network has a total of nN links, while a spanning tree has $N-1$ links in any network with N

nodes. Hence, it is possible to construct n edge-disjoint spanning trees in an nD torus network.

For a $k_1 \times k_2 \times \ldots \times k_n$ torus network, the construction method is as follows. First, randomly pick a node as the common root node for the n spanning trees. Then, for each i, 1 i n, call the algorithm shown in Fig. 1 to construct the ith spanning tree T_i. Here we use $\sigma_i(u)$ to indicate the index of node u in the ith dimension, while 0 $\sigma_i(u)$ $k_i - 1$.

```
Algorithm: Constructing the ith spanning tree Ti in k1 × k2 × … × kn torus networks
Input: root node r, integer i, 1   i   n
begin
  for each node u in the network do
    if σi(u) = σi(r) then
      mark node u as masked;
    else
      mark node u as unvisited;
    endif
  endfor
  add node r to spanning tree Ti and mark r as visited;
  put node r into the queue Q;
  while Q is not empty do
    delete the first node v in Q;
    for m = 0 to n – 1 do
      j = (m + i) mod n;
      let w be the adjacent node of v in the positive direction of the jth dimension.
      if (w is unvisited) or (w is masked and (σi(v) + 1) mod ki = σi(r)) then
        add node w to spanning tree Ti with v as its parent node and mark w as visited;
        put node w into the queue Q;
      endif
    endfor
  endwile
end
```

Fig. 1. The algorithm to construct edge-disjoint spanning trees

In order to construct spanning tree T_i, we first mask all the nodes that have the same indices as the root node in the ith dimension. Then, visit all other nodes in the network using a method similar with the breadth-first search, beginning from the root node. For each visited node, only those adjacent nodes in the positive direction of each dimension may be added to the tree. Note that the adjacent nodes are visited in an increasing dimensional order (modulo n if necessary) beginning from the ith dimension. The masked nodes are handled specially.

Fig. 2 shows the two edge-disjoint spanning trees applying the construction algorithm to an 8×8 torus network, while Fig. 3 shows the three edge-disjoint spanning trees constructed in a $4 \times 3 \times 3$ torus network.

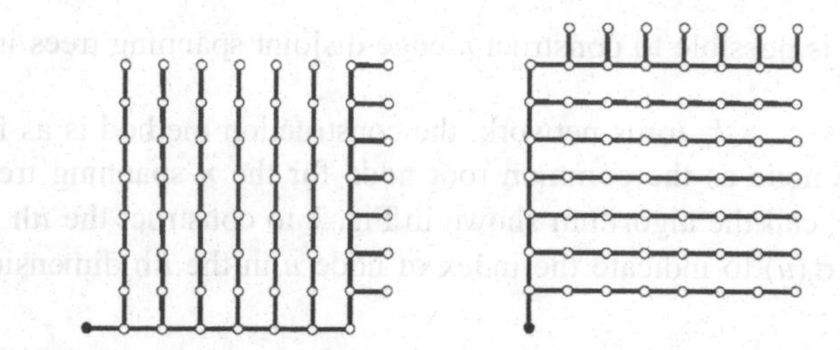

Fig. 2. The two edge-disjoint spanning trees in an 8 × 8 torus network

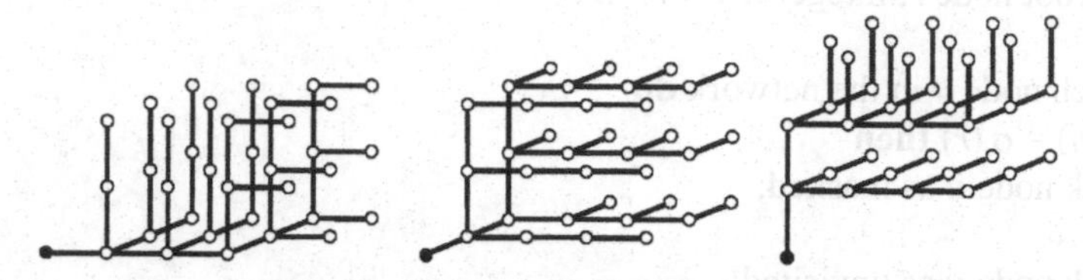

Fig. 3. The three edge-disjoint spanning trees in a 4 × 3 × 3 torus network

Once the edge-disjoint spanning trees are constructed, our routing algorithm for unicast and multicast messages works as follows. To save network resources, we pick the tree with the shortest path to transmit a unicast message. While for a multicast message, since the amount of resources that will be taken cannot be easily foreseen, a single tree is picked randomly. For both unicast and multicast messages, only tree links are used, no cross links are allowed. Since only tree links are used and an up-down order is followed, the messages are deadlock-free in each tree. To guarantee that the trees are truly edge-disjoint, n consumption channels are needed at each node for an nD torus network. Since the n trees are edge-disjoint, they will not deadlock with each other either. Therefore, STBM-torus algorithm is deadlock-free.

4 Multicast in Mesh Networks

A N-node $k_1 \times k_2$ mesh network has a total of $2N - k_1 - k_2$ links, while a spanning tree has $N - 1$ links in any network with N nodes. Therefore, it is impossible to construct two edge-disjoint spanning trees in arbitrary 2D mesh networks. However, we will show that the STBM-mesh algorithm is still deadlock-free although the two spanning trees have common links.

To construct the two spanning trees, we pick any corner node of the mesh network as the common root node of the two trees. For simplicity, we assume throughout the paper that the node (0, 0) is picked as the common root node. The first spanning tree, T_1, is constructed as follows. Starting from the root node (0, 0), visit all the nodes in the same row along $+x$ direction, and add nodes to the spanning tree as they are visited. Then, for each node that is already in the tree, visit each column along $+y$ direction, and add nodes to the tree as they are visited. The second spanning tree, T_2, is constructed similarly. Starting from the root node (0, 0), visit all the nodes in the same column along $+y$ direction, and add nodes to the spanning tree as they are visited.

Then, for each node that is already in the tree, visit each row along $+x$ direction, and add nodes to the tree as they are visited. The constructed two spanning trees in an 8×8 mesh network are shown in Fig. 4.

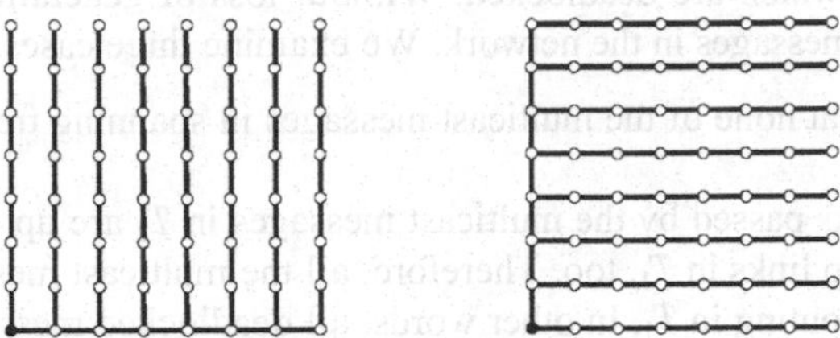

Fig. 4. The constructed two spanning trees in an 8×8 mesh network

Instead of restricting the route to only tree links, in STBM-mesh algorithm, both unicast and multicast messages route according to the SPAM algorithm. For unicast messages, any cross link can be used if possible. As a result, the unicast messages can route along shortest paths. For multicast messages, any up cross link can be used while the down cross links are prohibited. However, to avoid deadlock, we always pick the first spanning tree to transmit unicast messages. While for each multicast message, if the source node or any of the destination nodes exists in the first column (but it isn't the root node) of the network, then the second tree is picked; or a tree is picked randomly otherwise. So, for the multicast messages routing along the first spanning tree, the links in the first column of the network will never be used.

Following theorem will help us to prove the deadlock-freedom of STBM-mesh algorithm.

Theorem 1. *For any link c (tree link or cross link) in the network, if it is an up (down) link in one spanning tree, it must be an up (down) link in another spanning tree, too.*

Proof: Note that if link $c = <u, v>$ is an up (down) link in spanning tree T_1, then node u must be closer (farther) to the root of T_1 than node v. Since the two spanning trees have the common root node, then node u must also be closer (farther) to the root of T_2 than node v. That is, link c is an up (down) link in spanning tree T_2, too. The proof is similar if link c is an up (down) link in spanning tree T_2.

By Theorem 1, when we say an up link or a down link, we don't need to indicate which tree the link is relative to.

Definition 1. *For any multicast message m, if link $c = <u, v>$ is the first down link in the path from source node to the LCA of the destination set of m, then node u is referred to as the first down point of multicast message m. If no down links exist in the path, then LCA is the first down point of m.*

Observe that no down cross links are used by multicast messages. Hence, once a multicast message passes its first down point, all subsequent routing is restricted to down tree links.

We can label each node in the network such that node u has a less label than node v if u is visited before v when tree T_1 is traversed by breadth-first search. Each link $c = <u, v>$ is also labeled such that c has the same label as node v. Now we can prove the deadlock-freedom of STBM-mesh algorithm.

Theorem 2. *STBM-mesh algorithm is deadlock-free.*

Proof: Assume that STBM-mesh algorithm isn't deadlock-free. Let $m_1, m_2, \ldots, m_n$ be a set of messages which are deadlocked. Without loss of generality, we assume that these are the only messages in the network. We examine three cases.

Case 1: Assume that none of the multicast messages in spanning tree T_2 has passed its first down point.

Then all the links passed by the multicast messages in T_2 are up links. By Theorem 1, these links are up links in T_1, too. Therefore, all the multicast messages in T_2 can be regarded as those routing in T_1. In other words, all deadlocked messages are routing in T_1. However, by SPAM algorithm, the messages in a single spanning tree won't cause deadlock, a contradiction.

Case 2: Assume that none of the multicast messages in spanning tree T_1 has passed its first down point.

By the proof of Case 1, at least one multicast message in T_2 has passed its first down point. Let w be the maximum node (with respect to node label) at which some multicast message in T_2 passed its first down point and let m_i be any multicast message in T_2 that passed its first down point at w.

Let m_i be blocked at link $c = <u, v>$ by message m_j. Since m_i has passed its first down point, then c must be a down link. If m_j is a multicast message, then m_j must also have passed its first down point since m_j occupied the down link c. By the assumption of Case 2, m_j can't be a multicast message in T_1. Hence, m_j is either a multicast message in T_2 that has passed its first down point or a unicast message in T_1. If m_j is a multicast message in T_2 that has passed its first down point, then the first down point cannot be a descendant of w in T_2 since this would contradict the maximality of w. Therefore, the first down point is w or an ancestor of w in T_2, which means that the first down tree link on the path from w to u in T_2 is reserved by both m_i and m_j, a contradiction. Thus, m_j is a unicast message in T_1. In the same way we know that m_j cannot be blocked by a multicast message, otherwise the multicast message must be one in T_2 that has passed its first down point and again a contradiction occurs. That is, the unicast message m_j must be blocked by another unicast message. Repeating the deducing procedure, and we can conclude that all the messages block m_i directly or indirectly are unicast messages in T_1. By SPAM algorithm, these unicast messages are not deadlocked, contradicting the assumption that all messages in the network are deadlocked.

Case 3: Finally, we consider the case that at least one multicast message in spanning tree T_1 has passed its first down point.

Let w be the maximum node (with respect to node label) at which some multicast message in T_1 passed its first down point and let m_i be any multicast message in T_1 that passed its first down point at w. Let u be any descendant of w in T_1 at which m_i is blocked. In the subtree of T_1 with u as the root node, let $c = <x, y>$ be the maximum down tree link (with respect to link label) of T_1 at which a message is blocked, and let m_j occupy link c. There are two cases should be considered with respect to message m_j.

a) m_j is a message in T_1. If m_j is a multicast message, then m_j must have passed its first down point since m_j occupied the down link c. The first down point cannot be a descendant of w in T_1 since this would contradict the maximality of w. Therefore, the first down point is w or an ancestor of w in T_1, which means that the first down tree link on the path from w to u in T_1 is reserved by both m_i and m_j, a contradiction. Thus, m_j is a unicast message in T_1. Since c is a down tree link in T_1, then m_j must be blocked at some descendant of y in T_1. Hence, the link at which m_j is blocked must have a greater label than link c, which contradicts the maximality of c.

b) m_j is a multicast message in T_2. Observe that all the common links of T_1 and T_2 exist in the first row or column of the network. Since the multicast messages in T_1 will never use the links in the first column of the network, then link c must exist in the first row. As a result, node w and u must exist in the first row, too. Let l be the maximum down link (with respect to link label) in the first row at which a message is blocked, and let m_k occupy link l. By the discussion of Case 3.a, m_k cannot be a message in T_1. Thus, m_k is a multicast message in T_2 which has passed its first down point. If m_k has no other branches except for the path in the first row, then m_k must be blocked at a down link in the first row which has a greater label than link c, which contradicts the maximality of c. Hence, m_k has other branches and the root node of T_2 is its first down point. This means that the first down link on the path from w to u is reserved by both m_i and m_k, a contradiction.

Since each of the above three cases leads to a contradiction, the assumption that STBM-mesh algorithm isn't deadlock-free doesn't hold.

5 Simulation Results

We developed a flit-level simulation program to evaluate the performance of the STBM-torus and STBM-mesh algorithms. The evaluations were done in a 16 × 16 torus network for STBM-torus algorithm and in a 16 × 16 mesh network for STBM-mesh algorithm, respectively. For comparison, the SPAM algorithm was also simulated in both torus and mesh networks.

The following parameters are used for all the experiments reported in this section: 1 clock cycle to transmit a flit across a physical channel, 5 clock cycles for the header flit to make a routing decision and 1 clock cycle as the switching delay for successive flits. Since each of the three algorithms requires only one startup for each message, we did not consider the startup latency. Both the input and output buffers of the router can holds only one flit. Each node has two pairs of internal channels to consume independently the messages arriving over each tree. Each message comprises 128 data flits. A uniform distribution is used to select the source node and the destination node set of each message.

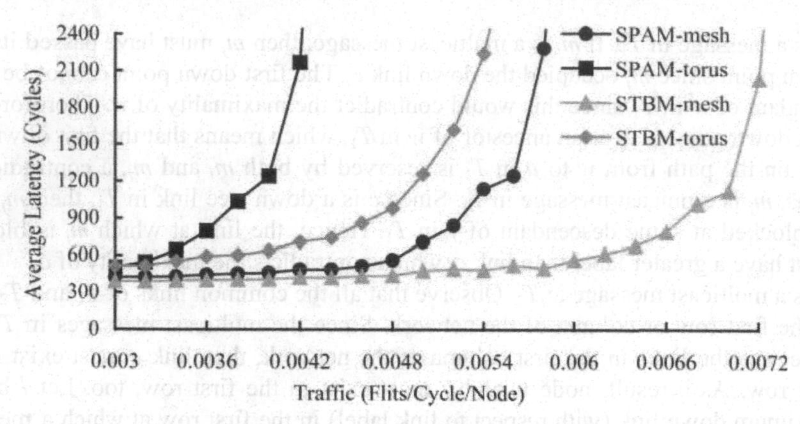

Fig. 5. Latency versus load with traffic comprised of 90% unicast and 10% multicast messages

In the first set of simulations, message latency was measured for mixed unicast and multicast traffic in which 90% of messages were unicast and 10% of messages were multicast. Each multicast message has 20 destinations. Fig. 5 shows the average message latency for the different multicast algorithms under different traffic loads. These results clearly show that the STBM-torus and STBM-mesh algorithms outperform SPAM significantly. This is because the traffic can be better balanced in STBM-torus and STBM-mesh since two spanning trees are used. STBM-mesh algorithm performs better than STBM-torus because unicast messages can use shortest paths in STBM-mesh algorithm.

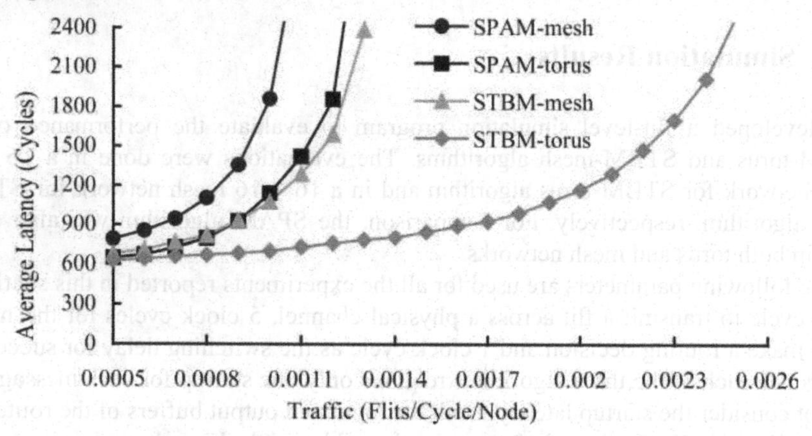

Fig. 6. Latency versus load with pure multicast traffic

In the second set of simulations, message latency was measured for pure multicast traffic with the number of destinations is still 20. As shown in Fig. 6, both STBM-torus and STBM-mesh perform better than SPAM under all the traffic loads. However, STBM-torus outperforms STBM-mesh when load is high. The reason is that the two spanning trees are edge-disjoint in STBM-torus, which causes the multicast messages in different trees not to interfere with each other.

6 Conclusion

In this paper, two tree-based multicast algorithms for wormhole-routed *n*D torus and 2D mesh networks are presented, respectively. Both of the two algorithms use multiple spanning trees with an up-down routing strategy to provide deadlock-free routing. In contrast with single spanning tree algorithms that use non-tree links only as shortcuts, the presented algorithms use all links (for *n*D tori, all but *n* links) as tree links and, therefore, better balance network traffic and resource utilization. For the algorithm in 2D mesh networks, shortcuts can still be used to shorten routing path for both unicast and multicast messages. Simulation results show that both of the proposed algorithms outperform the SPAM algorithm significantly.

An interesting open question is how well these approaches can be extended to other networks, including irregular topology networks. Since the vendors usually provide the routers or the network interface cards with a fixed number of ports, multiple spanning trees could then be constructed in an irregular network to better balance the network traffic.

References

[1] W. J. Dally and C. L. Seitz, "The torus routing chip," Journal of Distributed Computing, vol. 1, no. 3, pp. 187-196, 1986.

[2] R. Sivaram, D. Panda, and C. Stunkel, "Multicasting in irregular networks with cutthrough switches using tree-based multidestination worms," Proceedings of the 2nd Parallel Computing, Routing, and Communication Workshop, Jun. 1997.

[3] M. P. Malumbres, J. Duato, and J. Torrellas, "An efficient implementation of tree-based multicast routing for distributed shared-memory multiprocessors," Proceedings of the 8th IEEE Symposium on Parallel and Distributed Processing, pp. 186-189, Oct. 1996.

[4] M. D. Schroeder, et al., "Autonet: A high-speed, self-configuring local area network using point-to-point links," Technical Report 59, DEC SRC, Apr. 1990.

[5] M. Gelar, P. Palnati, S. Walton, "Multicasting protocols for high-speed, wormhole-routing local area networks," Computer Communication Review, vol. 25, no. 4, (ACM SIGCOMM'96 Conference, Aug. 1996), pp. 184-193, ACM, Oct. 1996.

[6] R. Libeskind-Hadas, D. Mazzoni, R. Rajagopalan, "Tree-based multicasting in wormhole-routed irregular topologies," Proceedings of the 13th International Parallel Processing Symposium and 10th Symposium on Parallel and Distributed Processing, pp. 244-249, 1998.

[7] R. Libeskind-Hadas, D. Mazzoni, R. Rajagopalan, "Tree-based multicasting in wormhole-routed irregular topologies," Technical Report HMC-CS-97-02, Harvey Mudd College, Aug. 1997.

[8] H. Wang and D. M. Blough, "Tree-based multicast in wormhole-routed torus networks," Proceedings of the 1998 International Conference on Parallel and Distributed Processing Techniques and Applications, pp. 702-709, 1998.

Parallel Algorithms for Identification of Basis Polygons in an Image

Arijit Laha[1], Amitava Sen[2], and Bhabani P. Sinha[2]

[1] National Institute of Management Calcutta
Alipore, Calcutta – 700027, India
arijitl@yahoo.com

[2] ACM Unit, Indian Statistical Institute
Calcutta – 700 108, India
amitavasen@hotmail.com
bhabani@isical.ac.in

Abstract. Given a set of n straight line segments each described by its two end points, we propose two novel algorithms for detecting all basis polygons formed by them. The algorithms, based on traversals along the sides of the basis polygons, detect the polygons in $O(n)$ time using n^2 processors. The first algorithm handles the simple scenes consisting of convex basis polygons only, while the second one deals with the general situation. These algorithms have been simulated and tested for a number of input sets of intersecting line segments.

1 Introduction

One of the fundamental problems associated with two-dimensional images consisting only of the staright line segments is that of detecting the straight lines. Most often this problem is tackled by using Hough transform [1], [2]. In recent years, a number of parallel algorithms for computing the Hough transform on different architectures have been presented by several authors [3], [4], [5]. Asano et al [6] proposed a scheme for detection of straight lines based on topological walk on an arrangement of sinusoidal curves defined by Hough transform. In [7], a novel parallel algorithm for identiying all straight line segments has been proposed, which overcomes several difficulties associated with the approach using the traditional Hough transform.

However, once the straight line segments in such an image are detected, an even higher level description of the image can be generated in terms of the polygons created by the constituent line segments. Such a description is more useful for syntactic pattern recognition and computer vision tasks. A number of interesting operations on polygons have been investigated by the researchers in computational geometry and computer graphics, once these polygons are identified. These include i) finding the convex hull of a polygon [8], [9], ii) testing the convexity of a polygon [10], iii) finding the intersection of two convex polygons [11], [12], [13], iv) finding the minimum vertex distance between two crossing convex polygons [14], v) triangulation of polygons [15], etc.

S. Sahni et al. (Eds.) HiPC 2002, LNCS 2552, pp. 302–312, 2002.

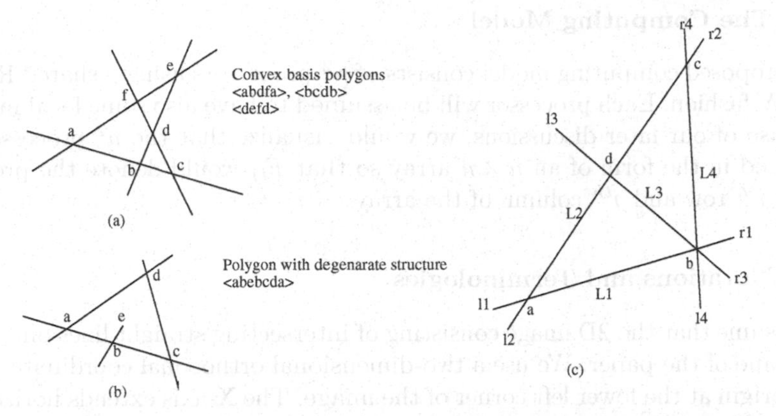

Fig. 1. Example of (a) convex basis polygons, (b) polygon with degenerate structures and (c) An image with four lines forming two basis polygons

Given the set of straight lines in a 2D image, the set of polygons formed by all these lines can best be described by the *basis polygons*, where a basis polygon is defined as one which would not enclose any other polygon. The maximum number of basis polygons that can be created by n intersecting straight line segments can be computed as

$$\sum_{k=2}^{n-1}(k-1) = \binom{n-1}{2} \approx O(n^2)$$

Again, onc polygon can have a maximum of n sides. Hence, in the worst case, the computational complexity for identifying all polygons can be as high as $O(n^3)$.

In this paper, we present two parallel algorithms for detecting all the basis polygons created by n intersecting straight line segments using $n \times n$ array of processors. Both these algorithms need $O(n)$ time to identify all the basis polygons using $O(n^2)$ processors in a CREW PRAM model. Thus, for the worst case scenarios when $O(n^2)$ basis polygons are formed by n intersecting straight lines, the proposed algorithms will be AT-optimal. The first one of the two algorithms presented here is developed under the assumption that all the basis polygons are convex (Fig. 1(a)) and no degenerate structures (e.g., a line ending inside a polygon (Fig. 1(b))) or nested polygons (in which one polygon is completely enclosed by another) exist. In Fig. 1(b), the polygon *abebcda* contains a degenerate structure *beb*. The second algorithm is designed to tackle the difficulties introduced by non-convexity, degenerate structures and nested polygons.

2 Basic Concepts

We describe here the computing model, notations and terminologies used.

2.1 The Computing Model

The proposed computing model consists of n^2 processors using a shared RAM in CREW fashion. Each processor will be assumed to have also some local memory. For ease of our later discussions, we would visualize that the n^2 processors are arranged in the form of an $n \times n$ array so that P_{ij} would denote the processor in the ith row and j^{th} column of the array.

2.2 Notations and Terminologies

We assume that the 2D image consisting of intersecting straight lines only, lies in the plane of the paper. We use a two-dimensional orthogonal coordinate system with origin at the lower left corner of the image. The X-axis extends horizontally rightward and the Y-axis extends vertically upward in the plane of the paper.

A point $p = (x, y)$ is denoted by a two-tuple of x and y coordinate values.

Definition 1 : An ordering among the points on the 2D plane is defined as follows: (1) a point $p_1 = (x_1, y_1)$ is equal to another point $p_2 = (x_2, y_2)$ if $x_1 = x_2$ and $y_1 = y_2$, (2) p_1 is on the left of p_2 (denoted as $p_1 < p_2$), if $x_1 < x_2$ or ($x_1 = x_2$ and $y_1 < y_2$), (3) p_1 is on the right of p_2 (denoted as $p_1 > p_2$), if $x_1 > x_2$ or ($x_1 = x_2$ and $y_1 > y_2$).

The input to our algorithms will be a set of n lines $\{\bar{L}_i : i = 1, 2, ..., n\}$, each line being expressed as an ordered 2-tuple of end points (p_l, p_r), where p_l is the left end point and p_r is the right end point, i.e., $p_l < p_r$.

To facilitate our later discussions, whenever we would refer to a point, this would exclusively mean either an end point of a straight line or an intesection point of a set of straight lines.

Definition 2 : A point p_i said to be a neighbor of another point p_j on a line $\bar{L}_k$ if (1) $p_i \neq p_j$ and (2) there is no other point on $\bar{L}_k$ between p_i and p_j.

Example 1: In Fig. 1(c), four lines (L_1, L_2, L_3 and L_4) with end points as $(l_1, r_1), (l_2, r_2), (l_3, r_3)$ and (l_4, r_4), respectively are shown. The point a has the points l_1 and b as its left and right neighbors, respectively on line 1. Point a is also on line 2 and it has points l_2 and c as left and right neighbors, respectively on line 2. However, l_1 has only a as its right neighbor.

Both $Intersection(i, j)$ and $Intersection(j, i)$ can be used to denote the intersection point of lines $\bar{L}_i$ and $\bar{L}_j$. The intersection point of more than two lines can be referred to using the indices of any two intersecting lines.

A polygon will be represented by a list of vertices $\{v_i : i = 1, 2, ..., s\}$ satisfying the conditions : (1) $v_1 = v_s$, (2) v_i is either an intersection point of two or more lines or a pure endpoint (such vertices may appear if degenerate structures (Fig. 1(b)) exist), (3) for $1 \leq i < s$, v_i and v_{i+1} are joined by a single line segment and (4) there is no other intersection point between v_i and v_{i+1} (otherwise, that point should have appeared between v_i and v_{i+1} in the vertex list).

Example 2: Fig. 1(a) contains three basis polygons described by vertex lists $< abdfa >$, $< bcdb >$ and $< defd >$. Fig. 1(b) depicts a polygon with degenerate structures which is represented by the vertex list $< abebcda >$.

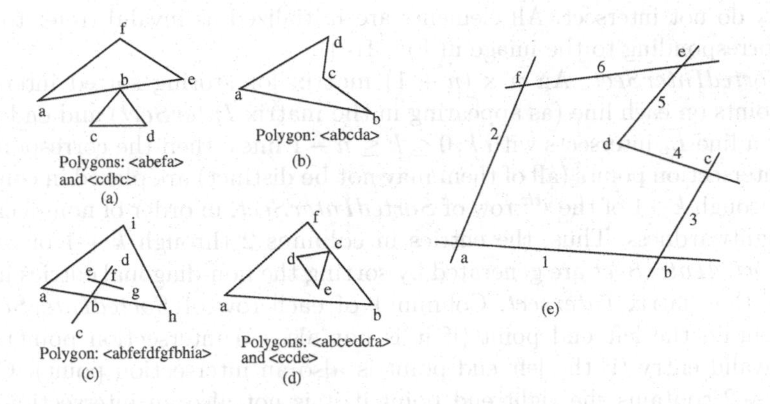

Fig. 2. The situations which can be faced in general cases (a) touching polygon, (b) non-convex polygon, (c) degenerate structure and (d) nested polygon, (e) depicts six line segments forming a non-convex polygon

Apart from the convex polygons as shown in Fig. 1(a) and Fig. 1(c) which are the simplest to work with, we would like to consider also the following polygonal structures : (1) **touching polygons**, e.g., polygons $< abefa >$ and $< bcdb >$ in Fig. 2(a). (2) **non-convex polygons** (Fig. 2(b)). (3) **polygons with degenerate structure** when one or more lines have their end point(s) within the polygon, e.g., the polygon $< abfefdfgfbhia >$ in Fig. 2(c) with one end point of line $\boldsymbol{cd}$ and a whole line $\boldsymbol{eg}$ within it. (4) **nested polygons**, e.g., the polygon $< ecde >$ (Fig 2(d)) contained within the outer polygon $< abcfa >$. A polygon with any combination of these four situations will be called a **general polygon**.

3 Algorithms

The proposed algorithms are based on ordered traversals along the sides of the potential basis polygons. Each traversal starts from an intersection point on a line, proceeds along this line towards the right neighbor of the point on the line. At that point the traversal chooses another neighboring point as the next vertex based on some criteria. Each intersection point has a special correspondence to a particular processor in that the processor P_{ij} would initiate a traversal from the $Intersection(i, j)$ along the line i.

3.1 Algorithm CP (for Convex Polygons)

We use the following data structures to describe this algorithm :

Data Structures:

(1) $InterSect$: An $n \times n$ matrix for storing intersection points. $InterSect_{ij}$ stores the intersection point of lines $\bar{L}_i$ and $\bar{L}_j$ or an *invalid* value if $\bar{L}_i$ and

$\bar{L}_j$ do not intersect. All elements are initialized as invalid (refer to Fig. 3 corresponding to the image in Fig. 1(c)).

(2) *SortedInterSect*: An $n \times (n+1)$ matrix for storing sorted intersection points on each line (as appearing in the matrix *InterSect*) and end points. If a line $\bar{L}_i$ intersects with $k, 0 \leq k \leq n-1$ lines, then the corresponding k intersection points (all of them may not be distinct) are placed in columns 2 through $k+1$ of the ith row of *SortedInterSect*, in order of non-decreasing rightwardness. Thus, the entries in columns 2 through $k+1$ of row i of *SortedInterSect* are generated by sorting the non-diagonal entries in row i of the matrix *Intersect*. Column 1 of each row of *SortedInterSect* will contain the left end point (if it is not also an intersection point), or an invalid entry (if the left end point is also an intersection point). Column $k+2$ contains the right end point if it is not also an intersection point; otherwise it contains an invalid entry. All columns from $k+3$ to $n+1$ contain invalid entries (see Fig. 3 corresponding to the image in Fig. 1(c)).

(3) *SortingIndex*: An $n \times (n+1)$ matrix of integer values, to create a positional correspondence between the sorted entries of the intersection points in *SortedInterSect* and their original (assorted) entries in *InterSect*. Thus, if $SortedInterSect_{ij} = InterSect_{i,k}$, then $SortingIndex_{ij} = k$. For non-distinct entries in *SortingIndex*, if $SortedInterSect_{ij} = SortedInterSect_{i,j+1}$, then the corresponding entries in $SortingIndex_{ij}$ and $SortingIndex_{i,j+1}$ are filled up with index values $k1$ and $k2$ (when $SortedInterSect_{ij} =$ $SortedInterSect_{i,j+1} = InterSect_{i,k1} = InterSect_{i,k2}$), such that $k1 < k2 < \cdots$. All entries in *SortingIndex* corresponding to pure end points and invalid values in *SortedInterSect* matrix are set to an invalid index value of -1.

(4) *InterSectIndex*: An $n \times n$ matrix of integer values, to create an inverse correspondence as done by *SortingIndex*. Thus, if $InterSect_{ij} = SortedInterSect_{ik}$ then $InterSectIndex_{ij} = k$. All entries corresponding to invalid values in *InterSect* matrix are set to an invalid index value of -1. In case of non-distinct entries in *SortedInterSect*, the corresponding entries in *InterSectIndex* are so entered that the index values are in ascending order from left to right.

The matrices *SortingIndex* and *InterSectIndex* corresponding to Fig. 1(c) are shown in Fig. 3. Before presenting the algorithm in a formal manner we first describe its basic idea with the help of the example of Fig. 1(c).

A traversal for detecting a basis polygon starts from an intersection point on one of the intersecting lines. This point is marked as the *starting vertex*, and put in the vertex list describing a basis polygon. The line on which this starting vertex lies, is termed (temporarily) as the *current line*. For example, point a on line L_1 may be the starting vertex and line L_1 will then be the current line. The next vertex is chosen as the closest intersection point on the current line on the right side of the starting vertex (**Start Rightward** strategy). This next vertex can easily be found with the help of the matrices *InterSect* and *SortedInterSect*.

InterSect:

Invalid	a	b	b
a	Invalid	c	d
b	c	Invalid	b
b	d	b	Invalid

SortedInterSect:

l1	a	b	b	r1
l2	a	c	d	r2
l3	c	b	b	r3
l4	b	b	d	r4

InterSectIndex:

−1	2	3	4
2	−1	3	4
3	2	−1	4
2	4	3	−1

SortingIndex:

−1	2	3	4	−1
−1	1	3	4	−1
−1	2	1	4	−1
−1	1	3	2	−1

Fig. 3. Data structures associated with the image shown in Fig. 1(c)

For example, the vertex on the right of the point a along line L_1 is b, which is then added to the vertex list of the polygon.

To continue the traversal from b, we need to (1) first identify all the lines (other than the current line) intersecting at b, (2) find the two neighboring points of b on each of these intersecting lines, and (3) finally select one of all these neighboring points for inclusion in the vertex list representing the basis polygon. The first two steps are straightforward. For the third step, we move to a point if the point is on the left of the current line. This is called **Move Left** strategy. Each of the intersecting lines in consideration can contribute at most one candidate point satisfying this condition. In our case c and d are those candidate points. Among these points, the one on the line that makes the smallest angle with the current line on the left side is chosen. This is known as **Move Left Sharpest** strategy. Since $\angle abc < \angle abd$, c is chosen as the next vertex and added to the vertex list and line L_3 is set as the current line.

Following in a similar manner, line L_2 can be found as intersecting line L_3 at point c, and hence, the point a on line L_2 will be chosen as the next one to be included in the vertex list for the basis polygon, according to the *move left strategy*. Since a is the start vertex, the algorithm ends successfully detecting the basis polygon *abca*.

The algorithm is formally presented in Fig. 4. The following notations have been used in describing the algorithms :

1) Operations on a collection of data elements (e.g., list, array) has been denoted as $collection\_name.operation(arguments)$, 2) Some functions compute a tuple of two values which have been represented in the form $[Value1, Value2] \leftarrow function\_name(arguments)$.

The functions $Intersection$, $LeftEnd$ and $RightEnd$ in Fig. 4 are self-explanatory. In the procedure **DetectPolygon**, as described in Fig. 5, data type

```
Algorithm CP:
Step 1: For i, j = 1 to n and i ≠ j
            All processors P_ij do in parallel
              InterSect_ij ← Intersection(L_i, L_j);
Step 2: For i = 1 to n
            Each row of processors P_i do in parallel
              [SortedInterSect_{i,2:n+1}, SortingIndex_{i,2:n+1}] ←ParallelSort(InterSect_i);
Step 3: For i = 1 to n
            All processors P_ii do in parallel
              If LeftEnd(L_i) ≠ SortedInterSect_{i,2}
                SortedInterSect_{i,2} ← LeftEnd(L_i);
                SortedInterSect_{i,1} ← Invalid;
              Else
                SortedInterSect_{i,1} ← LeftEnd(L_i);
              j ← 2;
              While (j ≤ n + 1)AND(SortedInterSect_{i,j} ≠ Invalid) do
                InterSectIndex_{i,SortingIndex_ij} ← j;
                j ← j + 1;
              Endwhile;
              If RightEnd(L_i) ≠ SortedInterSect_{i,j-1}
                SortedInterSect_{i,j} ←RightEnd(L_i);
Step 4: For i, j = 1 to n and i ≠ j
            All processors P_ij do in parallel
              If InterSect_ij ≠ Invalid
                DetectPolygon(i, j);
```

Fig. 4. Algorithm for basis polygon detection

Point is a 2-tuple of co-ordinate values, type *Identity* is a 2-tuple of index values (for processors) and type *LineNumber* is an integer identifying a line. The functions and several termination conditions used in DetectPolygon are explained below :

Position is a 2-tuple of index values corresponding to the entry of the *NextVertex* in *SortedIntersect*. The function *RowValue(Position)* extracts the row index part, which is the number of the line connecting the *CurrentVertex* and the *NextVertex* that becomes the *CurrentLine* for the next step of traversal. The column index for the entry of *NextVertex* in the *CurrentLine*-th row of *SortedInterSect* is extracted with the function *ColumnValue(Position)*.

The condition Terminate[1] allows only one processor (the processor having the lowest value of column index) to start traversal from a point along a particular line, when the *StartVertex* is an intersection of more than one lines, to avoid duplicate computation. The function *FindCandidateLines* ouputs a list containing the line numbers of all lines intersecting with the *CurrentLine* at *CurrentVertex*. The function *FindCandidatePoints* outputs the list of two tuples [*CandidatePoints, PointPositions*]. The function *FindNextVertex* takes the list of 2-tuples [*CandidatePoints, PointPositions*] as input, and it outputs the tuple [*NextVertex*, *Position*] to indicate the point where to move next. It first applies the **Move Left** strategy to select a candidate point. If more than one point qualify for left move, **Move Left Sharpest** strategy is used to select a single point. If no point qualifies for left move, an invalid value is returned for *NextVertex*. The condition Terminate[2] allows only that processor to complete the traversal for which the point it is starting from is the **leftmost** vertex of the polygon.

```
Procedure DetectPolygon(i, j)
List VertexList; /* List of vertices. */
List CandidateLines; /* List of line numbers. */
List CandidatePoints; /* List of points. */
List PointPositions; /* List of positions (two-tuples of indexes). */
Point StartVertex, CurrentVertex, NextVertex;
Identity Position;
LineNumber CurrentLine;

VertexList ← Empty;
StartVertex ← InterSect_ij;
VertexList.add(StartVertex);
    /* The above operation represents adding a vertex to the vertex list. */
CurrentLine ← i;
k ← InterSectIndex_ij;
If SortedInterSect_i,k+1 = StartVertex
    Terminate^1;
CurrentVertex ← SortedInterSect_i,k;
VertexList.add(CurrentVertex);
SortedColumn ← k;
While (CurrentVertex ≠ StartVertex) do
    CandidateLines ← FindCandidateLines(CurrentLine, SortedColumn);
    If CandidateLines is empty
        Terminate; /* CurrentVertex is a pure end point. */
  [CandidatePoints, PointPositions] ← FindCandidatePoints(CandidateLines, CurrentLine);
   [NextVertex, Position] ← FindNextVertex(CandidatePoints, PointPositions);
   If NextVertex = Invalid
        Terminate; /*No left move is possible. The traversal has to end.*/
   If NextVertex < StartVertex
        Terminate^2;
    CurrentLine ← RowValue(Position);
    SortedColumn ← ColumnValue(Position);
    CurrentVertex ← NextVertex;
    VertexList.add(CurrentVertex);
End While.
Report polygon.
End DetectPolygon
```

Fig. 5. The DetectPolygon procedure for convex basis polygons

Complexity of the Algorithm: Step 1 needs $O(1)$ time. Step 2 can be performed in $O(n)$ time if simple Odd-Even Transposition is used. Step 3 needs $O(n)$ time. In step 4, all the processors P_{ij} for which $InterSect_{ij}$ is a valid point, start a traversal for detecting a basis polygon with start vertex $InterSect_{ij}$ along line L_i. If a polygon has $s \leq n$ (since polygons are convex) sides the traversal that detects it has to cover s sides. If a vertex is intersection of $k \geq 2$ lines the algorithm has to choose one line out of $k-1$ lines in $O(k-1)$ time. But due to the convexity of the polygons, the $k-2$ lines rejected at this vertex can never be the sides of the same polygon. So the total number of computations for finding the next vertex is at most $O(n)$ for a polygon. Since one polygon is detected by one processor, step 4 of the algorithm also has the worst case complexity of $O(n)$. Thus the overall time complexity of the algorithm CP is $O(n)$. The algorithm uses four data structures each of size n^2, with a space complexity of $O(n^2)$.

3.2 Algorithm GP (for General Polygons)

We now modify the above algorithm to consider the most general case. This modified algorithm uses the same data structures and same steps of computationas as those used by Algorithm CP. But We make only two changes in the procedure **DetectPolygon** as explained below.

First, we change our strategy to find the next vertex as follows : Search for a possible left move from the current line. If not possible, then find the next neighboring vertex on the current line along the direction of traversal. If this is not possible, then find a possible **right move** from the current line. If there exist more than one candidate point for the right move (current vertex is the point of intersection of more than two lines), then choose, as the next vertex, the point in that line which makes the biggest angle with the current line at the current vertex (**Move Right Widest** strategy). If no such move is possible, then set the vertex prior to the current vertex as the next vertex (**Turn around** strategy).

Another change is concerned with the generation of the vertex list during the traversal of basis polygons. Each processor P_{ij} corresponding to a valid entry in $InterSect_{ij}$ starts a rightward traversal along line L_i as before. However, if at a vertex (say, b) during the traversal, the next vertex (say, c) is the right neighbor of the current vertex b, then the processor P_{ij} stops its traversal creating just a partial vertex list of a basis polygon. The partial vertex list of P_{ij} would then consist of the list of the traversed vertices upto the vertex (say, a) previous to the current vertex (i.e., the current vertex is not included in its partial list). Since c is the right neighbor of b, there must be another processor that has started a traversal along the same route (from b to c) to traverse the same basis polygon as that by P_{ij}. Let this processor be P_{kl} which can easily be detected by the processor P_{ij} by noting the line on which b and c lie. We call the processor P_{kl} as the successor of P_{ij} in this traversal. Conversely, P_{ij} will be termed as the predecessor processor of P_{kl}.

The processor P_{ij} at this stage also writes its own index values (i, j) at a location in the shared memory designated as the *Predecessor Index Buffer* of P_{kl}.

Example 3 : Referring to Fig. 2(e), the processor P_{12} would start traversing along line 1 from the intersection point a and then it finds the vertex b as the curent vertex. At b, it chooses a left move and c becomes the next vertex in the traversal. However, since $b < c$, it implies that the processor P_{31} also would start a traversal from b along line 3 towards c. So, P_{12} stops its traversal and keeps the partial vertex list computed by it before the current vertex (b here), i.e., $< a >$ and it writes its identity (1,2) in the predecessor index buffer of the processor P_{31}. Proceeding in this manner, the partial vertex lists and the predecessor index buffers of different processors will be as follows:

P_{12} : Vertex list = $< a >$, predecessor index = $(5, 4)$
P_{31} : Vertex list = $< bc >$, predecessor index = $(1, 2)$
P_{54} : Vertex list = $< def >$, predecessor index = $(3, 1)$

Since a polygon cannot have more than n vertices, it follows that the above process will be completed in $O(n)$ parallel time. Now the remaining task is to collect and merge these partial vertex lists to produce the full vertex list forming a basis polygon. We fix our idea that the processor with the lowest index value forming a chain of such partial vertex lists will collect all these partial lists and form the complete vertex list representing the basis polygon in consideration. At this stage every processor will start reading its own predecessor index buffer, follow the index value and read the predecessor index values of the successive processors in the chain, until it reads its own index value. By this, every processor in a chain forms an index list of all processors holding the partial vertex lists of a specific basis polygon. This requires $O(n)$ parallel time. Every processor will then find in parallel the minimum of all these (maximum n in number) index values in $\log n$ time.

Only the processor whose index is equal to this minimum index found by the above step, will continue with merging the partial vertex lists of the processors in the chain, as formed by the processor index buffer values of the successive processors. All other processors will exit. This requires $O(n)$ time again.

The modified version of the DetectPolygon procedure easily follows from these ideas (details are omitted due to brevity).

Time Complexity: In view of the above discussions, it follows that the modified version of the DetectPolygon procedure in algorithm GP would take $O(n)$ time. Thus the overall time complexity of the algorithm GP is also $O(n)$.

4 Conclusion

We have presented two algorithms for detecting basis polygons created by n straight line segments, the first one for convex polygons and the second one for the general case. Both algorithms use n^2 processors with $O(n^2)$ space complexity and $O(n)$ time complexity.

References

[1] P. V. C. Hough, "Methods and Means for Recognizing Complex Patterns," U. S. Patent 3069654, 1962. 302

[2] P. O. Duda and P. E. Hart, "Use of the Hough transformation to detect lines and curves in pictures," *Communications of the ACM,* vol. 15, no. 1, pp. 11-15, 1972. 302

[3] C. Guerra and S. Hambrusch, "Parallel Algorithms for line detection in a mesh." *Journal of Parallel and Distributed Computing,* vol. 6, pp. 1-19, February 1989. 302

[4] R. E. Cypher, J. L. C. Sanz and L. Snyder, "The Hough transform has $O(N)$ complexity on $N \times N$ mesh connected computers," *SIAM J. Computing,* vol. 19, pp. 805-820, October, 1990. 302

[5] P. Yi and H. Y. H. Chuang, "Parallel Hough transform algorithms on SIMD hypercube array," *Proc. International Conference on Parallel Processing,* August 1990, pp. 83-86. 302

[6] T. Asano, K. Obokata and T. Tokuyama, "On detecting digital line components in a binary image," *Proc. Workshop on Computational Geometry,* Calcutta, India, March 18-19, 2002. 302

[7] A. Sen, M. De, B.P. Sinha and A. Mukherjee, "A new parallel algorithm for identification of straight lines in images," *Proc. 8th International Conference on Advanced Computing and Communications*, December 14-16, 2000, pp. 152-159. 302

[8] D. McCallum and D. Avis, "A linear algorithm for finding the convex hull of a simple polygon," *Information Processing Letters,* Vol. 9, 1979, pp. 201-206. 302

[9] A.A. Melkman, "On-line construction of the convex hull of a simple polyline," *Information Processing Letters*, Vol. 25, 1987, pp. 11-12. 302

[10] P. Heckbert (ed.), *Graphics Gems IV*. Academic Press, 1994. 302

[11] G.T. Toussaint, "A simple linear algorithm for intersecting convex polygons," *The Visual Computer*, Vol. 1, 1985, pp. 118-123. 302

[12] J. O'Rourke, "A new linear algorithm for intersecting convex polygons," *Comput. Graph. Image Processing* Vol. 19, 1982, pp. 384-391. 302

[13] S. Kundu, "A new *O(n log n* algorithm for computing the intersection of convex polygons," *Pattern Recognition*, Vol. 20, 1987, pp. 419-424. 302

[14] G.T. Toussaint, "An optimal algorithm for computing the minimum vertex distance between two crossing convex polygons," *Computing*, Vol. 32, 1984, pp. 357-364. 302

[15] S. Sen Gupta and B.P. Sinha, "An *O(log n)* time algorithm for testing isomorphism of maximal outerplanar graphs', *Journal of Parallel and Distributed Computing*, Vol. 56, 1999, pp. 144-156. 302

Range Image Segmentation on a Cluster

Mary Ellen Bock[1] and Concettina Guerra[2]

[1] Dept of Statistics
Purdue University West-Lafayette, IN 47907
mbock@stat.purdue.edu
[2] Dip. Ingegneria dell'Informazione, Università di Padova
via Gradenigo 6a, 35131, Padova, Italy
guerra@dei.unipd.it

Abstract. We report on the implementation of a range image segmentation approach on a cluster using Message Passing Interface (MPI). The approach combines and integrates different strategies to find the best fitting planes for a set of three dimensional points. There are basically three distint modules for plane recovery; each module has a distinct method for generating a candidate plane and a distinct objective function for evaluating the candidate and selecting the "best" plane among the candidates. Parallelism can be exploited in two different ways. First, all three modules can be executed concurrently and asynchronously by distinct processes. The scheduling of the modules in the parallel implementation differs significantly from that of the sequential implementation. Thus, different output images can be obtained for the two implementations. However, the experiments conducted on several range images show that on average the quality of results is similar in comparison with ground truth images. Second, the computation within each module can be performed in parallel. A module chooses the best plane among a large set of randomly selected candidate planes; this is a highly parallel task that can be efficiently partitioned among a group of processors. The approach proposed in this paper for a multiprocessor environment has been implemented on a cluster of workstations using MPI. Preliminary results are presented.

1 Introduction

A number of vision tasks require substantial computation time and resources; for this reason there is a large body of literature on ways of making vision tasks more efficient by either designing special-purpose hardware or by exploiting the available systems in an effective way.

Often the parallelism that can be used in vision and image processing is rather obvious and the speed-up that can be obtained easily predictable for specific architectures or systems. Sometimes solving a problem on a specific architecture involves redesigning the algorithm and advanced analysis tools for evaluating its performance.

We consider the problem of segmenting a range image into planar regions, each region denoted by the parameters of a specific plane and the points in

S. Sahni et al. (Eds.) HiPC 2002, LNCS 2552, pp. 313–322, 2002.

the image associated with that plane. Our approach is to solve the problem in a multiprocessor environment by combining and integrating different plane detection strategies in a novel way. The results of this approach are not easily predictable from the sequential implementation both in terms of the output images and execution times; a careful analysis is therefore needed to evaluate the results.

The range image segmentation problem has received a lot of attention in the vision community. Recently, an experimental framework for evaluating different approaches to range image segmentation was initiated in [6] and continued in [7]. The average processing times of the segmentation algorithms on a set of range images ABW vary significantly ranging from a few hours to a few seconds. All experiments within this framework were conducted on conventional machines. The set of ABW test images consists of 30 range images available at http://marathon.csee.usf.edu/range/seg-comp/images.html

The multiprocessor approach we present combines three different modules to generate a list of different planes and their associated sets of range image elements that combine to make up the description of the image. The modules use different criteria to select the best fitting plane for sets of three dimensional points. In the sequential implementation, at each iteration to find the next plane to be added to the list, the three main procedures are invoked sequentially in a given order; the next one is executed only when the previous ones do not detect a significant plane (as defined later).

In the distributed implementation, based on a master/slave model, the three procedures are executed concurrently and asynchronously by different slaves; the master receives a plane sent by any slave in any order, and accepts it based on the significance threshold. Thus, a considerable speedup is obtained if all procedures return acceptable planes and if these plane correspond to distinct regions of the image. Crucial to the performance is that the three procedures may generate different planes even when applied to the same input data. This is generally true since they use different criteria for selecting the best plane. From the above considerations, it follows that the distributed implementation may produce results that are different from the ones generated sequentially. They may differ in the parameters of the planes detected, and/or in the order in which the planes are added to the list and in the procedure used to find a plane. The comparison with ground truth images shows that the quality of the results on average is the same as that of the sequential algorithm for the set of ABW images.

There is another way in which parallelism is exploited in the proposed range image segmentation algorithm. In each module several candidate planes are randomly selected and then evaluated on the basis of a given optimization criterion; the best of all the hypothesized planes is then sent to the master process. The task of generating candidate planes and then verifying them is highly parallel and can be performed efficiently by distinct slaves. The number of examined planes depends on the iteration ranging from thousands at the beginning of the processing, when a large fraction of the range image is still unlabeled, to few hundreds towards the end of the processing.

In the current version, we have implemented the concurrent execution of the three modules on different slaves. Of the three modules, only the most computation intensive is executed in parallel on a group of slaves. We report in this paper on the preliminary results obtained using MPI [4,8-9] on a cluster of workstations.

The paper is organized as follows. In section 2, we review the range image segmentation strategy. Then in section 3 we present the parallel algorithm. In section 4 we discuss an implementation using MPI and show preliminary results on ABW images. We conclude in section 5 with future work.

2 The PPU Segmenter

The PPU segmenter [3] combines several strategies for generating a list of different planes and their associated sets of range image elements that combine to make up the description of the image. The three main procedures are all based on random sampling [5] to reduce the computational complexity of the search. Each procedure has a distinct method for generating a candidate plane and a distinct objective function for evaluating the candidate and selecting the "best" plane among the candidates. (The number of candidate planes generated by a procedure is set by a parameter that changes as the size of the list increases.) Each plane on the list has an associated distinct set of range image elements labeled for the plane.

The first two procedures *CoplanarLines* and *LinePoint* depend on the image only through its edge points which are generated by an edge detector in a preprocessing phase. The parameters of candidate planes are found by randomly selecting a pair of proper subsets of the edge points: two coplanar lines, either parallel or not, for CoplanarLines, and a line and a point not belonging to the line, for LinePoint. In both cases, the objective function for a candidate plane is the number of edge points that belong to the plane, i.e. are within a given ϵ distance from the plane, where ϵ is a small constant set to 0.5 in our implementation.

The third procedure *AlternativeStrategy* generates a candidate plane that fits a small window of range points unlabeled by previously selected planes in the list. It uses as objective function the total number of unlabeled range points that are within ϵ distance from the candidate plane. This procedure tends to be computationally intensive due to the lengthy computation of the objective function. Thus, it is invoked only when the other two procedures produce "best" planes that are not significant, as described below.

At each iteration, to detect a new plane for the list the three main procedures are invoked sequentially in a given order; the next one executed only when the previous ones do not detect a significant plane. The significance of a plane selected as "best" by a procedure is expressed as the percentage of range points that are unlabeled and are within a given distance of the plane. When this percentage is above a certain threshold S, the detected plane is accepted for the list and a set associated with the new plane is assigned its new label; otherwise, the plane is discarded. The threshold S is updated dynamically during the processing. It may happen that none of the procedures recovers a significant plane even though there are still many unlabeled points. This may due to the fact that the threshold S is too high for this stage. Thus, the plane with the highest significance among the "best" ones detected by the three procedures is chosen and accepted as the new plane. Furthermore, the threshold S is updated and set to a value lower than the one corresponding to the significance of the new plane.

Once the parameters of the new plane for the list are found at a given iteration, the tentative set of range points associated with the new plane becomes all the points in the image which are within a given distance of the plane. Thus the plane is expanded over the entire image and all the fragments of the same surface in the image are labeled as belonging to the same plane. The fragments are generally due to occlusion but may also correspond to different faces of the same concave object or of different objects. Labeling conflicts may arise because some points in the tentative set may be close to another plane on the list and may also occur in the set of points labeled for the other plane. A conflict at a range point in the tentative set is resolved in favor of the label with the highest number of occurrences in a small window centered at the point. There are cases, however, where a conflict cannot be resolved within a small local window and a more global procedure is needed. Before proceeding with the next iteration to detect another plane, the list of the edge points is updated by removing the edge points within a given small distance from the new plane.

The need to combine three different strategies to solve the segmentation problem using the random sampling approach is supported by an extensive experimentation, as shown in [3]. We have experimentally determined that each procedure would perform poorly in a way or another if used in isolation. If CoplanarLines is the only one used, then the first few planes are found quickly and effectively, however towards the end of the processing the procedure fails because pairs of lines cannot be detected reliably from the few edge points left in the list of remaining edge points. AlternativeStrategy, on the other hand, is very effective both at the beginning of the processing to detect large regions (for instance, background or ground) as well as when most of the image points have already been labeled and only small patches are left. However, this procedure is too time consuming to be used in all stages. As for the last procedure LinePoint this is the fastest of the three; however the planes it recovers sometime fail to pass the significance test and therefore are rejected.

Following the comparison framework of Hoover et al. [6], [7], we compared the results of our method (PP) with approaches of the Universities of South Florida (USF), Bern (UB), Edinburgh (UE) and Washington State University (WSU). The evaluation of the different approaches is based on the number of correctly detected regions and other error metrics such as: over and under-segmented regions, angles differences between adjacent planes, and noise. A detailed description of these error metrics can be found in [6].

Table 1 shows the average results of different segmenters on all 30 ABW test images when compared with ground truth images.

Our approach has been implemented and the program written in C has been executed a Sun Sparc5. The average running time of the program on the 30 ABW images are is 352s. Average running times of the other approaches range from seconds to hours.

3 Implementation on a Cluster

The strategy is based on a *master/slave* model in which the master distributes data and collects the results and the slaves execute the three independent procedures, *LinePoint, CoplanarLines and AlternativeStrategy.* For the sake of simplicity, we first assume that each procedure is executed by a single slave process. However, the task of each procedure is highly parallelizable and therefore could be assigned to many slaves to speedup the computation even further. In our current implementation, only AlternativeStrategy is performed by a group of slaves.

Initially, the master performs edge detection on the input range image. It then distributes information to the slaves. It includes the range image data for the slave doing AlternativeStrategy, and the edge points for the slaves performing CoplanarLines and LinePoint.

Each slave asynchronously detects the best plane on the input data according to its own selection criteria and sends the obtained plane parameters to the master. The master upon receiving the plane parameters from a slave decides whether to accept it or not (based on the significance threshold S). If the plane is acceptable, then the master expands the plane over the entire image and labels the output image accordingly. It also resolves the conflicts with all previous planes. Finally, if more planes can be detected in the image, the master sends the updated information to the slave from which the new plane was received. The type of information sent depends on the slave itself. For the CoplanarLines and LinePoint slaves, it consists of the list of remaining edge points, i.e. the list of points after the removal of all the edge points that are at distance less than a small constant value from the new plane. For AlternativeStrategy, the information sent by the master is the labeled range image.

It is important to note that the three procedures, when applied to the same input data, generally return distinct planes since they are based on different criteria for choosing the candidate planes and selecting the best one among them. This observation is crucial for a successful distributed implementation. However, it may happen, both in the sequential and distributed implementations, that the same plane is returned more than once either by the same procedure or by distinct procedures. For this reason, for each received plane the master needs to check if such a plane already exists in the master list.

If the master receives a few unacceptable planes in a row (three in our current version) and there are still many unlabeled points in the image, the plane with the highest significance among them is chosen and accepted as the new plane.

Furthermore, the threshold S based on which the significance of a plane is judged is updated and set to a value lower than the one corresponding to the significance of the new plane. We next describe the three main modules of our method.

A plane from a line and a point (LinePoint)

The procedure *LinePoint* chooses a candidate plane defined by a line and a point outside the line. The line is the best fitting line for a collection of edge points. It is extracted from the set of remaining edge points, not associated to a previously detected plane. The line extraction in three dimensional space is based on a simple approach that selects the optimal line among those that are defined by pairs of points in the dataset. It does so by determining for each line passing through a pair of edge points the number of other remaining edge points lying within the expected measurement error ϵ from that line; the best line is the one that corresponds to the maximum computed value. The number of pairs $\frac{n \times (n-1)}{2}$ is generally very large for any reasonable value of n. However, in practice a number of pairs much less than the above is acceptable. A relatively small subset of pairs of points randomly chosen can yield results that are within a given bound from the optimal. The output of the line extraction is a pair of edge points belonging to the line. The pair of edge points for the line and a third point randomly selected in the set of remaining edges define a model plane. Several model planes are considered and the best one is chosen according to a figure of merit that takes into consideration all edge points currently assigned to other discovered planes as well as the remaining edge points. It is the number of edge points within ϵ distance from the plane.

It is easy to see that each plane selection and evaluation can be performed independently from the others. Thus, the load of this procedure can be uniformly distributed among several processes by assigning to each of them the same fraction of the total number of candidated planes to be examined. The best among all such planes can be determined by a max operation, at the end of the stage. Similarly, the determination of the best line in the set of edge points can be uniformly distributed to several slaves.

A plane defined by two coplanar lines (CoplanarLines)

The procedure *CoplanarLines* determines candidate planes defined by two coplanar lines. As described in the previous section, a line is the best fitting line for a set of edge points. Of the two lines, one at least is extracted from the set of remaining edge points and the second may also come from there or may belong to a previous plane.

Approximate coplanarity of the two lines is tested. To determine a plane associated with the two lines, three points are needed. The first two are the two edge points that determine the first line. The third point is one of the edge points in the ϵ neighborhood of the second line. Each of these third points determines a different candidate plane and the one finally chosen has the most number of edge points within ϵ of the two lines and also within ϵ of the candidate plane.

A plane defined by two coplanar lines detected in the scene does not necessarily correspond to an actual surface in the image. For instance, in a polyhedra with parallel faces, pairs of parallel lines not belonging to the same face generate planes that are not on the surface of the object. For a set of stairs in an image, the parallel lines that form the leading edge of two steps give an example. Thus, before making the final decision about the acceptance of a plane such cases have to be eliminated. This is done by drawing a line segment joining the two lines and checking that most of the points close to it lie in the candidate plane. If two parallel lines are very close they are ignored. Close parallel lines are generally present at occlusion boundaries and do not form a region in the image. The number of lines considered is an input parameters.

As in the previous procedure, several candidate planes are examined and the best one is chosen. Again, this computation can be efficiently distributed over separate processes.

A plane from a seed in the range image (AlternativeStrategy)

The procedure *AlternativeStrategy* randomly selects three uncolored range points $(i, j), (i, j+16), (i+14, j+8)$ (the vertices of an almost equilateral triangle) and forms a plane with them. It considers the smallest rectangular window enclosing the three points and counts the number of uncolored range points in the window that are within ϵ distance from the plane. If such count is below a given threshold, the plane is discarded because it is considered not significant. The threshold is initially high so that almost all points in the window need to be covered by the plane and is relaxed towards the end of the processing when few range points are still uncolored. If the plane is accepted according to its local significance, we estimate the goodness of the fit over the entire range image by counting the total number of uncolored range points close to the plane.

The above steps are repeated several times and the plane that best fits the range points over the entire image is selected. The number of iterations depends on stage of the processing and is set experimentally. Iterations can be partitioned among separate processes each working independently. At the end, the best plane is chosen among the candidate planes returned by all processes.

4 Results

The application has been programmed using the MPI libraries for interprocess communication. MPI is the method of choice for programming distributed memory parallel systems. For the experiments we have used a cluster of 8 sun-sparc5 workstations linked together by a 10Mb/s Ethernet network. Future experiments will be conducted on a faster dedicated cluster and/or on a grid.

In the sequential version, the order of execution of the three main procedures CoplanarLines, LinePoint and AlternativeStrategy has been decided experimentally based on many runs with different permutations. Except for the first two planes, the procedures are called in the following order: first CoplanarLines, second LinePoint and, finally, AlternativeStrategy. In the determination of the first two planes, generally corresponding to the background and the ground, the procedure CoplanarLines is not included. Recall that at given iteration, to find the i-th plane, a procedure is executed only if the previous one(s) fail to detect a significant plane.

We have compared the results obtained by the distributed implementation with ground truth images on a few test images within the framework described in [6]. Table 2 and 3 reports the results obtained on a few ABW test images, when executed with one slave and six slaves, respectively.

For a visual comparison, figure 1 shows the output of the sequential algorithm (left) and of the distributed algorithm (right) on the test image abw.test.10. The output images are almost identical. However, the planes in the two outputs were obtained by different procedures.
More precisely, in the sequential algorithm for the image abw.test.10 the planes were obtained by:

plane 0: LinePoint
plane 1: AlternativeStrategy
plane 2: CoplanarLines
plane 3: CoplanarLines
plane 4: CoplanarLines
plane 5: CoplanarLines
plane 6: CoplanarLines
while in the distributed version by:
plane 0: LinePoint
plane 1: AlternativeStrategy
plane 2: LinePoint
plane 3: CoplanarLines
plane 4: LinePoint
plane 5: CoplanarLines
plane 6: AlternativeStrategy

The running time of the program on the input image abw.test.10 went down from 77s to 34s on six processors.

5 Conclusions

We have designed a distributed range image segmentation algorithm that combines different modules and executes them asynchronously to improve the performance of the segmenter. The initial results obtained on a cluster of SUN-SPARC workstations are promising. Many of the existing segmenters do not necessarely benefit from a multiprocessor environment using some form of data parallelism. The natural way of using data parallelism in an image, that is by partitioning the image data into blocks and assigning each block to a distinct computing element, does not generally work for range image segmenters. This is especially true in our approach, since once the parameters of a plane are determined the plane is expanded over the entire image and all image points wich do not produce conflicts are labeled.

Further improvement of our approach can be expected from a more complete software implementation and from the use of a faster cluster of machines. The issue of scalability needs to be addressed in more depth. An implementation on a grid will also be considered within the DataGrid project [1].

References

[1] A. Apostolico, V. Breton, E. Cornillot, S. Du, L. Duret, C. Gautier, C. Guerra, N. Jacq, R. Medina, C. Michau, J. Montagnat, A. Robinson, M. Senger. DataGrid. requirements for grid-aware biology applications. Tech. report.

[2] M. E. Bock, C. Guerra. "A geometric approach to the segmentation of range images", *Proceedings of the Second International Conference on 3D-Digital Imaging and Modeling*, Ottawa, Canada, pp. 261-269, 1999.

[3] M. E. Bock, C. Guerra. "Segmentation of range images through the integration of different strategies", 6th Int. Work.Vision, Modeling, and Visualization, Stuttgart, Germany, 2001.

[4] J. Bruck, D. Dolev, C. T. Ho, M. Rosu, R. Strong, "Efficient message passing interface (MPI) for parallel computing on clusters of workstations", *Journal of Parallel and Distributed Computing*, v. 40, pp. 19-34, 1997.

[5] M. A. Fischler and R. C. Bolles, "Random sample consensus: a paradigm for model fitting with applications to image analysis and automated cartography," *Communications of the ACM* **24**, pp. 381–395, June 1981.

[6] A. Hoover, J. B. Gillina, X. Jiang, P. Flynn, H. Bunke, D. Goldgolf, K. Bowyer, D. Eggert, A. Fitzgibbon, and R. Fischer, "An experimental comparison of range image segmentation algorithms," *IEEE Trans. on Pattern Analysis and Machine Intelligence* **18**(7), pp. 673–689, 1996.

[7] X.Jiang, K.Bowyer, Y.Morioka, S.Hiura, K. Sato, S. Inokuchi, M.Bock, C.Guerra, R. E.Loke, J. M. H.duBuf. "Some Further Results of Experimental Comparison of Range Image Segmentation Algorithms", *15th Int. Conference on Pattern Recognition*, Spain, 2000.

[8] MPI Forum, A Message Passing Interface standard, www.mpi-forum.org

[9] P. J. Morrow and D. Crookes and J. Brown and G. McAleese and D. Roantree and I. Spence, "Efficient implementation of a portable parallel programming model for image processing", *Concurrency-Practice and Experience*, **11**, 11, pp. 671-685, 1999.

Table 1. Average results of segmenters on 30 range images abw.test.10 – abw.test.29

Research group	GT regions	Correct detection:	Angle: diff	Overseg:	Underseg:	Missed:	Noise:
USF	15.2	12.7	1.6	0.2	0.1	2.1	1.2
WSU	15.2	9.7	1.6	0.5	0.2	4.5	2.2
UB	15.2	12.8	1.3	0.5	0.1	1.7	2.1
UE	15.2	13.4	1.6	0.4	0.2	1.1	0.8
PP	15.2	12.7	3.2	0.7	0.5	0.8	0.5

Table 2. Results of the PP segmenter for a few ABW test images with 1 slave

Research group	Image num.	GT regions	Correct detection:	Angle: diff	Overseg:	Underseg:	Missed:	Noise:
PP	abw.9	9	7	1.1	0	1	1	0
PP	abw.10	7	7	2.3	0	0	0	0
PP	abw.11	9	7	0.6	0	1	0	0
PP	abw.15	17	16	3.5	0	0	1	0

Table 3. Results of the PP segmenter for a few ABW test images with 6 slaves

Research group	Image num.	GT regions	Correct detection:	Angle: diff	Overseg:	Underseg:	Missed:	Noise:
PP	abw.9	9	7	1.1	1	0	1	0
PP	abw.10	7	7	2.0	0	0	0	0
PP	abw.11	9	7	2.5	1	0	0	0
PP	abw.15	17	16	3.7	0	0	1	0

Fig. 1. The image abw.test.10 segmented with the sequential algorithm (left) and with the distributed algorithm (right)

Detection of Orthogonal Interval Relations

Punit Chandra and Ajay D. Kshemkalyani

Dept. of Computer Science, Univ. of Illinois at Chicago
Chicago, IL 60607, USA
{pchandra,ajayk}@cs.uic.edu

Abstract. The complete set $\Re$ of orthogonal temporal interactions between pairs of intervals, formulated by Kshemkalyani, allows the detailed specification of the manner in which intervals can be related to one another in a distributed execution. This paper presents a distributed algorithm to detect whether pre-specified interaction types between intervals at different processes hold. Specifically, for each pair of processes i and j, given a relation $r_{i,j}$ from the set of orthogonal relations $\Re$, this paper presents a distributed (on-line) algorithm to determine the intervals, if they exist, one from each process, such that each relation $r_{i,j}$ is satisfied for that (i, j) process pair. The algorithm uses $O(n \min(np, 4mn))$ messages of size $O(n)$ each, where n is the number of processes, m is the maximum number of messages sent by any process, and p is the maximum number of intervals at any process. The average time complexity per process is $O(\min(np, 4mn))$, and the total space complexity across all the processes is $\min(4pn^2 - 2np, 10mn^2)$.

1 Introduction

Monitoring, synchronization and coordination, debugging, and industrial process control in a distributed system inherently identify local durations at processes when certain application-specific local predicates defined on local variables are true. To design and develop such applications to their fullest, we require a way to *specify* how durations at different processes are related to one another, and also a way to *detect* whether specified relationships hold in an execution. The formalism and axiom system formulated by Kshemkalyani [5] identified a complete orthogonal set $\Re$ of 40 fine-grained temporal interactions (or relationships) between intervals to *specify* how durations at different processes are related to one another. This gives flexibility and power to monitor, synchronize, and control distributed executions. Given a specific orthogonal relation that needs to hold between each pair of processes in a distributed execution, this paper presents a distributed (on-line) algorithm to *detect* the earliest intervals, one on each process, such that the specified relation between each pair of intervals is satisfied.

The pairwise interaction between processes is an important way of information exchange even in many large-scale distributed systems. Examples of such systems are sensor networks, ad-hoc mobile networks, mobile agent systems, and on-line collaborative motion planning and navigation systems. The various

S. Sahni et al. (Eds.) HiPC 2002, LNCS 2552, pp. 323–333, 2002.

participating nodes can compute a dynamic global function (e.g., the classical distance-vector routing or AODV) or a dynamic local function such as the velocity of a mobile agent participating in a cooperative endeavor. Dynamic and on-line computation of local functions are used for centroidal Voronoi tessellations with applications to problems as diverse as image compression, quadrature, finite difference methods, distribution of resources, cellular biology, statistics, and the territorial behavior of animals [3].

To capture the pairwise interaction between processes, intervals at each process are identified to be the durations during which some application-specific local predicate is true. We introduce and address the following problem DOOR for the Detection of Orthogonal Relations.

Problem DOOR: Given a relation $r_{i,j}$ from $\Re$ for each pair of processes i and j, determine in an on-line, distributed manner the intervals, if they exist, one from each process, such that each relation $r_{i,j}$ is satisfied by the (i, j) pair.

The algorithm we propose uses $O(n \min(np, 4mn))$ messages of size $O(n)$ each, where n is the number of processes, m is the maximum number of messages sent by any process, and p is the maximum number of intervals at any process. The average time complexity per process is $O(\min(np, 4mn))$, and the total space complexity across all the processes is $\min(4pn^2 - 2np, 10mn^2)$.

A solution satisfying the set of relations $\{r_{i,j}(\forall i, j)\}$ identifies a global state of the system [2]. Note that a solution may exist in an execution only if the set of specified relations, one for each process pair, that need to hold, satisfies the axioms given in [5].

Section 2 gives the system model and preliminaries. Section 3 gives the theory used to determine when two given intervals at different processes can never be part of a solution set, and thus one of them can be discarded. Section 4 gives the data structures and local processing for tracking intervals at each process, and gives some tests used to determine the interaction type between a pair of intervals. Section 5 presents the distributed algorithm to solve problem **DOOR**. Section 6 gives concluding remarks.

2 System Model and Preliminaries

We assume an asynchronous distributed system in which n processes communicate by reliable message passing. Without loss of generality, we assume FIFO message delivery on the channels. A poset event structure model $(E, \prec)$, where $\prec$ is an irreflexive partial ordering representing the causality relation on the event set E, is used to model the distributed system execution. E is partitioned into local executions at each process. E_i is the linearly ordered set of events executed by process P_i. An event e executed by P_i is denoted e_i. The causality relation on E is the transitive closure of the local ordering relation on each E_i and the ordering imposed by message send events and message receive events [8]. This execution model is analogous to that in [5, 6, 9].

Table 1. Dependent relations for interactions between intervals are given in the first two columns [5]. Tests for the relations are given in the third column [7]

Relation r	Expression for $r(X,Y)$	Test for $r(X,Y)$
R1	$\forall x \in X \forall y \in Y, x \prec y$	$V_y^-[x] > V_x^+[x]$
R2	$\forall x \in X \exists y \in Y, x \prec y$	$V_y^+[x] > V_x^+[x]$
R3	$\exists x \in X \forall y \in Y, x \prec y$	$V_y^-[x] > V_x^-[x]$
R4	$\exists x \in X \exists y \in Y, x \prec y$	$V_y^+[x] > V_x^-[x]$
S1	$\exists x \in X \forall y \in Y, x \not\preceq y \bigwedge y \not\preceq x$	**if** $V_y^-[y] \not< V_x^-[y] \bigwedge V_y^+[x] \not> V_x^+[x]$ **then** $(\exists x^0 \in X$: $V_y^-[y] \not\leq V_x^{x^0}[y] \wedge$ $V_x^{x^0}[x] \not\leq V_y^+[x])$ **else** *false*
S2	$\exists x_1, x_2 \in X \exists y \in Y, x_1 \prec y \prec x_2$	**if** $V_y^+[x] > V_x^-[x] \bigwedge V_y^-[y] < V_x^+[y]$ **then** $(\exists y^0 \in Y : V_x^+[y] \not< V_y^{y^0}[y] \wedge$ $V_y^{y^0}[x] \not< V_x^-[x])$ **else** *false*

We assume vector clocks are available [4, 10]. The vector clock V has the property that $e \prec f \iff V(e) < V(f)$. The durations of interest at each process are the durations during which the local predicate is true. Such a duration, also termed as an interval, at process P_i is identified by the corresponding events within E_i.

Kshemkalyani showed in [5] that there are 29 or 40 possible mutually orthogonal ways in which any two durations can be related to each other, depending on whether the dense or the nondense time model is assumed. Informally speaking, with dense time, $\forall x, y$ in interval A, $x \prec y \implies \exists z \in A \mid x \prec z \prec y$. The orthogonal interaction types were identified by first using the six relations given in the first two columns of Table 1. Relations R1 (strong precedence), R2 (partially strong precedence), R3 (partially weak precedence), R4 (weak precedence) defined *causality conditions* whereas S1 and S2 defined *coupling conditions*.

Assuming that time is dense, it was shown in [5] that there are 29 possible interaction types between a pair of intervals, as given in the upper part of Table 2. The twenty-nine interaction types are specified using boolean vectors. The six relations R1-R4 and S1-S2 form a boolean vector of length 12, (six bits for $r(X,Y)$ and six bits for $r(Y,X)$). The nondense time model is significant because clocks which measure dense linear time use a nondense linear scale in practice. This model is also significant because actions at each node in a distributed system are a linear sequence of discrete events. This model permits 11 interaction types between a pair of intervals, defined in the lower part of Table 2, in addition to the 29 identified before. The interaction types are in pairs of inverses. For illustrations of these interactions and explanation of the table, the reader is requested to refer to [5]. The set of 40 relations is denoted as $\Re$.

Given a set of orthogonal relations, one between each pair of processes, that need to be detected, each of the 29 (40) possible independent relations in the dense (nondense) model of time can be tested for using the bit-patterns for the dependent relations, as given in Table 2. The tests for the relations $R1$, $R2$, $R3$,

Table 2. The 40 independent relations in $\Re$ [5]. The upper part of the table gives the 29 relations assuming dense time. The lower part of the table gives 11 additional relations if nondense time is assumed

Interaction Type	Relation $r(X,Y)$						Relation $r(Y,X)$					
	R1	R2	R3	R4	S1	S2	R1	R2	R3	R4	S1	S2
$IA(=IQ^{-1})$	1	1	1	1	0	0	0	0	0	0	0	0
$IB(=IR^{-1})$	0	1	1	1	0	0	0	0	0	0	0	0
$IC(=IV^{-1})$	0	0	1	1	1	0	0	0	0	0	0	0
$ID(=IX^{-1})$	0	0	1	1	1	1	0	1	0	1	0	0
$ID'(=IU^{-1})$	0	0	1	1	0	1	0	1	0	1	0	1
$IE(=IW^{-1})$	0	0	1	1	1	1	0	0	0	1	0	0
$IE'(=IT^{-1})$	0	0	1	1	0	1	0	0	0	1	0	1
$IF(=IS^{-1})$	0	1	1	1	0	1	0	0	0	1	0	1
$IG(=IG^{-1})$	0	0	0	0	1	0	0	0	0	0	1	0
$IH(=IK^{-1})$	0	0	0	1	1	0	0	0	0	0	1	0
$II(=IJ^{-1})$	0	1	0	1	0	0	0	0	0	0	1	0
$IL(=IO^{-1})$	0	0	0	1	1	1	0	1	0	1	0	0
$IL'(=IP^{-1})$	0	0	0	1	0	1	0	1	0	1	0	1
$IM(=IM^{-1})$	0	0	0	1	1	0	0	0	0	1	1	0
$IN(=IM'^{-1})$	0	0	0	1	1	1	0	0	0	1	0	0
$IN'(=IN'^{-1})$	0	0	0	1	0	1	0	0	0	1	0	1
$ID''(=(IUX)^{-1})$	0	0	1	1	0	1	0	1	0	1	0	0
$IE''(=(ITW)^{-1})$	0	0	1	1	0	1	0	0	0	1	0	0
$IL''(=(IOP)^{-1})$	0	0	0	1	0	1	0	1	0	1	0	0
$IM''(=(IMN)^{-1})$	0	0	0	1	0	0	0	0	0	1	1	0
$IN''(=(IMN')^{-1})$	0	0	0	1	0	1	0	0	0	1	0	0
$IMN''(=(IMN'')^{-1})$	0	0	0	1	0	0	0	0	0	1	0	0

$R4$, $S1$, and $S2$ in terms of vector timestamps are given in the third column of Table 1 [7]. V_i^- and V_i^+ denote the vector timestamp at process P_i at the start of an interval and at the end of an interval, respectively. V_i^x denotes the vector timestamp of event x_i at process P_i. The tests in Table 1 can be run by each process in a distributed manner. Each process P_i, $1 \leq i \leq n$, maintains information about the timestamps of the start and end of its local intervals, and certain other local information, in a local queue Q_i. The n processes collectively run a token-based algorithm to process the information in the local queues and solve problem **DOOR**.

We note some assumptions and terminology. (1) There are a maximum of p intervals at any process. (2) Interval X occurs at P_i and interval Y occurs at P_j. (3) For any two intervals X and X' that occur at the same process, if $R1(X, X')$, then we say that X is a predecessor of X' and X' is a successor of X.

1. When an internal event or send event occurs at process P_i, $V_i[i] = V_i[i] + 1$.
2. Every message contains the vector clock and *Interval Clock* of its send event.
3. When process P_i receives a message msg, then $\forall\ j$ do,

   if $(j == i)$ **then** $V_i[i] = V_i[i] + 1$,
   else $V_i[j] = \max(V_i[j], msg.V[j])$.

4. When an interval starts at P_i (local predicate ϕ_i becomes true), $I_i[i] = V_i[i]$.
5. When process P_i receives a message msg, then $\forall\ j$ do,
   $I_i[j] = \max(I_i[j], msg.I[j])$

Fig. 1. The vector clock V_i and *Interval Clock* I_i at process P_i

3 Conditions for Satisfying Given Interaction Types

A critical aspect of any distributed algorithm to solve problem **DOOR** is to design an efficient way to prune the intervals from the queues of the n processes. This section gives the condition for pruning an interval and the property which makes efficient pruning possible. We introduce the notion of prohibition function $S(r_{i,j})$ and relation $\lhd$ which give the condition for pruning of intervals. We also show that if the given relationship between a pair of intervals does not hold, then at least one of the intervals is deleted. This property makes the pruning and hence the algorithm efficient. Theorem 1 identifies this basic property.

For each $r_{i,j} \in \Re$, we define $S(r_{i,j})$ as the set of all relations R such that if $R(X,Y)$ is true, then $r_{i,j}(X,Y')$ can never be true for some successor Y' of Y. $S(r_{i,j})$ is the set of relations that prohibit $r_{i,j}$ from being true in the future.

Definition 1. Prohibition Function $S : \Re \rightarrow 2^{\Re}$ *is defined to be* $S(r_{i,j}) = \{R \in \Re \mid R \neq r_{i,j} \wedge\ if\ R(X,Y)\ is\ true\ then\ r_{i,j}(X,Y')\ is\ false\ for\ all\ Y'\ that\ succeed\ Y\}$.

Two relations R' and R'' in $\Re$ are related by $\lhd$ if the occurrence of $R'(X,Y)$ does not prohibit $R''(X,Y')$ for some successor Y' of Y.

Definition 2. $\lhd$ *is a relation on* $\Re \times \Re$ *such that* $R' \lhd R''$ *if (1)* $R' \neq R''$*, and (2) if* $R'(X,Y)$ *is true then* $R''(X,Y')$ *can be true for some* Y' *that succeeds* Y.

For example, $IC \lhd IB$ because (1) $IC \neq IB$ and, (2) if $IC(X,Y)$ is true, then there is a possibility that $IB(X,Y')$ is also true, where Y' succeeds Y.

Theorem 1. *For* $R', R'' \in \Re$*, if* $R' \lhd R''$ *then* $R'^{-1} \not\lhd R''^{-1}$*. (Proof is in [1].)*

Taking the same example, $IC \lhd IB \Rightarrow IV(= IC^{-1}) \not\lhd IR(= IB^{-1})$, which is indeed true. Note that $R' \neq R''$ in the definition of relation $\lhd$ is necessary; otherwise $R' \lhd R'$ leads to $R'^{-1} \not\lhd R'^{-1}$, a contradiction.

Lemma 1. *If* $R \in S(r_{i,j})$ *then* $R \not\lhd r_{i,j}$ *else if* $(R \notin S(r_{i,j})$ *and* $R \neq r_{i,j})$ *then* $R \lhd r_{i,j}$.

```
type Event_Interval = record
    interval_id : integer;
    local_event: integer;
end

type Process_Log = record
    event_interval_queue: queue of Event_Interval;
end

type Log = record
    start: array[1..n] of integer;
    end: array[1..n] of integer;
    p_log: array[1..n] of Process_Log;
end
```

Fig. 2. The *Event_Interval*, *Log*, and *Process_Log* data structures at P_i ($1 \leq i \leq n$)

Proof. If $R \in S(r_{i,j})$, using Definition 1, it can be inferred that $r_{i,j}$ is false for all Y' that succeed Y. This does not satisfy the second part of Definition 2. Hence $R \not\lhd r_{i,j}$. If $R \notin S(r_{i,j})$ and $R \neq r_{i,j}$, it follows that $r_{i,j}$ can be true for some Y' that succeeds Y. This satisfies Definition 2 and hence $R \lhd r_{i,j}$. □

$S(r_{i,j})$ for each of the interaction types in $\Re$ is given in [1]. The following lemmas are used to show the correctness of the algorithm in Figure 6.

Lemma 2. *If the relationship $R(X,Y)$ between intervals X and Y (belonging to processes P_i and P_j, resp.) is contained in the set $S(r_{i,j})$, then interval X can be removed from the queue Q_i.*

Proof. From the definition of $S(r_{i,j})$, we get that $r_{i,j}(X,Y')$ cannot exist, where Y' is any successor interval of Y. Hence interval X can never be a part of the solution and can be deleted from the queue. □

Lemma 3. *If the relationship between a pair of intervals X and Y (belonging to processes P_i and P_j respectively) is not equal to $r_{i,j}$, then interval X or interval Y is removed from its queue Q_i or Q_j, respectively.*

Proof. We use contradiction. Assume relation $R(X,Y)$ ($\neq r_{i,j}(X,Y)$) is true for intervals X and Y. From Lemma 2, the only time neither X nor Y will be deleted is when $R \notin S(r_{i,j})$ and $R^{-1} \notin S(r_{j,i})$. From Lemma 1, it can be inferred that $R \lhd r_{i,j}$ and $R^{-1} \lhd r_{j,i}$. As $r_{i,j}^{-1} = r_{j,i}$, we get $R \lhd r_{i,j}$ and $R^{-1} \lhd r_{i,j}^{-1}$. This is a contradiction as by Theorem 1, $R \lhd r_{i,j} \Rightarrow R^{-1} \not\lhd r_{i,j}^{-1}$. Hence $R \in S(r_{i,j})$ or $R^{-1} \in S(r_{j,i})$, and thus at least one of the intervals gets deleted. □

4 Tracking Intervals and Evaluating Relations

This section gives the operations and data structures to track intervals at each process. These are used by our algorithm given in the next section.

Each process P_i, where $1 \leq i \leq n$, maintains the following data structures. (1) V_i : array[1..n] of integer. This is the *Vector Clock* [4, 10]. (2) I_i : array[1..n] of integer. This is the *Interval Clock* which tracks the latest intervals at processes.

Start of an interval:
  $Log_i.start = V_i^-$. //Store the timestamp V_i^- of the starting of the interval.
On receiving a message during an interval: //Store the local component of
  if (change in I_i) **then** //vector clock and *interval_id* which caused
    for each k such that $I_i[k]$ was changed //the change in I_i
      insert $(I_i[k], V_i[i])$ in $Log_i.p\_log[k].event\_interval\_queue$
End of interval:
  $Log_i.end = V_i^+$ //Store the timestamp V_i^+ of the end of the interval.
  if (a receive or send occurs between start of previous interval and end of present interval) **then**
    Enqueue Log_i on to the local queue Q_i.

Fig. 3. The scheme for constructing *Log* at P_i $(1 \leq i \leq n)$

$I_i[j]$ is the timestamp $V_j[j]$ when ϕ_j last became true, as known to P_i. (3) Log_i: contains the information about an interval, needed to compare it with other intervals. Figure 1 shows how to update the vector clock and *Interval Clock.*

To maintain Log_i, the data structures are defined in Figure 2. The *Log* consists of vector timestamps *start* and *end* for the start and end of an interval, respectively. It also contains an array of *Process_Log*, where each *Process_Log* is a queue of type *Event_Interval. Event_Interval* consists of a tuple composed of *interval_id* and *local_event*. Let *local_event* be the local component of the clock value of a receive event at which the kth component of *Interval Clock* gets updated — then the tuple composed of the *local_event* and the kth component of *Interval Clock* is added into the *Process_Log* queue which forms the kth component of *p_log*. Log_i is constructed and stored on the local queue Q_i using the protocol shown in Figure 3. Note that not all the intervals are stored in the local queue. The *Log* corresponding to an interval is stored only if the relationship between the interval and all other intervals (at other processes) is different from the relationship which its predecessor interval had with all the other intervals (at other processes). In other words, if two or more successive intervals on the same process have the same relationship with all other intervals, then *Log* corresponding to only one of them needs to be stored on the queue. Two successive intervals Y and Y' on process P_j will have the same relationship if no message is sent or received by P_j between the start of Y and the end of Y'.

The *Log* is used to determine the relationship between two intervals. The tests in Table 1 are used to find which of $R1$, $R2$, $R3$, $R4$, $S1$, and $S2$ are true. Figure 4 shows how to implement the tests for $S1(Y, X)$ and $S2(X, Y)$ using the *Log* data structure.

$S2(X,Y)$:

1. // Eliminate from *Log* of interval Y (on P_j), all receives of messages //which were sent by i before the start of interval X (on P_i).
   (1a) **for** each $event\_interval \in Log_j.p\_log[i].event\_interval\_queue$
   (1b) **if** $(event\_interval.interval\_id < Log_i.start[i])$ **then**
   (1c) remove $event\_interval$

2. // Select from the pruned *Log*, the earliest message sent from X to Y.
   (2a) $temp = \infty$
   (2b) **if** $(Log_j.start[i] \geq Log_i.start[i])$ **then** $temp = Log_j.start[j]$
   (2c) **else**
   (2d) **for** each $event\_interval \in Log_j.p\_log[i].event\_interval\_queue$
   (2e) $temp = \min(temp, event\_interval.local\_event)$

3. **if** $(Log_i.end[j] \geq temp)$ **then** $S2(X,Y)$ is true.

$S1(Y,X)$:

1. Same as step 1 of scheme to determine $S2(X,Y)$.
2. Same as step 2 of scheme to determine $S2(X,Y)$.
3. **if** $(Log_i.end[j] < temp)$ and $(temp > Log_j.start[j])$ **then** $S1(Y,X)$ is true.

Fig. 4. Implementing the tests for $S1(X,Y)$ and $S2(Y,X)$

5 A Distributed Algorithm

5.1 Algorithm DOOR

To solve problem **DOOR**, defined in Section 1, recall that the given relations $\{r_{i,j} \mid 1 \leq i,j \leq n \text{ and } r_{i,j} \in \Re\}$ need to satisfy the axioms on $\Re$, given in [5]. Thus, it is possible for a solution to exist in some execution.

Algorithm Overview: The algorithm uses a token-based approach. Intuitively, the process P_i which has the token triggers at each other process P_j, the comparison of the interval at the head of P_i's queue with the interval at the head of P_j's queue. The comparison may result in either the interval at the head of P_i's queue or P_j's queue being deleted – the corresponding queue index gets inserted in *updatedQueues*, which is a part of the token at P_i. Once such a comparison is done with all other process queues, the token is then sent to some P_j whose index j is in *updatedQueues*. This allows comparison of the new interval at the head of P_j's queue with the interval at the head of the queue of each other process. A solution is found when *updatedQueue* becomes empty.

The Algorithm: Besides the token (T), two kinds of messages are exchanged between processes – *REQUEST* (*REQ*) and *REPLY* (*REP*). The data structures are given in Figure 5. The proposed algorithm is given in Figure 6. Each procedure is executed atomically. The token is used such that only the token-holder process can send *REQ*s and receive *REP*s. The process (P_i) having the

type *REQ* = **message**
  log : *Log*; //Contains the *Log* of the interval at the queue
end //head of the process sending the *REQ*
type *REP* = **message**
  updated: set of integer; //Contains the indices of queues updated after a test
end
type *T* = **token**
  updatedQueues: set of integer; //Contains the indices of all the updated queues
end

Fig. 5. The *REQ*, *REP*, and *Token* data structures

token broadcasts a *REQ* to all other processes (line 3b). The *Log* corresponding to the interval at the head of the queue Q_i is piggybacked on the *REQ* (line 3a). On receiving a *REQ* from P_i, each process P_j compares the piggybacked interval X with the interval Y at the head of its queue Q_j (line 4d). According to Lemma 3, the comparison between intervals on processes P_i and P_j can result in three cases. (1) $r_{i,j}$ is satisfied. (2) $r_{i,j}$ is not satisfied and interval X can be removed from the queue Q_i. (3) $r_{i,j}$ is not satisfied and interval Y can be removed from the queue Q_j. In the third case, the interval at the head of Q_j is dequeued and process index j is stored in *REP.updated* (lines 4g, 4h). In the second case, the process index i is stored in *REP.updated* (line 4e). Note that both cases (2) and (3) can occur after a comparison. P_j then sends *REP* to P_i. The *REP* carries the indices of the queues, if any, which got updated after the comparison. Once process P_i receives a *REP* from all other processes, it stores the indices of all the updated queues in set *updatedQueues* which is a part of the token currently at P_i. Process P_i then checks if its index i is contained in *updatedQueues*. If so, it deletes the interval at the head of Q_i (line 8f). A solution is detected when *updatedQueues* becomes empty. If *updatedQueues* is non-empty, then the token is sent to a randomly selected process from *updatedQueues* (line 8g).

5.2 Complexity Analysis

The complexity is analyzed in terms of the maximum number of messages sent per process (m) and the maximum number of intervals per process (p).

- Space overhead:
  - Worst case space overhead per process is $\min(4np-2p, 4mn^2+2mn-2m)$.
  - Total space overhead across all processes is $\min(4pn^2-2np, 10mn^2)$.
- Average time complexity (per process) is $O(\min(np, 4mn))$.
- Total number of messages sent is $O(n\min(np, 4mn))$. Total message space overhead is $O(n^2\min(np, 4mn))$.

The details of the complexity analysis are given in [1].

Note that in case of broadcast media, the number of *REQ*s sent for each *Log* is one because *REQ*s are broadcast by sending one message (line 3b). The message space complexity reduces to $O(n\min(np, 4mn))$, although the total number of messages sent stays at $O(n\min(np, 4mn))$.

(1) Initial state for process P_i
(1a) Q_i is empty

(2) Initial state of the token
(2a) $T.updatedQueues = \{1, 2...n\}$
(2b) A randomly elected process P_i holds the token.

(3) *SendReq* : Procedure called by process P_i to send *REQ* message
(3a) $REQ.log = Log_i$ at the head of the local queue
(3b) Broadcast request *REQ* to all processes

(4) *SendReply* : Procedure called by process P_j to send a *REP* message to P_i
(4a) $REP.updated = \phi$
(4b) Y=head of local queue Q_j
(4c) X=$REQ.log$
(4d) Determine $R(X, Y)$ using the tests given in Table 1 and Figure 4
(4e) **if** $(R(X, Y) \in S(r_{i,j}))$ **then** $REP.updated = REP.updated \cup \{i\}$
(4f) **if** $(R(Y, X) \in S(r_{j,i}))$ **then**
(4g) $REP.updated = REP.updated \cup \{j\}$
(4h) Dequeue Y from local queue Q_j
(4i) Send reply *REP* to P_i

(5) *RcvToken* : On receiving a token T at P_i
(5a) Remove index i from *T.updatedQueues*
(5b) **if** (Q_i is nonempty) **then** *SendReq*

(6) *IntQue* : When an interval gets queued on Q_i at P_i
(6a) **if** (number of elements in queue Q_i is 1) and (P_i has the token) **then** *SendReq*
(6b) **else**
(6c) **if** (number of elements in queue Q_i is 1) and (P_i has a pending request) **then**
(6d) *REQ* is not pending
(6e) *SendReply*

(7) *RcvReq* : On receiving a *REQ* at P_i
(7a) **if** (Q_i is nonempty) **then** *SendReply*
(7b) **else** *REQ* is pending

(8) *RcvReply* : On receiving a reply from P_i
(8a) $T.updatedQueues = T.updatedQueues \cup REP.updated$
(8b) **if** (reply received from all processes) **then**
(8c) **if** (*T.updatedQueues* is empty) **then**
(8d) Solution detected. Heads of the queues identify the intervals.
(8e) **else**
(8f) **if** $(i \in T.updatedQueues)$ **then** dequeue the head from Q_i
(8g) Send token to P_k where k is randomly selected from *T.updatedQueues*

Fig. 6. Distributed algorithm to solve problem **DOOR**

5.3 Optimizations

The following two modifications to the algorithm increase the pruning and decrease the number of messages sent. However, the order of space, time, and message complexities remains the same for both modifications.

– When procedure *SendReply* is executed by P_j in response to P_i's *REQ*, instead of comparing the interval at the head of Q_j with P_i's interval (piggybacked on *REQ*), multiple comparisons can be done. Each time the comparison results in the interval at the head of Q_j being deleted, the next interval on Q_j is compared with the piggybacked interval. A *REP* is sent back only when either the comparison results in a relation equal to $r_{i,j}$ or the relationship is such that the interval at the head of Q_i (the piggybacked interval) has to be deleted. Thus each *REQ* can result in multiple intervals being pruned from the queue of the process receiving the *REQ*.

– In procedure *RcvReply* (lines 8f-8g), if *T.updatedQueues* contains the index i, it means the interval at the head of queue Q_i needs to be deleted and that the token will be sent to process P_i again in the future. Hence, if index i is contained in *T.updatedQueues*, not only is the interval at the head of Q_i deleted but also the next token-holder is selected as P_i. This saves the extra message required to resend the token to P_i later.

6 Concluding Remarks

Pairwise temporal interactions in a distributed execution provide a valuable way to specify and model synchronization conditions and information interchange. This paper presented an on-line distributed algorithm to detect whether there exists a set of intervals, one at each process, such that a given set of pairwise temporal interactions, one for each process pair, holds for the set of intervals identified. Future work can be to explore how the orthogonal interaction types can formalize and simplify the exchange patterns for various applications.

Acknowledgements

This material is based upon work supported by the National Science Foundation under Grant No. CCR-9875617.

References

[1] P. Chandra, A. D. Kshemkalyani, Detection of orthogonal interval relations, *Tech. Report UIC-ECE-02-06*, Univ. of Illinois at Chicago, May 2002. 327, 328, 331

[2] K. M. Chandy, L. Lamport, Distributed snapshots: Determining global states of distributed systems, *ACM Transactions on Computer Systems*, 3(1): 63-75, 1985. 324

[3] Q. Du, V. Faber, M. Gunzburger, Centroidal Voronoi tessellations: applications and algorithms, *SIAM Review*, 41(4): 637-676, 1999. 324

[4] C. J. Fidge, Timestamps in message-passing systems that preserve partial ordering, *Australian Computer Science Communications*, 10(1): 56-66, February 1988. 325, 328

[5] A. D. Kshemkalyani, Temporal interactions of intervals in distributed systems, *Journal of Computer and System Sciences*, 52(2): 287-298, April 1996. 323, 324, 325, 326, 330

[6] A. D. Kshemkalyani, A framework for viewing atomic events in distributed computations, *Theoretical Computer Science*, 196(1-2), 45-70, April 1998. 324

[7] A. D. Kshemkalyani, A fine-grained modality classification for global predicates, *Tech. Report UIC-EECS-00-10*, Univ. of Illinois at Chicago, 2000. 325, 326

[8] L. Lamport, Time, clocks, and the ordering of events in a distributed system, *Communications of the ACM*, 558-565, 21(7), July 1978. 324

[9] L. Lamport, On interprocess communication, Part I: Basic formalism; Part II: Algorithms, *Distributed Computing*, 1:77-85 and 1:86-101, 1986. 324

[10] F. Mattern, Virtual time and global states of distributed systems, *Parallel and Distributed Algorithms*, North-Holland, 215-226, 1989. 325, 328

An Efficient Parallel Algorithm for Computing Bicompatible Elimination Ordering (BCO) of Proper Interval Graphs*

B.S. Panda[1] and Sajal K. Das[2]

[1] Department of Mathematics, Indian Institute of Technology, Delhi
Hauz Khas, New Delhi, 110 016, India
bspanda@maths.iitd.ernet.in

[2] Department of Computer Science and Engineering,
The University of Texas at Arlington
Arlington, TX 76019, USA
das@cse.uta.edu

Abstract. In this paper, we first show how a certein ordering of vertices, called *bicompatible elimination ordering* (BCO), of a proper interval graph (PIG) can be used to solve optimally the following problems: finding Hamiltonian cycle in a Hamiltonian PIG, the set of articulation points and bridges, and the single source or all pair shortest paths. We then propose an NC parallel algorithm (i.e., polylogarithmic-time employing a polynomial number of processors) to compute a BCO of a proper interval graph.

1 Introduction

A graph is *chordal* if every cycle in it of length at least four has a chord, i.e., an edge joining two non-consecutive vertices of the cycle. Chordal graphs arise in many real-life applications, including evolutionary trees [2], facility location [4] scheduling [14], archaeology [3], database systems [1], solutions of sparse systems of linear equations [16], and so on.

For a graph $G = (V, E)$, let $N(v) = \{w \in V | vw \in E\}$ be the set of neighbors of v and let $N[v] = N(v) \cup \{v\}$. Let $G[S]$, $S \subseteq V$, be the induced subgraph of G on S. If $G[N(v)]$ is a complete subgraph, then v is called a *simplicial vertex* of G. A *perfect elimination ordering* (PEO) of G of order n is an ordering $\alpha = (v_1, v_2, \ldots, v_n)$ of V such that v_i is a simplicial vertex of $G[\{v_i, v_{i+1}, \ldots, v_n\}]$, for $1 \leq i \leq n$. A graph G is *chordal* if and only if G has a PEO [6].

A PEO $\alpha = (v_1, v_2, \ldots, v_n)$ of a chordal graph G is a *bicompatible elimination ordering* (BCO) if $\alpha^{-1} = (v_n, v_{n-1}, \ldots, v_1)$, the reverse of α, is also a PEO of G.

* Part of this work was done while the first author was with the Dept of Computer and Information Sciences, Univ of Hyderabad and was visiting the Dept of Computer Science and Engineering at the Univ of Texas at Arlington. This work was partally supported by NASA Ames Research Center under Cooperative Agreement Number NCC 2-5359.

S. Sahni et al. (Eds.) HiPC 2002, LNCS 2552, pp. 334–343, 2002.

A graph G is called an *interval graph* for a family, F, of intervals in a linearly ordered set (like the real line), if there exists a one-to-one correspondence between the vertices of G and the intervals in F such that two vertices are adjacent if and only if the corresponding intervals intersect. If no interval of F properly contains another, set-theoretically, then G is called a *proper interval graph* (PIG, for short). A BCO characterizes PIGs in the same way as PEO characterizes chordal graphs, More specifically, it is known that a graph G has a BCO if and only if it is a PIG [11]. To the best of our knowledge, however, there exists no parallel algorithm to generate BCO of a PIG, which motivates our work.

A graph G is *Hamiltonian* if it has a Hamiltonian cycle containing all the vertices of G. The problem of deciding whether a graph is Hamiltonian is well known to be NP-complete [8].

Given a PIG, G, and its BCO, in this paper we design optimal sequential algorithms to solve such problems as finding (i) a Hamiltonian cycle if G is Hamiltonian, (ii) the set of all articulation points and bridges of G, and (iii) single source and all pair shortest paths. We also propose an NC parallel algorithm (i.e., polylogarithmic-time employing a polynomial number of processors) to compute BCO of a *PIG*. Precisely, the parallel algorithm requires $O(\log^2 n)$ time employing $O(n + m)$ processors where n is the number of vertices and m is the number of edges.

The rest of the paper is organized as follows. Section 2 presents some known results that will be used in this paper. Section 3 describes optimal sequential algorithms for various problems, while Section 4 proposes the NC parallel algorithm. Section 5 concludes the paper.

2 Preliminaries

In this section we state some known results and also derive some elementary results which will be used in the rest of this paper. Throughout, a graph is assumed to be connected with n vertices and m edges. If $G[C]$, $C \subseteq V$, is a maximal complete subgraphs of G, then C is called a *clique* of G. The classes of graphs characterized in terms of PEO and BCO are summarized in the following theorems.

Theorem 2.1 [6]: A graph G has a PEO if and only if it is chordal.

Theorem 2.2 [11]: A graph G has a BCO if and only if it is a PIG (proper interval graph).

The following two theorems characterize interval graphs and also which interval graphs are PIGs.

Theorem 2.3 [6]: A graph G is an interval graph if and only if its cliques can be linearly ordered such that, for every vertex x of G, the cliques containing x occur consecutively.

Theorem 2.4 [15]: A graph G is a PIG if and only if it is an interval graph containing no induced $K_{1,3}$ (star of 4 nodes).

That every BCO gives rise to a Hamiltonian path follows from the following result.

Lemma 2.5 [11]: If $v_1, v_2, \ldots, v_n$ is a BCO of a connected graph G, then $P = v_1, v_2, \ldots, v_n$ is a Hamiltonian path of G.

Let $G - C$ be disconnected for a given clique, C, resulting in the components $H_i = (V_i, E_i), 1 \leq i \leq r$ and $r \geq 2$. Then C is said to be a *separating clique* and $G_i = \mathrm{G}[(V_i \cup C)]$, is called a *separated graph* of G with respect to C. Let $\mathrm{W}(G_i) = \{v \in C \mid$ there is a $w \in V_i$ with $vw \in E(G)\}$. Cliques of G other than C which intersect C are called *relevant cliques* of G with respect to C. A *relevant clique* C_i of G_i for which $(C_i \cap C) = W(G_i)$ is called a *principal clique* of G_i. Let C(G) be the set of all cliques of G. For $v \in V(G)$, let $C_v(G)$ be the set of all cliques of G containing vertex v.

Walter [17], Gavril [7] and Buneman [2] have shown that chordal graphs are the intersection graph of subtrees of a tree. In fact, for every chordal graph G, there exists a tree T such that $V(T) = C(G)$ and $\mathrm{T}[C_v(G)]$ is a subtree of T for every $v \in V(G)$. Such a tree T is called a *clique tree* for G.

A chordal graph G is said to be an *undirected vertex* (UV) *graph* if there exists a clique tree T of G such that $\mathrm{T}[C_v(G)]$ is a path in T for every $v \in V(G)$. Such a clique tree is called a *UV clique tree* of G. Note that every UV graph is a chordal graph. The following lemma gives further structure to the clique tree of a UV graph.

Lemma 2.6 [13, Proposition 7′]: Let C be a clique in a UV graph G. If C is not a separating clique, then C is a leaf node in every UV clique tree T of G.

Let $C_1, C_2, \ldots, C_r$ be an ordering of the cliques of an interval graph G as in Theorem 2.3. If we construct a path T of these cliques, then T is a clique tree of G. It is known that

Theorem 2.7: Every interval graph G has a path T as a clique tree.

Now we derive the following lemma.

Lemma 2.8: Let G be a PIG. Then a clique C of G is a leaf node in every clique tree T of G if and only if C is a nonseparating clique of G.

Proof:

Necessity: Let C be a leaf node in every clique tree T of G. We will show that C is a nonseparating clique of G. If possible, let C be a separating clique of G and C be a leaf node of the clique tree T of G. Let G_1 and G_2 be two separated graphs of G with respect to C. Let C_i be a principal clique of G_i, for $i = 1, 2$. If $\mathrm{W}(G_1)$ and $\mathrm{W}(G_2)$ are comparable, set theoretically, then assume $\mathrm{W}(G_1) \subseteq \mathrm{W}(G_2)$. Let $x \in C_2 - C$, $x_1 \in C_1 - C$, $x_2 \in C_2 - C$, and $x_3 \in C_1 \cup C_2 \cup C$. Then, $G[\{x, x_1, x_2, x_3\}]$ is isomorphic to $K_{1,3}$, which is a contradiction. So, $W(G_1)$ and $W(G_2)$ are not comparable. Let $x \in (C_1 \cap C) - C_2$ and $y \in (C_2 \cap C) - C_1$. Now, $C_x(G)$ does not contain C_2 and $C_y(G)$ does not contain C_1. Since C is a leaf node of T, and $T[C_x(G)]$ and $T[(C_y(G)]$ are paths in T, either $C_1 \in C_x(G)$ or $C_2 \in C_y(G)$. So, we have a contradiction. Hence, C is a nonseparating clique of G.

Sufficiency: Let C be a nonseparating clique of G and T be any clique tree of G. We will show that C is a leaf node of T. Since every proper interval graph is a UV graph, by Lemma 2.6, C is a leaf node in every clique tree of G. Hence, C is leaf node of T.

□

Every PIG, G, being an interval graph, every proper interval graph G has clique tree T which is a path. Below, we show that this path is the unique clique tree of a PIG.

Lemma 2.9: Let G be a PIG having r cliques. Then P_r, the path of length r, is the unique clique tree of G.

Proof: By Theorem 2.3, a path on r vertices is a clique tree for G. Now, by Lemma 2.8, every nonseparating clique is a leaf node in every clique tree T of G. Hence G has exactly two nonseparating cliques. Also by Lemma 2.8, every separating clique corresponds to non leaf node in every clique tree of G. Since the path P_r is the only tree having exactly two leaf vertices, a path on r vertices is the unique clique tree of G. □

The following property of BCO will be used in establishing our result.

Lemma 2.10: Let $\alpha = (v_1, v_2, \ldots, v_n)$ be a BCO of a proper interval graph. If $v_iv_j \in E$, then $v_kv_j \in E$ for all k such that $i \leq k \leq j-1$.

Proof: As α is a BCO, the edge $v_iv_{i+1} \in E(G)$. Since v_i is a simplicial vertex of $G_i = G[\{v_i, v_{i+1}, \ldots, v_n\}]$ and $v_iv_j \in E(G)$, we conclude $v_{i+1}v_j \in E$. Again v_{i+1} is a simplicial vertex of $G_{i+1} = G[\{v_{i+1}, \ldots, v_n\}]$. So $v_{i+2}v_j \in E$. By the same argument, $v_kv_j \in E$ for all k, where $i \leq k \leq j-1$. Hence the lemma. □

3 Implications of BCO

Suppose a proper interval graph G and its BCO are given as input. In this section, we we design optimal algorithms for (i) constructing a Hamiltonian cycle, (ii) finding all the articulation points and bridges, and (iii) solving the single source shortest path and all pair shortest path problems.

3.1 Hamiltonian Cycle

Let us first show that a proper interval graph is Hamiltonian if and only if it has a BCO. Next, given a BCO, we construct a Hamiltonian cycle of a biconnected proper interval graph. To this end, we need the following lemma. For $H \subseteq G$, let $d_H(x) = |V(H) \cap N(x)|$ denote the degree of x in H.

Lemma 3.1: Let G be a biconnected PIG, and let $\alpha = (v_1, v_2, \ldots, v_n)$ be a BCO. Then $d_{G_i}(v_i) \geq 2$, where $1 \leq i \leq n-2$ and $G_i = G[\{v_i, v_{i+1}, \ldots, v_n\}]$.

Proof: If possible, let i be the largest index such that $d_{G_i}(v_i) < 2$. Then, since $v_iv_{i+1} \in E$, $d_{G_i}(v_i) = 1$. Again, $1 < i < n-1$. We claim that v_{i+1} is a cut vertex of G. Now $v_jv_k \notin E$, for $j \leq i$ and $k \geq i+2$, because if $v_jv_k \in E$, then by Lemma 2.10, $v_{j+s}v_k \in E$, for all s, $1 \leq s \leq k-j-1$. In particular, $v_iv_k \in E$. This contradicts $d_{G_i}(v_i) = 1$. So $v_jv_k \notin E$ for $j \leq i$ and $k \geq i+2$. So every path from v_j to v_k, where $j \leq i$ and $k \geq i+2$, passes through v_{i+1}. Hence our claim is true. This contradicts the biconnectedness of G and completes the proof. □

The following theorem characterizes a Hamiltonian PIG.

Theorem 3.2: A PIG G is Hamiltonian if and only if it is biconnected.
Proof: Since the necessity is trivial, we prove the sufficiency part only. Let $\alpha = (v_1, v_2, \ldots, v_n)$ be a BCO of G. Let i_1 be the largest index such that $v_1 v_{i_1} \in E$. Let i_2 be the largest index such that $v_{i_1-1}v_{i_2} \in E$. In general, let i_j be the largest index such that $v_{i_{j-1}-1}v_{i_j} \in E$. Let r be the smallest index such that $v_{i_r}v_n \in E$. As G is a biconnected PIG, $i_j - 1 > i_{j-1}$ and $i_{r-1} \leq n-2$. Let P_j be the path $v_{i_{j-1}-1}v_{i_j}$ consisting of only one edge, where $1 \leq j \leq r$. Let Q_j be the path from v_{i_j} to $v_{i_{j+1}-1}$ through consecutive vertices of the BCO α, $1 \leq j \leq r$. Let P_1 be the path $v_1 v_{i_1}$ consisting of only one edge. Let Q_1 be the path from v_1 to v_{i_1-1} through consecutive vertices. Let $v_{i_{r+1}} = v_n$ and Q_r be the path from v_{i_r} to v_n through consecutive vertices of the BCO, α. Then $C = P_1 \cup Q_1 \cup (\cup_{j=1}^{r} P_j) \cup_{j=1}^{r} Q_j)$ is a Hamiltonian cycle of G. Therefore, G is Hamiltonian. □

Given a biconnected PIG, G, and a BCO α of G, we next propose a linear time algorithm to construct a Hamiltonian cycle of G.

Algorithm Hamiltonian Cycle:
Input: $(G, \alpha = (v_1, v_2, \ldots, v_n))$, where G is a biconnected PIG and α is a BCO of G.
Output: A Hamiltonian cycle C of G. {
$C := v_1, v_2, \ldots, v_n$; $t := 1$; $r := 1$;
while ($t \neq n$) do {
Let l_r be the largest index such that $v_r v_{l_r} \in E$;
if ($l_r \neq n$) then
{ $C := C - \{v_{l_r-1}v_{l_r}\} \cup \{v_r v_{l_r}\}$;
$r := l_r - 1$; }
else { $C := C \cup \{v_r v_{l_r}\}; t = l_r$; }
} }

It is easy to see that the algorithm **Hamiltonian Cycle** runs in $O(n+m)$ time. The correctness of this algorithm follows from Theorem 3.2. So we have the following result.
Theorem 3.3: Given a biconnected PIG, G, and its BCO α, the Algorithm **Hamiltonian Cycle** correctly finds a Hamiltonian cycle of G in O(n+m) time.

3.2 Articulation Points and Bridges

This subsection shows that the set all articulation points and bridges of a PIG, G, can be computed in linear time once its BCO is given.

Let $\alpha = (v_1, v_2, \ldots, v_n)$ be a BCO of G. Let $High[i] = j$ if j is the largest index such that $v_i v_j \in E(G)$, for $1 \leq i \leq n$. Note that $High[i]$ can be computed for all i, $1 \leq i \leq n$, in linear time by scanning the adjacency list of G. If $d(v_1) = 1$, then $X_1 = \{v_1 v_2\}$. If $d(v_1) > 1$, then $X_1 = \phi$. If $d(v_n) = 1$, then $X_2 = \{v_{n-1}v_n\}$. If $d(v_n) > 1$, then $X_2 = \phi$. Let $X = X_1 \cup X_2$; $A = \{v_i, 1 < i < n | \; High[i-1] = i\}$; and $B = \{v_i v_{i+1}, i \leq n-2 | v_i, v_{i+1} \in A\} \cup X$. Here we claim that

Theorem 3.4: The sets A and B as defined above are the set of all articulation points and the set of all bridges of G, respectively. Moreover, they can be computed in linear time for a PIG, G, once a BCO of G is given.
Proof: Left to the reader. □

3.3 Shortest Paths

Let $\alpha = (v_1, v_2, \ldots, v_n)$ be a BCO of a PIG, G. For each i, $1 < i \leq n$, let $Low[i]$ be the minimum index $j < i$ such that $v_j v_i \in E(G)$. Note that $Low[i]$ can be computed by just scanning the adjacency list of v_i. So, $Low[i]$ for all i, where $1 < i \leq n$, can be computed in linear time. Since α is a BCO of G, implying α and α^{-1} are PEOs of G, by Lemma 2.10, $G[\{v_i, v_{i+1}, \ldots, v_{High[i]}\}]$ is a complete subgraph of G for $1 \leq i < n$. Similarly, $G[\{v_i, v_{i-1}, \ldots, v_{Low[i]}\}]$ is a complete subgraph of G for $1 < i \leq n$.

Let $High^k[i] = i$ if $k = 0$ and $High^k[i] = High[High^{k-1}[i]]$ for $k \geq 1$. Similarly we define $Low^k[i]$. Given a vertex v_i, $1 < i < n$, let r_1 and r_2 be the largest integers such that $High^{r_1}[i] < n$ and $Low^{r_2}[i] > 1$, respectively. Then, $d(v_i, v_j) = k$ for $j \in \{High^{k-1}[i] + 1, High^{k-1}[i] + 2, \ldots, High^k[i]\}, 1 \leq k \leq r_1$, and $d(v_i, v_j) = k + 1$ for $j \in \{High^k[i] + 1, High^k[i] + 2, \ldots, n\}$. Similarly, $d(v_i, v_j) = k$ for $j \in \{Low^{k-1}[i] - 1, Low^{k-1}[i] - 2, \ldots, Low^k[i]\}, 1 \leq k \leq r_2$, and $d(v_i, v_j) = k + 1$ for $j \in \{Low^k[i] - 1, Low^k[i] - 2, \ldots, 1\}$. For v_1, let r be the largest integer such that $High^r[1] \leq n$. Then, $d(v_1, v_j) = k$ for $j \in \{High^{k-1}[1] + 1, High^{k-1}[1] + 2, \ldots, High^k[1]\}, 1 \leq k \leq r$, and $d(v_1, v_j) = k+1$ for $j \in \{High^k[1] + 1, High^k[1] + 2, \ldots, n\}$. For v_n, let s be the largest integer such that $Low^s[n] \geq 1$. Then, $d(v_n, v_i) = k$ for $i \in \{Low^{k-1}[n] - 1, Low^{k-1}[n] - 2, \ldots, Low^k[n]\}, 1 \leq k \leq s$, and $d(v_n, v_j) = k+1$ for $j \in \{Low^s[n] - 1, Low^s[n] - 2, \ldots, 1\}$.

Thus, given any vertex v_i, we can find the distance of the shortest paths from v_i to all other vertices in linear time as the above expression can be computed in linear time. the computation of Low and $High$ arrays takes $O(n+m)$ time. However, once the Low and $High$ of each vertex are given, the distance calculation takes $O(n)$ time. So, d(v_i, v_j) for every pair (v_i, v_j) can be computed in $O(n^2)$ time. Also note that, the single-source as well as all-pair shortest paths can be computed in same time bound. This leads to the following theorem.
Theorem 3.5: Given a BCO, α, of a PIG having n vertices and m edges, the single-source and all-pair shortest path problems can be solved in $O(n+m)$ time and $O(n^2)$ time, respectively.

4 Parallel Algorithm for Generating BCO

In this section, we propose a parallel algorithm to generate a BCO of a PIG. For this purpose we prove some lemmas which will be useful for developing our algorithm.
Lemma 4.1: Let $G = (V, E)$ be a PIG. Then, every $v \in V$ belongs at at most three maximal cliques of G.

Proof: Omitted, due to space limitation. □

Let $C_1, C_2, \ldots, C_r$ be the cliques of G in the order of occurrence in the clique tree T, which is a path according to Lemma 2.11. Let X_i be the set of exclusive vertices of C_i; in other words, $x \in X_i$ implies that x belongs to the only clique C_i. Let $Y_i = (C_i \cap C_{i+1}) - (C_{i+2} \cup C_{i-1})$, for $2 \leq i \leq r-2$; $Y_1 = (C_1 \cap C_2) - C_3$; $Y_r = (C_{r-1} \cap C_r) - C_{r-2}$; and $Z_i = (C_{i-1} \cap C_i \cap C_{i+1})$, for $2 \leq i \leq r-1$

Lemma 4.2 If Z_i is non-empty, then X_i is empty for every i, $2 \leq i \leq r-1$
Proof: Now $G' = G[\{C_{i-1} \cup C_i \cup C_{i+1}\}]$ is connected and C_i is a separating clique of G'. Let G_1 and G_2 be the separated subgraphs of G' with respect to C_i. Let $x \in Z_i; x_1 \in C_{i-1} - C_i; and y_1 \in C_{i+1} - C_i$. If possible, let $z \in X_i$. Then $G[\{x, z, x_1, y_1\}]$ is isomorphic to a $K_{1,3}$, which is a contradiction. So, if Z_i is non-empty then X_i must be empty. □

A parallel algorithm presented in [10] generates a PEO of a chordal graph by taking the graph and its clique tree as input. In this approach, an arbitrary vertex C (clique of G) of T is chosen and a function $f : V \rightarrow Z^+$ is computed, where Z^+ is the set of non-negative integers, with the help of the following rule.
Rule 1: $f(v) = \min_{C' \in C_v(G)}\{d(C, C')\}$, where $C_v(G)$ is the set of cliques containing v and $d(C, C')$ is the distance between two clique vertices C and C'.

Note that f is not a total order on $V(G)$. Theapproach in [10] constructs a directed graph $D = (V', E')$ using the rule:
Rule 2: $V' = V$ and $E' = \{xy \in E|$. Then either $f(x) > f(y)$ or $f(x) = f(y)$ and $ord(x) > ord(y)\}$, where ord is an arbitrary ordering of the vertices of G.
Theorem 4.3 [10]: Let G be a chordal graph and T be a clique tree of G. Let $D = (V', F)$ be the directed graph constructed from G by using **Rule 1** and **Rule 2**. Then G is a directed acyclic graph (DAG) and every topological ordering of D is a PEO of G.

Following the above approach, we compute a function $g : V \rightarrow Z^+$ and show that g is one of the functions f that can be computed using **Rule 1**. We then construct a directed graph D following **Rule 2** and show that an ordering α of G is a BCO if and only if α is a topological ordering of D.

Let us now present our algorithm to construct a BCO of a proper interval graph. Assume that the vertices of the graph are numbered $1, 2, \ldots, n$.

Algorithm BCO_GEN
Input: A PIG, $G = (V, E)$;
Output: An ordering α of V such that α is a BCO of G. {

1. Compute $C(G)$, the set of all cliques of G.
2. Construct a clique tree T of G such that T is a path. Let $C_1, C_2, \ldots, C_r$ be the ordering of the cliques according to their occurrences in T.
3. Compute an array A such that $A[i] = j$ if vertex i belongs to j cliques of G. Computethe array M such that $M[i] = j$, where j is the smallest index of the clique containing i.
4. Compute the array X as follows:
   for all $i \in V$ do in parallel $X[i] = M[i]$;

5. Construct a DAG, $D = (V', E')$, as follows:
$V' = V$;
for each edge $ij \in E$ do in parallel {
if $(X[i] > X[j])$ then $E' = E' \cup \{ij\}$;
else if $(X[j] > X[i])$ then $E' = E' \cup \{ji\}$;
else if $(A[i] > A[j])$ then $E' = E' \cup \{ij\}$;
else if $(A[j] > A[i])$ then $E' = E' \cup \{ji\}$;
else if $(i > j)$ then $E' = E' \cup \{ij\}$;
else if $(j > i)$ then $E' = E' \cup \{ji\}$; }

6. Find a topological ordering α of D.
}

4.1 Proof of Correctness

Here we prove the correctness of the proposed algorithm *BCO_GEN*.

Theorem 4.4: The ordering α produced by Algorithm **BCO_GEN** is a BCO of G.

Proof: First, we claim that the directed graph D' constructed in step 5 is one of the DAGs that can be generated by using **Rule 1** and **Rule 2**. Let T be the clique tree constructed in Step 2. So, by Lemma 2.9, T is a path. Let $C = C_1$, a leaf node of T. From the computation of arrays X and A, it is clear that for each $i \in V$, $f(i) = X[i] - 1$, where f is as in **Rule 1**. So, $f(i) > f(j)$, if and only if $X[i] > X[j]$. An arc ij belongs to a directed graph constructed using **Rule 1** and **Rule 2**, if $f(i) > f(j)$, or $f(i) = f(j)$, and $ord(i) > ord(j)$, where ord is any arbitrary ordering of the vertices. Let us define $ord : V \to \{1, 2, \dots, n\}$ such that $ord(i) > ord(j)$ if and only if $A[i] > A[j]$, or $A[i] = A[j]$ and $i > j$. Then, clearly, $ij \in E'$ if and only if $f(i) > f(j)$ or $f(i) = f(j)$ and $ord(i) > ord(j)$. So, the digraph D constructed in Step 5 of algorithm **BCO_GEN** is constructed by applying **Rule 1** and **Rule 2**. By Theorem 5.3, every topological ordering of D' is a PEO of G. Hence, the ordering α generated in step 6, is a PEO of G.

Next, we construct a digraph D' and show that (i) α^{-1}, i.e., the reverse of α is a topological ordering of D' and (ii) D' can be constructed using **Rule 1** and **Rule 2**. So, α will be a BCO of G. Let $D' = (V'', E'')$ be constructed as follows:

$V'' = V$, and $E'' = \{ij | ji \in E'\}$. Let $ord^{-1} : V \to \{1, 2, \dots, n\}$ be defined by $ord^{-1}(i) > ord^{-1}(j)$ if and only if $A[i] > A[j]$ or $A[i] = A[j]$ and $i < j$. So, if $A[i] = A[j]$, then $ord(i) < ord(j)$ if and only if $ord^{-1}(j) < ord^{-1}(i)$. Let $f' : V \to Z^+$ be defined using Rule 1 and Rule 2 using the tree T and taking $C = C_k$, where $C_k \neq C_1$ is the leaf vertex of T. Next, we show that $ij \in E''$ implies $f'(i) > f'(j)$ or $f'(i) = f'(j)$ and $ord^{-1}(i) > ord^{-1}(j)$. Note that, $f'(i) = k - A[i] - f(i)$. Let $ji \in E''$. So, $ij \in E'$.

Case I: $A[i] = A[j]$.

Since $f'(i) = k - A[i] - f(i)$, $f(i) > f(j)$ if and only if $f'(i) < f'(j)$ in this case. Since $ij \in E'$, either $f(i) > f(j)$ or $i > j$. If $f(i) > f(j)$, then $f'(i) < f'(j)$. If $f(i) = f(j)$ and $i > j$, then $f'(i) = f'(j)$ and $j < i$. So, either $f'(i) < f'(j)$ or $ord^{-1}(i) < ord^{-1}(j)$.

Case II: $A[i] \neq A[j]$.

First assume that $A[i] > A[j]$. Since $f(i) \geq f(j)$, $-f(i) \leq -f(j)$, and $-A[i] < -A[j]$. So, $f'(i) < f'(j)$.

Next, assume that $A[i] < A[j]$. So, $ord(i) < ord(j)$. Since $ij \in E'$ and $ord(i) < ord(j)$, $f(i) > f(j)$. Let $A[i] = 1$ and $A[j] = 3$. If possible, let $f(i) = f(j) + 1 = r$. So, $X[i] = r + 1$ and $X[j] = r$. Hence $i \in C_{r+1}$ and $j \in C_r \cap C_{r+1} \cap C_{r+2}$. Since $A[i] = 1$, $i \in X_{r+1}$. Now, $j \in Z_{r+1}$ implies X_{r+1} and Z_{r+1} are simultaneously non-empty. This is a contradiction to Lemma 4.2. Hence, if $A[i] = 1$ and $A[j] = 3$, then $f(i) > f(j) + 1$. So, $f(i) + A(i) \geq f(j) + A[j]$. Hence, $k - f(i) - A[i] \leq k - f(j) - A[j]$, i.e. $f'(i) \leq f'(j)$ If $f'(i) = f'(j)$, then $ord^{-1}(i) < ord^{-1}(j)$. So, if $A[i] = 1$ and $A[j] = 3$, then either $f'(i) < f'(j)$ or $f'(i) = f'(j)$ and $ord^{-1}(i) < ord^{-1}(j)$.

Since, $A[i] < A[j]$, other choices of $A[i]$ and $A[j]$ are $A[i] = 1$ and $A[j] = 2$ or $A[i] = 2$ and $A[j] = 3$. In both the situations, $f(i) + A[i] \geq f(j) + A[j]$. So, $k - f(i) - A[i] \leq k - f(j) - A[j]$, i.e. $f'(i) \leq f'(j)$. If $f'(i) = f'(j)$, then $ord^{-1}(i) < ord^{-1}(j)$.

Hence, $ji \in E''$ implies $f'(j) > f'(i)$ or $f'(j) = f'(i)$ and $ord^{-1}(j) > ord^{-1}(i)$. Thus, the digraph D' defined above can be constructed by using **Rule 1** and **Rule 2**. Since D' is obtained by reversing the direction of each arc of D, the reverse of every topological ordering of D is a topological ordering of D'. Since, α, the ordering obtained in Step 6 of **Algorithm BCO_GEN** is a topological ordering of D, α^{-1} (i.e., the reverse of α), is a topological ordering of D'. So, by Theorem 4.3, α^{-1} is a PEO of G. Hence, α is a BCO of G. □

Theorem 4.5: The BCO of a proper interval graph (PIG) can be constructed in $O(\log^2 n)$ time using $O(n+m)$ processors on the CRCW (concurrent-read, concurrent-write) PRAM (parallel random access machine) model of parallel computation.

Proof: The set of all cliques of a chordal graph can be computed in $O(\log^2 n)$ time using $O(n+m)$ processors on the CRCW PRAM model [12]. Since PIGs are chordal graphs, Step 1 of Algorithm **BCO_GEN** takes $O(\log^2 n)$ time using $O(n+m)$ processors on the CRCW PRAM. A clique tree of a chordal graph can be constructed in the same processor and time bounds [12]. Since, by Theorem 2.9, a path is the unique clique tree of a PIG, Step 2 takes $O(\log^2 n)$ time using $O(n+m)$ processors. By Lemma 4.1, $A[i] \leq 3$ for each vertex i of a PIG; thus the array A can be computed as follows. Prepare a list consisting of all (i, j), where i is a vertex on clique j. The length of this list is $O(n+m)$ [9]. This list can be lexicographically sorted in $O(\log n)$ time using $O(n+m)$ processors by using a parallel merge sort [5]. From this sorted list, arrays A and M of Step 3 can be computed in constant time using $O(n+m)$ processors. Step 4 takes constant time using $O(n+m)$ processors. We need not construct the digraph in Step 5. We have to construct a topological ordering. Maintain the list of triplets $(X[i], A[i], i)$. A lexicographically non-decreasing ordering of this triplet gives rise to a topological ordering and hence a BCO of G. This takes $O(\log n)$ time using $O(n)$ processors [5]. Hence the theorem. □

5 Conclusion

In this paper, we show how to find optimally a Hamiltonian cycle, articulation points and bridges, single source and all pair shortest paths for a PIG once a BCO of the PIG is given as input. We also proposed an efficient parallel algorithm for generating a BCO of a PIG.

References

[1] C. Been, R. Fagin, D. maier , and M. Yannakakis, On the desirability of acyclic database schemes, J. ACM 30 (1983) 479-513.

[2] P.Buneman, A characterization of rigid circuit graphs, Discrete Mathe. 9 (1974) 205-212.

[3] P.Buneman, The recovery of trees from measures of dissimilarity, in: mathematics in the Archaeological and Historical Sciences, Edinburg University Press, Edinburg (1972) 387-395.

[4] R. Chandrasekharan and A. Tamir, Polylomially bounded algorithms for locating p-centers on a tree, Math. Programming 22 (1982) 304-315.

[5] R. Cole, Parallel merge sort, SIAM J. Comput. 17 (1988) 770-785.

[6] D. R. Fulkerson and O. S.Gross, Incidence matrices and interval graphs, Pacific J. Math. 15 (1965) 835-855.

[7] F.Gavril, The intersection graphs of subtrees in trees are exactly the chordal graphs, J. Combin. Theory, ser B 16 (1974) 47-56.

[8] M. R. Garey and D. S. Johnson, Computers and Intractability: A guide to the theory of NP-completeness, W. H. Freeman and Company (1979).

[9] M. C. Golumbic, Algorithmic graph theory and perfect graphs, Academic press, New York. (1980)

[10] Chin-wen Ho and R. C. T.Lee, Counting clique trees and computing perfect elimination schemes in parallel, IPL 31 (1989) 61-68.

[11] R. E. Jamison and R. Laskar, Elimination orderings of chordal graphs, in "Proc. of the seminar on Combinatorics and applications," (K. S. Vijain and N. M. Singhi eds. 1982) ISI, Calcutta, pp. 192-200.

[12] P. N.Klein, Efficient parallel algorithms for chordal graphs, SIAM J. Comput. 25 (1996) 797-827.

[13] C. L.Monma and V. K.Wei, Intersection graphs of paths in a tree, J. Combin. Theory, Ser. B 41 (1986) 141-181.

[14] C. Papadimitriou and M. Yannakakis, Scheduling interval ordered tasks, SIAM J. Comput. 8 (1979) 405-409.

[15] F. S. Roberts, Indifferences graphs, in:"Proof Techniques in Graph Theory", (F.Harary ed.) 1971 , pp.139-146, Academic Press.

[16] D. Rose, A graph theoretic study of the numerical solution of sparse positive definite systems of linear equations: in R. Read ed, Graph Theory and Computing (Academic Press, New York, 1972) 183-217.

[17] J. R. Walter, representation of chordal graphs as subtrees of a tree, J. Graph Theory 2 (1978) 265-267.

Router Handoff: An Approach for Preemptive Route Repair in Mobile Ad Hoc Networks

P. Abhilash, Srinath Perur, and Sridhar Iyer

K.R. School of Information Technology, IIT Bombay
Mumbai – 400076, India
abhilash,srinath,sri@it.iitb.ac.in

Abstract. Mobile Adhoc Networks (MANET) are distributed, mobile, wireless, multihop networks that operate without pre-existing communication infrastructure. Several routing protocols both reactive and proactive have been proposed to provide the self starting behavior needed for adhoc networks. Unreliable wireless links and node mobility may result in a large number of route breakages. The standard approach followed in reactive routing protocols in case of broken routes is to flag an error and re-initiate route discovery either at the source or at an intermediate node. Repairing these broken links increases routing overhead and the delay in delivering packets. In this paper, we propose an approach called Router Handoff which repairs routes preemptively, before they break, by using other nodes in the vicinity of a weak link. We have incorporated Router Handoff into the AODV routing protocol and have validated the idea through analysis and simulations. The results of the simulations indicate an increase in throughput under certain conditions. This improvement is a result of smaller overhead and delay.

1 Introduction

A Mobile Adhoc Network (MANET) [1] is a cooperative engagement of a collection of mobile devices, herein referred to as *nodes*, without the required intervention of any centralized access point or existing infrastructure. Each node is equipped with a wireless transmitter and a receiver. In order to facilitate communication within the network, each node acts as a router and a routing protocol is used to discover and maintain multi-hop routes between nodes.

Several routing protocols both reactive [2, 3] and pro-active [4] have been proposed for routing in adhoc networks. The performance of reactive protocols and their variants have been affected by *high routing overheads* and *delays* in the process of repairing broken routes. The current approach is to flag an error and re-initiate a route discovery either at the source or at an intermediate node where the route was broken, which in a reactive protocol typically involves flooding packets through the network. Several approaches like caching of learned routes, use of location information [5], and the use of virtual backbones [6, 7] have been proposed to reduce routing overheads. But in all these approaches route failure results in increased overheads and delays.

S. Sahni et al. (Eds.) HiPC 2002, LNCS 2552, pp. 347–357, 2002.

We present an approach called *Router Handoff* to preemptively repair routes that might break, using mobile nodes in the vicinity of the broken link. We have incorporated this idea into the AODV (Ad Hoc On-Demand Distance Vector) routing protocol. We also present a theoretical analysis of this approach where we compare AODV with Router Handoff, plain AODV, and AODV with Local Route Repair (LRR). The analysis shows remarkable improvement in terms of reduction in both routing overhead and delay. The simulations validate our claims to a large extent, showing an increase in throughput as a result of smaller overhead in repairing broken routes.

The main contribution of this paper is the preemptive technique of Router Handoff. Another contribution is the novel theoretical analysis of routing overhead and delays.

2 Router Handoff in AODV

Router Handoff is a preemptive approach to deal with route breaks. In Router Handoff, each node makes use of its *Neighbour Information Table* (NIT). This table contains information about the status of links with each neighbour. The central idea of Router Handoff is to find an alternate node in the vicinity of a potential link break, which can bypass the weak link. A node which finds that it is routing traffic on a link that is about to break *hands off* its routing information to a suitable node when the ratio between received power on the link and the threshold receiving power is less than a particular *Handoff THreshold* (HTH).

When movement of an intermediate node or the destination may cause a link to break, a node which uses the link as the next hop broadcasts a *Handoff REQuest* (HREQ). HREQ is a single hop packet and contains the next hop node and all the previous hop nodes that use the link. When a neighbor node which receives the HREQ is in a position to route packets from some of the previous hop nodes to the next hop node, a decision made by using the NIT, it sends a *Handoff REPly* (HREP). The node also updates its routing table. The previous hop nodes, which receive the HREP update their routing tables to make the node which sent the HREP as the next hop, thereby avoiding the broken link. The HREP is a single hop packet.

The advantage of this approach is that routes that may be about to break are repaired with just two packets HREQ and HREP. Since it tries to find an alternate route locally, before a route break, the delay involved is lower. Moreover more than one route can be fixed at a time.

2.1 Algorithm

The algorithm followed by each node in the network to perform Router Handoff is outlined below. Timers prevent multiple back-to-back HREQs for the same routes. The details regarding timers and handling of multiple HREPs are omitted here for ease of presentation. Here, `Received Packet` refers to data, routing or

Hello Message packets. Hello Messages are used by nodes to discover neighbors and maintain the Neighbor Information Table. For each node in the network:

```
Begin
          :
  if((Power of Received Packet/Threshold Power) < HTH)
  {
     Create Handoff Request Packet;
     Send Handoff Request Packet;
  }
  if(Received Packet == Handoff Request)
  {
     Check Neighbor Information Table;
     if(Next Hop Node in HREQ is a~Neighbor)
     {
        if(Any Previous Hop Node in HREQ is a~Neighbor)
        {
           Update Routing Table;
           Create Handoff Reply Packet;
           Send Handoff Reply Packet;
        }
     }
  }
  if(Received Packet == Handoff Reply)
  {
     if(Handoff Reply is for this Node)
     {
        Update Routing Table;
     }
  }
          :
End
```

2.2 Example

To make the concept of Router Handoff more concrete, consider the situation in Fig. 1. Let the route from source A to destination D pass through B and C, and the route from source E to destination D pass through . Now, if node C were to move, it could break either the link BC, or the link CD, or both.

Fig. 2 shows the scenario when the movement of C causes link CD to break. Before the the link CD is about to break, since node C has D as the next hop for some of the routes, it initiates a HREQ. HREQ invites responses from nodes that are within range of D and either B or E. The hop count of the HREQ is 1 and reaches only immediate neighbours. Node F sees from its Neighborhood

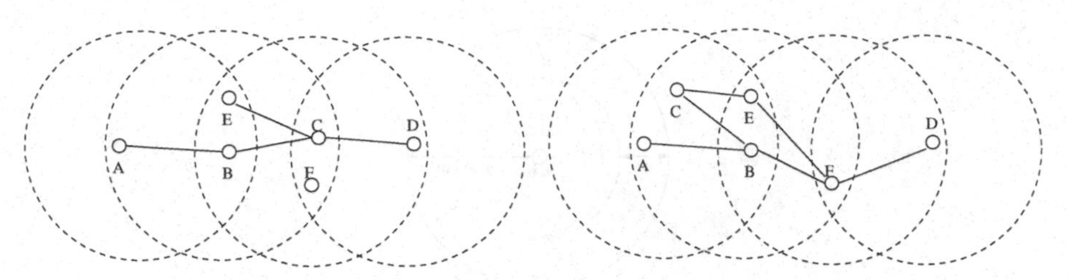

Fig. 1. Original routes **Fig. 2.** Link CD breaks

Information Table that it is within range of B, E and D, and sends a HREP packet. The HREP from node F indicates that it can route packets from B and E to D. The hop count of HREP is 1 and is received only by immediate neighbours of F. B and E on receiving the HREP update their routing tables so that F becomes the next hop for packets from sources *A* and source *E*.

2.3 Computation of Handoff Threshold (HTH)

The idea in router handoff is to hand over routes and associated information before the link breaks. This is done by performing handoff when the ratio of the received power (RxPr) from a next hop node and the receive threshold power (RxThresh) of the received packet is less than or equal to Handoff Threshold (HTH).

$$\frac{\mathrm{RxPr}}{\mathrm{RxThresh}} \leq \mathrm{HTH} \tag{1}$$

Let t be the time required for router handoff to take place, s be the maximum speed of the node and d be the distance that can be covered during which the handoff is to take place (refer figure 3). We know that:

$$\text{Received Power} \propto \frac{1}{\text{distance}^4}$$

$$\mathrm{RxThresh} \propto \frac{1}{\mathrm{R}^4} \tag{2}$$

$$\mathrm{RxPr} \propto \frac{1}{(\mathrm{R} - \mathrm{d})^4} \tag{3}$$

Substituting 2 and 3 in equation 1 we get:

$$\frac{\mathrm{R}^4}{(\mathrm{R} - \mathrm{d})^4} \leq \mathrm{HTH} \tag{4}$$

Substituting for d in equation 4 we have:

$$\frac{\mathrm{R}^4}{(\mathrm{R} - (\mathrm{s} * \mathrm{t}))^4} \leq \mathrm{HTH} \tag{5}$$

We use this result to compute HTH for a given network's degree of mobility and radio range.

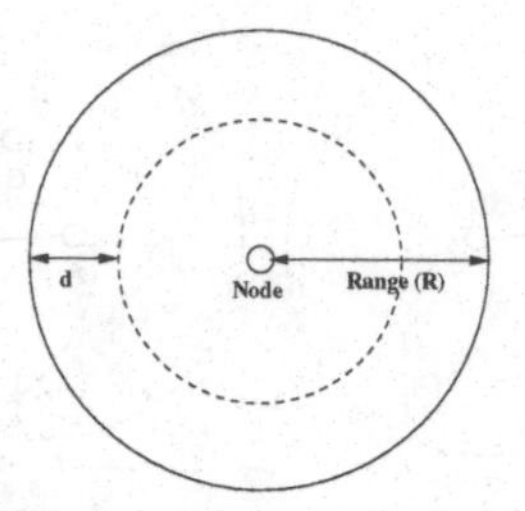

Fig. 3. A node in the annulus performs routing handoff

3 Theoretical Analysis

3.1 Network Model

Let A be the area of the network under consideration. N is the number of nodes uniformly distributed over the network. Assuming each node in the network has the same transmitting power, the corresponding range of transmission is R. We assume a random traffic pattern: each source node initiates packets to randomly chosen destinations in the network. The expected length $\overline{L}$ for such traffic, as derived in Sect. 4.1 of [8] is:

$$\overline{L} = \frac{2\sqrt{A}}{3} \quad (6)$$

When a link breaks, let ϕ be the number of routes affected. Between any source-destination pair, each of the links can break with equal probability. For our analysis, we do not take cached entries into account during route discovery. We also assume that that end-to-end delay on a route is proportional to the number of hops from source to destination.

3.2 Basic Results

Here we show some basic results which form the basis of our analysis. The results are based on the model stated in Subsection 3.1.

1. Number of packets involved in flooding the network:
   Here we assume that RREQ broadcasts reach all nodes in the network. Since each node forwards the RREQ packets only once, the number of broadcasts required is N. Hence the number of packets involved in flooding is N.
2. Number of hops H on the average to reach the destination:
   Since the expected path length is $\overline{L}$ and the transmission range is R, the number of hops required is:

$$H = \frac{\overline{L}}{R} = \frac{2\sqrt{A}}{3R} \quad (7)$$

3. Average number of hops (or time) to discover a route:
   From the above result the average number of hops to discover a route is $2H$.
4. Number of RERR broadcasts involved when a link breaks:
   Since our model assumes each of the H hops can break with equal probability, a link break one hop from the source results in 1 RERR broadcast and a link break one hop from the destination results in $H-1$ broadcasts, provided only one route is affected by the link breakage. Then the average number of RERR packets, per affected route.

$$k = \frac{1+2+3+\ldots+H-1}{H} = \frac{H-1}{2} \approx \frac{H}{2} = \frac{\sqrt{A}}{3R} \tag{8}$$

   But if ϕ routes are affected by a link breakage and the path from these sources to the point of link breakage do not overlap, the maximum number of RERR broadcasts required, K, is:

$$K = \phi k = \frac{\phi\sqrt{A}}{3R} \quad K < N \tag{9}$$

   Note the the value of K is bounded by number of nodes in the network N. Such a scenario arises when a RERR broadcast ends up flooding the network.

3.3 Analysis of AODV

In AODV, a link failure causes a RERR broadcast to the sources affected. We assume that the source on receiving a RERR, has a packet to send, and will initiate a route discovery.

1. Number of packets involved in repairing a broken route (PKT):
   PKT = RERR broadcast to the sources affected + flooding to discover the route for each route+ RREP unicast from the destination to the source

$$PKT = K + \phi N + \phi H = \frac{\phi\sqrt{A}}{3R} + \phi N + \phi\frac{2\sqrt{A}}{3R} \tag{10}$$

2. Delay involved in repairing a broken route (DEL):
   DEL = RERR broadcast to reach the source + RREQ to reach the destination + RREP to reach the source

$$DEL = k + H + H = k + 2H = \frac{\sqrt{A}}{3R} + \frac{4\sqrt{A}}{3R} = \frac{5\sqrt{A}}{3R} \tag{11}$$

3.4 Analysis of Local Route Repair

In AODV with local route repair we assume that the intermediate route discovery succeeds.

1. Number of packets involved in repairing a broken route (PKT):
   PKT = RERR broadcast + flooding to discover the route for each route + RREP unicast from destination to the intermediate node
$$PKT = K + \phi N + \frac{\phi\sqrt{A}}{3R} = \frac{\phi\sqrt{A}}{3R} + \phi N + \phi\frac{\sqrt{A}}{3R} \tag{12}$$
2. Delay involved in repairing a broken route (DEL):
   DEL = RREQ to reach the destination + RREP to reach the intermediate node
$$DEL = \frac{H}{2} + \frac{H}{2} = H = \frac{2\sqrt{A}}{3R} \tag{13}$$

3.5 Analysis of Router Handoff

For analysis of Router Handoff, we need a criterion to determine if there exists a suitable node for handoff to occur. Consider a section of the network as shown is Figure 4. When node C moves, it is not possible to find another node that will take the responsibility of routing packets from B to D (unless of course a new node moves to the original position of C). The point to note here is that a certain overlap of transmission ranges of B and D is required. This alone will not suffice. We also need nodes in the overlapping area that will take up the responsibility of routing packets from B to D as in Figure 5.

The maximum extent of overlapping between nodes B and D, should be less than shown in Figure 6, or we would not have used node C to relay the packets to D. The overlapping area will be less than $1.23R^2$. Since the nodes are uniformly distributed over the network, the number of nodes in the overlapping area:

$$\eta \leq \frac{1.23R^2N}{A} \quad \text{and} \quad \eta \geq 2 \tag{14}$$

$$N \geq \frac{\eta A}{1.23R^2} \tag{15}$$

Equation 15 provides us the condition under which nodes will be present in the overlapping area. But to ensure that the transmission ranges of B and D overlap, consider Figure 7 which is a representative snapshot of our model. For the transmission range of B and D to overlap:

$$R > \frac{\sqrt{A}}{\sqrt{N}} \quad \text{or} \quad N > \frac{A}{R^2} \tag{16}$$

which is essentially similar to equation 15.

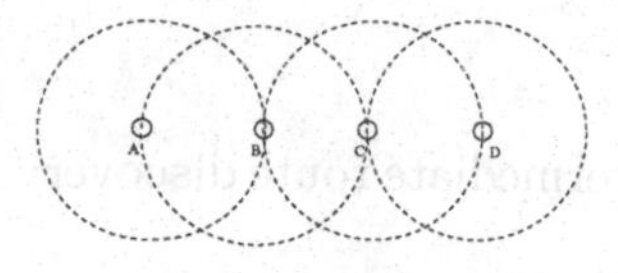

Fig. 4. No Overlap

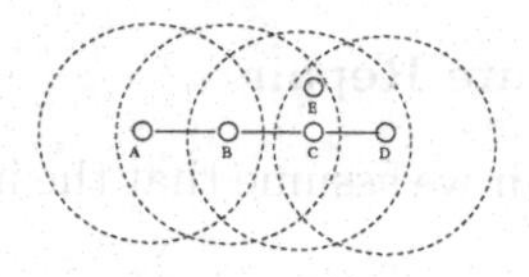

Fig. 5. With Overlap

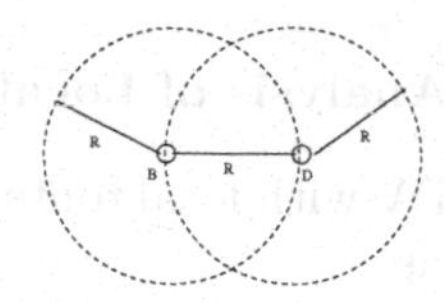

Fig. 6. Max. Overlap

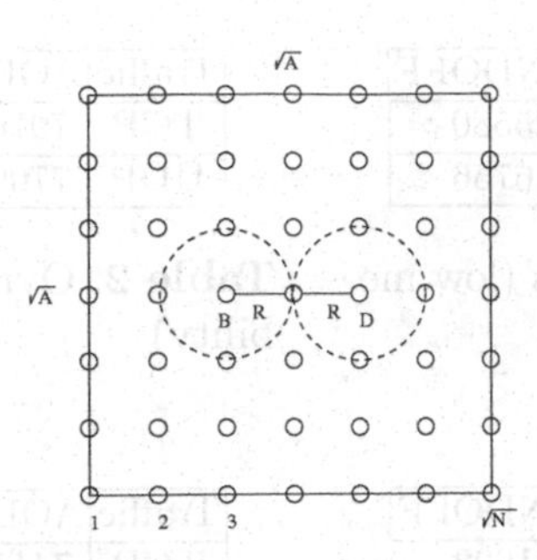

Fig. 7. Snapshot of our Network Model

1. Number of packets involved in repairing a broken link: PKT = HREQ + HREP = 1 + 1 = 2.
2. Delay involved in repairing a broken link: DEL = HREQ + HREP = 1 + 1 = 2.

These results of the analysis of Router Handoff apply only when a suitable node is available for handoff. In practice, such a node may not always be present.

4 Experiments

We used the ns-2.1b8a network simulator [9] for running simulations. The simulation results for AODV, AODV with Local Route Repair (LRR) and AODV with Router Handoff are presented here. We conducted the simulations on networks with 25, 50 and 75 nodes, for low and high mobility scenarios. The transmission range of each node was 250 m. and each simulation was run for a period of 200 seconds. Routing overhead for AODV and LRR is the sum of RREQ, RREP and RERR packets broadcast. Routing overhead for AODV with Router Handoff is the sum of RREQ, RREP, RERR, HREQ and HREP packets. Total data packets received by all nodes in a run is taken as a measure of throughput.

In the low mobility scenario, the minimum pause time is 5 seconds and maximum pause time is 10 seconds. The speed of the node varies between 20 m/s to 40 m/s. The value of Handoff Threshold (HTH) is computed according to equation 5 and is set to 1.5.

In the high mobility scenario, the minimum pause time is 1 second and maximum pause time is 5 seconds. The speed of the node varies between 40 m/s to 60 m/s. The value of Handoff Threshold (HTH) is computed according to equation 5 and is set to 2.

4.1 Results

For a network with 25 Nodes and satisfying the criterion in equation 16 we set the area to 950 × 950 sq. m. The network was subjected to traffic with 15 TCP

Traffic	AODV	LRR	HANDOFF
TCP	24607	23891	25580
UDP	18009	14695	16756

Table 1. T'put. 25 Nodes (low mobility)

Traffic	AODV	LRR	HANDOFF
TCP	79760	74753	75984
UDP	77098	80416	73842

Table 2. Ovrhd. 25 Nodes (low mobility)

Traffic	AODV	LRR	HANDOFF
TCP	27239	25060	21622
UDP	13704	12146	16397

Table 3. T'put. 25 Nodes (high mobility)

Traffic	AODV	LRR	HANDOFF
TCP	74115	81994	72293
UDP	84310	89169	78359

Table 4. Ovrhd. 25 Nodes (high mobility)

connections in one simulation and 15 UDP connections in another simulation for both low and high mobility scenarios. Similarly we repeated experiments for 50 and 75 nodes using areas of 1650 × 1650 sq. m. and 1950 × 1950 sq. m. with 30 and 45 connections respectively. We do not present results for the experiments with 75 nodes due to paucity of space.

Tables 1 to 8 indicate throughput and routing overheads in terms of number of packets, for plain AODV, LRR, and AODV with Router Handoff in various scenarios. We can conclude from the data that AODV with Router Handoff in general shows greater throughput and smaller routing overheads than both plain AODV and LRR.

5 Related Work

To the best of our knowledge, the only work related to the work presented in this paper is *Preemptive Routing* [10]. Preemptive routing keeps track of signal strengths and resorts to route repair procedures before a link breaks. The difference between Router Handoff and Preemptive Routing is that the latter does a normal route repair procedure involving flooding whereas Router Handoff tries to locally find an alternate node and hands off existing routing information to it using only two broadcasts. Of the two techniques, only Router Handoff reduces overheads of route repair but both Router Handoff and Preemptive Routing attempt to reduce delays due to route breakages.

6 Conclusion

In this paper we presented an approach called Router Handoff as a preemptive method of preserving routes in the presence of link failures. Simulation results show that AODV with Router Handoff performs better than plain AODV and

Traffic	AODV	LRR	HANDOFF
TCP	17166	16230	20949
UDP	13995	11034	15028

Table 5. T'put. 50 Nodes (low mobility)

Traffic	AODV	LRR	HANDOFF
TCP	89813	96044	101642
UDP	155261	175389	146763

Table 6. Ovrhd. 50 Nodes (low mobility)

Traffic	AODV	LRR	HANDOFF
TCP	18169	18897	21718
UDP	9611	9268	12035

Table 7. T'put. 50 Nodes (high mobility)

Traffic	AODV	LRR	HANDOFF
TCP	80187	89285	90902
UDP	159121	177173	154005

Table 8. Ovrhd. 50 Nodes (high mobility)

AODV with Local Route Repair when the the network satisfies conditions derived by theoretical analysis. This gain in performance is due to reduction in routing overheads and route repair delays. We believe that other reactive routing protocols could benefit from incorporating Router Handoff. This would be an interesting area for future work.

References

[1] *MANET Working Group Charter*, http://www.ietf.org/html.charters/manet-charter.html 347

[2] Charles Perkins and Elizabeth Royer, *Ad Hoc On-Demand Distance Vector Routing*, Proc. of the 2nd IEEE Workshop on Mobile Computing Systems and Applications, New Orleans. 347

[3] David Johnson and David Maltz, *Dynamic Source Routing in Ad Hoc Wireless Networks*, in *Mobile Computing* (1996), T.Imielinski and H.Korth, eds., Kluwer Academic Publishers. 347

[4] Charles Perkins and Pravin Bhagwat, *Highly Dynamic Destination Sequenced Distance Vector Routing*, ACM SIGCOMM, October 1994. 347

[5] Young-Bae Ko and Nitin H.Vaidya, *Location-Aided Routing (LAR) in Mobile Ad Hoc Networks*, Proc. of ACM/IEEE MobiCom (1998), Dallas, Texas, pp. 66–75. 347

[6] Raghupaty Sivakumar, Bevan Das and Vaduvur Bhargavan, *Routing in Ad Hoc Networks Using a Spine*, Proc. of ICCCN (1997), Las Vegas. 347

[7] Bo Ryu, Jason Erickson, Jim Smallcomb and Son Dao, *Virtual Wire for Managing Virtual Dynamic Backbone in Wireless Ad Hoc Networks*, Proc. of Dial-M (1999), Seattle. 347

[8] Jinyang Li, Charles Blake, Douglas S. J., De Couto, Hu Imm Lee and Robert Morris, *Capacity of Ad Hoc Wireless Networks*, Proc. of ACM MobiCom (2001), Rome, Italy. 351

[9] *Network Simulator, ns-2*, http://www.isi.edu/nsnam/ns/ 354

[10] T. Goff, N. B. Abu-Ghazaleh, D. S. Phatak and R.Kahveciogl u, *Preemptive Routing in Ad Hoc Networks*, Proc. of ACM MobiCom (2001), Rome, Italy. 355

A 2-D Random Walk Based Mobility Model for Location Tracking

Srabani Mukhopadhyaya and Krishnendu Mukhopadhyaya

Advanced Computing & Microelectronics Unit, Indian Statistical Institute
203 B. T. Road, Kolkata 700 108, India
{srabani_v, krishnendu}@isical.ac.in

Abstract. Performance analysis of different location tracking schemes needs a model that reflects the mobility of the mobile terminals in a realistic way. We propose a two-dimensional random walk based model with *inertia* for performance analysis of different location management schemes with specific update and paging mechanism. We used our mobility model to analyze the cost of movement-based location update technique with selective paging. We extend the model to have different values for inertia of motion and inertia of rest.

1 Introduction

In mobile communication system, tracking a mobile terminal (MT) is an important problem. As the number of mobile users and the areas covered expand rapidly, *paging* for an MT over the full area may be time consuming as well as costly. Communication channel is one of the costliest resources in this domain. If the MT frequently informs the base stations about its current location, cost of paging can be reduced substantially by directing the search using this knowledge. But these frequent location updates may add up to a lot of cost.

In dynamic location update schemes, updates take place depending on the movement of individual MTs and frequency of the incoming calls. In *distance-based update scheme* [4, 8, 12], an MT updates when it crosses a prefixed distance threshold from the place of last update. This scheme is difficult to implement and needs a prior knowledge of the cell map. In *movement-based update scheme* [4] an update is made when an MT crosses a certain number of cell boundaries. In *time-based update scheme* [1, 4] an MT updates its location after certain intervals of time. Distance-based update scheme consistently outperforms the movement-based and the time-based update schemes [4].

For finding an MT to deliver a call, it is paged. The last update of the MT is used for a starting position of the search. Different tracking mechanisms are designed to predict the cell location probability which estimates the next location of MT's movement [3, 11]. Several paging schemes like *cluster paging* [14], *selective paging* [1, 2, 8, 16] have been proposed in the literature. However, some of these schemes require significant amount of computing power. Implementation of such schemes may not be feasible in general.

S. Sahni et al. (Eds.) HiPC 2002, LNCS 2552, pp. 357–366, 2002.

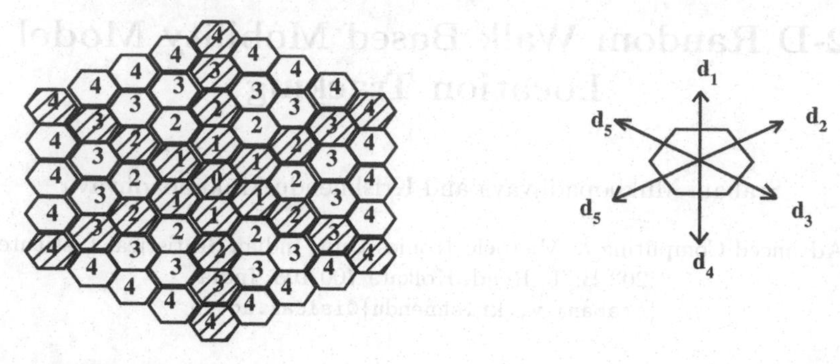

Fig. 1. Cell layout and the six directions of motion from a cell

One of the important issues in the performance analysis of the tracking schemes is to properly model the movement of the MTs. The model should be a reasonable approximation of the movement pattern. Most of the schemes discussed in the literature are based on simplified assumptions. The existing mobility models are the *fluid-flow model* [13, 15, 17] and the *random walk model* [2, 7, 8, 10]. Most of the works using the random walk model on a hexagonal cell-configuration (mesh configuration) assume that an MT in a cell can move to any of the six (four) neighboring cells with probability $\frac{1}{6}$ ($\frac{1}{4}$). This model, though easier to handle, does not reflect the actual movement patterns.

We propose a more realistic random walk based model for the mobility of the MTs. We assume that when an MT is moving, it is more likely to continue to move in the same direction. Similarly, a stationary MT is also more likely to remain in the same cell. A similar notion of direction of motion was used in [5] and [6]. But, in both the cases *one-dimensional* movement pattern has been considered. We assume a *two-dimensional* Markovian movement such that in a hexagonal cell configuration an MT can move to any of its six neighboring cells. The proposed mobility model can be used to analyze several location management schemes. We have chosen movement based dynamic location update scheme with selective paging technique [2] to analyze the location management cost under the proposed mobility model.

We consider hexagonal cell configuration. Cells are hexagonal in shape. Each cell has six neighbors. In Fig. 1, it can be seen that due to regularity of the shape of the cells, with respect to every cell, the whole area can be viewed as a collection of several concentric rings of cells. The innermost ring (ring 0) consists of that particular cell. We call this cell as the central cell. Total number of cells in ring i is $6i$, $i = 1, 2, 3, \cdots$. The cells in each ring can be classified into two categories. A cell of *type* A, has among its six neighbors one neighbor in ring $i-1$, two neighbors in ring i and three neighbors in ring $i+1$. On the other hand, a cell of *type* B has two neighbors in each of the three rings $i-1$, i, $i+1$. The positions of these two types of cells are shown in Fig. 1. In Fig.1, all shaded cells are of type A and rest are of type B. We shall refer to a cell of type A (B) in the ring i as cell i_A (i_B). In each ring i ($i = 1, 2, 3, \cdots$) there are exactly six cells of type A and $6(i-1)$ cells of type B. It can be noted that all the six cells in ring 1 are of type A. For the unique cell in ring 0, there is no type. For the sake of convenience of notation we shall denote it as type A.

2 Random Walk Model with Inertia

We assume that the time is slotted. That is, all events occur at discrete integer values of time. If the discrete points are close enough, the analysis thus produced would be close to continuous time. With respect to a cell we consider six directions $d_1, d_2, \cdots, d_6$ as shown in the Fig. 1. At any time instant we associate a direction with each MT. At time t_1 if an MT enters a cell, moving in the direction d_i then we say that the current (at the time t_1) direction of the MT is d_i. At time t_1+1, the MT may move to one of its six neighbors. The transition may or may not be in the same direction d_i. It is also possible that the MT may reside in the same cell for certain units of time. If the MT stays in the same cell, at the time t_1+1 the associated direction is defined to be d_0. Therefore, at any time instant there are seven possible directions associated with each MT. Let p denote the probability that at any time t an MT will move in the same direction (including d_0) as it has moved in the previous time instant. For example, at time t_1 if the direction of an MT is d_2, then at time t_1+1 the MT moves to a neighboring cell in the same direction d_2 with a probability p. Moreover, at time t_1 if an MT resides in the same cell then the probability that at the time t_1+1 it would again stay back in the same cell is p. In a way p represents the *inertia* of an MT. With probability p it continues doing what it did in the last time instant. The probability of an MT changing its direction is $1-p$. In that case there are six possible directions that the MT can take. Hence, an MT can move in each of these six directions with a probability $\frac{1-p}{6}$ ($= q$, say). In the above example, at the time t_1+1 the probability that the MT would move in the direction d_i ($i = 0, 1, 3, 4, 5, 6$) is q. Clearly, $p + 6q = 1$.

2.1 Cell Residence Time Distribution

Let X be the random variable representing the cell residence time of an MT.
Then, $\mathrm{P}(X=1) = 1-q$ and $\mathrm{P}(X=t) = q\,p^{t-2}(1-p),\ t > 1$.
The average cell residence time is:

$$E(X) = \sum_{i=1}^{\infty} i\,P(X = i) = (1-q) + \sum_{i=2}^{\infty} i\,q\,p^{i-2}(1-p) = 7/6\ [p + 6q = 1]$$

2.2 Transition Probability

An MT enters a cell C with an *entry direction*, say, d_i, $1 \leq i \leq 6$. It resides in the cell for some period of time and then moves to one of its six neighbors. In the latter move the MT may or may not change its direction. The direction in which the MT moves from that cell C to one of its six neighbors will be called the *transition direction* corresponding to the cell C. First we calculate the probability that the transition direction is same as entry direction. That is, the transition direction (TD) is also d_i. Let X denote cell residence time.

$$\begin{aligned} \mathrm{P}(TD = d_i) &= \mathrm{P}(TD = d_i \;\&\; X = 1) + \mathrm{P}(TD = d_i \;\& X > 1) \\ &= \mathrm{P}(TD = d_i | X = 1) \times \mathrm{P}(X = 1) + \mathrm{P}(TD = d_i | X > 1) \times \mathrm{P}(X > 1) \\ &= \tfrac{p}{p+5q} \times (1-q) + \tfrac{q}{6q} \times (1-(1-q)) = p + \tfrac{q}{6} = p' \text{ (say)} \end{aligned}$$

Similarly, if $\bar{d}_i$ is any direction other than d_i,

$$\mathrm{P}(TD = \bar{d}_i) = \tfrac{q}{p+5q} \times (1-q) + \tfrac{q}{6q} \times (1-(1-q)) = q + \tfrac{q}{6} = q' \text{ (say)}.$$

Note that $p' + 5q' = 1$.

We can model the mobility of an MT by a 2-D random walk. An MT moves from one cell to one of its neighbors in a direction same as the entry direction with a higher probability p' and it moves to any of the remaining five neighbors with a probability q' (assuming $p \geq q$). To describe the state of an MT we use a 3-tuple. An MT is in the state $(r,\ t,\ d)$ if it is currently residing in a cell of type t in ring r with an entry direction d. r can take a value from 0 to $K-1$. Since K is the movement threshold, being in ring K is actually being in ring 0 with respect to the next update. t can be either A or B and d can be any one of $d_1, d_2, \cdots, d_6$. Corresponding to the unique cell of ring 0, there are six possible states. Since all the cells in ring 1 are of type A, there can be six possible states corresponding to the ring 1. For the rest of the rings there can be 12 possible states. Hence the total number of possible states corresponding to the rings $0, 1, \cdots, K-1$ is $12(K-1)$. We name these states as $S_0, S_1, S_2, \cdots, S_{12(K-1)-1}$. The first 6 states (from S_0 to S_5) correspond to the ring 0 and the next 6 states (from S_6 to S_{11}) correspond to the ring 1. The states $S_{12(i-1)}, S_{12(i-1)+1}, S_{12(i-1)+2}, \cdots, S_{12i-1}$ correspond to the ring i.

If it is given that there will be no update, transitions form ring $K-1$ will be limited to cells in ring $K-1$ and ring $K-2$. So, there can be no transition to ring K. In the following discussion we assume that the MT will not have any update. That is, we are considering the time period between successive updates.

Transition matrix :

Suppose $P_K = [p_{ij}]_{N \times N}$, $N = 12(K-1)$, denotes the transition matrix. The (ij)th entry p_{ij} represents the probability that in one step (i.e., after a single cell boundary crossing) an MT moves from the state S_i to S_j.

To calculate the entries of the transition matrix P_K, we assume that the states S_i and S_j are represented by the 3-tuples (r, t, d) and (r', t', d') respectively. The entry p_{ij} therefore represents the probability that in one step the MT moves in the direction d' from a cell of type t in ring r with entry direction d to a cell of type t' in ring r'. Let us now calculate p_{ij} (or $p_{(r,t,d),(r',t',d')}$).

Since from the ring r after a single cell boundary crossing an MT can move only to the ring $r-1$ or to the ring $r+1$ or to some other cell in the same ring r, in case $|r - r'| > 1$, $p_{(r,t,d),(r',t',d')} = 0$.

Case 1 : $r = 0$.

$$\begin{aligned} p_{(0,A,d),(1,A,d')} &= p' \text{ if } d = d' \\ &= q' \text{ if } d \neq d' \end{aligned}$$

Case 2 : $r = 1$.

$$\begin{aligned} p_{(1,A,d),(0,A,d')} &= \tfrac{p'}{6} \text{ if } d = d' \\ &= \tfrac{q'}{6} \text{ if } d \neq d' \end{aligned} \qquad \begin{aligned} p_{(1,A,d),(1,A,d')} &= \tfrac{p'}{3} \text{ if } d = d' \\ &= \tfrac{q'}{3} \text{ if } d \neq d' \end{aligned}$$

$p_{(1,A,d),(2,A,d')} = \frac{p'}{6}$ if $d = d'$ $\quad p_{(1,A,d),(2,B,d')} = \frac{p'}{3}$ if $d = d'$
$= \frac{q'}{6}$ if $d \neq d'$ $\quad = \frac{q'}{3}$ if $d \neq d'$

Case 3 : $K - 1 > r > 1$

Subcase 3a : $t = A$.

The MT may move to a cell in ring $(r-1)$ or r or $(r+1)$. The six neighbors of a cell of type A in ring r are: i) one cell of type A in ring $(r-1)$, ii) one cell of type A in ring $(r+1)$, iii) two cells of type B in ring r, and iv) two cells of type B in ring $(r+1)$.

When $d' = d$:

$$
\begin{aligned}
p_{(r,A,d),(r',t',d)} &= 0 && \text{if } (r' = r \text{ and } t' = A) \text{ or } (r' = (r-1) \text{ and } t' = B)\\
&= \tfrac{p'}{6} && \text{if } r' = (r-1) \text{ or } (r+1) \text{ and } t' = A\\
&= \tfrac{p'}{3} && \text{if } r' = r \text{ or } (r+1) \text{ and } t' = B
\end{aligned}
$$

When $d' \neq d$:

$$
\begin{aligned}
p_{(r,A,d),(r',t',d')} &= 0 && \text{if } (r' = r \text{ and } t' = A) \text{ or } (r' = (r-1) \text{ and } t' = B)\\
&= \tfrac{q'}{6} && \text{if } r' = (r-1) \text{ or } (r+1) \text{ and } t' = A\\
&= \tfrac{q'}{3} && \text{if } r' = r \text{ or } (r+1) \text{ and } t' = B
\end{aligned}
$$

Subcase 3(b) : $t = B$.

Similarly, when $d' = d$

$$
\begin{aligned}
p_{(r,B,d),(r',t',d)} &= 0 && \text{if } r' = (r+1) \text{ and } t' = A\\
&= \tfrac{2p'}{6(r-1)} && \text{if } r' = (r-1) \text{ or r and } t' = A\\
&= \tfrac{2(r-1)p'}{6(r-1)} && \text{if } r' = (r+1) \text{ and } t' = B\\
&= \tfrac{2(r-2)p'}{6(r-1)} && \text{if } r' = r \text{ or } (r-1) \text{ and } t' = B
\end{aligned}
$$

When $d' \neq d$

$$
\begin{aligned}
p_{(r,B,d),(r',t',d')} &= 0 && \text{if } r' = (r+1) \text{ and } t' = A\\
&= \tfrac{2q'}{6(r-1)} && \text{if } r' = (r-1) \text{ or } (r) \text{ and } t' = A\\
&= \tfrac{2(r-1)q'}{6(r-1)} && \text{if } r' = (r+1) \text{ and } t' = B\\
&= \tfrac{2(r-2)q'}{6(r-1)} && \text{if } r' = r \text{ or } (r-1) \text{ and } t' = B
\end{aligned}
$$

Case 4 : $r = K - 1$

If transitions to ring K were allowed, the total probability of an MT in a cell of type t in ring $K-1$ moving to any cell in ring K is 0.5 if $t = A$ and $\frac{1}{3}$ if $t = B$. But since it is given that there would be no more update within the time period we are considering, there would be no transition to ring K. The transitions are to the cells in ring $(K-1)$ and $(K-2)$. Using the rules of conditional probability, the probabilities for these events under the conditional setup, can be found by dividing the unconditional transition probabilities by $(1 - .05)$ and $(1 - \frac{1}{3})$ for $t = A$ and B respectively.

3 Location Update and Terminal Paging Costs

Under the proposed mobility model we now estimate the average cost for location update and terminal paging. We assume movement based location update scheme and selective paging [2] based on SDF (shortest distance first) scheme. In movement based update schemes an update is made when an MT crosses a certain number of cell boundaries, called *movement threshold.* So, there is a limit to the distance an MT may travel since its last update.

In selective paging, one assumes a maximum allowable paging delay of η ($\eta \geq 1$). *Residing area* is defined as the collection of cells that are within a maximum distance of $K-1$ from the location of the last update. Movement based location update scheme, with movement threshold K, guarantees that the MT is located within the residing area. Selective paging partitions the residing area of the called MT into $l = \min(\eta, K)$ number of sub areas and then polls each subarea, one after another, until the MT is found. The sub area A_j, $1 \leq j \leq l$, contains one or more rings. The partitioning scheme assures that each subarea has approximately an equal number of rings.

Let,

$$P_K^{(n)} = \begin{cases} P_K, & n = 1 \\ P_K \times P_K^{(n-1)}, & n > 1 \end{cases}$$

and $\beta(r, s)$ = Probability that the MT is in some cell in the ring r after s cell boundary crossings since last update, $r, s < K$.

Note that $P_K^{(n)}$ gives the transition probabilities from a state to another in n steps, i.e., after n cell boundary crossings. Being in a cell in ring r implies being in one of the several possible states corresponding to the ring r. Hence taking the sum over those,

$$\beta(r, s) = \begin{cases} 0; \ \text{if } s < r \\ \frac{1}{6} \sum_{i=0}^{5} \sum_{j=6r}^{6r+5} p_{ij}^{(s)}; \ \text{if } 1 \geq r \geq 0 \\ \frac{1}{6} \sum_{i=0}^{5} \sum_{j=12(r-1)}^{12r-1} p_{ij}^{(s)}; \ \text{if } K > r > 1 \end{cases}$$

We now compute the probability of crossing a certain number of cell boundaries in a given time period. Let, $\gamma(t, s)$ = Probability that the MT crosses s number of cell boundaries in time t.

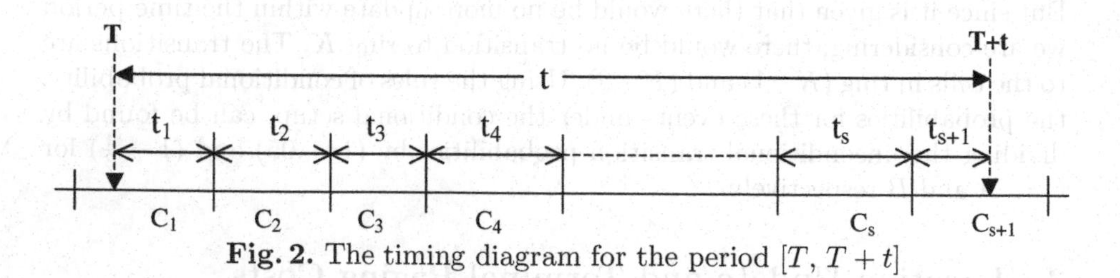

Fig. 2. The timing diagram for the period $[T, T+t]$

First we consider the case when $s > 0$. Consider the timing diagram shown in Fig. 3. Suppose the mobile terminal is in some cell C_1 at time T and it spends t_1 more units of time in the cell C_1 before leaving it. Afterwards the MT visits s more cells and resides t_j units of time in the jth cell $C_j, j = 2, 3, \cdots, s$. Moreover, we assume that the MT spends t_{s+1} units of time in the $(s+1)$th cell C_{s+1} before the time period expires. Clearly, $t_1 + t_2 + t_3 + \cdots t_s + t_{s+1} = t$ and $t_i \geq 1 \; \forall \, i, 1 \leq i \leq s$ and $t_{s+1} \geq 0$.

Suppose X_i denotes the total time that the MT spends in the cell C_i, $2 \leq i \leq s+1$. These X_i's $\forall \, i, 2 \leq i \leq s$ are independent and identically distributed random variables. Let X_1 denote the time spent by the MT in the cell C_1, after time T. Now, suppose, $\forall \, i, \; 2 \leq i \leq s$, $\text{Prob}(X_i = t_i) = P_i$, where,
$P_i = \begin{cases} (1-q), & \text{when } t_i = 1 \\ qp^{t_i - 2}(1-p), & \text{when } t_i \geq 2. \end{cases}$

For the starting cell C_1, let $\text{P}(X_1 = t_1) = P_1$. Now, at any time instant the direction associated with the MT can be any of the seven directions d_0, d_1, d_2, d_3, d_4, d_5 and d_6. For an MT, at any time instant the chances of having any one of these directions are same, i.e., $\frac{1}{7}$ (see section 5 for a formal proof).

Case 1: $t_1 = 1$.

P(After time T, the MT stays in the cell C_1 for exactly one unit of time)
= P(At time T, the direction of the MT is d_0 and after T the MT stays in C_1 for exactly one unit of time) + P(At time T, the direction of the MT is other than d_0 and after T the MT stays in C_1 for exactly one unit of time)
= P(after time T the MT resides in the cell for 1 unit of time | at the time T the direction associated with MT is d_0) $\times$ P(at the time T the associated direction of the MT is d_0) + P(after the time T the MT resides in the cell for 1 unit of time | at the time T the direction associated with MT is other than d_0) $\times$ P(at the time T the associated direction of the MT is other than d_0)
$= (1-p) \times \frac{1}{7} + (1-q) \times \frac{6}{7} = \frac{6}{7}$ (using $p + 6q = 1$)

Case 2: $t_1 > 1$.

Proceeding in a similar way as in the above case, we get
$P_1 \;= \text{P(after time } T\text{, the MT resides in } C_1 \text{ for exactly } t_1 \text{ units of time)}$
$\quad\; = p^{t_1 - 1}(1-p) \times \frac{1}{7} + qp^{t_1 - 2}(1-p) \times \frac{6}{7} = \frac{1}{7}p^{t_1 - 2}(1-p)$

Let, $P_{s+1} =$ P(the MT spends exactly t_{s+1} units of time in C_{s+1} before the time interval ends)
Clearly, $P_{s+1} = P(X_{s+1} > t_{s+1}) = 1 - P(X_{s+1} \leq t_{s+1}) = qp^{t_{s+1} - 1}$, if $t_{s+1} > 0$.
If $t_{s+1} = 0$, then $P_{s+1} = 1$.

We can express $\gamma(t, s)$ in terms of the above probabilities as follows:
Suppose, $P(t_1, t_2, t_3, \cdots, t_s, t_{s+1}) = \Pi_{i=1}^{s+1} P(X_i = t_i)$. Then,

$$\gamma(t, s) = \sum_{t_1 + t_2 + \cdots + t_s + t_{s+1} = t} P(t_1, t_2, t_3, \cdots, t_s, t_{s+1})$$

The above sum is taken over all possible $(s+1)$ partition of t such that $t_i \geq 1$, $\forall i$, $1 \leq i \leq s$ and $t_{s+1} \geq 0$.

Though the above sum, has a very large number of terms, all of them need not be computed separately. The terms may be divided into several groups so that the value in each group is constant.

Now, if $s = 0$, proceeding as before, $\gamma(1, 0) = \frac{1}{7}p + \frac{6}{7}q = \frac{1}{7}$.
For $t > 1$, $\gamma(t, 0) = \frac{1}{7}p^t + \frac{6}{7}q(1-p)^{t-1} = \frac{1}{7}p^{t-1}$.

Let, $\alpha(s)$ = Probability that there are s cell boundary crossings between two call arrivals.

We assume that call arrival to an MT follows geometric distribution with parameter λ. Then, the probability that the interval between two consecutive phone calls will be t is $\lambda(1-\lambda)^{t-1}$. Hence, $\alpha(s) = \sum_{t=s}^{\infty} \gamma(t,s) * \lambda(1-\lambda)^{t-1}$.

Suppose, cost for polling a cell is V and that for a single location update is U. Let K be the movement threshold for the location update scheme. Then the expected location update cost C_U per call arrival can be calculated as in [2]:

$$C_U = U \sum_{i=1}^{\infty} i \sum_{j=iK}^{(i+1)K-1} \alpha(j)$$

The expected terminal paging cost per call arrival, C_V, is given by

$$C_V = V \sum_{j=0}^{l-1} \rho_j \omega_j$$

Where, ρ_j is the probability that the MT is residing in the Sub area A_j when a call arrives and ω_j is the number of cells polled before the terminal is successfully located, given that the terminal is residing in the subarea A_j. As discussed in [2], ρ_j and ω_j can be calculated as follows:

$\rho_j = \sum_{\text{ring } i \in A_j} \pi(i)$, where $\pi(i)$ is the probability that the MT is located in a ring i cell when a call arrival occurs and is given by $\pi(i) = \sum_{k=0}^{\infty} \alpha(k)\beta(i, k \bmod K)$.

If $N(A_j)$ denotes the number of cells present in the Sub area A_j and $g(i)$ is the total number of cells constituting ring i, then

$$\omega_j = \sum_{r=0}^{j} N(A_r) = \sum_{r=0}^{j} \sum_{\text{ring } i \in A_r} g(i)$$

Hence, expected total cost for location update and terminal paging is

$$C_T = C_U + C_V$$

4 Inertia of Rest and Inertia of Mobility

We extend our model to allow two different values for inertia of motion and rest. Let,

Inertia of rest (p_1) = Probability that an MT would remain in the same cell in the next time instant, when the current direction is d_0,

q_1 = Probability that an MT would move in the direction d_i (other than d_0) at the next time instant, when the current direction is d_0,

Inertia of motion (p_2) = Probability that an MT would move in the same direction d_i at the next time instant, when the current direction is d_i, which is not d_0, and

q_2 = Probability that an MT would change its direction to $\bar{d}_i$ at the next time instant, when the current direction is d_i, which is not d_0.

Clearly, $q_1 = \frac{1-p_1}{6}$ and $q_2 = \frac{1-p_2}{6}$

Let X denote the cell residence time.

$P(X = 1) = 1 - q_2$, $P(X = t) = q_2 p_1^{t-2}(1-p_1)$, $t > 1$.

$E(X) = 1 + \frac{q_2}{(1-p_1)}$

Suppose an MT enters a cell C with an entry direction d_i.

P(Transition direction $= d_i) = p' = p_2 + \frac{q_2}{6}$

P(Transition direction $= \bar{d}_i) = q' = q_2 + \frac{q_2}{6}$.

With these transition probabilities p' and q', the transition matrix P_K and $P_K^{(n)}$ can be calculated in a similar fashion as in section 3.2. The expression for $\beta(r,s)$ also remains unchanged.

In the process of finding the probability $\gamma(t,s)$, the constituent probabilities need to be calculated differently in this case. Consider the timing diagram shown in Fig. 3. Following the same notations used in the previous sections we calculate the following probabilities:

First we assume $s > 0$

(i) P_i can be calculated in a similar way as in section 4, $\forall i,\ 2 \leq i \leq s$.

$$P_i = P(X_i = t_i) = \begin{cases} (1-q_2), & \text{when } t_i = 1 \\ q_2 p_1^{t_i-2}(1-p_1), & \text{when } t_i \geq 2. \end{cases}$$

(ii) To calculate P_1 we need to calculate the following two probabilities first:

π_1: probability that at any time instant the direction of an MT is d_0

π_2: prob. that at any time instant the direction of an MT be d_i, other than d_0.

Using standard results in Stochastic Processes [9], we get $\pi_1 = \frac{q_2}{1-p_1+q_2}$ and $\pi_2 = \frac{1-p_1}{1-p_1+q_2}$. Note that if $p_1 = p_2$ then $\pi_1 = \frac{1}{7}$ and $\pi_2 = \frac{6}{7}$, which is nothing but the previous case. With these probabilities, one can compute the values of $P(X = t)$ and hence $\gamma(t,s)$ in a similar way.

When $s = 0$, $\gamma(1,0) = \frac{q_2}{1-p_1+q_2}$ and $\gamma(t,0) = \frac{q_2 p_1^{t-1}}{1-p_1+q_2}$.

After finding $\gamma(t,s)$, $\alpha(s)$ can be calculated accordingly.

5 Conclusion

We have proposed a new 2-D random walk based model with inertia for mobility of MTs. The model is intuitively more realistic in representing user mobility. Under this model, a moving MT is more likely to continue to move in the same direction; a stationary MT is also more likely to remain in the same cell. We assume discrete time. Under this model we have found expressions for update and paging costs when movement-based update scheme with selective paging is used. We have also extended the model to allow different values for the inertia of motion and inertia of rest.

References

[1] I. F. Akyildiz and J. S. M. Ho, "Dynamic mobile user location update for wireless PCS networks," *ACM-Baltzer J. Wireless Networks,* vol. 1, no. 2, pp. 187-196, July 1995. 357

[2] I. F. Akyildiz, J. S. M. Ho, and Y.-B. Lin, "Movement-based location update and selective paging for PCS networks," *IEEE/ACM Transactions on Networking,* vol. 4, no. 4, pp. 629-638, August 1996. 357, 358, 361, 364

[3] A. Bhattacharya and S. K. Das, "Lezi-Update: an information-theoretic approach to track mobile users in PCS networks," in *Proc. of ACM/IEEE MobiCom '99,* pp. 1-12, August 1999. 357

[4] A. Bar-Noy, I. Kessler, and M. Sidi, "Mobile users: To update or not to update," *ACM-Baltzer J. Wireless Networks,* vol. 1, no. 2, pp. 175-185, July 1995. 357

[5] A. Bar-Noy, I. Kessler, and M. Sidi, "Topology-based tracking strategies for personal communication networks," *ACM-Baltzer J. Mobile Networks and Applications (MONET),* vol. 1, no. 1, pp. 49-56, 1996. 358

[6] Y. Birk and Y. Nachman, "Using direction and elapsed-time information to reduce the wireless cost of locating mobile units in cellular networks," *Wireless Networks,* vol. 1, pp. 403-412, 1995. 358

[7] P. G. Escalle, V. C. Giner and J. M. Oltra, "Reducing location update and paging costs in a PCS network," *IEEE Trans. Wireless Communications,* vol. 1, no. 1, pp. 200-209, January 2002. 358

[8] J. S. M. Ho and I. F. Akyildiz, "Mobile user location update and paging under delay constraints," *ACM-Baltzer J. Wireless Networks,* vol. 1, no. 4, pp. 413-425, December 1995. 357, 358

[9] L. Kleinrock, *Queueing Systems Volume 1: Theory.* New York: Wiley, 1975. 365

[10] Y.-B. Lin, "Reducing location update cost in a PCS networks," *IEEE/ACM Trans Networking,* vol. 5, pp. 25-33, February 1997. 358

[11] T. Liu, P. Bhal and I. Chlamtac, "Mobility modeling, location tracking and trajectory prediction in wireless ATM networks," *IEEE J. on Selected Areas in Communications,* vol. 16, no. 6, pp. 389-400, August 1998. 357

[12] U. Madhow, M. L. Honig, and K. Steiglitz, "Optimization of wireless resources for personal communication mobility tracking," *IEEE/ACM Transactions on Networking,* vol. 3, no. 6, pp. 698-707, December 1995. 357

[13] J. G. Markoulidakis, G. L. Lyberopoulos, and M. E. Anagnostou, "Traffic model for third generation cellular mobile telecommunication systems," *ACM-Baltzer J. Wireless Networks,* vol. 4, pp. 389-400, August 1998. 358

[14] D. Munoz-Rodriguez, "Cluster paging for travelling subscribers," in *Proc. IEEE Vehicular Technology Conference,* 1990. 357

[15] G. Wan and E. Lin, "Cost reduction in location management using semi-realtime movement information," *ACM-Baltzer J. Wireless Networks,* vol. 5, no. 4, pp. 245-256, 1999. 358

[16] W. Wang, I. F. Akylidiz, G. Stuber and B-Y. Chung, "Effective paging schemes with delay bounds as QoS constraints in wireless systems," *Wireless Networks,* vol. 7, pp. 455-466, 2001. 357

[17] H. Xie, S. Tabbane, and D. Goodman, "Dynamic location area management and performance analysis," in *Proc. 42nd IEEE Vehicular Technology Conference,* pp. 536-539, May 1993. 358

Data Placement in Intermittently Available Environments

Yun Huang and Nalini Venkatasubramanian

Dept. of Information & Computer Science, University of California, Irvine
Irvine, CA 92697-3425, USA
{yunh, nalini}@ ics.uci.edu

Abstract. In this paper, we address the problem of data placement in a grid based multimedia environment, where the resource providers, i.e. servers, are intermittently available. The goal is to optimize the system performance by admitting maximum number of users into the system while ensuring user Quality of Service (QoS). We define and formulate various placement strategies that determine the degree of replication necessary for video objects by using a cost-based optimization procedure based on predictions of expected requests under various time-map scenarios and QoS demands. We also devise methods for dereplication of videos based on changes in popularity and server usage patterns. Our performance results indicate the benefits obtained the judicious use of dynamic placement strategies.

1 Introduction

Global grid infrastructures [1, 11] enable the use of idle computing and communication resources distributed in a wide-area environment. Multimedia applications are resource intensive and can effectively exploit idle resources available on a grid. Many multimedia applications, e.g. streaming media, must ensure continuous access to information sources to maintain Quality-of-Service requirements. However, systems on a computational grid are not continuously available. Our objective is to ensure application QoS and effective resource utilization in "intermittently available" environments. We define "intermittently available" systems as those in which servers and service providers may not be available all the time. Effective load management in such an intermittently available environment requires: (1) Resource discovery and scheduling mechanisms that ensure the continuity of data to the user [9]; (2) Data placement mechanisms to ensure data availability. In this paper, we focus on the second problem.

Generally, a placement policy for a multimedia (MM) system will address: (1) how many replicas are needed for each MM object; (2) which servers the replicas should be created on; (3) when to replicate. Placement decisions directly affect requests acceptance ratios for different MM objects. A bad placement policy will deteriorate the system performance, since it causes unused replicas to occupy premium storage resources. An

S. Sahni et al. (Eds.): HiPC 2002, LNCS 2552, pp. 367-376, 2002.

ideal placement policy must be capable of adjusting the mapping of replicas on the servers according to the run-time request pattern. Previous work has addressed issues in data placement for servers that are continuously available [4]. In this paper, we propose placement strategies that consider the intermittent availability of servers in addition to the request patterns and resource limitations. Basically, we determine the degree of replication of an object by using a cost-based optimization procedure based on predictions of expected requests for that object. To enforce the replication decisions, the placement and dereplication strategies proposed take into consideration the time maps of server availability in addition to object popularity and server utilization.

We illustrate the system architecture in Section 2. Section 3 proposes a family of placement strategies for intermittently available environment. Section 4 introduces the time-aware predictive placement algorithm for dynamic object placement. We evaluate the performance of the proposed approaches in Section 5 and conclude in Section 6.

2 System Architecture

The envisioned system consists of clients and multimedia servers distributed across a wide area network (see Fig 1). The resources provided by the distributed servers include high capacity storage devices to store the multimedia data, processor, buffer memory, and NIC (network interface card) resources for real-time multimedia retrieval and transmission. Server availability is specified using a server time-map that indicates specific times when a server is available. The availability of resources on servers can vary dynamically due to request arrivals and completions; the stored data on a server also changes dynamically due

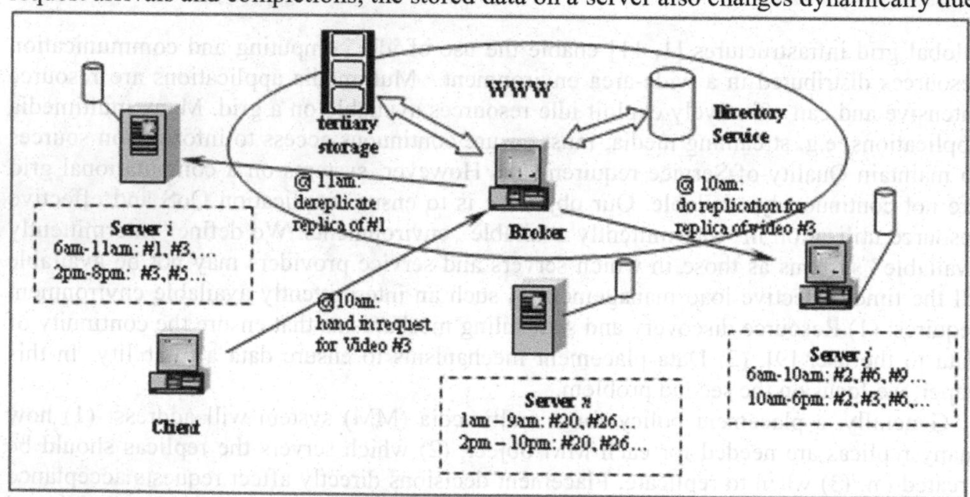

Fig. 1. System Architecture

to replication and dereplication of MM objects. To accommodate a large number of video objects, the environment includes tertiary storage. The key component of this architecture is a *brokerage service.* Specifically, the broker: determines an initial placement of video objects on servers; discovers the appropriate set of resources to handle an incoming request; coordinates resource reservation and schedules these requests on the selected resources; and initiates replication and dereplication of multimedia objects to cater to changes in request pattern and load conditions. Information required for effective data placement and resource provisioning include server resource availabilities, server time maps, replica maps, network conditions etc. This information is held in a directory service (DS) that is accessed and updated suitably by the broker.

Client requests for MM objects are routed to the broker that determines whether or not to accept the request based on current system conditions and request characteristics. Using state information in the DS, the broker determines a candidate server or a set of servers that can satisfy the request. Once a solution for the incoming request (i.e. scheduled servers and times) has been determined, the broker will update the directory service to reflect the allocated schedule. The goal is to improve the overall system performance and increase the number of accepted requests.

3 Placement Strategies

Given the time map and server resource configurations, an effective placement strategy will determine the optimal mapping of replicas to servers, so that the overall system performance is improved and accepted requests will be guaranteed QoS. Specifically, a placement policy for a MM system will provide decisions on which objects to replicate, how many replicas are required, where and when to replicate. Placement decisions can be made statically (in advance) or dynamically changed at runtime. We propose and compare a family of static and dynamic placement policies that can be used in intermittently available environments. Specifically, we devise and evaluate a Time-aware Predictive Placement (TAPP) algorithm for dynamic placement in a multimedia grid.

Static Placement Strategies: are determined at system initialization time; this placement is not altered during the course of request execution. Static placement policies may be popularity based where information on request popularity is taken into account when determining the replication degree of a MM object. We propose deterministic and non-deterministic policies to place the replicas on servers. Three static placement policies have been studied in this paper: (1) **SP1:** *Cluster-based static placement* – We cluster the servers into groups so that the total available service time of each group covers the entire day. Each video object is associated with exactly one group, so that every server in the group has a replica of this video object. (2) **SP2:** *Popularity-enhanced deterministic placement* – Here, we classify video objects into two groups – *very-popular* and *less-popular*. A replica of a very-popular video object is placed on every server, assuming resource availability. Less-popular video objects are evenly placed in the remaining storage space. (3) **SP3:** *Popularity-based random placement* – Here, we choose the number of replicas for a video object based on its popularity, however, these replicas are randomly distributed among the feasible servers, i.e those that have available disk storage.

Dynamic Placement Strategies consider the available disk storage, the current load on the servers, and the changing request patterns for access to video objects so as to dynamically reconfigure the number of replicas for each video object and their placement. We model the request R from a client as: R: < VID_R, ST_R, ET_R, QoS_R>, where VID_R corresponds to the requested video ID; ST_R is the request start time; ET_R is the end time by which the request should be finished; QoS_R represents the resources required by the request, such as: the required disk bandwidth (R_{DBW}), required memory resource (R_{MEM}), required CPU (R_{CPU}), and the required network transfer bandwidth (R_{NBW}); and the duration for which these resources are required (D_v). In order to deal with the capacity of each server over time in a unified way and represent how much a request will affect the server during the requested period of time, we define a *Load Factor*. Initially, we define a *Load Factor* (*R, S, t*) for a request r on server s at a particular time unit *t*, as:

$$\begin{aligned} &Load\ Factor\ (R, S, t) = \mathrm{Max}\ [\ CPU_a, MEM_a, NBW_a, DBW_a\ ] \\ &CPU_a = R_{CPU} / S_{AvailCPU}(t),\ MEM_a = R_{MEM} / S_{AvailMEM}(t), \\ &NBW_a = R_{NBW} / S_{AvailNBW}(t),\ DBW_a = R_{DBW} / S_{AvailDBW}(t). \end{aligned} \quad (1)$$

Thus, the *Load Factor* of a server *S* at time *t*, LF(*R, S, t*), is determined by the bottleneck resource at time *t*. However, the duration of the requested period from ST_R to ET_R may cover multiple time units; we therefore use the average *Load Factor* over all time units (between ST_R and ET_R) during which the server is available. For example, if the granularity of the time units in a day is 24 (24 hours/day), and ST_R is 5am and ET_R is 10am, we consider

$$LF(S) = \mathrm{Average}\ (LF_5, LF_6, LF_7, LF_8, LF_9, LF_{10}). \quad (2)$$

Dynamic placement strategies can be initiated on-demand to satisfy an individual request, or issued in advance based on predicted request arrivals. The efficacy of the on-demand technique depends on (a) the startup latency, i.e. how long a user will wait for a request to start and (b) the available tertiary storage bandwidth for replication. With low startup latencies, large amounts of server bandwidth and tertiary storage bandwidth are required to create replicas; hence the on-demand strategy may not always be feasible. In the following section, we propose a time-aware predictive placement approach that integrates server time-map information with request access histories to design a data placement mechanism for intermittently available environments.

4 The Time-Aware Predictive Placement (TAPP) Algorithm

The generalized *Time-Aware Predictive Placement* (TAPP) algorithm (Figure 2) has three main steps - *Popularity-Estimation, Candidate-server-selection, Pseudo-Replication*. An additional *Pseudo-Dereplication* phase may be added if sufficient storage space is not available[1]. We assume that the TAPP algorithm executes periodically with a predefined prediction-period.

Time-Aware Predictive Dynamic Placement Algorithm:

1 Obtain a snapshot of replica-maps and server state informaiton
  / (2) to (5) below execute on the snapshot information and do not modify the actual data.*/*
2 *Initialize*
  */*set* candidate_servers, add_replicas, del_replicas, replication_set, dereplication_set = null */
3 ***Popularity-Estimation*** /* *update* add_replicas and del_replicas */
4 candidate_servers = ***Candidate-Server-Selection( )***
5 **if** (*storage-biased* **and** candidate_servers == null) **then**
  dereplication_set += ***Pseudo-Dereplication*** (*system_wide*) */* update candidate_servers*/*
6 **while** (candidate_servers != null **and** add_replicas!= null) **do**
    replication_set = *Pseudo-Replication()*
    if (replication_set != null) **then**
      for each combination (V_i, S_j) in replication_set **do**
        if (S_j does not have enough storage for a replica of V_i) **then**
          dereplication_set += *Pseudo-Dereplication* (*server_specific*)
      /* update replication_set, add_replicas and candidate_servers appropriately*/
7 **if** (replication_set != null **and** dereplication_set != null) **then**
    do dereplication of video objects from appropriate real servers
    do replication of video objects on appropriate real servers

Fig. 2. Time-aware Predictive Placement Algorithm (TAPP)

Popularity-Estimation decides which video objects need more replicas based on a *rejection-popularity (RP)* factor defined for each video object V_i. Let #rejection(i) be the number of rejections of V_i in the last time period, #rejections be the total number of rejections for all videos within last time period, and #requests(i) be the number of requests for V_i in the last time period. Then $RP(i)$ is defined as:

$$RP(i) \quad \frac{\#rejections(i)}{\#rejections} \quad \#requests(i)\,. \tag{3}$$

Hence, the larger the *RP* is, the more problematic this video object is; which implies the necessity to add replicas. Furthermore, we group videos with rejection ratio > 5% into *add_replicas* list maintained in decreasing order of the *RP*. We also group videos with rejection ratio < 0.5% into a list *del_replicas,* maintained in ascending order of the RP.

Candidate-Server-Selection determines which servers will be considered for the creation of new replicas. It may be implemented using two approaches. A *Bandwidth – biased* approach chooses candidate servers using the current load and service time availability as primary criteria and ignores information about storage availability on servers. The *Storage – biased* approach uses the storage availability as a primary criterion

[1] Replication and dereplication decisions are not executed until a final mapping has been achieved – hence the names pseudo replication and pseudo dereplication.

in choosing candidate servers for new replicas. If there are no servers with sufficient disk storage, we choose servers that have replicas on the *del_replicas* list and attempt to dereplicate the less problematic replicas using the *Pseudo-Dereplication* process described later.

Table 1. Placement Cost Matrix (PCM)

	S_1	S_2	S_3
V_1	Min (N_1 ,1/LF(R_1, S_1)*T_1)	Min (N_1 ,1/LF(R_1, S_2)*T_2)	Min (N_1 ,1/LF(R_1, S_3)*T_3)
.	...	...	...
V_n	Min (N_n ,1/LF(R_n, S_1)*T_1)	Min (N_n ,1/LF(R_n, S_2)*T_2)	Min (N_n ,1/LF(R_n, S_3)*T_3)
	Max(PCM(V_n, S_1))	Max(PCM(V_n, S_2))	Max(PCM(V_n, S_3))

Pseudo-Replication: Using the set of servers from Step 2 (*candidate server selection*) and the *add_replicas* list obtained from Step 1 (*Popularity-Estimation)* we build a placement cost matrix, PCM (Table 1) in order to derive a mapping of video objects to data servers. The matrix represents the relative costs of servicing subscriber requests from each of the data servers. The columns represent data servers and rows represent video objects.

If server S_j already has a replica for video *Vi*, then the placement matrix entry is set as null, that is, this combination will not be considered for replication. Othewise, 1/ LF(R_i, S_j) represents the average number of requests similar to R_i that data server S_j can service per time-unit. To account for the intermittent availability of servers, we introduce a factor T_j to represent the duration of time for which server j is available in the next prediction-period. Then, the value (1/LF(R_i, S_j) * T_j) represents the average number of concurrently executing requests for video *i* that a server *j* can accept during the next prediction-period. N_i represents the number of rejections in the last period. We therefore set $PCM(V_i, S_j) \quad Min(N_i, (1/LF(R_i, S_j) * T_j)$ that represents the benefit that can accrue from allocating V_i to S_j.

After calculating the entries in the placement matrix, we choose the maximum value Max(PCM(V_i, S_j)). This gives us the object-server combination that is expected to improve request acceptance in the next prediction-period. We shrink the matrix by either deleting a column (if N_i >= 1/LF(R_i, S_j)*T_j) correspondingly decreasing the value of N_i by 1/LF(Ri, S_j)*tj; or deleting a row (if N_i <= 1/LF(R_i, S_j)*T_j) appropriately decreasing the available network bandwidth of S_j. We then repeat the pseudo-replication process until there are no more columns or rows left. At the end of the *Pseudo-Replication*, we have an ordered list of replication decisions that the broker can initiate.

Pseudo-Dereplication marks video objects in the *del_replicas* list (obtained during Popularity Estimation) as dereplicatable if they are not currently in use and/or have not been reserved for future use. Dereplication decisions can be made on a system wide basis or on a specific server. *System_wide dereplication* may be invoked prior to *Pseudo-Replication*. The goal is to dereplicate the least popular replica from lightly loaded servers with more service time available. For each video V_k in *del_replicas*, we order the servers that have a copy of V_k in ascending order of their load-factors. We then pick the most beneficial replica for dereplication. The number of replicas that are selected for dereplication can be tailored to meet the storage needs of popular objects. *Server_specific dereplication* may be issued to ensure replication decisions made by the *Pseudo-Replication* process. In order to provide enough space for a new replica on a specific server, we may need to dereplicate appropriate number of videos from the *del_replicas* list that are not in use or have not been reserved for future use on that server.

5 Performance Evaluations

In this section, we evaluate the performance of the proposed placement strategies under various time-maps scenarios and server resource configurations.

Simulation Environment: We characterize incoming multimedia requests using a Zipfian distribution [6,13,14], with the request arrivals per day for each video V_i given:

$$\text{Pr. } (V_i \text{ is requested}) = \frac{K_M}{i}, \text{ where } K_M = \left(\sum_{i=1}^{M} \frac{1}{i}\right)^{-1}. \quad \textbf{(4)}$$

We compute the probability of request arrival in an hour j to be: $p_j = c/(j^{1-\theta})$ for $1 \le j \le 24$, where θ is the degree of skew and assumed to be 0.8 and $c = 1/(\sum (1/j^{1-\theta}))$, $1 \le j \le 24$. Hence, the number of requests that arrive in each hour for each video V_j are computed.

The basic video server configuration used in this simulation includes 20 data servers each with 100 GB storage and 100Mbps network transfer bandwidth. For simplicity, the CPU and memory resources of the data servers are assumed not to be bottlenecks; besides the 100Mbps network bandwidth available for video streaming, a fraction of server bandwidth is reserved for replications into the server. In the following simulations, we set the duration of each video to be three hours; each video replica requires 2 GB disk storage, and 2 Mbps for network transmission bandwidth. In fact, these parameters will be changed for different purposes during the simulation.

In such an intermittently available system, specifically for each server, we use the service *time map* to keep the information of when the servers will be available during the span of a day. We study three approaches to model the time map of each server: (1) *T1*: *Uniform availability* – All the servers are available for an equal amount of time; and the servers are divided into groups, so that within each group, the time distribution covers the entire 24-hour day. (2) *T2*: *Random availability* – How long and when the servers are available are all randomly determined. (3) *T3: Total availability* – All the servers are available all the time. In the remainder of this paper, we will use T1, T2 and T3 to identify the three time map strategies.

Performance Results: Given the time map information and resource configuration of the video server system, we study the system performance by applying the scheduling polices for discovering intermittently available resources (DIAR) described in [9] and use the number of rejections (i.e. success of the admission control process) as the main metric of evaluation. In order to focus on the placement problem, we present our simulation results by using a scheduling policy where requests are initiated immediately (albeit startup latency) and continuously serviced without interruption by (possibly) multiple servers.

Overall performance of static placement under different time map patterns: Intuitively, when servers are always available (T3), the multiple server cases should always have a better acceptance rate, as in Fig 3. This is because, in general, the single server case is a constrained version of the multiple server case and therefore has fewer options for resource selection. With time-maps T2 and T1, we observe that the number of rejections of policy SP3 is much smaller than SP2 or SP1. Although both SP2 and SP3 take request popularity into account, SP3 is better than SP2, because the random distribution enlarges the possibility for a replica to be created on more servers with different time maps. Although SP1 (cluster-based placement) does not take the request popularity into account, it works well in conjunction with the T1 time-map since the clustering technique meshes well with the T1 time-map.

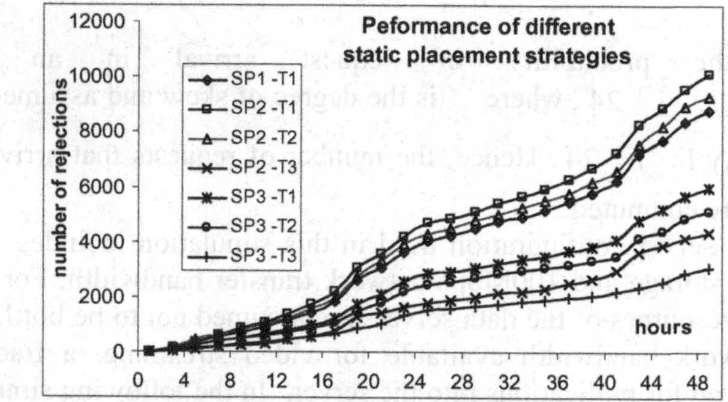

Fig. 3. Performance of different static placement strategies. The caption of "SPx-Ty" implies the SPx static placement and Ty time map model used.

Comparing dynamic placement strategies with static approaches: In order to avoid the influence of randomness, we choose the SP1 and T1 configurations. From Fig 4a, we observe that all the dynamic placement policies outperform SP1 with less number of rejections. The bandwidth biased TAPP seems to perform similar to the storage biased variant. However, comparing with Fig 4b, we observed that with more network bandwidth, the bandwidth-biased TAPP performs marginally better than storage-biased. We attribute this to the fact that when bandwidth is sufficient, the impact of effective storage utilization is less, since more effective load balance is achieved.

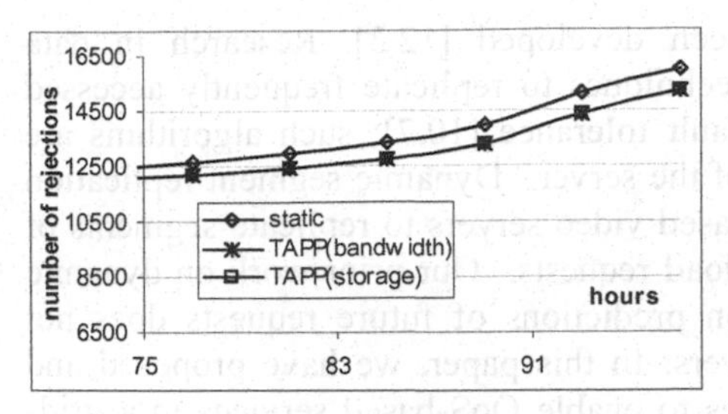

Fig. 4a. Placement strategies under limited network bandwidth resources.

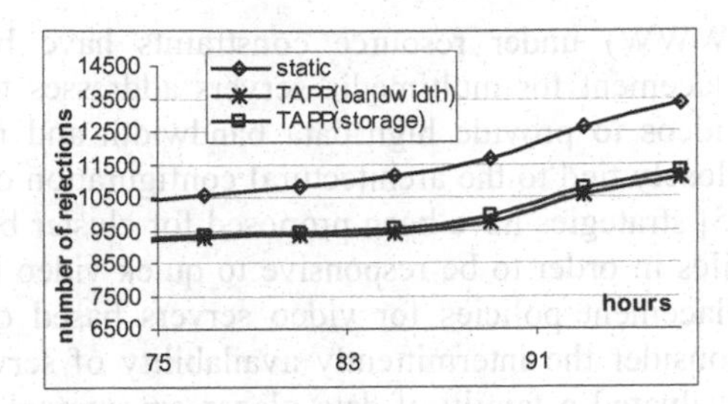

Fig. 4b. Placement strategies under large network bandwidth resource.

We also observe that more storage provides greater possibility of replicating popular videos yielding better acceptance ratios. However, our experiments indicate that beyond a point, merely increasing storage is insufficient to improve performance significantly. Other factors such as network bandwidth and server time-maps are bottlenecks to performance at high storage levels. We studied several factors that may influence the system performance such as time-map continuity, the number of data servers, etc. The TAPP policy exhibited low sensitivity to variations in time-map continuity; we believe that this policy is well suited to grid environments where systems exhibit randomized availability patterns. Furthermore, dynamic placement exhibits better performance with larger numbers of servers (as Fig 5) indicating the enhanced benefits that can be obtained as the size of the grid increases. We conclude that effective data placement can significantly improve system performance. Furthermore, dynamic placement is particularly effective when network bandwidth is limited.

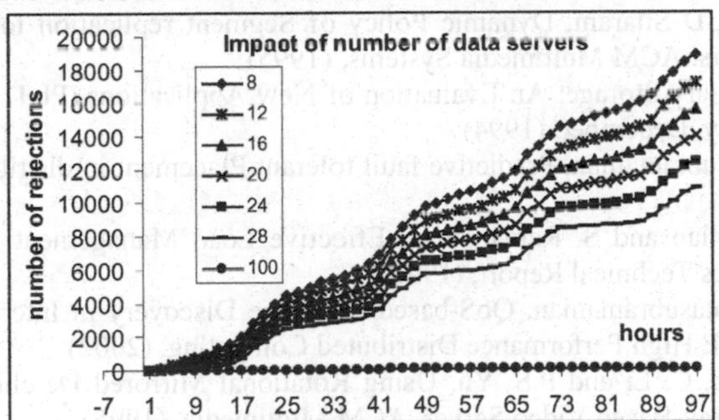

Fig. 5. The impact of number of data servers for TAPP (bandwidth-biased)

6. Related work and Future Research Directions

Data placement has been studied extensively in the context of distributed systems. [2] proposes a dissemination tree for replica placement algorithms which reduces the number of replicas deployed. Static data management policies for general purpose applications

(WWW) under resource constraints have been developed [12,3]. Research in data placement for multimedia servers addresses techniques to replicate frequently accessed videos to provide high data bandwidth and fault tolerance [10,7]; such algorithms are closely tied to the architectural configuration of the server. Dynamic segment replication [5] strategies have been proposed for cluster based video servers to replicate segments of files in order to be responsive to quick video load requests. Our prior work on dynamic placement policies for video servers based on predictions of future requests does not consider the intermittently availability of servers. In this paper, we have proposed and evaluated a family of data placement strategies to enable QoS-based services in a grid-based environment. We intend to conduct further performance studies with heterogeneous servers and combinations of request patterns. We also plan to explore effective middleware support in situations where server availability is not known ahead of time.

References

1. R.J. Allan. Survey of Computational Grid, Meta-computing and Network Information Tools, report, Parallel Application Software on High Performance Computers. (1999)
2. Y. Chen, R. H. Katz, J. Kubiatowicz, Dynamic Replica Placement for Scalable Content Delivery, 1st International Workshop on Peer-to-Peer Systems. (2002)
3. I. Baev and R. Rajaraman. Approximation algorithms for data placement in arbitrary networks. 12th ACM-SIAM Symposium on Discrete Algorithms (SODA), (2001)
4. C. Chou, L. Golubchik, and J.C.S. Lui, A Performance Study of Dynamic Replication Techniques in Continuous Media Servers, ACM (1999)
5. A. Dan, M Kienzle, D Sitaram, Dynamic Policy of Segment replication for load-balancing in video-on-demand servers. ACM Multimedia Systems, (1995)
6. A.L.Chervenak. Tertiary Storage: An Evaluation of New Applications, Ph.D. Thesis, University of California at Berkeley, December, (1994)
7. X. Wei, N. Venkatasubramanian, Predictive fault tolerant Placement in distributed video servers, ICME (2001)
8. N. Venkatasubramanian and S. Ramanathan, Effective Load Management for Scalable Video Servers, HP Laboratories Technical Report, (1996)
9. Y. Huang, N. Venkatasubramanian. QoS-based Resource Discovery in Intermittently Available Environments. 11th IEEE High Performance Distributed Computing, (2002)
10. M Chen, H. Hsiao, C. Li and P.S. Yu, Using Rotational Mirrored De-clustering for Replica Placement in a Disk-Array-Based Video Server, ACM Multimedia, (1995)
11. I. Foster, C. Kesselman. The Grid: Blueprint for a New Computing Infrastructure, book, preface, (1998)
12. C. Krick, H. Räcke, M. Westermann Approximation Algorithms for Data Management in Networks. 13th ACM Symposium on Parallel Algorithms and Architectures. 237-246, (2001)
13. A. Dan and D.Sitaram. An online video placement policy based on bandwidth to space ration (bsr). In *SIGMOD* '95, pages 376-385, (1995)
14. A. Dan, D. Sitaram, P. Shahabuddin. Scheduling Policies for an On-Demand Video Server with Batching, 2th ACM Multimedia Conference and Exposition, (1994).

RT-MuPAC: Multi-power Architecture for Voice Cellular Networks*

K. Jayanth Kumar, B.S. Manoj, and C. Siva Ram Murthy**

Department of Computer Science and Engineering
Indian Institute of Technology Madras, Chennai 600 036, India
kjk@dcs.iitm.ernet.in
bsmanoj@cs.iitm.ernet.in
murthy@iitm.ernet.in

Abstract. We have considered the problem of providing greater throughput in cellular networks. We propose a novel cellular architecture, RT-MuPAC, that supports greater throughput compared to conventional cellular architectures. RT-MuPAC (Real-time Multi-Power Architecture for Cellular Networks) is based on two fundamental features not present in today's cellular networks: *usage of multiple hops* and *power control* (power control is used only in a limited fashion to reduce interference in today's networks). These features, we believe, will become increasingly important in next generation cellular systems as heterogeneous networks will operate in synergy. We show using detailed simulations that RT-MuPAC is indeed a significant improvement over conventional networks. RT-MuPAC can evolve from the existing infrastructure and offer advantages to both the service provider and the users. RT-MuPAC also serves as a proof of concept for the use of multi-hop architectures in cellular networks.

1 Introduction

Wireless applications have begun putting increasing demands on the bandwidth owing to the advent of powerful processors. This work proposes RT-MuPAC, a novel variable-power alternative for the Cellular Architecture, one of the most common wide-area mobile communication architecture today. RT-MuPAC can provide greater throughput with the same number of base stations, thus allowing the network operator to support a greater number of users with the same infrastructure.

Currently, the Single-hop Cellular Network (SCN) architecture is extensively used in the 2G and 3G networks. In the SCN networks, the base station (also referred to as base in this work) can be reached through a single hop. The entire service area of the operator (the service provider) is divided into cells and one

* This work was supported by Infosys, Bangalore, and the Department of Science and Technology, New Delhi, India.

** Author for correspondence.

S. Sahni et al. (Eds.) HiPC 2002, LNCS 2552, pp. 377–387, 2002.

base station placed per cell. Nodes (mobile hosts) establish calls by requesting the base for a channel, which is subsequently used to send packetized voice data.

Future Fourth Generation (4G) systems [1] are expected to support increasing data services, multimedia traffic, efficient resource utilization and provide greater bandwidth. The authors of [1] also discuss the use of multihop architectures for 4G cellular networks. Increasing attention is being focused on the use of multiple hops in cellular data networks: Multi-hop Cellular Network (henceforth referred to as MCN) in [2] and [3], the Integrated Cellular and Ad hoc Relaying System (iCAR) architecture in [4], and the A-GSM architecture in [5]. These architectures attempt to introduce some of the desirable features of Ad hoc networks into traditional cellular networks. It is in this context that one can appreciate the need for multiple hops in *voice* cellular networks since using multiple hops has already been illustrated as a useful technique for *data* networks. RT-MuPAC combines the idea of using multiple hops along with power control techniques. RT-MuPAC is one of the first architectures to incorporate the use of multiple hops in a full-fledged fashion for *voice* cellular networks.

2 Related Work

In order to support lower transmission power and higher throughput *without* additional base stations, Lin and Hsu [2] have proposed Multi-hop Cellular Networks (MCN) as an alternative to SCN. In the MCN architecture, nodes use a transmission power that is a fraction $1/k$ of the cell radius. The mobile nodes use multiple hops to reach the base (and each other). This architecture has been extended in [3]. Power control and real-time traffic support are not considered in [2, 3].

Unlike the previously discussed work which have been proposed for *packet data* cellular networks, [4] and [5] suggest the use of multiple hops in *voice* cellular networks, though in a limited fashion. In [4], the authors propose the Integrated Cellular and Ad hoc Relaying System (iCAR) architecture for the purpose of load balancing. In the iCAR architecture, nodes in congested cells establish calls through Ad hoc Relaying Stations (ARSs), placed at the boundary of cells, to the adjoining possibly lightly loaded base stations. Multiple hops are used in [4] only for the purpose of load balancing and relaying is done through ARSs, which requires additional infrastructure. The A-GSM architecture in [5] suggests the use of relaying through mobile nodes to allow nodes in dead spots (where the signal of the base station can not reach due to geographical features) to reach the base. However, extending service to dead spots is the objective in [5], not throughput enhancement.

3 Real-Time MuPAC

In this section, we discuss RT-MuPAC (Real-time Multi-Power Architecture for Cellular Networks) for real-time communication in cellular networks in detail.

By real-time communication, we mean the voice traffic supported in the current day, SCN based, cellular networks.

3.1 Basic Operation in RT-MuPAC

There are three main features in RT-MuPAC:

1. Multiple hops are used by nodes to send their packetized voice data to the base. The use of multiple hops allows the use of a lower transmission range than the cell radius and hence, greater spatial reuse, and greater throughput.
2. Power control is used to maximize spatial reuse. Each transmission is kept at a power that is calculated based on the distance between the communicating nodes. Power control has been employed in the SCN architecture only in a limited fashion with the objective of saving battery power and reducing interference (for example, in reuse partitioning). RT-MuPAC, on the other hand, employs power control in a more extensive fashion because of the use of multiple hops.
3. Control messages are sent over a single hop to the base through a dedicated control channel while data packets are sent over multiple hops.

In RT-MuPAC, similar to SCN, the bandwidth is divided into a control channel and a number of duplex voice channels. Control messages are sent over a single hop using the control channel (the transmission range over the control channel is R where R is the cell radius). Voice packets are sent, possibly through multiple hops, over the duplex voice channels. These voice channels are not divided over the cells in a cluster: they are maintained in a common pool that can be used in any cell provided this usage does not lead to interference in other on-going transmissions. One of the distinctive design features of RT-MuPAC is that the *control* plane of RT-MuPAC is similar to that of SCN, while on the *data* plane, RT-MuPAC resembles an Ad hoc network. This means that RT-MuPAC can support greater throughput because of the use of multiple hops on the data plane, and control messages can be sent over a single hop to the base ensuring that the node can get in contact with the base without the need for any route discovery process.

When a node makes a call (by sending a *Call Request* packet over the control channel to its base), a path (possibly over multiple hops) is computed by the base. The base now reserves channels over every hop on this path (marks these channels as reserved, to be exact). The channel allocation and the address of the relaying nodes for this call are now broadcast over the control channel, so that the intermediate nodes in the path can receive packets over one channel and forward it to the next node in the path over another channel. Multiple powers are used in RT-MuPAC as follows: over each hop, the power used is a function of the distance between the nodes communicating over that hop. Thus, each channel operates independently at its own power: the power used by a node over a channel is purely a local decision made at the node. The base only specifies the channel to be used over every hop in the path: the nodes decide what power

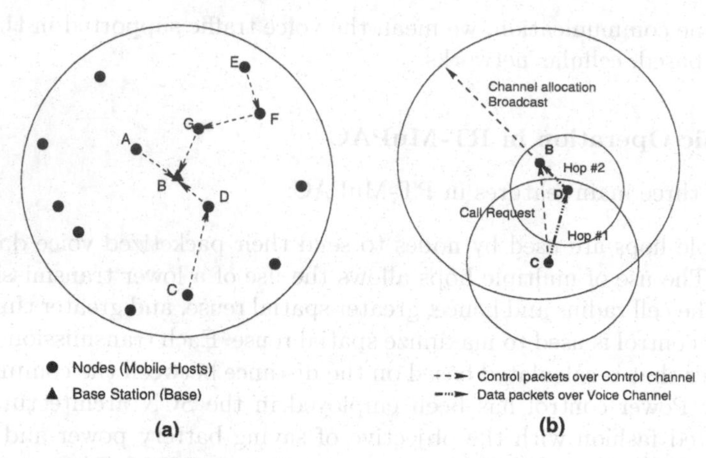

Fig. 1. Operation of RT-MuPAC

to use over this channel based on the power of the *Hello* beacons from the next hop. These *Hello* beacons are transmitted by all nodes periodically so that each node knows its neighbors and can decide what transmission range to use in a transmission with each neighbor.

Figure 1 (a) shows the use of multiple hops in RT-MuPAC. Nodes A, C, and E transmit packets to the base B over multiple hops along the paths $A - B$, $C - D - B$, and $E - F - G - B$ respectively. These paths are computed by the base B when the nodes A, C, and E make a *Call Request* to the base.

Figure 1 (b) shows the power control in RT-MuPAC. Node C makes a *Call Request* to the base B over the control channel and the base broadcasts the channel allocation over the multiple hop path $(C - D - B)$. Thus, node C is relayed through node D to the base B. Note that since the $C - D$ distance is greater than the distance $D - B$, the transmission range over the first hop (from C to D) is greater than that over the second hop (from D to B).

The base requires information about the topology of the nodes in its cell in order to compute paths from nodes to their base. For this purpose, we have used the following mechanisms: beaconing and neighbor table update mechanism. All nodes periodically transmit beacons with a transmission range of $R/2$ (using TDMA). This is used by nodes to keep track of their neighbors. Each node maintains a list of its neighbors along with the reception power of the *Hello* beacon from each neighbor. This list is periodically sent to its base, so that the base can construct the complete topology graph of all nodes in its cell. We detail the other features of RT-MuPAC in the subsequent sections.

3.2 Power Control

In this section, we describe the power control scheme used by the nodes. Let node x use channel C to communicate with node y. The transmission range r

it uses over this channel C is varied by the nodes as αd where α is a tunable parameter in RT-MuPAC and d is the distance between these nodes. The distance d is estimated by the nodes based on the power of the *Hello* beacons of x received by y (or vice-versa). Every time a *Hello* beacon is received from the next hop, each node re-calculates the distance d of the next hop and adjusts its transmission range accordingly. These *Hello* beacons are transmitted with a transmission range of $R/2$ with a frequency of 1 per second (this frequency is sufficient at even high mobility conditions as the simulations in Section 4 show — if required this can be changed). Note that this means that the maximum possible transmission range is $\alpha R/2$ since only nodes which can hear each others beacons will be selected as two consecutive nodes in the path. Since the maximum transmission range has been restricted to $\alpha R/2$, nodes in isolated regions of low node density may not be able to find neighboring nodes to relay to the base. However, a transmission radius of $\alpha R/2$ ensures a very low probability of network partition at typical node densities. At low node densities, when relaying nodes are not available, channels can be directly setup to the base (over a single hop) with an appropriate transmission range.

Therefore, in RT-MuPAC the distance d is used to calculate the transmission range. This is possible because of the use of multiple voice channels: each channel operates at different powers. Since the channel is reserved for operation for transmission over a particular hop, the power of transmission can be decided by the nodes involved in this transmission (provided interference is not caused at other nodes which also use the same channel). We describe the role of the parameter α: it might not always be possible to estimate the distance d accurately based on the reception power of the *Hello* beacons. Hence, α is used as a "safety" factor in the transmission power calculation.

3.3 Path Computation

The nodes use beacons to keep track of neighbors and send this neighbor table information (list of neighbors along with the reception power of their *Hello* beacons) to the base periodically. The base computes the path from the calling node to the base using this information. For this purpose, firstly, the base constructs a graph where each node in its cell is represented by a vertex. Vertex A (in the graph) is connected by an edge to vertex B if the corresponding node A is within the (maximum) transmission range of node B. The base now uses Dijkstra's algorithm (or any other minimum weight path algorithm) to compute a path from the node to the base. During this path computation, we have assigned weights to the edges as follows: $w(x, y) = 1$ if x can hear y's *Hello* beacons or vice-versa. The base now allocates channels to every hop on this path and broadcasts this on the control channel. Note that any given hop might be used for more than one call.

3.4 Channel Assignment

Implementing a channel assignment scheme is more complicated in RT-MuPAC than in SCN for two reasons: a) Unlike SCN, the base is not involved in every hop b) Nodes independently choose their transmission power depending on the distance of the next hop. For RT-MuPAC, we have chosen to use dynamic channel allocation because of the the additional complexity of using multiple powers and multiple hops by relaying through mobile nodes.

To accomplish channel allocation, the base needs to know which channels are being used by which nodes and the topology information. Note that in this process, it also needs to consult adjacent base stations. Recall that the nodes send their neighbor table information to the base: this includes the received power of their neighbor's *Hello* beacons. In RT-MuPAC, the distance d is used directly in the transmission range computation, hence this neighbor update is periodic (we have used a frequency of 1 per second, which is the same as the beaconing frequency). Thus the base can also estimate the distance d between nodes, in the same way the nodes calculate it. Since the base has knowledge of the currently setup calls, the path these calls use, the relaying nodes in this path and the channels used, it can find out the different nodes using the channels and their transmission ranges over this channel (αd).

For this purpose, the base maintains the following table entry for every channel: $< C, N(C) >$ where C is the channel identifier, and $N(C) = \{ (x, y, r) \mid$ *node* x *and node* y *use channel* C *with transmission range* $r\}$. Thus, $N(C)$ is a set recording all usages for the channel C. Entries are added to this table when a new call is setup and entries are deleted when a call is torn down. The transmission range r is updated every time a neighbor table update is received from x or y. The base uses these table entries to assign channels to different hops in the path.

Let the path from the node to the base be denoted by $P = x_1, x_2, \cdots, x_n$ where x_1 is the node and x_n is the base. A channel C can be assigned for the hop x_i, x_{i+1} if and only if (1) this usage does not interfere with nodes already using the channel C and (2) the nodes already using the channel C do not cause interference at x_i, x_{i+1}. Note that condition (1) does not imply (2) or vice-versa because of the usage of multiple powers. Note that by "interference", we mean unacceptable signal quality.

This condition is checked by the base as follows: If channel C is to be used for the hop x_i, x_{i+1}, these nodes will use a transmission power of $r' = \alpha d(x_i, x_{i+1})$ where $d(x_i, x_{i+1})$ is the distance between the nodes x_i and x_{i+1} as estimated using the neighbor table information sent by the nodes. Thus, channel C can be permitted for use in the hop x_i, x_{i+1}, if and only if for every entry (x, y, r) in $N(C)$, the following conditions hold:

$$
\begin{aligned}
d(x, x_i) > \max(r, r'), d(x, x_{i+1}) > \max(r, r') \\
d(y, x_i) > \max(r, r'), d(y, x_{i+1}) > \max(r, r')
\end{aligned} \tag{1}
$$

It is easy to see that these conditions ensure that these two transmissions will not interfere with each other. We have used a randomized first fit channel allocation policy: the channels are ordered in a linear order $C_1, C_2, \cdots$ (this ordering is chosen randomly every time a channel needs to be assigned) and the first channel C_k satisfying the constraints specified earlier, is used for the hop x_i, x_{i+1}. Thus, the base allocates channels to each hop in the path P and broadcasts this information over the control channel. Then the intermediate nodes establish required state (their two neighbors in the path P and channels used to communicate with them) and the calling node begins transmitting voice packets over the allocated channels. If the base can not find any channels to allocate over some hop in the path P, the call is said to be *blocked* and the calling node is informed of the same over the control channel.

3.5 Path Re-configuration

In this section, we discuss the control protocol operation when channel re-configuration or path re-configuration is required. Channel re-configuration is required when two different transmissions over the same channel come too close to one another and leads to interference to at least one of these transmissions. Path re-configuration (and channel re-configuration) is required when two nodes communicating over a single hop move away from each other beyond a distance $R/2$ and so a new path and accompanying channel allocation need to be found.

In both of the above cases, the node which detects interference or which discovers that his next hop has moved away, informs the base of the same over the control channel. The base frees up channel resources, if necessary, (deleting table entries from the channel usage table), and finds a new path, if necessary. It also assigns a new set of channels to this path (and correspondingly adds entries to the channel usage table) and broadcasts this information along with the path over the control channel. This triggers formerly intermediate nodes to delete their forwarding state and the new relaying nodes to establish forwarding state. In case, a new path can not be found or channels could not be found, the call is said to be *dropped* and the calling node is informed of the same (this also triggers formerly relaying nodes to delete their forwarding state for that call). A similar procedure occurs when interference is detected over the channel (a *Channel Error* message is sent to the base in this case).

3.6 Dropping-Blocking Trade-off

In this section, we describe a technique to control the dropping-blocking trade-off in RT-MuPAC. Traditionally, such techniques have been known for the SCN architecture, the most popular one among them is perhaps the technique of guard channels. A similar technique is also necessary for RT-MuPAC, perhaps more so. Besides allowing us to control the trade-off, such a technique would be able to limit the dropping probability: the use of multiple hops by relaying through mobile nodes (unlike relaying through a stationary base station) can otherwise increase the dropping probability.

In Equation (1), we have indicated the conditions under which a channel C may be allowed for transmission between nodes x_i, x_{i+1}. We introduce an additional parameter $\beta\,(>1)$ to control the dropping probability. Instead of ensuring that the nodes which are part of different transmissions are at a distance d $(d > \max(r, r'))$, we impose a tighter constraint $d > \beta \max(r, r')$. Thus the new constraints are:

$$d(x, x_i) > \beta \max(r, r'), d(x, x_{i+1}) > \beta \max(r, r')$$
$$d(y, x_i) > \beta \max(r, r'), d(y, x_{i+1}) > \beta \max(r, r') \quad (2)$$

Note that this constraint is checked only during call admission — when a new call is made in the system. When an existing call is reconfigured (due to a path break or channel interference), the constraint in Equation (1) is checked. The rationale behind this scheme is that during call admission, tighter constraints are checked: only if the potentially interfering transmissions are a "safe" distance $(\beta \max(r, r'))$ apart, the call is accepted by the system. It is easy to see that as the parameter β increases, the dropping probability decreases and the blocking probability increases. Thus, the parameter β gives the network provider control over the trade-off between blocking and dropping in the system.

4 Comparative Simulation Results of RT-MuPAC and SCN

In this section, we will present simulation results (using GlomoSim [7]) comparing the performance of RT-MuPAC against SCN for voice traffic (an analytical comparison of RT-MuPAC against SCN has been presented in [6]). We used 60 voice channels and for RT-MuPAC, the beaconing frequency was 1 per second. The parameter α in RT-MuPAC has been chosen to be 1.2 and we have simulated RT-MuPAC with parameter $\beta = 1.0, 1.25,$ *and* 1.5 (to illustrate the dropping-blocking trade-off). We have used a call generation scheme wherein the inter-arrival times are obtained using a exponential distribution with a mean fixed at 30 seconds. The holding time has been varied (from 10 seconds to 20 seconds) to vary the load on the system. We have used a node density of 40 nodes in a cell of radius 500 meters. We used the random waypoint model where a node chooses a random direction, moves for a given travel time and waits for a given pause period before moving again. The speed is chosen randomly from [*min_speed*, *max_speed*]. In our simulations, the nodes travel 80% of the time using a travel time of 8s, and a pause time of 2s: the minimum speed was set to 0 m/s and we have varied the maximum speed to obtain different mobility conditions. The simulation was run over seven cells and statistics are collected only in the central cell. Statistics are measured for 300s and all simulations have been averaged over 15 runs. *Blocking Probability* and *Dropping Probability* were used as the comparison metrics.

4.1 Blocking Probability

Blocking probability is one of the key indices of performance of a real-time cellular system. We experimented with RT-MuPAC and SCN by measuring the blocking probability over a range of holding times varying from 10 seconds to 20 seconds and the results are plotted in Figure 2 (a). RT-MuPAC performs significantly better than SCN and as expected, greater the value of β greater the blocking probability. At high load, the blocking probability is brought down by about 90% for $\beta = 1$ and by about 50% for $\beta = 1.25$. To illustrate the effect of mobility, the blocking probability has been plotted against the maximum speed at high load in Figure 2 (b). In both these scenarios, RT-MuPAC performs better than SCN. Under high load, the blocking probability is nearly independent of the mobility, as expected.

4.2 Dropping Probability

The dropping probability has been plotted for varying load in Figure 2 (c). Again, RT-MuPAC performs significantly better than SCN with the dropping probability brought down to less than 0.1% at high load. As expected, the dropping probability is greater for RT-MuPAC with greater β. Also notice that the blocking probability is much more sensitive to the value of β as compared to the dropping probability. The dropping probability has been plotted against mobility in Figure 2 (d). At high load, the dropping probability in RT-MuPAC is much lower compared to SCN. In this graph, the effect of β on the dropping probability is clearly seen: at greater β, the dropping probability decreases. Despite relaying through mobile nodes, the techniques of using a transmission range of αd and further controlling the blocking-dropping trade-off through a tunable parameter β, have resulted in lower dropping probability. Thus, RT-MuPAC provides superior throughput (lower blocking probability) and Quality of Service as compared to SCN.

5 Conclusions

We have proposed a novel cellular architecture, RT-MuPAC, based on using multiple hops and power control. RT-MuPAC resembles SCN on the *control* plane, and Ad hoc networks on the *data* plane, combining the resilient routing in the former and the greater throughput in the latter. This architecture provides better performance compared to SCN and offers advantages to the network operator (who can support a greater number of users and earn greater revenue without additional infrastructure) and the users (who can enjoy greater throughput). The superiority of RT-MuPAC has been illustrated using analysis (in [6]) and simulation. RT-MuPAC supports lower call blocking probability and greater total call time supported, and lower dropping probability (in most cases), as compared to SCN networks. RT-MuPAC is an ideal choice for next generation cellular networks which will need to inter-operate with a wide variety of wireless networks

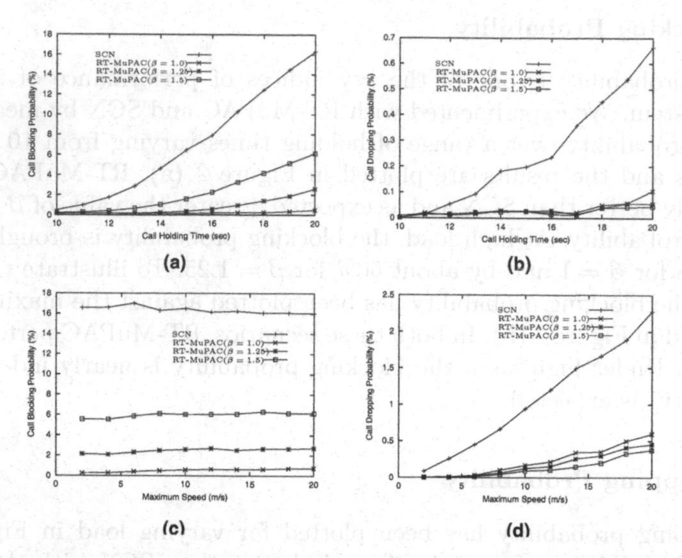

Fig. 2. Simulation Results of RT-MuPAC and SCN

supporting a variety of devices with different transmission power capability. RT-MuPAC also substantiates the fact that multi-hop networks can offer similar or better Quality of Service (in terms of dropping probability) as single-hop networks. Though the complexity of the nodes in RT-MuPAC is greater compared to SCN (greater buffer space, for example), we believe that wireless bandwidth, rather than processor memory or computing power, is the limiting factor in wireless networks.

References

[1] M. Frodigh, S. Parkvall, C. Roobol, P. Johansson, and P. Larsson, "Future-Generation Wireless Networks", in *IEEE Personal Communications Mag.*, Vol. 8, No. 5, pp. 10-17, October 2001. 378

[2] Y. D. Lin and Y. C. Hsu, "Multi-Hop Cellular: A New Architecture for Wireless Communications", in *Proc. IEEE INFOCOM 2000*, Tel Aviv, Israel, March 2000. 378

[3] R. Ananthapadmanabha, B. S. Manoj, and C. Siva Ram Murthy, "Multi-hop Cellular Networks: The Architecture and Routing Protocols", in *Proc. IEEE PIMRC 2001*, San Diego, USA, October 2001. 378

[4] H. Wu, C. Qiao, S. De, and O. Tonguz, "Integrated Cellular and Ad hoc Relaying Systems: iCAR", in *IEEE Journal on Selected Areas in Communications 2001*, Vol. 19, No. 10, pp. 2105-2115, October 2001. 378

[5] G. Aggelou and R. Tafazolli, "On the Relaying Capacity of Next-Generation GSM Cellular Networks", in *IEEE Personal Communications Mag.*, Vol. 8, No. 1, pp. 40-47, February 2001. 378

[6] K. Jayanth Kumar, "Multi-hop Packet Data and Voice Cellular Networks: Architectures, Protocols, and Performance Analysis", *BTech Thesis*, Dept. of Computer Science and Engineering, Indian Institute of Technology Madras, May 2002. 384, 385

[7] X. Zeng, R. Bagrodia, and M. Gerla, "GloMoSim: A Library for Parallel Simulation of Large-scale Wireless Networks", in *Proc. PADS-98*, Banff, Canada, May 1998. 384

Asynchronous Transaction Processing for Updates by Client: With Elimination of Wait-for State

Subhash Bhalla

Database System Laboratory, The University of Aizu
Aizu-wakamatsu city, Fukushima, Japan
bhalla@u-aizu.ac.jp

Abstract. In a distributed database system, time-critical transactions need to complete their processing, within a time limit. This is especially true in the case of a real-time database systems, and also in the case of mobile database systems. This study considers server level enhancements. By adopting transaction classification, few changes can be accommodated within 2-phase locking at a low cost that enable the database update by time-critical Clients. A use of an instant priority based execution scheme, based on transaction classification, can reduce delays caused by resource conflicts with ordinary transactions. We further investigate a procedure that performs critical functions asynchronously.

1 Introduction

Transaction updates by mobile clients are a desirable feature for many applications. Most existing research efforts consider a limited case, of the read-only support for mobile clients. Few other studies consider relaxing the criteria of serializability, or propose a prolonged execution sequence. In a disconnection prone system, prolonged execution of transactions is undesirable. We consider, an environment based on transaction classification, in which the server is assumed to have a high capacity and also supports requests of mobile clients for updates. The other transactions at the server end are considered to be short. It is assumed that these can be easily restarted. The mobile client's transactions on the other hand are considered instant execution requests of highest (real-time) priority. Conflicts among two mobile client transactions are dealt with by detection of conflicts as soon as these occur. Firstly the lock grant/denial messages carry the wait-for information. Secondly, the conventional deadlock and blocking related processing has been modified. The transaction manager, on occurrence of a blocking informs all the connected data mangers, about the wait-for information. Under the proposed asynchronous organization a number of conflicts can be detected without exchange of additional messages or delays.

In order to preserve serializability, the conventional systems depend on 2 phase locking (2PL) protocol [2]. Whereas the 2PL protocol enforces a two phase

S. Sahni et al. (Eds.) HiPC 2002, LNCS 2552, pp. 388–398, 2002.

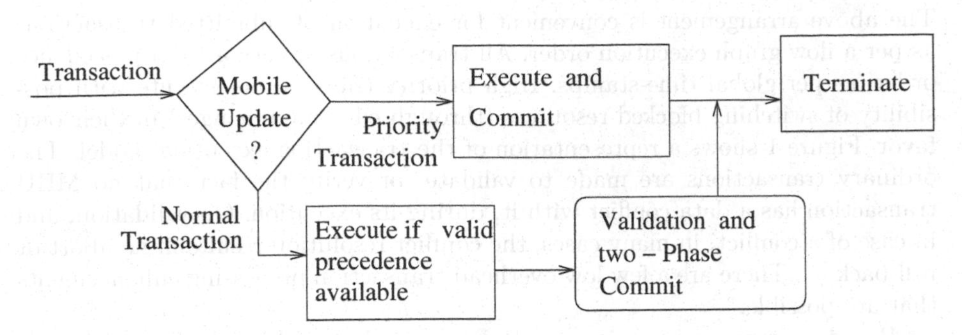

Fig. 1. Execution of MH update transactions in isolation

disciple, the criteria of serializability does not dictate the order in which a collection of conflicting transactions need to execute [2]. This option provides an opportunity to make a modified system that follows 2 PL protocol at the TM's level, but can be flexible at the data manager's (DM's) level. It can permit a interference free and 'non-blocked' execution for MH transactions. This change necessitates maintaining locking information in the form of site level graphs. Although this is the first effort (to the best of our knowledge) to use the technique for mobile databases, many graph based techniques have been studied earlier by [6, 12]. It is proposed to execute a mobile host update (MHU) transaction in a special priority fashion. It may need to wait for another low-priority transaction, only if, that transaction has completed and local DM is participating in the 2nd phase of commit processing.

The introduction of these possibilities integrates well with the existing transaction execution models. Earlier efforts at separating read-only transactions and update transactions exist [2]. The present study is an effort that proposes an implementation strategy for isolation of Serializable MHU transactions, for execution, that is free from interference by other transactions (Figure 1). Similar efforts have been made to isolate mobile transaction activities [11].

The data flow graph based approaches have been studied by Eich [6], and Katoh [7] for the centralized environment. In a distributed system, scheduling can be carried out by constructing partial graphs at different sites for individual transactions [14]. These graphs are used to substitute the lock tables managed by the (local) site data managers (DMs). This change permits increased interaction between local managers (TMs and DMs) for removal of blocking on account of distributed deadlocks. Such a data flow graph is referred to as local access graph (LAG) or data access graph (DAG). Based on the approach, a lock request (LR_i) of transaction (T_i) is sent to the concerned sites. At the site, the LAG is formed. A LAG of T_i at site S_k contains the edges ($< T_j, T_i >$) of all transactions T_j such that, both T_i and T_j have a conflict on some data items resident at S_k .

The possibility of transaction blocking is removed by interchanging local wait-for edges (transactions) that seek incorrect wait for precedences. Many of these interchanges can be performed locally, without any effect on serializability [13].

The above arrangement is convenient for execution of submitted transactions as per a flow graph execution order. All transactions are accorded a precedence order, as per global time-stamps. High priority transactions execute with possibility of switching blocked resources (through edge interchanges) in their own favor. Figure 1 shows a representation of the transaction execution model. The ordinary transactions are made to validate, or verify the fact that no MHU transaction has a data conflict with it, during its execution. On validation, and in case of a conflict, in many cases, the conflict resolution results in no abort or roll-back [2]. There are a few low overhead transaction processing enhancements that are possible.

Based on some assumptions a predetermined precedence order can be defined. Thus, a transaction Ti can precede Tj, and as both transactions soon detect the deadlock on their own, a decision to eliminate the deadlock can also be done locally (with no exchange of inter-site messages). In addition, access order can be sorted for any other item (say) z, that may be sought by both the transactions, while (say) Tj is still waiting.

Given a mobile computing system (MCS), the transactions need to execute as per the allotted priorities. Assuming that T1 is a MH update (MHU) transaction with a higher priority, its request will be processed by DM2 by revoking grant of T2, unless it is participating in a 2 phase commit for T2.

Thus, in the paper, we present a modified generalized framework for exchange of precedence among n transactions executing at m sites with few MHU transactions executing with instant precedence order (priorities) [13]. Research studies on MHU transactions is an active area of research [4, 5, 8, 10, 16].

The body of the paper is organized as follows. In the next Section, we define a system model, along with some definitions. In section 3, the background of asynchronous processing operations is examined. Section 4 considers events in the distributed system for sake of ordering as per the concurrency control criterion. In section 5, an algorithm has been presented. Section 6 considers performance study based on the message exchange technique. The last section consists of summary and conclusions.

2 The System Model

Based on the models of 2 phase locking and real-time computational environment with no slack time [3, 17, 9], a set of assumptions for executing transactions are organized. It is assumed that a 2 phase locking discipline is followed and the transaction execution is based on the criteria of serializability.

Ideally, the MHU transactions should be able to do the following :

- over ride conventional delays;
- integrate with existing modes of transaction executions. The two phases of the 2 PL protocol must execute with no blocking;
- a critical transaction may proceed without interference from other transactions.

- execute and commit, i.e., if phase 1 is completed, then phase 2 needs to complete.

In the following section, a scheme to execute transactions as per a precedence order is described.

2.1 Definitions: Mobile Database System

Mobile database system (MDS) consists of a set of data items (say set 'D'). The MDS is assumed to be based on a collection of (fixed) servers that are occasionally accessed by mobile hosts. Our assumptions are similar to earlier examples [15, 1]. Each site supports a transaction manager (TM) and a data manager (DM). The TM supervises the execution of the transactions. The DMs manage individual databases. Each mobile host supports a TM, that interacts with a fixed MSS. That performs other TM functions of interaction with other DMs. The network is assumed to detect failures, as and when these occur. When a site fails, it simply stops running and other sites detect this fact.

2.2 The Transaction Model

We define a transaction as a set of atomic operations on data items. The system contains a mixture of instant priority real-time transactions (MHU, or MH reads) and ordinary transactions. We assume that the ordinary transactions can be aborted, in case of a data conflict with the real-time transactions.

3 Background

On a centralized system, a deadlock is detected by the combined effort of a transaction manager (TM) and a data manager (DM). The DM refers to its lock table in order to determine the existence of cycles in a transaction-wait-for (TWF) graph. In the case of a distributed system, the possibility of a similar cooperative activity can be determined by considering the following four example cases.

Consider that the earlier example of a distributed system with two sites.

- Case 1: Both data items x, and y may reside at either one of the two sites. The deadlock detection can be achieved by a local activity as in the case of a centralized system.
- Case 2: Both the transactions T1, and T2 may occur at the same site. The deadlock detection can be achieved by a local exchange of messages.
- Case 3: Data items and transactions are located at different sites. Inter-site exchange of messages and the locally available information can be combined to detect the deadlock as shown, without incurring additional inter-site message overheads.
- Case 4: Data items and transactions are located at different sites. Messages and the locally available information is not sufficient. Deadlock is detection is achieved by additional exchange of messages.

We examine the permutations in which the data or transactions can be at the same site to support conflict detection by local computations. The probability of no message exchange cases in case of a 2 level deadlocks is given by P_n, given n sites.

$$P_n = \frac{n(n-1)^2 + (n-1)(n-2)(n-3)}{2n^3} \quad (1)$$

3.1 Asynchronous Operations

For the early detection of a deadlock, asynchronous operations are supported as advanced wait-for information messages, sent by waiting TM. Each transaction has an associated set of DMs that have granted or considering lock grant requests from the site. It is proposed that as soon as a transaction faces a wait-for condition, it informs the other DM sites about its wait-for condition by virtue of a message. In contrast, conventional techniques wait for delays to occur and then start a deadlock detection process. We have studied all possibilities of message exchanges that are possible. It is possible to detect the global occurrence of a deadlock with no additional messages. The following asynchronous exchanges of messages are studied in this report.

1. Odd-Messages : A blocked transaction, checks if the waiting order is proper (that is, it is waiting for an older transaction). Otherwise, it informs the site, holding the data item about its wait-for condition.
2. Even-Messages : A blocked transaction, checks if the waiting order is proper (it is waiting for an older transaction). It informs the site, holding the data item about its wait-for condition.
3. All-Messages : A blocked transaction, informs the site, holding the data item about its wait-for condition.

4 Ordering of Transactions in a Distributed System

In the proposed approach, the transactions are ordered by constructing local access graphs (LAGs) for access requests. In this section, we define a LAG.

Definition 1 : A directed graph G consists of a set of vertices $V = V_1, V_2, \dots$, a set of edges $E = E_1, E_2, \dots$, and a mapping function Ψ that maps every edge on to some ordered pair of vertices $< V_i, V_j >$. A pair is ordered, if $< V_i, V_j >$ is different from $< V_j, V_i >$. A vertex is represented by a point, and an edge is represented by a line segment between V_i and V_j with an arrow directed from V_i to V_j.

Insertion of an edge $< V_i, V_j >$ into the graph G =(V,E) results in graph G' =(V',E'), where $V' = V \cup \{V_i, V_j\}$ and $E' = E \cup \{< V_i, V_j >\}$. The union of two graphs $G_1 = (V_1, E_1)$ and $G_2 = (V_2, E_2)$ is another graph G_3 (written as $G_3 = G_1 \cup G_2$), whose vertex set is $V_3 = V_1 \cup V_2$ and the edge set is $E_3 = E_1 \cup E_3$.

Let, there be a partial ordering relation $\ll_T$ defined over T (the collection of executing transactions $T_1, \ldots, T_n$), to indicate a precedence order among transactions, based on criteria of serializability.

Definition 2 : An access graph of T_i (AG_i) is a graph $AG_i(V, E)$, where $V \subseteq T$, and $E = \{< T_j, T_i > | LV_j \cap LV_i \neq \phi$ and $T_j \ll_T T_i\}$.

Example 1: Consider the transactions T_{aRT} a real-time transaction (MHU transaction), and T_b, T_c, T_d and T_e as shown below. Let X,Y,Z be data items. Also,

- **(real-time)** $T_{aRT} = r_a(X)r_a(Y)w_a(X)w_a(Y)$.
- $T_b = r_b(X)r_b(Y)r_b(Z)w_b(X)w_b(Y)w_b(Z)$.
- $T_c = r_c(Z)w_c(Z)$.
- $T_d = r_d(X)r_d(Z)w_d(X)w_d(Z)$.
- $T_e = r_e(Y)r_e(Z)w_e(Y)w_e(Z)$.

Consider a situation, where X,Y and Z are located at one site. The execution of above transactions' operations can follow any one of the sequences as per the criteria of serializability [2]. For each execution (equivalent to a serial execution), the AGs of transactions are different. If we consider the arrival pattern of transactions in the order T_b, T_c, T_d and T_e, T_{aRT}, then, the corresponding AGs of above transactions are shown in Figure 2. Noting that the T_{aRT} is a real-time transaction. In this, $T_i \rightarrow^x T_j$ indicates, T_j is waiting for data item X which will be released after completion of T_i.

Definition 3 : A local access graph (LAG) of T_i at S_k, is a graph $LAG_{ik}(V, E)$, where, $V \subseteq T$, and $E = \{< T_j, T_i > | LV_{jk} \cap LV_{ik} \neq \phi$ and $T_j \ll_T T_i\}$. In this expression, T_j has previously visited site S_k, and LV_{ik} denotes the part of LV_i, resident at S_k.

When a locking request LR_i (for T_i) is sent to S_j, a LAG_{ij} is constructed at S_j (Figure 3).

In example 1, the equivalent serial order obtained in this way is indicated by: $T_{aRT} \ll_T T_b \ll_T T_c \ll_T T_d \ll_T T_e$.

As shown above, the AG_i and LAG_{ij} differ from each other. An AG_i includes edges ($< T_j, T_i >$), of all active and conflicting T_j. But, constructing such a graph in a distributed system introduces significant inter-site message overheads. In order to eliminate the overhead, LAG_{ij} are used. To summarize, based on the definitions of AG, and LAG:

AG_a	AG_b	AG_c	AG_d	AG_e
$.T_{aRT}$	$T_{aRT} \longrightarrow T_b(x,y)$	$T_b \longrightarrow T_c(z)$	$T_{aRT} \longrightarrow T_d(x)$	$T_{aRT} \longrightarrow T_e(y)$
			$T_b \longrightarrow T_d(x,z)$	$T_b \longrightarrow T_e(y,z)$
			$T_c \longrightarrow T_d(z)$	$T_c \longrightarrow T_e(z)$
				$T_d \longrightarrow T_e(z)$

Fig. 2. Access Graphs (AGs) of transactions

	$LAG_{a,site}$	$LAG_{b,site}$	$LAG_{c,site}$	$LAG_{d,site}$	$LAG_{e,site}$
Site S_1	$.T_{aRT}$	$T_{aRT} \longrightarrow T_b(x)$		$T_{aRT} \longrightarrow T_d(x)$ $T_b \longrightarrow T_d(x)$	
Site S_2	$.T_{aRT}$	$T_{aRT} \longrightarrow T_b(y)$			$T_{aRT} \longrightarrow T_e(y)$ $T_b \longrightarrow T_e(y)$
Site S_3		$.T_b$	$T_b \longrightarrow T_c(z)$	$T_b \longrightarrow T_d(z)$ $T_c \longrightarrow T_d(z)$	$T_b \longrightarrow T_e(z)$ $T_c \longrightarrow T_e(z)$ $T_d \longrightarrow T_e(z)$

Fig. 3. Local Access Graphs (LAGs) of active transactions

Observation 1 : Let LV_i be stored at sites $S_1, \ldots, S_m$. And, LAG_{ij} be the LAG of T_i at S_j. Then,
$AG_i \;=\; \cup_{j=1}^{m} LAG_{ij}$

5 An Algorithm to Construct LAG

5.1 Informal Description of the Algorithm

In this algorithm, whenever a transaction needs to access data items at S_i, its LR_{ik} are prepared and are sent to each concerned site S_k. The LAG_{ik} is updated at these sites. At any site S_k, if LAG_{ik} contains odd edge $< T_j, T_i >$, then it is an indication of possible blocking or delay. It is proposed that the MHU transactions exchange the precedence with the lock holding transaction, to generate a normal precedence (even edge). In all cases, the odd edge is nullified by exchange of precedence to revoke the grant. This is called the confirmation of the edge. Thus, a time-critical transaction is permitted to execute a revoke grant (if necessary) for the conflicting item, in order to cancel an odd edge.

5.2 Deadlocks among Mobile Host Updates

On account of a higher priority, the MHU avoid all interference from executing ordinary transactions. In all cases, an odd edge is formed by a waiting MHU transaction. It results in edge confirmation.

Conflicts among MHU transaction that execute in parallel, are prevented by sending asynchronous wait-for messages. Advance despatch of the wait-for information enables at least one of the DMs to detect the deadlock as soon as it occurs. In case of a conflict, one of the transaction needs to wait for completion of the other transaction.

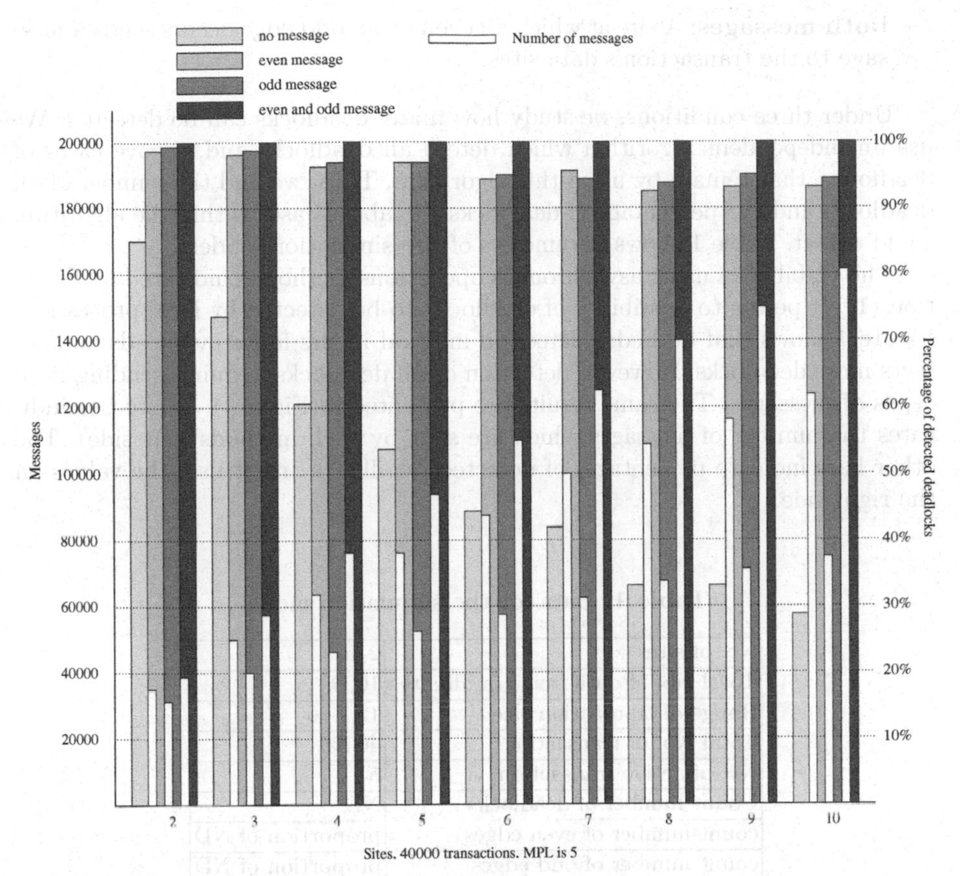

Fig. 4. Execution of MH update transactions in isolation

6 Performance Considerations

6.1 Simulation Model

A comparison of performance the proposed approach with respect to conventional techniques has been presented in [13]. Intuitively, many transactions gain quick response on account of local processing. Many advanced steps carried out on account of asynchronous computing prevent delays for mobile transactions. Our simulation study in this report considers the performance of the following types of asynchronous messages (separately).

- No messages are sent, or
- **even message**: A site in which an even edge occurs sends a message to the transaction's data sites (e.g. when T2 → T1, the site (TM of T2) sends messages to data sites of T2),
- **odd message**: A site in which an odd edge occurs sends a message to the transaction's data sites,

– **both messages**: A site at which an even or an odd edge occurs sends a message to the transaction's data sites.

Under three conditions, we study how many deadlocks can be detected. We use an independent algorithm which detect all deadlocks and remove cases of deadlocks, that remain by using this algorithm. Thus, we find the number of all deadlocks and the percentage of deadlocks the above cases within the algorithm could detect. Table 1 shows parameters of the simulation model.

The viability of using asynchronous operations method is indicated by Equation (1). It points to possibility of deadlocks to be detected by local processing. Figure 4 shows that odd edge detection method incurs fewer overheads and detects more deadlocks. However, detection of all deadlocks, requires sending both types of messages. The same results are presented in Figure 4. White bar indicates the number of messages which are sent by each methods (left side). The other bars indicate percentages of detected deadlocks and it uses the values on the right side.

Table 1. Data for the simulation model

No. of sites	2-10
Total No. of data items in all sites	1000
Range of transaction size	12 - 20
Total No. of transaction	40000
No. of active transaction	5
Count number of deadlocks	ND
count number of even edges	proportion of ND
count number of odd edges	proportion of ND
count number of all edges	proportion of ND
count number of messages	NM

7 Summary and Conclusions

Transaction rejects and delays are two of the main problems that concern MHU transaction processing activity. In this study, a procedure has been identified that shows a possibility of execution of critical transactions under serializability conditions. As a result, the MH update transactions do not undergo blocking and roll-back due to ordinary transactions. The technique avoids deadlocks among MHU transactions by virtue of asynchronous processing. The level of concurrency that is achieved, is higher as a result of removal of excessive blocking.

References

[1] Acharya S, Franklin, M. and Zdonik, S., " Disseminating Updates on Broadcast Disks. In Proc. of 1996 VLDB conf. 391

[2] P. A.Bernstein, V.hadzilacos and N.Goodman, Concurrency control and recovery in database systems, Addison-Wesley,1987. 388, 389, 390, 393

[3] Cheng S., J. A. Stankovic, and K. Ramamritham, "Scheduling Algorithm for Hard Real Time Systems - A Brief Survey", Tutorial : Hard Real Time Systems, Computer Society Press, 1988. 390

[4] P. K. Chrysanthis, "Transaction Processing in Mobile Computing Environment", in Proceedings of *IEEE workshop on Advances in Parallel and Distributed Systems*, 1993, pp. 77-82. 390

[5] M. H. Eich and A. Helal, "A Mobile Transaction Model That Captures Both Data and Movement Behaviour", *ACM/Baltzer Journal on Special Topics on Mobile Networks and Applications*, vol. 2, no. 2, pp. 149-162. 390

[6] M. H.Eich and S. H. Garad, The performance of flow graph locking, IEEE Transactions on Software Engineering, vol.16, no.4, pp.477-483, April 1990. 389

[7] N.Katoh, T .Ibaraki and T. Kameda, Cautious transaction schedulers with admission control, ACM Transactions on Database Systems, vol.10, no.2, pp.205-229, June 1985. 389

[8] J. Kistler and M. Satyanarayanan, "Disconnected Operations in Coda File System", *ACM Transactions on Computer Systems*, vol. 10, No. 1, pp. 3-25, 1992. 390

[9] Korth H. F., E. Levy, and A. Silberschatz, "Compensating Transactions: a New Recovery Paradigm," in Proc. 16th Intl. Conf. Very Large Databases, Brisbane, Australia, 1990, pp. 95-106. 390

[10] E. Pitoura and B. Bhargava, "Maintaining Consistency of Data in Mobile Computing Environments", in Proceedings of 15th International Conference on Distributed Systems, June 1995, extended version in *IEEE Transactions on Knowledge and Data Engineering*, 2001. 390

[11] Qi Lu, and M. Satyanarayanan, "Improving Data Consistency in Mobile Computing Using Isolation Only Transactions", Proceedings of 5th Workshop on Hot Topics in Operating Systems, May 4 - 5, 1995. 389

[12] P. K. Reddy, and S. Bhalla, A Non-Blocking Transaction Data Flow Graph Based Protocol for Replicated Databases, IEEE Transactions on Knowledge and Data Engineering, vol. 7, No. 5, pp. 829-834, October 1995. 389

[13] P. K. Reddy, and S. Bhalla, "Asynchronous Operations in Distributed Concurrency Control", to appear in "IEEE Transactions on KDE", accepted for publication February 2002. 389, 390, 395

[14] S. C.Shyu, V. O. K.Li, and C. P.Weng, An abortion free distributed deadlock detection/resolution algorithm, Proc. IEEE 10th International Conference on Distributed Computing Systems, pp.1-8, June 1990. 389

[15] J. Shanmugasundaram, A. NitharKashyap, J. Padhye, R. SivaShankaran, M. Xiong, and K. Ramamritham, "Transaction Processing in Broadcast Disk Environments, in Adv. Transaction Models and Architectures, Kluwer Academic, eds. S. Jajodia, and Larry Kerschberg. 391

[16] G. D. Walbborn and P. K. Chrysanthis,"Supporting Semantics Based Transaction Processing in Mobile Database Applications", in Proceedings of 14th IEEE Symposium on Reliable Distributed Systems, 1995, pp. 31-40. 390

[17] Zhao W., K. Ramamritham, and J. A. Stankovic, "Scheduling Tasks with Resource Requirements in Hard Real Time Systems", IEEE Transactions on Software Engineering, Vol.SE-13, No.5, May 1987. 390

Active File Systems for Data Mining and Multimedia

S. H. Srinivasan and Pranay Singh

Satyam Computer Services Ltd
Applied Research Group
14 Langford Avenue, Bangalore, INDIA – 560 025
{SH_Srinivasan, Pranay_Singh}@Satyam.com

Abstract. Data mining and multimedia applications require huge amounts of storage. These applications are also compute-intensive. Active disks make use of the computational power available in the disk to reduce storage traffic. Many of the file system proposals for active disks work at the block level. In this paper we argue for the necessity of filtering at application level. We propose two file systems for active disks: active file system (ACFS) which binds files and filters at the file system level and active network file system (ANFS) which extends ACFS over networks. These file systems preserve the familiar Unix file system semantics to a large extent. We present an implementation of the file systems which makes minimal changes to the existing file system code in Linux.

1 Introduction

Storage has been one of the less popular areas of computer science, but advances in storage technology have been spectacular in the last few years. This is reflected in the sizes of current storage systems. In addition to capacity, there has been advances in storage systems architectures also. Two important developments in storage architectures are: active disks and storage networks. These two developments are motivated by different reasons.

The availability of cheap memory and processing power has opened a new paradigm called active disks [1]. Active disks are storage devices with some capability for code execution. Using active disks it is possible to perform some operations on the disk itself. The server can use the results of these operations or perform additional operations on the results. Thus part of compute-intensive operations can be moved to storage devices.

With the increase in network speeds and reliability, storage can be connected to the server via a network instead of a bus. This move is indeed required since local storage is fraught with problems of scalability, sharing, and management. The new architectures are: Storage Area Networks (SAN) and Network Attached Storage (NAS). While SAN works at the SCSI level, NAS abstracts UNIX file system calls. There has been extensive work in this area [2] [3] [4].

S. Sahni et al. (Eds.) HiPC 2002, LNCS 2552, pp. 398–407, 2002.

Two applications which have large storage and computing requirements are data mining and multimedia. Data mining attempts identify patterns and establish relationships implicit in datasets. Datamining algorithms – classification, correlation/association, forecasting, etc. – are compute-intensive and are amenable to parallel implementation. Because of dataset sizes, any architecture which permits sharing will speed up datamining operations. Since SANs permit sharing, storage bottlenecks can be removed by deploying SANs. The compute-intensive nature of data mining can eased by performing part of the computations on storage devices themselves. Multimedia retrieval applications also have similar requirements.

Applications seldom concern themselves with storage devices or blocks. Applications use file system calls which abstract details of file system implementation. Hence we require a file system abstraction for SANs containing active disks. There is some research work in this direction [4] [5] [6] [7]. The previous works illustrate modifying the user and the client side systems [6] [7]. These file systems work at the block level. But in some applications, the filters need application-level information [7] [8] [9] [10]. The following example uses indexing to illustrate this.

Consider nearest neighbor search implemented on an active disks. If this is implemented at the block level, the filter has to

1. Interpret the data blocks to get size (height, width, etc.) of the table. Note that this interpretation is application dependent. (Some applications may use integers to store this information where as other may use long or ultra long integers.)
2. Calculate the distance between the query vector and each of the vectors in the dataset. This also requires information at the application level since the elements of vectors can be binary, integer, float, etc.
3. Return the row with minimum distance.

It can be seen that the filter requires information at the file system and application level. If the filters work at block-level, getting this information securely requires other mechanisms. In this paper, we propose running the filter infrastructure at the file system level. This is implemented in two parts. The file system (called ACFS – Active File System) keeps track of the binding between a filter and a file. The filter infrastructure (called filter monitor) handles registering and deployment of filters. We extend ACFS for use over networks. The resulting file system is called Active Network File System (ANFS).

The paper is organized as follows. Section 2 discusses various aspects of active disks and proposals for their file systems. Section 3 discusses the details of active disk file system. The design and implementation issues are discussed in sections 4 and 5. ANFS is outlined in section 6. We close the paper with conclusions.

2 Active Disks

Disks contain controllers which perform a variety of functions like head scheduling. Because of advances in VLSI, additional computing power is becoming available at little or no cost. Active disks [1] make this computing power available to applications: the disk performs application-dependent filtering before returning data. In addition to reducing load on servers, such "preprocessing" reduces the storage traffic.

Storage devices exist one level below file systems. While file systems contain metadata like storage layout, the code executing on disks has no information at the file system level: only block level information is available. This limits the type of computations that can be performed by active disks. The examples considered in [1] are: nearest neighbor search, histogram computation, image edge detection, and image registration.

2.1 File Systems

The active disk file system [11] is a distributed file system using CORBA. The design is object oriented. Servers control access and lookup of objects.

In [7], a *sequential* stream-based programming model is presented for active disks. Applications can deploy *disklets* which execute in a sandbox. In the scenarios presented in [7], disklets run to completion. Hence only one application can deploy disklets at any time.

The Multi-View Storage System [5] works at the block level. The files having the filtered content are called as virtual files. They are made of virtual blocks. This approach uses the auxiliary table for allocating the data blocks to the virtual file. This table also maintains the filter-file binding information or the meta data information. The MVSS implementation provides sequential support for the filter operations, lacking the indexed approach. The approach is compatible with Unix file system.

In this paper we propose a file system which supports application level filtering. The file system is compatible with Unix file system to a large extent. The file system provides the traditional stream interface with random access and record access.

3 ACFS API

ACFS has two aspects: API and implementation. The API corresponds to the system call interface and the implementation to inodes and disk blocks. The important Unix file system calls are: *open, close, read, write, lseek*, etc. The following calls are provided in ACFS. Since the calls are very similar to Unix system calls (section 4.1), the actual arguments are not shown.

1. *open*: The parameters to this function include a file and a filter. Hence, in addition to usual parameters, the filters associated with the read or write request are passed as parameters. This call creates a binding between the file and the filter.
2. *read*: This routine reads the filtered data output by the filter associated with the read operation.

3. *write*: The data supplied by the application is input to the filter and the filter output is written to the output file.
4. *stat*: This function returns the attributes of a file. In ACFS, it is possible to associate filters with files. A directory, for example, can be associated with an encryption filter. All files in that directory are encrypted and stored. The function is modified to return associated filters in addition to usual file system attributes.
5. *mkdir*: The directory is created and the filters associated with a directory are added along with the other attributes supported in the Unix implementation. The filters are associated with read and write operations on files created under the directory.

Apart from the above operations the following new procedures are available.

1. *aseek(fileDescriptor, cue, buffer)*: The aim of this function is to provide "associative seek". The *cue* string is handed to the filter. The filter interprets the cue in an application dependent manner. The filter returns result in *buffer*. Associative memories can be implemented using this call.
2. *flookup(filtSpec)*: This looks up the presence of filters having the characteristics specified by "filtSpec". The characteristic includes but is not limited to filter names.
3. *close(fileDescriptor)*: The close procedure is required to free the filter resources at the server side. See section 4.4.

4 ACFS Design Considerations

We now address some issues which need to be considered when implementing ACFS.

4.1 Composition Format

The expected format of a read system call for active file open is

open (filename, access flag, mode, filter)

In many UNIX implementations, all file systems are subsumed by VFS and hence this system call is routed through VFS. We want to maintain compatibility to this framework as much as possible. To achieve compatibility, we propose the following new format for the ACFS calls called the *composition format* or *composition string.*

The operation of a filter on a file is represented as

$acfs://filter \circ fileName$

Instead of passing a plain file name and a filter name to *open* system call, we pass the composition string as shown below.

open("acfs://filter∘fileName", flag, mode)

Note that this format is compatible with the traditional *open* call. For file names in the composition string format, VFS first parses the composition string to filter and file name parts. The file name is then looked up in the usual manner.

For the last component, VFS associates the filter name to create a *minimal composition string* which is dispatched to the ACFS. The remaining processing is described in section 5.

In the composition format, pure filter names are represented as

acfs://filter∘

Filters can also be composed using the following syntax

acfs://filter$_2$∘ *filter*$_1$∘

to denote the application of "*filter*$_2$" to the output of "*filter*$_1$".

Finally any string without "*acfs://*" represents a pure file.

Note: We use the symbol ∘ here to denote function application and composition. In ANFS implementation, the symbol # is used. The filter names are forbidden to have this character.

4.2 Incremental and Random Accesses

Consider a filter performing a transcoding. We can permit access to transcoded data after the transcoding operation is complete. In this case, the application needs to wait/block till transcoding is over. For large files, the wait can be long. If we permit incremental reads, a fast application can overrun the filter.

The same issue arises in another context. The *open* call associates a filter with a file. The *read* and *write* system calls have the following arguments:

fileDescriptor: File descriptor created by the ACFS *open*.
offset: The offset at which the operation is to be performed
count: The number of bytes to be read/written

Applications can perform *lseek* to read from arbitrary offsets. For example, an application can open a file using a filter and immediately start reading from a large offset. The *read* call has to block till data is created at that offset (synchronous read) or return a "data not available yet" code (asynchronous read).

We implement this functionality by adding a new pid field to Unix inode structure. This field contains the pid of the process writing to the file. The field is set to zero when the writing process terminates. For any read request, if the offset is more than the current file length and if the pid field is nonzero, the request blocks.

4.3 Record Accesses

Unix file system provides the abstraction of a random access stream. For some applications, it is desirable to provide record-oriented access. The application performs read/write in terms of records with the file system preserving record boundaries. While this can be done by ACFS, we do not include this functionality.

On the other hand, record-oriented accesses are provided by *aseek*. The cue string argument of *aseek* is delivered to the filter and the data returned by the filter is delivered to the application. Thus *aseek* can be used to interact with the filter. The filter can return records. Applications using *aseek* can also use *read/* or *write* system calls. These calls interact with the ACFS file. Thus both stream and record access are possible.

4.4 Garbage Collection

Temporary files need to be removed when applications no longer need them. All ACFS files are temporary files. Once the filter process completes execution, the binding is no longer needed. The files themselves are not needed once the corresponding applications terminate. This is effected through the procedure *close*. Applications can close the files they no longer need. This terminates the filter process if it is running and removes the temporary (ACFS) file. If applications crash without a issuing *close*, a timeout mechanism is used to terminate filters and remove temporary files.

5 ACFS Implementation

In this section we consider implementing ACFS in Linux. ACFS is implemented on top of VFS. (See [12] for an overview of VFS.)

Consider the *open* call. Because of the composition format, the system call interface also remains unchanged. VFS recognizes file names starting with "*acfs://*". It separates the filter and file name parts. After several lookups, VFS calls ACFS with the minimal composition string. ACFS binds the file and the associated process as shown in figure 1. The binding itself is the result of the

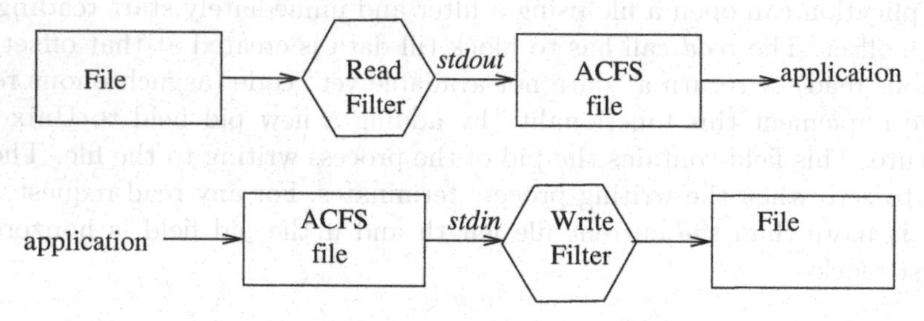

Fig. 1. ACFS binding. ACFS is used to create a binding between a file and a process. The process is the source of data for the file. The ACFS inode carries a field for the pid of the associated process. When the associated process completes execution, the pid field is set to zero. Reads on the ACFS file can be blocking. When the application issues a read request *open("acfs://filterR∘file", "r")*, the filter monitor creates the ACFS file and redirects the *stdout* of *filterR* to the ACFS file. The file descriptor returned to the application corresponds to the ACFS file. For the write request, *open("acfs://filterW∘file", "w")*, the application writes to the ACFS file which is read by the filter. The output of the filter is sent to the actual output file

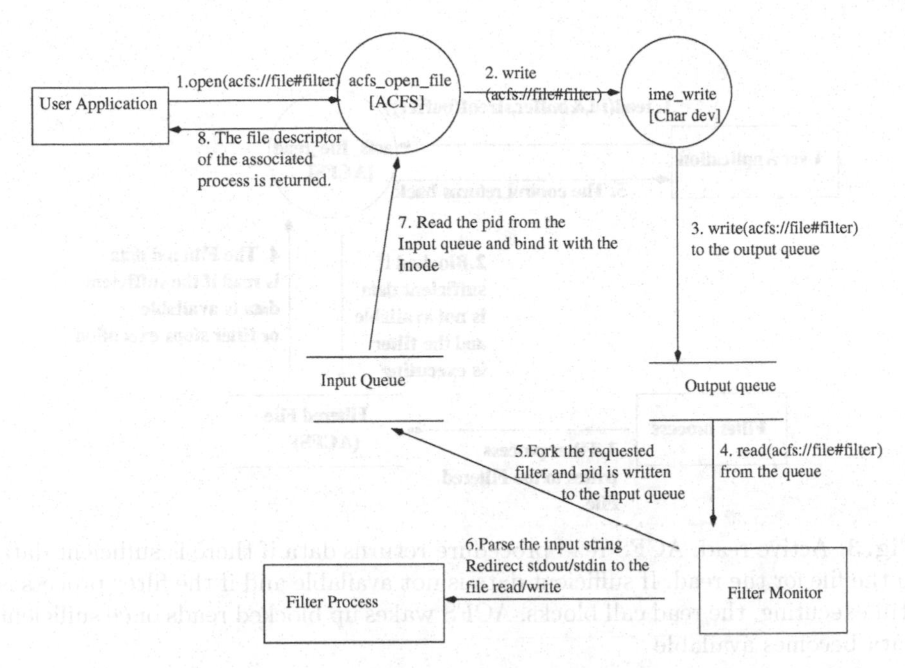

Fig. 2. ACFS *open*. User space operations are shown in rectangles while kernel space operations are shown in circles. The character device is currently used as a kernel-user space communication mechanism. This can be integrated into ACFS itself. ACFS creates an inode for the output file. This inode has a field for pid of the process which writes to the file. For read requests, this id is that of filter process pid. For write requests, this is the pid of the application. The filter monitor is responsible for running the filter. The filter monitor returns the pid of the filter process to ACFS (step 7) which sets the pid field of the inode. The filter monitor is also responsible for redirecting the standard output of the filter process to the ACFS file for read requests (step 6). For write requests, the ACFS file is redirected to standard input of the filter process

interplay between several user and kernel level operations which are detailed in figure 2.

The application gets a file descriptor after as a result of active *open* call. Reads/writes are now identical to conventional system calls. The processing is transparent to the application. Figure 3 shows the processing for active *read*.

As can be seen the above calls preserve the random-access stream nature of file access. In some applications, applications need to interact with the filter for record-oriented random accesses. An example of this is dictionary lookup. Consider a database of user preferences. The application may want the preferences of a particular user. The ACFS call *aseek* provides this functionality. The user supplied argument of *aseek* is delivered to the filter process and the filter process

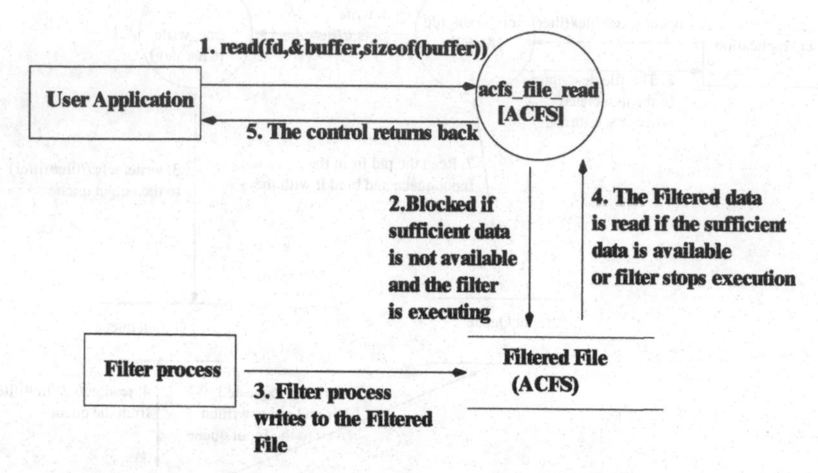

Fig. 3. Active read. ACFS read procedure returns data if there is sufficient data in the file for the read. If sufficient data is not available and if the filter process is still executing, the read call blocks. ACFS wakes up blocked reads once sufficient data becomes available

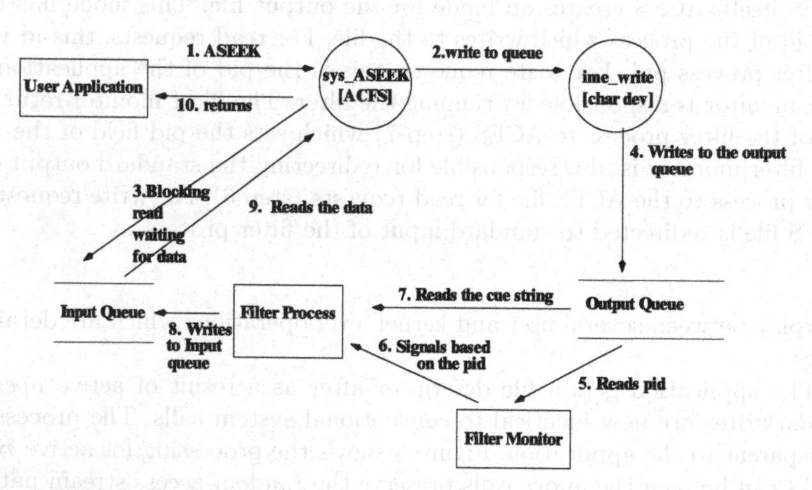

Fig. 4. ACFS *aseek*. *aseek* is used to implement record-oriented associative accesses. The application issues an aseek with a cue string. This information is read by the filter monitor which passes it to the filter through an *ioctl* call. The filter writes the output to a queue using the new system call *awrite*. ACFS returns the result to the application

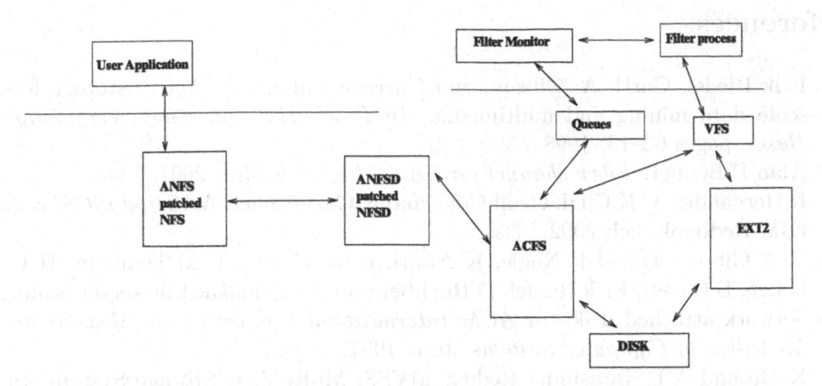

Fig. 5. ANFS architecture. ANFS implementation requires patches to NFS client and server codes. The server side requires ACFS. Hence client is minimally affected. NFS is based on the idea "servers are dumb and clients are smart". For ANFS the opposite is true

output is returned to the application as a result of *aseek*. The application blocks till the result is returned. Figure 4 shows the details of this operation.

6 ANFS

NFS provides file system access over networks. ANFS provides similar abstraction for ACFS. Because of the use of composition string, both ANFS client and server are minimally affected. Bulk of the processing is performed by ACFS and filter monitor on the server side. The client and server still communicate using file handles. This also preserves the spirit of NFS.

The overall scheme is shown in figure 5. ANFS API corresponds very closely to ACFS API. The important functions are: ANFSPROC_GETATTR, ANFSPROC_READ, ANFSPROC_WRITE, ANFSPROC_CREATE, ANFSPROC_MKDIR, ANFSPROC_ASEEK, ANFSPROC_FLOOKUP, and ANFSPROC_CLOSE. These correspond closely to the system calls discussed in section 3.

7 Conclusion

Active file systems are required in case of large scale data mining and multimedia applications. ACFS and ANFS can be used with SANs and NASs containing active disks respectively. Our implementation of ACFS and ANFS makes minimal changes on VFS using the notion of composition strings. All state-dependent structures are maintained by the filter monitor. Efforts are underway to optimize the implementation and to provide asynchronous call and NFS version 4 semantics.

References

[1] Erik Riedel, Garth A. Gibson, and Christos Faloutsos. Active storage for large-scale data mining and multimedia. In *Proc. 24th Int. Conf. Very Large Data Bases*, pages 62–73, 1998. 398, 400

[2] Alan F Benner. *Fiber channel for SANs*. McGraw-Hill, 2001. 398

[3] R Hernandez, C K Chal, Geoff Cole, and K Carmichael. *NAS and iSCSI Solutions*. IBM Redbook, Feb 2002. 398

[4] G A Gibson, David F Nagle, K Amiri, F W Chang, E M Feinberg, H Gobioff, C Lee, B Ozceri, Erik Reidel, D Rochberg, and J Zelenka. File server scaling with network attached disk. In *ACM International Conference on Measurement and Modelling of Computer Systems*, June 1997. 398, 399

[5] X Ma and A L Narasimha Reddy. MVSS: Multi View Storage System. In *Proc. of ICDCS*, Apr 2001. 399, 400

[6] H Lim, V Kapoor, C Wighe, and David Du. Active disk file system: A distributed, scalable file system. In *Proc. of the Eighteenth IEEE Symposium on Mass Storage Systems*, pages 101–116, Apr 2001. 399

[7] A Acharya, M Uysal, and J Saltz. Active disks programming model, algorithms and evaluation. In *International Conference on Architectural Support for Programming Languages and Operating Systems*, Oct 1998. 399, 400

[8] S Berchtold, C Bohm, and H P Kriegel. Improving the query performance of high-dimensional index structures by bulk load operations. In *Proc. of the Int. Conf. on Extending Database Technology*, pages 216–230, Mar 1998. 399

[9] S Berchtold, C Bohm, B Braunmuller, D A Keim, and H P Kriegel. Fast parallel similarity search in multimedia databases. In *Proc ACM SIGMOD Int. Conf. on Management of Data*, pages 1–12, 1997. 399

[10] Rakesh Agrawal and Ramakrishnan Srikant. Fast algorithms for mining association rules. In *Proc. 20th Int. Conf. Very Large Data Bases*, 1994. 399

[11] Hyeran Lim, Vikram Kapoor, Chirag Wighe, and David H.-C Du. Active disk file system: A distributed scalable file system. In *IEEE Symposium on Mass Storage Systems*, 2001. 400

[12] Uresh Vahalia. *Unix internals: The new frontiers*. Prentice-Hall, 1996. 403

Simulating DNA Computing

Sanjeev Baskiyar*

Department of C.Sci. and Soft. Eng.
Auburn University, Auburn, AL 36849
baskiyar@eng.auburn.edu

Abstract. Although DNA (deoxy-ribo nucleic acid) can perform 10^{22} computations per second, it is time intensive and complex to set up input and output of data to and from a biological computer and to filter the final result. This paper, discusses how to simulate DNA computing on a digital computer to solve the Hamiltonian path problem using Adleman's model. The simulation serves as an educational tool to teach DNA computing without the elaborate bio-experiments. As an aside, it also digitally verifies Adleman's notion of DNA computing to solve the Hamiltonian path problem. Future work will involve a parallel implementation of the algorithm and investigation of the possibility of construction of simple regular *VLSI* structures to implement the basics of the model for fixed-sized problems.

Keywords: Parallel Computing, DNA, Hamiltonian path, Simulation, Educational tool.

1 Introduction

The idea of viewing living cells and molecular complexes as computing components dates back to the late 1950s, when Feynman [9] described "sub-microscopic computers." Recently, Adleman [1] showed that random highly parallel searches could be performed using DNA-manipulation techniques.

DNA is a nucleic acid that carries the genetic information in cells and is capable of self-replication. It consists of two long chains of nucleotides twisted into a double helix and joined by hydrogen bonds between the complementary bases Adenine (A) and Thymine (T) or Cytosine (C) and Guanine (G). The two linear chains of nucleotides must have complementary base sequences and opposite chemical polarities to form a partial or a full duplex.

Massively parallel computations can be performed by recombinant bio-operations on many DNA molecules simultaneously. Parallel molecular machines can be constructed by encoding a problem using DNA strands. The core of a DNA computer is a collection of selected DNA strands whose combinations result in solutions to problems. Technology is currently available to select and filter strands.

* Assistant Professor

S. Sahni et al. (Eds.) HiPC 2002, LNCS 2552, pp. 411-419, 2002.

Theoretically, DNA computing has the potential to outperform electronic computers. Adleman's model [2] shows that the number of operations per second ($1.2*10^{18}$) performed by a DNA computer is approximately *12* million times of that of the current fastest supercomputer. Also, DNA computers are energy efficient—a DNA computer [1] performs $2*10^{19}$ operations per joule, whereas the fastest supercomputer executes 10^9 operations per joule. Information can be stored in DNA molecules at a density of *1* bit/nm^3, whereas electronic storage density is approximately *1* bit/10^{12} nm^3 [5]. Thus DNA computing may offer excellent improvements in speed, energy consumption and information density.

However, the main obstacles to creating a practical DNA computer remain. Such obstacles are encountered in dealing with the complexity of the experimental setup and errors. DNA computing is efficient in solving large intractable problems such as the directed Hamiltonian path problem, the satisfiability problem (SAT) and for deciphering codes. However, for simple Boolean and arithmetic operations, electronic computers outperform DNA computers in speed. Thus problems, whose algorithms can be highly parallelized, may become the domain of DNA computers, whereas others, whose algorithms are inherently sequential, may remain the specialty of electronic computers.

The organization of this paper is as follows. In Section 2, we review previous work in DNA computing. In Section 3, we develop a digital computer simulation of Adelman's DNA computing technique to solve the Hamiltonian path problem. In Section 4, we conclude by discussing the results of the simulation and suggesting directions for future work.

2 Background

Research in DNA computing has incorporated both experimental and theoretical aspects. Theoretically, it has been shown that using DNA one can construct a Turing machine [12]. Also, mathematical models of the computing process have been developed and their in-vitro feasibility examined. On the other hand, experiments have been performed to solve various problems such as: the satisfiability problem [19], breaking the data encryption standard [7], expansions of symbolic determinants [18], matrix multiplication [22], graph connectivity and knapsack problems using dynamic programming [4], the road coloring problem [13], exascale computer algebra problems [26], and simple Horn clause computation [17]. Other studies include: Kaplan's [16] replication of Adleman's experiment, the partial progress of a Wisconsin team of computer scientists and biochemists in solving a *5*-variable instance of the *SAT* problem using a surface-based approach [21] and addition using horizontal chain-reaction [10].

Various aspects of the ability to implement DNA computing have been experimentally investigated. Notable studies are: the impact of encoding techniques on simplifying Adleman's experiment [8], the complications in using the Polymerase Chain Reaction (PCR) [15], the use of self-assembly of DNA [27], the experimental gap in the design and assembly of unusual DNA structures [25], joining and rotating data with molecules [3], evaluating simple Boolean formulas [11] and the use of ligation [14].

2.1 DNA Computing Models

The various models with complementary features illustrate the versatility of DNA as a computing device. Below we outline Adleman's, Lipton's, Sticker and PAM models and discuss their advantages and disadvantages.

2.1.1 Adleman's Model

Adleman [1] used the ability to encode information in a DNA sequence to solve an instance of the directed Hamiltonian path problem by performing simple bio-operations. Adleman's model has a formal mathematical proof. A V-vertex E-edge directed graph, $G(V,E)$ with designated vertices *vin* and *vout* is said to have a Hamiltonian path iff there exists a path that begins at *vin,* ends at *vout* and enters every vertex exactly once. The following non-deterministic algorithm [2] was used by Adleman to solve the Hamiltonian path problem in a molecular biology lab.

1. Generate random paths through the graph, $G = (V, E)$.
2. Delete paths that do not begin at *vin* and end at *vout.*
3. Delete paths that do not have exactly $|V|$ vertices.
4. Delete paths that visit any vertex more than once.
5. If any path remains, return *true*, otherwise *false*.

Fig. 1. Adleman's algorithm to solve the Hamiltonian path problem

A disadvantage of Adleman's model is that it involves extensive manipulation of test tubes with possibilities of errors.

2.1.2 Lipton's Model

Lipton ([19],[20]) extended Adleman's idea to solve the *SAT* problem. The *SAT* problem consists in finding values to n literals $x_1,\ldots,x_n$ such that $F = C_1 \quad C_2 \quad \ldots \quad C_m$ evaluates true, where C_i , $1 \quad i \quad m$, is a clause of the form $x_j \quad x_{j+1} \ldots \quad x_k$, $1 \quad j,k \quad n$. Lipton's solution uses the same initial set of DNA strands for every instance of the problem making it more general than Adleman's. The computation can be executed in a number of steps proportional to the number of literals in F.

2.1.3 Sticker Model

In the above models, the presence of a certain strand in the test tube sets a particular value to a particular bit. Thus, it is necessary to associate two DNA strands to a Boolean variable.

The Sticker model [24] avoids using enzymes, which are expensive and have a short life. It employs two types of DNA strands called memory strands and stickers. Memory strands are long strands that can represent any binary string. They are n $(=mk)$ base sequences which are a concatenation of k non-overlapping regions, where each region is m bases long and refers to a bit position of the encoded k-bit string. Stickers are m-base DNA strands that are the Watson-Crick complementary (A-T and C-G) of the regions of the memory strands to which they can bind. If a region is single stranded, it means that the corresponding bit value is 0. Instead if a sticker is bonded to the region, the bit value is 1. A partially bonded memory strand is called a complex. The information density is $(1/m)$ bits/base, which is comparable to other models. Such a representation implements Boolean operations efficiently. Heating the solution containing DNAs can reset all bits.

2.1.4 PAM Model

Reif's [23] Parallel Associative Memory (PAM) model extends Adleman's and Lipton's models to allow dynamic changes in memory. It uses a parallel associative match operation similar to the string join operation used to encode strings in the Sticker model. Reif formalized another model called the Recombinant DNA model to implement DNA computing at molecular level, as opposed to the system level PAM model.

2.2 Implementation

We chose to simulate Adleman's model as it is easy to implement and since other models are extensions of this model. Also, unlike other models, Adleman's model provides a clear algorithm for solving the Hamiltonian path problem. In Figure 1 we briefly outline the experimental implementation of the algorithm—a detailed explanation appears in [1].

To implement *Step 1* of the algorithm, each vertex of the graph was encoded as a random *20*-nucleotide strand or a *20*-letter sequence of DNA. Then, for each edge of the graph, a DNA sequence was created whose second half encoded the start vertex and the first half the end vertex. By using complements of the vertices as splints, DNA sequences corresponding to compatible edges were ligated, forming DNA molecules that encoded random paths through the graph.

To implement *Step 2*, the output of *Step 1* was amplified by PCR. Thus, only those molecules encoding paths that began with *vin* and ended in *vout* were amplified.

For implementing *Step 3*, a technique called *gel electrophoresis* was used, which separated DNA strands by length. The molecules were placed on top of a wet gel, to which an electric field was applied, drawing larger molecules slower than the smaller. Thus, after a certain time, the molecules striated according to size.

Step 4 was accomplished by iteratively using a process called *affinity purification*. This process permitted strands with a given subsequence that encoded a vertex v of the graph to be filtered out. After synthesizing strands complementary to v and attaching them to magnetic beads, the solution was passed over the beads. Strands containing v, which annealed to the complementary sequence, were retained whereas others filtered through.

In *Step 5*, the presence of a molecule encoding a Hamiltonian path was checked.

The run-time of a molecular algorithm is proportional to the number of operations on test tubes. An important complexity measure is the "solution volume size" or the maximum number of strings in the test tube. Adleman has speculated that molecular computing multiplicities with a solution volume of size 2^{70} might be possible [6].

3 Digital DNA Computer System

We simulate DNA computing on a single processor system to determine a Hamiltonian path using Adleman's algorithm, as explained in Figure 1. We call the simulation system Digital DNA Computing System (*DDCS*).

3.1 Software

The *DDCS* has three modules: *Graph Encoder*, *DNA Combiner* and *Path Filter* as shown in Figure 2. The *Graph Encoder* has two software modules: the *Parser/ DNA Packet Creator* and the *DNA Replicator*. The *Parser/ DNA Packet Creator* parses the input file and creates an array of DNA packets with each packet representing a node in the input graph. It outputs the *graph size,* V , to the *DNA Combiner* and the array of DNA packets to the *DNA Replicator*.

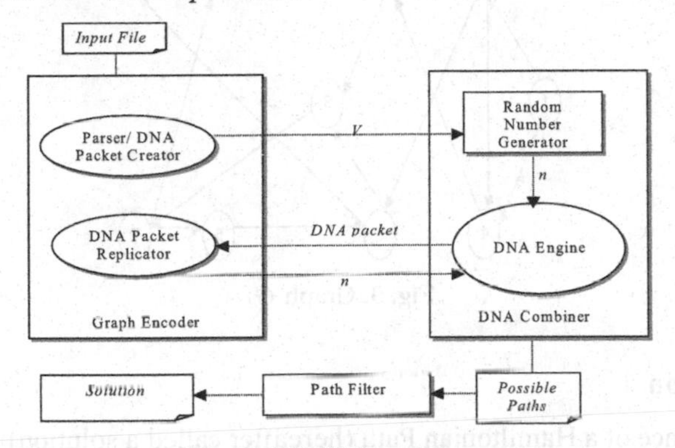

Fig. 2. *DDCS* flow diagram

The *DNA Combiner* consists of two software modules: the *Random Number Generator* and the *DNA Engine*. The *Random Number Generator* generates a random number, *n*, within the bounded set *(0.. V -1)*. When the *DNA Engine* requests a DNA packet having a particular node-id *n*, the *DNA Replicator* indexes into the DNA packet array, and sends a copy of the DNA packet with the requested node-id, *n*. The *DNA Engine* imitates DNA operations to combine DNA packets and generates the *Possible Paths* file some of which may be Hamiltonian paths. The *Path Filter* filters paths that have duplicate indices, whose lengths are not V and which do not start at *vin* and end in *vout*.

3.2 Sample Run

To illustrate how *DDCS* solves the Hamiltonian path problem, below we present an example run on the graph of Figure 3. The entry node, *vin,* is node *1* and the exit node, *vout*, is node *10*. By observation, a valid Hamiltonian path for the graph would be *1-2-4-3-5-7-8-6-9-10*. Let there be an empty path list P. The inset of a node, *i*, consists of the nodes with edges incident on *i*.

Step 1. Let the first randomly selected node be *6*. Let this node be put in path *P*.

Step 2. Let the second selected node be *8*. Since node *8* belongs to the *InSet* of path *P*, it is placed at the head of *P*. Let the *InSet* of *P* be replaced by the *InSet* of node *8*. The *OutSet* of *P* is unchanged.

Step 3. Let the next selected node be *7*. Since node *7* is in the *InSet* of *P*, the procedure adopted for node 8 in *Step 2* is repeated for node *7*.

Step 4. Let the next selected node be *2*. Since node *2* is neither in the *InSet* nor the *OutSet* of the path, it is discarded and another node is randomly chosen.

The above process continues until the number of nodes in the linked list becomes $|V| = 10$. However, it is possible that the path length may never reach $|V|$. If the size of the path does not increase after a certain number of trials, it is discarded assuming that the path is not the answer.

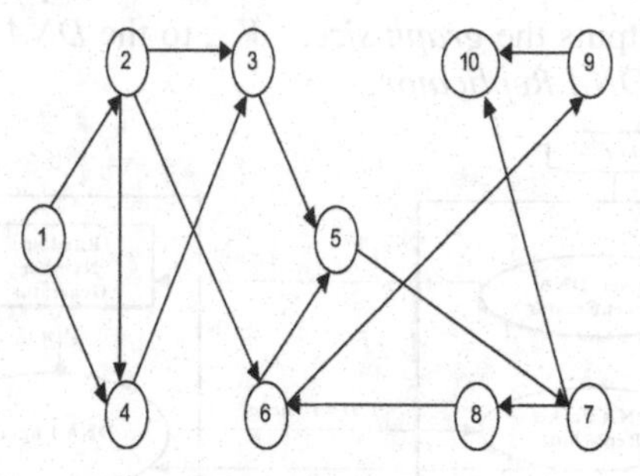

Fig. 3. Graph G_1

3.3 Discussion

Since the existence of a Hamiltonian Path (hereafter called a solution) is checked after a set of possible paths is generated, it is necessary to specify the size of the set that has a high likelihood of generating a solution. How can we specify this size?

We hypothesized that for any graph, the number of possible paths required to be generated before a solution is hit is related to the complexity of the graph, as measured either by the number of nodes or by the ratio of the number of edges to that of nodes. To verify this hypothesis, two statistical experiments were performed on graphs having *8* and *12* nodes with different number of edges.

The first experiment was to generate different number of paths and record the number of solutions over several runs of the algorithm. The second experiment was to generate paths until a solution appeared. The results of both experiments were averaged over a hundred runs. Figure 4 shows an average of the number of solutions generated vs. average number of paths. The results (all not displayed here) show that the number of paths that needs to be generated, before a solution is hit, grows exponentially with the ratio of the number of edges to number of nodes. Also, for a fixed number of possible paths generated, the average number of solutions increases with the decrease in the number of edges in the input graph.

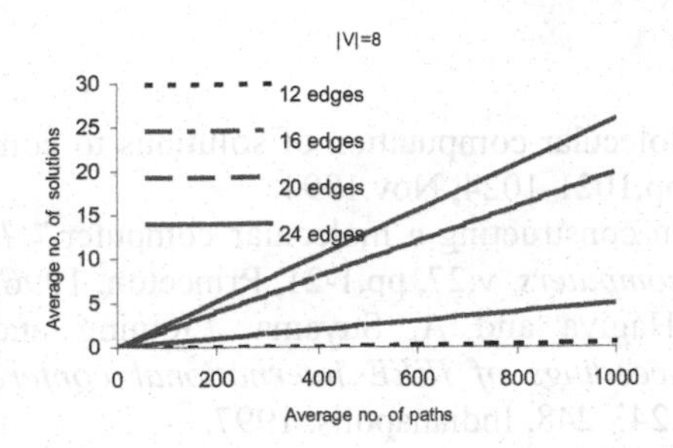

Fig. 4. Ave. no. of solutions vs. Ave. no. of paths

4 Conclusions

This paper reports the development of a DNA computing simulator to solve the Hamiltonian path problem. The simulator is an educational tool to teach DNA computing to students who do not have access to a bio-lab or would like to avoid the elaborate and time intensive input, output and filtering of data in a biological computer. We observe that, to find a Hamiltonian path, on the average a large number of paths must be generated for graphs with high edge to node ratios. Future work will involve a parallel implementation of the algorithm and investigation of the possibility of construction of simple regular *VLSI* structures to implement the basics of the algorithm for fixed-sized problems. Future research in DNA computing must address development of error-detection and error-correction in-vitro techniques to solve classes of problems for which robust DNA algorithms do not exist.

Acknowledgements

The author wishes to thank and acknowledge numerous students who participated in some form or the other during the development of this work; in particular thanks are due to students D. Caughlan, C. Tidwell, J. Crawford and Z. Maung in an undergraduate senior design student project for preliminary tests, to Dr. R. O. Chapman who taught the senior design project, to the students in my Real-time Embedded Computing class particularly, D. Cipperly, Ser-Geon Fu, J. Clark, A. Abdel-Rahim, Y.Li and H.Liang for verifying the simulations, and to research assistants N. Meghanathan, H. Lin and H. Qin, for background material searches and word processing in the early stages.

References

[1] L. Adleman. "Molecular computation of solutions to combinatorial problems," *Science,* v.266, pp.1021-1024, Nov.1994.

[2] L. Adleman, "On constructing a molecular computer," *1st DIMACS workshop on DNA based computers,* v.27, pp.1-21, Princeton, 1996.

[3] M. Arita, M. Hagiya and A. Suyama, "Joining and rotating data with molecules," *Proceedings of IEEE international conference on evolutionary computation*, pp.243-248, Indianapolis, 1997.

[4] E. Baum and D. Boneh, "Running dynamic programming algorithms on a DNA computer," *2nd DIMACS workshop on DNA based computers,* pp.141-147, Princeton, 1996.

[5] E. Baum, "Building an associative memory vastly larger than the brain," *Science*, v.268, pp.583-585, Apr. 1995.

[6] R. Beigel and B. Fu, "Solving intractable problems with DNA computing," *Proceedings of the 13th Annual Conference on Computational Complexity*, pp. 154-168, 1998.

[7] D. Boneh, C. Dunworth and R.Lipton, "Breaking DES using a molecular computer," *1st DIMACS workshop on DNA based computers,* pp.37-65, v. 27, Princeton, 1996.

[8] R. Deaton, R. Murphy, J. Rose, M. Garzon, D. Franceschetti and S. Stevens. "A DNA based implementation of an evolutionary search for good encodings for DNA computation," *Proc. of the IEEE international conference on evolutionary computation,* pp. 267-271, Indianapolis, 1997.

[9] R. P. Feynman, "There's plenty of room at the bottom," *Miniaturization*, Reinhold. pp.282-296, 1961.

[10] F. Guarnieri, M. Fliss and C. Bancroft, "Making DNA add," *Science,* pp.220-223, v.273, July 1996.

[11] M. Hagiya, M. Arita, D. Kiga, K. Sakamoto and S. Yokoyama, "Towards parallel evaluation and learning of Boolean formulas with molecules," *DNA Based Computers III, DIMACS Series in Discrete Mathematics and Theoretical Computer Science*, pp.57-72, v.48, 1999.

[12] http://www.ugcs.caltech.edu/~pwkr/dna_comp.html

[13] N. Jonoska and S. Karl, "A molecular computation of the road coloring problem," *2nd DIMACS workshop on DNA based* computers, pp.148-158, Princeton, 1996.

[14] N. Jonoska and S. Karl, "Ligation experiments in computing with DNA," *Proceedings of IEEE International conference on evolutionary computation,* pp.261-266, Indianapolis, 1977.

[15] P. Kaplan, G. Cecchi and A.Libchaber, "DNA based molecular computation: template-template interactions in PCR," *2nd DIMACS workshop on DNA based computers*, pp.159-171, Princeton, 1996.

[16] P. Kaplan, G. Cecchi and A.Libchaber, "Molecular computation: Adleman's experiment repeated," *Technical report*, NEC Research Institute, 1995.

[17] S. Kobayashi, T. Yokomori, G. Sampei and K. Mizobuchi, "DNA implementation of simple horn clause computation," *Proceedings of IEEE international conference on evolutionary computation*, pp.213-217, Indianapolis, 1977.

[18] T. Leete, M. Schwartz, R. Williams, D. Wood, J. Salem and H.Rubin, "Massively parallel DNA computation: Expansion of symbolic determinants," *2nd DIMACS workshop on DNA based computers*, pp.49-66, Princeton, 1996.

[19] R. Lipton, "DNA solution of hard computational problems," *Science*, v.268, pp.542-545, April 1995.

[20] R. Lipton, "Using DNA to solve SAT," http://www.cs.princeton.edu/~rjl.

[21] Q. Liu, Z. Guo, A. Condon, R. Corn, M. Lagally and L. Smith, "A surface based approach to DNA computation," *2nd DIMACS workshop on DNA based computers,* pp.206-216, Princeton, 1996.

[22] J. Oliver, "Computation with DNA: matrix multiplication," *2nd DIMACS workshop on DNA based computers*, pp.236-248, Princeton, 1996.

[23] J. Reif, "Parallel Molecular Computation," 7th annual ACM symposium on parallel algorithms and architectures, Santa Barbara, 1995.

[24] R. R. Roweis, E. Winfree, R. Burgoyne, N. V. Chelyapov, M. F. Goodman, P. W. K. Rothemund and L.Adleman, "A sticker based architecture for DNA computation," *Proc. of the second annual meeting on DNA-based computers,* Princeton, 1998.

[25] N. Seemen et al, "The perils of polynucleotides: the new experimental gap between the design and assembly of unusual DNA structures," *2nd DIMACS workshop on DNA based computers,* pp.191-205, Princeton, 1995.

[26] R. Williams and D. Woods, "Exascale computer algebra problems interconnect with molecular reactions and complexity theory," *2nd DIMACS workshop on DNA based computers,* pp.260-268, Princeton, 1996.

[27] E. Winfree, X. Yang and N. Seemen, "Universal computation via self assembly of DNA: some theory and experiments," *2nd DIMACS workshop on DNA based computers,* pp.172-190, Princeton, 1996.

Parallel Syntenic Alignments

Natsuhiko Futamura, Srinivas Aluru*, and Xiaoqiu Huang**

Iowa State University, Ames, IA 50011, USA
{nfutamur,aluru,xqhuang}@iastate.edu

Abstract. Given two genomic DNA sequences, the syntenic alignment problem is to compute an ordered list of subsequences for each sequence such that the corresponding subsequence pairs exhibit a high degree of similarity. Syntenic alignments are useful in comparing genomic DNA from related species and in identifying conserved genes. In this paper, we present a parallel algorithm for computing syntenic alignments that runs in $O\left(\frac{mn}{p}\right)$ time and $O\left(m + \frac{n}{p}\right)$ memory per processor, where m and n are the respective lengths of the two genomic sequences. Our algorithm is time optimal with respect to the corresponding sequential algorithm and can use $O\left(\frac{n}{\log n}\right)$ processors, where n is the length of the larger sequence. Using an implementation of this parallel algorithm, we report the alignment of human chromosome $12p13$ and its syntenic region in mouse chromosome 6 (both over 220,000 base pairs in length) in under 24 minutes on a 64-processor IBM xSeries cluster.

1 Introduction

Sequence alignments are fundamental to many applications in computational biology, and comprise one of the best studied and well understood problem areas in this discipline. Much of the early pioneering work concentrated on two types of alignments – 1) global alignments, which are intended for comparing two sequences that are entirely similar [7][14][15], and 2) local alignments, which are intended for comparing sequences that have locally similar regions [10][18]. In general, these problems can be solved in time proportional to the product of the lengths of the sequences and in space proportional to the sum of the lengths of the sequences. Research has also been conducted in developing parallel algorithms for solving global and local alignment problems [1][5][6][9][13].

It is widely recognized that evolutionary processes tend to conserve genes. Along a chromosome, genes are interspersed by large regions known as 'junk DNA'. A gene itself is comprised of alternating regions known as *exons* and *introns*, and the introns are intervening regions that do not participate in the translation of a gene to its corresponding protein. Homologous DNA sequences from related organisms, such as the human and the mouse, are usually similar over the exon regions but different over other regions. Because the different

* Research supported by NSF Career under CCR-0096288 and NSF EIA-0130861.
** Research supported by NIH Grants R01 HG01502-05 and R01 HG01676-05 from NHGRI.

S. Sahni et al. (Eds.) HiPC 2002, LNCS 2552, pp. 420–430, 2002.

regions are much longer than similar regions, conserved sequences cannot be identified through global alignment. This results in the problem of aligning two sequences where an ordered list of subsequences of one sequence is highly similar to a corresponding ordered list of subsequences from the other sequence. We refer to this problem as the *syntenic alignment* problem.

A number of fast comparison algorithms have been developed for comparing syntenic genomic sequences [3][4][12][16]. Generally, these methods perform fast identification of significant local similarities, and narrow further consideration to such regions. Because of this, such methods tend to work well on sequences with highly similar regions. Recently, Huang [11] developed a dynamic programming based solution to the syntenic alignment problem. This method guarantees finding an optimal solution and is capable of detecting weak similarities. However, the run-time of this scheme is quadratic, making it difficult to apply over long sequences.

In this paper, we present a parallel syntenic alignment algorithm based on the sequential dynamic programming solution developed by Huang [11]. The algorithm runs in $O\left(\frac{mn}{p}\right)$ time and $O\left(m+\frac{n}{p}\right)$ space, where m and n are the lengths of the two genomic sequences ($m \leq n$). The algorithm is time optimal with respect to the sequential algorithm and can use up to $O\left(\frac{n}{\log n}\right)$ processors. We implemented the parallel algorithm using C and MPI, and demonstrate its scalability using an IBM xSeries cluster. Using this software, we report the alignment of human chromosome 12p13 and its syntenic region in mouse chromosome 6 (both over 220kb in length) in 23.32 minutes using 64-processors.

The rest of the paper is organized as follows: In Section 2, we describe the syntenic alignment problem and describe a dynamic programming solution for solving it. In Section 3, we present our parallel algorithm. Experimental results are presented in Section 4. Section 5 concludes the paper.

2 Problem Formulation

An alignment of two sequences $S = s_1s_2\dots s_k$ and $T = t_1t_2\dots t_l$ over an alphabet Σ is obtained by inserting gaps in chosen positions and stacking the sequences such that each character in a sequence is either matched with a character in the other sequence or a gap. The quality of an alignment is computed as follows: A scoring function $f : \Sigma \times \Sigma \rightarrow I\!R$ specifies the score for matching a character in one sequence with a character in the other sequence. Gaps are penalized by using an affine gap penalty function that charges a penalty of $h + gr$ for a sequence of r maximal gaps. Here, h is referred to as gap opening penalty and g is referred to as gap continuation penalty. An optimal alignment of S and T is an alignment resulting in the maximum possible score over all possible alignments. Let $score(S,T)$ denote the score of an optimal alignment. In line with the tradition in molecular biology, we use *sequence* to mean string and *subsequence* to mean substring.

Let $A = a_1a_2 \ldots a_m$ and $B = b_1b_2 \ldots b_n$ be two sequences. A subsequence A' of A is said to precede another subsequence A'' of A, written $A' \prec A''$, if the last character of A' occurs strictly before the first character of A'' in A. An ordered list of subsequences of A, $(A_1, A_2, \ldots, A_k)$ is called a chain if $A_1 \prec A_2 \prec \ldots A_k$. The syntenic alignment problem for sequences A and B is to find a chain $(A_1, A_2, \ldots, A_k)$ of subsequences in A and a chain $(B_1, B_2, \ldots, B_k)$ of subsequences in B such that the score

$$\left\{\sum_{i=1}^{k} score(A_i, B_i)\right\} - (k-1)d$$

is maximized (see Figure 1).

The parameter d is a large penalty aimed at preventing alignment of short subsequences which occur by chance and not because of any biological significance. Intuitively, we are interested in finding an ordered list of matching subsequence pairs that correspond to conserved exons. One can think of unmatched subsequence pairs that lie between consecutive matched subsequences, i.e. the subsequence between A_i and A_{i+1} and the subsequence between B_i and B_{i+1}. The penalty d can be viewed as corresponding to an unmatched subsequence pair. For a small alphabet size, given a character in an unmatched subsequence, there is a high probability of finding the same character in the corresponding unmatched subsequence. In the absence of the penalty d, using these two characters as another matched subsequence pair would increase the score of the syntenic alignment. The penalty d serves to avoid declaring such irrelevant matching subsequences as part of the syntenic alignment, and its value should be chosen carefully.

Based on the problem definition, the syntenic alignment of two sequences $A = a_1a_2 \ldots a_m$ and $B = b_1b_2 \ldots b_n$ can be computed by dynamic programming. Basically, we compute the syntenic alignment between every prefix of A and every prefix of B. We compute 4 tables C, D, I and H of size $(m+1) \times (n+1)$. Entry $[i, j]$ in each table corresponds to the optimal score of a syntenic alignment between $a_1a_2 \ldots a_i$ and $b_1b_2 \ldots b_j$, subject to the following conditions: 1) In C, a_i is matched with b_j, 2) In D, a_i is matched with a gap, 3) In I, gap is matched with b_j, and 4) In H, either a_i or b_j is part of an unmatched subsequence. It

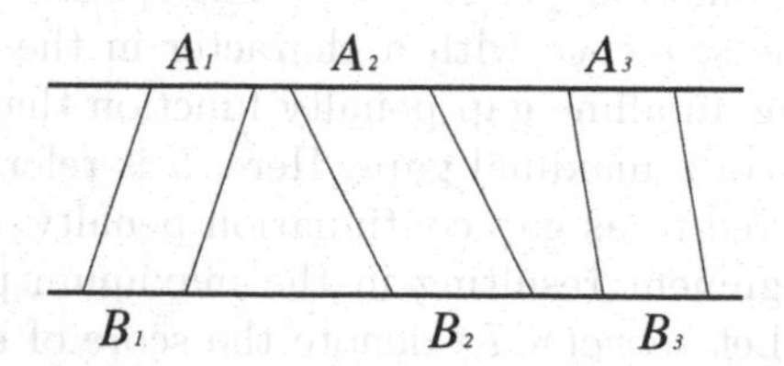

Fig. 1. An illustration of the syntenic alignment problem

follows from these definitions that the tables can be computed using the following recurrence equations:

$$\begin{aligned}
C[i,j] &= f(a_i,b_j) + \max\{C[i-1,j-1], D[i-1,j-1], \\
&\qquad\qquad I[i-1,j-1], H[i-1,j-1]\} \\
D[i,j] &= \max\{C[i-1,j]-g', D[i-1,j]-g, I[i-1,j]-g', H[i-1,j]-g'\} \\
I[i,j] &= \max\{C[i,j-1]-g', D[i,j-1]-g', I[i,j-1]-g, H[i,j-1]-g'\} \\
H[i,j] &= \max\{C[i-1,j]-d, I[i-1,j]-d, C[i,j-1]-d, D[i,j-1]-d, \\
&\qquad H[i-1,j], H[i,j-1]\}
\end{aligned}$$

where $g' = (g+h)$.

Prior to computation, the top row and left column of each table should be initialized. These initial values can be directly computed. After computing the tables, the optimal score of a syntenic alignment is given by the maximum score in $C[m,n]$, $D[m,n]$, $I[m,n]$, or $H[m,n]$. Thus, the problem can be solved in $O(mn)$ time and space. If we draw links from each table entry to an entry which gives the maximum value in equation (1), (2), (3) or (4), the optimal syntenic alignment can be retrieved by tracing backward in the tables starting from the largest $[m,n]$ entry and ending at $C[0,0]$. Using the now standard technique of space-saving, introduced originally by Hirschberg [8], the space required can be reduced to $O(m+n)$, while increasing the run-time by at most a factor of 2.

3 Parallel Syntenic Alignment Algorithm

Let p denote the number of processors, with *id*'s ranging from 0 to $p-1$. Without loss of generality, assume that $m \leq n$. We compute the four tables C, D, I and H together in parallel. We use a columnwise decomposition to partition the tables to the processors. For simplicity, assume m and n are multiples of p. Processor i receives columns $i\frac{n}{p}+1, \ldots, (i+1)\frac{n}{p}$ of each table, and is responsible for computing the table entries allocated to it. The tables are computed one row at a time, in the order C, D, H and I.

Consider computing the ith row of the tables. The recurrence relation for D uses entries from the already computed $(i-1)^{th}$ row and in the same column. These are readily available on the same processor. In computing C, entries that are in the previous row and previous column are needed. These are available on the same processor, except in the case of the first column assigned to each processor. After computing the $(i-1)^{th}$ row, each processor sends the last entry it computed in each of the four tables to the next processor. This is sufficient to compute the next row of C, and requires communicating just four entries per processor irrespective of the problem size. Next, we compute the ith row of H. Let

$$v[j] = max\{C[i-1,j]-d,\ I[i-1,j]-d,\ C[i,j-1]-d,\ D[i,j-1]-d,\ H[i-1,j]\}$$

Because the ith rows of C and D are already computed, the vector v can be computed directly in parallel using the information available within each processor. Then, $H[i,j]$ can be written as $\max\{v[j], H[i,j-1]\}$. It is easy to see that the computation of $H[i,j]$ can be done using the *parallel prefix*[1] operation with 'max' as the binary associative operator.

Now, let us turn to the computation of the ith row of table I.

Let $\quad w[j] = \max\{C[i,j-1], D[i,j-1], H[i,j-1]\} - g'$
Then, $\quad I[i,j] = \max\{w[j], I[i,j-1] - g\}$

Let

$$\begin{aligned} x[j] &= I[i,j] + gj \\ &= \max\{w[j] + gj, I[i,j-1] + gj - g\} \\ &= \max\{w[j] + gj, I[i,j-1] + g(j-1)\} \\ &= \max\{w[j] + gj, x[j-1]\} \end{aligned}$$

Let $\quad z[j] = w[j] + gj$
Then, $\quad x[j] = \max\{z[j], x[j-1]\}$

Since the $z[j]$'s can be easily computed from the ith row of C, D, and H, $x[j]$'s can be computed using parallel prefix with 'max' as the binary associative operator. In turn, $I[i,j]$ can be computed from $x[j]$ by simply subtracting gj from it.

As mentioned before, processor i is responsible for computing columns $i\frac{n}{p}+1$ through $(i+1)\frac{n}{p}$ of the tables C, D, I and H. Distribution of sequence B is trivial because b_j is needed only in computing column j. Therefore, processor i is given $b_{i\frac{n}{p}+1} \ldots b_{(i+1)\frac{n}{p}}$. Each a_i is needed by all the processors at the same time when row i is being computed. Sequence A is stored in each processor. It remains to describe how the traceback procedure is performed in parallel to retrieve the optimal syntenic alignment. However, we defer this as the scheme presented so far cannot be used directly due to the unreasonably large amount of memory required.

Run-time and Space Analysis: Each processor computes $\frac{n}{p}$ entries per row of each of the four tables. The run-time is dominated by parallel prefix, which takes $O\left(\frac{n}{p} + \log p\right)$ time. To achieve optimal $O\left(\frac{n}{p}\right)$ run-time, the number of processors used should be $O\left(\frac{n}{\log n}\right)$. To enable using as large a number of processors as possible, and more importantly because practical efficiencies are better when

[1] Given $x_1, x_2, \ldots, x_n$ and a binary associative operator $\otimes$, parallel prefix is the problem of computing $s_1, s_2, \ldots, s_n$, where $s_i = x_1 \otimes x_2 \otimes \ldots \otimes x_i$ (or equivalently, $s_i = s_{i-1} \otimes x_i$). This is a well-known primitive operation in parallel computing, and is readily available on most parallel computers. For example, the function MPI_Scan computes parallel prefix.

the problem size per processor is large, we choose the larger sequence to represent the columns of the table (i.e., $n \geq m$). The parallel run-time for computing all the tables is $O\left(\frac{mn}{p}\right)$, optimal with respect to the sequential algorithm. The space required is also $O\left(\frac{mn}{p}\right)$.

The space required by the algorithm presented can be prohibitively large for syntenic alignments. For example, consider the alignment of two sequences of length one million each, on 100 processors. Assuming each entry of the table requires two memory words (one for the value and one for the pointer), the space required per processor can be estimated as $\frac{1}{100} \times 10^6 \times 10^6 \times 2 \times 4 \approx 80GB$!

Note that complete storing of the tables is required only because of the necessity to traceback to retrieve the optimal alignment. If only the optimal score is required, we only need the entry $[m, n]$ in each of the four tables. In computing row i of the tables, only the previous row of the tables is required. Thus, one can compute the tables by keeping track of at most two rows at a time (this can be actually reduced to one row plus constant storage). This will immediately reduce the storage to $O\left(\frac{n}{p}\right)$, but it would not allow traceback.

Parallel Space-Saving Algorithm

Define p special columns C_k $(0 \leq k \leq p-1)$ of a table to be the last columns of the parts of the tables allocated to each processor, except for the last processor, i.e., $C_k = (k+1) \times \frac{n}{p}$. If the intersections of an optimal path with the special columns are identified, the problem can be split into p subproblems (see Figure 2), to be solved one per processor using the sequential space-saving algorithm. The solutions of the subproblems are then concatenated to get the total alignment. Each subproblem receives exactly $\frac{1}{p}^{th}$ of sequence B but an undetermined portion of sequence A. Since the total length of the sequence A is m, each subproblem can be solved sequentially in $O\left(\frac{mn}{p}\right)$ time and $O\left(m + \frac{n}{p}\right)$ space. This memory requirement can be easily satisfied even if the input sequences span entire chromosomes. Memory permitting, multiple special columns per processor can be used, resulting in smaller subproblems and decreased overall run-time. Thus, there is a memory vs. run-time tradeoff.

It remains to be described how the intersection of an optimal path with the special columns can be computed. We only store information on the special columns of a table. In addition, we store the most recently computed row of a table in order to compute the next row using parallel prefix. This gives a space bound of $O\left(m + \frac{n}{p}\right)$. For each entry of a table, the 'value' of the entry and the ⟨*table number*, *row number*⟩ tuple of the entry in the closest special column to the left that lies on an optimal path from $C[0, 0]$ to the entry are computed. Call such a tuple a *pointer* to the previous special column. This essentially gives the ability to perform a traceback through special columns, without considering other columns. The values in a row of each table are computed as before. The pointers for tables C and D can be copied from the entry in the previous rows

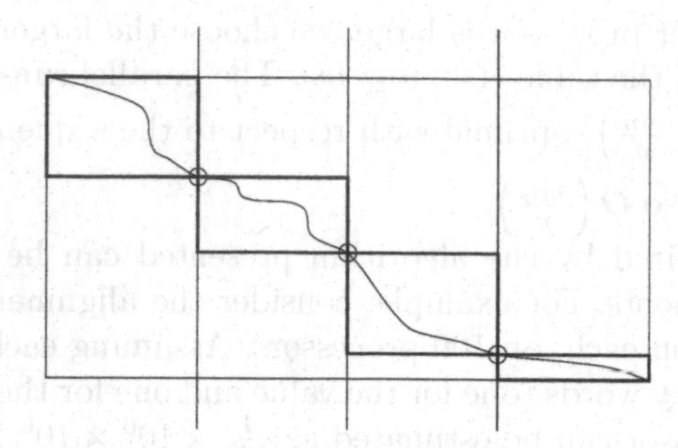

Fig. 2. Problem decomposition in parallel space-saving algorithm

of the four tables that is responsible for the value chosen by the max operator. For H and I, the pointer is similarly known if it results from one of the known entries but is not known if it results from the previous entry in the same row of the table being computed. Therefore it is initially set to u (undefined), unless $j-1$ is a special column. If so, $\langle H, i\rangle$ (or $\langle I, i\rangle$ when computing I) is taken to be the pointer. The undefined entries can then be filled using parallel prefix and the following operation:

$$x \oplus y = \begin{cases} x, & y = u \\ y, & y \neq u \end{cases}$$

In fact, the parallel prefix for establishing the pointers can be avoided altogether. This is because the last column of the table allocated to each processor is a special column and the pointer value in an entry next to the special column is already known. Therefore, a sequential prefix computation within each processor is enough to determine the pointers.

A sequential traceback procedure along the special columns can be used to split the problem into p subproblems in $O(p)$ time. This does not significantly affect the run-time of $O\left(\frac{mn}{p}\right)$ provided $p^2 = O(mn)$. While this is a reasonable assumption in practice, time-optimality can be retained even if this is not true.

The idea is to parallelize the traceback procedure itself using parallel prefix. Each element on a special column contains a pointer to the element on the previous special column in one of the four tables. It is required to establish a pointer from each element on the last special column of each table to an element on every other special column in one of the four tables following the chain of pointers leading to it. Consider the special columns $C_{p-1}, C_{p-2}, \ldots, C_0$ of all the tables. To operate on the special columns on the four tables at once, special columns with the same column numbers are concatenated together and considered as an array of size $4(m+1)$, and pointer tuples $\langle \textit{table number}, \textit{row number}\rangle$ stored at each special column be adjusted accordingly. Define the operator $\oplus$ such that

$$(A \oplus B)[i] = B[A[i]]$$

where A, B and $A \oplus B$ are arrays of length $4(m+1)$ representing array of pointers on special columns. Partial sums $s_{p-1}, s_{p-2}, \ldots, s_0$, where

$$s_k = C_{p-1} \oplus C_{p-2} \oplus \ldots \oplus C_{k+1}$$

can be computed using parallel prefix. As applying this operator takes $O(m)$ time, such a parallel prefix takes $O(m \log p)$ time. As we can take $O\left(\frac{mn}{p}\right)$ time, up to $O\left(\frac{n}{\log n}\right)$ processors can be utilized.

It remains to be described how the data required for the subproblems is moved to the respective processors. Sequence B is already distributed appropriately. Distribute sequence A uniformly across all the processors. While better methods can be designed, the following suffices to prove the required time complexity. Perform p circular shift operations on sequence A such that the entire sequence passes through each processor. Each processor retains as much of sequence A as it needs. If there is sufficient memory on each processor, data movement can be avoided by storing the entire sequence A throughout the computation, without violating our space bound of $O\left(m + \frac{n}{p}\right)$.

4 Experimental Results

We implemented the parallel syntenic alignment algorithm in C and MPI and experimentally evaluated its performance using an IBM xSeries cluster. The cluster consists of 64 Pentium processors each with a clock rate of 1.26GHZ and 512MB of main memory, connected by Myrinet, supporting bidirectional communication rates of 2Gb/sec. The parallel syntenic alignment algorithm consists of a problem decomposition stage, followed by a local computation stage: In the decomposition stage, the tables are computed in parallel, storing entries only on the special columns. This is followed by a traceback procedure to split the problem into p subproblems. In the local computation stage, the subproblems are solved independently on each processor. The time spent in the decomposition stage depends only on the size of the tables. Even though the time spent in the local computation stage is worst-case optimal ($O\left(\frac{mn}{p}\right)$), the actual time spent depends upon how evenly the problem splits into subproblems, which in turn depends upon the structure of the optimal alignment. In the best case where all subproblems have equal size, the run-time for the local computation stage is only $O\left(\frac{mn}{p^2}\right)$. The worst-case corresponds to a single processor receiving a problem of size $m \times \frac{n}{p}$, which translates to all conserved exons confined to $\frac{1}{p}^{th}$ of an input sequence allocated to the same processor. This situation is highly unlikely, and the actual performance is expected to be closer to the best-case.

To study the scalability of the algorithm, the program is run using sequences of the same length and varying the number of processors. Note that the communication required in computing a row depends only on the number of processors and is independent of the problem size. Thus, it is interesting to determine the

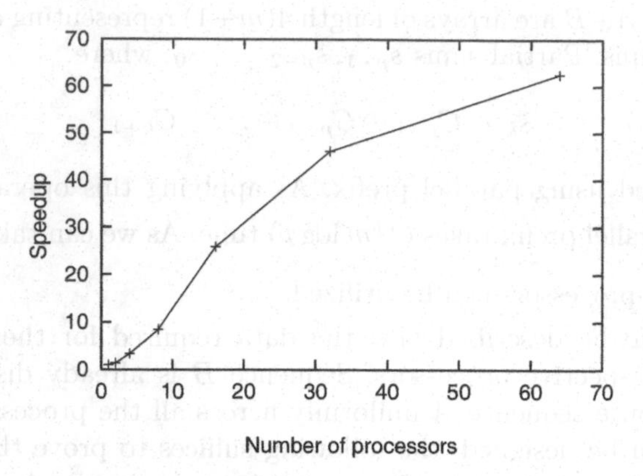

Fig. 3. Speedup as a function of the number of processors for syntenic alignment of two sequences of length 30,000

smallest problem size per processor (*grain-size*) that gives good scaling results. This can be used to calculate the largest number of processors that can be beneficially used to solve a given problem. On the IBM cluster, we determined that the grain-size required for efficient parallel execution is about $500 - 1000$ per processor.

The speedups as a function of the number of processors for a syntenic alignment of two sequences of length 30,000 are shown in Figure 3. Notice that superlinear speed up is observed in several cases. Apart from the typical beneficial effect due to better caching, this is due to the fact that increasing the number of processors causes a proportionate increase in the number of special columns, which reduces total work. Based on an approximate calculation (ignoring lower order terms etc.), our parallel algorithm requires computing $6mn$ entries per table in the problem decomposition stage. Note that parallel prefix on a row of a table requires processing each table entry twice, and only two of the four tables need parallel prefix. If the subproblems are perfectly balanced, each processor computes $\frac{8mn}{p^2}$ entries in the local computation stage. Thus, the total work done by our parallel algorithm is $6mn + \frac{8mn}{p}$, including the idle time on processors (taking work to be the product of parallel run-time multiplied by the number of processors). This is a major reason for the superlinear speedup observed. Table 1 shows the total run-time and the time spent in the problem decomposition and local computation stages as the number of processors is varied. The local computation stage can scale anywhere between linearly and quadratically (run-time reduces by a factor of 4 for a twofold increase in the number of processors).

The effect of caching is the key factor in the superlinear speedup observed in the problem decomposition stage, when the number of processors is increased from 8 to 16. On 16 processors, each processor has an approximate row size of 2,000 entries per table. We need to store 4 tables, 2 rows per table, and need 3 memory words (12 bytes) per entry. Thus, the memory required in the problem

Table 1. Run-time (in seconds) spent in the problem decomposition and local computation stages for $m = n = 30,000$, as the number of processors is varied

Number of processors	Problem decomposition stage	local computation stage	Total time
1	1497	409	1906
2	1220	166	1386
4	507	67	574
8	200	28	228
16	61	13	74
32	35	6	41
64	27	3	30

decomposition stage is 192KB per processor, which will nicely fit into the 256KB cache. On 8 processors, the rows will have to be continually swapped between cache and main memory, causing significant slowdown.

The program is used to compare two syntenic human and mouse sequences containing 17 genes [2]. The human sequence is of length 222,930 bp (GenBank Accession U47924) and the mouse sequence is of length 227,538 bp (GenBank Accession AC002397). The following parameters are used based on our prior experiences with standard alignment programs: match = 10; mismatch = −20; gap opening penalty, $h = 60$; gap continuation penalty, $g = 2$. A value for the parameter d was selected on the basis of internal exon lengths, often of length at least 50 bp. The score of 50 matches at 10 per match is 500. The value of 300 was used for the parameter d. The human and mouse sequences were screened for repeats with RepeatMasker [17]. The masked sequences are then used as input. The program produced a syntenic alignment of the two sequences in 23.32 minutes on 64 processors. The alignment consists of 154 ordered subsequence pairs separated by unmatched subsequences. The alignment fully displays the similar regions but omits most of the dissimilar regions. The 154 similar regions are mostly coding exon regions and untranslated regions. Gaps occur much more frequently in alignments of untranslated regions than in alignments of coding exon regions. The total length of the 154 similar regions is 43,445 bp and their average identity is 79%. The 154 similar regions constitute about 19% of each of the two sequences.

5 Conclusions

In this paper, we developed a parallel algorithm and its implementation for comparing sequences with intermittent similarities. The proposed method allows fast computation of syntenic alignments of long DNA sequences. It enables the computation of long genomic regions with weak similarities.

References

[1] S. Aluru, N. Futamura and K. Mehrotra, Biological sequence comparison using prefix computations, *Proc. International Parallel Processing Symposium* (1999) 653-659. 420

[2] M. A. Ansari-Lari, J. C. Oeltjen, S. Schwartz, Z. Zhang, D. M. Muzny, J. Lu, J. H. Gorrell, A. C. Chinault, J. W. Belmont, W. Miller and R. A. Gibbs, Comparative sequence analysis of a gene-rich cluster at human chromosome 12p13 and its syntenic region in mouse chromosome 6, *Genome Research, 8* (1998) 29-40. 429

[3] S. Batzoglou, L. Pachter, J. P. Mesirov, B. Berger and E. S. Lander, Human and mouse gene structure: comparative analysis and application to exon prediction, *Genome Research, 10* (2000) 950-958. 421

[4] A. L. Delcher, S. Kasif, R. D. Fleischmann, J. Peterson, O. While and S. L. Salzberg, Alignment of whole genomes. *Nucleic Acids Research, 27* (1999) 2369-2376. 421

[5] E. W. Edmiston and R. A. Wagner, Parallelization of the dynamic programming algorithm for comparison of sequences, *Proc. International Conference on Parallel Processing* (1987) 78-80. 420

[6] E. W. Edmiston, N. G. Core, J. H. Saltz and R. M. Smith, Parallel processing of biological sequence comparison algorithms, *International Journal of Parallel Programming, 17(3)* (1988) 259-275. 420

[7] O. Gotoh, An improved algorithm for matching biological sequences. *Journal of Molecular Biology, 162* (1982) 705-708. 420

[8] D. S. Hirschberg, A linear space algorithm for computing maximal common subsequences, *Communications of the ACM, 18(6)* (1975) 341-343. 423

[9] X. Huang, A space-efficient parallel sequence comparison algorithm for a message-passing multiprocessor, *International Journal of Parallel Programming, 18(3)* (1989) 223-239. 420

[10] X. Huang, A space-efficient algorithm for local similarities, *Computer Applications in the Biosciences,6(4)* (1990) 373-381. 420

[11] X. Huang and K. Chao, A generalized global alignement algorithm, manuscript in preparation. 421

[12] N. Jareborg, E. Birney, and R. Durbin, Comparative analysis of noncoding regions of 77 orthologous mouse and human gene pairs, *Genome Research, 9,* (1999) 815-824. 421

[13] E. Lander, J. P. Mesirov and W. Taylor, Protein sequence comparison on a data parallel computer, *Proc. International Conference on Parallel Processing* (1988) 257-263. 420

[14] E. W. Mayers and W.Miller, Optimal alignments in linear space, *Computer Applications in the Biosciences, 4(1)* (1988) 11-17. 420

[15] S. B. Needleman and C. D. Wunsch, A general method applicable to the search for similarities in the amino acid sequence of two proteins, *Journal of Molecular Biology, 48* (1970) 443-453. 420

[16] S. Schwartz, Z. Zhang, K. Frazer, A. Smit, C. Riemer, J. Bouck, R. Gibbs, R. Hardison, and W. Miller, PipMaker–A web server for aligning two genomic DNA sequences, *Genome Research, 10* (2000) 577-586. 421

[17] A. Smit and P. Green, `http://ftp.genome.washington.edu/RM/RepeatMasker.html`, 1999. 429

[18] T. F. Smith and M. S. Waterman, Identification of common molecular subsequences, *Journal of Molecular Biology, 147* (1981) 195-197. 420

XS-systems: eXtended S-Systems and Algebraic Differential Automata for Modeling Cellular Behavior*

Marco Antoniotti[1], Alberto Policriti[2], Nadia Ugel[1], and Bud Mishra[1,3]

[1] Courant Institute of Mathematical Sciences, NYU
New York, NY, USA
[2] Università di Udine, Udine, (UD) Italy
[3] Watson School of Biological Sciences
Cold Spring Harbor, NY, USA

Abstract. Several biological and biochemical mechanisms can be modeled with relatively simple sets of differential algebraic equations (DAE). The numerical solution to these differential equations provide the main investigative tool for biologists and biochemists. However, the set of numerical *traces* of very complex systems become unwieldy to wade through when several variables are involved. To address this problem, we propose a novel way to query large sets of numerical traces by combining in a new way well known tools from numerical analysis, temporal logic and verification, and visualization.

In this paper we describe *XS-systems*: computational models whose aim is to provide the users of S-systems with the extra tool of an *automaton* modeling the *temporal evolution* of complex biochemical reactions. The automaton construction is described starting from both numerical and analytic solutions of the differential equations involved, and parameter determination and tuning are also considered. A temporal logic language for expressing and verifying properties of XS-systems is introduced and a prototype implementation is presented.

1 Introduction

In this paper we reason about issues related to the construction of tools aiming at helping biologists and biochemists who perform simulation of complex biochemical pathways in conjunction with their experimental activities. The content of this paper is a work in progress, whose aim is to create a framework where to bring together several disciplines in a focused way.

Several biological and biochemical mechanisms can be modeled with sets of relatively simple differential algebraic equations (DAE). The numerical solution

* The work reported in this paper was supported by grants from NSF's Qubic program, DARPA, HHMI biomedical support research grant, the US department of Energy, the US air force, National Institutes of Health and New York State Office of Science, Technology & Academic Research.

S. Sahni et al. (Eds.) HiPC 2002, LNCS 2552, pp. 431–442, 2002.

to these differential equations provide the main investigative tool for biologists and biochemists. The simple (canonical) forms of the differential equations may contrast with the actual "complexity" of the system being modeled, as measured by the number of *variables* involved. The set of numerical *traces* of very complex systems rapidly becomes unwieldy to wade through when several variables are involved. To cope with this problem, we propose a novel way to query large sets of numerical traces by combining in a new way well known tools from numerical analysis, temporal logic and verification, and visualization.

Our starting reference points are the S-systems, described in [15, 16], and the idea that a natural completion for that approach would be an *automaton summarizing* the phases along which the simulated biochemical system passes during its evolution in time. The automata we are proposing will allow the user to view and manipulate a, detectably specified, representation of the set of states through which the system evolves. The approach will mainly serve the following two purposes:

- Explicitly render the significant changes in the values of substances involved in the biochemical reactions during the *in silico* experiment, thereby providing a better control on the physical plausibility of the latter.
- Provide a precise language for controlling sets of possible experiments, based on different values of the parameters involved.

The automaton construction is based on the computation of the approximate numerical solution of the S-system and is performed in two steps: first, starting from any given *time step* and from the corresponding approximate numerical solution for the DAEs, a synchronous automaton whose states are collection of values for the dependent values is determined. Then, a qualitative analysis (based on the first derivatives $\dot{X}_i$'s of the functions expressing concentrations) is carried out to the effect of collapsing states modeling non-significant variations in the evolution of the system.

The second operation simply corresponds to finding a set of "linear approximations" of the function flow as represented in the computed or sampled trace. We note here some correspondences with some of the work done in the analysis of Hybrid Systems (*cf.* [2]), which we will explore in a future work.

The automaton obtained after the second of the above two steps is "asynchronous" (untimed, *cf.* [2]), as the values of the temporal intervals connecting collapsed states are ignored, but suitable for a verification analysis of temporal properties. To this end a (temporal) language suited for such a kind of analysis is proposed and studied. Moreover, the equivalence relation that collapses states at different stages of the temporal evolution can be either global (unique), or local, and is itself a function of time, providing the ability of refining the qualitative analysis in sensitive regions of the metabolic pathway.

Tools based on temporal specification and verification are widely recognized as crucial in the realm of embedded reactive systems, and the approach we present here has many similarities with ideas very much exploited in that area; modeling the evolution of a biochemical system as we are proposing consists, in fact, in seeing the system's state-sequence as the analogue of a computation, and different computations as (simulations of) experiments differing for some given parameters' values. However, a special feature of the biochemical-systems' modeling arena is its stronger focus on one (or a restricted family of) experiment(s). This must be contrasted with the case of formal verification of temporal properties of embedded systems, in which the focus must stand much more on the careful consideration of *all* possible computations (in order to be able exclude the bad one!). In fact, our automata construction can produce different values as the parameters involved (i.e. time-step, rate constants, kinetic-order exponents, etc.) are varied at the start. E.g., as a limit to the choice of a smaller time-step in the construction, we could consider the analytic description of the solution to the DAE governing the system. The, more abstract, study of the issues related with families of possible numerical solutions (and hence automaton constructions), is going to be our next goal.

One of the main features of our approach is the intermixing of both quantitative and qualitative modeling, In particular, the qualitative modeling of the system is supposed to be specified *after* a quantitative description has been determined and (numerically) solved. The automaton is in fact obtained by "gluing" together different representations of possible evolution of the system. This sort of *bottom-up* automata construction is one of the characterizing aspects of our proposal, the other being the idea that the notion of state of the system should be less restrictive than the one usually employed in formal verification and strictly based on variables' values. Notice, finally, that the proposal we are putting forward could be discussed in the context of regulatory pathways modeling as well. We will discuss the details of this application in a future work.

Preliminaries

In the following we will build on ideas introduced in [15, 16] (from which we will borrow also most of the notation) and, e.g., [5].

S-systems. The basic ingredients of an S-system are n dependent variables to be denoted $X_1, \ldots, X_n$, m independent variables $X_{n+1}, \ldots, X_m$ and let $D_1, \ldots, D_{n+m}$ be the domains where the $n + m$ variables take value. We augment the form described in [15, 16] with a set of *algebraic constraints* which serve to characterize the conditions under which a given set of equations is derived from a set of maps. The justification for this construction is beyond the scope of this paper and it appears elsewhere.

The basic differential equations constituting the system take the general form:

$$\dot{X}_i(t) = V_i^+(X_1(t), \ldots, X_m(t)) - V_i^-(X_1(t), \ldots, X_m(t)), \tag{1}$$

for each dependent variable X_i (see [16] for a complete discussion and justification of the assumptions underlying the format of the above equation). The set of algebraic constraints take the form

$$\{C_j(X_1(t), \ldots, X_m(t)) = 0\} \tag{2}$$

The above equations take, in general, the following *power law* form:

$$\dot{X}_i = \alpha_i \prod_{j=1}^{n+m} X_j^{g_{ij}} - \beta_i \prod_{j=1}^{n+m} X_j^{h_{ij}} \tag{3}$$

$$C_j(X_1(t), \ldots, X_m(t)) = \sum \left(\gamma_j \prod_{k=1}^{n+m} X_k^{f_{jk}} \right) = 0 \tag{4}$$

where the α_i's and β_i's are called *rate constants* and govern the positive or negative contributions to a given substance (represented by X_i as a function of time) with other variables entering in the differential equation with exponents to be denoted as g_{ij}'s and h_{ij}'s. The γ_j are called *rate constraints* acting concurrently with the exponents f_{jk} to delimit the evolution of the system over a specified manifold embedded in the $n+m$-dimensional surface. Note that we have $\alpha_i \geq 0$ and $\beta_i \geq 0$ for all i's.

A system of differential equations (power-laws) such as the one above can be integrated by either symbolically – in particularly favorable cases – or by numerical approximation. The particular S-system "canonical" form allows for very efficient computations of both the function flows X_i and the derivative field $\dot{X}_i$ [11]. In the following we will concentrate on the "numerical" case and we will also exploit the special nature of the S-system traces for our Temporal Logic"query language". We will address the "symbolic" case in a much more general way in a future work. When a numerical approximation is involved, the notion of *time-step* "**step**" becomes central to our considerations.

Example 1. Consider the following example consisting of the pathway of a so-called *repressilator system* [7]. The metabolic map corresponding to the above system is shown in Figure 1 (a). The repressilator system metabolic map involves six variables $X_1, \ldots, X_6$, the first three of which (namely X_1 ,X_2 and ,X_3) are independent while the remaining are dependent.

The following is the S-system corresponding to the above metabolic map:

$$\begin{cases} \dot{X}_1 = \alpha_1 X_3^{g_{13}} X_4^{g_{14}} - \beta_1 X_1^{h_{11}}; \\ \dot{X}_2 = \alpha_2 X_1^{g_{21}} X_5^{g_{25}} - \beta_2 X_2^{h_{22}}; \\ \dot{X}_3 = \alpha_3 X_2^{g_{32}} X_6^{g_{36}} - \beta_3 X_3^{h_{33}}, \end{cases}$$

with X_4, X_5, and X_6 as controls (independent variables).
Figure 1(b) shows the oscillatory trace of the system (*cf.* [7] for a discussion of the numerical and analytical analysis of the system).

Related Works. Many interesting researches are dealing with more or less the general themes treated in this work.

For example, in [6] the problem of modeling and simulating qualitatively complex genetic regulatory networks of large dimension is studied. That work, as well as others along the same line, aims at dealing with situations in which

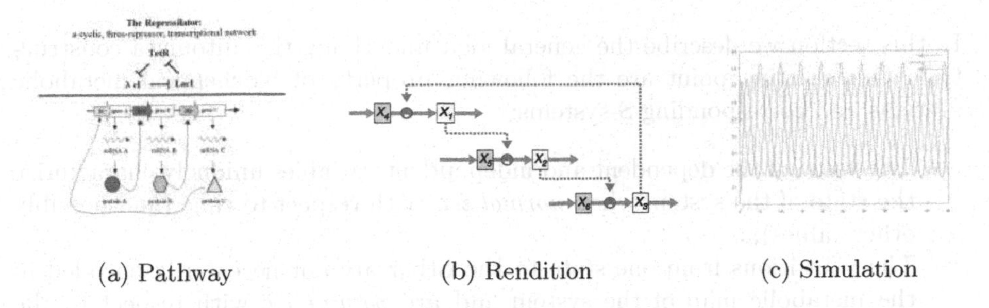

(a) Pathway (b) Rendition (c) Simulation

Fig. 1. The repressilator system metabolic map (a) (reprinted from [7]), its rendition in "cascade" form (b) and its oscillatory trace (c)

the lack of quantitative information forces simulation in a qualitative way, an assumption that marks the difference with the situation we study here and with kind of qualitative modeling we propose.

The problem of constructing an automaton from a given mathematical model of a complex system has also been previously considered in the literature. In particular, in the control literature, it has been deeply studied by Brockett in [4]. Our approach here is certainly at a lower level of generality as it deals with specific mathematical models (S-systems) and, moreover, tries to integrate the numerical determination of a solution for the system of differential equations involved with the automata construction.

The kind of formalization and tools we are proposing in this paper could, in general, be used to study on a more systematic way hypothesis on properties of complex systems of biochemical reactions. The research by Bhalla et al. in [3], for example, aims at proving that a sort of "learned behaviour" of biological systems is in fact stored within the mechanisms regulating intracellular biochemical reactions constituting signaling pathways. For this kind of studies, following [9], both qualitative and quantitative features of the system under study should be taken into account and we hope the system we propose can turn out useful on the ground of its ability to capture and compare temporal evolution of the system under study.

In [13] Cellerator is presented, a Mathematica package for biological modeling that bears many similarities with our project here.

Using a Temporal Logic query language to analyze continuous systems is investigated also in [14], as an extension of the *Qualitative Reasoning* approach (*cf.* [12]).

Finally, in [1] is reported a use of Hybrid Systems in modeling properties of systems of biochemical reactions. It is very interesting the idea of using the discrete component of the (hybrid) automaton to switch from one mode to another when (for example) the number of molecules grows over a certain threshold. In our framework the same effect should be captured in a sort of bottom-up fashion, by "gluing" different simulations determined with different sets of parameters.

2 XS-systems: S-Systems Extended with Automata

In this section we describe the general idea underlying the automata construction. Our starting point are the following property of biochemical metabolic systems and corresponding S-systems:

- The value of the dependent and independent variables uniquely characterize the state of the system when *normalized* with respect to time (and possibly other values);
- The transitions from one state to the other are not necessarily encoded in the metabolic map of the system and are *parametric* with respect to the value of constants in the S-system.

The idea behind the automata definition and construction we are going to define is to start with snapshots of the system variables' values that will constitute the possible states of the automaton. Transitions will be inferred from *traces* of the system variables' values evolution.

On the ground of the above observations we define:

Definition 1. *Given an S-system $\mathcal{S}$, the* S-system automaton $\mathcal{A_S}$ *associated to* $\mathcal{S}$ *is 4-tuple* $\mathcal{A_S} = (S, \Delta, S_0, F)$*, where* $S \subseteq D_1 \times \cdots \times D_{n+m}$ *is a (finite or infinite) set of states,* $\Delta \subseteq S \times S$ *is the transition relation, and* $S_0, F \subset S$ *are the initial and final states, respectively.*

Final states will be those states in which the simulation reaches a recognizable end and are supposed to represent "equilibrium" points for the pathway being modeled.

Definition 2. *A* trace *of an* S-system automaton $\mathcal{A_S}$ *is a (finite or infinite) sequence* $s_0, s_1, \ldots, s_n, \ldots$*, such that* $s_0 \in S_0$*,* $\Delta(s_i, s_{i+1})$ *for* $i \geq 0$*. A trace can also be defined as:*

$$\mathsf{trace}(\mathcal{A_S}) = \langle\langle X_1(t) \ldots X_n(t)\rangle \mid t \in \{t_0 + k\,\mathbf{step}\ :\ k \in \mathbb{N}\}\rangle,$$

is called the trace of $\mathcal{A_S}$

Clearly, from now on, if the analysis of the system is to be carried out in a finite interval of time $[0, t]$, a k such as the one in the above definition of trace will vary in $\{1, \ldots \lceil \frac{t}{\mathbf{step}} \rceil\}$.

Notice that, for fixed values of the independent variables, a unique trace is obtained when a simulation is performed. Moreover, while studying a trace of a system, it can be useful to concentrate (i.e. *project*) on one or more variables, which justifies the following definition:

Definition 3. *Given any set of variables* $U \subseteq \{X_1, \ldots, X_{n+m}\}$*, the sequence:*

$$\mathsf{trace}(\mathcal{A_S}|_U) = \langle\langle X_i(t) \mid X_i \in U\rangle\ :\ t \in \{t_0 + k\,\mathbf{step}\ :\ k \in \mathbb{N}\}\rangle,$$

is called the trace of U*. If* U *consists of a single variable* X_i *the trace is called the* trace of X_i.

Multiple traces arise as we start varying the values in the primary parameter sets. Collection of such traces, in general, allows one to study the different instances of the simulated metabolic pathway evolution. Such a collection will give rise to the automaton with corresponding transitions when a suitable *equivalence relation* on states is defined.

Construction of a *Collapsed* Automaton. We can easily construct an (in general unreduced) automaton $\mathcal{A}_S$ by simply associating a different state to each tuple $\langle X_1(t) \ldots X_n(t) \rangle$ as time grows according to a given time step. In this case there is a unique trace for the obtained automaton that is therefore called *linear* automaton.

Consider the function $X_i(t)$ at times $t_i, t_{i+1}, \ldots t_{i+5}$ as depicted in Figure 2 (a). In this case we have $\mathbf{step} = t_{i+1} - t_i$ as a result of (fixed) sampling or numerical integration. We associate the automata $\mathcal{A}_S$ to the trace of X_i simply by taking into account each time step. Note that this is not much different than what it is done by "untiming" a Timed Automata.

These automata are in some way "synchronous", in the sense that the time elapsed while passing from one state to the other is known and fixed (with respect to the numerical trace obtained from a source sampled at "fixed" intervals or from an integration algorithm employing a "fixed" step size).

Eventually, given a collection of linear automata (traces), we will propose to "glue" them together in a unique automaton capable of modeling various possible behaviors of the system. However, before that, we propose a method to collapse states of a linear automaton into equivalence classes capturing *qualitatively* the behavior of the system (along a single trace).

A solution to a given S-system is, in general, determined by a numerical approximation once the values for the independent (constant) variables are given. Given a numerical approximation to a solution of our S-system, for any given time step and any given (dependent) variable X_i, a linear automaton corresponding to the trace $\mathsf{trace}(X_i)$ could be reduced by using the following equivalence relation R_{i,δ_i}, which depends on the parameter δ_i: R_{δ_i} holds between two states $X_i(t + k\,\mathbf{step})$ and $X_i(t + (k+j)\,\mathbf{step})$ for $j > 0$, if and only if

$$| \dot{X}_i(t + k\,\mathbf{step}) - \dot{X}_i(t + (k-1)\,\mathbf{step}) | \leq \delta_i.$$

Notice that the first derivatives of functions expressing the variations of dependent variables, involved in establishing the validity of the above condition, are available during the numerical computation carried out to solve the underlying S-system. Moreover, notice that if the above collapsing is performed on an independent variable (assumed to be constant), the construction trivializes and a unique state is obtained. This construction is extended to the full collection of variables as follows:

Definition 4. *The relation $R_{\boldsymbol{\delta}}$ holds between two states $s_k = \boldsymbol{X}(t+k\,\mathbf{step})$ and $s_{k+j} = \boldsymbol{X}(t+(k+j)\,\mathbf{step}))$ with $j > 0$, if and only if, for each $i \in \{1, \ldots, n+m\}$,*

$$|\dot{X}_i(t + k\,\mathbf{step}) - \dot{X}_i(t + (k+j)\,\mathbf{step})| \leq \delta_i.$$

The collection $\{\delta_i \mid 1 \leq i \leq n+m\}$ is denoted by $\boldsymbol{\delta}$.

The (simple) idea is to choose as representative in each equivalence class, the element corresponding to the minimum time in the class. The following pseudo-algorithm explains how we compute the set of states of the collapsed automata:

Algorithm 1 COLLAPSE_STATES_INCREMENTALLY($\boldsymbol{\delta}$, **step**, t)

---

let $s_0 = \langle X_1(t_0) \ldots X_n(t_0) \rangle \in S_0$;
STATE := 0; *–initialize the current state*
REPR := 0; *–initialize the representative of the equivalence class*
while STATE $< \lceil \frac{t}{\mathbf{step}} \rceil$ **do**
  STATE := STATE +1;
  ... *–simulation proceeds using e.g. Euler's method*
  if $\exists \delta_i \in \boldsymbol{\delta} \left(|X_i(t_0 + \text{REPR}\,\mathbf{step}) - X_i(t_0 + \text{STATE}\,\mathbf{step})| \geq \delta_i \right)$ **then**
    REPR := STATE;
  end if
end while

If the guard of the **if**-statement in the above algorithm is weakened (e.g.) restricting the set of variables for which the condition is checked, the effect will be to "concentrate" on the restricted set of variables and, in general, to producing *less* states.

Consider again the function $X_i(t)$ described in Figure 2 (a). In Figure 2 (b) the effects of applying the collapsing algorithm are shown. With respect to $X_i(t)$ we obtain an automata $\mathcal{A}_{\mathcal{S}i}$ which has fewer states

$$\begin{aligned}\mathsf{states}(\mathcal{A}_{\mathcal{S}i}) = \langle \ldots &(t_i, X_j(t_i), \dot{X}_j(t_i)),\\ &(t_{i+2}, X_j(t_{i+2}), \dot{X}_j(t_{i+2})),\\ &(t_{i+5}, X_j(t_{i+5}), \dot{X}_j(t_{i+5})), \ldots \rangle\end{aligned}$$

Now suppose to have a different function $X_k(t)$. We associate to $X_k(t)$ the collapsed automata $\mathcal{A}_{\mathcal{S}k}$, such that

$$\mathsf{states}(\mathcal{A}_{\mathcal{S}k}) = \langle \ldots (t_i, X_k(t_i), \dot{X}_k(t_i)),\ \ (t_{i+4}, X_k(t_{i+4}), \dot{X}_k(t_{i+4})), \ldots \rangle$$

i.e. the "landmark" times are t_i and t_{i+4} in this case. In order to construct a useful automata for the analysis tool we construct the *merged* automata $\mathcal{A}_{\mathcal{S}jk}$ such that

$$\begin{aligned}\mathsf{states}(\mathcal{A}_{\mathcal{S}k}) = \langle \ldots\ &(t_i, X_j(t_i), \dot{X}_j(t_i)),\\ &(t_{i+2}, X_j(t_{i+2}), \dot{X}_j(t_{i+2})),\\ &(t_{i+4}, X_k(t_{i+4}), \dot{X}_k(t_{i+4})),\\ &(t_{i+5}, X_j(t_{i+5}), \dot{X}_j(t_{i+5})), \ldots \rangle\end{aligned}$$

i.e. automata $\mathcal{A}_{\mathcal{S}jk}$ is an *ordered merge* of the two automata $\mathcal{A}_{\mathcal{S}j}$, $\mathcal{A}_{\mathcal{S}k}$.

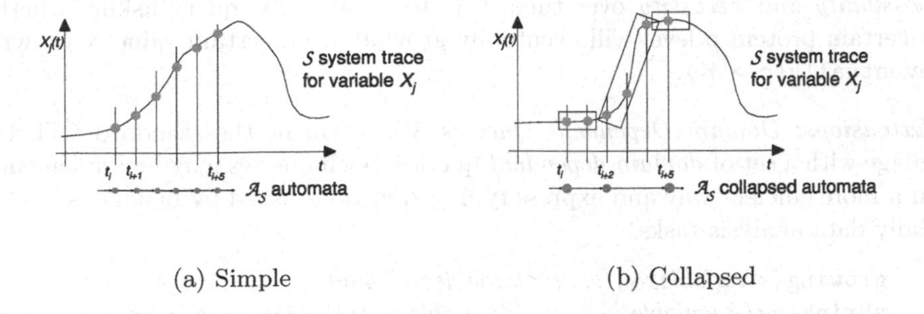

(a) Simple (b) Collapsed

Fig. 2. (a) Simple one-to-one construction of the "trace" automata $\mathcal{A}_\mathcal{S}$ for a S-system $\mathcal{S}$, and (b) the effects of the *collapsing* construction of the "trace" automata $\mathcal{A}_\mathcal{S}$ for a S-system $\mathcal{S}$

Normalizing and Projecting. According to the previous remark, in order to capture state-equivalence *modulo* normalization we begin with the following definition:

Definition 5. *Given a subset V of the set $\{X_1, \ldots, X_{n+m}\}$ of variables, we define the set of states* normalized *with respect to V as the following set of tuples:*

$$S_{\backslash V} = \left\{ \left\langle \frac{X_i(t_0 + k\,\mathbf{step})}{\nu_i(V, t_0 + k\,\mathbf{step})} \right\rangle \;:\; k \in \mathbb{N} \right\},$$

with the ν_i's are normalizing functions.

More complex forms of normalization can be obtained when other variable contribute into play. Moreover, notice that when we normalize with a collection of normalizing functions $\{\nu_i\}$ defined as $\nu_i(U,t) = X_i$ if $X_i \notin U$, or $\nu_i(U,t) = 1$ otherwise. Then the normalization corresponds to projecting with respect to the set of variables U.

3 Pathways Simulation Query System

In this section we briefly outline a language that can be used to inspect and formulate queries on the simulation results of XS-systems. The language is called ASySA (*Automata S-systems Simulation Analysis* language) and is used for expressing and verifying temporal properties of XS-systems and ADA systems. ASySA is essentially a *Temporal Logic* language (*cf.* [8]) with a specialized set of predicate variables whose aim is to make it easy to formulate queries on numerical quantities. The full rendition and semantics of ASySA is beyond the scope of this paper. Suffice to say that the standard CTL operators are available in

English-ized form[1]. The main operators in ASySA (and CTL) are used to denote *possibility* and *necessity* over time. E.g. to express the query asking whether a certain protein p level will eventually grow above a certain value K we write eventually(p > K).

Extensions: Domain Dependent Queries We augment the standard CTL language with a set of *domain dependent* queries. Such queries may be implemented in a more efficient way and express typical questions asked by biologists in their daily data analysis tasks.

- growing(<*variable*>$_1$, ..., <*variable*>$_k$) and shrinking(<*variable*>$_1$, ..., <*variable*>$_k$): the growing special operator is a *state formula* saying that all the variables mentioned are *growing* (*shrinking*) in a given state.
- represses(<*variable*>$_1$, <*variable*>$_2$) and activates(<*variable*>$_1$, <*variable*>$_2$): this special predicate is to be interpreted as a *path formula* stating that <*variable*>$_1$ (informally interpreted as a "gene product") *represses* (*activates*) the production of <*variable*>$_2$.

A Simple Example. Suppose we have a system like the well known *repressilator* system [7], coded as an *S*-system, as displayed in Figure 1. One of the proteins in the system is *LacI*. We can easily formulate and compute the following query oscillates(lacI_low, lacI_high) which can be translated into a regular Temporal Logic formula stating

```
eventually (not (always(lacI_low)) or (always(lacI_high))))
```

where lacI_low ≡ (lacI_low ≤ *low*) and lacI_high ≡ (lacI_high ≥ *high*), for appropriate values of *low* < *high*. The query asks the system whether the main property of the repressilator system holds over the length of the trace.

Implementation. We have implemented a prototype system embodying the concepts we have described in the previous sections. The system we describe here is the analysis component of the larger Simpathica (*Sim*ulation of *Path*ways and *I*ntegrated *C*oncurrent *A*nalysis). Figure 3 shows the main windows of the Simpathica pathway simulation tool and of the XSSYS analysis tool with the repressilator trace loaded and several queries performed.

4 Conclusions and Future Work

We have presented a "work in progress" that aims to construct a useful analysis tool by combining in a novel way several tools and techniques from Computer Science, Engineering and Biology/Biochemistry. By combining different

[1] We are providing an English form to the standard operators, in order to make the content of resulting language easier to manipulate for the intended audience, who has not been exposed to the notation used in Temporal Logic. We also note that, technically, we are missing EG, since we are only providing always as a rendition of AG.

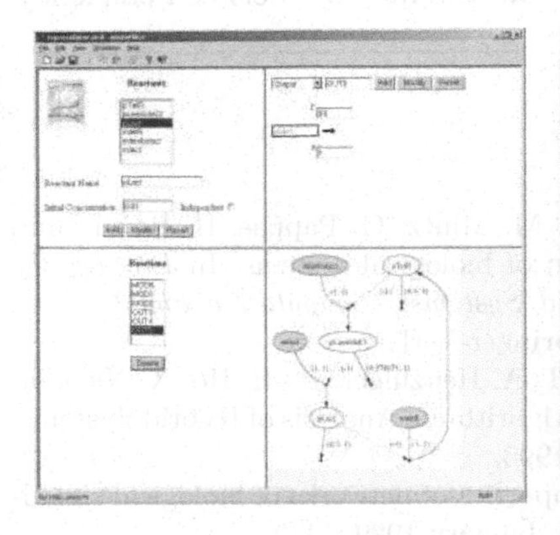

(a) Simpathica Main Window

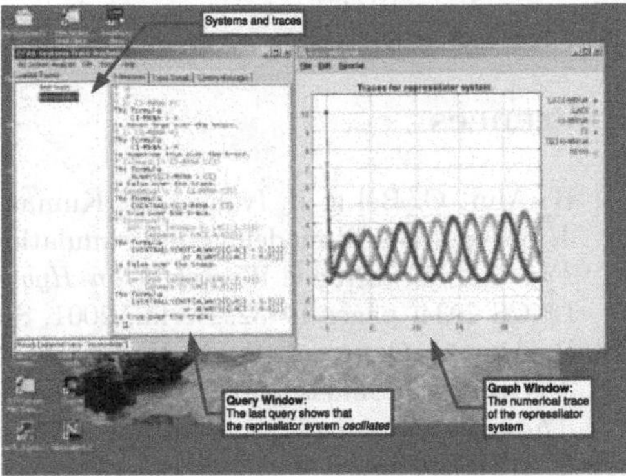

(b) XSSYS Interaction

Fig. 3. (a) The Simpathica main window. Reactants are entered on the upper left side and single reactions are entered in the top right side. Their list and graphical rendition appears in the bottom quadrants. (b) A view of the main window of the XSSYS XS-system analysis tool

and relatively simple components we obtain a synergistic effect that allows us to construct an efficient and effective analysis tool. We motivated our approach by analyzing a simple, yet interesting synthetic biological system: the "repressilator" [7].

Future Work. There are several topics we will investigate in our future work. First of all, since our collapsed automata relies on a "linear" approximation of a system trace, we will investigate the connections with the body of work on the analysis of Hybrid Systems (*cf.* [2]).

We are aware that our approach is be generalizable in several ways (e.g. see [4]). We will explore ideas from *Signal Processing* in our future work. The "model checking" algorithm we have implemented so far is extremely simple minded (yet extremely efficient), and is exploits the *linear* and *finite* nature of the XS-systems traces. This may turn out to be insufficient when comparing different traces produced under different conditions. Nevertheless, we conjecture that because of the special format of the XS-system traces we will be able to construct simple and efficient algorithms even in that case, without resorting to the full complexity of general Model Checkers and Theorem Provers.

Finally, the study of the XS-systems' canonical form poses two sets of problems. First, the use of XS-systems automata as *semantics* of S-systems in connection with the modular design of large maps simulating biochemical systems is still not completely resolved. Secondly, the *symbolic* manipulation of the DAE, algebraic constraints and collapsed automata, along with novel property checking algorithms is still open. We conjecture that the specialized and constrained form of XS-systems will allow us to successfully walk the fine line between efficiency, efficacy and expressiveness.

References

[1] R. Alur, C. Belta, F. Ivančić, V. Kumar, M. Mintz, G. Pappas, H. Rubin, and J. Schug. Hybrid modeling and simulation of biological systems. In *Proc. of the Fourth International Workshop on Hybrid Systems: Computation and Control*, LNCS 2034, pages 19–32, Berlin, 2001. Springer-Verlag. 435

[2] R. Alur, C. Courcoubetis, N. Halbwachs, T. A. Henzinger, P. -H. Ho, X. Nicollin, A. Olivero, J. Sifakis, and S. Yovine. The Algorithmic Analysis of Hybrid Systems. *Theoretical Computer Science*, 138:3–34, 1995. 432, 441

[3] U. S. Bhalla and R. Iyengar. Emergent properties of networks of biological signaling pathways. *SCIENCE*, 283:381–387, 15 January 1999. 435

[4] R. W. Brockett. Dynamical systems and their associated automata. In U. Helmke, R. Mennicken, and J. Saurer, editors, *Systems and Networks: Mathematical Theory and Applications—Proceedings of the 1993 MTNS*, volume 77, pages 49–69, Berlin, 1994. Akademie-Verlag. 435, 441

[5] A. Cornish-Bowden. *Fundamentals of Enzyme Kinetics.* Portland Press, London, second revised edition, 1999. 433

[6] H. de-Jong, M. Page, C. Hernandez, and J. Geiselmann. Qualitative simulation of genetic regulatory networks: methods and applications. In B. Nebel, editor, *Proc. of the 17th Int. Joint Conf. on Art. Int.*, San Mateo, CA, 2001. Morgan Kaufmann. 434

[7] M. Elowitz and S. Leibler. A synthetic oscillatory network of transcriptional regulators. *Nature*, 403:335–338, 2000. 434, 435, 440, 441

[8] E. A. Emerson. Temporal and Modal Logic. In J. van Leeuwen, editor, *Handbook of Theoretical Computer Science*, volume B, chapter 16, pages 995–1072. MIT Press, 1990. 439

[9] D. Endy and R. Brent. Modeling cellular behavior. *Nature*, 409(18):391–395, January 2001. 435

[10] R. Hofestädt and U. Scholz. Information processing for the analysis of metabolic pathways and inborn errors. *BioSystems*, 47:91–102, 1998.

[11] D. H. Irvine and M. A. Savageau. Efficient solution of nonlinear ordinary differential equations expressed in S-System canonical form. *SIAM Journal on Numerical Analysis*, 27(3):704–735, 1990. 434

[12] B. Kuipers. *Qualitative Reasoning.* MIT Press, 1994. 435

[13] B. E. Shapiro and E. D. Mjolsness. Developmental simulation with cellerator. In *Proc. of the Second International Conference on Systems Biology (ICSB)*, Pasadena, CA, November 2001. 435

[14] B. Shults and B. J. Kuipers. Proving properties of continuous systmes: qualitative simulation and temporal logic. *Artificial Intelligence Journal*, 92(1-2), 1997. 435

[15] E. O. Voit. *Canonical Nonlinear Modeling, S-system Approach to Understanding Complexity.* Van Nostrand Reinhold, New York, 1991. 432, 433

[16] E. O. Voit. *Computational Analysis of Biochemical Systems A Practical Guide for Biochemists and Molecular Biologists.* Cambridge University Press, 2000. 432, 433

A High Performance Scheme for EEG Compression Using a Multichannel Model

D. Gopikrishna and Anamitra Makur

Dept. of Electrical Communication Engineering, Indian Institute of Science
Bangalore, India.
gopikrishna@protocol.ece.iisc.ernet.in,
amakur@ece.iisc.ernet.in

Abstract. The amount of data contained in electroencephalogram (EEG) recordings is quite massive and this places constraints on bandwidth and storage. The requirement of online transmission of data needs a scheme that allows higher performance with lower computation. Single channel algorithms, when applied on multichannel EEG data fail to meet this requirement. While there have been many methods proposed for multichannel ECG compression, not much work appears to have been done in the area of multichannel EEG compression. In this paper, we present an EEG compression algorithm based on a multichannel model, which gives higher performance compared to other algorithms. Simulations have been performed on both normal and pathological EEG data and it is observed that a high compression ratio with very large SNR is obtained in both cases. The reconstructed signals are found to match the original signals very closely, thus confirming that diagnostic information is being preserved during transmission.

1 Introduction

Electroencephalogram (EEG), the manifestation of brain's electrical activity as scalp potentials, remains as one of the commonly used noninvasive techniques for understanding brain functions in health and disease. The amount of data contained in these EEG records is quite massive and hence a large bandwidth of transmission is needed if the data is to be transmitted along a communication channel. Since there are many situations where EEG data is transmitted over a communication channel, like transmission of ambulatory EEG to hospital, the problem of reducing the bandwidth of transmission of EEG data is of great importance. Also, the huge amount of data places a constraint on memory when the data needs to be stored in hospital database systems for future references.

Developing algorithms for the compression of EEG data without loosing relevant diagnostic information could solve the above problem. As there are many situations where online transmission of EEG data is required (for example in transmitting

S. Sahni et al. (Eds.) HiPC 2002, LNCS 2552, pp. 443-451, 2002.

ambulatory EEG to hospital in cases of head trauma, where time is very critical), there is a constraint on the computational complexity of the compression algorithm. If we decrease the computational complexity, we may not achieve low bit rate (channel bandwidth constraint) or lose diagnostic information. Thus, a scheme that allows higher performance with lower computation is the need of the hour.

There have been numerous efforts towards achieving lossless and near-lossless compression of EEG. They can be broadly classified into two categories viz. single channel and multichannel compression.

In single channel compression, each channel of EEG is considered as a separate time series and the compression algorithm devised for a single time series is applied iteratively to all channels. Some of the attempts in this direction are Huffman coding, repetition count compression, linear predictive coding and adaptive linear prediction [1]. Compression schemes based on neural network predictor [2], context based bias cancellation [3], sub-band coding [4], [5] and chaotic modeling [6] have also been proposed.

Single channel compression is sub-optimal, as the correlation between channels [7] is not utilized. With all the channels emanating from the same process, there is bound to be some mutual information between them [8]. Hence multichannel compression techniques have been used for effective compression. Ke Chu Yi, et al used an adaptive bit assignment algorithm to achieve EEG compression [9]. Agarwal et al have proposed a long-term prediction (LTP) for EEG compression [10]. This attempt at EEG-LTP is more a scheme of optimal representation/display of information that would assist the doctor to see the huge data at once than a 'compression' scheme, which would allow the reconstruction of the signal. In the light of the fact that no substantial work on multichannel EEG compression is reported in literature, we will look into some of the multichannel schemes reported for electrocardiogram (ECG).

Mammen and Ramamurthy have used m-AZTEC and classified vector quantization (VQ) in [11]. Cetin et al have used multirate signal processing and transform coding in [12] for compressing multichannel ECG. But the increase in computational complexity due to the method is not justified by the small improvement in performance compared to iterative application of single channel algorithm to multichannel data. Paggetti et al have used a template matching algorithm in [13] for multichannel ECG compression. But for a random-like signal like EEG, this method is not encouraging. Cohen et al have used long term ECG prediction [14], which uses pseudo-periodic nature of ECG and makes it unsuitable for EEG compression. Carl Taswell et al have used near-best wavelet packet transform in [15] and [16] for compression of multichannel multimodal signals. The method is more useful for quality-controlled compression in applications like telemedicine where multimodal signals are transmitted and is nowhere near an optimal scheme for multichannel compression. To overcome the mismatch problem of the VQ scheme in [11], Shaou-Gang Miaou et al have used Adaptive vector quantization [17]. But for some of the reasons given in [17], this method does not give a significant improvement in performance compared to the methods outlined above. Prieto et al have proposed a model based compression scheme using FIR filter identification in [18].

As seen above, there have been many methods proposed for multi-channel ECG compression, but negligible work has taken place in the arena of multichannel EEG compression. As EEG is a very different signal from that of ECG with its random like nature, the above methods do not automatically apply for EEG and there is a need to develop independent methods to compress multichannel EEG data.

In this paper, we have used a new model for the multichannel EEG signals that captures the inter-channel correlation and gives a higher performance with respect to compression, than single channel methods and other multichannel methods at an affordable complexity. Wavelets and ARX (autoregressive with exogenous input) model have been used effectively to obtain high compression ratios with improved SNR.

2 Compression Method

EEG is a multi-channel signal, in the sense that in any standard of recording EEG, different electrodes are placed on different locations on the scalp and signals recorded. Hence the signals are emanating from the same process, but sensed at different locations of the system. Therefore, an assumption is made that there exists a model or a filter $H_i(z)$ whose input is one of the channels x(n), and the output another channel $y_i(n)$. m such models, for m channels $y_1(n)$ to $y_m(n)$, can be fitted into a compression scheme; better the models, better the compression system. Thus the problem is to identify the models/filters in question.

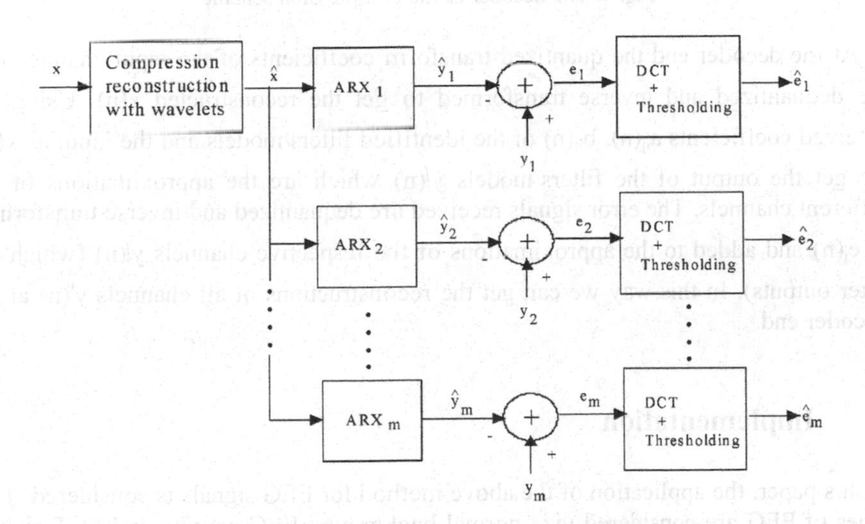

Fig. 1. The encoder of the compression scheme

In the general scheme of the encoder, the input channel x(n) to all the filters/models is compressed and reconstructed using Wavelet transform. This reconstructed signal $\hat{x}(n)$ is used as the input to the models or IIR filters $H_i(z)$, whose coefficients/parameters $a_{ij}(n)$ (denominator) and $b_{ij}(n)$ (numerator) are to be found. This is done because only the reconstructed signal $\hat{x}(n)$ is available at the decoder end. The outputs of the filters at the encoder $\hat{y}_i(n)$ are subtracted from that of the original signal $y_i(n)$ to get an error signal $e_i(n)$. This error signal is transformed with Discrete Cosine Transform (DCT), quantized and transmitted. Also the co-efficients of the identified filters/models are transmitted without loss using an entropy code. The quantized transform coefficients of the input channel $x(n)$ are also transmitted to the decoder.

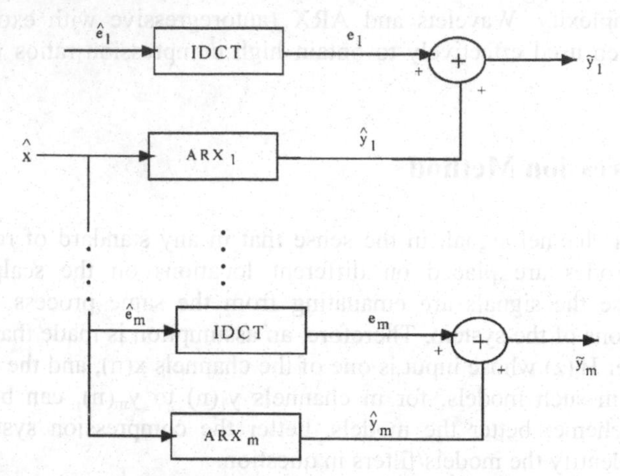

Fig. 2. The decoder of the compression scheme

At the decoder end the quantized transform coefficients of the input channel $x(n)$ are dequantized and inverse transformed to get the reconstructed $x'(n)$. Using the received coefficients $a_{ij}(n)$, $b_{ij}(n)$ of the identified filters/models and the input as $x'(n)$, we get the output of the filters/models $y_i'(n)$ which are the approximations of the different channels. The error signals received are dequantized and inverse transformed to $e_i(n)$ and added to the approximations of the respective channels $y_i'(n)$ (which are filter outputs). In this way we can get the reconstructions of all channels $y_i''(n)$ at the decoder end.

3 Implementation

In this paper, the application of the above method for EEG signals is considered. Two types of EEG are considered viz., normal background EEG activity and ECT signals (spike and wave activity). ECT stands for Electro-convulsive therapy, wherein electric current is passed through the brain to induce seizure. This signal consists of spike and wave complexes and it represents a pathological EEG waveform.

If FIR filters are employed as $H_i(z)$, then a very high order (typically 100 [18]) has to be used, which increases the computational complexity and make the method unsuitable for online applications; also implementation on a VLSI chip becomes costly. Hence we have used an ARX model (IIR of both numerator and denominator order 10), which is given by the equation

$$y_i(n)= \;_{j}\, b_{ij}\, x(n-j) - \;_{j}\, a_{ij}\, y_i'(n-j) + e_i(n)$$

where $y_i(n)$ is the channel i , $x(n)$ the input to $H_1(z)...H_m(z)$ and $y_i'(n)$ the output of $H_i(z)$. a_{ij} and b_{ij} are the denominator and numerator coefficients of $H_i(z)$ respectively.

The performance, as is shown later in the paper, is as good as or even better than using an FIR filter. Thus the reduced order of the model is an improvement, which gives higher performance with lower complexity.

As the reconstruction of the input channel is extremely important for the performance of the codec, wavelet compression is used instead of DCT. A 3-level wavelet decomposition using Daubechies-3 wavelet is obtained and only the first approximation and two detail levels are retained and bits are optimally allocated to achieve compression (Refer [19] for further details). Nevertheless, for coding the error signal, DCT is used as using wavelets in that case does not provide any improvement in performance.

4 Results and Discussion

We present the results in two parts. First, we consider the results obtained by applying the method on ECT data.

4.1 Application to ECT Data

ECT signal that we have considered is a 4-channel signal consisting of the right frontal (RF), left frontal (LF), left temporal (LT) and right temporal (RT) signals. A plot of 1000 points of a typical channel (original and reconstructed) in mid-seizure looks as shown in fig.3.

The quantitative measure of the quality of compression is given by the Signal to Noise Ratio (SNR) and multichannel compression ratio (CR) which are defined as given below

$$\text{SNR} \quad 10*\log 10 \frac{\text{Variance (signal)}}{\text{Variance (reconstruction error)}}$$

$$\text{CR} \quad \frac{\text{Total bits to represent the signal points}}{\text{Totbits}}$$

where Totbits = Bits used to represent the wavelet decomposition of the first channel +
Bits used to represent the DCT coefficients of the errors of each channel +
Bits used to represent the model parameters

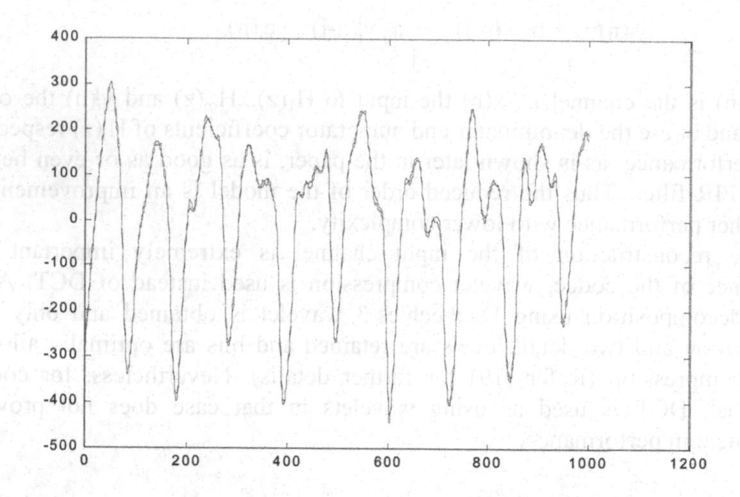

Fig. 3. Original and reconstructed signal of a typical channel of ECT (1000 pts) showing mid-seizure spike and wave complex

For the 4-channel signal mentioned above, we get an SNR (Signal to Noise Ratio) of 24.25 dB, 26.5 dB, 20.1 dB and 21.6 dB for RF, LF, LT and RT respectively with RF taken as the input channel. The multi-channel compression ratio got is 13.35. To judge the performance of the system vis a vis the single channel compression system, we define a parameter Improvement Factor (IF) as the ratio of the compression ratio obtained by applying the single channel coding on each of the channels iteratively to the compression ratio got by applying the given multi-channel algorithm, keeping the SNR constant. This parameter gives a fair idea of the improvement we get by using the multi-channel algorithm instead of the single channel algorithm. The value of IF we get here is 1.67, which means that for the same SNR, using single channel algorithm to compress the 4-channel data would have yielded a compression ratio of just 7.9 instead of 13.35. This indeed is a tremendous improvement.

4.2 Application to Background EEG

We have considered 4 channels of background EEG activity. This looks like a random signal and does not posses the structure of ECT signals. Nevertheless the ARX model is able to capture the inter-channel correlation as shown below in the good reconstruction of the signal. The typical channel and its reconstruction are given in figure 4.

For the 4-channel signal (X1, X2, X3 and X4) mentioned above, the SNR Vs CR characteristic of channel-2 is as given in fig.5.

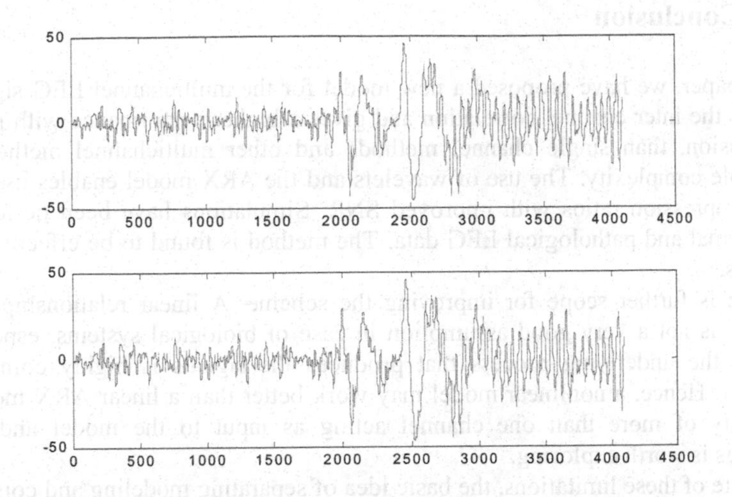

Fig. 4. Original and reconstruction of a typical channel of background EEG

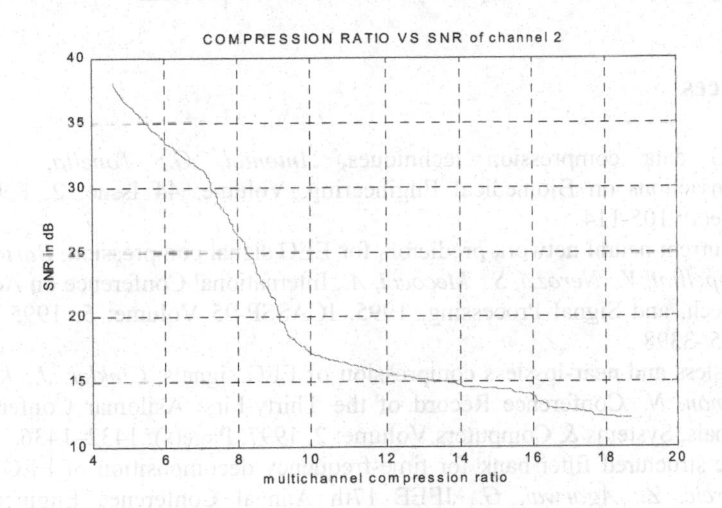

Fig. 5. CR vs SNR characteristic for EEG channel 2

We get similar curves for the other channels with an IF of 1.57. Also as seen from the figures, the original and the reconstructed signals are almost indistinguishable. Even though our SNR is finite, all the diagnostic information is preserved. This type of coding is called near-lossless coding.

It is observed that the use of wavelets and ARX model has not only reduced the order of the model by a factor of 5 (from 100 to 20), but also improved the compression ratio for the same SNR. Thus there is an overall improvement in the performance of the system by using wavelets with an ARX model instead of DCT with FIR filters. Thus the proposed method has achieved a higher performance with lower computation.

5 Conclusion

In this paper, we have proposed a new model for the multichannel EEG signals that captures the inter-channel correlation and gives a higher performance with respect to compression, than single channel methods and other multichannel methods at an affordable complexity. The use of wavelets and the ARX model enables us to obtain high compression ratios with improved SNR. Simulations have been performed on both normal and pathological EEG data. The method is found to be effective in both the cases.

There is further scope for improving the scheme. A linear relationship between channels is not a very good assumption in case of biological systems, especially in EEG as the underlying process that produces the signals is highly complex and nonlinear. Hence, a nonlinear model may work better than a linear ARX model. The possibility of more than one channel acting as input to the model and channel groupings is worth exploring.

In spite of these limitations, the basic idea of separating modeling and compression is illuminating. As we need high performance computation for compression, a performing model with low complexity would solve the problem.

References

[1] EEG data compression techniques, *Antoniol, G.; Tonella, P.*, IEEE Transactions on Biomedical Engineering, Volume: 44 Issue: 2, Feb. 1997, Page(s): 105-114

[2] Recurrent neural network predictors for EEG signal compression, *Bartolini, F.; Cappellini, V.; Nerozzi, S.; Mecocci, A.*, International Conference on Acoustics, Speech, and Signal Processing, 1995. ICASSP-95 Volume: 5, 1995 Page(s): 3395 -3398 .

[3] Lossless and near-lossless compression of EEG signals, *Cinkler, J.; Kong, X., Memon, N.* ,Conference Record of the Thirty-First Asilomar Conference on Signals, Systems & Computers.Volume: 2, 1997, Page(s): 1432–1436.

[4] Tree structured filter bank for time-frequency decomposition of EEG signals, *Sijercic, Z.; Agarwal, G.*, IEEE 17th Annual Conference Engineering in Medicine and Biology Society 1995, Volume: 2, 1995, Page(s): 991–992.

[5] EEG signal compression with ADPCM subband coding, *Sijercic, Z.; Agarwal, G.C.; Anderson, C.W.*, IEEE 39th Midwest symposium on Circuits and Systems, Volume: 2, 1996, Page(s): 695–698.

[6] Use of chaotic modeling for transmission of EEG data , *Kavitha, V.; Narayana Dutt, D.*, ICICS Proceedings of 1997 International Conference on Information, Communications and Signal Processing, 1997. Volume: 3, 1997, Page(s): 1262 –1265.

[7] Spatio-temporal EEG information transfer in an episode of epilepsy, A.M. Albano et al Nonlinear Dynamics and Brain Functioning (Editors-N.Pradhan, P.E.Rapp and R.Sreenivasan), Nova Science Publishers, Newyork, 199, Page(s): 411-434.

[8] Entropy of brain rhythms: normal versus injury EEG, *Thakor, N.V.; Paul, J.; Tong, S.; Zhu, Y.; Bezerianos, A.*, Proceedings of the 11th IEEE Signal Processing Workshop on Statistical Signal Processing, 2001, Page(s): 261–264

[9] A lossless compression algorithm for multichannel EEG, *Ke Chu Yi; Mingui Sun; Ching Chung Li; Sclabassi, R.J.*, Proceedings of the First Joint BMES/EMBS Conference, 1999. Volume: 1, 1999, Page(s): 429

[10] Long-term EEG compression for intensive-care settings, *Agarwal, R.; Gotman, J.* IEEE Engineering in Medicine and Biology Magazine, Volume: 20 Issue: 5, Sept.-Oct. 2001, Page(s): 23 –29

[11] Vector quantization for compression of multichannel ECG, *Mammen, C.P.; Ramamurthi, B.* ,IEEE Transactions on Biomedical Engineering, Volume: 37 Issue: 9, Sept. 1990, Page(s): 821 –825.

[12] Multichannel ECG data compression by multirate signal processing and transform domain coding techniques ,*Cetin, A.E.; Koymen, H.; Aydin, M.C.*, IEEE Transactions on Biomedical Engineering, Volume: 40 Issue: 5 , May 1993,Page(s): 495 -499

[13] A multichannel template based data compression algorithm ,*Paggetti, C.; Lusini, M.; Varanini, M.; Taddei, A.; Marchesi, C.*, Computers in Cardiology 1994, Page(s): 629 -632

[14] Compression of multichannel ECG through multichannel long-term prediction *Cohen, A.; Zigel, Y.*, IEEE Engineering in Medicine and Biology Magazine, Volume: 17 Issue: 1 , Jan.-Feb. 1998, Page(s): 109 -115

[15] Near-best WPT compression of polysomnograms, *Niederholz, J.; Taswell, C.*,Proceedings of the First Joint BMES/EMBS Conference, Volume: 2, 1999, Page(s): 961.

[16] Quality controlled compression of polysomnograms ,*Taswell, C.; Niederholz, J.*,Proceedings of the First Joint BMES/EMBS Conference, 1999, Volume: 2, 1999, Page(s): 944.

[17] Multichannel ECG compression using multichannel adaptive vector quantization , *Shaou-Gang Miaou; Heng-Lin Yen,* IEEE Transactions on Biomedical Engineering, Volume: 48 Issue: 10, Oct. 2001, Page(s): 1203 -1207

[18] Multichannel ECG data compression method based on a new modeling method *Prieto, A.; Mailhes, C.* ,Computers in Cardiology, 2001, Page(s): 261 -264

[19] Gilbert Strang and Truong Ngugen. : Wavelets and Filterbanks, Cambridge University Press, 1996.

Scalability and Performance of Multi-threaded Algorithms for International Fare Construction on High-Performance Machines

Chandra N. Sekharan[1], Krishnan Saranathan[2], Raj Sivakumar[2], and Zia Taherbhai[2]

[1]Department of Computer Science, Loyola University of Chicago
6525 N. Sheridan Road, IL 60626
chandra@cs.luc.edu
[2]Information Sciences Division- WHQKB, United Airlines
P.O.Box 66100, Chicago, IL 60666-0100

Abstract. We describe the design, implementation and performance of a project for constructing international fares at United Airlines. An efficient fare construction engine allows an airline to simplify, and automate the process of pricing fares in lucrative international markets. The impact of a fare engine to the revenues of a large airline has been estimated at between $20M and $60M per year. The goal is to design an efficient software system for handling 10 Gb of data pertaining to base fares from all airlines, and to generate over 250 million memory-resident records (fares). The software architecture uses a 64-bit, object-oriented, and multi-threaded approach and the hardware platforms used for benchmarking include a 24-CPU, IBM S-80 and 32-CPU, Hewlett-Packard Superdome. Two competing software designs using (i) dynamic memory and (ii) static memory are compared. Critical part of the design includes a scheduler that uses a heuristic for load balancing which is provably within a constant factor of optimality. Performance results are presented and discussed.

1 Introduction and Terminology

In general, a pricing department in an airline is responsible for setting price levels and fare rules for the airline's fare products. Fare rules involve conditions for which travel may be permitted or the circumstances under which the given fare may be applicable; such conditions may include for instance, Saturday night stay, 21-day advance purchase, specific travel dates etc. The process of arriving at fares is based on analyzing the competitors' fare actions and initiatives and the airline's own proactive pricing moves. It is quite obvious that pricing is an important part of an airline's business model and a driver for profits. Although pricing is essential in both domestic and international markets, it has a special significance in the international markets owing to its potential for generating higher revenues. By using marketplace knowledge and high volumes of fare data from other airlines, pricing analysts can price fare products effectively. Pricing is a bootstrapped process wherein new fares are determined by applying changes to previous fares, which in turn, were determined by applying changes to the fares that had existed before and so on. To facilitate the collection and

S. Sahni et al. (Eds.) HiPC 2002, LNCS 2552, pp. 452-460, 2002.

distribution of fares and fare related data for the airlines and travel industry, over 20 international airlines have set up and manage an organization called the Airline Tariff Publishing Company (ATPCO). ATPCO [1] collects fare information from over 500 airlines and distributes it to global distribution systems such as Galileo International and Sabre and computer reservation systems (CRS). The fare related data is sent to subscribers three times a day and the data contains all the changes to fares from the previous transmission for all airlines along with rules and footnotes. The reverse transmission of fare data involves an airline sending new, adjusted fares to ATPCO, based on pricing analysts' determination.

1.1 Terminology

For an airline, a *fare* is characterized by an origin city, a destination city, carrier name, certain rules and footnotes, cost, and several other attributes. A *domestic fare* is defined to be a fare where the origin and destination are US cities. An *international fare* is defined to be a fare where either the *origin city* is a US city and the *destination city* is a foreign city or vice versa. We call a fare associated with an origin city *A* and a destination city *B,* as *specified* if either cities *A* and *B* are US cities or *A* and *B* are major gateway cities in different countries. In the first scenario, the fares are called domestic specified fares and in the second, international specified fares. Examples of specified international fares include those associated with Chicago to London (United Airlines), New York City to Frankfurt (United Airlines), and Detroit to Frankfurt (Northwest Airlines). An important characteristic of specified fares is that these are fares that can actually be sold to passengers and hence readily available for look-up. In contrast to specified fares, there are *add-on* or *arbitrary* fares which are not saleable, only exist in conjunction with specified international fares, and are associated with travel between a gateway city and an interior city in the same country. For instance, with United Airlines as the carrier, in the itinerary St. Louis Chicago London Manchester, the fares associated with St. Louis Chicago and London Manchester are add-on fares where Chicago and London are gateways and the others are interior cities. The fares associated with Chicago London are specified international fares. In [2] a version of the international pricing problem has been formulated as a linear program, which minimizes the total penalty across the fare network. The whole problem of constructing fares is fairly involved and hence we will not be able to describe it in complete detail here. The purpose of this work is to outline an architectural overview of the software, the design issues at a high-level of abstraction, and benchmarking on high-performance parallel machines.

The rest of the paper is organized as follows. In Section 2, we will cover the basics of international fare construction, data sizes, and requirements. Section 3 will address design issues in the algorithms, scheduling, static and dynamic memory implementations. In Section 4 we will talk about some basic architectural reviews of the hardware platforms used in benchmarking, performance results, and theoretical results. We then present some conclusions.

2 International Fare Construction

International fare information from ATPCO consists of separate flat files for arbitrary fares, specified fares, tariff tables etc. Loosely speaking, fare construction involves putting together specified fares and add-on fares, based on certain criteria and forming new fares. As an example, an airline may have a Chicago to London fare (specified), but not have a fare from St. Louis to London. One solution is to create an add-on fare for St. Louis to Chicago segment and combine it with the known specified fare from Chicago to London forming a St. Louis Chicago London *constructed* fare. It may be noted that the add-on fare is typically way below the cost of an equivalent domestic specified fare or even negative. We will call such a constructed fare obtained by combining an add-on with a specified, as of type *AS*. Analogously, we could have a constructed fare of type *SA (*by combining a specified with an add-on), and of type *ASA* (by combining an add-on, specified, and an add-on in that order). Figure 1 shows the different combinations that could potentially be generated during fare construction. In Figure 1, as an example, there are four add-on fares from origin city to gateway city **A**, three fares from gateway city **B** to the destination, and one specified from **A** and **B.**

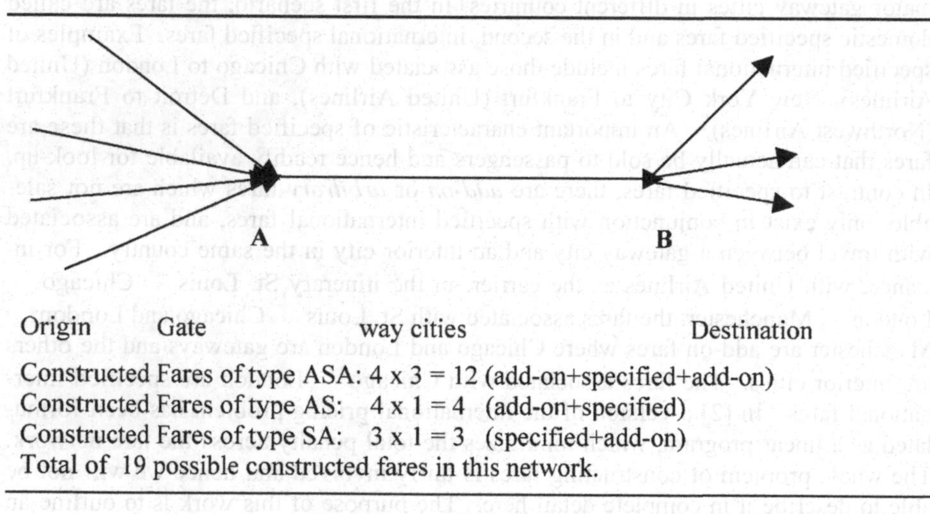

A
B
Origin
Gate way cities
Destination
Constructed Fares of type ASA: 4 x 3 = 12 (add-on+specified+add-on)
Constructed Fares of type AS: 4 x 1 = 4 (add-on+specified)
Constructed Fares of type SA: 3 x 1 = 3 (specified+add-on)
Total of 19 possible constructed fares in this network.

Fig. 1.

In actuality, not all combinations of constructed fares are possible as shown in Figure 1 which only shows a worst-case scenario. This is owing to the fact that there are several compatibility criteria that limit an add-on fare combining with a specified and vice-versa. Without going into too much detail, the fare combinability criteria include checks for compatible tariff codes, fare classes, travel dates, zone codes, carrier name, one way or round trip fares, and gateway city compatibility. To get an idea of the total number of constructed fares let us briefly discuss the sizes of various data that are needed for fare construction. The number of specified fares is typically between 10 and 15 million and the number of add-ons is between 1 and 2 million. Other tables such as for zone, city codes etc., take up to 1 million records. The fare construction needs to be done for all airlines. The data changes three times daily and hence fast construction of fares is needed. The number of constructed fares has been estimated to be around 250 million, each fare occupying about 120 bytes of data. The software requirements by the pricing analysts include the following:

Ability to identify all the markets (origin city-destination city pair) which are affected by the incoming transmission of changed fares.
Generation of all constructed fares in a given market.
Identification of all markets that will be affected by a user-defined change in the fare records.

To accommodate all of these requirements, the design of the software comprises two phases. The first phase is called *Link Formation* and the second the *Fare Construction.* The next section will describe the Link Formation phase in great detail.

3 Link Formation: Design and Implementation

The link formation phase determines, in essence, the arbitrary records that are compatible with the specified records and vice versa. To do this in parallel, we need to ascertain the right amount of granularity for each processor considering the huge numbers of arbitrary and specified records. Each fare can be considered to have an *indexed part* and a *non-indexed part.* The indexed-part of the fare is the tuple *[city,country,carrier,tariff,one way/round trip]* and the non-indexed part is the tuple that consists of all remaining attributes of the fare record [zone,dates, fare class, link#, …]. The records for which the indexed-part is identical is called a **block**. The idea of a block is an equivalence relation that partitions the specified and arbitrary non-index tables into specified blocks and arbitrary blocks, the sizes of which are roughly 1 million and 15,000 respectively for a typical master fare file. It may also be noted that each block of data from either table is only compatible with exactly one block from the other table. This is owing to the fact, the indexed-part should be the same for AS or SA type of compatibility. As a consequence, we need to compare roughly 15,000 blocks of specified table with 15,000 of arbitrary table for performing link formation of one type (AS or SA). When we find a pair of conforming arbitrary-specified blocks, then we compare each record of one block with each record of the other within the blocks for compatibility. This is called the *cross-product* operation in the link formation phase. Hence for both AS and SA we would need to do about 30,000 block cross-products. The problem is we do not know which arbitrary block goes with which specified block (or vice-versa) for combining records. Although the number of blocks actually involved in the cross-product is only 30,000 there is a tremendous amount of imbalance in the computational load. The size of each block varies from 1 to several 100, 000's. Hence the block-to-block cross-product suffers from an inherent load imbalance and the granularity is uneven. Furthermore, the generation of enormous in-memory links creates crucial bottlenecks for both run-time and memory requirements.

The design choices are as follows (the reasons for which would be elaborated in the full paper).

1. We form a linked-list of compatible records for each record in the Specified and Arbitrary table respectively for both AS and SA types but not ASA type. This has the effect of reducing the memory requirement to a great degree.
2. We postpone complete compatibility checking to when fares are constructed and not when links are formed. This choice greatly increases the amount of memory needed and to an unpredictable level. Full compatibility checking would whittle down the number of links formed but the user requirements made it a necessity to do this in the second phase.
3. AS and SA links are kept track of in both specified and arbitrary tables. This has the effect of doubling the memory needed.
4. Each arbitrary block vs specified block cross-product operation is performed by a thread.

There are several data structures that are used in the link formation phase which include the specified and arbitrary tables, the specified index and arbitrary index arrays that index into the specified and arbitrary tables, the specified changed fares table, arbitrary changed fares table, zone code table, and fare class table. We know that the work involved is roughly, 30,000 arbitrary block versus specified block cross-products. But the key problem is that these are widely varying in terms of computational work-loads. For instance, the largest cross-product involves around 220 million iterations of (partial) compatibility checking and the smallest involves just 1 iteration and there are any number of work-loads that are between the two extremes. Basically, a thread's work is proportional to the size of |Arbitrary block| x |Specified block|. Hence we designed a simple Scheduler that schedules the threads using a greedy strategy and determines the arbitrary versus specified block mapping in parallel. We will discuss the Scheduler later in this section. As for links formed, we have two possible approaches: one is to create the links dynamically (using *New[]* call in C++ or the *malloc* call in C) or to use static arrays to store the links generated. There is a time-space tradeoff here and hence we will discuss the implementation and performance of both strategies. We used C++ and POSIX thread library to implement our algorithms.

3.1 Link Formation: Implementation

Dynamic Memory: Initially we allocate a chunk of *new[]* say, of size 50 (a variable parameter in the actual implementation) for each of the records for all the four possible cases, namely, AS Specified Links, SA Specified Links, AS Arbitrary Links, and SA Arbitrary Links. Then we perform the cross-product using multiple threads and as and when compatible links between records are found, we store the link information using dynamic memory. If the number of links generated turns out to be bigger than the size allocated, we resize allocated storage with an increment (also parametrized). The size of in-memory storage for the links is roughly of the order of 6-8 Gb necessitating a 64-bit implementation. There are two serious drawbacks associated with link storage using dynamic memory. The first is memory fragmentation which resulted in about 1 to 2 Gb of extra memory over what is really needed. The second is scalability due to heap management using a single heap. However, this is overcome in a straightforward manner by using a separate heap for each thread supported by most Unix operating systems. The initial run took about an hour for generating the links using 16 threads. Using a variety of code optimization and tuning techniques, we benchmarked the results on two high-performance machines, namely, the IBM S-80, a pure symmetric multi-processing machine (SMP) and HP Superdome, a CC-NUMA machine.

3.2 Performance Benchmarks

The link formation phase and the data structures needed to construct fares are all coded up in C++ using the POSIX thread library. The hardware involved an IBM S-80 and a HP Superdome. The IBM S-80 is an SMP system that can have up to 24 CPUs (Power 3, 450 MHz chips) with up to 64 Gb of RAM and an aggregate memory bandwidth of 24 Gb/second. The operating system used is AIX version 4.3.3. Figure 2 shows the results of the benchmarking with up to 18 threads on S-80. The HP Superdome is a cell-based hierarchical cross bar system. Each cell consists of 4 CPUs configured as an SMP and up to 16 cells can be interconnected by a layered cross-bar system. The Superdome that was used for benchmarking had PA-8600, 550 MHz chips with 2Gb of RAM per cell and 32 CPUs for a total of 64 Gb of distributed memory. The operating system used is HP-UX version 11. The results are shown in Figure 3. We used a threads-to-processor mapping ratio of 1:1 without processor binding. In both of the systems, the link formation showed good scalability up to about 16 threads and then flattens out in performance. This due to the fact that the scalability of the solution is ultimately limited to computing the largest block-pair cross-product (about 220 million iterations).

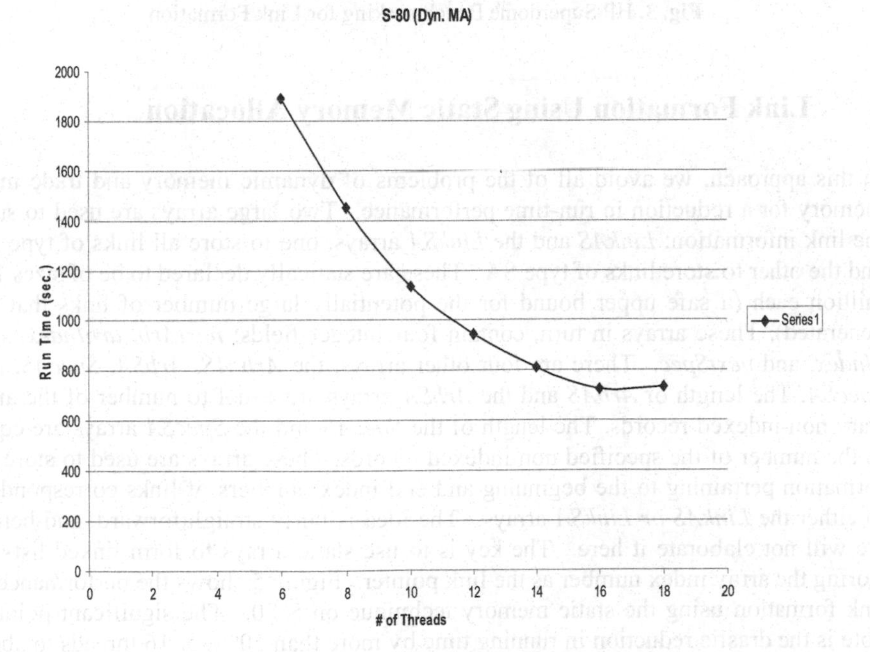

Fig. 2. IBM S-80 benchmarking for Link Formation

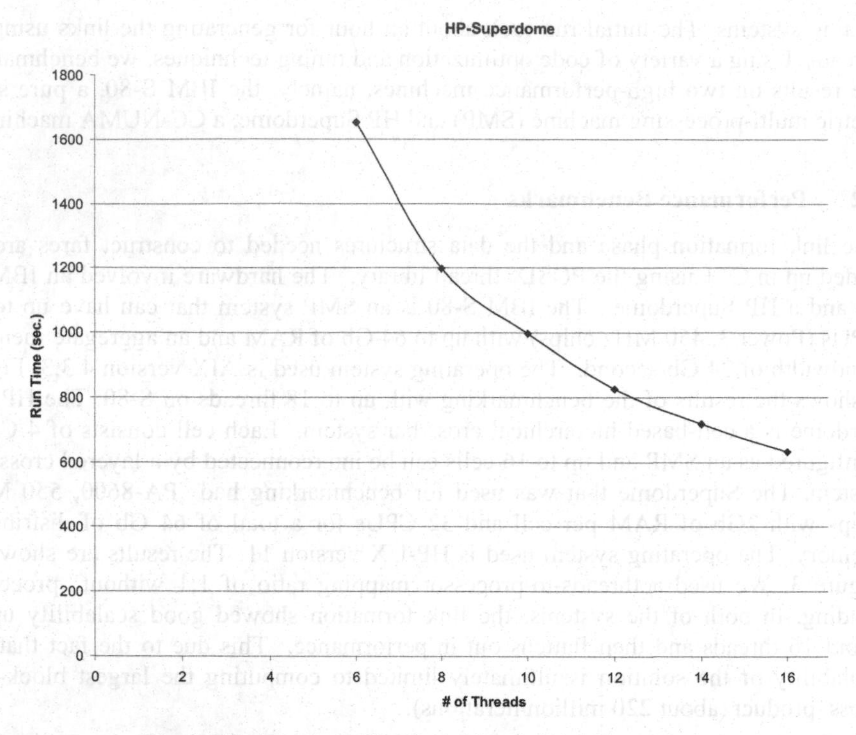

Fig. 3. HP Superdome Benchmarking for Link Formation

4 Link Formation Using Static Memory Allocation

In this approach, we avoid all of the problems of dynamic memory and trade more memory for a reduction in run-time performance. Two large arrays are used to store the link information: *LinkAS* and the *LinkSA* arrays, one to store all links of type AS and the other to store links of type SA. These are statically declared to be of sizes 140 million each (a safe upper bound for the potentially large number of links that are generated). These arrays in turn, contain four integer fields: *nextArb, arbIndex, specIndex*, and *nextSpec*. There are four other arrays, the *Arb AS*, *ArbSA*, *SpecAS*, and *SpecSA*. The length of *ArbAS* and the *ArbSA* arrays are equal to number of the arbitrary non-indexed records. The length of the *SpecAS* and the *SpecSA* arrays are equal to the number of the specified non-indexed records. These arrays are used to store information pertaining to the beginning and end index numbers of links corresponding to either the *LinkAS or LinkSA* arrays. The idea is fairly straightforward, and hence, we will not elaborate it here. The key is to use static arrays to form linked-lists by storing the array index number as the link pointer. Figure 5 shows the performance of link formation using the static memory technique on S-80. The significant point to note is the drastic reduction in running time by more than 50% for 16 threads to about 5 minutes. The total memory needed goes up to about 12 Gb which appears to be a great trade-off, considering that memory has become inexpensive lately.

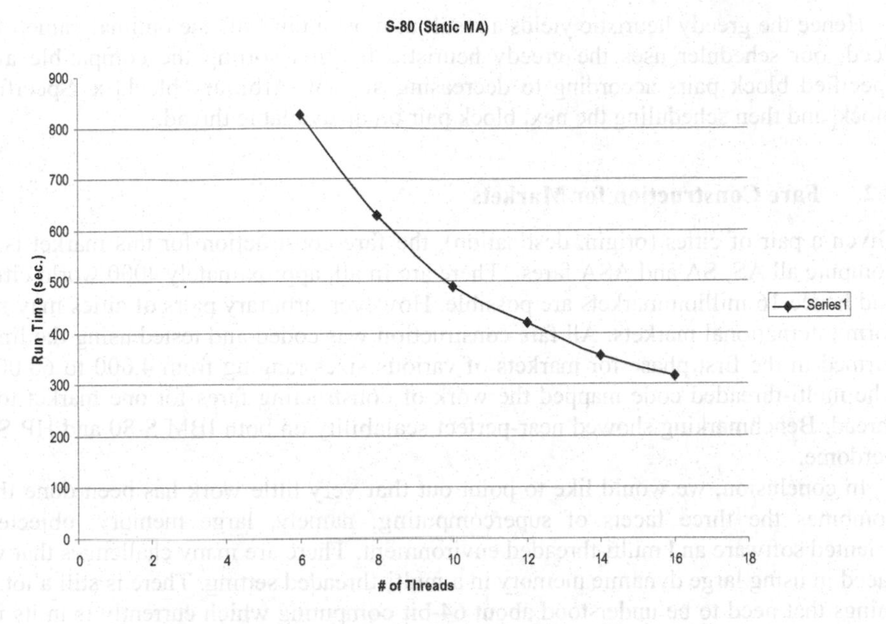

Fig. 4. Performance of Static memory allocation on S-80

4.1 Scheduler Heuristic and Analysis

We now provide theoretical estimates of the efficiency of the heuristic used in our scheduler algorithm. We need some terminology first. Basically, the cross-product of block-pair is the task given to each thread and we know that the computational load is widely varying across various cross-products (all 30,000 of them). It is reasonable to assume that the load is represented by the number |Arbitrary block| x |Specified block|, given a specific pair of compatible arbitrary and specified blocks. In essence, there are 30,000 such tasks (with their corresponding load numbers) to be scheduled among a fixed pool of threads (processors) so as to minimize the overall running time. Formally, we are given a set of n tasks with designated processing times P_j to be scheduled on m identical processors. An assignment of tasks to the processors is a schedule so that each processor is scheduled for a certain total time. The maximum time that any machine is scheduled for is called the *span* of the schedule. The objective is to find a schedule with minimum span, denoted by *OPT(I,m)* where *I* is the set of processing times. This problem is known to be NP-complete [4] and hence only heuristics can be used to solve the problem reasonably fast for large number of tasks. There are two simple heuristics for scheduling the tasks whose performance are indicated in the following lemmas.

Lemma 1: [5] The heuristic is to pick the next task arbitrarily to run on an idle cpu. In this case, the *span* *(2-(1/m))*OPT(I,m)*.

Lemma 2: [5] The heuristic approach uses a greedy paradigm where the next task scheduled is the one with the longest processing time. In this case the *span* *(4/3 – (1/3m))*OPT(I,m)*.

Hence the greedy heuristic yields a result at most a third off the optimal value. Indeed, our scheduler uses the greedy heuristic by first sorting the compatible arb-specified block pairs according to decreasing size of |Arbitrary block| x |Specified block| and then scheduling the next block pair on an available thread.

4.2 Fare Construction for Markets

Given a pair of cities (origin, destination), the fare construction for this market is to compute all AS, SA and ASA fares. There are in all, approximately 4000 world cities and hence 16 million markets are possible. However, arbitrary pairs of cities may not form international markets. All fare construction was coded and tested using the links formed in the first phase for markets of various sizes ranging from 4,000 to 60,000. The multi-threaded code mapped the work of constructing fares for one market to a thread. Benchmarking showed near-perfect scalability on both IBM S-80 and HP Superdome.

In conclusion, we would like to point out that very little work has been done that combines the three facets of supercomputing, namely, large memory, objected-oriented software and multi-threaded environment. There are many challenges that we faced in using large dynamic memory in a multi-threaded setting. There is still a lot of things that need to be understood about 64-bit computing which currently is in its infancy.

(Disclaimer: It is to be noted that the benchmarks we have provided were generated on high-performance machines available from IBM and HP towards the end of the year 2000 and hence perhaps not indicative of performance of newer and more recent machines from both manufacturers.)

5 References

[1] "Airline Tariff Publishing Company", at URL: www.atpco.net
[2] Sarac, S. Rayaprolu, K. Saranathan, and S. Ramaswamy. "Automated Fare Response: An International Pricing Problem", INFORMS, Aviation Applications, Miami, 2001.
[3] Garey, J and Johnson, D. "Theory of NP-Completeness", Freeman Press, 1976.
[4] Hochbaum, D. Shmoys, D. "Using Dual Approximation Algorithms for Scheduling Problems: Theoretical and Practical Results", Journal of ACM, Vol. 34, No.1, January 1987, pp.144-162.
[5] Graham, R. "Bounds on Multi-processing Timing Anomalies", SIAM J. Applied Mathematics, Vol. 17, 1969, pp. 416-429.

A Resource Brokering Infrastructure for Computational Grids

Ahmed Al-Theneyan[1], Piyush Mehrotra[2], and Mohammad Zubair[1]

[1] Computer Science Department, Old Dominion University
Norfolk, VA 23529 USA
{theneyan,zubair}@cs.odu.edu
[2] NAS Division, M/S T27A-1, NASA Ames Research Center
Moffett Field, CA 94035 USA
pmehrotra@arc.nasa.gov

Abstract. With the advances in the networking infrastructure in general, and the Internet in specific, we can build grid environments that allow users to utilize a diverse set of distributed and heterogeneous resources. Since the focus of such environments is the efficient usage of the underlying resources, a critical component is the brokering module that mediates the discovery, access and usage of these resources. One of the major tasks of the brokering module is brokering of resources. With the consumer's constraints, provider's rules, distributed heterogeneous resources and the large number of scheduling choices, the brokering module needs to decide where to place the user's jobs and when to start their execution in a way that yields the best performance to the user and the best utilization to the resource provider. In this paper we present the design and implementation of a flexible, extensible and generic policy-based resource brokering infrastructure for computational grids following a layered façade design pattern and using XML as the underlying specification language. We also describe a testbed environment and our efforts at integrating it with several grid systems.

1 Introduction

The increasing availability of cheap high-speed computational resources is making it feasible for engineers and scientists to address large-size simulations and computational problems. For example, multi-disciplinary applications such as the design and optimization of aerospace vehicles require heterogeneous computational resources that are distributed geographically. With the advances in the networking infrastructure, many groups [4],[7],[8],[14], both research and commercial, are attempting to build grid infrastructures which allow users to utilize distributed heterogeneous resources to solve their problems. Among these efforts, the Globus research project is noteworthy which enables software applications to integrate

S. Sahni et al. (Eds.) HiPC 2002, LNCS 2552, pp. 463-473, 2002.

instruments, displays, computational and information resources that are managed by diverse organizations in widespread locations [8].

The Globus middleware is open and extendable, making it attractive to build higher-level services and components to support end user needs. One such component is the resource brokering environment, which given tasks requirements, constraints and status of resources, and optimization objectives maps tasks to resources. This may include co-allocation in which multiple resources may need to be simultaneously allocated to subtasks of a job and advanced reservations wherein resources may need to be reserved for use at a future time to satisfy some real-time constraints. A resource brokering environment typically utilizes scheduling heuristics to achieve the optimization objectives. The optimization objective of a resource brokering environment can be system-centric, for example as in Condor [4], where overall performance of the system is the key optimization objective. On the other hand, there are application-centric systems, such as AppLeS [5]. However, most of these systems are narrowly focused on one or the other objective without providing support for both application and system centric optimization objectives.

In this paper, we discuss a Policy-based ResOurce Broker Environment (PROBE) [2], which is modular with well-defined APIs and can interface with multiple grid environments such as Globus and Sun Grid Engine. PROBE consists of a set of extensible and replaceable modules that define the services for brokering resources. The main module of the proposed PROBE is a *Resource Broker*, which can support a variety of underlying scheduling heuristics. The design of the *Resource Broker* is based on façade design pattern approach and uses XML as the underlying specification language. This approach provides support for plug-and-play of any scheduling algorithm and application problem the user might provide.

We have integrated PROBE with Globus and Sun Grid Engine [14], the most popular grid systems that have wide acceptance in the grid community. We also implement some static and dynamic scheduling algorithms for Directed Acyclic Graph (DAG) applications based on the classic Critical Path Method (CPM) [11]. We use Pathfinder, an aircraft Multidisciplinary Design Optimization (MDO) problem, as a real test application. This provides a testbed for our experiments to evaluate our system with respect to the ease of use and deployment.

The rest of this paper is organized as follows. In Section 2, we review several related projects focusing on their resource brokering environments and scheduling components. Section 3 describes the approach that we have followed in designing our resource brokering infrastructure. Section 4 gives details of the implementation issues. Section 5 then describes the experimental testbed and the evaluation experiments that we carry out with respect to ease of use and deployment. Finally, in Section 6, we conclude by summarizing the work that we have described in this paper and brief our future research.

2 Related Work

The problem of managing a distributed heterogeneous collection of shared resources has been an active area of research [4],[7],[8],[14].

Globus [8], at Argonne National Laboratory and University of Southern California, is a system that provides the basic software infrastructure for computations that use geographically distributed computational and information resources. A central element of the Globus system is the Globus grid toolkit, which defines the basic services and capabilities necessary to construct a computational grid. The main focus of the resource management infrastructure in Globus is to provide a uniform and scalable mechanism for naming and locating computational resources. Brokering is partially supported where the main focus is to provide interfaces to other underlying batch-queuing systems and support site autonomy and security.

Other efforts such as Legion, Condor, and Sun Grid Engine (formerly known as Codine) are in the same spirit. Legion [7], at the University of Virginia, is an object-based metacomputing system, which allows users to access a large collection of heterogeneous resources unified into a single coherent system. Condor [4], at the University of Wisconsin-Madison, is a high-throughput computing system, which runs on a cluster of workstations to harness wasted CPU cycles. Sun Grid Engine [13] is a resource management tool for Grid Computing that accepts jobs submitted by users and schedules them for execution on appropriate resources in the Grid based upon the provided resource management policies.

There are also efforts to build resource brokering environments that can work with grid computing environment such as Globus. AppLeS [5], at the University of California in San Diego, is a system that provides an application-centric approach to efficient scheduling of distributed supercomputing applications. A recent effort within the AppLeS project is the development of AppLeS templates, stand-alone classes that can be re-used to automatically schedule applications of similar structure. Nimrod [1], at Griffith University in Australia, is a system for managing the execution of parameterized simulations on distributed workstations including a scheduling component that manages the scheduling of individual parametric experiments onto set of idle resources in a local area network. Nimrod/O, is an extension of Nimrod, which employs number of different optimization algorithms. The work is continued in Nimrod/G, which runs on top of Globus.

3 Approach

We have designed and implemented a Policy-based ResOurce Broker Environment (PROBE) [2], as shown in Figure 1. The aim is to develop a modular and fully integrated resource brokering framework with well-defined APIs flexible enough to be utilized in any grid environment. PROBE has been divided into a set of extensible and replaceable modules, where each module implements a specific function. In this subsection, we describe the approach that we have chosen in order to implement the prototype of our system.

3.1 Overview of PROBE

PROBE, as illustrated in Figure 1, consists of several modules interacting with each other to achieve the overall required functionality. The *Client Interface Module* provides an interface to interact with different clients including to other PROBE deployments. The *Resource Repository* maintains an up-to-date information and historical performance information about all the available resources. The *Resource Broker* is the core component of the PROBE, which allocates resources based on the client requirements as we explain in the next subsection. The *Policy Enforcement Manager* works with the *Resource Broker* in finding resources and is responsible for enforcing policies. The *Resource Monitor* keeps track of the current status of the resources and updates the *Resource Repository* periodically. The *Job Monitor* monitors the execution of the jobs that occupy the managed resources while the Jobs *Repository* keeps information about all the currently running jobs.

PROBE infrastructure has been implemented using Jini technology [9], which provides a plug-and-play networking environment. One issue with Jini is that it cannot be used across networks that do not support multicasting. To address this limitation, we enhanced Jini with a tunneling service that propagates Jini multicast messages across such networks [3]. Each module of PROBE has been implemented as a Jini service. We have also utilized Jini's distributed event notification mechanism to keep track of the allocated jobs.

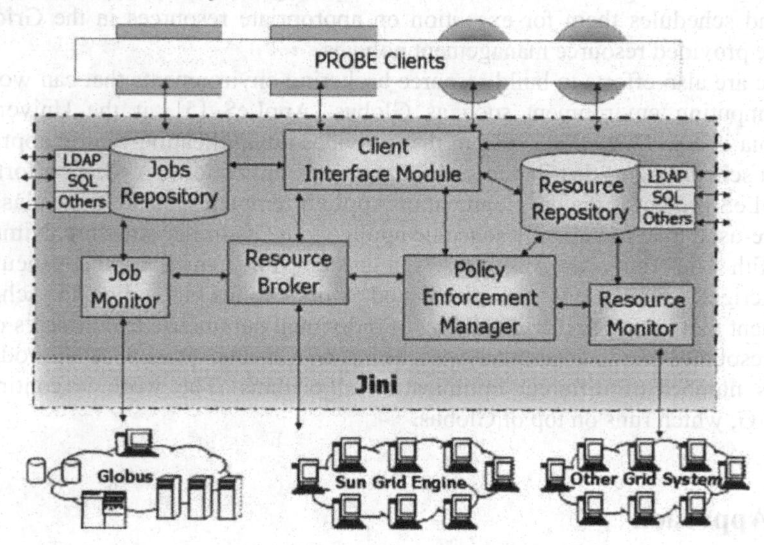

Fig. 1. Architecture of PROBE

3.2 Resource Broker: An Overall Architectural View

As illustrated in Figure 1, the *Resource Broker* module is the one in charge of the brokering tasks. The *Resource Broker* needs to be flexible and generic enough not only to handle the different kinds of user tasks but also to handle the different kinds of scheduling techniques the system is going to incorporate. We have designed and implemented a resource brokering infrastructure for computational grids that can be easily utilized by grid systems. As shown in Figure 2, we have divided our *Resource Broker* into two flexible, extensible, and replaceable agents, where each agent implements an individual function. These agents define the basic services and capabilities required to construct a distributed resource brokering system. Dividing into agents provides flexibility and ease of replacement making it easier to satisfy users' requirements in the future. Also, scalability and high availability can be achieved by replicating those agents.

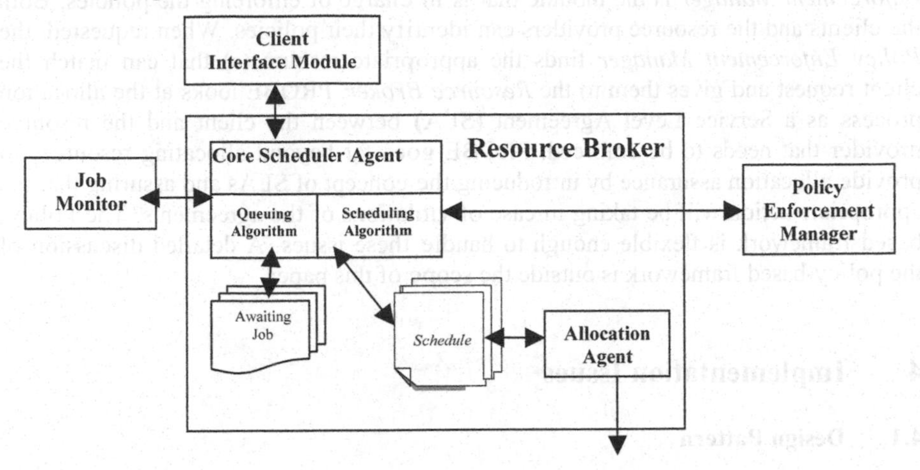

Fig. 2. Overall Architecture of the Resource Broker

The *Core Scheduler Agent* is the heart of our *Resource Broker* and the first point of contact for the user's job. Based on the underlying scheduling algorithm, the user's job and the matched sub-set of resources provided by the *Policy Enforcement Manager*, the *Core Scheduler Agent* is going to construct a near optimal active schedule object and pass it to the *Allocation Agent* where it is going to be implemented. The schedule is an active object that has an order and placement of tasks that need to be allocated. It gets created by the *Core Scheduler Agent* based on the underlying scheduling algorithm and then manipulated by the different components of our *Resource Broker* as necessary. A unique job ID is assigned for each job at the time of creating the schedule by the *Core Scheduler Agent*. The *Core Scheduler Agent* maintains an internal queue of jobs currently in the system and that have not been scheduled yet including those that failed and need to be rescheduled. The *Core Scheduler Agent* uses a queuing algorithm to select the next job to schedule. The approach we follow allows the users to plug in their scheduling and queuing algorithms as needed. The design approach makes these algorithms look like black boxes to the *Core Scheduling Agent*.

The *Allocation Agent* is responsible for implementing the created schedule that is launching the tasks on the designated resources. The *Allocation Agent* will notify the *Job Monitor*, which will in turn keep on monitoring the allocated jobs. The *Job Monitor* then updates the *Core Scheduling Agent* as necessary about the significant changes in the job status (FINISHED, FAILED, SUSPENDED, etc). The *Core Scheduler Agent* in such a case might need to re-schedule some of the associated tasks based on the underlying scheduling algorithm.

3.3 Policy-Based Brokering Approach

PROBE adopts a policy-based approach for resource brokering. The *Policy Enforcement Manager* is the module that is in charge of enforcing the policies. Both the clients and the resource providers can identify their policies. When requested, the *Policy Enforcement Manager* finds the appropriate resource(s) that can match the client request and gives them to the *Resource Broker*. PROBE looks at the allocation process as a Service Level Agreement (SLA) between the client and the resource provider that needs to be enforced. PROBE goes far beyond allocating resources to provide allocation assurance by introducing the concept of SLAs and assuring that the appropriate action will be taking in case of violations of the agreements. The Policy-based framework is flexible enough to handle these issues. A detailed discussion of the policy-based framework is outside the scope of this paper.

4 Implementation Issues

4.1 Design Pattern

We need to decouple our *Resource Broker* from a specific queuing algorithm, scheduling algorithm and the type of jobs that it is going to deal with. One way to address this is to use a facade object that provides a single, simplified interface to more general facilities of a subsystem. We follow the facade design pattern [13] for the objects being used by the *Resource Broker*. This shields our *Resource Broker* from the Queuing Algorithms, Scheduling Algorithms and Job Types. Our *Resource Broker* sees them as black boxes.

An example of the use of the façade approach is the job types. *Job* is an abstract class and needs to be implemented by the job type. The Resource Broker and the Scheduling Algorithm have a unified interface to a set of Job Types. This makes the design independent of any job type. Initially, we support *Single*, *Aggregated* and *Direct Acyclic Graph (DAG)* jobs. A *Single Job* is the basic job type in our framework that represents the executable portion of an application. An *Aggregated Job* is where a group of tasks are combined to form a unified job such as: *CoAllocation Job* that requires that a given set of resources are available for use simultaneously; and *Parametric Job* where the same program is repeatedly executed with different initial conditions. A *DAG Job* represents an application program that consists of a collection of heterogeneous modules (application codes from different disciplines). A typical distributed application requires these modules to be executed in some order and possibly on different machines.

Adding new job types to the system does not require modification to the code nor its recompilation. We just need to create a class inheriting Job and implement the abstract methods. The same approach is used for the scheduling algorithm and the queuing algorithm. This gives our system the flexibility to plug and play any one of them based on the requirements of the overall system.

```
<!--Request.dtd-->
<!DOCTYPE Request [
<!ENTITY % operator "AND|OR|NOT">
<!ENTITY % JobType "Single|Aggregated|DAG">
<!ENTITY % comparison "EQ|NEQ|GR|GREQ|LS|LSEQ">
<!ENTITY % aggregationType "CoAllocation|Parametric">
<!ENTITY % CoAllocationTiming "SameTime|DifferentTime">
<!ELEMENT Request ((%JobType;))>
<!ELEMENT Single (Rule?,AdditionalInfo*)>
<!ATTLIST Single
        Name            CDATA #IMPLIED
        Executable      CDATA #IMPLIED
        RunDirectory    CDATA #IMPLIED
        Arguments       CDATA #IMPLIED>
<!ELEMENT Aggregated (Single+,Rule?,AdditionalInfo*)>
<!ATTLIST Aggregated
        Name            CDATA #IMPLIED
        Type            (%aggregationType;)    #IMPLIED
        Timing          (%CoAllocationTiming;)    #IMPLIED>
<!ELEMENT DAG ((%JobType;)+,Dependency*,Rule?,AdditionalInfo*)>
<!ATTLIST DAG
        Name            CDATA #IMPLIED>
<!ELEMENT Dependency EMPTY>
<!ATTLIST Dependency
        From            CDATA #IMPLIED
        To              CDATA #IMPLIED>
<!ELEMENT Rule ((Condition)|(%operator;))>
<!ELEMENT AND ((Condition)|(%operator;))*>
<!ELEMENT OR ((Condition)|(%operator;))*>
<!ELEMENT NOT ((Condition)|(%operator;))>
<!ELEMENT Condition EMPTY>
<!ATTLIST Condition
        Entity          CDATA #IMPLIED
        Operator        (%comparison;) #IMPLIED
        Value           CDATA #IMPLIED>
<!ELEMENT AdditionalInfo EMPTY>
<!ATTLIST AdditionalInfo
        Name            CDATA #IMPLIED
        Value           CDATA #IMPLIED>
]>
```

Fig. 3. Flexible Job Language (FJL)

4.2 Specification Language

The underlying language used to specify the user's request is based on XML. This allows our system to interact with external systems and exchange jobs information. We have designed a Flexible Job Language (FJL) that can be used to express the user's request. Our FJL can be easily extended to satisfy complicated user's requirements. Figure 3 illustrates the schema that specifies how the request can be generated.

5 Integration and Experimental Evaluation

In this section, we describe our efforts of PROBE with different grid systems, in particular, Globus [8] and Sun Grid Engine [14]. We also describe a testbed on which we have evaluated our system using Pathfinder, an aircraft preliminary Multidisciplinary Design Optimization application that demonstrates the methodology for multidisciplinary communications and couplings between several engineering disciplines.

5.1 PROBE Resource Daemon – Gateway to Grid Systems

PROBE supports a daemon that is started on each resource under the control of their administrative domains. This daemon is implemented as a Jini service and acts as gateways between the PROBE and the managed resource. As illustrated in Figure 4, the daemon has four components: the *Core Daemon* implements the infrastructure necessary for the daemon to be a Jini service and for managing the interactions among the other components; the *Data Collector* handles the collection of statistical data about the resource and passes it to the *Resource Monitor*; the *Execution Monitor* keeps track of the allocated jobs within the resource and updates the *Job Monitor* about their status (implemented using the distributed event notification mechanism in Jini); the *Local Policy Enforcer* keeps track of the policies associated with the resource. It also performs some optimization by normalizing the associated policies before updating the *Policy Enforcement Manager*; and the *Platform Specific Adapter* maps the data collection and job execution/monitoring requests to the specific platform (such as Globus, Sun Gird Engine, etc).

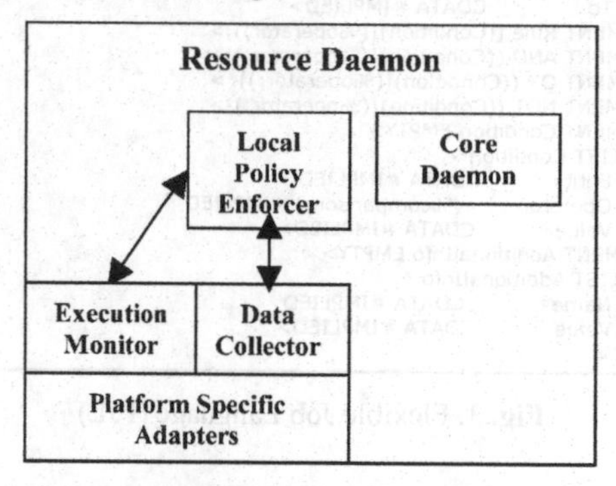

Fig. 4. PROBE Resource Daemon

5.2 Integration with Globus and Sun Grid Engine

We have integrated PROBE with Globus 2.0 using the Java Commodity Grid (CoG) Kit 0.9.13 [10]. PROBE acts as client for the Globus GRAM and generates RSL on the fly for each job being submitted to a resource managed by Globus. We have used the *RSL package* to manipulate the translated RSL request and check its validity; the *GRAM package* to create, submit and monitor jobs with the RSL being created by the RSL package; the *MDS package* to query and collect data about the status of the resources being manages by Globus; and the *GSI package* to enable secure access to the resources.

The Sun Grid Engine is an open source community effort which sponsored by Sun Microsystems and compatible with the Sun Grid Engine. Its main objective is to extend Sun's Grid Engine. We have built a pure JNI adaptor that allows PROBE to interact with Sun Grid Engine 5.3.

5.3 Testbed

Our testbed environment is made up of the three local administrative domains. In the first domain, we have installed Globus Tool Kits 2.0 on a 733 MHz PIII PC with Redhat 7.2 Linux. This administrative domain has eight 600 MHz PIII PCs and a system wide policy that resources are not to be accessed between 9 AM and 5 PM. In the second administrative domain, we have installed Sun Grid Engine 5.3 on 10 Sun ULTRAstation-10 workstations with Solaris 2.7. Some of the resources have a resource policy stating that the resource can not be used when its load exceed 50%. In the third administrative domain, we have installed PROBE on 1 GHz PIV PC with Windows 2000 on it. This administrative domain has 5 PIV PCs some with Windows 2000 and some with NT. Some of the resources have a resource policy stating that the resource cannot be used when number of interactive users exceeds 5.

5.4 Scenario

In this subsection, we describe a typical scenario that illustrates how PROBE handles a DAG job. Using the command-line client interface of PROBE, we submit an FJL-based problem description for a DAG job. For our experiment, we use the DAG representing the Pathfinder problem [12]. We also specify the constraint on the type of resources we would like to use, for example a computing resource with a CPU of 700 MHz and 128 MB of Memory.

The *Client Interface Module* on receiving the job description creates a *Job* object and passes the request to the *Resource Broker*. After generating the global unique IDs for the job and all its sub-tasks, the *Core Scheduler Agent* consults with the *Policy Enforcement Manager*. Out of all resources (in our experiment 23) that are available in the system, the *Policy Enforcement Manager* matches a subset of 4 appropriate resources and forwards the set to the *Core Scheduler Agent*. Given these set of 4 resources, the *Core Scheduler Agent* constructs a near-optimal schedule. The underlying algorithm used for scheduling the DAG is a static CPM-based, which first assigns high priority tasks to the required resources [11] that we have integrated as part of the overall system. As each sub-task in the DAG gets allocated onto the designated resources, the *Job Monitor* is informed so that it can keep track of the job. After the successful completion of the last sub-task, the schedule is terminated by the *Core Scheduling Agent*.

6 Conclusion and Future Work

Computational grids are evolving and becoming a basic infrastructure for the future of high performance and distributed computing. In this paper, we have described a framework of a brokering infrastructure for Computational Grid that is flexible enough to be utilized on various grid systems. We have also described a testbed for our experiments to evaluate our system with respect to ease of use and deployment.

One of the recent directions is applying economics to resource management and scheduling. This has been an active area of research recently. Buyya [6] has proposed an economic-based model for the grid. Our *Resource Broker* can be easily extended to adopt economic-based scheduling policies. The *Policy Enforcement Manager* is flexible enough to handle this issue. It can be also easily integrated with other models.

Also, for efficient scheduling of resources, it is more useful for the *Resource Broker* to use an estimate of the performance in the near future rather than current performance. Based on some historical performance information, the Resource Broker should be able to predict the performance each resource is going to deliver at the time of the allocation. This could result in a more efficient scheduling of the resources. We are planning to extend our model in the near future to handle prediction. A new agent, called *Predictor*, will be introduced for that purpose. Our *Predictor* is going to keep historical performance information and predicts future performance based on that. Work also will continue in interfacing with NWS [15], a Distributed Resource Performance Forecasting Service for Metacomputing by the University of Tennessee.

References

[1] Abramson, D., Sosic, R., Giddy, J., Hall, B.: Nimrod: A Tool for Performing Parametised Simulations using Distributed Workstations. The 4th IEEE Symposium on High Performance Distributed Computing, Virginia, August 1995.

[2] Al-Theneyan, A., Mehrotra, P., Zubair, M.: PROBE: A Policy-based Resource Brokering Environment for the Grid. Under preparation.

[3] Al-Theneyan, A., Mehrotra, P., Zubair, M.: Enhancing Jini for Use Across Non-Multicastable Networks. Proceedings of the First Saudi Technical Conference and Exhibition, Volume II, pp. 18-23, Riyadh, Saudi Arabia, November 2000.

[4] Basney, J., Livny, M.: Managing Network Resources in Condor. Proceedings of the Ninth IEEE Symposium on High Performance Distributed Computing (HPDC9). Pittsburgh, Pennsylvania, August 2000.

[5] Berman, F., Wolski, R.: The AppLeS Project: A Status Report. Proceedings of the 8th NEC Research Symposium, Berlin, Germany, May 1997.

[6] Buyya, R.: Economic-based Distributed Resource Management and Scheduling for Grid Computing. Ph.D.Thesis, School of Computer Science and Software Engineering, Monash University, Melbourne, Australia, April 2002.

[7] Chapin, S., Karpovich, J., Grimshaw, A.: The Legion Resource Management System. Proceedings of the 5th Workshop on Job Scheduling Strategies for Parallel Processing (JSSPP '99), San Juan, Puerto Rico, April 1999.

[8] Czajkowski, K., Foster, I., Karonis, N., Kesselman, C., Martin, S., Smith, W., Tuecke, S.: A Resource Management Architecture for Grid Systems. Proceedings of the IPPS/SPDP '98 Workshop on Job Scheduling Strategies for Parallel Processing, 1998.

[9] Keith, W.: Core Jini. Prentice Hall, ISBN 013014469X, 1999.

[10] Laszewski, G., Foster, I., Gawor, J., Lane, P.: A Java Commodity Grid Kit. Concurrency and Computation: Practice and Experience, Volume 13, Issue 8-9, pp. 643-662, 2001.

[11] Liu, G.: Two Approaches to Critical Path Scheduling for a Hetrogeneous Environment. *M.S.Thesis*, Department of Computer Science, Old Dominion University, Norfolk, VA, USA, October 1998.
[12] Multidisciplinary Optimization Branch (MDOB) at NASA Langley Research Center. Available from http://fmad-www.larc.nasa.gov/mdob/MDOB/index.html.
[13] Schmidt, D., Stal, M., Rohnert, H., Buschmann, F.: Pattern-Oriented Software Architecture: Patterns for Concurrent and Networked Objects. Wiley & Sons, ISBN 0-471-60695-2, 2000.
[14] Sun Microsystems: Sun Grid Engine Software. Available from http://wwws.sun.com/software/gridware/.
[15] Wolski, R., Spring, N. T., Hayes, J.: The Network Weather Service: A Distributed Resource Performance Forecasting Service for Grid. The Journal of Future Generation Computing Systems, 1999.

On Improving Thread Migration: Safety and Performance*

Hai Jiang[1] and Vipin Chaudhary[2]

[1] Institute for Scientific Computing, Wayne State University
Detroit, MI 48202 USA
haj@cs.wayne.edu

[2] Institute for Scientific Computing, Wayne State University
and Cradle Technologies, Inc.
vipin@wayne.edu

Abstract. Application-level migration schemes have been paid more attention recently because of their great potential for heterogeneous migration. But they are facing an obstacle that few migration-unsafe features in certain programming languages prevent some programs from migrating. Most application-level migration schemes declare or assume they are dealing with "safe" programs which confuse users without explanation. This paper proposes an application-level thread migration package, *MigThread*, to identify "unsafe" features in C/C++ and migrate this kind of programs with correct results. Therefore, users need not worry if their programs are qualified for migration as they experienced before. Besides the existing characteristics like scalability and flexibility, *MigThread* improves transparency and reliability. Complexity analysis and performance evaluation illustrate the migration efficiency.

1 Introduction

Recent improvements in commodity processors and networks have provided a chance to support high-performance parallel applications within an everyday computing infrastructure. As high-performance facilities shift from supercomputers to Networks of Workstations (NOWs), migration of computing from one node to another will be indispensable. Thread/process migration enables dynamic load distribution, fault tolerance, eased system administration, data access locality and mobile computing [1, 2, 7].

Thread migration can be achieved at kernel, user, or application level. Kernel level thread migration is a part of the operating system. Threads are moved around among processors if they are on multi-processors, such as SMPs, or among workstations by distributed operating systems. Kernel-level migration is complicated, but efficient. User-level approaches move migration functionality from the kernel into user space and typically yield simpler implementations,

* This research was supported in part by NSF IGERT grant 9987598, NSF MRI grant 9977815, NSF ITR grant 0081696, US Army Contract DAEA-32-93-D-004, Ford Motor Company Grants 96-136R and 96-628R, and Institute for Scientific Computing.

S. Sahni et al. (Eds.) HiPC 2002, LNCS 2552, pp. 474–485, 2002.

but suffer too much from reduced performance and less transparency. User-level migration is targeted for long-running threads with few OS requirements, less transparency, and a limited set of system calls. Programs will need to be re-linked with certain library to enable migration feature.

Traditional application-level migration is implemented as part of an application. It achieves simplicity by sacrificing transparency and reusability. But it has an attractive potential for heterogeneous migration feature. As internet is popular and grid computing is emerging, heterogeneous migration will be indispensable. We have proposed an application-level thread migration scheme, *MigThread*, to improve transparency and reusability for migration [3]. A big impediment to thread migration is due to "migration-unsafe" features within C/C++. If the programmer uses these "unsafe" features in their programs, the migration leads to errors. To ensure the correctness, most application-level or language level migration schemes declare they only work on "migration-safe" programs. There are two problems with this. First, the "unsafe" features are not well defined. Second, such restriction greatly reduces the domain where migration can be utilized.

In this paper, we make the following contributions:

- Determine and overcome "migration-unsafe" features in programs to widen the applicability of thread migration.
- Improve existing thread migration scheme[3].
  - Speed up source-to-source transformation at compile time.
  - Handle pointers and pointer arithmetic efficiently.
  - Support better memory segment management.
- Provide complexity comparison and analysis, and performance evaluation on real applications.

The remainder of this paper is organized as follows: Section 2 describes the performance improvements to the existing one in [3]. Section 3 identifies and overcomes some migration-unsafe features in C. In section 4, we compare the complexity of our scheme with existing implementations and show experiment results on benchmark programs. Section 5 gives an overview of related work. We wrap up with conclusions and continuing work in Section 6.

2 Optimizing Thread Migration Using *MigThread*

In this section we present some optimization to the migration scheme in [3].

2.1 *MigThread*

MigThread is an portable and scalable application-level thread migration package. It takes thread state out of kernel or libraries, and moves it up to the language level. *MigThread* consists of two parts: preprocessor and runtime support module. At compile time, the preprocessor scans the source code and collects related thread state information into two data structures which will be integrated

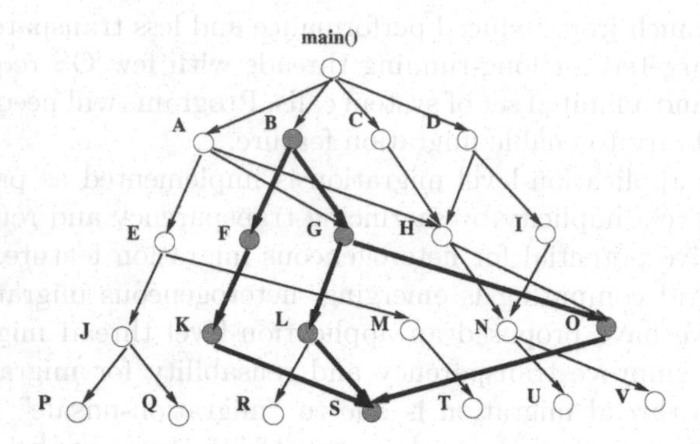

Fig. 1. Function call graph to reduce compile time overhead

into the thread state at runtime. Local variables, function arguments, and Program Counters (PC) are all parts of thread state. Dynamically allocated memory is also supported. Thus data in heap are migrated with the thread state and restored remotely. Since destination nodes might use different address spaces, pointers referencing stack or heap might be invalid after migration. *MigThread* detects and marks pointers at language level so that at runtime it just accesses predefined data structures to update most of them precisely. Adaptation points where thread migration can take place are detected, labelled, and pointed by **switch** statement [3].

At runtime, *MigThread* maintains a thread control area (TCA) which holds a record for each thread containing references to a stack, a control block for memory segments in the heap and a pointer translation table. During migration, thread state is constructed, transferred, and restored. After updating pointers, *MigThread* resumes computation at the right place. Since the physical thread state is transformed into a logical form, *MigThread* has great potential to be used in heterogeneous environments without relying on any type of thread libraries or operating systems. More design and implementation details are in [3].

2.2 Reducing Compile Time Overhead

To reduce the overhead, *MigThread* only transforms the related functions invoked by the migrating threads. It creates a call graph starting out of **main()** to detect the thread starting function and the migration function. These two and all other functions between them should be transformed into the migration-enabled code. For example, in Fig. 1, **B** and **S** are the thread starting and migration functions respectively. There are three paths between them: **BFKS**, **BGLS** and **BGOS**. The program execution has to take one of them based on the runtime situation. Therefore, at compile-time, *MigThread* uses breadth first search to identify and transform related functions on these possible paths. Since only a fraction of the entire program is transformed, compile time overhead is greatly reduced (see Fig. 6 in Section 4).

2.3 Generalizing Pointer Handling

The scheme in [3] identifies pointer variables at language level and collects them into a data structure *sr_ptr*. If some structure type variables in *sr_var* contain pointer fields, they need to be referenced by new pointer variables in the other data structure *sr_ptr*. On the destination node, it scans the memory area of *sr_ptr* to translate most pointers. This strategy is more efficient than reporting pointers one-by-one as in Porch[6] and SNOW[7]. *MigThread* extends this model further to handle more complicated cases in dynamically allocated memory. If some pointer variables in *sr_ptr* contains pointer type subfields, the preprocessor just reports their offsets in base units and the runtime support module will detect other dynamic pointer fields by pointer arithmetic. *MigThread* does not trace pointers if programs are "migration-safe". No matter how pointers are manipulated, only the current values of variables and pointers hold the correct thread state. This "ignore-strategy" makes *MigThread* efficient. *MigThread* traces pointers only when "unsafe" features are involved as in Section 3.

2.4 Memory Management Optimization

Unlike the linked-list structure in [3], *MigThread* maintains a red-black tree of memory segment records, traces all dynamically allocated memory in local or shared heap, and provides the information for pointer updating. Each segment record consists of the address, size, and type of the referenced memory block, with an extra linked list of offsets for inner pointer subfields. In user applications, when **malloc()** and **free()** are invoked to allocate and deallocate memory space, the preprocessor inserts **STR_mig_reg()** and **STR_mig_unreg()** accordingly to let *MigThread* create and delete memory segment records at runtime. Since memory blocks are maintained in order, the insertion, deletion or searching of one node in the red-black tree takes $O(logN)$ time. Again, the dynamically allocated memory management is moved up to the application level [3].

3 Handling Migration-Unsafe Features

Application level migration schemes rely on programming style to ensure the correctness of the resumed computation after migration. Most migration schemes declare that they only work on "safe" programs to avoid those "unsafe" features in C and obtain the correct thread state. *MigThread* can detect and handles some such "unsafe" features, including pointer casting, pointers in unions, library calls, and state-carrying instructions.

3.1 Pointer Casting

Pointer manipulations can cause problems with migration. Pointer casting is one of them. It does not mean the cast between different pointer types, but the cast to/from integral types, such as integer, long, or double. The problem is

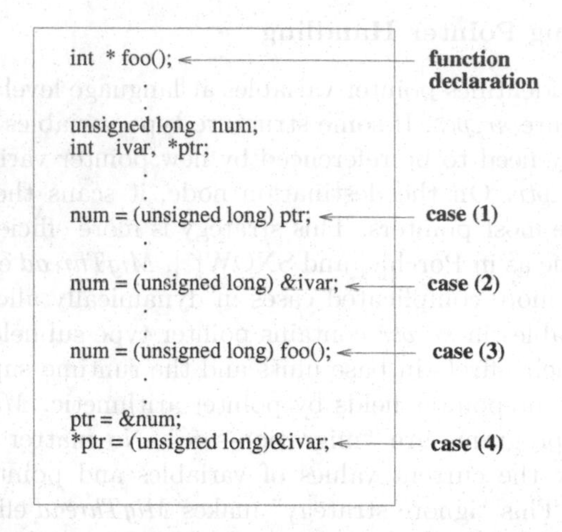

Fig. 2. Four cases of hiding pointers in integral type variables

that pointers might hide in integral type variables. Application level migration schemes identifies pointer values (or memory addresses) by pointer names or even types if they are in dynamically allocated memory segments. If pointers are cast into integral type variables, migration schemes might miss updating them when address space changes during migration. So the central issue is to detect those integral variables containing pointer values (or memory addresses) so that they could be updated during state restoration. Casting could be direct or indirect. There are four ways to hide pointers in integral type variables (shown in Fig. 2):

1. Cast pointers directly or indirectly. In Fig. 2, case (1) only shows the direct cast. If *num* is assigned to another integral type variable, indirect cast happens and it also can cause problems.
2. Memory addresses are cast into integral type variables directly.
3. Functions' returning values are cast in.
4. Integral variables are referenced indirectly by pointers or pointer arithmetic and their values are changed by all the above three cases.

To avoid dangerous pointer casting, *MigThread* investigates pointer operations at compile time. The preprocessor creates a pointer-group by collecting pointers, functions with pointer type return values, and integral variables that have already been cast in pointer values. When the left-hand side of an assignment is an integral type variable, the preprocessor checks the right-hand side to see if pointer casting happens. If members of pointer-group exist without changing their types, the left-hand side variable should also be put into pointer-group for future detection and reported to the runtime support module for possible pointer update during migration. The preprocessor ignores all other cases.

The preprocessor is insufficient for indirect access and pointer arithmetic as case (4) in Fig. 2. The preprocessor inserts primitive **STR_check_ptr(mem1, mem2)** to request the runtime support module to check if *mem1* is actually

```
union u_type {
    struct s_type {
        int  idx;
        int  *first;   <---- exist
        int  *second;  <---- together
    }    a;
    int *b;
    int c;
};
```

Fig. 3. Pointers in Union

an integral variable's address (not on pointer trees) and *mem*2 is an address (pointer type). If so, *mem*1 will be registered as a pointer which could also be deregistered later. Here *mem*1 is the left-hand side of assignment and *mem*2 is one member of right-hand side components. If there are multiple components on the right-hand side, this primitive will be called multiple times. Frequently using pointer arithmetic on the left-hand side can definitely cause heavy burden on tracing and sacrifice performance. This is a rare case since normally pointer arithmetic is applied more on the right-hand side. Thus, computation is not affected dramatically. During the migration, registered pointers will be updated no matter if their original types are pointer ones or not. *MigThread*'s preprocessor and runtime support module work together to find out memory addresses hidden in integral variables and update them for migration safety.

3.2 Pointers in Union

Union is another construct where pointers can evade updating. In the example of Fig. 3, using member *a* means two pointers are meaningful; member *b* indicates one; and member *c* requires no update. Migration schemes have to identify dynamic situations on the fly. Application-level migration schemes have advantages over kernel- and user-level ones. When a **union** variable is declared, the compiler automatically allocates enough storage to hold the largest member of the **union**. In the program, once the preprocessor detects that a certain member of the **union** variable is in use, it inserts primitive **STR_union_upd()** to inform the runtime support module which member and its corresponding pointer fields are in activation. The records for previous members' pointer subfields become invalid because of the ownership changing of the **union** variable. We use linked list to maintain these inner pointers and get them updated after migration.

3.3 Library Calls

Library calls bring difficulties to all migration schemes since it is hard to figure out what is going on inside the library code. Application-level migration schemes work on the source code. Without the source code of libraries, problems can

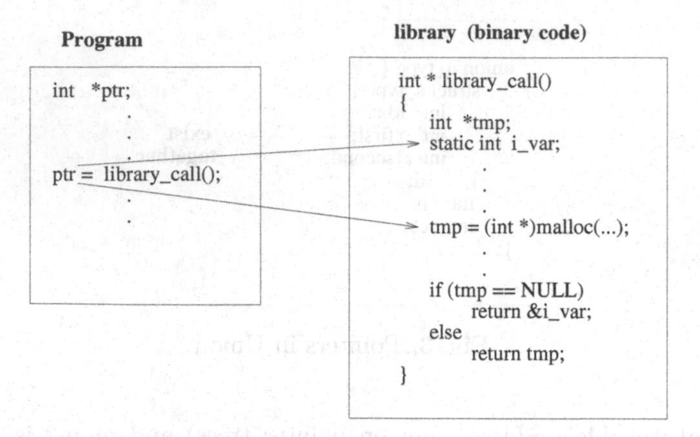

Fig. 4. Pointers dangling after library calls

occur when pointers are involved. For *MigThread*, the major concerns are static local variables and dynamically allocated memory. In the example of Fig. 4, the pointer *ptr* might be pointing to an address of static local variable *i_var* for which compilers creates permanent storage or a dynamically allocated memory block. Both of them are invisible to *MigThread*. Pointers pointing to these unregistered locations are also called "dangling pointers", as those pointing to de-allocated memory blocks. This phenomena indicates that *MigThread* is unable to catch all memory allocations because of the "blackbox" effect. The current version of *MigThread* can inform users of the possible danger of "memory leakage" so that programmers can register these memory blocks by hands if they know the library calls well. This is one workaround solution whereas all other migration schemes have to face the same problem. For **malloc()** wrappers, another option is for users to specify the syntax so that the preprocessor can know how to insert proper primitives for memory management.

3.4 State-Carrying Instructions

Since *MigThread* works at language level, the adaptation points can only be inserted at this level. Thus, there is at least one C language statement between two adaptation points. It seems that the migration can only happen between statements. But functions break this rule and enable migration to take place

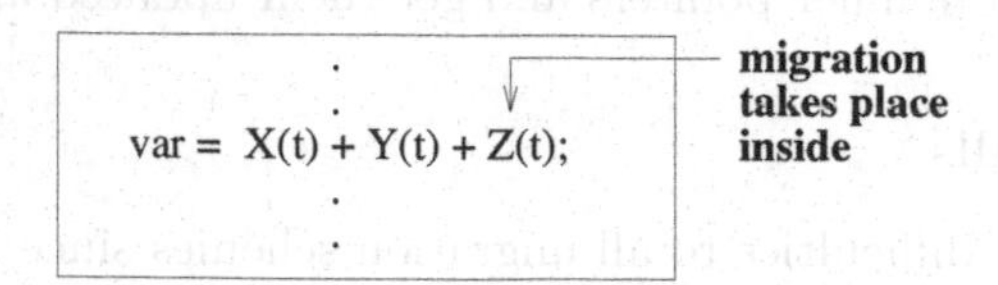

Fig. 5. Migration happens inside of single complex statement

Table 1. Complexity comparison in data collecting

System	Collect Variables	Collect Pointers	Collect Memory Blocks	Save Variables	Save Pointers	Allocate Memory Blocks
Porch	$O(N_{var})$	$O(N_{ptr})$	$O(N_{mem})$	$O(N_{var})$	$O(N_{ptr})$	$O(N_{mem} * logN_{mem})$
SNOW	$O(N_{var})$	$O(N_{ptr})$	$O(N_{mem} * logN_{mem})$	$O(N_{var})$	$O(N_{ptr})$	$O(N_{mem})$
MigThread	1	1	$O(N_{mem})$	0	0	$O(N_{mem} * logN_{mem})$

Table 2. Complexity comparison in data restoration

System	Restore Variables	Restore Pointers	Update Pointers	Re-allocate Memory Blocks	Delete Memory Blocks
Porch	$O(N_{var})$	$O(N_{ptr})$	$O(N_{ptr} * logN_{mem})$	$O(N_{mem})$	$O(N_{mem} * logN_{mem})$
SNOW	$O(N_{var})$	$O(N_{ptr})$	$O(N_{ptr} * N_{mem})$	$O(N_{mem})$	$O(N_{mem}^2)$
MigThread	1	1	$O(N_{ptr} * logN_{mem})$	$O(N_{mem})$	$O(N_{mem} * logN_{mem})$

within single statement as the example in Fig. 5. The right-hand side is a summation of three functions' results. Suppose thread migration takes place inside the third function $Z(t)$. We assume that the functions are executed in order. Before migration, the compiler saves the results of the first two functions in some temporary storages which are useful only if they are under *MigThread*'s control. To achieve this, the statement should be broken down and temporary variables are introduced to save temporary results. The advantage of this is that we increase the adaptation points. But this brings minor changes to the program's structure.

To avoid modification of programs, the first two functions could be rerun to retrieve their return values. This means they should be "re-entrant" and deliver the same result with the same inputs. *MigThread* has to detect "state-carrying" functions and make them "stateless" and label the position right before each **return** statement. During the re-running of the first two functions after migration, their **switch** statements dispatch computation directly to their last escape points. Therefore, no actual computation goes through function bodies and functions become stateless.

4 Complexity Analysis and Performance Evaluation

Besides *MigThread*, there are two other application level migration systems, Porch [6] and SNOW [7] which collect and restore variables one-by-one explicitly at each adaptation point in time $O(N)$. This makes it hard for users to insert adaptation points by themselves. Our *MigThread* only registers variables once in time $O(1)$ and at adaptation points the programs only check for condition variables. Therefore, *MigThread* is much faster dealing with thread state.

For memory blocks, Porch and *MigThread* have similar complexity because they both maintain memory information in red-black trees. SNOW uses a mem-

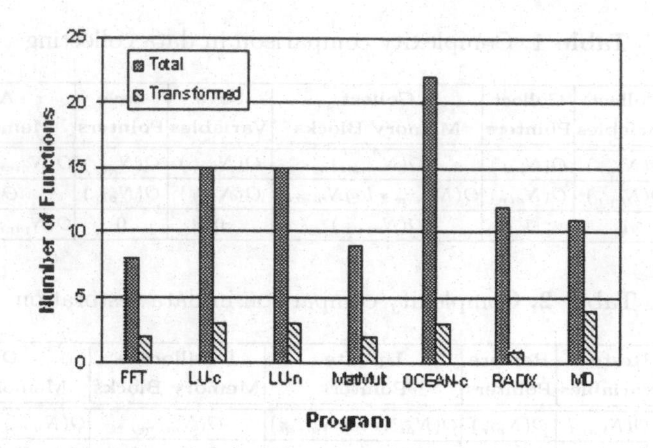

Fig. 6. Functions transformed by Preprocessor

ory space representation graph, which is quick to create a memory node, but extremely slow for other operations because searching for a particular node in a randomly generated graph is time-consuming. Also, SNOW traces all pointers and slows down much for pointer-intensive programs. *MigThread* virtually only cares about results ("ignore process") and is therefore less dependent on the types of programs. We summarize the complexity of these three systems, and list the results in Table 1 and 2. The N_{var}, N_{ptr} and N_{mem} represent numbers of variables, pointers and dynamically allocated memory blocks. From these, we can see that *MigThread* is very efficient.

The migration platform is a software Distributed Shared Memory (DSM) System [5] over SMPs (SUN UltraEnterprise 3000s) connected by fast Ethernet. Each SMP contains four 330Mhz UltraSparc processors. The parallelized programs are running on two SMP machines with one thread on each. The communication layer is UDP/IP. Since the inserted primitives do not cause any noticeable slowdown when no migration happens, we only focus on the migration cost and compare it with pure execution time on two SMP nodes. We use several applications from the SPLASH-2 application suite, matrix multiplication, and Molecular Dynamics (MD) simulation to evaluate the thread migration cost.

MigThread's preprocessor scans and transforms C programs automatically. The function call graph eliminates unnecessary functions so that preprocessor only transforms a fraction of functions to reduce the compile-time cost. Only about 10-20% functions require to be transformed (see Fig. 6). This one-time transform procedure takes about 1-8 seconds for our benchmark programs.

The runtime overheads are shown in Table 3. As mentioned before, no noticeable overhead is seen when no migration happens. For most applications, the thread states range from 100 to 184 bytes, and their migration time is around 2.4 ms. Even though the thread state of OCEAN-c is increased to 432 bytes, its migration time does not change. Only thread states of RADIX and MD are big enough to make difference. Since shared data are in DSM's global shared regions

Table 3. Migration Overhead in real applications

Program	Input Size	State Size (bytes)	Transform Time (sec)	Execution Time (ms)	Migration Time (ms)	Migr./Exec. Rate (%)
FFT	64 Points	160	5.87	85	2.42	2.85
	1024 Points	160	5.87	112	2.46	2.20
LU-c	16 x 16	184	4.19	77	2.35	3.05
	512 x 512	184	4.19	7,699	2.41	0.03
LU-n	16 x 16	176	4.17	346	2.34	0.68
	128 x 128	176	4.17	596	2.37	0.40
MatMult	16 x 16	100	1.34	371	2.32	0.63
	128 x 128	100	1.34	703	2.47	0.35
OCEAN-c	18 x 18	432	7.98	2,884	2.45	0.08
	258 x 258	432	7.98	14496	2.40	0.02
RADIX	64 keys	32,984	2.86	688	5.12	0.74
	1024 keys	32,984	2.86	694	5.14	0.74
MD	5,286 Atoms	7,040,532	2.45	38,067	83.65	0.22

which do not need to be migrated with threads, thread state sizes are invariant to problem sizes in Table 3. Compared to programs' execution time, migration cost is so small (mostly less than 1% and at most 3%) for benchmark programs.

The chosen programs are popular, but all array-based. Fortunately, *MigThread* does not slow down particularly for pointer-intensive applications because pointers are not traced all the time. Definitely more memory blocks incur bigger overhead, which is inevitable.

5 Related Work

The major concern in thread migration is that the address space could be totally different on various machines and internal self-referential pointers may no longer be valid. There are three approaches to handle the pointer issue. The first approach is to use language and compiler support to identify and update pointers[4, 8], such as in Emerald[9] and Arachne[8]. But they rely on new languages and compilers. The second approach requires scanning the stacks at run-time to detect and translate the possible pointers dynamically, as in Ariadne[10]. Since some pointers in stack are probably misidentified, the resumed execution can be incorrect. The third approach is popular, such as in Millipede[11] and necessitates the partitioning of address spaces and reservation of unique virtual addresses for the stack of each thread so that the update of internal pointers becomes unnecessary. This faces severe scalability and portability problem [11, 4].

Application-level implementation achieves the heterogeneity feature. The Tui system [2] is a heterogeneous process migration package which modifies a compiler (ACK) to provide runtime information via debugging code and relies on Unix ptrace to obtain the state of processes. SNOW [7] is another heterogeneous process migration scheme which only work on "migration-safe" programs.

Its memory representation model implies a pointer-sensitive design which slows down the migration dramatically. The Porch system [6] reports pointers individually to create state as in SNOW. This might cause flexibility and efficiency problems for complex applications. The thread migration approach in [3] is similar to *MigThread* but has limitations. *MigThread* can handle pointer arithmetic, memory management, and "migration-unsafe" features efficiently.

6 Conclusion and Future Work

MigThread is shown to be generic in its scope. It handles four major "migration-unsafe" features in C/C++. Under *MigThread*, more programs become migratable and programmers do not need to worry if they are coding in "migration-safe" style. Thread state is constructed efficiently at runtime. More adaptation points can be inserted into programs to improve sensitivity of dynamic environment without sacrificing performance. As an application-level approach, *MigThread* places no restriction on thread types and operating systems. Experiments on real applications indicate that the overhead of *MigThread* is minimal.

We are currently porting *MigThread* to multiple platforms to exploit its heterogeneity potential. More work is being conducted on transferring process state and communication state for a complete thread migration package.

References

[1] Milojicic, D., Douglis, F., Paindaveine, Y., Wheeler, R. and Zhou, S.: Process Migration, ACM Computing Surveys (2000) 474

[2] Smith, P. and Hutchinson, N.: Heterogeneous process migration: the TUI system, Tech rep 96-04, University of British Columbia (1996) 474, 483

[3] Jiang H. and Chaudhary, V.: Compile/Run-time Support for Thread Migration, Proc. of 16th Int. Parallel and Distributed Processing Symposium (2002) 475, 476, 477, 484

[4] Thitikamol, K. and Keleher, P.: Thread Migration and Communication Minimization in DSM Systems, Proc. of the IEEE (1999) 483

[5] Roy, S. and Chaudhary, V.: Design Issues for a High-Performance Distributed Shared Memory on Symmetrical Multiprocessor Clusters, Cluster Computing: The Journal of Networks, Software Tools and Applications No. 2 (1999) 482

[6] Strumpen, V.: Compiler Technology for Portable Checkpoints, submitted for publication (http://theory.lcs. mit.edu/ strumpen/porch.ps.gz) (1998) 477, 481, 484

[7] Chanchio, K. and Sun, X. H.: Data Collection and Restoration for Heterogeneous Process Migration, Proc. of Int. Conf. on Distributed Computing Systems (2001) 474, 477, 481, 483

[8] Dimitrov, B. and Rego, V.: Arachne: A Portable Threads System Supporting Migrant Threads on Heterogeneous Network Farms, IEEE Transactions on Parallel and Distributed Systems 9(5), 1998. 483

[9] Jul, E., Levy, H., Hutchinson, N. and Blad, A.: Fine-Grained Mobility in the Emerald System, ACM Transactions on Computer Systems, Vol. 6, No. 1, 1998. 483

[10] Mascarenhas, E., and Rego, V: Ariadne: Architecture of a Portable Threads system supporting Mobile Processes, CSD-TR 95-017, Purdue University (1995) 483
[11] Itzkovitz, A., Schuster, A., and Wolfovich, L.: Thread Migration and its Applications in Distributed Shared Memory Systems, Journal of Systems and Software, vol. 42, no. 1, 1998. 483

Improved Preprocessing Methods for Modulo Scheduling Algorithms

D.V. Ravindra and Y.N. Srikant

Computer Science & Automation, Indian Institute of Science, India
{ravindra,srikant}@csa.iisc.ernet.in

Abstract. Instruction scheduling with an automaton-based resource conflict model is well-established for normal scheduling. Such models have been generalized to software pipelining in the modulo-scheduling framework. One weakness with existing methods is that a distinct automaton must be constructed for each combination of a reservation table and initiation interval. In this work, we present a different approach to model conflicts. We construct one automaton for each reservation table which acts as a compact encoding of all the conflict automata for this table, which can be recovered for use in modulo-scheduling. The basic premise of the construction is to move away from the Proebsting-Fraser model of conflict automaton to the Müller model of automaton modelling issue sequences. The latter turns out to be useful and efficient in this situation. Having constructed this automaton, we show how to improve the estimate of resource constrained initiation interval. Such a bound is always better than the average-use estimate. We show that our bound is safe: it is always lower than the true initiation interval. This use of the automaton is orthogonal to its use in modulo-scheduling. Once we generate the required information during pre-processing, we can compute the lower bound for a program without any further reference to the automaton.

1 The Problem

Instruction scheduling based on finite state automata is an elegant way of capturing the resource conflicts which can occur in current processors. The resource usage model in such a formulation is based on reservation tables and their compositions. In normal scheduling, we wish to answer the following query efficiently: *Given a set of currently active operations in the schedule, what are the possible offsets at which a new operation can be issued*? There are dual views about what the automaton should model:

- The issue automaton records all the active issues in a state. A new issue at a permitted latency adds to the resource usage and takes us to another state. An issue which retires at some point will take us back to a previous state. We will call this the Müller style automaton. This automaton is too fine-grained for scheduling.

S. Sahni et al. (Eds.) HiPC 2002, LNCS 2552, pp. 485–494, 2002.

– The conflict automaton records the set of permitted offsets and reservation tables as a state. An issue in a state at a permitted offset takes us to a new state with some of the offsets being invalidated due to resource conflicts. We will call this the Proebsting-Fraser (PF) style automaton. The conflict information captured by this model is just sufficient for scheduling.

In a generalization of automaton-based scheduling to modulo-scheduling, we are also interested in answering the following query efficiently for each reservation table r: *What are the permitted offsets at which r can be issued in a software pipeline with an initiation interval of* **II** *cycles* ? Note that we are talking of issuing a single reservation table repeatedly, but within a cyclic schedule which will wrap the reservation table around at **II** cycles (modelling an overlap between issues of r in different iterations). Our focus in this work is on constructing an automaton which can answer this query efficiently at compile-time for a given **II**.

2 Background and Existing Solutions

The practical basis for conflict modelling are those proposed by Proebsting and Fraser [3] (and extended in [1]) and Müller [2]. The former generates a minimal automaton to capture all the possible issues in normal (acyclic) scheduling. For a given initiation interval i, a modulo-scheduling algorithm keeps track of the resource usage in a *modulo reservation table* (MRT). This is a $n \times i$ matrix where n is the number of resources in the machine. When a reservation table is issued at t, its effect is reflected by composing the MRT with r at all offsets $t + ki$ modulo i. This repeated composition reflects the issue of this operation at every i cycles in the software pipeline. The extension of the PF model to modulo-scheduling is based on the *cyclic reservation tables* proposed in [4] in the framework of co-scheduling. The CRT is a $n \times i$ usage matrix where i is the initiation interval. A reservation table r issued at an offset t causes r to be composed at $t + ki, k \geq 0$ modulo i. The usage matrix in this case is always of size i reflecting the kernel. Valid initiation sequences of r under an initiation interval i are now obtained by a method similar to the PF-automaton construction except that an offset is valid iff there is no conflict in the CRT. Improvements to the basic method in pre-computing valid offset sequences for different initiation intervals have been presented in [5]. The automaton so generated is acyclic and each path denotes a valid offset sequence of issue of r so that the resulting schedule is valid for an initiation interval i.

3 Outline of Our Solution

The problem with the conflict model of resources is that the automaton structure is heavily dependent on the initiation interval. Our intent is to show the applicability of the Müller-style automata in this restricted domain. Once the automaton structure is independent of the initiation interval, the automata we generate can be used for other purposes. We show one such application in improving the estimate of the resource constrained initiation interval. The automata can be pre-processed to get this information so that using this estimate is no more expensive than the usual average-use estimate.

Notation: We use r to denote a reservation table, u to denote a usage matrix, t to denote an offset, s to denote a state. Each such symbol may be subscripted to specialize its meaning (as in t_i or t_c) or to enumerate (t_1, t_2). Tuples, sequences and sets are delimited by $\langle\rangle$, $[]$ and $\{\}$ respectively. The "size" of an object x is denoted by $|x|$. For a sequence v, $v[i{:}j]$ denotes the sub-sequence from i to j inclusive. A usage matrix is considered as a sequence of resource usage vectors. We consider the usage vectors as elementary units. We never select rows out of a usage matrix. Composition of (a usage matrix) u with (a reservation table) r at (an offset) t is denoted $u \oplus \langle r, t\rangle$. If the offset is illegal (causes a resource overuse), the result will be $\perp$ else it will be a new usage matrix. A term of the form k^* indicates zero-or-more repetitions of k. We usually write the reservation tables as $[r_1; r_2; \cdots]$ where r_i corresponds to the usage pattern of the resource i.

4 Construction of the M-automaton

The construction of the automaton is a two-stage process:

1. We construct a Müller style automaton with a form of redundancy removal applied to decrease the number of states (M-automaton). It contains two kinds of transitions, I-transitions for a new issue and C-transitions for a "clock" at which an active issue finishes.
2. In this stage, we construct a reduced automaton which contains only the issue transitions (M_i-automaton). It makes sense only if the M-automaton is exact.

One M-automaton is constructed for each reservation table. It essentially records the resource usages defined by repeated issues of r starting at $[\,]$. The development of the automaton for [00010; 10001; 01100] is illustrated in Fig. 1. Edges and nodes are labelled by an index. The second part of the edge label(after the colon) is due to the transitions.

The states of the automaton are characterized by a tuple $\langle u, c\rangle$. c is an increasing sequence of offsets at which an active issue retires. We call this the *clock sequence* of the state while u is a usage matrix. The initial state is $\langle[\,],[\,]\rangle$. For instance, the state (3) is associated with $\langle$[00010010; 10011001; 01101100], [5, 8]$\rangle$. Two instances of r are due to retire at clocks 5 and 8 respectively. The state (7) is the initial state in the example. From each state $\langle u, c\rangle$, there are two kinds of transitions in the automaton:

- There is one *clock transition* which corresponds to the finish of the earliest finishing operation in this state (which is also the earliest issue in the state). The resulting state will be u with its first c[0] columns shifted out. The clock sequence is c without c[0] with each element decremented by c[0].

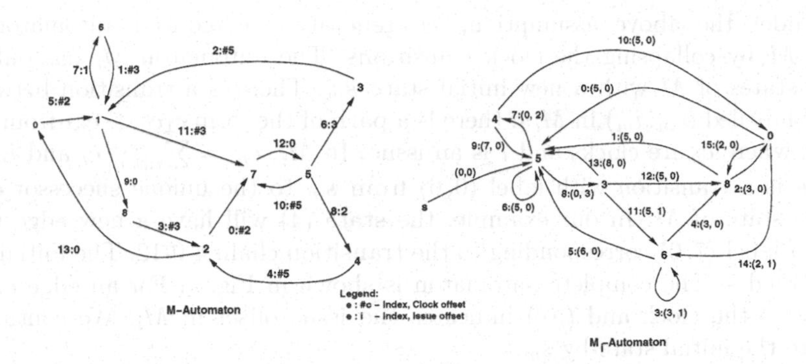

Fig. 1. M-automata for [00010; 10001; 01100]

– For every permitted offset between 0 to $c[0]-1$ (or at 0 in the initial state), there is an *issue transition.* The restriction on the issue offsets ensures that all operations in the state are active at the time of issue. For an issue at t, the resulting state is $u \oplus \langle r, t \rangle$. The clock sequence is c extended with $t + |r|$ and sorted to restore order.

The automaton is constructed by applying the transition rule repeatedly with the initial state $\langle [\,], [\,] \rangle$ as the seed. Henceforth, we will refer to the M-automaton as M with an implicit reservation table r. The timespan of a state s is the timespan of its usage matrix and is also donoted $|s|$.
The following observations regarding the structure and use of M are important:

1. Traversing a clock edge increments the next reference point of issue and traversing an issue edge adds one more issue to the sequence at the appropriate offset from the last clock. In our example, the issue sequence at offsets $[0, 2, 7, 10, 13]$ from state 7 corresponds to the edge sequence 12-8-4-0-12-6-2-7 which takes us to state 6. The clock is advanced as 0-5-7-12 during the clock traversals.
2. During the construction of the M-automaton, we check for states which have already been generated. In our implementation, we approximate state equivalence to the usage matrix equivalence. We ignore the clock sequence during this comparison. It can be verified that a reservation table of the form (1000) will cause heavy aliasing under this scheme.
3. For every state, $|s| < 2|r|$. If (s_1, s_2) is a clock transition, then $|s_2| < |s_1|$ and $|s_2| < |r|$. If (s_1, s_2) is an issue transition, then $|s_2| \geq |s_1|$ and $|s_2| \geq |r|$. The states with in-clock transitions are called clock states and the rest as issue states. There exists no state which is both an issue state and a clock state. The clock states in our example are $\{7, 1, 2\}$.

Aliasing of states cannot occur between an issue state and clock state. We can state something stronger: *If there are no idle columns in r, then no aliasing can take place between the states of the automaton.*

Under the above assumption, we generate our reduced automaton, M_i from M by collapsing the clock transitions. The automaton M_i has only the issue states of M with a new initial state s_m. There is a transition between s and t labelled (c_m, i_m) in M_i if there is a path of the form $c_1c_2 \cdots c_n i$ from s to t in M where c_i are clocks and i is an issue. In M_i, $c_m = \sum_{i:[1,n]} c_i$ and $i_m = i$. There is a transition with label $(0, 0)$ from s_m to the unique successor of the initial state of M. In our example, the state (4) will have a new edge to (5) with a label $\langle 7, 0\rangle$ corresponding to the transition chain 4-0-12. The initial state is marked s. The complete automaton is shown in Fig. 1. For an edge e, $t_c(e)$ indicates the clock and $t_i(e)$ indicates the issue offset in M_i. We continue to denote the initial state by s_m.

The traversal of M_i to trace issues of r is similar to M. An important property of M_i is that no two edges out of a state have the same $t_i(e) + t_c(e)$. Every sequence of issue offsets from an initial state is either invalid or has exactly one path in the automaton. For a path $p : s_1, s_2, \cdots, s_n$ in the automaton, we define

$$C_p = \sum_{i=1}^{n-1} t_c(s_i, s_{i+1}) \ . \tag{1}$$

$$I_p[k] = C_{p'} + t_i(s_k, s_{k+1}) \text{ where } p' = p[s_1 : s_k] \ . \tag{2}$$

$$T_p = I_p[n-1] + |r| \ . \tag{3}$$

C_p records the base time-step for the issue offset in s_n, T_p records the size of the schedule implied by the traversal through p and I_p denotes the issue sequence of p, the time-steps at which an issue is made. For example, for the path s_m-5-3-8-0-6, $I_p = [0, 3, 5, 8, 11]$, $C_p = 10$ and $T_p = 16$.

Consider a state s and a sequence of valid offsets I which traces a path p in the automaton. We can surely issue this sequence with an initiation interval of T_p. But, it is possible that we can actually issue it at an offset somewhere between $I_p[n-1]$ and T_p. We can determine the validity of an offset T by tracing the path p issued every T cycles in M_i. A legal offset will eventually induce a periodic cycle while an illegal offset will miss an issue at some point in the trace.

5 Generation of Valid Issue Sequences

Using M_i, we show how to solve our motivating problem:

Problem 1. Find all the legal offsets for repeated issue of a reservation table r under a software pipeline with an initiation interval i under modulo-scheduling.

The basic idea is to unroll the paths in the automaton till an issue at a time less than i is not possible. Then, the issues which caused this usage table are checked for validity for an initiation interval not more than i. The systematic algorithm, **genpath**, for doing this is using M_i follows. We use the function **kernel_size** to find the least kernel size.

1: **algorithm** genpath: {*params*: valid, s, p, t_p}
2: extend $\leftarrow$ **false**
3: **for** $e : (s, s') \in M_i$ **do**
4: **let** $|p|' \leftarrow t_c(e) + t_i(e) + t_p$ **in**
5: **if** $|p|' \leq i$ **then**
6: genpath (valid, s', $[p; t_p + t_i(e) + t_c(e)]$, $t_p + t_c(e)$)
7: extend $\leftarrow$ **true**
8: **if not** extend **and** kernel_size(p) $\leq i$ **then**
9: Add p to *valid*

The toplevel invocation is `genpath` ($\emptyset$, initial(M_i), [0], 0). The set of paths is accumulated in `valid`. The development of the paths to generate the valid set of offsets for $i = 11$ is illustrated in Fig. 2. Some invalid extensions are shown underlined. The leaves correspond to the maximal valid sequences. The fourth component here is the actual size of the table developed which might be larger than i.

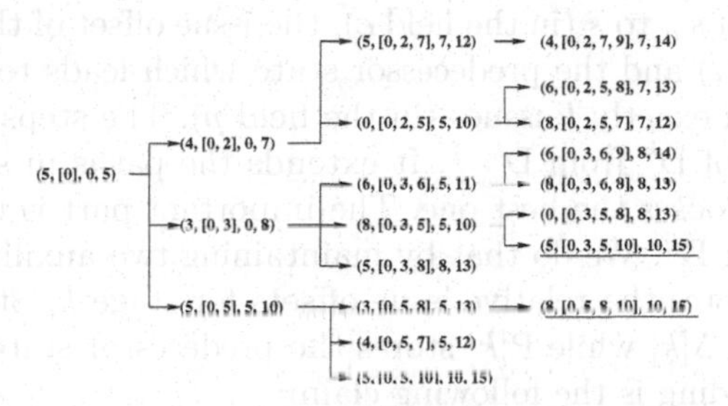

Fig. 2. Unfolding of M_i for **II** = 11

6 Estimation of Resource Constrained II

There are two components contributing to the estimate of an initiation interval: the dependence induced bound which depends on the program and the resource bound which depends on the architecture. Usually, $\lambda_r = max(\frac{P_r}{N_r})$ where P_r is the number of times resource r is used in the program and N_r is the number of units of that resource in the architecture. The actual initiation interval $\lambda \geq max(\lambda_d, \lambda_r)$.

Problem 2. Find a safe estimate (which is $\leq \lambda$) of the architecture-influenced initiation interval which is better (numerically higher) than λ_r.

Let $K_m = k_0, k_1, k_2 \cdots$ be the sequence where k_i is the minimum size of a kernel required to hold i copies of r. Our goal is to find a compact encoding of K_m so that we can find that bound for any number of copies of r. Consider M_i for a reservation table r with an initial state s_m. Our first step is to find a compact encoding for the (infinite) sequences D_s where $D_s[k] = \min(C_p$, p is a path between s_m and s with exactly k issues). We construct D_s in stages $D^0, D^1 \cdots$ where D^k is a *stage* of size $|M_i|$ and $D^k[s]$ for a state s is $D_s[k]$:

1: Initialize $\mathrm{D}^0[succ(s_m)]$ as $\langle c:0, i:0, p:s_m\rangle$[1] and rest as $\langle c:\infty, i:\infty, p:-\rangle$
2: $\Delta \leftarrow \emptyset$ and $\mathrm{S} \leftarrow \emptyset$
3: $k \leftarrow 1$
4: **loop**
5:    **for** $s \in M_i$ **do**
6:       let $s' \in pred(s)$ such that $\mathrm{D}^{k-1}[s'].c + t_c(s', s) + t_i(s', s)$ is minimum
7:       $\mathrm{D}^k[s] \leftarrow \langle c : \mathrm{D}^{k-1}[s'].c + t_c(s', s), i : t_i(s', s), p : s'\rangle$
8:    $t_{min} \leftarrow \min_{s \in \mathrm{D}^k} (\mathrm{D}^k[s].c + \mathrm{D}^k[s].i)$
9:    $\Delta[k] \leftarrow [\mathrm{D}^k[s].c + \mathrm{D}^k[s].i - t_{min} \mid s \in M_i]$
10:    $\mathrm{P}[k] \leftarrow [\mathrm{D}^k[s].p \mid s \in M_i]$
11:    **if** $\exists j < k$, $\mathrm{P}[j] = \mathrm{P}[k]$ and $\Delta[j] = \Delta[k]$ **then**
12:       **return**
13:    $k \leftarrow k + 1$

The information we store in $\mathrm{D}^k[s]$ while developing the profile is the C_p along the best path from s_m to s (in the field c), the issue offset of the last issue in this path (in the field i) and the predecessor state which leads to the shortest path from s_m to s with exactly k issues (in the field p). The steps (5)–(7) constitute the development of D^k from D^{k-1}. It extends the paths in stage $k-1$ by one more issue and chooses the best one. The important part is to determine when to stop generating D^k. We do that by maintaining two auxiliary data Δ and S. The term $\Delta[k]$ stores the relative issue offsets for stage k. Steps (8), (9) in the algorithm update $\Delta[k]$ while $\mathrm{P}[k]$ stores the predecessor states for stage k. The basis for our encoding is the following claim:

Claim. There exist indices i and j, $i > j$ such that $\mathrm{P}[i] = \mathrm{P}[j]$ and $\Delta[i] = \Delta[j]$.

Proof. Each element of Δ_k is between 0 and $|r|$ (after $|r|$ cycles, we can always issue an new instance of r) and the number of states is finite. So, the number of combinations the two sequences together can take is finite. □

A consequence of this is that the profile looking from i is identical to the profile at j except for a change in the base cost ($\min(D^k[s].c + D^k[s].i)$). The sequence D_s for a given state s (which is $D^k[s].c + D^k[s].i$ for $k = 0, 1, 2 \cdots$) is monotonically increasing with its difference sequence $[D_s[n] - D_s[n-1]]$ of the form $appp\cdots$. Hence, D_s is completely defined by simply storing a and p. We can generate any term of D_s in constant time from a and p: Let $S_a = \{0, \mathrm{a}[0], \mathrm{a}[0] + \mathrm{a}[1], \cdots\}$, $\mathrm{A} = S_a[|S_a| - 1]$ $S_p = \{\mathrm{A} + \mathrm{p}[0], \mathrm{A} + \mathrm{p}[0] + \mathrm{p}[1], \cdots\}$ and $\mathrm{P} = \mathrm{p}[0] + \mathrm{p}[1] + \mathrm{p}[2] + \cdots$. Then, the ith term of the sequence can be found as $S_a[i]$ if $i < |S_a|$ and $S_p[r] + qP$ where q and r are the quotient and remainder obtained when $i - |S_a|$ is divided by $|S_p|$.

The sequence K_m is bounded above by the sequence $K^+ = [D_{s_m}^{(k)}]$ and below by the sequence $K^- = [\min(D_u^{(k)} + i_m(u))]$ where $i_m(u)$ is the least offset at

[1] $\langle \ldots x_i : v_i \ldots \rangle$ denotes a record with fields x_i and values v_i.

which u has an issue. Both these sequences have the same structure as D_n and hence can be efficiently encoded. We use K^- as our lower bound. It is a safe estimate on the minimum kernel size.

In our example, the profile will be (the first component is the clock and issue while the second component is the previous state):

State ⇒	6	5	4	3	0	8
1 :	$(-,-)/-$	$(0,0)/7$	$(-,-)/-$	$(-,-)/-$	$(-,-)/-$	$(-,-)/-$
2 :	$(-,-)/-$	$(5,0)/5$	$(0,2)/5$	$(0,3)/5$	$(-,-)/-$	$(-,-)/-$
3 :	$(5,1)/3$	$(7,0)/4$	$(5,2)/5$	$(5,3)/5$	$(5,0)/4$	$(5,0)/3$
4 :	$(7,1)/0$	$(10,0)/0$	$(7,2)/5$	$(7,3)/5$	$(8,0)/8$	$(7,0)/0$
5 :	$(10,1)/6$	$(12,0)/8$	$(10,2)/5$	$(10,3)/5$	$(10,0)/8$	$(10,0)/6$
6 :	$(12,1)/0$	$(15,0)/0$	$(12,2)/5$	$(12,3)/5$	$(13,0)/8$	$(12,0)/0$

It can be seen that the delta cost and state sequence at Stage 6 is identical to that at Stage 4. Hence, the sequences, D_n, can be partitioned at the indices (1-2-3)(4-5)*. In our example, i_m is [0:2, 3:5, 4:5, 5:2, 6:3, 8:3]. Hence, the sequence $K^- = [2,5,7,10,12,15]$ which is decomposed as $S_a = [0,2,5,7]$ and $S_p = [10,12]$ and $P = 5$.

We can now use the sequences $[K_r]$ (K_m for each r) generated for each reservation table r to improve the estimate of λ_r. Let resources in the program be denoted e_i. Let P_e denote the number of times a resource e is used in the program and A_e denote the number of units of resource e available in the machine. The conventional estimate is

$$\lambda_r = \max_{e \in E(P)} \left\lceil \frac{P_e}{A_e} \right\rceil$$

where $E(P)$ denotes the resources used in the program P. All our estimates will be compared with this estimate. Using K^- instead of K_r in what follows is perfectly valid since the estimate λ_r models a busy schedule of single-cycle operations packed as compactly as the machine would allow. Our assumption of no-idle columns in the reservation tables implies that the $i_m(s) \geq 1$ for a state s.

Suppose there are n copies of r used in the program, then an obvious lower bound on the size of the kernel is $K_r[n]$. So, we get our first estimate $\lambda_r^{(1)} = \max_{r \in R(P)} K_r[n_r]$ where $R(P)$ is the set of reservation tables used in the program and n_r is the number of copies of r in P. We note that if there is only type of reservation table in use in the program, then $\lambda_r^{(1)} \geq \lambda_r$. In our example, if there are ten instances, $\lambda_r = 10$ since each resource is used exactly once while $\lambda_r^{(1)} = 10 + 3 \times 5 = 25$ which is a much better estimate. But, this does not scale up when we have more than one reservation table used in the program since it does not account for the contribution to the usage of e from the other reservation tables.

Consider a resource e and let R_e be the set of reservation tables in the program which contribute to the use of e. Our next estimate is

$$\lambda_r^{(2)} = \max_{e \in E(P)} \max_{r \in R_e} \left(K_r[n_r] + \frac{P_e - (K_r[n_r] - N_r)}{A_e} \right)$$

where N_r indicates the number of slots in K_r at which the resource e is used. This is independent of the schedule (but dependent on r, the sum of N_r for a given e will be P_e). The idea is to build the tightest possible kernel respecting the reservation table-resource association followed by a naive busy scheduling for the rest of the usages of the resource. It can be shown that $\lambda_r^{(2)} \geq \lambda_r$.

As an example, consider an architecture with three resources {A, B, C} and the reservation tables: {R1: B C C A B, R2: A B B C, R3: B A C A B A}. The kernel sequences for the tables can be determined to be: { [2, 5, 7, 10, 12, 15], [2, 4, 6, 8, 10], [1, 6, 7, 12, 13, 18]}. The pattern of the sequence is quite obvious. Consider a program: {R1:10, R2:4, R3:6} (indicating the number of occurrences of each table.) Then, $\lambda_r = \max(28, 40, 30) = 40$. $\lambda_r^{(1)} = \max(25, 8, 18) = 25$ and $\lambda_r^{(2)}$ can be computed to be $\max(32, 40, 32) = 40$. This example illustrates the necessity of $\lambda_r^{(2)}$ to handle programs which have a balanced mix of reservation tables. If only R1 or R2 dominate, then $\lambda_r^{(1)}$ alone would suffice as we saw earlier (R1 in our running example.)

7 Experimental Studies on Existing Processors

In this section, we give our preliminary results on the size of M and M_i on some processors: Alpha 21264, MIPS R4000 and the external issue logic of IA-64.

It turns out that we do not need to consider all the reservation tables since the automata generated are identical in some cases. Consider two reservation tables of the form `IF+ID MUL+ADD SUB` and `IF+ID SUB+DIV ADD`. Even though the reservation tables look sufficiently different, they will have the same abstract structure of M and M_i under the mapping `MUL-SUB, ADD-DIV, SUB-ADD`. Most processors admit this equivalence. On such processors, we can use λ_1 instead of λ_2 as an improved estimate of λ_r.

The second optimization makes the issue based model feasible for pipelines of the form S1-S2-S3-S4-$\cdots$, i.e. a linear pipeline which can support one issue per cycle and flows smoothly. The Alpha model is a perfect example of such functional units. Such a linear pipeline results in state explosion for the M-automaton. We avoid this by considering a linear pipeline as a single indivisible unit. This does not affect the automaton which depends on the *rate* of the pipeline. But, the estimates of λ_r must account for this. Table 1 shows the three largest automata sizes (the column *Table* corresponds to the indices of the reservation tables in our test data which is omitted here).

In the cases of Alpha and R4000, the reduction outlined earlier and the architecture constraints on repetition rates effectively make almost all reservation tables isomorphic. We note that the automaton sizes are fairly small, but are highly dense. The uniformity of the architecture is also reflected in $K^- = 1\ 1\ 1\ 1\ 1\ 1\ 7\ 8\ 9\ 10\ 11\ 12\ \cdots$. i.e. in the absence of constraints, we can do no better than predict that an issue is made every cycle.

The R4000's automaton is influenced by the processor mandating a particular repeat rate. In the absence of this constraint, the tables are reasonable since the

Table 1. Size of M-automaton on some processors

Arch	Table	$\|N(M)\|$	$\|E(M)\|$	$\|N(M_i)\|$	$\|E(M_i)\|$	D
R4000	9	5	7	4	7	(5, 1)
	1	2	2	2	2	(3, 1)
Alpha	1	449	1313	386	3176	(10, 1)
Itanium	9	13	23	10	30	(6, 1)
	3	9	14	7	17	(6, 2)
	5	5	7	4	7	(5, 1)

processor is not entirely pipelined. Some non-pipelined units in the execution path keep the automaton size small. The IA-64 exhibits substantial isomorphism in its dispersal logic (For example, the use of I-units and the M-units).

Finally, the last column indicates the length of the encoding of K^-, the initial prefix and the repeating sequence. They are very short and hence can be used directly to improve estimates of λ_r.

Conclusion

The Müller style automaton which in the general case are too huge to be useful are suitable when we model resource usage information for modulo-scheduling. Independent of modulo-scheduling, we have shown how preprocessing the resource conflicts can improve the estimates of the resource constrained initiation interval. They are very compact and the limit generated from them can be formally shown to be a safe estimate greater than the naive bound.

References

[1] Vasanth Bala and Norman Rubin. Efficient instruction scheduling using finite state automata. In *28th Annual ACM/IEEE International Symposium on Microarchitecture*, pages 46–56, 1995. 486

[2] Thomas Müller. Employing finite automata for resource scheduling. In *26th Annual ACM/IEEE International Symposium on Microarchitecture*, pages 12–20, 1993. 486

[3] Todd W. Proebsting and Christopher Fraser. Detecting pipeline structural hazards quickly. In *21st Annual ACM Symposium on Principles of Programming Languages*, pages 280–286, 1994. 486

[4] Govindarajan R, Erik R. Altman, and Guang R. Gao. Co-scheduling hardware and software pipelines. In *2nd IEEE Symposium on High-Performance Computer Architecture*, pages 52–61, 1996. 486

[5] Chihong Zhang, Govindarajan R, Sean Ryan, and Guang R. Gao. Efficient state-diagram construction methods for software pipelining. In *Proc. of International Conference on Compiler Construction (CC-99)*, pages 153–167, 1999. 486

Dynamic Path Profile Aided Recompilation in a JAVA Just-In-Time Compiler*

R. Vinodh Kumar[1], B. Lakshmi Narayanan[2], and R. Govindarajan[3]

[1] Cisco Systems, Bangalore, India
vinodh@cisco.com

[2] School of Computer Science and Engineering, College of Engineering-Guindy
Anna University, Chennai 600025, India
blnarayanan@hotmail.com

[3] Department of Computer Science and Automation and Supercomputer Education and Research Centre, Indian Institute of Science
Bangalore 560 012, India
govind@csa.iisc.ernet.in

Abstract. Just-in-Time (JIT) compilers for Java can be augmented by making use of runtime profile information to produce better quality code and hence achieve higher performance. In a JIT compilation environment, the profile information obtained can be readily exploited in the same run to aid recompilation and optimization of frequently executed (hot) methods. This paper discusses a low overhead path profiling scheme for dynamically profiling JIT produced native code. The profile information is used in recompilation during a subsequent invocation of the hot method. During recompilation tree regions along the hot paths are enlarged and instruction scheduling at the superblock level is performed. We have used the open source LaTTe JIT compiler framework for our implementation. Our results on a SPARC platform for SPEC JVM98 benchmarks indicate that (i) there is a significant reduction in the number of tree regions along the hot paths, and (ii) profile aided recompilation in LaTTe achieves performance comparable to that of adaptive LaTTe in spite of retranslation and profiling overheads.

1 Introduction

A Java JIT compiler is a component of the Java Virtual Machine (JVM) [20], which translates Java's bytecode methods into native code prior to execution, so that the Java program under execution runs as a real executable. Since the

* This work was done when the first author was a graduate student at the Department of Computer Science and Automation, Indian Institute of Science, Bangalore. It was extended when the second author was a Summer Research Fellow of Jawaharlal Nehru Centre for Advanced Scientific Research, Bangalore, at the Supercomputer Education and Research Centre, Indian Institute of Science, Bangalore. The last author acknowledges the research funding received from the Department of Science and Technology, India, which partly supported this work.

S. Sahni et al. (Eds.) HiPC 2002, LNCS 2552, pp. 495–506, 2002.

machine code is *cached* in the JVM, the translated native code is used for subsequent executions of the method. The execution of the translated code avoids the inefficiencies associated with the interpretation of bytecode. However, such a translation incurs runtime overheads and hence it cannot apply aggressive compiler optimizations to obtain efficient code. To overcome this, adaptive compilation techniques, which apply aggressive compiler optimizations only on frequently executed methods, have been proposed in the recent past [2, 11, 19]. Such adaptive compilation facilitates the exploitation of runtime program behavior through profile information for better performance.

Programs typically spend most of their time in a small set of paths and optimization along these paths definitely yields performance benefits. In the static mode of compilation, path profiles from preparatory runs of the program are used for off-line re-compilation and optimization of the program along its frequently executed paths. Several profiling methods [5, 6, 7] have been proposed to reduce the time incurred in obtaining the profile information. In a dynamic compilation environment, programs can be monitored for some time and the parts of program which are expected to run frequently for the remaining part of the execution can be optimized. This form of adaptive compilation of programs exists in modern JIT compilers such as HotSpot compiler from *Sun* [14], *Intel* Vtune JIT compiler for Java [1], and LaTTe [19]. These compilers retranslate and optimize methods that are invoked more frequently (hot methods).

In our work, we plan to aid the retranslation of the hot methods in a Java JIT compiler using intra-procedural (i.e., intra-method) path profiling information gathered from a low overhead path profiling strategy so that the methods can be better optimized for execution along the hot paths. The technique adopted for profiling must have low (profile) instrumentation overheads and low profiling overheads, since these constitute a part of the running time of the program. As a consequence, sophisticated low profile overhead algorithms such as the Ball and Larus' efficient path profiling algorithm [6] or the efficient edge profiling algorithm [5], which take significant time to identify profile points in the program, cannot be used in dynamic compilation.

We use a profiling strategy called *bit tracing* [5] for generating the path profiles dynamically. Using the profile information, we enlarge tree regions, which are the units of optimization in LaTTe, along the hot path. For this purpose, we clone basic blocks along the hot path which have more than one incoming edge in the control flow graph to ensure that all the basic blocks along the hot path are within a single region. This allows the optimizations to be more effective on the hot paths in the program. The profile information is then further used to perform instruction scheduling on superblocks [10] which are present along the hot paths. Thus the major contributions of this paper are: (i) implementing an efficient profile-assisted aggressive recompilation method for hot paths in an adaptive JIT compiler, (ii) expanding regions to include all basic blocks of a hot path, and (iii) superblock instruction scheduling in a JIT compiler framework. Our initial results show that region expansion along the hot paths significantly reduces the number of regions in a method, by a factor of 4 to 5, in all SPEC

JVM98 benchmarks. This facilitates aggressive optimization along the hot path. Our experiments with superblock instruction scheduling reveal that profile aided recompilation in LaTTe achieves performance comparable to that of adaptive LaTTe in spite of the retranslation and profiling overheads. This suggests that if retranslation overheads, when retranslating again from bytecode to native code, can be mitigated, profile assisted aggressive compilation methods could yield further performance improvements.

Section 2 presents the necessary background on LaTTe. In Section 3, we discuss the path profiling technique used in our implementation. Section 4 deals with the recompilation strategy. Section 5 focusses on region expansion and superblock scheduling performed in our implementation. In Section 6 we present our performance results. Section 7 discusses related work. In Section 8 we provide concluding remarks.

2 LaTTe JIT Compiler

LaTTe is a Java virtual machine [19] that includes a JIT compiler targeted at RISC machines, specifically the Ultra SPARC processor [17]. The JIT compiler generates good quality code through a mapping of the Java Stack to registers, incurring very little overhead. It also performs some traditional optimizations such as common subexpression elimination and loop invariant code motion [19]. In LaTTe, the basic unit of optimization is a tree region (or simply, region). The CFG of pseudo SPARC code is partitioned into regions which are single entry, multiple-exit sub-graphs shaped like trees. Tree regions start at the beginning of a method or at other join points and end at start of the other regions.

In LaTTe, translation of bytecode to native code is done in 4 phases. The details of these phases can be found in [19].

BYTECODE_ANALYSIS: LaTTe identifies all control join points and subroutines in the bytecode via a depth first traversal.

CFG_GEN: The bytecode is translated into a control flow graph (CFG) of pseudo SPARC instructions with symbolic registers. If speculative virtual method inlining is enabled, it is done during this phase.

OPT: In the third stage, LaTTe performs region-based traditional optimizations, along with region-based register allocation.

CODE_GEN: In the final phase, the CFG is converted into SPARC code, which is allocated in the heap area of the JVM and is made ready for execution.

Adaptive retranslation in a JIT compiler is used to perform the costly runtime optimizations selectively on hot methods based on the program behavior. In the adaptive version of LaTTe, methods are selected for optimizations based on method run counts [15]. This is achieved as follows. When a method is called for the first time, it is translated without any optimization and method inlining. Each method has an associated method run counter which is initially loaded with a threshold value and decremented each time the method is invoked. When the method count becomes zero, i.e., when the number of times this method is invoked exceeds a certain threshold, the method is retranslated with all optimizations and conditional inlining enabled.

3 Path Profiling Technique

In this work, we use a low overhead path profiling technique called bit tracing [5] which is described below. The input to the path profiler is the CFG of a method. The nodes of the CFG are basic blocks. We identify the nodes that are headers of the method or loop headers (i.e., nodes that have an incoming back edge) or exit nodes (nodes that exit out of the method) as *store nodes.* A *store node* stores an encoding of the path starting from the previous store node to the current store node. The store node then initializes the path string by inserting a '1' into it. Code to left shift bits into the path string based on the branch outcome are introduced along the branch edges as depicted in Figure 1.

In Figure 1, block A is the store node since it has an incoming back edge (loop). Node A initializes the path string to '1'. So now path *ABEIJ* has the encoding '101' since the branch outcomes at *A* and *B* cause the edges *AB* and *BE* to left shift a '0' and '1' respectively, in order.

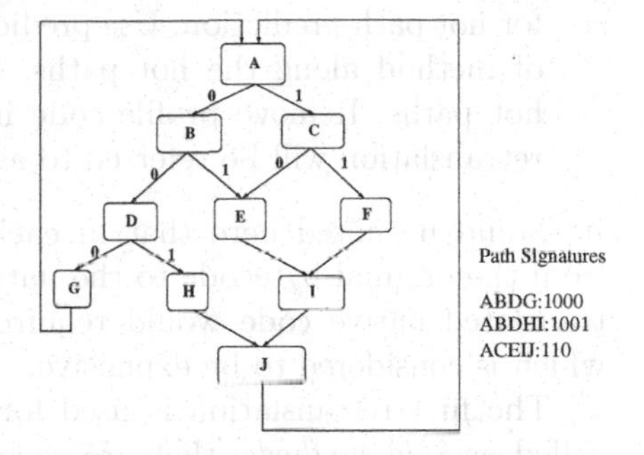

Fig. 1. Path Profiling strategy

The profiling technique employs two registers: one to hold the path encoding so far (referred to as shift register) and the other to hold the address where the path encoding has to be stored. We have used two SPARC registers **g6** and **g7** for these purposes and these registers are not used by LaTTe for register allocation or any other purposes. However since these registers are global we need to store and restore them during each method invocation. This profiling strategy has minimal overhead, as the instrumentation can be added to the CFG in one pass and the profiling information gathered needs minimal processing for hot path prediction. Lastly, this profiling method gives more accurate information about paths than edge profiling methods [5].

Associated with each store node is a circular queue of size n to store the path encodings of the last n recently executed paths. The choice of n depends on the amount of history needed for prediction. In this context, we shall refer to this strategy as n-MRET (n-Most Recently Executed Tails strategy). Apart from the circular queue, a counter is associated with each store node which is incremented each time execution reaches it. If the counter exceeds a particular threshold, referred to as *hotness threshold*, then it implies that the paths originating from the store node have become hot enough to be optimized. A point to be noted is that the hotness threshold may differ from n in that the threshold determines when to predict the hot path, while n determines the amount of profile history seen for making the prediction. A store node stores the previous path encoding reaching it and then increments its counter. It then loads the store register with the address where the encoding of the next path originating from it is to be stored subsequently.

4 Recompilation Strategy

Our recompilation strategy is as follows:

1. During the first translation, when a method is invoked for the first time, the method is compiled with no optimization.
2. Select hot methods for retranslation using method run counts. Compile these hot methods with all optimizations including method inlining. Instrument the method for gathering path profiles. This retranslation will be referred to as the *first retranslation*.
3. Another retranslation is invoked once enough profile information is available for hot path prediction. Use predicted hot path information for optimization of method along the hot paths. Perform instruction scheduling along the hot paths. Remove profile code introduced in the first retranslation. This retranslation will be referred to as the *second retranslation*.

It should be noted here that in each retranslation, the translation is always from the original bytecode to the native code. To retranslate from a previously translated native code would require retaining many intermediate structures which is considered to be expensive.

The first retranslation is used for two purposes: (i) filtering out methods, called as *cold methods*, that are executed only a few times[1] and (ii) allowing for paths through the method to mature. In our implementation, the first retranslation of a method is invoked whenever the method is invoked more than certain retranslation threshold. The retranslation threshold is a product of hotness threshold and a retranslation factor (referred to as first retranslation factor). In the first retranslation, our instrumentation for performing path profiling is done in the CODE_GEN phase (refer to Section 2), before the SPARC code is generated.

The second retranslation is triggered based on a combination of both method run count and hot path information stored in the store node counter. More specifically, the second retranslation of a method is enabled when either the method run count or the profile count of a hot path in the method exceeds a threshold value. The second retranslation threshold value is the product of hotness threshold and a second retranslation factor. During the second retranslation, the template of the newly generated CFG must match the CFG that was used for instrumentation in the first retranslation. If there is a CFG template mismatch, then the profiling information gathered would be rendered useless. Such a situation could arise in the presence of virtual method inlining. We omit the details here due to space constraints. The reader is referred to [18] for details.

[1] However, there may be many cold methods.

5 Path-Based Optimization

5.1 Region Expansion

A tree region is a unit of local optimizations in LaTTe. Tree regions have a single entry and multiple exits. For example, in Figure 2(a), basic blocks *A, B, C* and *E* fall in a single region. Similarly *D* and *F* are the other two regions in the CFG. In LaTTe, optimizations such as redundancy elimination, common subexpression elimination, constant propagation, loop invariant code motion, as well as local register allocation are performed for tree regions. Hence, bigger the tree regions, greater is the scope of optimization. So, we perform region expansion based on the predicted hot paths.

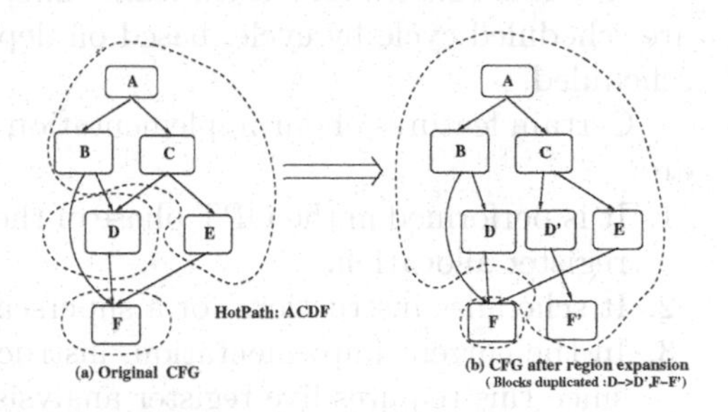

Fig. 2. Region Expansion

In our approach, using the path profile gathered, we duplicate basic blocks that occur in the hot path, to ensure that there are no incoming edges into the hot path. Thus, the region gets expanded and the scope for optimizations is increased. We illustrate this with an example. Consider the CFG in Figure 2(a). Let us assume that path profiling of this CFG yields the information that $ACDF$ is the most frequently executed path through it. Since LaTTe does region based optimizations and basic blocks *D* and *F* fall in different regions from nodes *A* and *C*, the path is not optimized completely with respect to the optimizations performed.

In our approach, we clone the blocks *D* and *F* to create two duplicate basic blocks D' and F'. Now in the modified CFG shown in Figure 2(b), the region with header as node A contains the nodes A, B, C, D, E, D' and F'. Since now the hot path lies entirely within the region, the path will be optimized completely with respect to the optimizations performed. In order to avoid excessive code duplication, we use a threshold called *cloning threshold*. Only if the store node associated with the hot path has executed more number of times than the cloning threshold, cloning is initiated.

5.2 Superblock Scheduling

The steps involved in superblock scheduling are superblock formation and list scheduling [10]. It is in superblock formation that the profile information is put to use. The basic blocks that are along a hot path are used to form a superblock. The branch targets, which are predicted to be *taken* based on profile information, are placed on the fall-through path of the branch. Then, the superblock is said to be the sequence of instructions from one join point (a point where control enters through more than one edge) up to the next join point. Superblock formation has an added advantage in that the predicted branch targets (along the hot path) are not likely to miss on the instruction cache, thereby improving cache performance. List scheduling [16] has been chosen as the scheduling mechanism because it is efficient and at the same time, has a low time overhead. Instructions are scheduled cycle by cycle, based on dependences, till all the instructions are scheduled.

Certain features of our implementation of the superblock scheduler in LaTTe are:

1. It is performed in the OPT phase of the translation (refer to Section 2) after register allocation.
2. It schedules instructions for a superscalar processor.
3. In the current implementation, instructions are not moved above a branch, since this requires live register analysis, which is expensive in terms of run-time overhead.

6 Experimental Results

This section describes our experimental setup and reports the performance of our profile-aided recompilation method. We consider the performance of our technique both with and without superblock scheduling. We compare these with the performance of the original LaTTe JVM (without our profile aided retranslation mechanism.)

6.1 Experimental Setup

We have used the open source LaTTe JIT compiler version 0.9.0 for our implementation [19]. Our test machine is a Sun Ultra 1 workstation having a UltraSPARC I 167 MHz processor with 32 KB L1-cache (16 KB I-cache and 16 KB D-cache), 512 KB L2-cache and 64 MB RAM and running SunOS 5.7. We used 5 benchmarks from SPEC JVM98.

We set the threshold values in our implementation as follows. The hotness threshold, used in the first and second retranslations, of a method is inversely proportional to the number of branches (bytecode branches) in the method. Setting the hotness threshold this way facilitates early retranslation for methods that have larger number of branches or paths. This threshold value is kept the same for our profile-aided retranslation (for both first and second retranslations) and for the original adaptive LaTTe. The first and second retranslation factors (discussed in Section 4) range from 0.1 to 1.0. By tuning these factors, we can control the thresholds for methods to become hot and for paths within the methods to become hot enough for prediction.

6.2 Experimental Results

First we present the reduction in the number of regions along the hot path when profile-aided retranslation and region expansion were performed. Table 1 reports the number of regions across which the basic blocks in the hot paths were spread over before and after region expansion. In this experiment, we have chosen the multiplication factors for method run count threshold in the first and second retranslation to be 0.1 and 0.5 respectively. We used a hotness threshold of 20 and cloning threshold of 15. After region expansion, a hot path lies entirely within a single region. The table indicates that our region expansion technique significantly decreases the number of regions in which the hot paths lie by a factor four or five times. Thus region expansion exposes more scope for aggressive optimizations on hot regions.

Next we present the execution times of SPEC JVM programs in LaTTe with and without profile aided recompilation. We refer to LaTTe with profile aided recompilation as P-LaTTe. In this experiment, we have chosen 0.2 as the multiplication factor for the first retranslation and 1.0 as the second retranslation factor in P-LaTTe, as against adaptive LaTTe which uses the multiplication factor 0.2 in its first retranslation. Further, in P-LaTTe, the hotness threshold for the store nodes is chosen to be 20 and the cloning threshold is set to be 15. With these parameters, more time is given for profiling, and the second retranslation is triggered only after "enough" profile is gathered.

The results of this experiment are mixed. We find that the execution times under LaTTe and P-LaTTe (with and without instruction scheduling) are comparable (refer to Table 2), in spite of the retranslation, profiling, and cloning overheads. First we note that the comparison between original LaTTe with and without adaptive retranslation also gives mixed results. More specifically, in 2 out of the 5 benchmarks, original LaTTe with adaptive retranslation performs poorly compared to original LaTTe without adaptive retranslation. Our P-LaTTe with instruction scheduling performs better only in one benchmark, namely `_222_mpegaudio`, and relatively poorly in the other four. Comparing the execution times of the programs under P-LaTTe with and without instruction scheduling, we observe that the benefits due to instruction scheduling compensates fairly well, though they are not significantly high to result in better performance compared to the original LaTTe. One reason for the improvement in performance due to instruction scheduling being small is the fact that our instruction scheduler is a post-pass scheduler, i.e., it performs scheduling after

Table 1. Number of regions containing the hot paths before and after region expansion

Benchmark	Before region expansion	After Region expansion
_202_jess	466	104
_201_compress	165	44
_209_db	181	44
_222_mpegaudio	192	31
_228_jack	506	105

Table 2. Performance comparison with execution times: Original LaTTe, Original Adaptive LaTTe, P-LaTTe, and P-LaTTe with Instruction Scheduling

Benchmark	Original LaTTe	Original Adaptive LaTTe	P-LaTTe	P-LaTTe with instruction scheduling
_201_compress	88.07 s	85.94 s	90.30 s	90.01s
_202_jess	53.71 s	52.22 s	54.97 s	55.48s
_209_db	85.56 s	91.59 s	92.75 s	91.99s
_222_mpegaudio	76.60 s	75.50 s	77.06 s	75.04s
_228_jack	58.82 s	63.11 s	66.05 s	66.40s

register allocation. As in any postpass scheduling, register allocation introduces anti- and output-dependences which somewhat restricts the parallelism exposed by the instruction scheduler. Thus, we conclude that our results are encouraging in the sense that the performed optimization results in execution times which are comparable to that of the original LaTTe, despite the additional profiling and retranslation overhead. To obtain greater benefits, additional profile-based optimizations could be tried.

In order to estimate the overheads involved in our method, we measured the time taken for the second retranslation. Table 3 presents the total retranslation overheads with and without instruction scheduling. The (second) retranslation overhead without instruction scheduling is significantly higher in two of the benchmarks, viz., _202_jess and _228_jack. Note that, in these benchmarks, P-LaTTe performed relatively worse than adaptive LaTTe. The considerable retranslation overhead comes from the fact that both first and second retranslations still translate code from bytecode to native code, rather than from native code to native code. Due to this, the performance gain achieved from our optimizations is lost. We also observe that the additional overhead for performing superblock scheduling is very low.

Table 3. Retranslation time for profiling only, profiling with instruction scheduling

Benchmark	Profiling only	Profiling with instruction scheduling
_201_compress	0.30 s	0.37 s
_202_jess	2.21 s	2.59 s
_209_db	0.47 s	0.58 s
_222_mpegaudio	0.54 s	0.78 s
_228_jack	3.02 s	3.80 s

7 Related Work

Several recent research projects have focused on aspects of dynamic optimization [4, 8, 9, 12]. Dynamo [4], a dynamic optimization software, attempts to run a statically optimized binary faster. This system uses the Most Recently Executed Tail (MRET) strategy [11], in which the path that is executed after a start of trace node has become hot is predicted as the hot path. Our n-MRET strategy reduces the probability of noise paths (infrequently executed paths) being selected as the hot path which is more probable in MRET. Similar work on profile-driven dynamic recompilation for Scheme makes use of Ball and Larus edge profiling algorithm [8]. We have chosen path profiling over edge profiling because of more accurate information available in the former. The theoretic and algorithmic results, which may be used to determine when an edge profile is a good predictor for hot paths and when it is a poor predictor, are discussed in [7]. The efficient path profiling algorithm by Ball and Larus has a high instrumentation overhead though it has the least profiling overhead of all path profiling algorithms.

For choosing hot methods, LaTTe relies on method run counts. Alternatively, time based and sample based profiling can also be used [3]. The IBM Jalapeno optimizing compiler [9] for Java uses sample based profiling for determining hot methods for retranslations. In this case, the program counter (PC) is sampled at regular intervals and this gives a more accurate measure than mere run counts, since long-running methods with less number of invocations are also accounted for in this strategy.

8 Conclusions and Future Work

In this work we have implemented a dynamic path profiling strategy for aiding recompilation in the LaTTe JIT compilation framework. Our profile aided recompilation mechanism and region expansion reduces the number of regions significantly, by a factor of 4 to 5, which increases the scope for performing aggressive compiler optimization on hot methods and hot paths. With profile assisted compilation and instruction scheduling, our implementation performs comparable to the original adaptive LaTTe in spite of the retranslation and profiling overheads. Future work could concentrate on avoiding a complete retranslation by retaining certain structures from the previous translation of the method. Another future work could be on developing heuristics that vary the cloning threshold for different paths, instead of a fixed (static) threshold value. Last, using the profile information to perform other optimizations along the hot paths could be explored.

References

[1] A.-R. Adl-Tabatabai, M. Cierniak, G-Y. Lueh, V. M. Parikh, and J. M. Stichnoth. Fast effective code generation in a Just-In-Time Java compiler. In *Proc. of the ACM SIGPLAN '98 Conf. on Programming Language Design and Implementation*, June 1998. 496

[2] M. Arnold, S. J. Fink, D. Grove, M. Hind, P. F. Sweeney. Adaptive optimization in the Jalapeno JVM. In *Proc. of the ACM SIGPLAN Conf. on Object Oriented Programming, Systems, Languages and Applications (OOPSLA'00)*, Oct. 2000. 496

[3] M. Arnold, M. Hind and B. G. Ryder. An empirical study of selective optimization. In *Proc. of the 13th Intl. Workshop on Languages and Compilers for Parallel Computing*, 2000. 504

[4] V. Bala, E. Duesterwald, S. Banerjia. Dynamo: A transparent dynamic optimization system. In *Proc. of the SIGPLAN '98 Conf. on Programming Language Design and Implementation*, 2000. 503

[5] T. Ball and J. R.Larus. Optimally profiling and tracing programs. In *Proc. of the 19th Symp. on Principles of Programming Languages*, Jan. 1992. 496, 498

[6] T. Ball and J. R.Larus. Efficient path profiling. In *Proc. of 29th Symp. on Microarchitecture*, Dec. 1996. 496

[7] T. Ball, P. Mataga and M. Sagiv. Edge profiling versus path profiling: The showdown. In *Proc. of the 25th ACM SIGPLAN-SIGACT Symp. on Principles of Programming Languages*, Jan. 1998. 496, 504

[8] R. G. Burger and R. K. Dybvig. An infrastructure for profile-driven dynamic recompilation. In *Proc. of Intl. Conf. on Computer Languages (ICCL'98)*, May 1998. 503, 504

[9] M. G.Burke, D. Choi, S. Fink, D. Grove, M. Hind, V. Sarkar, M. J. Serrano, V. C. Sreedhar, H. Srinivasan, and J. Whaley. The Jalapeno dynamic optimizing compiler for Java. In *Proc. of 1999 ACM Java Grande Conference*, June 1999. 503, 504

[10] W. Y. Chen, S. A. Mahlke, N. J. Warter, S. Anik, W. W. Hwu. Profile-assisted instruction scheduling. *International Journal of Parallel Programming*, 1994. 496, 500

[11] E. Duesterwald and V. Bala. Software profiling for hot path prediction: Less is more. In *Proc. of the 9th Intl. Conf. on Architecture Support for Programming Languages and Operating Systems*, 2000. 496, 503

[12] B. Grant, M. Mock, M. Philipose, C. Chambers and S. J. Eggers. DyC: An expressive annotation-directed dynamic compiler for C. Technical Report UW-CSE-97-03-03, Dept. of Computer Science and Engineering, University of Washington, Seattle, WA, 1997. 503

[13] W. W. Hwu, S. A. Mahlke, W. Y. Chen, P. P. Chang, N. J. Warter, R. A. Bringmann, R. G. Ouellette, R. E. Hank, T. Kiyohara, G. E. Haab, J. G. Holm, and D. M. Lavery. The Superblock: An effective structure for VLIW and superscalar compilation. *Journal of Supercomputing*, Feb. 1993.

[14] *Java HotSpot Performance Engine*, http://java.sun.com/products/hotspot/, 1999. 496

[15] J. Lee, B-S. Yang, S. Kim, S. Lee, Y. C. Chung, H. Lee, J. H. Lee, S-M. Moon, K. Ebicioglu, E. Altman. Reducing virtual call overheads in a Java VM Just-in-Time compiler. In *Proc. of 1999 Workshop on Interaction between Compilers and Computer Architectures*, Jan. 2000. 497

[16] S. S. Muchnick. *Advanced Compiler Design and Implementation*. Morgan Kaufmann Publishers Inc., 1997. 501

[17] *The SPARC Architecture Manual Version 8*. 497

[18] R. Vinodh Kumar. Dynamic path profile aided recompilation in a Java Just-In-Time compiler M. E. Dissertation Project, Indian Institute of Science, Dept. of Computer Science & Automation, Bangalore, 560 012, India, Jan. 2000. 499

[19] B-S. Yang, S-M. Moon, S. Park, J. Lee, S. Lee, J. Park, Y. C. Chung,S. Kim, K. Ebcioglu, and E. Altman. LaTTe: A Java VM Just-In-Time compiler with fast and efficient register allocation. In *Proceedings of the 1999 Intl. Conf. on Parallel Architectures and Compilation Techniques, Oct. 1999.* 496, 497, 501

[20] F.Yellin and T.Lindholm, *The Java Virtual Machine Specification*, Addison-Wesley, 1996. Hall, 1999. 495

Exploiting Data Value Prediction in Compiler Based Thread Formation

Anasua Bhowmik[1] and Manoj Franklin[2]

[1] Computer Science Department
[2] ECE Department and UMIACS
University of Maryland, College Park, MD 20742

Abstract. Speculative multithreading (SpMT) is an effective execution model for parallelizing non-numeric programs, which tend to use irregular and pointer-intensive data structures, and have complex flows of control. An SpMT compiler performs program partitioning by carefully considering the data dependencies present in the program. However, at run-time, the data dependency picture changes dramatically if the SpMT hardware performs data value prediction. Many of the data dependencies, which guided the compiler's partitioning algorithm in taking decisions, may lose their relevance due to successful data value prediction. This paper presents a compiler framework that uses profile-based value predictability information when making program partitioning decisions. We have developed a Value Predictability Profiler (VPP) that generates the value prediction statistics for the source variables in a program. Our SpMT compiler utilizes this information by ignoring the data dependencies due to variables with high prediction accuracies. The compiler can thus perform more efficient thread formation. This SpMT compiler framework is implemented on the SUIF-MachSUIF platform. A simulation-based evaluation of SPEC programs shows that the speedup with 6 processing elements increases up to 21% when utilizing value predictability information during program partitioning.

Keywords: data dependency, data value prediction, parallelization, profiling, speculative multithreading (SpMT), thread-level parallelism (TLP)

1 Introduction

Reducing the completion time of a single computation task has been an important challenge for the last several decades. A prominent technique used to meet this challenge—besides clock speed increase and memory latency reduction—is *parallel processing*. Parallelization started being successful for non-numeric applications with the advent of the *speculative multithreading* (SpMT) model in the last decade. Hardware support for speculative thread execution makes it possible for an SpMT compiler to parallelize sequential applications without worrying about data dependencies and control dependencies. The most important task in SpMT compilation is partitioning a program into separate threads

S. Sahni et al. (Eds.) HiPC 2002, LNCS 2552, pp. 506–516, 2002.

of execution. Several researchers have developed compilers for different SpMT execution models; some of the notable ones are [1, 4, 7, 8].

Program partitioning needs to consider several factors such as data dependencies, control dependencies, thread granularity, and load balancing [1, 8]. Among these, data dependencies are perhaps the most important factor; if there are many inter-thread data dependencies, very little speedup will result, unless the SpMT hardware uses data value prediction. Past research has shown that data value prediction can provide good speedup in SpMT processors, by temporarily ignoring data dependencies and executing instructions based upon the predicted values [5]. However, if data value prediction is employed, then the data dependency modeling done by the compiler becomes inaccurate. If the compiler ignores data dependency information altogether, to model the effect of data value prediction, then also the partitioning is likely to be poor, because some instructions are not very predictable. The compiler must make an educated guess to decide which dependencies should be honored and which ones can be ignored.

In this paper we study the potential of exploiting value predictability information at compile time to make program partitioning decisions. To do this, we map the predictable instructions to the actual source variables they operate on. Some work [2] had been done earlier to use value predictability information to do better compile-time scheduling for ILP processors. However, to the best of our knowledge, no SpMT compilers have so far used value predictability information to guide thread formation. Using our compiler framework, we have compiled large non-numeric applications from the SPEC benchmark suites. Our studies with two different data dependency models have led to the following conclusions:

- The use of value predictability information in thread formation results in better threads, and gives better speedups in SpMT processors.
- When using value predictability information for thread formation, the two data dependency models tend to have similar performance; this permits the simpler model to be used.

Section 2 provides background information on SpMT, and motivates the use of value predictability information during thread formation. Section 3 details our SpMT compiler framework and program partitioning algorithm. Section 4 presents the simulation results. Section 5 presents the conclusions.

2 Speculative Multithreading (SpMT)

The SpMT execution model executes in parallel speculative threads belonging to the same program. It envisions a strict sequential ordering among the threads. In case of mis-speculation, the results of the mis-speculated thread and subsequent threads are discarded. The control flow of the sequential code imposes an order on the threads and we can use the terms *predecessor* and *successor* to qualify the relation between any given pair of threads. This means that inter-thread communication between any two threads (if any) is strictly in one direction, as dictated by the sequential thread ordering. No explicit synchronization operations are necessary, as the sequential semantics of the threads guarantee proper synchronization. This relaxation allows us to "parallelize" non-numeric applications without using explicit synchronization, even if there is a potential inter-thread data dependency.

2.1 Data Dependency Modeling

The exact places in a program where threads begin and end impact SpMT performance in a major way. Inter-thread data dependencies are a major factor in this decision, as they affect inter-thread data communication and the amount of thread level parallelism exploited. It is not possible to detect all data dependencies at compile time because of ambiguous memory dependencies. It is also not possible to accurately determine the *impact* of data dependencies, because of uncertainties like branch outcomes and cache misses. The compiler can use profile information and heuristics to estimate the relative distance between dependent instructions. It can also perform intra-thread scheduling to reduce the wait time due to a data dependency. Two different metrics can be used to quantify the data dependencies between adjacent threads: *data dependency count* and *data dependency distance* [1].

Data Dependency Count (DDC): The *data dependency count* is a weighted count of the number of data dependency arcs coming into a thread from the previous threads. This models the extent of data dependency this thread has on other threads. If the dependency count is small then this thread is more or less data independent from other threads. While counting the data dependency arcs, the compiler can give less weights to the arcs coming from distant threads and those coming from the less likely paths.

Data Dependency Distance (DDD): The *data dependency distance* between two threads models the maximum time that the instructions in the successor thread will stall for instructions in the predecessor thread to complete, if the two are executed in parallel. It is not beneficial to parallely execute threads having large data dependency distances. In order to decide whether to start a new thread at a particular point in the program, the compiler calculates the data dependency distance that will result if a new thread is started at that point. If the distance is small, then a new thread can be started at that point.

2.2 Importance of Exploiting Value Predictability Information

Data value prediction has been found to speed up the execution of SpMT programs [1, 5]. We shall use an example to illustrate the importance of exploiting value predictability information at compile time. The code fragment and CFG shown in Figure 1 is taken from a frequently executed function in `mcf`, a SPEC2000 integer benchmark program. The CFG has 3 basic blocks, B1, B2, and B3. The straight arrows show the flows of control through these blocks, and

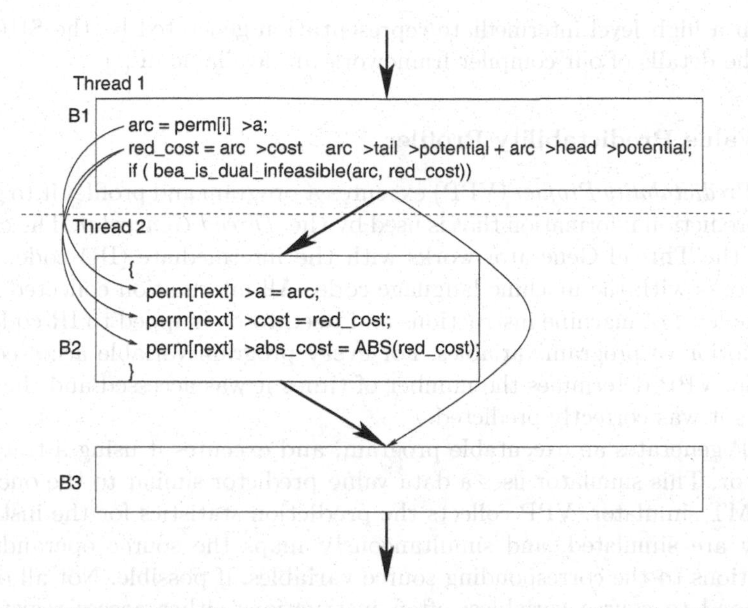

Fig. 1. An Example Code Fragment to Show the Importance of Data Value Profiling

the arcs show the data dependencies between the statements. Most of the time, control goes to B2 after executing B1, as indicated by the thick arrows. The compiler gets this information from the profiler. Normally, the compiler groups B1 and B2 in a single thread, because there are many dependencies between B1 and B2 through variables *arc* and *red_cost*.

However, data value prediction statistics show that the variables *arc* are *red_cost* are correctly predicted more than 80% and 50% of the time, respectively. Therefore, more than 50% of the time, it is actually beneficial to execute B1 and B2 in parallel. Thus, by using the prediction statistics of variables, the compiler can ignore the dependencies due to these variables and assign these basic blocks into two separate threads to be executed in parallel.

3 SpMT Compiler Framework

The compiler framework is implemented on the SUIF-MACHSUIF platform. [3] and has three major components: *Path Profiler*, *Value Predictability Profiler (VPP)*, and *Thread Generator*. The Thread Generator is the core component in the framework; it analyzes the program and partitions it into threads. The Path Profiler generates the information on the most likely path through the program and the Value Predictability Profiler generates the data value prediction information. Information generated by these profilers is used by the Thread Generator to do the thread formation. Program analysis and partitioning are done on a high level intermediate representation genereted by the SUIF front-end. The details of our compiler framework are available in [1].

3.1 Value Predictability Profiler

Value Predictability Profiler (VPP) executes a program and profiles it to generate value prediction information that is used by the *Thread Generator*. The challenge is that the Thread Generator works with the intermediate (IR) code, whereas VPP works with the machine language code. All information collected by VPP is at the level of machine instructions and need to be mapped to IR code, which has a notion of program variables. For every program variable accessed during profiling, VPP determines the number of times it was accessed and the number of times it was correctly predicted.

VPP generates an executable program, and executes it using a trace-driven simulator. This simulator uses a data value predictor similar to the one used in the SpMT simulator. VPP collects the prediction statistics for the instructions as they are simulated, and simultaneously maps the source operands of the instructions to the corresponding source variables, if possible. Not all operands correspond to source variables; often instructions either access registers that hold intermediate results or load from memory locations, values that spilled from registers. The symbol table information present in the executable (when compiled with the `-g` option) aids this reverse mapping. VPP dynamically maps the load instructions to the source variable with the help of the symbol table information and the operand values. For example, when loading a global variable, the variable can be determined by matching the memory address with the elements of the symbol table. For local variables, the variable is identified by determining its position in the current stack frame. For structure variables, VPP maps the source operand to the particular field of the structure it corresponds. For array variables, VPP maps the source operand to the array variable and does not narrow it down further to the particular index. However, the SpMT compiler later uses the prediction statistics of the array variable along with array index analysis to determine whether the dependency due to the array variable can be ignored. Similarly, for accesses to an array of structures, VPP does not map the source operand to the specific field within the structure.

3.2 Using Value Prediction Information in Thread Formation

In this subsection we describe how the value prediction information generated by VPP is used by the Thread Generator while doing the program partitioning. The Thread Generator works in multiple passes. In the first pass it builds the CFG and determines the data dependence information by analyzing the program. The outline of the program partitioning algorithm is given in Figure 2. The program partitioning is done by traversing the CFG for every subroutine of the program, and calling the procedure *partition_thread* (of the partitioning algorithm) with appropriate basic blocks.

```
partition_thread ( start_block, postdom_block, curr_thread)
 {
   path = find_most_likely_path(start_block, postdom_block);
   new_curr_thread = add_path(curr_thread, path);
   probable_future_thread = build_next_likely_thread(postdom_block);
   max_delay = compute_delay(new_curr_thread, probable_future_thread);
   make_partition(start_block, postdom_block, curr_thread, max_delay);
 }
compute_delay(new_curr_thread, probable_future_thread)
 {
   delay = maxdelay = 0;
   for every instruction  instr  in probable_future_thread
     for every variable  var  read by  instr
        if ( prediction accuracy of  var < PREDICTION_THRESHOLD )
           delay = compute_instruction_delay ( ins, var,new_curr_thread, probable_future_thread );
           if ( delay > max_delay ) max_delay = delay;
        }
   return max_delay;
 }
```

Fig. 2. Outline of the Program Partitioning Algorithm that uses Value Predictability

The procedure *partition_thread* partitions the program segment between the basic blocks *start_block* and *postdom_block*, where the latter is the first basic block that is control independent of the *start_block*. The procedure first finds the most likely path from *start_block* to *postdom_block*, and then augments the current thread with this path to build a possible new thread. Also, a probable future thread is built starting from *postdom_block*. Now, the maximum delay due to data dependency is computed by calling the procedure *compute_delay*. Based on the data dependency, control dependency, current thread size (i.e., size of *curr_thread*) and the size of the program segment under consideration, *partition_thread* can do one of the following: (i) start a new thread from *postdom_block*, (ii) include the whole program segment under consideration as a part of *curr_thread*, (iii) partition the program segment into further threads. A detailed description of the function *partition_thread* can be found in [1].

The procedure *compute_delay* computes the maximum delay between two threads if they are executed in parallel. For a variable read by an instruction in the future thread, its prediction accuracy, as generated by the VPP, is consulted. If it is lower than a threshold value, the probable delay suffered by this instruction due to this variable is computed using one of the data dependency models described in section 2.1.

For our experimental evaluation, we have used the prediction accuracy threshold of 50%. VPP generates the predictability statistics on a per location basis, i.e., if the same variable is accessed by different static instructions (different program counter values), VPP generates separate statistics for each of these different locations. However, our experiments have shown that generally the prediction accuracies of a particular variable tend to be similar even when accessed from different locations. This is because the prediction accuracy of a variable is not only dependent on the dynamic predictor used, but also is an inherent property of the program. So, instead of considering the prediction accuracy of the variable at a particular location, the compiler uses the overall prediction accuracy of the variable, in order to decide the partitioning.

4 Experimental Evaluation

To study the effectiveness of utilizing value predictability information for program partitioning, we conducted a simulation-based evaluation. This section presents the simulation framework and the simulation results obtained.

4.1 Experimental Setup

To search through a large space of thread formation schemes effectively, we built an SpMT simulator on top of a trace-driven simulator. The SpMT simulator models multiple processing elements (PEs), each of which can sequence through a thread. The number of PEs, issue width per PE, etc., are parameterized. The simulator uses the Alpha ISA. For simplicity, all functional units (except the memory unit) have a single-cycle latency. When encountering a conditional branch instruction in a thread, the PE consults a branch predictor for making a prediction. The simulator models a hybrid data value predictor [9] for predicting the results of instructions whose operands are unavailable at fetch time. It also models a shared L1 d-cache.

For our simulations we use a PE issue width of 4 instructions per cycle, and the PEs use out-of-order issue. Each PE has an instruction window of 128 instructions. The L1 cache size is 256 Kbytes. There is a 2-cycle overhead in assigning a thread to a PE. Thread pre-emption incurs a 2-cycle penalty. Furthermore, there is a 2-cycle latency for forwarding register values across multiple PEs. The L1 d-cache has 1 cycle access latency, and a miss latency of 10 cycles.

For benchmarks, we use a SPECINT95 program (`ijpeg`) and 5 programs from the SPEC2000 suite, all written in C. We have used a "fast forward" mode to skip the initialization phases of the programs, before beginning the actual simulation. Each benchmark is simulated for 300 million instructions after fast forwarding, as we found the performance numbers to be reaching steady state by about 200 million instructions. The code executed in the supervisor mode is unavailable to the simulator, and is therefore not taken into account in the parallelism studies. The library code is not parallelized, as we use the standard libraries in our experiments. The library code therefore executes in serial mode, providing a conservative treatment to our parallelism values. Table 1 shows the number of source variables, the overall prediction accuracies of the variables, and the number of variables with prediction accuracies higher than 50%.

Table 1. Prediction Statistics of the Source Variables

Program Name	No. of Source Variables	Overall Pred. Accuracies of the Source Variables	No. of Variables with Pred Accuracies greater than 50%
ijpeg	1992	42.45%	124
crafty	1077	34.19%	187
equake	168	75.49%	51
mcf	155	51.62%	55
twolf	1727	41.34%	162
vpr	502	48.00%	502

Table 2. Thread Statistics for Different Partitioning Schemes

Program Name	Model Type	With Using Value Predictability			Without Using Value Predictability		
		No. of Threads		Avg. Dyn. Thread Size	No. of Threads		Avg. Dyn. Thread Size
		Static	Dynamic		Static	Dynamic	
ijpeg	DDD	1167	4933169	60.8	882	2664627	112.6
	DDC	1211	6038371	49.7	941	5628660	53.3
crafty	DDD	1150	4873757	61.6	1061	3678551	81.6
	DDC	1237	5539624	54.2	1208	5196837	57.8
equake	DDD	230	10798639	27.8	230	10717271	28.0
	DDC	229	10836695	27.7	219	10455248	28.7
mcf	DDD	99	8251294	29.7	83	7277269	33.2
	DDC	105	8129892	30.1	87	5999878	40.3
twolf	DDD	1495	9256135	32.4	1256	8966135	33.5
	DDC	1574	9353547	32.1	1403	9287968	32.3
vpr	DDD	804	3889689	77.1	732	3581348	83.8
	DDC	806	4075973	73.6	763	3528688	85.0

4.2 Experimental Results and Analysis

Figure 3 presents the result of 4 different schemes. Bars marked *a* and *b* present the speedups of the programs compiled without and with value predictability information for *Data Dependency Distance* based modeling, respectively. Similarly, bars *c* and *d* present the data for *Data Dependency Count* based modeling when compiled without and with value predictability information, respectively. We also tested the scalability of the partitioning by varying the number of PEs from 2 to 4 to 6.

The results shown in Figure 3 can be analyzed with the help of Tables 1 and 2. Table 2 shows the number of static threads and dynamic threads, and the average size of the dynamic threads. Comparing bar *a* with *b*, and *c* with *d* in Figure 3, we find that `mcf` shows an 18.7% increase in speedup for *Data Dependency Distance* (DDD) based modeling and 14.4% increase for *Data Dependency Count* (DDC) based modeling, when using 6 PEs. The reason behind this large improvement was already explained in section 2.2. Moreover, in Table 1, we see that more than

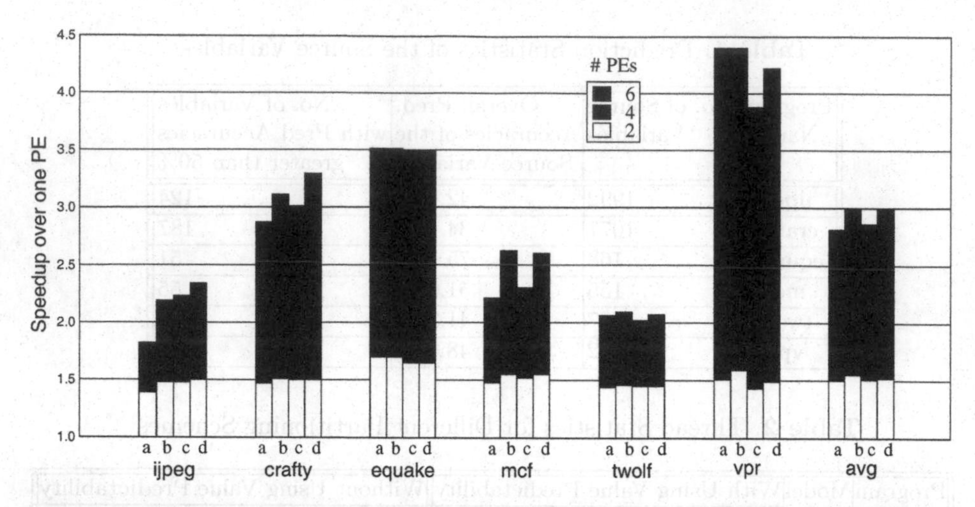

Fig. 3. Speedups Obtained with Threads Formed by Different Partitioning Schemes. **a:** Data Dependency Distance modeling without using value predictability; **b:** Data Dependency Distance modeling with using value predictability; **c:** Data Dependency Count modeling without using value predictability; **d:** Data Dependency Count modeling with using value predictability

one third of the variables in `mcf` have prediction accuracies greater than 50% and the overall prediction accuracies for the variables are also above 50%. This means that the compiler is able to disregard many of the data dependencies and perform aggressive partitioning. This is also evident from the increased static and dynamic thread count presented in Table 2.

For `ijpeg`, there is an improvement of 20.4% in speedup with 6 PEs when DDD modeling is used, but for DDC modeling the speedup is only 4.7%. The large improvement for `ijpeg` with DDD modeling is because the compiler did a conservative job with this modeling in the first place. This is apparent from the low speedup found in bar *a*. Although a very small fraction of the variables in `ijpeg` have prediction accuracies over 50%, the compiler has been able to ignore them effectively and achieve a much better partitioning. It seems these variables played a critical role in the decision making under DDD modeling. Although `ijpeg` shows better speedup with DDC modeling even without using value predictability information, the difference in speedups between the two models is reduced by a significant amount when using predictability information. The speedup characteristics for `vpr` are similar to those of `ijpeg`.

`crafty` shows reasonably good improvements in speedup with both modelings, although its overall prediction accuracy is not high. However, `crafty` shows significant increase in the number of static and dynamic thread counts when compiled using value predictability information for both models. This justifies the significant improvements in speedup.

twolf shows little improvement in speedups when using value predictability information. From Table 2, we see that in twolf, the static thread count has increased significantly for both models when using value predictability information, but the dynamic thread counts have not increased proportionally. This suggests that the program spends less time on those parts that got affected by the new partitioning scheme.

Finally, we see that, although equake has a high overall value prediction accuracy, it does not show any change in speedups under any scheme. Also, we can see from Table 2 that there are very few changes in the static and dynamic thread counts for equake. This indicates that the variables that have high prediction accuracies in equake do not play a critical role in thread formation. In other words, the dependencies caused by these variables do not create any obstacles in creating the threads under the original schemes. For example, if a variable is written and then read inside the same basic block, then this data dependency is not considered for deciding the partitioning. Similarly, the dependency coming from a distant thread also does not have a major influence on the partitioning.

In Figure 3, we also see that for all benchmarks, the partitioning shows good scalability for both models when value predictability information is used. Hence we can expect that with more PEs, we can achieve even better improvements by using the predictability information. Also, in general, the programs that originally had lower speedups, such as ijpeg, crafty, and mcf, show more improvement than the ones which have comparatively higher speedups, such as vpr and equake. Furthermore, the difference in speedups for the two dependency models becomes less significant when utilizing the value predictability information. This can also be seen from the fact that the average speedup achieved by the two models are the same when using the predictability information.

5 Conclusions

Proper thread formation by the compiler is crucial to obtaining good speedup in a speculative multithreading (SpMT) processor. Among various program characteristics, inter-thread data dependencies play a major role in SpMT performance, and in compile-time thread formation. The data dependency picture changes dramatically at run-time, however, if the SpMT hardware performs data value prediction. Many of the data dependencies, which guided the compiler's partitioning algorithm in taking decisions in a particular manner, may lose their relevance when the hardware makes successful data value predictions. This paper presented a compiler framework that uses profile-based value predictability information when making thread formation decisions. It uses a Value Predictability Profiler (VPP) that generates the prediction statistics for the source variables in a program. Our SpMT compiler uses that information to form better threads This SpMT compiler framework is implemented on the SUIF-MachSUIF platform. A simulation-based evaluation of the generated threads using value predictability information shows that an improvement of up to 21% in the speedup can be obtained with 6 processing elements compared to a thread formation scheme that does not use value predictability information. Moreover, the speedups of the programs become less sensitive to the data dependency modeling.

Acknowledgements

This work was supported by the U.S. National Science Foundation (NSF) through a CAREER grant (MIP 9702569) and a regular grant (CCR 0073582).

References

[1] A. Bhowmik and M. Franklin, "A General Compiler Framework for Speculative Multithreading," *Proc. 14th ACM Symp. on Parallel Algorithms and Architectures (SPAA 2002)*. 507, 508, 510, 511

[2] C-Y. Fu, M. D. Jennings, S. Larin, and T. M. Conte, "Value Speculation Scheduling for High Performance Processor," *Proc. ASPLOS-VIII*, 1998. 507

[3] M. W. Hall, et al, "Maximizing Multiprocessor Performance with the SUIF Compiler," *IEEE Computer*, December 1996. 509

[4] O. C. Maquelin, H. H. J. Hum, and G. R. Gao, "Costs and Benefits of Multithreading with Off-the-Shelf RISC Processor," *Proc. 1st Int'l EURO-PAR Conf.*, 1995. 507

[5] P. Marcuello, J. Tubella and A. Gonzalez, "Value Prediction for Speculative Multithreaded Architectures," *Proc. 32nd Int'l Symp. on Microarchitecture*, 1998. 507, 508

[6] K. Olukotun, et al, "A Chip-Multiprocessor Architecture with Speculative Multithreading," *IEEE Transactions on Computers*, September 1999.

[7] J-Y. Tsai and P-C. Yew, "The Superthreaded Architecture: Thread Pipelining with Run-Time Data Dendence Checking and Control Speculation," *Proc. Int'l Conf. on Parallel Architectures and Compilation Techniques (PACT)*, 1996. 507

[8] T. N. Vijaykumar and G. S. Sohi, "Task Selection for a Multiscalar Processor," *Proc. 31st Int'l Symp. on Microarchitecture (MICRO-31)*, 1998. 507

[9] K. Wang and M. Franklin, "Highly Accurate Data Value Prediction using Hybrid Predictors," *Proc. 30th Int'l Symp. on Microarchitecture (MICRO-30)*, pp. 281-290, 1997. 512

High Performance Computing of Fluid-Structure Interactions in Hydrodynamics Applications Using Unstructured Meshes with More than One Billion Elements

S. Aliabadi[1], A. Johnson[2], J. Abedi[1], and B. Zellars[1]

[1]Department of Engineering, Clark Atlanta University
223 James P. Brawley Dr., Atlanta, GA 30314, USA
aliabadi@cau.edu
[2]Network Computing Services, Inc., Army HPC Research Center
1200 Washington Ave. S., Minneapolis, MN 55415, USA
ajohn@networkcs.com

Abstract. A parallel finite element fluid-structure interaction solver is developed for numerical simulation of water waves interacting with floating objects. In our approach, the governing equations are the Navier-Stokes equations written for two incompressible fluids. An interface function with two distinct values serves as a marker identifying the location of the interface. The numerical method is based on writing stabilized finite element formulations in an arbitrary Lagrangian-Eulerian frame. This allows us to handle the motion of the floating objects by moving the computational nodes. In the mesh-moving schemes, we assume that the computational domain is made of elastic materials. The linear elasticity equations are solved to obtain the displacements. In order to update the position of the floating object, the nonlinear rigid body dynamics equations are coupled with the governing equations of fluids and are solved simultaneously. The mooring forces are modeled using nonlinear cables and linear spring models.

1 Introduction

In the past, there have been many scientific works devoted to the development of numerical algorithms capable of simulating free-surface flow applications. Unfortunately, due to hardware and software limitations, most of the algorithms could solve simple 2D problems, with many simplifications and assumptions. Recently, there has been significant progress in 3D simulations of free-surface flows [1-2], especially the in marine industry [1-4]. Our free-surface flow simulator is a powerful and fully realistic tool, which takes into account the wind, waves, water depth, banks, and even moving objects [2].

S. Sahni et al. (Eds.) HiPC 2002, LNCS 2552, pp. 519-533, 2002.

In this article, we describe a new finite element technique for simulation of free-surface flow problems interacting with moored floating objects. In this approach, we keep track of the free-surface using the IS-GMC (Interface-Sharpening/Global Mass Conservation) [5] interface-capturing methods [1-2,5], while we absorb the motion of the physical boundaries resulting from translation and rotation of floating objects using mesh-moving methods [6-8].

2 Governing Equations and Finite Element Formulations

We consider the governing equations for two interacting fluids in the spatial domain and its boundary . Here we assume that the spatial domain and its boundary are both functions of time, *t*. The two fluids are incompressible (e.g. air-water) and are separated with an interface. Along the interface, the traction force is continuous (surface tension is negligible). The governing equations of two fluids are the Navier-Stokes equations written in the Arbitrary Lagrangian-Eulerian (ALE) frame [9-10]. These equations are:

$$\left[\left.\frac{u}{t}\right| \;\; (u \;\; u_{mesh}) \;\; u \;\; g\right] \;\; 0, \tag{1}$$

$$u \;\; 0, \tag{2}$$

where

$$pI \;\; 2 \;\; (u), \quad \frac{1}{2}(\;\; u \quad u^T). \tag{3}$$

Here u, u_{mesh}, p, , **g**, and are the fluid velocity, mesh velocity, pressure, density, gravitational force, and dynamic viscosity, respectively. The strain tensor is denoted by and **I** represents the identity tensor. Equations (1-2) are completed by an appropriate set of boundary and initial conditions. The stabilized finite element formulations [1,5-6] for equations (1-2) can be written as:

$$\begin{aligned}
&\int \mathbf{w}^h \;\; [\frac{u^h}{t} \;\; (u^h \;\; u^h_{mesh}) \;\; u^h \;\; g]d \qquad \int \;\; (\mathbf{w}^h): \;\; (p^h, \mathbf{u}^h)d \\
&\int q^h_p \;\; \mathbf{u}^h d \qquad \sum_{e\;1}^{ne} \int_e \frac{m}{}[\;\; (u^h \;\; u^h_{mesh}) \;\; \mathbf{w}^h \qquad (q^h_p, \mathbf{w}^h)] \\
&\left[[\frac{u^h}{t} \;\; (u^h \;\; u^h_{mesh}) \;\; u^h \;\; g] \qquad (p^h, \mathbf{u}^h)\right]d \\
&\sum_{e\;1}^{ne} \int_{e} {}_{c} \;\; \mathbf{w}^h \qquad \mathbf{u}^h d \qquad \int_{h_u} \mathbf{w}^h.hd \;.
\end{aligned} \tag{4}$$

Here, **w** and q are linear test functions for the velocity and pressure, respectively. In this formulation, the first three integrals together with the right hand side term are the Galerkin finite element formulation. The first element-level integral is the SUPG [11-12] and PSPG [5-6] stabilizations. The second element-level integral is the least-square stabilization of the continuity equation, which enhanced the robustness of the finite element formulation at high Reynolds numbers. The coefficients $_{m}$ and $_{c}$ are defined in [6].

The interface function, , has two distinct values (0,1) and is used to differentiate between the two fluids [13]. A time-dependent advection equation transports this function throughout the computational domain with the fluid velocity as:

$$\frac{}{t}\Big| \quad (u \quad u_{mesh}) \qquad 0 . \tag{5}$$

Using , the density and viscosity can be calculate as:

$$_{A} \; (1 \quad) \; _{B}, \qquad\qquad _{A} \; (1 \quad) \; _{B}, \tag{6}$$

where the subscripts *A* and *B* denote the fluid *A* and fluid *B*. The artificial diffusion finite element formulation for Equation (5) leads to:

$$\int \; ^{h}\left[\frac{^{h}}{t} \quad (u^{h} \quad u^{h}_{mesh}) \qquad ^{h}\right] d \qquad\qquad\qquad , \qquad \sum_{e\;1}^{ne} \int_{e} \; _{i} \quad ^{h} \quad ^{h} d \qquad 0 \tag{7}$$

where is a linear test function for the interface function. Here the first integral is the Galerkin finite element formulation and the second integral is the artificial diffusion stabilization. The artificial diffusion stabilization technique is used for the interface function for over stabilization [1,5]. This feature allows us to enforce the conservation of mass not only locally, but also globally.

There is no limit on the number of cables that can be attached to the floating objects. The equation governing the dynamics of the nonlinear cables can be written as:

$$_{c}\left(\frac{^{2}x}{t^{2}} \quad g\right) \qquad\qquad , \tag{8}$$

$$x\Big|_{t\;0} \quad X, \quad \frac{x}{t}\Big|_{t\;0} \qquad 0, \tag{9}$$

where **x** is the position vector, **X** is the initial position, $_{c}$ is the cable density, and **T** is the Cauchy stress tensor. Locally, the Green strain tensor has only one component, which can be defined as:

$$E_{11} \quad \frac{1}{2}\left(\left\|\frac{x}{s}\right\|^{2} \quad \left\|\frac{X}{s}\right\|^{2}\right), \tag{10}$$

where *s* is a tangent vector in the original configuration. The Cauchy stress tensor, **T**, and the 2nd Piola-Kirchoff stress tensor, **S**, are related through geometry deformation. Under the assumption of small strain, but large geometry displacements, the only component of **S** in the *s* direction can be written as:

$$S_{11} \quad E_c \, E_{11} \, . \tag{11}$$

Here E_c is the Young's modulus of elasticity of the cable. The finite element formulation for the Equation (8) is derived from the principle of virtual work leading to:

$$\begin{aligned} &\int_0 H^h \left(\frac{^2 x^h}{t^2} \quad \frac{x^h}{t} \quad g \right) d \\ &\frac{1}{2} E_c \int_0 \left(\left\| \frac{x}{s} \right\|^2 \quad \left\| \frac{X}{s} \right\|^2 \right) \frac{H}{s} \, \frac{x}{s} d \quad 0 \end{aligned} \tag{12}$$

where **H** is the test function. Here we assume that the fluid does not affect the cable motion directly. Instead, a numerical damping, , is introduced to dampen the oscillations in time. In this formulation, all the integrations are carried out in the original domain rather than current deformed domain. In some applications, we use a simple spring model for the cables. This is valid if the cable density is comparable to the density of water. For spring models, we have

$$S_{11} \quad E_c \frac{L_f - L_i}{L_i} , \qquad L_f - L_i \begin{cases} L_f - L_i & L_f \quad L_i \\ 0 & L_f \quad L_i \end{cases} . \tag{13}$$

Here, L_f and L_i are the final and initial length of the cable, respectively.

The motions of the floating objects are absorbed by moving the computational nodes. In this approach, the mesh connectivity does not change and only the nodes are displaced to new positions. In our mesh-moving scheme, we assume that the computational domain is made of elastic material [6-7]. We solve linear elasticity equations to obtain the displacements for every computational node. These equations are:

$$_1(\ d)I \ 2 \ _2 \ \ (d) \quad 0, \qquad \frac{1}{2}(\ d \quad d^T), \tag{14}$$

where d is the displacement, is the strain tensor, and $_1$ and $_2$ are the linear elasticity coefficients. The finite element formulation for these equations is the Galerkin formulation written as:

$$\int (\ ^h) : \left[_1(\ d^h)I \ 2 \ _2 \ \ (d^h) \right] d \quad 0 \, , \tag{15}$$

where is the test function for the displacements.

The six degrees of freedom nonlinear rigid body dynamics are solved to locate the new position and orientation of the floating objects. Here, we consider two coordinate systems, one attached to the computational domain, X, and the other attached to the floating object, Y. The rotation matrix, Q, transfers components of any arbitrary vector from the X coordinate system to the Y coordinate system. The nonlinear rigid body dynamics equations for the center of gravity of the object are:

$$F_X \quad m g_X \quad m a_X , \tag{16}$$

$$M_Y \quad J_Y \quad _Y , \tag{17}$$

where F_X and M_Y are the total force and angular momentum exerted on the center of gravity of the object by the fluids, respectively. The linear and angular accelerations are a_X and $\quad _Y$. The mass of the object is *m*, and its moment of inertia is J_Y. Here the subscribed "X" and "Y" denotes the coordinate system where components of the vectors and matrices are evaluated.

3 Interface Sharpening/Global Mass Conservation (IS-GMC)

The finite element formulation in Equation (7) introduces numerical diffusion for the interface function . In the IS-GMC algorithm, we recover the sharpness of the interface function by replacing with $_{new}$ as following:

$$_{new} \begin{cases} \left(\frac{}{}\right), & 0 \\ 1\text{-}(1\text{-} \ )\left(\frac{1}{1}\right), & 1 \end{cases} \tag{18}$$

where 1.2 1.5 is a sharpening parameter [1], and 0 1 is a point satisfying the global conservation of mass for each fluid.

To determine , we satisfy the mass conservation at a given time, *t*, using $_{new}$. Therefore:

$$_A\int \ _{new} d \quad m_A \quad _A\int_t \int \ u.n \, d \ dt , \tag{19}$$

$$_B\int (1 \quad _{new}) d \quad m_B \quad _B\int_t \int (1 \quad) u.n \, d \ dt , \tag{20}$$

where m_A and m_B are the initial mass of Fluid A and Fluid B, respectively. Note that we only need to satisfy either Equation (19) or (20). Combining Equations (18) and (19) and assuming that the parameter is given and constant, we obtain:

$$M \ ^{(1\text{-} \ )} \quad N(1\text{-} \ )^{(1\text{-} \ )} \quad K , \tag{21}$$

where M, N, and K are all functions of . This nonlinear equation is solved using a Newton-Raphson algorithm. Typically, with the initial guess of 0.5, the algorithm converges in three iterations [1].

4 Iterative Solution Strategy

The discretization of the finite element formulations results in a series of coupled, nonlinear systems of equations that need to be solved at every time step. The nonlinear system of equations in vector form can be written as:

$$F(\dot{s},s) \quad L, \tag{22}$$

where the vector F is the function of nodal unknowns, s, and its time derivative, $\dot{s}$. Here L is the known right-hand-side vector. After linearization using the Newton-Raphson algorithm, we need to solve a series of first order linear differential equation systems. These systems are also solved iteratively using the GMRES update algorithm [14]. For very large systems of equations, we use a matrix-free iteration strategy [11]. This element-vector-based computation totally eliminates the need to form any matrices, even at the element-level.

5 Parallel Implementation

The computations of 3D free-surface flow applications are large-scale. Parallel supercomputers with hundreds of fast processors, such as CRAY T3E and IBM SP are used to reduce the computing time [7,11]. In the parallel implementation, we use a message-passing computing paradigm, making the cross-processor communication explicit. This is accomplished by using the message passing interface (MPI) libraries. Prior to the computation, the finite element mesh is partitioned into contiguous subdomains using METIS [15], and these subdomains are assigned to individual processors.

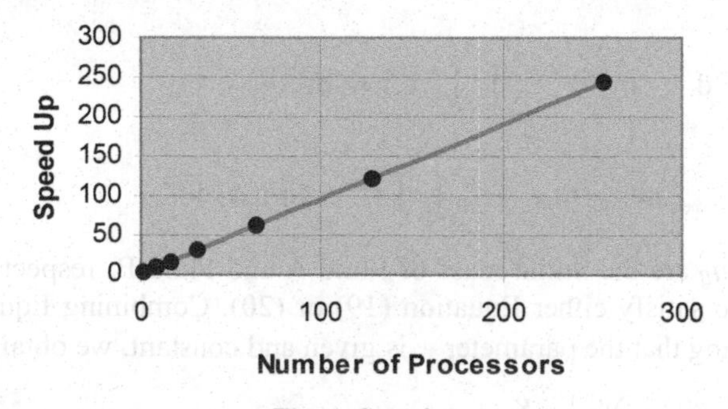

Fig. 1. Speedup

To ensure load balancing for each processor, each subdomain contains approximately the same number of elements. Element-level computations are carried out independently for each processor. Data transfer between the elements and nodes is accomplished in two steps. First, data is gathered or scattered between the elements and the nodes residing on the same processor. This step does not involve any communication. At the second step, the gather and scatter operations are performed to exchange data across the processors only for those nodes residing on the boundary of subdomains. Figure 1 shows the speed up versus the number of processors. Here, and unstructured finite element mesh with approximately one million tetrahedral elements is used.

6 Automatic Mesh Generation

Our automatic mesh generation tools have been under development at the Army High Performance Computing (HPC) Research Center for quite some time, and have been used in a wide variety of applications. The automatic mesh generator is not a single application, but consists of a series of three applications including a 3D geometric modeler, an automatic surface mesh generator, and an automatic 3D volume mesh generator. For details, see [7-16].

7 Parallel Mesh Multiplication

In one specific application, we use an unstructured finite element mesh with almost one billion tetrahedral elements. Generating such a large mesh with our automatic mesh generator created unique challenges due to the fact that the largest mesh that could be created on a workstation is roughly 3 million elements. Due to this constraint, we had to develop a parallel mesh multiplication program in order to turn an unstructured mesh into a billion-element mesh. We believe that such large meshes, which are becoming more common for modern HPC resources, should be created with such a mesh multiplication technique. An automatic mesh generator is capable of handling complex geometry and making high quality meshes, but not necessarily useful at just adding more nodes in order to increase the number of elements to reach a desired size. A two-step technique, as was used in some of the simulations here, is a better approach to generating extremely large unstructured meshes [7]. The mesh multiplication technique is based on dividing each tetrahedral element into 8 new sub-elements (Figure 2). The entire mesh size is increased by a factor of 8 in total element size. Of course, this algorithm is rather simple to implement on a serial (i.e. single processor) computer, but the size of the new (i.e. divided) mesh becomes very large rather quickly, and will usually require more memory than a single processor workstation has available. To overcome the memory limitation, we implemented this algorithm in parallel using MPI so that we can have access to the large memory usually available on large distributed memory computing platforms.

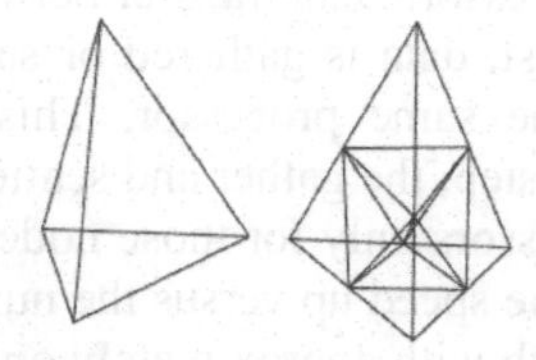

Fig. 2. A single tetrahedral element divided into eight sub-elements

8 Large-Scale Data Visualization

The visualization of such large data sets such as the ones used here posed unique challenges due to its raw size. A typical workstation does not have nearly the memory or computing power needed to process the data set alone. Furthermore, the transport and storage of this large data set (up to 300 Gigabytes in our applications) from the parallel computing server to the users desktop system can become a very time-consuming and possibly impossible task. To overcome this bottleneck, we used the Presto visualization program, developed at the Army HPC Research Center to perform the data visualization used in the examples presented here.

This visualization program is written in a client-server framework. The server part of the program runs in parallel (MPI-based) on the computer where the data is located. This server program is responsible for loading the large 3D data set, processing it, and extracting any visualization constructs requested by the client program. The client program runs on a user's desktop system and is responsible for all user interaction and displaying all 3D geometry sent to it by the server. The two programs communicate with standard Internet protocols. The program can visualize the 3D data set through surface (i.e. boundary) shadings, cross-sections, iso-surfaces, and streamlines, and that is the key to the effectiveness of this client-server model. The visualization constructs (i.e. the geometry that actually gets shown on the computer screen) are all based on "2D" or flat surface geometry, and the size of these surfaces is orders of magnitude less than the raw 3D simulation data that the parallel server program is processing. This surface data can more easily be sent across a moderately configured Internet connection and be displayed on the client workstation with the OpenGL 3D graphics library.

9 Numerical Example

9.1 Contraction Channel

To measure the accuracy of the method, we simulate flow in a contraction channel at super-critical condition and compare the results with available experimental data. The contraction channel walls are composed of two equal circular arcs each having a radius of 75 inches (see Figure 3). Water at high velocity enters the 24-inch wide channel at the speed of 85.3 inch/s and passes the narrow section of the channel, which is 12 inches wide. The Froude number with respect to the entering water elevation of 1 3/16 inch is 4.0. Here, we use a mesh consisting of 2,715,171 nodes and

2,640,000 hexahedral elements. The number of elements in the axial, vertical and cross-flow directions are 401, 61 and 111, respectively. The channel lengths before and after the contraction section are 19 11/16 inch and 136 inch, respectively.

The computation starts with an initial uniform velocity field equal to the entering water velocity. The time step is set to 0.00346 s. As computations continue, water waves are formed in the contraction section and reflected back into the narrow section of the channel from the channel walls. The two pictures in Figure 4 show the comparison of the computed water elevation along the channel wall and center to experimental data. All the flow features observed from the experiments are also captured in the computation. Here the computed results are very comparable with the experimental data.

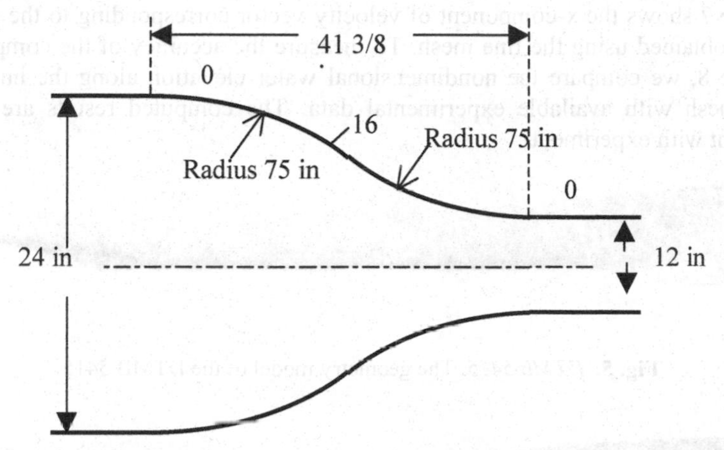

Fig. 3. *Contraction Channel.* Geometry of the channel

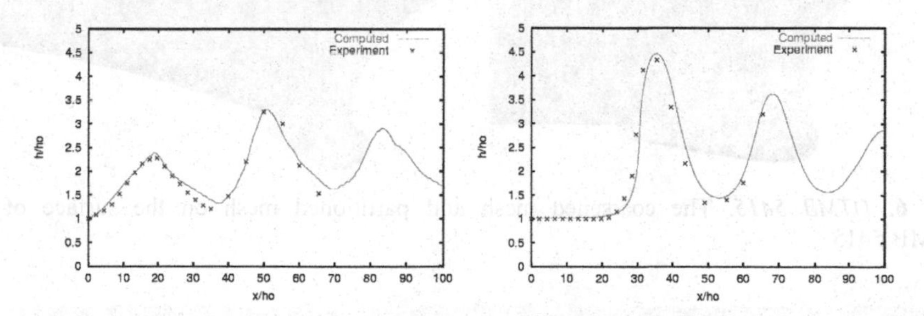

Fig. 4. *Contraction Channel.* Comparison of the computed water elevation along the channel wall (left picture) and the channel center (right picture) to experimental data

9.2 DTMB 5415

Here we apply our numerical techniques to simulate water flow around the US Navy Combatant, DTMB 5415. Model 5415 was conceived as a preliminary design for a Navy surface combatant ca. 1980 [17]. This model was also selected in the Gothenburg 2000 workshop [17]. Figure 5 shows the geometric model of the DTMB

5415. For the scaled model, the experimental data already exists for two speeds, 4 and 6 knots. The hull geometry includes both a sonar dome and a transom stern. The length of the ship is 18 7/10 feet long. To validate our flow solver, the boundary conditions are exactly the same as the ones used in experiments.

The Froude number is set to 0.28, which corresponds to the speed of 4 knots. We carried out simulations on the Cray T3E-1200 with 256 processors. Two unstructured finite element meshes are used in this simulation. The coarse mesh has more than 25 million tetrahedral elements and the fine mesh has more than one billion tetrahedral elements. The steady-state solutions are obtained for these two meshes. The pictures in Figure 6 show the coarsened mesh and the element processor distribution on the surface of the DTMB 5415.

Figure 7 shows the x-component of velocity vector corresponding to the seat-state solution obtained using the fine mesh. To measure the accuracy of the computations, in Figure 8, we compare the nondimensional water elevation along the hull for the coarse mesh with available experimental data. The computed results are in good agreement with experiment.

Fig. 5. *DTMB 5415.* The geometry model of the DTMB 5415

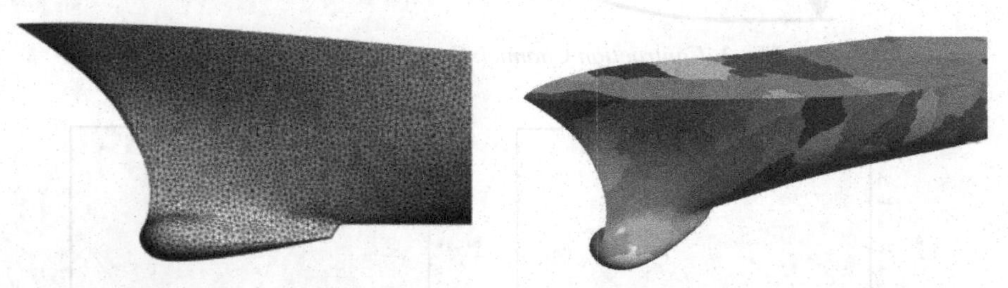

Fig. 6. *DTMB 5415.* The coarsened mesh and partitioned mesh on the surface of the DTMB 5415

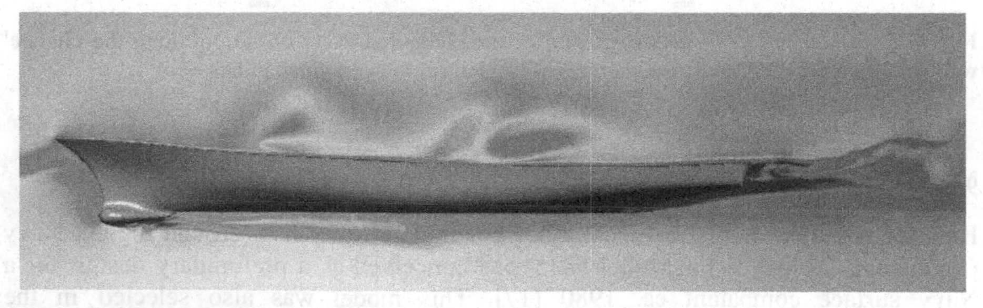

Fig. 7. *DTMB 5415.* The colors represent the x-component of velocity. The solution is obtained using a fine mesh with more than one billion tetrahedral elements

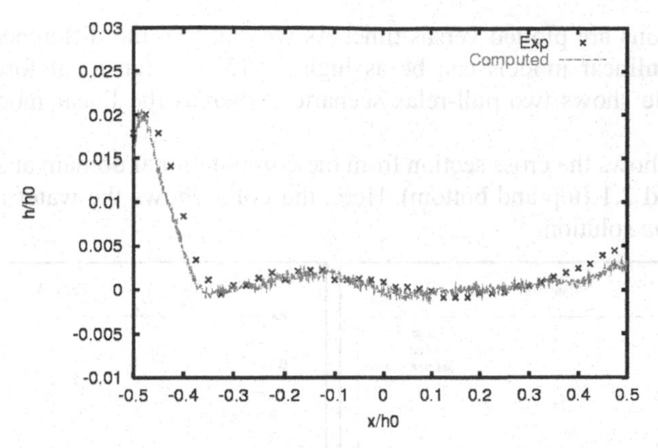

Fig. 8. *DTMB 5425.* Comparison between the computed and measured water elevation along the hull

9.3 Comparison of Linear and Nonlinear Cable Models

Here we simulate the buoyancy force exerted on a cubic object completely under water. The computational domain covers the volume $-0.25 \leq x \leq 4.0$, $0.0 \leq y \leq 1.25$ and $-0.75 \leq z \leq 0.75$, and the cubic box covers the volume of $1.5 \leq x \leq 2.5$, $0.35 \leq y \leq 0.65$ and $-0.25 \leq z \leq 0.25$. Here the numbers are nondimensional. The density of the cubic object is half of the water. The water elevation (in y direction) is 0.75. A single cable is attached in one end to the ground at location x =0.0, y = 0.0, and z = 0.0 and in other end to the object at x = 1.5, y = 0.35 and z = 0.0. The density of the cable is 1.5 times the density of water. Since the cable remains completely under water, the effective density of the cable is half of the water (due to buoyancy forces). The Young's modulus of elasticity and the cross section area of the cable are 1000 and 0.001, respectively (nondimensional).

The simulations are carried out using two levels of meshes. The coarse mesh has 199,666 nodes and 1,214,076 tetrahedral elements. The fine mesh is obtained by simply subdividing each element of coarse mesh into 8 elements.

The cable is modeled using both the Equation (8) (nonlinear) and Equation (13) (linear). Here we present 4 solutions labeled as *Nonlinear-Coarse* for coarse mesh using Equation (8), *Linear-Coarse*, for coarse mesh using Equation (13), *Nonlinear-Fine* for fine mesh using Equation (8) and *Linear-Fine* for fine mesh using Equation (13).

As the cable rises to the free-surface of water, tensional force is generated in the cable, which causes the floating object to move toward left (-x direction) and rotate counterclockwise.

Figure 9 shows the position of the cable at time equal to 0.5 and 1.0 (nondimensional). In this figure, the graphs on the left correspond to the coarse mesh and the graphs on the right correspond to the fine mesh. The difference between the solutions obtained using both meshes is small. From these graphs, we can also see that the linear and nonlinear models for the cable result in almost identical solutions. However, this is misleading. The differences between the solutions of these two models can be seen in more detail in Figure 10. In this figure, the tensional forces for all four solutions are plotted versus time. As we can see, the difference between the linear and nonlinear models can be as high as 15% in tensional forces. Also, the nonlinear cable shows two pull-relax scenarios whereas the linear model experience only one.

Figure 11 shows the cross section from the computational domain at z = 0.0 and for time at 0.1 and 2.1 (top and bottom). Here, the color shows the water and air for the *Nonlinear-Fine* solution.

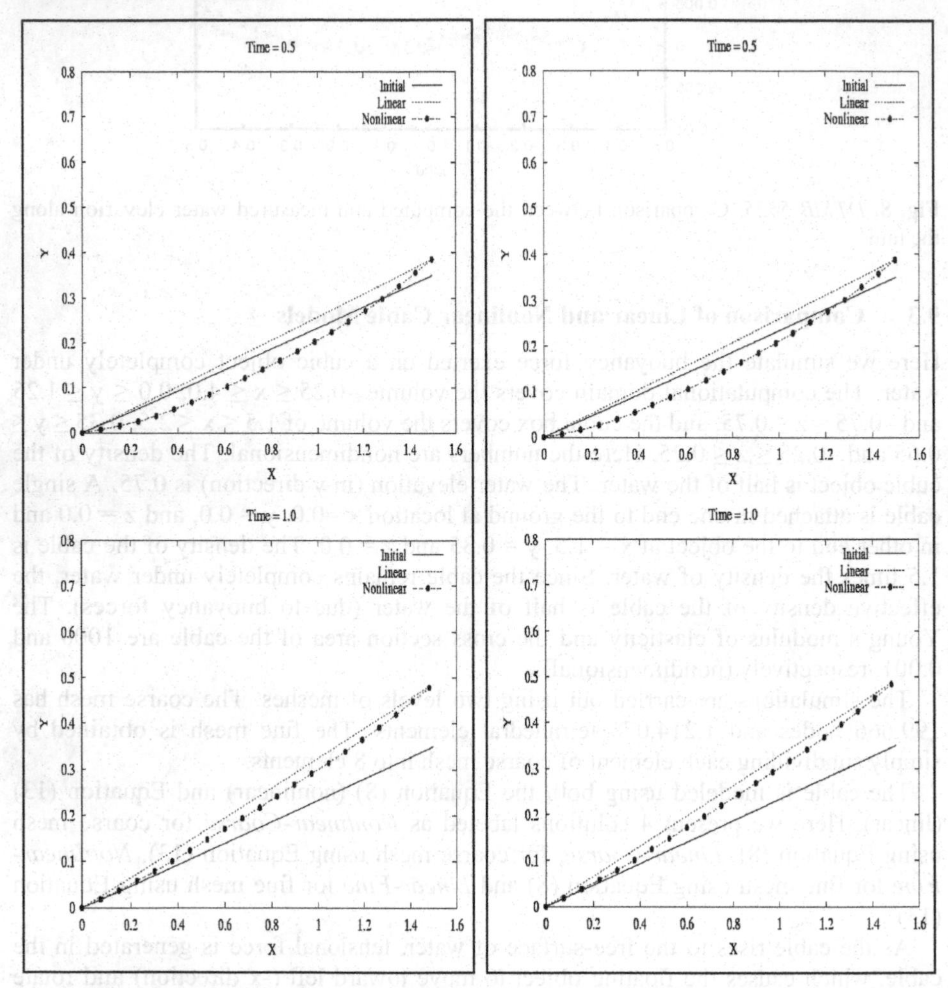

Fig. 9. *Comparison of Linear and Nonlinear Cable Models.* The graphs show the position of the cable at time equal to 0.5 and 1.0 (nondimensional). In this figure, the graphs on the left correspond to the coarse mesh and the graphs on the right correspond to the fine mesh

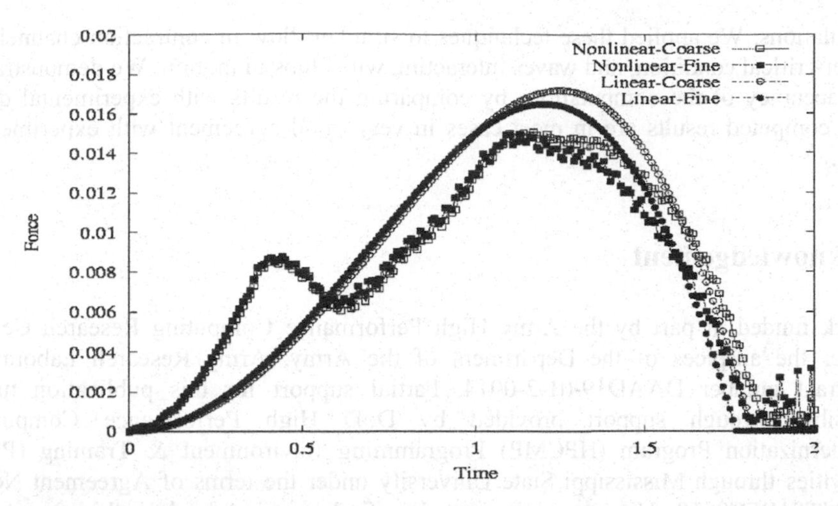

Fig. 10. *Comparison of Linear and Nonlinear Cable Models.* The tensional forces for all four solutions versus time

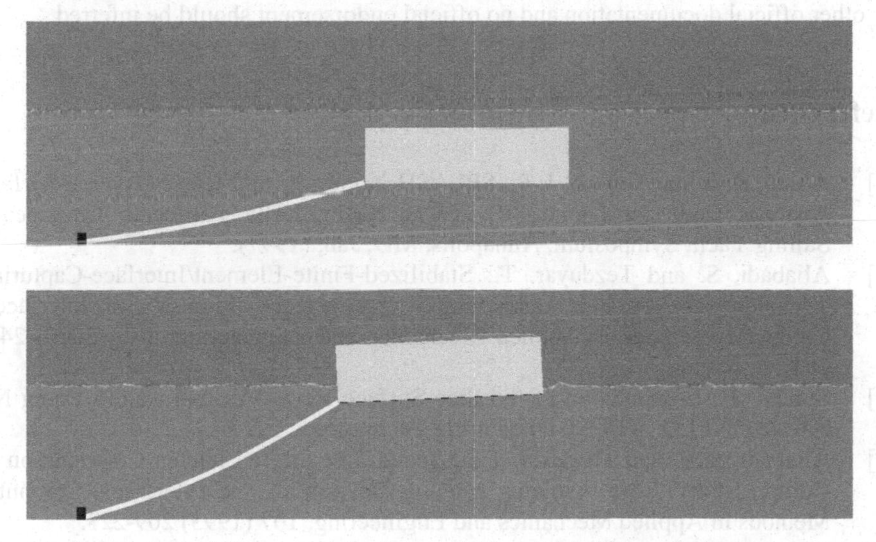

Fig. 11. *Comparison of Linear and Nonlinear Cable Models.* Figure shows the cross section from the computational domain at z = 0.0 and for time at 0.1 (top) and 2.1 (bottom). Here, the color shows the water and air for the *Nonlinear-Fine* solution

10 Concluding Remarks

We developed high performance computing tools and techniques to simulate waves interacting with ships in motion. We outlined the details of the techniques we use to solve applications on unstructured meshes. We used our client-server based Presto Visualizer to interactively visualize the very large data sets generated by these simulations. We applied these techniques to simulate flow in contraction channels at super-critical condition, and waves interacting with ships in motion. We demonstrated the accuracy of the computations by comparing the results with experimental data. The computed results are in most cases in very good agreement with experimental data.

Acknowledgement

Work funded in part by the Army High Performance Computing Research Center under the auspices of the Department of the Army, Army Research Laboratory contract number DAAD19-01-2-0014. Partial support for this publication made possible through support provided by DoD High Performance Computing Modernization Program (HPCMP) Programming Environment & Training (PET) activities through Mississippi State University under the terms of Agreement No. # GS04T01BFC0060. Views, opinions, and/or findings contained in this report are those of the author(s) and should not be construed as an official Department of the Army and Department of Defense position, policy, or decision unless so designated by other official documentation and no official endorsement should be inferred.

References

[1] Rosen, B. S. and Laiosa, J. P., SPLASH Nonlinear and Unsteady Free-Surface Analysis Code for Grand Prix Yacht Racing. The Thirteenth Chesapeake Sailing Yacht Symposium, Annapolis, MD, Jan. (1997).

[2] Aliabadi, S. and Tezduyar, T.: Stabilized-Finite-Element/Interface-Capturing Technique for Parallel Computation of Unsteady Flows with Interfaces. Computer Methods in Applied Mechanics and Engineering, 190 (2000) 243-261.

[3] Sundel, T., Computation of the Free-Surface Flows Around a Ship Using NS Solver FINFLO. VTT Manufacturing Technology, 1997.

[4] Aliabadi, S. K. and Tezduyar, T.E.: Space-time Finite Element Computation of Compressible Flows Involving Moving Boundaries and Interfaces. Computer Methods in Applied Mechanics and Engineering, 107 (1993) 209-223.

[5] Hughes, T. J. R. and Brooks, A. N., A multi-dimensional upwind scheme with no crosswind diffusion. In: Hughes, T. R. (Ed.): Finite Element Methods for Convection Dominated Flows, ASME, New York, AMD-Vol. 34 (1979) 19-35.

[6] Donea, J.: An Arbitrary Lagrangian-Eulerian Finite Element Method for Transient Fluid-Structure Interactions, Computer Methods in Applied Mechanics and Engineering Computational Mechanics, 33 (1982) 689-723.

[7] Johnson, A. and Tezduyar, T., Advanced Mesh Generation and Update Methods for 3D Flow Simulations. Computational Mechanics, 23 (1999) 130-143.

[8] Aliabadi, S., Johnson, A., Zellars, B., Abatan, A., and Berger, C.: Parallel Simulation of Flows in Open Channels. Journal of Future Generation Computer Systems, Vol. 18/5 (2002) 627-637.

[9] Aliabadi, S. and Shujaee, S.: Free Surface Flow Simulations Using Parallel Finite Element Method. Simulation, Volume 76, No. 5, ISSN 0037-5497/01 (2001) 257-262.

[10] Aliabadi, S., Abedi, J., Zellars, B., and Bota, K.: New Finite Element Technique for Simulation of Wave-Object Interaction. AIAA Paper 2002-0876 (2002).

[11] Johnson, A. and Aliabadi, S., Application of Automatic Mesh Generation and Mesh Multiplication Techniques to Very Large Scale Free-Surface Flow Simulations. Proceeding of the 7th International Conference on Numerical Grid Generation in Computational Field Simulations, Whistler, British Columbia, Canada, September 2000.

[12] Aliabadi, S., and Tezduyar, T.: Parallel Fluid Dynamics Computations in Aerospace Applications, International Journal for the Numerical Methods in Fluids, 21 (1995) 783-805.

[13] Hirt, W. and Nichols, B. D., Volume of Fluid (VOF) Method for the Dynamics of Free Boundaries. Journal of Computational Physics, 39 (1981) 201-225.

[14] Farhat, C., Lesoinne, M., and Maman, N., Mixed Explicit/Implicit Time Integration of Coupled Aeroelastic Problems: Three-Field Formulation, Geometric Conservation and Distributed Solution. International Journal for the Numerical Methods in Fluids, 21 (1995) 807-835.

[15] Karypis, G. and Kumar, V., Parallel Multilevel k-Way Partitioning Scheme for Irregular Graphs. SIAM Review, 41 (1999) 278-300.

[16] Saad, Y. and Schultz, M., GMRES: Generalized Minimal Residual Algorithm for Solving Nonsymmetic Linear Systems. SIAM Journal of Scientific and Statistical Computing, 7 (1986) 856-896.

[17] http://www.iihr.uiowa.edu/gothenburg2000/5415/combatant.html

An Efficient and Exponentially Accurate Parallel *h-p* Spectral Element Method for Elliptic Problems on Polygonal Domains – The Dirichlet Case

S.K. Tomar[1], P. Dutt[2], and B.V. Rathish Kumar[2]

[1] Faculty of Mathematical Sciences, University of Twente
P.O. Box 217, 7500 AE, Enschede, The Netherlands
s.k.tomar@math.utwente.nl

[2] Department of Mathematics
IIT Kanpur, 208016, UP, India
{pravir,bvrk}@iitk.ac.in

Abstract. For smooth problems spectral element methods (SEM) exhibit exponential convergence and have been very successfully used in practical problems. However, in many engineering and scientific applications we frequently encounter the numerical solutions of elliptic boundary value problems in non-smooth domains which give rise to singularities in the solution. In such cases the accuracy of the solution obtained by SEM deteriorates and they offer no advantages over low order methods. A new Parallel *h-p* Spectral Element Method is presented which resolves this form of singularity by employing a *geometric mesh* in the neighborhood of the corners and gives exponential convergence with asymptotically faster results than conventional methods. The normal equations are solved by the *Preconditioned Conjugate Gradient* (PCG) method. Except for the assemblage of the resulting solution vector, all computations are done on the element level and we don't need to compute and store *mass* and *stiffness* like matrices. The technique to compute the preconditioner is quite simple and very easy to implement. The method is based on a parallel computer with distributed memory and the library used for message passing is *MPI.* Load balancing issues are discussed and the communication involved among the processors is shown to be quite small.

1 Introduction

Many important real life problems, for example, shock waves in compressible flow or crack problems in structural mechanics, encounter irregular geometries which give rise to singularities in the solution. In such cases we require the solution of elliptic boundary value problems on non-smooth domains.

Current formulations of the spectral methods to solve elliptic boundary value problems on polygonal domains allow us to recover only algebraic convergence [4, 8]. Using conformal mapping of the form $z = \xi^{\alpha}$ relatively fast convergence can

S. Sahni et al. (Eds.) HiPC 2002, LNCS 2552, pp. 534–544, 2002.

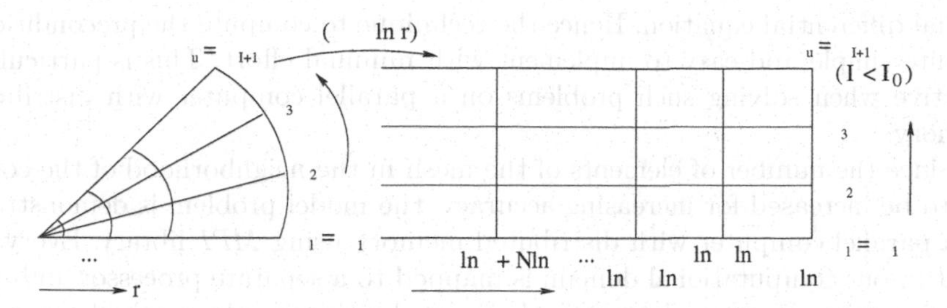
Fig. 1. Mapping $z = \ln \xi$

be achieved but exponential convergence can not be fully recovered [8]. A method for obtaining a numerical solution with exponential accuracy to elliptic boundary value problems with analytic coefficients on curvilinear polygons with piecewise analytic boundary was first proposed by Babuska and Guo [1, 2] within the framework of the finite element method. They were able to resolve the singularities which arise at the corners by using a *geometric mesh.* The method we present here solves the same class of problems to exponential accuracy within the framework of spectral methods, and gives asymptotically faster results than conventional methods.

In a neighborhood of the corner A_k we also use a *geometric mesh,* but the important difference is the mapping, which is of the form $z = \ln \xi$ and this enables us to obtain the solution with exponential accuracy. In this neighborhood we switch to new variables (τ, θ) where $\tau = \ln r$ and (r, θ) are the usual polar coordinates with origin at A_k. In doing so the geometric mesh is reduced to a quasi-uniform mesh in a sectoral neighborhood of the corners. Away from these sectoral neighborhoods of the corners we retain (x, y) variables for our coordinate system.

With this mesh we seek a solution which minimizes a functional associated with the problem. *Differentiability estimates* and a *stability estimate* with respect to these new variables have been derived in [9]. A *Parallel preconditioner* and *error estimates* for the solution of the *minimization problem* are then obtained using the stability estimate.

The preconditioner is of block diagonal form and allows the solutions for different elements to *decouple* completely, which plays an important role in solving the problem *efficiently* on parallel computers. Moreover, the preconditioner is nearly optimal as the *condition number* of the preconditioned system is polylogarithmic in N, the number of elements in the radial direction.

The *normal equations* resulting from the minimization problem are then solved by a PCG method. Since we only need *matrix-vector products* in this procedure, the *assemblage* is to be performed *only* on the resulting vector. The matrix-vector product can be carried out at element level and hence we do not need to compute and store *mass* and *stiffness* matrices [7].

The preconditioner is obtained in the same way as the residuals in the normal equations but with homogenous boundary data and the homogenous form of the partial differential equation. Hence the technique to compute the preconditioner is quite simple and easy to implement with minimal effort. This is particularly effective when solving such problems on a parallel computer with distributed memory.

Since the number of elements of the mesh in the neighborhood of the corner are to be increased for increasing accuracy, the model problem is demonstrated on a parallel computer with distributed memory using *MPI* library. Every element in our computational domain is mapped to a separate processor and since the dimension of every element is chosen to be the same on each element, the load on individual processors is perfectly balanced. The communication involved is also shown to be quite small [9].

For the purely Dirichlet problem our spectral element functions are fully nonconforming and hence there are no *common boundary values* to solve for. This turns out to be computationally more efficient than conventional methods [9]. The spectral element functions for the mixed Dirichlet and Neumann case are discussed in [9]. If the data is analytic then the error is shown to be exponentially small in N [9].

The outline of the paper is as follows. In Section 2 we describe the mesh strategy and the *stability estimate*. The numerical scheme based on this stability estimate is then presented. In Section 3 we have discussed the computation of a parallel preconditioner and the load balancing issues together with the overall complexity of the method. Finally, in Section 4 computational results are provided which verify the asymptotic estimates we have obtained.

2 The Problem and the Numerical Scheme

Here we examine the solution of the problem

$$Lu = f \quad \text{for } (x, y) \in \Omega, \tag{1}$$

with Dirichlet boundary conditions $u = g_j$ for $(x, y) \in \Gamma_j$ where $\partial\Omega = \bigcup \Gamma_j$. Here L denotes the Laplacian operator.

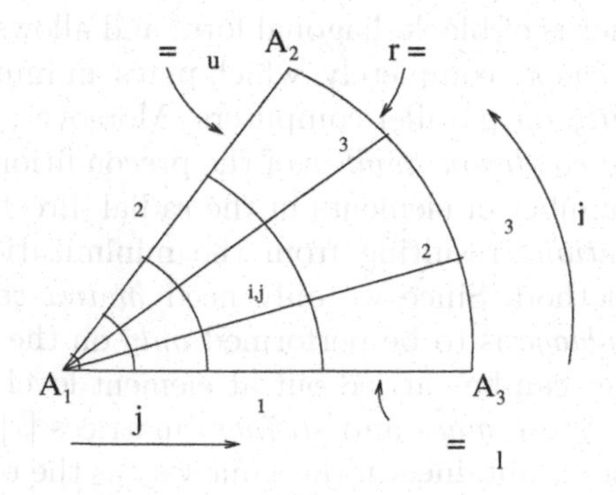

Fig. 2. The mesh at sectoral domain

For simplicity we present here only the sectoral domain case which consists of only one vertex, namely A_1 and its sectoral neighborhood. For the treatment of a polygonal domain, as well as the mixed Dirichlet and Neumann boundary conditions case, the reader is referred to [9].

On a sectoral domain with vertex A_1, as shown in Fig. 2, we define a geometric mesh with ratio $0 < \mu < 1$, as has been done in [2]. Let (r, θ) denote polar coordinates with center at A_1 and $\tau = \ln r$. The sector S_ρ (in this case Ω itself), may be represented as $S_\rho = \{(x, y) : 0 < r < \rho, \psi_l < \theta < \psi_u\}$.

- Let $\{\psi_i\}_{i=1,\ldots,I+1}$ be an increasing sequence of points such that $\psi_1 = \psi_l$ and $\psi_{I+1} = \psi_u$. Let $\Delta\psi_i = \psi_{i+1} - \psi_i$. Let $\sigma_1 = 0$, and $\sigma_j = \rho(\mu)^{N+1-j}$ for $2 \le j \le N+1$. Finally we define $\eta_j = \ln \sigma_j$ for $1 \le j \le N+1$.
- Let us denote the index sets $\mathcal{I} = \{i | 1 \le i \le I\}$ and $\mathcal{J} = \{j | 1 \le j \le N\}$. Let $\Omega_{i,j} = \{(x, y) : \sigma_j < r < \sigma_{j+1}, \psi_i < \theta < \psi_{i+1}\}$ for $i \in \mathcal{I}$ and $j \in \mathcal{J}$. Let B_ρ is the circular arc which bounds Ω.
- Now let $\tau = \ln r$. Define $\widetilde{\Omega}_{i,j} = \{(\tau, \theta) : \eta_j < \tau < \eta_{j+1}, \psi_i < \theta < \psi_{i+1}\}$ for $i \in \mathcal{I}$ and $j \in \mathcal{J}$. Also let $\widetilde{B}_\rho$ denotes the representation of B_ρ in (τ, θ) coordinates, i.e. $\widetilde{B}_\rho = \{(\tau, \theta) : \tau = \ln \rho, \psi_l < \theta < \psi_u\}$. Similarly let $\widetilde{\Gamma}_k$, $\widetilde{\Gamma}_{k+1}$ denote the representation of the sides Γ_k and Γ_{k+1} in (τ, θ) coordinates. Let γ_l be a side of $\Omega_{i,j}$ for some i and j and let $\widetilde{\gamma}_l$ denotes its representation in (τ, θ) coordinates.

Let $u_{i,j}(\tau, \theta)$ be a set of nonconforming elements, defined on $\widetilde{\Omega}_{i,j}$, given by

$$u_{i,j}(\tau, \theta) = \sum_{n=0}^{N} \sum_{m=0}^{N} a_{m,n} \tau^m \theta^n$$

for $j > 1$. Since we are assuming that the data g is analytic and compatible at the vertex A_1, the value of u at the vertex A_1 is well defined. Thus if we subtract from u an analytic function which assumes this value at the vertex A_1 then the difference would satisfy (1) with a modified set of analytic data and the Dirichlet boundary data would assume the value zero at A_1. Hence without loss of generality we may assume $g(A_1) = 0$. We shall thus choose $u_{i,1} \equiv 0$ for all i. Let us denote the jump in u across inter-element boundaries as follows

$$[u_{i,j}](\eta_{j+1}, \theta) := (u_{i,j+1} - u_{i,j})(\eta_{j+1}, \theta),$$
$$[u_{i,j}](\tau, \psi_{i+1}) := (u_{i+1,j} - u_{i,j})(\tau, \psi_{i+1}).$$

We now state the stability result on which our numerical scheme and preconditioner are based. For the proof the reader is referred to [9].

Theorem 1. *For the sectoral domain $\widetilde{\Omega}$ the following stability estimate holds.*

$$\begin{aligned}
&\sum_{j=2}^{N}\sum_{i=1}^{I}\|u_{i,j}(\tau,\theta)\|^2_{2,\widetilde{\Omega}_{i,j}} \qquad (2)\\
&\leq C(\ln N)^2\left\{\sum_{j=2}^{N}\sum_{i=1}^{I}\|Lu_{i,j}(\tau,\theta)\|^2_{0,\widetilde{\Omega}_{i,j}}\right.\\
&+\sum_{\widetilde{\gamma}_l\subseteq\widetilde{\Omega}}\left(\|[u]\|^2_{0,\widetilde{\gamma}_l}+\|[u_\tau]\|^2_{1/2,\widetilde{\gamma}_l}+\|[u_\theta]\|^2_{1/2,\widetilde{\gamma}_l}\right)\\
&+\sum_{\widetilde{\gamma}_l\subseteq\widetilde{B}_\rho}\left(\|u\|^2_{0,\widetilde{\gamma}_l}+\|u_\theta\|^2_{1/2,\widetilde{\gamma}_l}\right)\\
&\left.+\sum_{m=1}^{2}\sum_{\widetilde{\gamma}_l\subseteq\partial\widetilde{\Omega}\bigcap\widetilde{\Gamma}_m}\left(\|u\|^2_{0,\widetilde{\gamma}_l}+\|u_\tau\|^2_{1/2,\widetilde{\gamma}_l}\right)\right\}.
\end{aligned}$$

Here $u_\tau=\frac{\partial u}{\partial\tau}$, $u_\theta=\frac{\partial u}{\partial\theta}$, and $\|.\|_{s,\widetilde{\gamma}_l}$ denotes the fractional Sobolev norm when s is not an integer.

In [9] it has been shown that an exponentially accurate solution in the $H^1(\Omega)$ norm is obtained by using the above estimate and trace theorems for Sobolev spaces.

Now to find $\{u_{i,j}(\tau,\theta)\}_{i\in\mathcal{I},j\in\mathcal{J}}$ which minimizes a functional

$$\mathfrak{R}^N\left(\{v_{i,j}(\tau,\theta)\}_{i\in\mathcal{I},j\in\mathcal{J}}\right)$$

closely related to the right hand side of the stability estimate stated above, we need to solve the normal equations of the least-squares problem corresponding to collocating the partial differential equation and boundary conditions at an over-determined set of collocation points [5, 6]. We can then obtain a solution by using PCG techniques for solving the normal equations.

Now

$$\mathfrak{R}^N(U+\varepsilon V)=\mathfrak{R}^N(U)+2\varepsilon V^t(SU-TG)+O\left(\varepsilon^2\right)$$

for all V, where U is a vector assembled from the values of

$$\left\{\left\{u_{i,j}\left(\tau^{N_j}_{j,l},\theta^{N_j}_{i,m}\right)\right\}_{0\leq l,m\leq N_j}\right\}_{2\leq j\leq N,i\in\mathcal{I}}.$$

The vector V is similarly assembled and G is assembled from the data. Here S and T denote matrices. For the least-squares discretization we now need to solve the linear system $SU-TG=0$; which is symmetric and positive definite and hence the solution can be obtained efficiently by the PCG method. This requires an efficient computation of $SV-TG$ during the iterative process.

For the details of element wise computations and how to solve the corresponding linear system without having to compute and store mass and stiffness matrices, the reader can refer to [9].

3 Parallel Computations

3.1 Preconditioner

We define the quadratic form $\mathcal{W}^N\left(\{v_{i,j}(\tau,\theta)\}_{i,j}\right)$ as:

$$\mathcal{W}^N = \sum_{j=2}^{N}\sum_{i=1}^{I} \|v_{i,j}(\tau,\theta)\|^2_{2,\tilde{\Omega}_{i,j}}\ . \tag{3}$$

In the same way we may define the quadratic form $\mathcal{V}^N\left(\{v_{i,j}(\tau,\theta)\}_{i,j}\right)$ as the right hand side of the Theorem 1. Then, using the stability estimate (2) and the trace theorems for Sobolev spaces it can be been shown that the two quadratic forms $\mathcal{W}^N$ and $\mathcal{V}^N$ are spectrally equivalent and the constant of equivalence is $O(\ln N)^2$ [9]. Hence, if we use $\mathcal{W}^N$ as a preconditioner the condition number of the preconditioned system is $O(\ln N)^2$.

Now since $\mathcal{W}^N$ is of block diagonal form, this is an important benefit for computations on a parallel computer. Each block can be computed and stored on a respective element and the computation of $\mathcal{W}^N$ for each element is completely decoupled from other elements. Here each block corresponds to the H^2 norm of the spectral element function defined on a particular element which is mapped onto the master square S $([-1,1]\times[-1,1])$. Next we show how this can be computed.

Let $v_{i,j}(\xi,\eta)$ be the spectral element function defined on the square S to which the domain $\overline{\Omega}_{i,j}$ is mapped. Then $v_{i,j}(\xi,\eta)$ is determined by its values at the points $\{\xi_l,\eta_m\}_{0\le l\le N, 0\le m\le N}$. Dropping sub and superscripts we order the values of $v(\xi_l,\eta_m)$ in lexicographic order and denote them as v_n for $1\le n\le (N+1)^2$. Now consider the bilinear form $\mathcal{S}^N(u,v)$ induced by the H^2 norm on S, i.e. $\mathcal{S}^N(u,u)=\|u\|^2_{H^2(S)}$. Then there is a matrix A such that

$$\mathcal{S}^N(u,v) = \sum_{i=1}^{(N+1)^2}\left(\sum_{j=1}^{(N+1)^2} A_{i,j}u_i\right)v_j .$$

The matrix A can be determined by its columns Ae_i where e_i is a unit vector with a one in its ith place and zero everywhere else.

Now using integration by parts Ae_i can be computed in $O\left(N^3\right)$ operations in exactly the same way as we compute the residual in the normal equations. For this we need to consider homogenous form of the partial differential equation and boundary data. Hence it requires minimal extra effort to compute and store the preconditioner.

If we distribute the $(N+1)^2$ columns among the N_B processors then the matrix A can be computed in $O\left(N^4\right)$ operations on a parallel computer since $N_B = O(N)$. Moreover the L-U factorization of A can be performed in $O\left(N^5\right)$ operations and stored on every processor. Once this has been done the *action of the inverse* of the matrix on different right hand sides can be computed in $O\left(N^4\right)$ operations on every processor.

3.2 Load Balancing

The dimension of our spectral element function is $(N_j+1)^2$ for $2 \leq j \leq N$ where $\alpha j \leq N_j \leq N$, α is a positive constant. We choose an upper bound i.e. $N_j = N$ for all $j \geq 2$ defined on $\widetilde{\Omega}_{i,j}$ and map each $u_{i,j}(\tau, \theta)$ onto separate processor. By doing so we are able to achieve perfect load balancing among individual processors, but at the cost of making some of the processors do extra computational work which would not increase the accuracy of the numerical solution substantially. Alternatively, if we were to assign each $u_{i,j}(\tau, \theta)$ with varying N_j onto different processors this would cause a severe imbalance in the loads assigned to different processors and we will additionally need to employ the load balancing techniques. We should also point out that the latter strategy has the drawback that the degree of the polynomials $N_j \sim \alpha j$ is data dependent since α is determined by the data. The strategy of choosing a uniform N as an upper bound for N_j for every element is thus more robust and hence is to be preferred as it would apply to the most general class of data. The numerical results with this upper bound perfectly match with the theoretical estimates. This affirms the robustness of the proposed method.

Further, to compute the residuals in the normal equations we need to enforce continuity across inter-element boundaries. In doing so we only need to exchange the function and its derivative's values at inter-element boundaries between neighboring processors. During the PCG steps we also need to compute two global scalars to update the approximate solution and the search direction which can be achieved by a simple *MPI_ALLREDUCE* call with *MPI_SUM* operation. The inter-processor communication is thus quite small.

Finally, since we would need to perform $O(N \log N)$ iterations to obtain the solution to exponential accuracy and every iteration requires $O(N^4)$ operations, the total operations required to compute the solution would be $O(N^5 \log N)$, which for the same class of problems and on a parallel computer with N processors requires $O(N^6 \log N)$ operations with an *h-p* finite element method.

4 Computational Results

We consider Poisson equation as a model problem with Dirichlet boundary conditions. We consider only a sectoral domain as shown in Fig. 3 and show that the geometric mesh with Dirichlet boundary data gives exponential convergence. We choose our data so that there is a singularity only at the vertex O and hence we need to impose a geometric mesh only for the vertex O, as has been done in [2].

The solution thus obtained is nonconforming and we can make a correction to the solution so that the corrected solution is conforming. The error between the exact solution and the corrected solution in the $H^1(\Omega)$ norm is exponentially small in N. The details of how to do this can be found in [9].

We present the results of our numerical simulations for a sector with sectoral angle $\omega = \frac{3\pi}{2}$ and radius $\rho = 1$. We choose our data so that the solution has

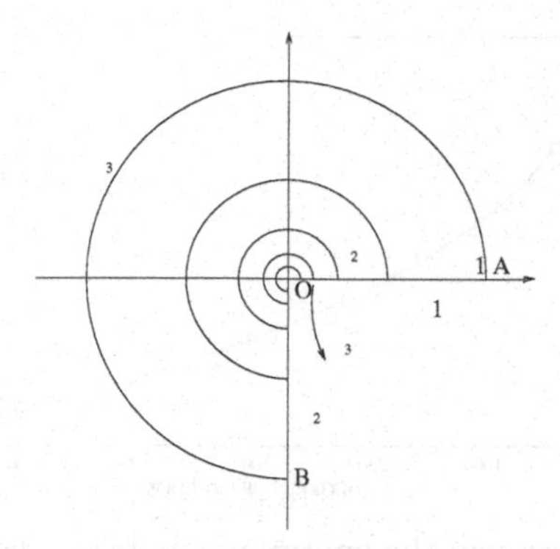

Fig. 3. The Geometric mesh at sector

the form of the leading singular solution $u = r^{\alpha} \sin(\alpha\theta)$ where $\alpha = \frac{\pi}{\omega}$. Then u vanishes along the sides OA and OB and is analytic along the curvilinear side AB. Hence the Dirichlet data is analytic and continuous at the vertices and moreover $\Delta u = 0$ in Ω. The solution of this problem is analytic in Ω except that it has a singularity at the vertex O. We divide the sector into three equal subsectors and choose the geometric ratio $\mu = .15$, which gives optimal convergence [3]. Let N be the number of spectral elements in the radial direction and the number of degrees of freedom of each variable in every element. The total degrees of freedom thus will be $3N^3$. Table 1 shows the relative error $\|e\|_{ER}$ against degrees of freedom, where the relative error $\|e\|_{ER}$ is defined as $\|e\|_{ER} = \|e\|_E / \|u\|_E$ and $\|.\|_E$ stands for the energy norm. Fig. 4 shows the error on the scale $\ln \|e\|_{ER}$ against total degrees of freedom.

We next choose the angle of the sector as $\omega = \frac{3\pi}{8}$ (with no division in θ direction) so that the solution is sufficiently regular in the neighborhood of the

Table 1. Percent of relative error against $\|e\|_{ER}$ the number of degrees of freedom

DOF	$\|e\|_{ER}$ %
24	.7462E+01
81	.3709E+01
192	.1407E+01
375	.5151E+00
648	.1736E+00
1029	.5720E-01
1536	.1830E-01
2187	.5779E-02
3000	.1798E-02
3993	.5547E-03
5184	.1696E-03

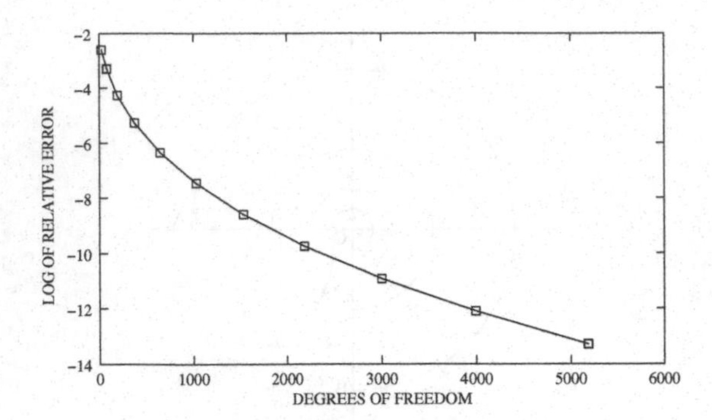

Fig. 4. Log of relative error in the energy norm $\|e\|_{ER}$ vs. the degrees of freedom

Table 2. Relative error $\|e\|_{ER}$ in percent against the number of iterations

Iterations	$\|e\|_{ER}$ %
10	.7921E-01
20	.3121E-02
30	.1157E-03
40	.8448E-05
50	.2035E-06
61	.4187E-08

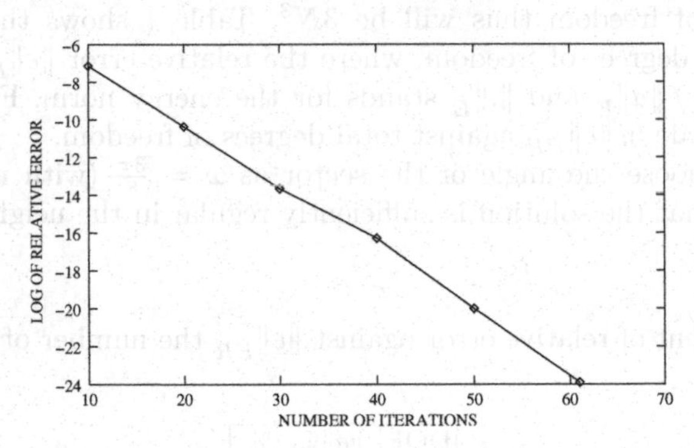

Fig. 5. Log of relative error in the energy norm $\|e\|_{ER}$ vs. the number of iterations

vertex O. We have achieved an accuracy of order e^{-10} in $H^1(\Omega)$ norm with 1350 degrees of freedom.

Table 2 shows how the relative error $\|e\|_{ER}$ depends on number of iterations. Fig. 5 shows the error on the scale $\ln \|e\|_{ER}$ against the number of iterations and the relationship is almost linear.

5 Conclusions

We have described a least-squares approach to solve elliptic boundary value problems on a sectoral domain with exponential convergence within the framework of spectral methods. The numerical scheme has a computational complexity which is less than that of finite element methods. Moreover, the construction of the preconditioner is simple and easy to implement with minimal effort. For the mesh strategy, various estimates, numerical scheme and parallelization strategies for the polygonal domain, the reader can refer to [9]. This includes additional terms on elements away from the sectoral neighborhood and is a straight forward extension of what has been presented here. The elliptic boundary value problems with mixed Dirichlet and Neumann boundary conditions on polygonal domains are solved in [9]. To solve the associated Schur Complement matrix the technique we have presented there is also computationally more efficient than *h-p* FEM. All these results are valid for elliptic problems with mixed boundary conditions on domains with curvilinear boundaries which satisfy the usual conditions [9].

Acknowledgment

The financial support provided by the CSIR-India (Project no. 9/92(123)/95-EMR-I) and ARDB-India (Project no. 95255) for this work is gratefully acknowledged. First author also gratefully acknowledges the financial support and computational resources from the Faculty of Mathematical Sciences, University of Twente, The Netherlands.

References

[1] I. Babuska and B. Q. Guo (1988): Regularity of the Solution of Elliptic Problems with Piecewise Analytic Data, Part- I, SIAM J. Math. Anal. 19, 172-203. 535
[2] I. Babuska and B. Q. Guo (1988): The h-p Version of the Finite Element Method on Domains with Curved Boundaries, SIAM J. Num. Anal., 25, 837-861. 535, 537, 540
[3] I. Babuska and M. Suri (1994): The *p* and *h-p* versions of the Finite Element Method, Basic Principles and Properties, SIAM Review, 36, 4, 578-632. 541
[4] C. Canuto, M. Y. Hussaini, A. Quarteroni and T. A. Zang (1988): Spectral Methods in Fluid Dynamics, Series in Computational Physics, Springer-Verlag. 534
[5] P. Dutt and S. Joshi (2001): Spectral Methods for Hyperbolic Initial Boundary Value Problems on Parallel Computers, Jour. Comp. Appl. Math., 134, No. 1-2, 165-190. 538
[6] P. K. Dutt and A. K. Singh (1994): The Galerkin-Collocation Method for Hyperbolic Initial Boundary Value Problems, Jour. Comp. Phys., 112, No. 2. 538
[7] Bo-nan Jiang (1998): The Least-Squares Finite Element Method - Theory and Applications in Computational Fluid Dynamics and Electromagnetics, Series in Scientific Computation, Springer-Verlag. 535
[8] G. Karniadakis and Sherwin J. Spencer (1999): Spectral/hp Element Methods for CFD, Oxford University Press. 534, 535

[9] S.K. Tomar (2001): *h-p* Spectral Element Methods for Elliptic Problems on Non-smooth Domains using Parallel Computers, Ph.D. thesis, IIT Kanpur, India. Reprint available as *Tec. Rep. no. 1631*, Faculty of Mathematical Sciences, University of Twente, The Netherlands. *http://www.math.utwente.nl/publications* 535, 536, 537, 538, 539, 540, 543

Fast Stable Solver for Sequentially Semi-separable Linear Systems of Equations

S. Chandrasekaran[1*], P. Dewilde[3], M. Gu[2**], T. Pals[1*], and A.-J. van der Veen[3]

[1] University of California, Santa Barbara
shiv@ece.ucsb.edu
[2] University of California, Berkeley
mgu@math.berkeley.edu
[3] DIMES, Delft

1 Introduction

In this paper we will present a fast backward stable algorithm for the solution of certain structured matrices which can be either sparse or dense. It essentially combines the fast solution techniques for banded plus semi-separable linear systems of equations of Chandrasekaran and Gu [4] with similar techniques of Dewilde and van der Veen for time-varying systems [12].

We will also use the proposed techniques to suggest fast direct solvers for a class of spectral methods for which there had been no known fast direct solvers (not even unstable ones). This will illustrate the usefulness of the algorithms presented in this paper. This is the spectral method by Kress [11] for solving the integral equations of classical exterior scattering theory in two dimensions.

To be more specific, let A be an $N \times N$ (possibly complex) matrix satisfying the matrix structure. Then there exist n positive integers $m_1, \cdots, m_n$ with $N = m_1 + \cdots + m_n$ to block-partition A as $A = (A_{i,j})$, where $A_{ij} \in \mathbf{C}^{m_i \times m_j}$ satisfies

$$A_{ij} = \begin{cases} D_i, & \text{if } i = j, \\ U_i W_{i+1} \cdots W_{j-1} V_j^H, & \text{if } j > i, \\ P_i R_{i-1} \cdots R_{j+1} Q_j^H, & \text{if } j < i. \end{cases} \quad (1)$$

Here we use the superscript H to denote the Hermitian transpose. The sequences $\{U_i\}_{i=1}^{n-1}$, $\{V_i\}_{i=2}^{n}$, $\{W_i\}_{i=2}^{n-1}$, $\{P_i\}_{i=2}^{n}$, $\{Q_i\}_{i=1}^{n-1}$, $\{R_i\}_{i=2}^{n-1}$ and $\{D_i\}_{i=1}^{n}$ are all matrices whose dimensions are defined in Table 1. While any matrix can be represented in this form for large enough k_i's and l_i's, our main focus will be on matrices of this special form that have relatively small values for the k_i's and l_i's (see Section 3). In the above equation, empty products are defined to

* Partially supported by grant B521537 from Lawrence Livermore national Laboratory and NSF Career Award CCR-9734290.

** This research was supported in part by NSF Career Award CCR-9702866 and by Alfred Sloan Research Fellowship BR-3720.

S. Sahni et al. (Eds.) HiPC 2002, LNCS 2552, pp. 545–554, 2002.

Table 1. Dimensions of matrices in (1). k_i and l_i are column dimensions of U_i and P_i, respectively

Matrix	U_i	V_i	W_i	P_i	Q_i	R_i
Dimensions	$m_i \times k_i$	$m_i \times k_{i-1}$	$k_{i-1} \times k_i$	$m_i \times l_i$	$m_i \times l_{i+1}$	$l_{i+1} \times l_i$

be the identity matrix. For $n = 4$, the matrix A has the form

$$A = \begin{pmatrix} D_1 & U_1 V_2^H & U_1 W_2 V_3^H & U_1 W_2 W_3 V_4^H \\ P_2 Q_1^H & D_2 & U_2 V_3^H & U_2 W_3 V_4^H \\ P_3 R_2 Q_1^H & P_3 Q_2^H & D_3 & U_3 V_4^H \\ P_4 R_3 R_2 Q_1^H & P_4 R_3 Q_2^H & P_4 Q_3^H & D_4 \end{pmatrix}.$$

We say that the matrix A is **sequentially semi-separable** if it satisfies (1). In the case where all W_i and R_i are identities, A reduces to a block-diagonal plus semi-separable matrix, which can be handled directly using techniques in Chandrasekaran and Gu [4]. It is shown in [12] that this class of matrices is closed under inversion and includes banded matrices, semi-separable matrices as well as their inverses as special cases.

It should be noted that the sequentially semi-separable structure of a given matrix A depends on the sequence m_i. Different sequences will lead to different representations. Through out this paper we will assume that the D_i's are square matrices. The methods in this paper can be generalized to non-square representations too, but that matter will not be pursued here.

2 Fast Backward Stable Solver

In this section we describe a recursive and fast backward stable solver for the linear system of equations $Ax = b$, where A satisfies (1) and b itself is an unstructured matrix

We assume that the sequentially semi-separable matrix A is represented by the seven sequences $\{U_i\}_{i=1}^{n-1}$, $\{V_i\}_{i=2}^{n}$, $\{W_i\}_{i=2}^{n-1}$, $\{P_i\}_{i=2}^{n}$, $\{Q_i\}_{i=1}^{n-1}$, $\{R_i\}_{i=2}^{n-1}$ and $\{D_i\}_{i=1}^{n}$ as in (1). We also partition $x = (\, x_i \,)$ and $b = (\, b_j \,)$ such that x_i and b_i have m_i rows. As in the 4×4 example, there are two cases at each step of the recursion.

Case of $n > 1$ and $k_1 < m_1$: Elimination. Our goal is to do orthogonal eliminations on both sides of A to create an $(m_1 - k_1) \times (m_1 - k_1)$ lower triangular submatrix at the top left corner of A.

We perform orthogonal eliminations by computing QL and LQ factorizations

$$U_1 = q_1 \begin{pmatrix} 0 \\ \hat{U}_1 \end{pmatrix} \begin{matrix} m_1 - k_1 \\ k_1 \end{matrix} \quad \text{and} \quad (q_1^H D_1) = \begin{matrix} \\ m_1 - k_1 \\ k_1 \end{matrix} \begin{matrix} m_1 - k_1 \quad k_1 \\ \begin{pmatrix} D_{11} & 0 \\ D_{21} & D_{22} \end{pmatrix} \end{matrix} w_1,$$

where q_1 and w_1 are unitary matrices. To complete the eliminations, we also need to apply q_1^H to b_1 and w_1 to Q_1 to obtain

$$q_1^H b_1 = \begin{matrix} m_1 - k_1 \\ k_1 \end{matrix} \begin{pmatrix} \beta_1 \\ \gamma_1 \end{pmatrix} \quad \text{and} \quad w_1 Q_1 = \begin{matrix} m_1 - k_1 \\ k_1 \end{matrix} \begin{pmatrix} Q_{11} \\ \hat{Q}_1 \end{pmatrix}.$$

Equations (1) have now become

$$\begin{pmatrix} q_1^H & 0 \\ 0 & I \end{pmatrix} A \begin{pmatrix} w_1^H & 0 \\ 0 & I \end{pmatrix} \begin{pmatrix} w_1 & 0 \\ 0 & I \end{pmatrix} x = \begin{pmatrix} q_1^H & 0 \\ 0 & I \end{pmatrix} b - \begin{pmatrix} 0 \\ P_2 \\ P_3 R_2 \\ P_4 R_3 R_2 \\ \vdots \\ P_n R_{n-1} \cdots R_2 \end{pmatrix} \tau, \tag{2}$$

We now orthogonally transform the unknowns x_1 and solve the $(m_1 - k_1) \times (m_1 - k_1)$ lower triangular system of equations. Let $\begin{matrix} m_1 - k_1 \\ k_1 \end{matrix} \begin{pmatrix} z_1 \\ \hat{x}_1 \end{pmatrix} = w_1 x_1$. Then the first $m_1 - k_1$ equations of (2) has been simplified to $D_{11} z_1 = \beta_1$. Hence we compute $z_1 = D_1^{-1} \beta_1$ by forward substitution.

We further compute $\hat{b}_1 = \gamma_1 - D_{21} z_1$. This in effect subtracts the D_{21} portion of the columns from the right-hand side. Finally we compute $\hat{\tau} = \tau + Q_{11}^H z_1$. This simple operation merges the previous pending subtraction at the right-hand side and the subtraction of the first $m_1 - k_1$ columns (those corresponding to z_1) from the new right-hand side.

At this stage, we discard the first $m_1 - k_1$ equations and are left with a new linear system of equations

$$\hat{A}\hat{x} = \hat{b} - \begin{pmatrix} 0 \\ P_2 \\ P_3 R_2 \\ P_4 R_3 R_2 \\ \vdots \\ P_n R_{n-1} \cdots R_2 \end{pmatrix} \hat{\tau}$$

with exactly the same form as (1). To see this, we note that among the seven sequences $\{U_i\}_{i=1}^{n-1}$, $\{V_i\}_{i=2}^{n}$, $\{W_i\}_{i=2}^{n-1}$, $\{P_i\}_{i=2}^{n}$, $\{Q_i\}_{i=1}^{n-1}$, $\{R_i\}_{i=2}^{n-1}$ and $\{D_i\}_{i=1}^{n}$, everything remains the same except that U_1, Q_1, and D_1 have been replaced by $\hat{U}_1$, $\hat{Q}_1$, and D_{22}. Among the partitioned unknown subvectors x_i's and right hand side subvectors b_i's, the only changes are that x_1 and b_1 have been replaced by $\hat{x}_1$ and $\hat{b}_1$, respectively. Of course, the new linear system of equations has a strictly smaller dimension, hence we can indeed proceed with this recursion. After we have computed the unknowns x_2 to x_n and the transformed unknowns $\hat{x}_1$, we can recover x_1 using the formula

$$x_1 = w_1^H \begin{pmatrix} z_1 \\ \hat{x}_1 \end{pmatrix}.$$

Case of $k_1 \geq m_1$: Merge. We perform merging in this case. In the case $n > 1$ and $m_1 \leq k_1$, we cannot perform eliminations. Instead we merge the first two block rows and columns of A while still maintaining the sequentially semi-separable structure.

We merge the first two blocks by computing

$$\hat{D}_1 = \begin{pmatrix} D_1 & U_1 V_2^H \\ P_2 Q_1^H & D_2 \end{pmatrix}, \quad \hat{U}_1 = \begin{pmatrix} U_1 W_2 \\ U_2 \end{pmatrix}, \quad \text{and} \quad \hat{Q}_1 = \begin{pmatrix} Q_1 R_2^H \\ Q_2 \end{pmatrix}.$$

We merge x_1 and x_2 into $\hat{x}_1$, and we merge the right hand sides by computing

$$\hat{b}_1 = \begin{pmatrix} b_1 \\ b_2 - P_2 \tau \end{pmatrix} \quad \text{and} \quad \hat{\tau} = R_2 \tau.$$

Let $\hat{A}$ and $\hat{b}$ denote the matrix A and the vector b after this merge. We can rewrite (1) equivalently as

$$\hat{A}\hat{x} = \hat{b} - \begin{pmatrix} 0 \\ P_2 \\ P_3 R_2 \\ P_4 R_3 R_2 \\ \vdots \\ P_{n-1} R_{n-2} \cdots R_2 \end{pmatrix} \hat{\tau}.$$

Clearly $\hat{A}$ is again a sequentially semi-separable matrix associated with the seven hatted sequences except that we have reduced the number of blocks from n to $n-1$.

To complete the recursion, we observe that if $n = 1$, the equations (1) become the standard linear system of equations and can therefore be solved by standard solution techniques.

2.1 Flop Count

The total flop count for this algorithm can be estimated as follows. For simplicity we assume that compression and merging steps always alternate. We also assume without loss of generality that b has only one column. Then we can show that the leading terms of the flop count are given by

$$2 \sum_{i=1}^{n} (m_i + k_{i-1}) k_i^2 + (m_i + k_{i-1})^3 + (m_i + k_{i-1})^2 l_{i+1}$$
$$+ k_i^2 m_{i+1} + k_i l_{i+1} (m_{i+1} + l_i + l_{i+2}).$$

To get a better feel for the operation count we look at the important case when $m_i = m$, $k_i = k$ and $l_i = l$. Then the count simplifies to

$$2n\left(m^3 + m^2(3k+l) + m(3k+l) + m(3kl+5k^2) + 2k^3 + k^2 l + 2kl^2\right).$$

We observe that the count is not symmetric in k_i and l_i. Therefore sometimes it is cheaper to compute a URV^T factorization instead. This matter is also covered in [4]. When $k = l$, the count simplifies further to

$$2n(m^3 + 4m^2 k + 8mk^2 + 5k^3).$$

If we make the further assumption that $m = k$ then we get the flop count $36nk^3$. Note that the constant in front of the leading term is not large.

2.2 Experimental Run-Times

We now report the run-times of this algorithm on a PowerBook G4 running at 400 MHz with 768 MB of RAM. We used the ATLAS BLAS version 3.3.14 and LAPACK version 3 libraries. For comparison we also report the run-times of the standard dense solvers from LAPACK and ATLAS BLAS. All timings are reported in Table 2. The columns are indexed by the actual size of the matrix, which range from 256 to 8192. The horizontal rows are indexed by the value of m_i which is set equal to k_i and l_i for all i and ranges from 16 to 128. These are representative for many classes of problems (see [9]). In the last row we report the run-times in seconds of a standard dense (Gaussian elimination) solver from the LAPACK version 3 library running on top of the ATLAS BLAS version 3.3.14. These are highly-tuned routines which essentially run at peak flop rates.

From the table we can see the expected linear dependence on the size of the matrix. The non-quadratic dependence on m_i (and k_i and l_i) seems to be due to the dominance of the low-order complexity terms. For example we observe a *decrease* in run-time when we increase m_i from 64 to 128 for a matrix of size 256! This is because at this size and rank the matrix has no structure and essentially a dense solver (without any of the overhead associated with a fast solver) is being used. There is also a non-linear increase in the run-time when we increase the size from 256 to 512 for $m_i = k_i = l_i = 128$. This is due to the lower over-heads associated with standard solver.

Restricting our attention to the last two rows in Table 2 where $m_i = k_i = l_i = 128$ for all i, we observe that the fast algorithm breaks even with the dense solver for matrices of size between 512 and 1024. (The estimated flop count actually predicts a break-even around matrices of size 940.) For matrices of size 4096 we have speed-ups in excess of 17.2401. Since the standard solver becomes unusually slower for matrices of size 8192 (possibly due to a shortage of RAM) we get a speed-up of 130 at this size. The speed-ups are even better for smaller values of m_i's.

We could further speed up the fast algorithm by using Gaussian elimination with partial pivoting instead of orthogonal transforms. This approach would still be completely stable as long as the dimensions of the diagonal blocks remain small.

Table 2. Run-times in seconds for both the fast stable algorithm and standard solver for random sequentially semi-separable matrices with $m_i = k_i = l_i$ for all i

	size					
$m_i = l_i = k_i$ for all i	256	512	1024	2048	4096	8192
16	0.04	0.08	0.16	0.36	0.67	1.34
32	0.08	0.19	0.42	0.83	1.66	3.44
64	0.18	0.48	1.12	2.36	4.8	9.87
128	0.15	1.01	2.73	6.09	12.91	26.9
Standard Solver (GEPP)	0.15	0.72	4.57	30.46	222.57	3499.46

3 Constructing Sequentially Semi-separable Matrices

In this section we consider the problem of computing the sequentially semi-separable structure of a matrix given the sequence $\{m_i\}_{i=1}^n$ and a low-rank representation of some off-diagonal blocks. The second assumption is to allow for the efficient computation of the sequentially semi-separable representation of matrices possessing some other structure. The method presented can be applied to any unstructured matrix, thus proving that any matrix has a sequentially semi-separable structure (of course, k_i and l_i will usually be large in this case, precluding any speed-ups).

3.1 General Construction Algorithm

Let A represent the matrix for which we wish to construct a sequentially semi-separable representation corresponding to the sequence $\{m_i\}_{i=1}^n$, where $\sum m_i = N$, the order of the matrix. Our procedure is similar to that of Dewilde and van der Veen [12]. Since the upper triangular part and lower triangular parts are so similar, we will only describe how to construct the sequentially semi-separable representation of the strictly block upper triangular part of A. The basic idea is to recursively compress off-diagonal blocks into low-rank representations.

Let H_i denote the off-diagonal block

$$H_i = \begin{pmatrix} U_1 W_2 \cdots W_i V_{i+1}^H & \cdots & U_1 W_2 \cdots W_{n-1} V_n^H \\ \vdots & \vdots & \vdots \\ U_i V_{i+1}^H & \cdots & U_i W_{i+1} \cdots W_{n-1} V_n^H \end{pmatrix}, \tag{3}$$

and let $H_i \approx E_i \Sigma_i F_i^H$ denote a low-rank (also called economy) SVD of H_i. That is, we assume that the matrix of singular values Σ_i, is a square invertible matrix, all of whose singular values below a certain threshold have been set to zero. Therefore, E_i and F_i have an orthonormal set of columns, but they may not be unitary. Following Dewilde and van der Veen [12] we will call H_i the ith Hankel block. Each H_i is a $\mu_i \times \nu_i$ matrix with $\mu_i = m_1 + \cdots + m_i$ and $\nu_i = m_{i+1} + \cdots + m_n$.

Observe that we can obtain H_{i+1} from H_i by dropping the first m_{i+1} columns of H_i and then appending to the resulting matrix the last m_{i+1} rows of H_{i+1}. We will discuss the details of computing the SVD of H_{i+1} from that of H_i shortly.

For now, we want to compute the representation of H_i in (3) using the SVDs. Partition the SVD of $H_i \approx E_i \Sigma_i F_i^H$ as:

$$E_i = \begin{matrix} \mu_{i-1} \\ m_i \end{matrix} \begin{pmatrix} E_{i,1} \\ E_{i,2} \end{pmatrix} \quad \text{and} \quad F_i = \begin{matrix} m_{i+1} \\ \nu_{i+1} \end{matrix} \begin{pmatrix} F_{i,1} \\ F_{i,2} \end{pmatrix}. \tag{4}$$

Observe that we can pick $U_i = E_{i,2}$ and $V_{i+1} = F_{i,1} \Sigma_i^H$ (of course $\Sigma_i^H = \Sigma_i$).

How do we pick W_i? Observe that H_{i+1} and H_i share a large block of the matrix. It follows from the sequentially semi-separable representation that we should pick W_{i+1} such that $E_i W_{i+1}$ will give a column basis for the upper portion of H_{i+1}. It follows that we should pick W_{i+1} to satisfy the following requirement, $E_i W_{i+1} = E_{i+1,1}$. We can solve this easily to obtain $W_{i+1} = E_i^H E_{i+1,1}$. The proof that these formulas work can be seen by substituting them back into the sequentially semi-separable representation beginning with H_1.

However, we are still not done. To compute the sequentially semi-separable representation efficiently it is important to compute the SVD of H_{i+1} quickly. To do that we need to use the SVD of H_i. As we mentioned earlier, H_{i+1} is obtained from H_i by dropping the first m_{i+1} columns of H_i and then appending to the resulting matrix the last m_{i+1} rows of H_{i+1}, which we will call Z. Hence we can rewrite H_{i+1} in the notation of (4) as

$$H_{i+1} \approx \begin{pmatrix} E_i \Sigma_i F_{i,2} \\ Z \end{pmatrix} = \begin{pmatrix} E_i & 0 \\ 0 & I \end{pmatrix} \begin{pmatrix} \Sigma_i F_{i,2} \\ Z \end{pmatrix}.$$

Hence we compute the low-rank SVD $\begin{pmatrix} \Sigma_i F_{i,2} \\ Z \end{pmatrix} \approx \tilde{E} \tilde{S} \tilde{F}^H$ and obtain the low-rank SVD of H_{i+1} as follows:

$$H_{i+1} \approx \left(\begin{pmatrix} E_i & 0 \\ 0 & I \end{pmatrix} \tilde{E} \right) \tilde{S} \tilde{F}^H.$$

Finally, we note that the sequentially semi-separable representation for the lower triangular part of A can be computed by applying exactly the same procedure above to A^H. The computational costs are similar as well.

This algorithm takes $O(N^2)$ flops, where the hidden constants depend on m_i, k_i and l_i. The algorithm can be implemented to require only $O(N)$ memory locations. This is particularly important in those applications where a large dense structured matrix can be generated (or read from a file) on the fly. Many computational electromagnetics problems involving integral equations fall in this class.

We can replace the use of singular value decompositions with rank-revealing QR factorizations (QR factorizations with column pivoting) quite easily. This may result in some speed ups with little loss of compression. The only difficulty might be the lack of easily available software.

A totally different alternative is to use the recursively semi-separable (RSS) representation presented in the paper by Chandrasekaran and Gu [4]. This is usually easier to compute efficiently, but may be less flexible.

In many important applications the sequentially semi-separable representation needs to be computed only once for a fixed problem size and stored in a file. In such cases the cost of the exact algorithm is not important. Such cases include computing the sequentially semi-separable structure of spectral discretization methods of Greengard and Rokhlin [23, 29] for two-point boundary value problems and that of Kress [11] for integral equations of classical potential theory in two dimensions.

4 Two-Dimensional Scattering

For two-dimensional exterior scattering problems on analytic curves for acoustic and electro-magnetic waves, Kress' method of discretization of order $2n$ will lead to a $2n \times 2n$ matrix of the form

$$A = I + R \odot K_1 + K_2,$$

where K_1 and K_2 are low-rank matrices and

$$R_{ij} = -\frac{2\pi}{n} \sum_{m=1}^{n-1} \frac{1}{m} \cos \frac{m|i-j|\pi}{n} - \frac{(-1)^{|i-j|}\pi}{n^2}.$$

From the results in [9] we see that it is sufficient to verify that R is a sequentially semi-separable matrix of low Hankel-block ranks. It would then follow that A is a sequentially semi-separable matrix of low Hankel-block ranks. In Table 3 we exhibit the peak Hankel block ranks of R. The rows are indexed by the (absolute) tolerance we used to determine the numerical ranks of the Hankel blocks. In particular we used tolerances of 10^{-8} and 10^{-12} that are useful in practice. The columns are indexed by the size of R.

As can be seen the ranks seem to depend logarithmically on the size N, of R. This implies that the fast algorithm will take $O(N \log^2 N)$ flops to solve linear systems involving A. We observe that the sequentially semi-separable representations of R for different sizes and tolerances need to be computed once and stored off-line. Then using the results in [9] we can compute the sequentially semi-separable representation of A rapidly on the fly.

Acknowledgements

The authors are grateful to Robert Sharpe and Daniel White of Lawrence Livermore National Laboratory for their suggestions and interests in this research.

Table 3. Peak Hankel block ranks for the spectral method of Kress, Martensen and Kussmaul for the exterior Helmholtz problem.

	size					
tolerance	256	512	1024	2048	4096	8192
1E-8	28	32	34	37	38	40
1E-12	40	46	52	58	62	66

References

[1] J. J. Dongarra, J. Du Croz, S. Hammarling, and I. Duff, *Algorithm 679: A Set of Level 3 Basic Linear Algebra Subprograms: Model Implementation and Test Programs*, ACM Transactions on Mathematical Software, 16 (1990), pp. 18-28.

[2] F.X. Canning and K. Rogovin, *Fast direct solution of moment-method matrices*, IEEE Antennas and Propagation Magazine, 40 (1998).

[3] J. Carrier, L. Greengard, and V. Rokhlin, *A fast adaptive multipole algorithm for particle simulations*, SIAM J. Sci. Stat. Comput., 9 (1988), pp. 669–686.

[4] S. Chandrasekaran and M. Gu, *Fast and stable algorithms for banded plus semi-separable matrices*, submitted to SIAM J. Matrix Anal. Appl., 2000. 545, 546, 549, 552

[5] S. Chandrasekaran and M. Gu, *A fast and stable solver for recursively semi-separable systems of equations*, in *Structured matrices in mathematics, computer science and engineering, II*, edited by Vadim Olshevsky, in the Contemporary Mathematics series, AMS publications, 2001.

[6] S. Chandrasekaran and M. Gu, *Fast and Stable Eigendecomposition of Symmetric Banded plus Semi-separable Matrices*, 1999, *Linear Algebra and its Applications*, Volume 313, Issues 1-3, 1 July 2000, pages 107-114.

[7] S. Chandrasekaran and M. Gu, *A Divide-and-Conquer Algorithm for the Eigendecomposition of Symmetric Block-Diagonal Plus Semiseparable Matrices*, 1999, accepted for publication in *Numerische Mathematik*.

[8] S. Chandrasekaran, M. Gu, and T. Pals, *A fast and stable solver for smooth recursively semi-separable systems.* Paper presented at the SIAM Annual Conference, San Diego, CA, 2001, and SIAM Conference of Linear Algebra in Controls, Signals and Systems, Boston, MA, 2001.

[9] S. Chandrasekaran, P. Dewilde, M. Gu, T. Pals, and A.-J. van der Veen, *Fast and stable solvers for sequentially recursively semi-separable linear systems of equations.* Submitted to SIAM Journal on Matrix Analysis and its Applications, 2002. 549, 552

[10] D. Colton and R. Kress, *Integral Equation Methods in Scattering Theory*, Wiley, 1983.

[11] D. Colton and R. Kress, *Inverse acoustic and electromagnetic scattering theory*, Applied Mathematical Sciences, vol. 93, Springer-Verlag, 1992. 545, 552

[12] P. Dewilde and A. van der Veen, *Time-varying systems and computations.* Kluwer Academic Publishers, 1998. 545, 546, 550

[13] Y. Eidelman and I. Gohberg, *Inversion formulas and linear complexity algorithm for diagonal plus semiseparable matrices*, Computers and Mathematics with Applications, 33 (1997), Elsevier, pp. 69–79.

[14] Y. Eidelman and I. Gohberg, *A look-ahead block Schur algorithm for diagonal plus semiseparable matrices*, Computers and Mathematics with Applications, 35 1998), pp. 25–34.

[15] Y. Eidelman and I. Gohberg, *A modification of the Dewilde van der Veen method for inversion of finite structured matrices,* Linear Algebra and its Applications, volumes 343–344, 1 march 2001, pages 419–450.

[16] L. Greengard and V. Rokhlin, *A fast algorithm for particle simulations,* J. Comp. Phys., 73 (1987), pp. 325–348.

[17] L. Gurel and W. C. Chew, *Fast direct (non-iterative) solvers for integral-equation formulations of scattering problems,* IEEE Antennas and Propagation Society International Symposium, 1998 Digest, Antennas: Gateways to the Global Network, pp. 298–301, vol. 1.

[18] W. Hackbusch, *A sparse arithmetic based on $\mathcal{H}$-matrices. Part-I: Introduction to $\mathcal{H}$-matrices*, Computing 62, pp. 89–108, 1999.

[19] W. Hackbusch and B. N. Khoromskij, *A sparse $\mathcal{H}$-matrix arithmetic. Part-II: application to multi-dimensional problems*, Computing 64, pp. 21–47, 2000.

[20] W. Hackbusch and B. N. Khoromskij, *A sparse $\mathcal{H}$-matrix arithmetic: general complexity estimates,* Journal of Computational and Applied Mathematics, volume 125, pp. 79–501, 2000.

[21] D. Gope and V. Jandhyala, *An iteration-free fast multilevel solver for dense method of moment systems,* Electrical Performance of Electronic Packaging, 2001, pp. 177–180.

[22] R. Kussmaul, *Ein numerisches Verfahren zur Lösung des Neumannschen Aussenraumproblems für die Helmholtzsche Schwingungsgleichung,* Computing 4, 246–273, 1969.

[23] June-Yub Lee and L. Greengard, *A fast adaptive numerical method for stiff two-point boundary value problems,*SIAM Journal on Scientific Computing, vol.18, (no.2), SIAM, March 1997, pp. 403–29. 552

[24] E. Martensen, *Über eine Methode zum räumlichen Neumannschen Problem mit einer Anwendung für torusartige Berandungen,* Acta Math. 109, 75–135, 1963.

[25] N. Mastronardi, S. Chandrasekaran and S. van Huffel, *Fast and stable two–way chasing algorithm for diagonal plus semi-separable systems of linear equations, Numerical Linear Algebra with Applications*, Volume 38, Issue 1, January 2000, pages 7–12.

[26] N. Mastronardi, S. Chandrasekaran and S. van Huffel, *Fast and stable algorithms for reducing diagonal plus semi-separable matrices to tridiagonal and bidiagonal form, BIT*, volume 41, number 1, March 2001, pages 149–157.

[27] T. Pals. Ph. D. Thesis, Department of Electrical and Computer Engineering, University of California, Santa Barbara, CA, 2002.

[28] V. Rokhlin, *Applications of volume integrals to the solution of PDEs*, J. Comp. Phys., 86 (1990), pp. 414–439.

[29] P. Starr, *On the numerical solution of one-dimensional integral and differential equations*, Thesis advisor: V. Rokhlin, Research Report YALEU/DCS/RR-888, Department of Computer Science, Yale University, New Haven, CT, December 1991. 552

Dynamic Network Information Collection for Distributed Scientific Application Adaptation*

Devdatta Kulkarni and Masha Sosonkina

Department of Computer Science
University of Minnesota - Duluth, MN 55812 USA
{kulk0015,masha}@d.umn.edu

Abstract. An application for collecting local dynamic network information during the course of distributed scientific application execution is presented. The technique used is light weight and provides adaptive capabilities to the application.

1 Introduction

Emergence of high capacity networks has led to increased usage of distributed resources for scientific application development. Technologies such as MPI [11] have proved to be useful in the development of distributed scientific applications. Resources such as memory and link capacity should be available for the timely execution of the distributed scientific application. It is difficult to guarantee the availability of such resources on a loosely coupled system such as a cluster of workstations due to the presence of dynamic users. Cluster of workstations provide a less expensive means of computation and are quite affordable. On the one hand utility and availability of cluster makes it a choice for many scientific application developers, however loose coupling of resources and relatively low control of the usage makes it challenging to use. Given the abilities of the current networking infrastructures and the growing propensity of the scientific community to use cluster of workstations for application development, some kind of a helper application is required that can assist the distributed application to glean dynamic network information easily and use that information advantageously. Some of the requirements of such a helper application can be stated as follows: non-intrusiveness, ease of use, flexibility, scalability and range of applicability. Different architectures for such helper applications have been proposed [7, 1]. The ability is provided to the application to know about its resource network environment through [7]. The improvement in the network route between the remote service and the application is achieved in [1] by collecting the information about network connections between the same source and destnation through an operating system module. CANS infrastructure [3] achieves the

* This work was supported in part by NSF under grants NSF/ACI-0000443 and NSF/INT-0003274, and in part by the Minnesota Supercomputing Institute.

S. Sahni et al. (Eds.) HiPC 2002, LNCS 2552, pp. 555–562, 2002.

application adaptation through monitoring the system and application events and re-planning the data paths through the use of semantics preserving modules of code. Another popular line of research is exemplefied by NWS [14] which provides an infrastructure for measuring and predicting network information to be used for making scheduling decisions. In our system we consider network monitoring to a given distributed scientific application *non-intrusively* and without scientific application code modification. In [13] we gave an initial design and experimental evaluation of our tool, Network Information Collection and Application Notification (NICAN). Here we detail the modified design of NICAN which has the above stated properties.

2 Design of NICAN

The objective of NICAN is to collect local dynamic network information and allow distributed applications to use this information advantageously. NICAN will be useful in the environment in which the scientific distributed applications have to share their compute and communication resources with competing applications resulting in the delay in execution of the scientific application. Application adaptation can be considered as an advantageous usage and NICAN strives to provide such adaptive capabilities to the application. Broadly speaking two types of adaptations are possible, viz. static adaptation and dynamic adaptation. NICAN infrastructure is built such that it allows for both static as well as dynamic adaptation. NICAN system can be divided into two main parts, the NICAN front end and the NICAN back end. Distributed application can specify the network characteristics to be monitored through the NICAN front end to the NICAN back end. NICAN back end, a thread of the application, is essentially an information collector with the ability to inform the application about any changes in the application specified network parameters.

The static adaptation capabilities are provided by *a priori* mapping the distributed application on less loaded nodes. Heuristics of process migration and dynamic changing of virtual computing environment cannot be addressed with the simple strategy of statically mapping the application. Nevertheless for the scientific applications built using MPI, which are tightly coupled with the machines at the start of the execution, this simple strategy is often effective. The dynamic adaptation is provided by presenting the dynamic network information to the distributed application. The collected network status information is made available to an application in a non-intrusive manner as and when the information becomes critical.

We accomplish the goal of decoupling network information collection from application development by using call-back mechanisms, with which the application incurs no idle time in receiving the performance information. The network information collector should have negligible overhead and not compete with application for resource usage. This design requirement is especially vital since we target high performance distributed applications which often demand full capacity of computer resources. The use of call-back mechanism serves this design objective. An application is supplied with the network information only *if this information becomes critical*, i.e., when the values for the network parameters to be observed fall outside some feasible bounds. The feasibility is determined by an application and is conveyed to the network information collector as a parameter. This selective notification approach is rather advantageous both when there is little change in the dynamic network characteristics and when the performance is very changeable. In the former case, there is no overhead associated with processing unnecessary information. In the latter, the knowledge of the network may be more accurate since it is obtained more frequently.

In Figure ??, the simultaneous collection of latency and effective throughput for the application is sketched. Note that for simplicity NICAN does not attempt to perform a combined parameter analysis: each network parameter is monitored and analyzed separately from others. The modular design, shown in Figure ??, enables an easy augmentation of the collection process with new options, which ensures its applicability to a variety of network interconnections.

Figure ?? sketches the design of NICAN. NICAN front end prepares the environment for application run by mapping the scientific application on less loaded nodes. The node load information is found out by considering the number of processes in the run queue of nodes for the previous one minute. This is static adaptation of NICAN. An application starts the NICAN back end thread and passes the request to it. The solid arrows indicate the information request passage to the NICAN back end and its subsequent information collection and data analysis phase. The dashed arrow indicates the selective notification of the application and its subsequent adaptation phase. If the adaptation is not invoked then the application continues as before. If the adaptation is invoked then the application adapts seamlessly since adaptations are transparent to the application as the adaptations are enclosed in the application adaptation handlers provided by NICAN front end.

3 Application Adaptation

Scientific applications using MPI are mapped on a set of nodes. Some nodes are logical neighbors which participate in some kind of data exchange. Thus for MPI applications, the information about the network characteristics of the neighboring node is as important as the local network information. Applications can make adaptation decisions based not only on the local conditions but also depending on the network conditions of the neighboring nodes. The application execution behavior is affected by the network characteristics of the neighboring nodes. There has to be some way by which the information about the prevalent network conditions of the neighboring nodes is supplied to the nodes. NICAN back end distributes such network information to all the neighbors of a given node.

An adaptation handler is a place where different applications can state application specific adaptations to be carried out. The application developer will be required to write the application specific adaptation handlers. Adaptations might involve changing the limit on some critical application parameter. Such parameters are shared with the NICAN front end. This allows adaptation handler to dynamically change the value of these parameters according to network conditions thus effecting the change in the execution of the application. In this design of NICAN, the adaptation handler is invoked as a part of the occurence of user specified network condition only.

Figure ?? shows the NICAN architecture. This is a node based view. The solid boxes show the independence between NICAN and the application and the dashed boxes show the interaction between them. The dotted box shows the scope of NICAN with respect to the application. The application starts the NICAN back end thread. Application passes the request for the network parameters to be monitored to the NICAN back end thread. Each network entity to be monitored is collected using a separate thread of the NICAN back end. For collecting the throughput information SNMP is used and for collecting the latency information *ping* and *traceroute* utilities are used. NICAN has been tested for a network of solaris SPARC workstations and IBM SP supercomputer.

3.1 Exchanging Critical Information

For distributed applications the network information of the neighbor nodes is equally important. NICAN provides a facility for carrying out such information exchange. This information exchange is carried out using the UDP sockets. The NICAN back end thread starts a thread, called an exchange server thread, which waits for any incoming requests. The NICAN front end thread supplies the NICAN back end thread with a list of nodes with which NICAN back end can exchange information. NICAN front end maps the MPI ranks to the IP addresses and forms a list of nodes with which this node can exchange information. NICAN back end thread contacts the exchange server thread running on each of the nodes from this list and supplies its network information to them. The exchange server on this node collects the network information from all the nodes on which this node is listed in the list supplied to the NICAN back end thread from the front end thread on those nodes.

Figure ?? shows the operation of the exchange server. In this figure solid arrows indicate passing of information from a particular node to different nodes and dashed arrows show passing of information from different nodes to a particular node.

4 Experiments and Evaluation of NICAN

Here we show the flexibility and ease in utilizing NICAN. Currently, NICAN has throughput and latency monitoring capabilities. Monitoring conditions are to be specified by the user. Thus flexibility is offered to the application with respect to the use of throughput or latency monitoring. Ease of NICAN usage is exemplefied by the fact that the application developer has to instrument the application code only to introduce the call to the NICAN back end thread and specify the monitoring request. The adaptation handlers provided by the NICAN are to be written by the application developer but the place holders to write the adaptation handlers are provided within the NICAN code itself, so the instrumentation of the application code is not required. NICAN has been successfully integrated with NetPIPE [2] and pARMS [4]. NetPIPE [10] and pARMS [6] are two very different applications. NetPIPE is a network benchmarking application whereas pARMS is a scientific distributed application. The integration of NICAN with two such different applications shows the range of applicability of this concept.

An experiment with the latency as the network parameter to be monitored was conducted. The application used was pARMS. (Complete experimental analysis showing the use of NICAN for pARMS on different computing platforms can be found in [4]). The problem is as follows [12]: We solve a system of convection-diffusion partial differential equations (PDE) on rectangular regions with Dirichlet boundary conditions, discretized with a five-point centered finite-difference scheme. If the number of points in the x and y directions (respectively) are m_x and m_y, excluding the boundary points, then the mesh is mapped to a virtual $p_x \times p_y$ grid of nodes, such that a sub-rectangle of m_x/p_x points in the x direction and m_y/p_y points in the y direction is mapped to a node. Each of the sub-problems associated with these sub-rectangles is generated in parallel. This problem is solved by FGMRES(100) using a domain decomposition pARMS preconditioning. The combining phase uses Additive Schwartz procedure. The experiment was conducted in the Computer Science department at the University of Minnesota Duluth. The network is a 10Mbps LAN running Ethernet protocol. pARMS was run on 4 workstations selected *a priori*. On two of these workstations NetPIPE was run. NetPIPE finds out the link bandwidth and protocol overhead by sending successively larger messages. Thus it can be considered as a competing network intensive application.

pARMS is a distributed iterative application which goes through alternating computation and communication phases. To start with, the problem is divided among different nodes. Each node solves its local system and then waits for the data exchange phase with the neighbors. In each iteration, local computations are alternated with the data exchange phase among all neighboring nodes following the pattern of sparse matrix-vector multiplication. This pattern is preserved if a domain decomposition type preconditioner (see e.g., [9]) is used. For such a preconditioner, it is possible to change the amount of local computations in each node depending on local sub-problem or computing platform characteristics. For varying sub-problem complexity, this issue has been considered in [8] and extended to encompass unequal node loads in [12].

The data exchange phase of pARMS is affected by the competing network intensive application like NetPIPE. Since a network intensive application makes the link heavily loaded hence there is an increase in the latency in the data acquisition phase of pARMS from the neighbor. A node which faces such a latency will have to wait for the data transfer phase to complete. In this scenario NICAN helps pARMS in two ways. Firstly it gives information about the latency to pARMS which will be difficult to obtain otherwise. Secondly the adaptation handler provided by the NICAN helps pARMS in arriving at the convergence earlier.

To guage the non-intrusiveness of NICAN, pARMS was run with and without NICAN. Without NICAN the convergence time of pARMS was 102.94 seconds. With NICAN there was not much change in this and the convergence time was 101.75 seconds. When NetPIPE was run on two nodes on which pARMS was mapped, the solution time increased to 163.89 sec. The increased network traffic affects the data exchange phase of pARMS resulting in increased latency to obtain the required data at each iteration. With NetPIPE running NICAN was introduced to see the effect of adaptation. NICAN without adaptation gave a solution time of 163.53 sec. This is comparable to the time when NetPIPE is running but in the absence of NICAN. With adaptation, however, the solution time decreased to 154.92 sec. The adaptation consisted of adjusting the number of inner iterations of pARMS based on the latency information. In this preliminary experiment, with latency as the network parameter to be monitored, the adjustment of the local inner iterations was done according to a rather simple approach. The number of inner iterations was increased by one in the face of the latency value with the neighbor crossing pARMS specified range. Without NICAN pARMS has no way of knowing about the competing NetPIPE. With NICAN pARMS can not only know about the existing NetPIPE but also can adapt its behavior. Latency information can be used to control the adjustment of the number of inner iterations more accurately as shown below. Let L_{nom} be the nominal latency between the node and its neighbor. Introduction of NetPIPE between the two nodes increases this latency value to a new value say, L_{new}. The difference between these two values, L_{diff}, gives the extra time spent in communicating the data in the face of competing NetPIPE. The adjustment of the local iteration number is similar to the strategy proposed in [8]. In the (next) jth outer iteration of the iterative process

$$n_j^i = n_{j-1}^i + \Delta_j^i,$$

where Δ_j^i, the number of iterations that node i can fit into the time to be wasted in idling otherwise at the jth outer iteration, is determined as follows:

$$\Delta_j^i = \frac{L_{max}}{T_s}$$

where T_s is the time for a single inner iteration and L_{max} is the maximum of L_{diff} from the neighbors of the node. Thus this ensures that the nodes on which NetPIPE is not running will not wait idlly but will be engaged in more local iterations. More local iterations result in increased locally accurate solution which is propagated in the subsequent communication phase and results in a decrease in the overall iterations and solution time. The number of inner iterations n_j^i can be updated provided that the limit $n_{\lim}$ on the number of inner iterations is not reached. This is not a unique strategy. As shown in [5] it is possible to use throughput information to adjust the number of inner iterations as well. Currently the difference between these two strategies is being investigated.

5 Summary

To conclude, modified design and implementation of a middleware, NICAN, to dynamically supply local network information to distributed scientific applications was presented. The application provides adaptive capabilities, is non-intrusive, flexible, easy to use and extensible. The distributed scientific application can use this tool for information collection to its advantage without substantial code instrumentation.

References

[1] D. Andersen, D. Bansal, D. Curtis, S. Seshan, and H. Balakrishnan. System support for bandwidth management and content adaptation in internet applications. In *Proceedings of 4th Symposium on Operating Systems Design and Implementation San Diego,* CA. USENIX Association.:213-226, October 2000. 555

[2] G. Chen. Providing dynamic network information to distributed applications. May 2001. Computer Science Deparment, University of Minnesota Duluth. Master's Thesis. 559

[3] Xiaodong Fu, Weisong Shi, Anatoly Akkerman, and Vijay Karamcheti. Cans: Composable, adaptive network services infrastructure. In *USENIX Symposium on Internet Technologies and Systems (USITS),* March 2001. 555

[4] D. Kulkarni and M. Sosonkina. Using dynamic network information to improve the runtime performance of a distributed sparse linear system solution. Technical Report UMSI-2002-10, 2002. 559

[5] Devdatta Kulkarni and Masha Sosonkina. A framework for integrating network information into distributed iterative solution of sparse linear systems. In *Proceedings of the 5th International Meeting of Vector and Parallel Processing, VECPAR'2002.* 560

[6] Z. Li, Y. Saad, and M. Sosonkina. pARMS: A parallel version of the algebraic recursive multilevel solver. Technical Report UMSI-2001-100, Minnesota Supercomputer Institute, University of Minnesota, Minneapolis, MN, 2001. 559

[7] B. Lowekamp, N. Miller, D. Sutherland, T. Gross, P. Steenkiste, and J. Subhlok. A resource query interface for networkaware applications. In *Cluster Computing,* volume 2, pages 139-151, 1999. 555

[8] Y. Saad and M. Sosonkina. Non-standard parallel solution strategies for distributed sparse linear systems. In A. Uhl P. Zinterhof, M. Vajtersic, editor, *Parallel Computation: Proc. of ACPC'99,* Lecture Notes in Computer Science, Berlin, 1999. Springer-Verlag. 559, 560

[9] B. Smith, P. Bjorstad, and W. Gropp. *Domian Decomposition: Parallel Multilevel Methods for Elliptic Partial Differential Equations.* Cambridge University Press New York, 1996. 559

[10] Q. Snell, A. Mikler, and J. Gustafson. NetPIPE: A network protocol independent performance evaluator. In *LASTED International Conference on Intelligent Information Management and Systems,* June 1996. 559

[11] M. Snir, S. Otto, S. Huss-Lederman, D. Walker, and J. Dongarra. MPI - *The complete Reference,* volume 1. The MIT Press, second edition, 1998. 555

[12] M. Sosonkina. Runtime adaptation of an iterative linear system solution to distributed environments. In *Applied Parallel Computing, PARA'2000,* volume 1947 of *Lecture Notes in Computer Science,* pages 132-140, Berlin, 2001. Springer-Verlag. 559

[13] M. Sosonkina and G. Chen. Design of a tool for providing network information to distributed applications. In *Proceedings PACT2001,* Lecture Notes in Computer Science. Springer-Verlag, 2001. 556
[14] Richard Wolski. Dynamically forecasting network performance using the network weather service. *Cluster Computing,* 1(1):119-132, 1998. 556

Adaptive Runtime Management of SAMR Applications*

Sumir Chandra[1], Shweta Sinha[1], Manish Parashar[1], Yeliang Zhang[2], Jingmei Yang[2], and Salim Hariri[2]

[1] The Applied Software Systems Laboratory
Rutgers, The State University of New Jersey
Piscataway, NJ 08854, USA
{sumir,shwetas,parashar}@caip.rutgers.edu
[2] Dept. of Electrical and Computer Engineering
University of Arizona
Tucson, AZ 85721, USA
{zhang,jm_yang,hariri}@ece.arizona.edu

Abstract. This paper presents the design, prototype implementation, and evaluation of a runtime management framework for structured adaptive mesh refinement applications. The framework is capable of reactively and proactively managing and optimizing application execution using current system and application state, predictive models for system behavior and application performance, and an agent based control network. The overall goal of this research is to enable large-scale dynamically adaptive scientific and engineering simulations on distributed, heterogeneous and dynamic execution environments such as the computational "grid".

Keywords: Adaptive runtime management; Structured adaptive mesh refinement; Dynamic applications; Heterogeneous distributed computing; Performance characterization.

1 Introduction

Next-generation scientific and engineering simulations of complex physical phenomena will be built on widely distributed, highly heterogeneous and dynamic, networked computational "grids". These simulations will provide new insights into complex systems such as interacting black holes and neutron stars, formations of galaxies, subsurface flows in oil reservoirs and aquifers, and dynamic response of materials to detonation. However, configuring and managing the execution of these applications to exploit the underlying computational power in spite of its heterogeneity and dynamism, presents many challenges. The overall goal of this research is to realize an adaptive runtime framework capable of

* Support for this work was provided by the NSF via grants numbers ACI 9984357 (CAREERS), EIA 0103674 (NGS) and EIA-0120934 (ITR), DOE ASCI/ASAP (Caltech) via grant number PC295251, and the DOE Scientific Discovery through Advanced Computing (SciDAC) program via grant number DE-FC02-01ER41184.

S. Sahni et al. (Eds.) HiPC 2002, LNCS 2552, pp. 564–574, 2002.

reactively and proactively managing and optimizing application execution using current system and application state, predictive models for system behavior and application performance, and an agent based control network. Its overarching motivation is enabling very large-scale, dynamically adaptive scientific and engineering simulations on distributed, heterogeneous and dynamic execution environments such as the computational "grid".

This paper presents the design, prototype implementation, and evaluation of a proactive and reactive system-sensitive runtime management framework for Structured Adaptive Mesh Refinement (SAMR) applications. System capabilities and current state are obtained using the NWS (Network Weather Service) [9] resource monitoring tool and used to appropriately distribute and load-balance the dynamic AMR computation domain. Performance prediction functions hierarchically combine analytical, experimental and empirical performance models to predict the performance of the application, and to determine when the overheads of dynamic load balancing are justified and if it is beneficial to redistribute load. An active control network combines sensors, actuators and application management agents and provides the mechanism to adapt the application at runtime.

The research presented in this paper extends our prior work on system-sensitive runtime management and integrates the different runtime approaches (reactive system-sensitive partitioning and proactive management using performance functions) within a single adaptive and automated framework.

2 Enabling Realistic Simulations Using AMR

The design of the adaptive runtime framework is driven by specific problems in enabling realistic simulations using AMR techniques. In this paper, we use the 3-D Richtmyer-Meshkov (RM3D[1]) instability encountered in compressible fluid dynamics. The RM instability occurs when a plane shock interacts with a corrugated interface between two fluids of different densities. As a result of such an interaction, interface perturbation starts to grow because the transmitted shock is converging at the wave peak and diverging at the valley. Converging shock increases pressure and accelerates perturbation peak into the second fluid. RM instabilities occur over a wide range of scales, from nearly microscopic objects, such as laser fusion pellets, to objects of astronomical size, such as supernovae.

A key challenge in such a simulation is that the physics exhibits multiple scales of length and time. If one were to employ zoning, which resolves the smallest scales, the required number of computational zones would be prohibitive. One solution is to use Adaptive Mesh Refinement (AMR) with multiple independent timesteps, which allows the grid resolution to adapt to a local estimate of the error in the solution. With AMR, the number of zones along with their location in the problem space is continuously changing. Besides dynamic communication and storage requirements, another challenge is that the local physics may change significantly from zone to zone as fronts move through the system.

[1] RM3D has been developed by Ravi Samtaney as part of the virtual test facility at the Caltech ASCI/ASAP Center (`http://www.cacr.caltech.edu/ASAP`).

Distributed implementations of these simulations lead to interesting challenges in dynamic resource allocation, data-distribution and load balancing, communications and coordination, and resource management. Furthermore, the complexity and heterogeneity of the environment make the selection of a "best" match between system resources, application algorithms, problem decompositions, mappings and load distributions, communication mechanisms, etc., nontrivial. System dynamics coupled with application adaptivity makes application configuration and runtime management a significant challenge. In this paper, we address dynamic system-sensitive partitioning and load-balancing.

3 Adaptive Runtime Framework: Design Overview

The runtime management framework is composed of three key components: a system characterization and abstraction component, a performance analysis module, and an active control network module, described as follows.

3.1 System Characterization and Abstraction

The objective of the system characterization/abstraction component is to monitor, abstract and characterize the current state of the underlying computational environment, and use this information to drive the predictive performance functions and models that can estimate its performance in the near future. Networked computational environments such as the computational "grid" are highly dynamic in nature. Thus, it is imperative that the application management system be able to react to this dynamism and make runtime decisions to satisfy application requirements and optimize performance. These decisions include selecting the appropriate number, type, and configuration of the computing elements, appropriate distribution and load-balancing schemes, the most efficient communication mechanism, as well as the right algorithms and parameters at the application level. Furthermore, proactive application management by predicting system behavior will enable a new generation of applications that can tolerate the dynamics of the grid and truly exploit its computational capabilities.

3.2 Performance Analysis Module

The performance analysis module is built on *Performance Functions.* Performance Functions (PF) describe the behavior of a system component, subsystem or compound system in terms of changes in one or more of its attributes. Using the PF concept, we can characterize the operations and performance of any resource in a distributed environment. Once the PFs of each resource used by an application are defined, we compose these PFs to generate an overall end-to-end PF that characterizes and quantifies application performance.

Our PF-based modeling approach includes three steps. First, we identify the attributes that can accurately express and quantify the operation and performance of a resource (e.g., Clock speed, Error, Capacity). The second step is to use experimental and analytical techniques to obtain the PF that characterizes and quantifies the performance of each system component in terms of these attributes. The final step is to compose the component PFs to generate an overall PF that can be used during runtime to estimate and project the operation and performance of the application for any system and network state. This composition approach is based on the performance interpretation approach for parallel and distributed applications [6, 7].

3.3 Active Control Network

The underlying mechanisms for adaptive runtime management of SAMR applications are realized by an active control network of sensors, actuators, and management agents. This network overlays the application data-network and allows application components to be interrogated, configured, and deployed at runtime to ensure that application requirements are satisfied. Sensors and actuators are embedded within the application and/or system software and define interfaces and mechanisms for adaptation. This approach has been successfully used to embed and deploy sensors and actuators for interactive computational steering of large, distributed and adaptive applications [4].

3.4 Adaptive Application Management

The key goal of the runtime management framework is to develop policies and mechanisms for both "application sensitive" and "system sensitive" runtime adaptations of SAMR applications. The former is based on current application state while the latter is driven by current system state and system performance predictions. Application sensitive adaptations [1] use the current state of the application to drive the runtime adaptations. The abstraction and characterization of the application state is used to drive the resource allocation, partitioning and mapping of application components onto the grid, selection and configuration of partitioning and load-balancing algorithms, communication mechanisms, etc. System sensitive application management [8] uses current and predicted system state characterization to make application adaptation decisions. For example, the information about the current load and available memory will determine the granularity of the mapping of application components to processing nodes, while available communication bandwidths will determine the communication strategy to be used. Similarly, application level algorithms may be selected based on the type, specifications, and status of the underlying architecture. Finally, the availability and "health" of computing elements on the grid may determine the nature (refined grid size, aspect ratios, etc.) of refinements to be allowed.

4 Runtime Management Framework: Evaluation

We have developed and deployed a prototype runtime management framework that uses current system state and predictive performance functions to proactively and reactively manage the distribution and load-balancing of SAMR applications. The prototype framework has been integrated into the GrACE (Grid Adaptive Computational Engine) [5] infrastructure's adaptive runtime system. GrACE is a data-management framework for parallel/distributed AMR and is being used to provide AMR support for varied applications including reservoir simulations, computational fluid dynamics, seismic modeling, and numerical relativity. This section presents the implementation and experimental evaluation of this prototype using the RM3D CFD kernel.

4.1 Reactive System Sensitive Partitioning and Load Balancing

The adaptive runtime framework reacts to system capabilities and current system state to select and tune distribution parameters by dynamically partitioning and load balancing the SAMR grid hierarchies. Current system state is obtained at runtime using the NWS resource monitoring tool. System state information along with system capabilities are then used to compute relative computational capacities of each of the computational nodes. These relative capacities are used by the "system-sensitive" partitioner for dynamic distribution and load-balancing. NWS periodically monitors and dynamically forecasts the performance delivered by the various network and computational resources over a given time interval. Measurements include the fraction of CPU time available for new processes, the fraction of CPU available to a process that is already running, end-to-end TCP network latency, end-to-end TCP network bandwidth, free memory, and the amount of space unused on a disk.

This system information provided by NWS is used to compute a relative capacity metric for each processor as follows [8]. Let us assume that there are K processors in the system among which the partitioner distributes the workload. For node k, let $\mathcal{P}_k$ be the percentage of CPU available, $\mathcal{M}_k$ the available memory, and $\mathcal{B}_k$ the link bandwidth. The available resource at k is first converted to a fraction of total available resources, i.e.

$$P_k = \mathcal{P}_k / \sum_{i=1}^{K} \mathcal{P}_i \quad \text{and} \quad M_k = \mathcal{M}_k / \sum_{i=1}^{K} \mathcal{M}_i \quad \text{and} \quad B_k = \mathcal{B}_k / \sum_{i=1}^{K} \mathcal{B}_i \tag{1}$$

The relative capacity C_k of a processor is then defined as the weighted sum of these normalized quantities, i.e.

$$C_k = w_p P_k + w_m M_k + w_b B_k \tag{2}$$

where w_p, w_m, and w_b are the weights associated with the relative CPU, memory, and link bandwidth availabilities, respectively, such that $w_p + w_m + w_b = 1$. The weights are application dependent and reflect its computational, memory, and communication requirements. Note that $\sum_{k=1}^{K} C_k = 1$. If L is the total work to be assigned to all the processors, then the work L_k assigned to the kth processor can be computed as $L_k = C_k L$. The overall operation is shown in Figure 1.

The system sensitive adaptive partitioner is evaluated using the RM3D CFD kernel on a Linux-based workstation cluster. The kernel used 3 levels of factor 2

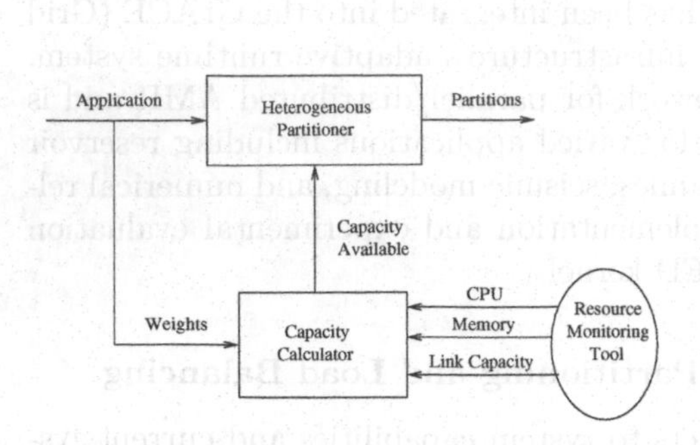

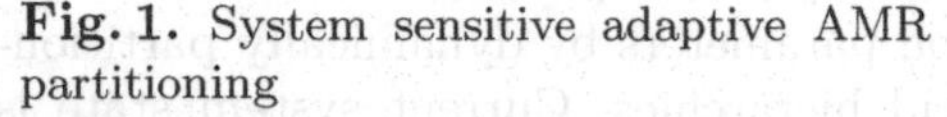
Fig. 1. System sensitive adaptive AMR partitioning

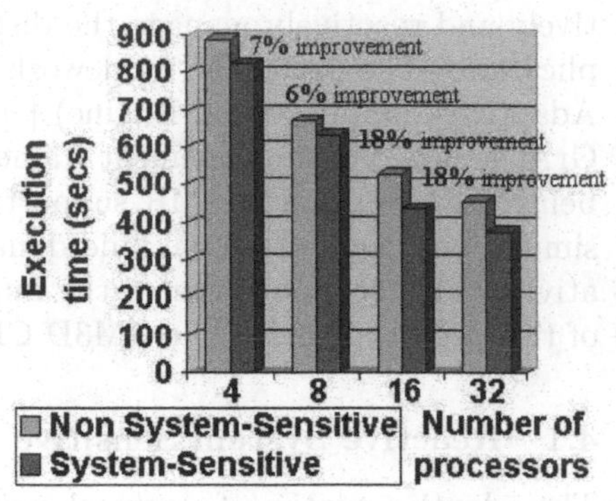

Fig. 2. Improvement in execution time due to system sensitive partitioning

space-time refinements on a base mesh of size 128*32*32. The cluster consisted of 32 nodes interconnected by fast Ethernet (100MB). The experimental setup consisted of a synthetic load generator (for simulating heterogeneous loads on the cluster nodes) and an external resource monitoring system (i.e. NWS). The evaluation comprised of comparing the runtimes and load balance generated for the system sensitive partitioner with those for the default partitioning scheme provided by GrACE. This latter scheme assumes homogeneous processors and performs an equal distribution of the workload on the processors.

The improvement in application execution time using the system sensitive partitioner as compared to the default non-system sensitive partitioner is illustrated in Figure 2. System sensitive partitioning reduced execution time by about 18% in the case of 32 nodes. We believe that the improvement will be more significant in the case of a cluster with greater heterogeneity and load dynamics.

The adaptivity of the system sensitive partitioner to system dynamics are evaluated for a cluster with 4 nodes, with the relative capacities C_1, C_2, C_3, and C_4 computed as 16%, 19%, 31%, and 34% respectively. The three system characteristics, viz. CPU, memory, and link bandwidth, are assumed to be equally important, i.e. $w_p = w_m = w_b = 1/3$, and the application regrids every 5 iterations. The load assignment for the GrACE default and the system sensitive (*ACEHeterogeneous*) partitioners are plotted in Figures 3 and 4 respectively. For the kth processor, the load imbalance I_k is defined as

$$I_k = \frac{|W_k - L_k|}{L_k} \times 100 \quad \% \tag{3}$$

As expected, the GrACE default partitioner generates large load imbalances as it does not consider relative capacities. The system sensitive partitioner produces about 45% smaller imbalances. Note that the load imbalances in the case of the system sensitive partitioner are due to the constraints (minimum box size and aspect ratio) that have to be satisfied while breaking boxes.

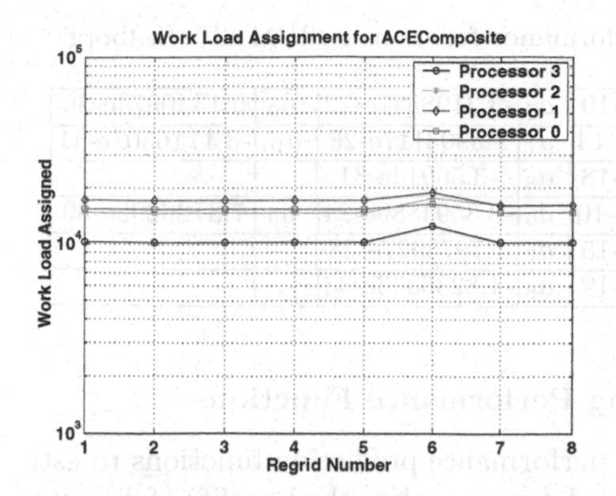

Fig. 3. Workload assignments for default partitioning scheme (Relative capacities of processors 0 – 3 are 16%, 19%, 31%, 34%)

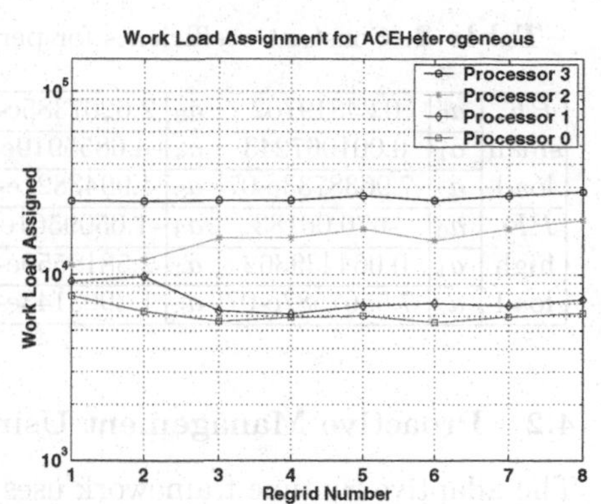

Fig. 4. Workload assignments for ACE-Heterogeneous partitioning scheme (Relative capacities of processors 0 – 3 are 16%, 19%, 31%, 34%)

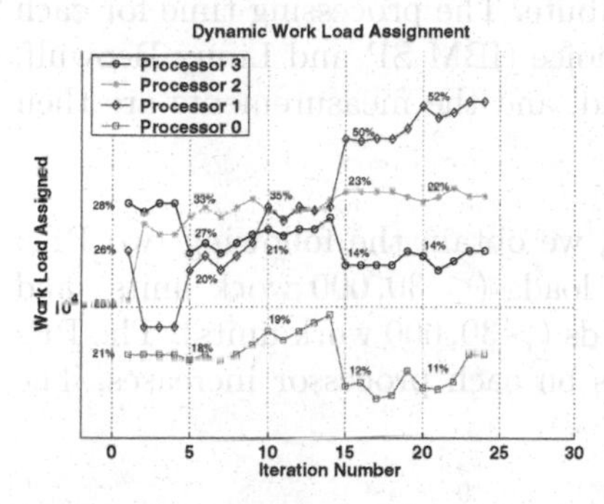

Fig. 5. Dynamic load allocation for a system state sensing frequency of 20 iterations

Table 1. Comparison of execution times using static (only once) and dynamic sensing (every 40 iterations)

Number of Processors	Execution time with Dynamic Sensing (secs)	Execution time with Sensing only once (secs)
2	423.7	805.5
4	292.0	450.0
6	272.0	442.0
8	225.0	430.0

In order to evaluate the ability of the system-sensitive partitioner to adapt to load dynamics in the cluster, the synthetic load generator was used on two processors to dynamically vary the system load. The load assignments at each processor were computed for different sensing frequencies. Table 1 illustrates the effect of sensing frequency on overall application performance. Dynamic runtime sensing improves application performance by as much as 45% compared to sensing only once at the beginning of the simulation. Figure 5 shows the relative processor capacities and load assignments for a sensing frequency of 20 iterations. The frequency of sensing depends on the load dynamics and can affect application performance. In our experimental setup, the best application performance was achieved for a sensing frequency of 20 iterations.

Table 2. Constant coefficients for performance functions on IBM SP "Seaborg"

PF_s	a_0	0.24819702	a_3	2.6207385e-10	a_6	-3.3198527e-22	a_9	5.1340974e-36
small	a_1	0.001067243	a_4	-4.6855919e-14	a_7	1.3903117e-26	a_{10}	-3.1716301e-41
load	a_2	-7.9638733e-07	a_5	4.9943821e-18	a_8	-3.56914e-31		
PF_h	a_0	-670.06183	a_3	-1.0509561e-10	a_6	-8.3594889e-24	a_9	7.3129295e-39
high	a_1	0.064439362	a_4	2.5819559e-15	a_7	1.7978927e-28		
load	a_2	-7.5567587e-07	a_5	1.3903143e-19	a_8	-1.8236343e-33		

4.2 Proactive Management Using Performance Functions

The adaptive runtime framework uses performance prediction functions to estimate application execution times and to determine when the benefits of dynamic load redistribution exceed the costs of repartitioning and data movement. The performance functions (PF) model the execution of the SAMR-based RM3D application and describe its overall behavior with respect to the desired metric. In this experiment, we use the computational load as the metric and model application execution time with respect to this attribute. The processing time for each application component on the machine of choice (IBM SP and Linux Beowulf, in our case) is measured in terms of the load, and the measurements are then used to obtain the corresponding PF.

IBM SP "Seaborg"[2]: For our evaluation, we obtain the following two PFs: PF_s denotes the PF associated with small loads ($\leq 30,000$ work units) and PF_h denotes the PF associated with high loads ($> 30,000$ work units). The PFs increase linearly as the number of processes on each processor increases. The PFs for small and high loads are as follows:

$$PF_s = \sum_{i=0}^{10} a_i * x^i \quad \text{and} \quad PF_h = \sum_{i=0}^{9} a_i * x^i \tag{4}$$

where a_i (i=0,1,...,10) are constants and x is the computational load. The coefficients for the PFs for small and large loads are listed in Table 2.

Linux Beowulf "Discover"[3]: The performance function evaluation on the Linux cluster yields a single PF as follows:

$$PF = \sum_{i=0}^{10} b_i * x^i \tag{5}$$

where b_i (i=0,1,...,10) are coefficients (listed in Table 3) and x is the workload.

Tables 4 and 5 show that the error incurred in modeling the execution time based on the PF modeling approach is low, roughly between 0-8% for the IBM SP

[2] The National Energy Research Scientific Computing Center (NERSC) IBM SP RS/6000, named seaborg.nersc.gov, is a distributed memory parallel supercomputer with 2,944 compute processors among 184 compute nodes.

[3] Discover is a 16-node Beowulf cluster at Rutgers University.

Table 3. Constant coefficients for performance function on Linux Beowulf "Discover"

PF	b_0	6.0826507	b_3	2.4465223e-13	b_6	-3.6882935e-29	b_9	7.8471222e-47
	b_1	0.00048341426	b_4	-2.10405e-18	b_7	7.8747669e-35	b_{10}	-2.5582951e-53
	b_2	-1.5930319e-08	b_5	1.1051294e-23	b_8	-1.0441739e-40		

Table 4. Accuracy of performance functions on IBM SP "Seaborg"

Workload (grid units)	Actual time (sec)	PF derived time (sec)	Error rate (%)
13824	0.4420	0.4088	7.5
18432	0.5140	0.5089	0.99
23040	0.5735	0.5623	1.9
35072	0.7088	0.7089	0

Table 5. Accuracy of performance functions on Linux Beowulf "Discover"

Workload (grid units)	Actual time (sec)	PF derived time (sec)	Error rate (%)
74880	9.4823	9.9852	5.3
97344	9.8688	9.8859	0.17
123656	10.3557	9.7612	5.74
173056	11.0533	10.4147	5.77
274560	11.2683	11.2393	0.26
430496	14.9336	15.5478	4.11

and between 0-6% for the Beowulf cluster. Details about the PF-based approach for modeling large-scale distributed systems are found in [2, 3].

This PF-based model is used by the adaptive framework to determine when the benefits of dynamic load redistribution exceed the costs of repartitioning and data movement. For N processors in the system, let $GlbLoad$ denote the global workload for the structured dynamic grid hierarchy and $LocLoad_k$ denote the local load for processor k. The ideal workload per processor is given by $IdlLoad = GlbLoad/N$. Using the PF-based approach, execution time estimates are obtained for ideal and local workloads for each processor, denoted by $PFtimeIdl$ and $PFtimeLoc$ respectively. Load redistribution is typically expensive for small load variations; however, it is justified when the workload imbalance exceeds a certain threshold, defined by

$$Thresh = \frac{PFtimeIdl - PFtimeLoc}{PFtimeLoc} \quad (6)$$

A threshold of 0 indicates regular periodic load redistribution regardless of the load-balancing costs. A high threshold represents the ability of the application hierarchy to tolerate workload imbalance and determines when the overheads of dynamic load balancing are justified and if it is beneficial to redistribute load.

The RM3D evaluation on the Beowulf cluster analyzes the effect of dynamic load balancing on application recompose time in order to achieve better performance. The experimental setup consists of the RM3D application executing on 8 processors for redistribution thresholds of 0 and 1. The application uses 3 levels of factor 2 space-time refinements on a base mesh of size 64*16*16 with regriding every 4 time-steps. Threshold of 1 considers the costs and benefits of redistributing load and results in recompose time being reduced by half (improvement of almost 100%) as compared to when a threshold of 0 is used.

5 Conclusions

In this paper, we presented the design, prototype implementation, and evaluation of an adaptive runtime framework capable of reactively and proactively managing and optimizing application execution using current system and application state, predictive models for system behavior and application performance, and an active control network. The overarching motivation for this research is to enable large-scale dynamically adaptive scientific and engineering simulations on distributed, heterogeneous and dynamic execution environments such as the computational "grid". Experimental results using a distributed and adaptive Richtmyer-Meshkov CFD kernel are presented. We are currently extending this evaluation to larger systems and more heterogeneous configurations and to different application domains.

References

[1] S. Chandra, J. Steensland, M. Parashar, and J. Cummings. An Experimental Study of Adaptive Application Sensitive Partitioning Strategies for SAMR Applications. Proceedings of the *2nd Los Alamos Computer Science Institute Symposium* (also best research poster at *Supercomputing Conference 2001*), October 2001. 567

[2] S. Hariri and et al. A Hierarchical Analysis Approach for High Performance Computing and Communication Applications. Proceedings of the *32nd Hawaii International Conference on System Sciences*, 1999. 572

[3] S. Hariri, H. Xu, and A. Balamash. A Multilevel Modeling and Analysis of Network-Centric Systems. Special Issue of *Microprocessors and Microsystems Journal*, Elsevier Science on Engineering Complex Computer Systems, 1999. 572

[4] S. Kaur, V. Mann, V. Matossian, R. Muralidhar, and M. Parashar. Engineering a Distributed Computational Collaboratory. Proceedings of the *34th Hawaii International Conference on System Sciences*, January 2001. 567

[5] M. Parashar and J. Browne. On Partitioning Dynamic Adaptive Grid Hierarchies. Proceedings of the *29th Hawaii International Conference on System Sciences*, January 1996. 568

[6] M. Parashar and S. Hariri. Interpretive Performance Prediction for Parallel Application Development. *Journal of Parallel and Distributed Computing*, vol. 60(1), pp. 17-47, January 2000. 567

[7] M. Parashar and S. Hariri. Compile-Time Performance Interpretation of HPF/Fortran 90D. *IEEE Parallel and Distributed Technology*, Spring 1996. 567

[8] S. Sinha and M. Parashar. Adaptive Runtime Partitioning of AMR Applications on Heterogeneous Clusters. Proceedings of the *3rd IEEE International Conference on Cluster Computing*, pp. 435-442, 2001. 567, 568

[9] R. Wolski. Forecasting Network Performance to Support Dynamic Scheduling using the Network Weather Service. Proceedings of the *6th IEEE Symposium on High Performance Distributed Computing*, 1997. 565

Mobile Agents – The Right Vehicle for Distributed Sequential Computing

Lei Pan, Lubomir F. Bic, Michael B. Dillencourt, and Ming Kin Lai

Information and Computer Science, University of California
Irvine, CA 92697-3425
{pan,bic,dillenco,mingl}@ics.uci.edu

Abstract. Distributed sequential computing (DSC) is computing with distributed data using a single locus of computation. In this paper we argue that computation mobility—the ability for the locus of computation to migrate across distributed memories and continue the computation as it meets the required data—facilitated by mobile agents with strong mobility is essential for scalable distributed sequential programs that preserve the integrity of the original algorithm.

Keywords: distributed sequential computing (DSC), computation mobility, mobile agents, distributed code building block (DBlock), scalability, algorithmic integrity, paging, Crout factorization

1 Introduction

Sequential computing will never disappear, even in distributed environments. There are at least three reasons for this. First, only very few algorithms (referred to as "embarrassingly parallel") can be perfectly parallelized. In fact, Amdahl's law tells us that the sequential portions in a parallel program often become bottlenecks for speedup, so we need to pay special attention to them. Second, when solving problems on distributed-memory systems, programmers may choose to decompose a problem into coarse-grained sequential sub-tasks that run in parallel, even if fine-grained data parallelism is an option [1]. Third, although parallel compilers can sometimes do a good job of translating sequential code to parallel code, not all auto-translations are viable and in many cases human intervention is unavoidable [2]. Re-implementing an algorithm in parallel can require a major programming effort. In some cases, significant performance improvement can be achieved without this additional programming effort by running the sequential algorithm in a distributed environment to solve large problems and avoid the performance penalty from disk thrashing [3, 4, 5]. We define *distributed sequential computing* (DSC) as computing with distributed data using a single locus of computation.

The key to scalability in distributed high performance computing is reducing the amount of data being moved. The classical distributed shared memory (DSM) systems cannot address this issue efficiently and nicely, therefore they cannot fulfill the initial expectations after almost 15 years of research [6].

S. Sahni et al. (Eds.) HiPC 2002, LNCS 2552, pp. 575–584, 2002.

Not willing to give up the convenience of "reading and writing remote memory with simple assignment statements" [7], researchers tried to exploit data locality in later improved DSM-based systems or compilers, such as HPF [8, 9] and UPC [7, 10], by following the general principle of "owner-computes" [11]. The principle states that computations should happen on the processor that owns the value that is being assigned to. The use of this principle in HPF or UPC is somewhat limited: a programmer can only exercise "owner-computes" by first specifying data distribution pattern and then using new language constructs such as **forall**. Strict adherence to the principle of "owner-computes" as articulated above does not always minimize data movement. For example, in an assignment, if the right-hand-side (RHS) objects are large in size and distributed, instead of all the RHS objects being sent to the owner of the left-hand-side (LHS) object for computation, the result could instead be computed on the home processor of the RHS objects and then sent to the owner of the LHS object for assignment. This would reduce the amount of communication required. In this paper, we will reuse the term "owner-computes" but extend the meaning to be "owner process/thread of large sized data computes" and the scope to be code building blocks that can have one or more assignments.

The goal of "owner-computes" in our extended sense is to minimize communication cost. The message passing (MP) approach, as exemplified by MPI, follows the rule strictly, which is why it is efficient and scalable, and hence popular in the high-performance commercial software market [12]. However, MP has moved a step too far, which makes it hard to use. MP is "too local" in two ways. First, MP programmers usually have only a "local view" of the distributed data, rather than a global view from which the "original algorithm" (the sequential or PRAM [13] algorithm) is developed [1]. This is because the data reference (e.g., array indexing) is local to a distributed data piece that is owned by a process. Second, moving the locus of computation from one "local process" to another in MP is cumbersome since it usually requires artificial constructs and synchronization. The inconsistency in data view and the additional programming required to support the transfer of computation locus can cause the MP implementation to look dramatically different from the original algorithm.

Moving computation locus across memory boundaries is unavoidable if we follow the rule of "owner-computes" in distributed memory. This observation is an immediate consequence of the following two basic facts: (1) in order for a computation to be performed on a data item, the data item and the locus of computation need to be together; and (2) the "owners" of distributed data pieces change across memory boundaries. We define *computation mobility* as the ability for the locus of computation to migrate across distributed memories and continue the computation as it meets the required data. This migration is controlled by a programmer either explicitly or implicitly through data distribution. Mobile agents that provide strong mobility [14] are a means to facilitate computation mobility. If the computation mobility in a distributed sequential program is implemented using a mobile agent system, we call this type of computing mobile-agent-based DSC. One subtle but important point is that mobile-agent-based computation mobility does not necessarily mean code has to move. In fact, a careful implementation of the underlying agent infrastructure allows code to either be loaded from a shared disk or, in a non-shared file system, to be sent across the network at most once, irrespective of how many times the locus of computation moves across the network [15]. This is crucial to performance. Strong mobility in our mobile agent system means that execution state but not always code is allowed to migrate.

One immediate application of DSC is to utilize the power of a network of workstations to improve the performance of data-intensive sequential programs without re-programming them. This is based on the observation that under certain circumstances partitioning the data onto different machines and reducing the disk paging overhead by using the collective memory of a network of workstations can result in considerable performance increase, without converting the underlying algorithm to a parallel implementation. One way of utilizing the distributed memories is the "remote memory paging" or "network RamDisk" approach [4, 5], a special case of DSM, in which a process runs on a single machine and accesses data remotely. A major disadvantage of this approach is its nonscalability because the rule of "owner-computes" is clearly violated here. Another approach is to use the mobile-agent-based DSC approach [3]. The data is distributed over the workstations in the network just as in the "remote memory paging" approach. The difference is that rather than having the program run on a single machine and remotely access the data, the computation, using a mobile agent as a vehicle, moves to the data. As computation locus, carried by a mobile agent, moves to each piece of the distributed data, it dynamically becomes the owner of large sized data and therefore the rule of "owner-computes" is always followed.

In this paper, we argue that computation mobility facilitated by mobile agents with strong mobility is not only *a good way* to implement DSC, but also *the right way*. Of course, anything that can be done with mobile agents can be done with MP: after all, at low level mobile agents are ultimately streams of bytes and hence messages. But mobile-agent-based DSC provides a new layer of abstraction that helps to improve programmability in two ways. First, a DSC program implemented as a mobile agent preserves a global view of the problem data through shared variable programming [16]. The distributed data is bridged by mobile agents at the application level. Second, mobile-agent-based DSC eliminates the need of manually adding artificial auxiliary threads, and handling state transfer and synchronization among the real and the auxiliary threads. This observation, that mobile agents simplify the programming task by eliminating the necessity of explicitly maintaining the state of the processes or threads, has been previously articulated elsewhere [17].

The rest of the paper is organized as follows. Section 2 compares, using a simple and abstract example, the different ways mobile agents, MP, and DSM would bring computation locus and data together sequentially in distributed memory. Section 3 describes a real application, Crout factorization, and its different implementations. The last section contains some final remarks.

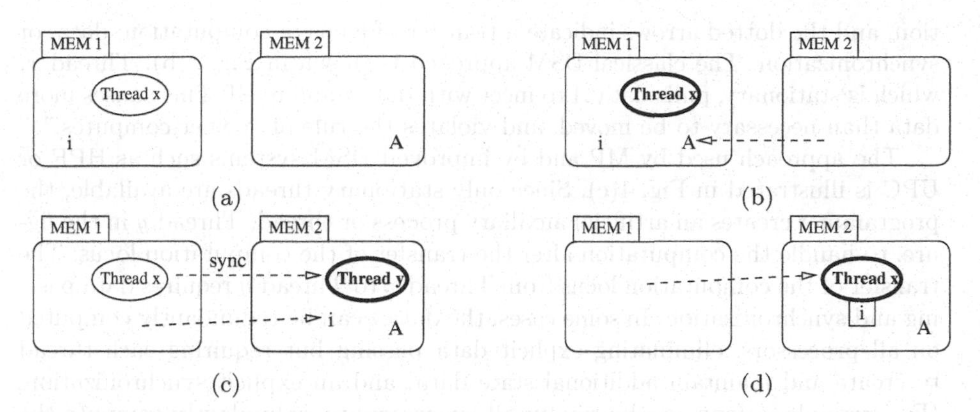

Fig. 1. Data and computation rendezvous. (a) The problem. (b) The classical DSM approach. (c) The MP or improved DSM approach. (d) The mobile agent approach

2 Bringing Together Data and Computation

A sequential program consists of code building blocks, which may be nested or may follow one another sequentially. We use the notation $\mathcal{B}_{\mathrm{T}}(\mathcal{D})$ to denote a code building block of type T that performs its computation on a collection of data pieces denoted by $\mathcal{D}$. The type T of a code building block can be any of the basic programming constructs such as a loop, an **if** statement, a multi-way conditional statement (e.g., a **switch/case** block), or a sequence of statements. The collection of data pieces $\mathcal{D}$ represents the data (input, output, intermediate data) used in the execution of the code building block. When the data pieces in the collection $\mathcal{D}$ are distributed over disjoint memories, $\mathcal{B}_{\mathrm{T}}(\mathcal{D})$ is called a *distributed code building block*, or a *DBlock*. DBlocks are the natural constructs to consider when performing sequential computation using distributed data; indeed, converting code building blocks into DBlocks is the fundamental problem that must be solved when turning a sequential program into a distributed sequential program. As the problem scales up, more code building blocks in a program become DBlocks, and the data in DBlocks are distributed across more memories. Hence the ability to handle DBlocks in an efficient and consistent way is the key to scalable distributed sequential computing.

Consider a DBlock $\mathcal{B}_{\mathrm{T}}(i, A)$, where i is a relatively small data piece and A is large. A and i could be from different variables (e.g., if i is the data stored in a loop index variable and A is a matrix or a portion thereof), or they could be from the same variable (e.g., if i is a column of a matrix and A is a portion of the same matrix that consists of a large number of columns). Suppose that the data pieces i and A are distributed as shown in Fig. 1(a). $\mathcal{B}_{\mathrm{T}}(i, A)$, executed by Thread x, needs to bring together these two data pieces to continue the computation. There are several ways of doing this. These are shown in Fig. 1(b)-(d), where in each case the highlighted thread is the one that continues the computation, and the dotted arrows indicate a transfer of locus of computation, data, or synchronization. The classical DSM approach is shown in Fig. 1(b). Thread x, which is stationary, pulls data A to meet with data i and itself. This causes more data than necessary to be moved, and violates the rule of "owner-computes."

The approach used by MP and by improved DSM systems such as HPF or UPC is illustrated in Fig. 1(c). Since only stationary threads are available, the programmer creates an artificial auxiliary process or thread, Thread y in the figure, to handle the computation after the transfer of the computation locus. The transfer of the computation locus from Thread x to Thread y requires data passing and synchronization. In some cases, the data i can be redundantly computed on all processors, eliminating explicit data passing but requiring each thread to create and maintain additional state data, and an explicit synchronization. (For example, a loop can be run on all processors to redundantly compute the value i, and an artificial mask, in the form of an **if** statement, can be used to identify the owner process of data A. This is the SPMD programming style; the **if** statements are the way SPMD creates auxiliary threads.) In other cases, synchronization can be performed implicitly via explicit data passing. It is worth noting that the **forall** construct in HPF or UPC is designed for data parallel loops (i.e., loops in which iterations can be done concurrently) rather than for constructing a sequential loop that spans distributed memories.

The mobile agent approach to implementing a DBlock is shown in Fig. 1(d). The thread performing the computation "bundles" the data i into a local agent variable, which it then carries to MEM 2 to perform the computation. This is not only efficient (because the agent thread becomes the owner of the large data A before it computes, without moving this data) but also natural to a programmer, because the rendezvous of computation locus and data is seamless.

At the level of abstraction appropriate for algorithm design, a DBlock is no different from a non-distributed block: the data is manipulated by the single computation locus. The requirement that computation and data be brought together is imposed by the distributed environment and the size of the problem being solved, since data is distributed across disjoint memories as the problem scales up. Thus the integrity of the algorithm is best preserved by bringing computation and data together via intra-thread data carrying rather than inter-thread data passing. Mobile agents with strong mobility provide a layer of abstraction that helps this to happen in a seamless and consistent fashion. At a low level, mobile agent code is generally compiled down to stationary threads passing data between them. By programming at the mobile agent level rather than the MP level and letting a compiler do the translation, the programmer avoids a tedious, time-consuming, and error-prone translation task. In the next section, we will provide an example illustrating how programming at this level of abstraction can help with a real-world application.

```
(1)  For j = 1 to N
(2)     For i = 2 to j - 1
(3)        K_ij ← K_ij - Σ_{l=1}^{i-1} K_li K_lj
(4)     End For
(5)     For i = 1 to j - 1
(6)        T ← K_ij
(7)        K_ij ← T / K_ii
(8)        K_jj ← K_jj - T K_ij
(9)     End For
(10) End For
```

(a)

```
(1)    For j = 1 to N
(1.1)     load column j
(2)       For i = 2 to j - 1
(2.1)        hop to column i
(2.2)        load K_ii
(3)          K_ij ← K_ij - Σ_{l=1}^{i-1} K_li K_lj
(4)       End For
(4.1)     hop to column j
(4.2)     unload column j
(5)       For i = 1 to j - 1
(6)          T ← K_ij
(7)          K_ij ← T / K_ii
(8)          K_jj ← K_jj - T K_ij
(9)       End For
(10)   End For
```

(b)

Fig. 2. Pseudocode for Crout factorization. (a) The sequential implementation. (b) The mobile-agent-based DSC implementation

3 Distributed Sequential Crout Factorization

In this section, we describe one example, Crout Factorization, for which performance data is provided in our other paper [3]; here the focus is on why, once we have decided to use DSC, the mobile-agent-based DSC approach is better than MP. In essence, Crout Factorization is a method of factoring a symmetric positive-definite $N \times N$ matrix K into the product of three matrices $K = U^T DU$, where U is an upper triangular matrix with unit diagonal entries and D is a diagonal matrix. Typically, K is a sparse banded matrix, meaning that entries that are more than a fixed distance b, called the *half-bandwidth*, from the diagonal are 0. Figure 2(a) shows the pseudocode for Crout factorization. In line (3), the summation over l corresponds to a dot product of two sub-vectors of columns i and j. These are the two shaded vectors in Fig. 3(a). The computation of column j depends on previously computed columns. The "working set" of matrix entries required to compute column j is shown shaded in Fig. 3(b).

When the size of the working set exceeds the size of the main memory on a single workstation, extensive paging overhead occurs. This thrashing can be eliminated by using the mobile-agent-based DSC implementation of the algorithm [3]. The idea is to split the matrix into pieces, where each piece is a contiguous set of columns. The size of the piece is chosen so that each piece can fit into the main memory of one workstation. The algorithm runs on P workstations, where P is the number of pieces that comprise K. Figure 3(c) shows an

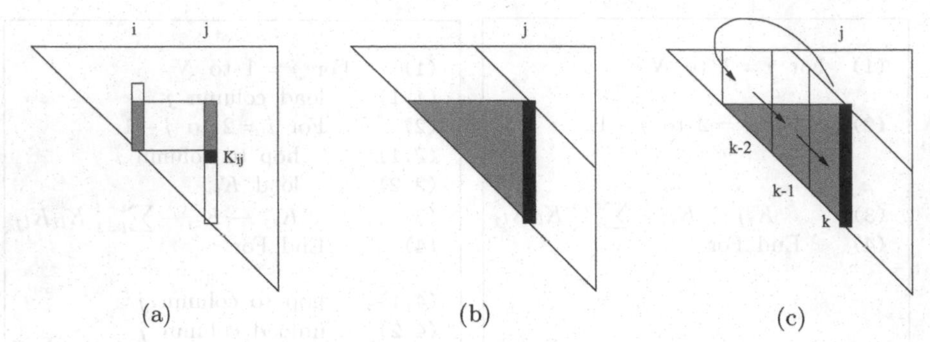

Fig. 3. Crout factorization: (a) Computing of K_{ij} requires the dot product of the two shaded vectors. (b) Working set for column j. (c) Working set decomposition

example for which the working set is subdivided into three pieces. The arrows indicate how an agent, carrying column j which it is computing, would move through the pieces of the working set.

When the size of the working set, which is problem dependent, is too large for a single workstation, pulling the entire working set to a single stationary process, as done in the "remote memory paging" or DSM approach, would not only require much more data to be transferred, but also cause "remote memory thrashing" instead of "local disk thrashing," if "least recently used" is the underlying protocol for handling paging. This is because the data access pattern, shown with arrows in Fig. 3(c), is such that columns are paged out of the (local) main memory right before they are going to be used.

The mobile-agent-based DSC implementation of Crout factorization is shown in Fig. 2(b). The only difference between this code and the sequential code is that two hop statements and three load/unload statements are inserted. These statements are navigational annotations telling the computation which node to hop to (given a column index) and what data to load or unload. They do not modify the structure of the existing sequential algorithm; in other words, they preserve algorithmic integrity. Although there are a large number of hops, most of them will be local and hence will be no-ops with negligible cost. The load statements involve copying a single column (at line (1.1)) or a single matrix entry (at line (2.2)) into agent variables. Once the new values of column j have been computed, by the agent visiting the nodes that contain the pieces of the working set, they are unloaded, or copied back into the appropriate location on the node storing column j by the unload statement at line (4.2).

There are two DBlocks that we focus on in our implementation shown in Fig. 2(b). The first one is $\mathcal{B}_{\text{for}}(\{j\}, \{K_{lj}\})$ which includes lines (1)–(10). In this DBlock, the data $\{K_{lj}\}$ is column j of matrix K which is much larger in size than data $\{j\}$ which holds the loop index variable. The variable j is an agent variable, thus the data $\{j\}$ can follow the computation locus and meet column j wherever the column resides. The second DBlock is $\mathcal{B}_{\text{for}}(\{i\}, \{j\}, \{K_{lj}\}, \{K_{ii}\}, \{K_{li}\})$ which consists of lines (2)–(4) (the loop that computes *dot product* of columns j

```
(1)  For k = 1 to P                          (17)        Recv (k, col j)
(2)    If (μ == k) Then                      (18)        For i = I_{k-2} to I_{k-1} - 1
(3)      For j = I_k to I_{k+1} - 1          (19)          K_ij ← K_ij - Σ_{l=1}^{i-1} K_li K_lj
(4)        Send (k - 2, col j)               (20)        End For
(5)        Recv (k - 1, col j, {K_dd})       (21)        Send (k - 1, col j, {K_dd})
(6)        For i = I_k to j - 1              (22)      End For
(7)          K_ij ← K_ij - Σ_{l=1}^{i-1} K_li K_lj
                                             (23)    Else If (μ == k - 1) Then
(8)        End For                           (24)      For m = I_k to I_{k+1} - 1
(9)        For i = 1 to j - 1                (25)        Recv (k - 2, col j, {K_dd})
(10)         T ← K_ij                        (26)        For i = I_{k-1} to I_k - 1
(11)         K_ij ← T / K_ii                 (27)          K_ij ← K_ij - Σ_{l=1}^{i-1} K_li K_lj
(12)         K_jj ← K_jj - T K_ij            (28)        End For
(13)       End For                           (29)        Send (k, col j, {K_dd})
(14)     End For                             (30)      End For
(15)   Else If (μ == k - 2) Then             (31)    End If
(16)     For m = I_k to I_{k+1} - 1          (32)  End For
```

Fig. 4. Pseudocode for Crout factorization using MP or improved DSM

and i). In this DBlock, the data $\{i\}$ is the value of the loop index variable i, an agent variable similar to variable j. The data $\{K_{lj}\}$ is a copy of column j in an agent variable represented by K_{ij} and K_{lj} (at line (3)). The data $\{K_{ii}\}$ is the diagonal entries of K loaded in an agent variable K_{ii} (at line (2.2)). The data $\{K_{li}\}$ is the working set of column j. It is the largest sized data in this DBlock, and is distributed in node variable K_{li}. In the second DBlock, the data pieces $\{i\}$, $\{j\}$, $\{K_{lj}\}$, and $\{K_{ii}\}$ are carried to meet with $\{K_{li}\}$. Lines (5)–(9) (the loop that does *scaling* of column j with diagonal entries $\{K_{ii}\}$) make a code building block that would run on one workstation, for each value of j, with all data local, and hence they do not make a DBlock. In this code building block, K_{ij} and K_{jj} are node variables, and K_{ii} is the same agent variable as the one in line (2.2).

It is of course possible to implement Crout factorization using MP-based DSC. Figure 4 shows the pseudocode. Notice that this pseudocode is only at a high level, and some details, e.g., the boundary cases of k, are left out. In the pseudocode, μ is the process ID, which is defined as the index of a matrix piece that a processor owns. I_k is the index of the first column that piece k owns. And $\{K_{dd}\}$ is a vector of diagonal entries. If we compare the MP implementation shown in Fig. 4, with the original algorithm shown in Fig. 2(a), the differences are considerable. Producing the MP version requires carefully analyzing and explicitly handling the "roles," or states of various processes using artificial **if** masks. These masks artificially cut a DBlock into sub- code building blocks that belong to different processes. These sub-blocks are each code building blocks, but no longer DBlocks, because they only work on data that is local (the received messages are buffered locally). In Fig. 4, code lines (6)–(8), (18)–(20), and (26)–(28) used to belong to the same code building block in the original algorithm (lines (2)–(4) in Fig. 2(a)), or the agent code (lines (2)–(4) in Fig. 2(b)), but they are broken up by MP into different sub-blocks, as the original code building block grows with the problem size to become DBlocks. In contrast, mobile-agent-based DSC implementation handles this block transition seamlessly and consistently through intra-agent data carrying. A local view of distributed data pieces is used in MP, which is reflected by the fact that each process only runs loops over the columns it owns, as it updates only its own data. Both the local data view and broken DBlocks in MP implementation contribute to the change of the high-level structure of the original algorithm, which significantly complicates the task of new code development as well as old code maintenance.

4 Final Remarks

We have demonstrated in this paper that DSC is a natural fit with mobile agents and a very poor fit with MP, which is a lower level approach to distributed computing. As mentioned in the introduction, one reason why DSC is important is that a parallel problem using distributed data can be decomposed into a collection of cooperating subtasks, each implemented using DSC. An example of this is presented in a companion paper [18].

One advantage of DSM is that it supports incremental parallelization of sequential programs [1]. This is because sequential programs can be ported to a DSM system without much efforts. Once this initial porting is complete, the programmer can incrementally re-implement portions in parallel. Our mobile-agent-based distributed computing offers the same advantage for essentially the same reason: because of algorithmic integrity, transforming a sequential program to a DSC program is straightforward.

The introduction of DBlocks in Section 2 gives some insight into the nature of the programming tasks using DSM, MP, or our mobile-agent-based approach. With classical DSM, DBlocks are handled completely transparently as if they were not distributed blocks at all. This is extremely convenient, but it comes at a steep price: scalability is lost because the "owner-computes" rule is violated. In MP or improved DSM, DBlocks are broken into code building blocks belonging to different processes. The "owner-computes" rule is followed, but the original code structure is changed significantly. In our approach, blocks that are not distributed are coded exactly as they are in the original code, while DBlocks are annotated with navigational commands but otherwise preserve their code structure. As the problem size increases, more code building blocks are turned into DBlocks, and more navigational commands are therefore inserted. This provides a natural migration path along which programs can evolve to solve increasingly large problems.

Acknowledgment

The authors wish to thank Prof. Keith Marzullo and the anonymous reviewers for their incisive comments on an earlier draft of this paper.

References

[1] Leopold, C.: Parallel and Distributed Computing: A Survey of Models, Paradigms, and Approaches. John Wiley & Sons, Inc. (2001) 575, 576, 583

[2] Berthou, J. Y., Colombet, L.: Which approach to parallelizing scientific codes—that is the question. Parallel Computing **23** (1997) 165–179 575

[3] Pan, L., Bic, L. F., Dillencourt, M. B.: Distributed sequential computing using mobile code: moving computation to data. In: ICPP2001: 30th International Conference on Parallel Processing, Valencia, Spain, IEEE (2001) 77–86 575, 577, 580

[4] Dramamitos, G., Marktos, E. P.: Adaptive and reliable paging to remote main memory. Journal of Parallel and Distributed Computing **58** (1999) 357–388 575, 577

[5] Flouris, M. D., Markatos, E. P.: The network ramdisk : Using remote memory on heterogeneous NOWs. Cluster Computing **2** (1999) 281–293 575, 577

[6] Tanenbaum, A. S., Van Steen, M.: Distributed Systems Principles and Paradigms. Prentice Hall, Upper Saddle River, NJ 07458 (2002) 575

[7] Carlson, W. W., Draper, J. M., Culler, D. E., Yelick, K., Brooks, E., Warren, K.: Introduction to UPC and language specification. Technical Report CCS-TR-99-157, IDA Center for Computing Sciences (1999) 576

[8] Merlin, J., Hey, A.: An introduction to High Performance Fortran. Scientific Programming **4** (1995) 87–113 576

[9] Schreiber, R. S.: An introduction to HPF. Lecture Notes in Computer Science **1132** (1996) 27–44 576

[10] El-Ghazawi, T., Chauvin, S.: UPC benchmarking issues. In: 2001 International Conference on Parallel Processing (ICPP '01), Valencia, Spain, IEEE (2001) 365–372 576

[11] Rogers, A., Pingali, K.: Compiling for distributed memory architectures. IEEE Transactions on Parallel and Distributed Systems **5** (1994) 281–298 576

[12] Gropp, W. D.: Learning from the success of MPI. Lecture Notes in Computer Science **2228** (2001) 81–92 576

[13] Keller, J., Kebler, C. W., Traff, J. L.: Practical PRAM Programming. Wiley, New York (2000) 576

[14] Fuggetta, A., Picco, G. P., Vigna, G.: Understanding Code Mobility. IEEE Transactions on Software Engineering **24** (1998) 342–361 576

[15] Gendelman, E., Bic, L. F., Dillencourt, M. B.: Fast file access for fast agents. In: MA 2001: 5th International Conference on Mobile Agents, Atlanta, Georgia (2001) 88–102 577

[16] Pan, L., Bic, L. F., Dillencourt, M. B.: Shared variable programming beyond shared memory: Bridging distributed memory with mobile agents. In: IDPT2002: The Sixth International Conference on Integrated Design and Process Technology, Pasadena, CA (2002) 577

[17] Chess, D., Harrison, C., Kershenbaum, A.: Mobile agents: Are they a good idea? In: 2nd Int. Workshop on Mobile Object Systems, Springer LNCS 1222 (1997) 25–47 577

[18] Pan, L., Bic, L. F., Dillencourt, M. B., Huseynov, J. J., Lai, M. K.: Distributed parallel computing using navigational programming: Orchestrating computations around data. In: PDCS2002: 2002 International Conference on Parallel and Distributed Computing and Systems, Cambridge, MA (2002) 583

Using Dataflow Based Context for Accurate Branch Prediction

Renju Thomas[1] and Manoj Franklin[2]

[1] ECE Department, University of Maryland
College Park, MD 20742
renju@eng.umd.edu

[2] ECE Department and UMIACS, University of Maryland
College Park, MD 20742
manoj@eng.umd.edu

Abstract. Contexts formed only from the outcomes of the last several instances of a static branch instruction or that of the last several dynamic branches do not always encapsulate all of the information required for correct prediction of the branch. Complex interactions between data flow and control flow change the context in ways that result in predictability loss for a significant number of dynamic branches. For improving the prediction accuracy, we use contexts derived from the predictable portions of the data flow graph. That is, the predictability of hard-to-predict branches can be improved by taking advantage of the predictability of the easy-to-predict instructions that precede it in the data flow graph. We propose and investigate a run-time scheme for producing such an improved context from the predicted values of preceding instructions. We also propose a novel branch predictor that uses *dynamic dataflow-inherited speculative context (DDISC)* for prediction. Simulation results verify that the use of dataflow-based contexts yields significant reduction in branch mispredictions, ranging up to 40%. This translates to an overall branch prediction accuracy of 89% to 99.5%.

Keywords: Data value based branch prediction, dataflow inherited branch history, dynamic branch prediction, speculative execution.

1 Introduction

Processor pipelines have been growing deeper and issue widths wider over the years. As this trend continues, branch misprediction penalty will soar because of more and more speculatively executed instructions. Better branch predictors and misprediction recovery mechanisms are essential to cut down the number of cycles lost due to mispredictions. For a branch predictor to provide good prediction accuracy, the history used for predicting a dynamic branch's outcome should encapsulate all of its predictability information. We investigate a scheme to generate such a history context for branches.

Our approach to improving branch predictability is motivated by an earlier study for improving value predictability [7]. In that earlier work, it was found

S. Sahni et al. (Eds.) HiPC 2002, LNCS 2552, pp. 587–596, 2002.

that a value unpredictable instruction can be made more predictable by using a context derived from the closest predicted values in the instruction's backward slice in the dataflow graph. In the proposed branch prediction scheme, the history pattern of a branch-under-prediction is derived from the closest predictable values in the branch's backward slice in the dataflow graph in a conceptually similar fashion. Thus, our methodology of providing the "right history pattern" involves correlating the branches-under-prediction to their closest value-predictable producer instructions along with the operations applied on this set of predictable values enveloping the dynamic dataflow graph (DFG).

A data value predictor can reveal points in the data flow path leading to the branch-under-prediction, where such closest predictable instructions can be identified to form the branch's context. We use a predictor called *dynamic dataflow-inherited speculative context (DDISC) predictor* that uses the dataflow-inherited context for predicting branches. Such a context also adapts more closely, tracking the "actual context" in which the branch to be predicted is in. The DDISC predictor can provide the prediction earlier than branch execution by saving the dataflow latency and most of the fixed pipeline latency.

Mainly two classes of branch predictors have been proposed so far - control flow based and data value based. In control flow based branch predictors, the global or local control flow history of a branch is used for its prediction. Data values have been previously used to improve branch prediction. Heil, *et al* [3] proposed a branch prediction scheme that derives the context for prediction from the closest previous instance of the branch-under-prediction with known operands. Notice that this scheme does not directly use dataflow information relevant to the branch-under-prediction. In contrast, Gonzalez, *et al* [2] used data value prediction to anticipate the predicted branch's source operands which is then used to pre-execute the branch. Notice that the value predictions used in this scheme need not be accurate. The work described in this paper uses predicted values from the predictable portions of the dataflow graph, which may not necessarily be the source operands of the branch to be predicted.

The rest of this paper is organized as follows. Section 2 derives the required background for understanding the relevance of the DDISC predictor, explains the concepts of this novel predictor and discusses its hardware considerations. Section 3 gives the simulation results. We conclude the paper in Section 4.

2 Improving Branch Prediction Accuracy Using DDISC

In this section, we explain the rationale and mechanics of the DDISC predictor. It was seen in the introduction that the new predictor uses data value predictability of instructions belonging to the backward slice of the branch. To utilize data value predictability, value prediction-eligible dynamic instructions are classified into two categories: *predictable instructions* and *unpredictable instructions.* A predictable instruction is one for which the conventional data value predictor provided a prediction because of high confidence, whereas an unpredictable instruction is one for which there was no prediction because of low confidence.

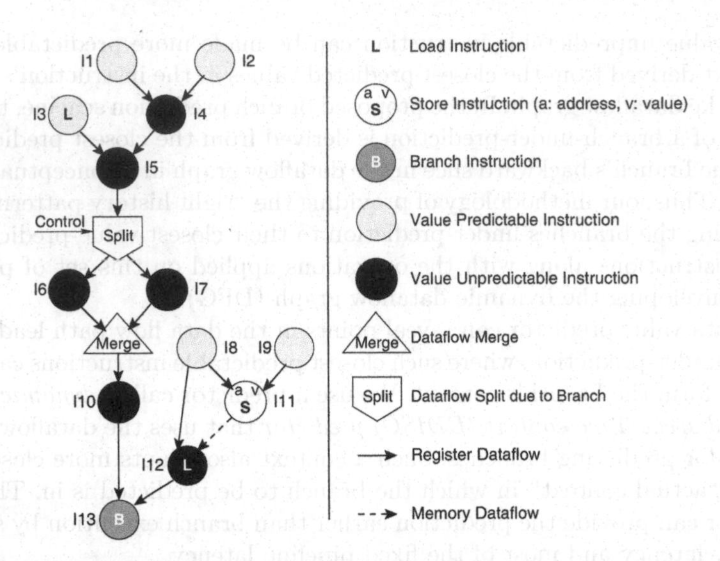

Fig. 1. An example dataflow graph illustrating value predictable instructions, value unpredictable instructions, and branch-under-prediction

Figure 1 shows an example dataflow graph depicting producer-consumer relationships between several instructions, without explicitly representing the control flow. In the figure, dataflow through registers are indicated by straight lines with arrows, and dataflow through memory are indicated by dotted lines with arrows. The figure has one store instruction (I9), marked by an unshaded circle with the letter **S** inside. The figure has the branch-under-prediction (I13), marked by a lightly shaded circle with the letter **B** inside. The unshaded instructions—I1, I2, I3, I8, and I9—are value predictable. Instructions marked by dark circles—I4, I5, I6, I7, I10, and I12—are value unpredictable. Among these, I12 is an unpredictable load instruction, for which the load value is not predictable. We will use this example throughout this paper.

2.1 Dynamic Dataflow-Inherited Speculative Context (DDISC)

The DDISC can be obtained from the DFG as a set of predictable values that affect a branch's outcome from the closest points to the branch-under-prediction along with the identifiers of the instructions that operate on those values. For example, in Figure 1, for branch I13, the predictable values mentioned above are those of instructions {I1, I2, I3, I8, I9}, and the instructions that operate on them are {I4, I5, I6 or I7, I10, I12, I13}. Note that the same set can be obtained by taking the union of the set for I10 and I12 (the operand producers of I13). Thus, we can say that in general, the DDISC of an instruction can be formed as a combination of the context(s) of its source operand producer instructions. Because the DDISC can be a large set, we *compress* the contexts when deriving the context for an instruction from its source operands.

2.2 Systematic Generation of DDISC: The Concept of Signatures

In this section, we propose one way of systematically deriving contexts from the DFG. The context is determined by assigning a *signature* to each node in the dataflow graph. For signature formation, we classify the prediction-eligible dynamic instructions into the following three classes, from the value predictor's point of view: (i) predictable instructions, (ii) unpredictable non-load instructions, and (iii) unpredictable load instructions.

Signature of Predictable Instructions: We define the signature of a predictable instruction to be its value predicted by the value predictor. For instance, in Figure 1, the signatures of I1, I2, I3, I8, and I9 are their predicted values themselves.
Signature of Unpredictable Non-load Instructions: We let the signature of an unpredictable non-load instruction to be inherited from the signatures of its operand producers. When there are multiple operands, the signatures of the producers need to be compressed to keep their lengths manageable; which can cause aliasing. One way of doing this compression is to take the EXOR of the producers' signatures[1]. Further, to incorporate information about the dependence order of nodes in the DFG, we shift this EXORed value left by one bit. To represent the operation performed by the instruction, the PC of the instruction is also EXOR'ed in to form the signature. For instance, the signature of instruction I4 in Figure 1 is formed by taking the EXOR of the signatures of its operand producers, I1 and I2, shifting left by one bit and EXORing the PC of I4.
Signature of Unpredictable Load Instructions: For each unpredictable load instruction, we can think of an immediately preceding store instruction that wrote the value into the same memory location (provided the load is not loading a program input data value). The signature of the load is inherited from the signature of the producer of the register value of this store instruction. For instance, the instruction I12 in Figure 1 is an unpredictable load. Note that its signature is not related to the signature of I8, its address register producer instruction, contrary to what might seem to be from a first look at Figure 1. Instead, I12's signature is inherited from that of I11's store value producer, I9. However, when a load does not have a corresponding store, its signature is inherited from the signature of the address register producer instruction. The PC of the load instruction is also EXORed in to form the signature.

2.3 Hardware Structures for Signature Calculation

The signature to be used for predicting each dynamic branch has to be dynamically generated and recorded in an efficient manner. We therefore need an efficient hardware structure to store the signatures of past instructions. The main issues are: (i) how to limit the storage space (i.e., discard signatures when no longer needed), and (ii) how to link the producers and consumers efficiently?

[1] If two of the signatures to be EXORed are the same, then instead of taking their EXOR, one of the identical signatures is taken.

Signature Register File For storing the signatures of register-result producing instructions, and to effectively link register value producers and consumers, we propose to use a *signature register file (SRF)*, which is very similar to a register file. The SRF has as many entries as the number of registers defined in the instruction set architecture. Figure 3 shows the structure of the SRF within a DDISC predictor. Each SRF entry stores the signature of the latest dynamic instruction that will produce its result into the corresponding register name.

Signature Memory Buffer As discussed in Section 2.2, we need a hardware mechanism to act as the interface between stores and loads that have a producer-consumer relation, so as to speculatively communicate the signature values of stores to dependent unpredictable load instructions. The main issue is how to establish the producer-consumer relation between stores and loads when the load is in the pipeline front-end. One of the ways this can be done is using memory renaming. [1] [6] [8]. The hardware structure we use for store-load signature communication is called a *signature memory buffer (SMB)*.

2.4 Signature Generation Algorithm and Example

The formal algorithm for generating the signature for each branch instruction, register-value producing instruction and memory-value producing instruction (i.e., store) is given in Figure 2 for a specific instruction I.

Consider again the example dataflow graph given in Figure 1. For simplicity of explanation, assume that instruction Ix in Figure 1 writes its result into register Rx. Let SRF[x] denote the SRF entry corresponding to register Rx. The signature determination for the example code of Figure 1 is given in Table 1.

```
if (I is a branch) /* Branch if Ry f Rz */
    SOB ← SRF[y] ⊕ SRF[z] ⊕ I's PC;
else if (I is a store) /* Mem[f(Ra)] ← Rx */
    SMB[*] ← SRF[a];
else if (I is predictable) /* Rx ← ... */
    SRF[x] ← Predicted value of I;
else if (I is a non-load) /* Rx ← Ry f Rz */
    SRF[x] ← ((SRF[y] ⊕ SRF[z]) ≪ 1) ⊕ I's PC;
else if (I is a load) /* Rx ← Mem[f(Ra)] */
    SRF[x] ← (SMB[*] ≪ 1) ⊕ I's PC;
```

*: depends on SMB implementation

SOB: Signature of Branch

Fig. 2. Signature Generation Algorithm

Table 1. Illustration of Signature Determination for the Example given in Figure 1

Inst	Type	Update of Signature Register File
I1	P	SRF[1] $\leftarrow$ Predicted value of I1
I2	P	SRF[2] $\leftarrow$ Predicted value of I2
I3	P	SRF[3] $\leftarrow$ Predicted value of I3
I4	U	SRF[4] $\leftarrow$ ((SRF[1] $\oplus$ SRF[2]) $\ll$ 1) $\oplus$ I4's PC
I5	U	SRF[5] $\leftarrow$ ((SRF[3] $\oplus$ SRF[4]) $\ll$ 1) $\oplus$ I5's PC
I6	U	SRF[6] $\leftarrow$ (SRF[5] $\ll$ 1) $\oplus$ I6's PC
I7	U	SRF[7] $\leftarrow$ (SRF[5] $\ll$ 1) $\oplus$ I7's PC
I8	P	SRF[8] $\leftarrow$ Predicted value of I8
I9	P	SRF[9] $\leftarrow$ Predicted value of I9
I10	U	SRF[10] $\leftarrow$ ((SRF[6] or SRF[7]) $\ll$ 1) $\oplus$ I10's PC
I11	S	SMB[*] $\leftarrow$ SRF[9]
I12	UL	SRF[12] $\leftarrow$ (SMB[*] $\ll$ 1) $\oplus$ I12's PC
I13	B	SOB $\leftarrow$ SRF[10] $\oplus$ SRF[12] $\oplus$ I13's PC

P: Predictable instruction, U: Unpredictable non-load instruction,
UL: Unpredictable load instruction, B: Unpredictable branch instruction,
*: Depends on SMB implementation

2.5 DDISC Branch Predictor Organization

Figure 3 shows a block diagram of the DDISC branch predictor. The DDISC table is a table of 2-bit saturating counters. It is indexed by using the *Signature of Branch (SOB)* in folded form. The DDISC table saturating counter updates are done similar to that in a conventional branch predictor at commit time.

Because the DDISC predictor may not be able to predict all the branches that are predictable by the conventional branch predictor, we combine the two into a hybrid predictor to take advantage of their complementary prediction behavior. A table of 2-bit saturating counters is used as the hybrid selector and is indexed using the conventional branch predictor history. Note that complementary prediction behavior can also be captured using more storage efficient hybrids such as the Rare Event Predictor [3].

The DDISC predictor's predictability can be affected by the predictability and prediction accuracy of the data value predictor. The DDISC predictor's predictability can also be affected by other factors such as aliasing in the DDISC table and the signature processing using limited hardware resources.

The additional hardware cost for the DDISC branch predictor includes a data value predictor, a DDISC table, the SRF, and the SMB.

3 Experimental Analysis

In this section, we perform an experimental evaluation of the DDISC branch predictor that we proposed in the previous section. We do not pursue a detailed microarchitectural evaluation in this paper, as this paper focuses on concepts and ideas, and not on bottom-line, configuration-dependent performance numbers.

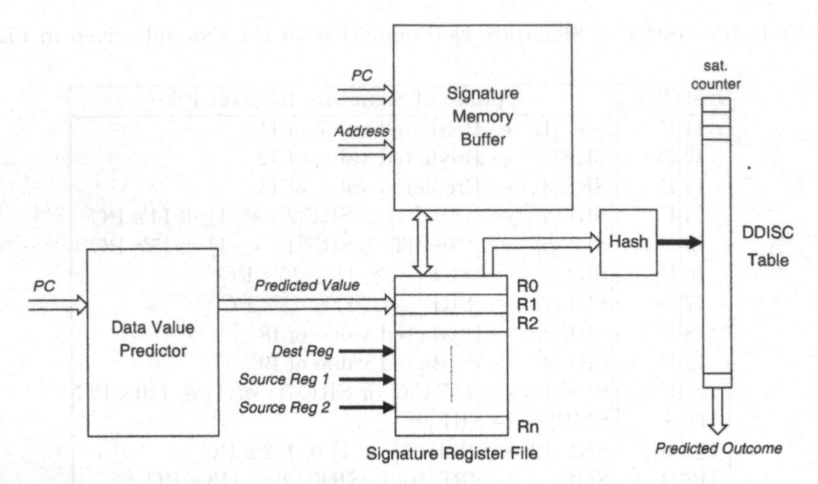

Fig. 3. Hardware Components of a DDISC Branch Predictor

3.1 Experimental Setup and Hardware Configurations

We use simplescalar v3.0 (sim-outorder) using Alpha ISA for the simulations in this paper. The measurements reported in this paper are taken for 13 benchmarks from the SPEC95 and SPEC2000 integer benchmark suites. The benchmarks used are **`bzip, eon, gap, gcc, go, gzip, ijpeg, mcf, parser, perlbmk, twolf, vortex and vpr`**. For all benchmarks except `ijpeg`, the initial 2 billion instructions are skipped and then the measurements are taken for next 500 million instructions. For `ijpeg`, only initial 100 million instructions are skipped (in order not to fast forward all instructions in the benchmark) and then measurements are taken for next 500 million instructions. We predict all conditional branches in the Alpha ISA.

All single register result producing instructions are considered value prediction-eligible except for `compare` instructions. `Compare` instructions are not value predicted because when a `compare` instruction is the value producer of a branch, it also has a binary outcome and hence is equivalent to predicting the outcome of the branch instruction itself using the value predictor.

DDISC Configuration Simulated: We implemented a DDISC branch predictor in our simulation framework. It uses a Signature Register File (SRF) with number of entries equal to the number of architectural registers in the Alpha ISA. The SRF entry fields are kept 32 bits wide, in line with the 32-bit word size of the Alpha ISA. For the Signature Memory Buffer (SMB), we assume perfect store-load linking when such a link exists. This is because an efficient SMB implementation is heavily dependent on the specific microarchitecture chosen, and will have to be accordingly tuned based on the hardware configurations used. The data value predictor for the DDISC branch predictor consists of a hybrid between a 4K-entry last value predictor (LVP) with a 2-bit saturating counter confidence estimator along with each entry and a 4K-entry stride predictor.

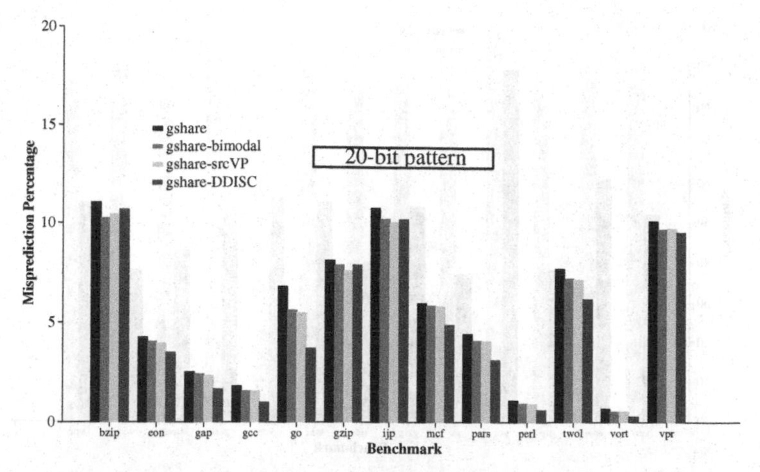

Fig. 4. Comparison of Mispredictions between gshare, gshare-bimodal hybrid, gshare-srcVP hybrid and gshare-DDISC hybrid

The value predictors are indexed using the lower 12 bits of the PC. The last value predictor predicts if the 2-bit saturating counter of the confidence estimator is greater than or equal to 2 and it has priority over the stride predictor's prediction. The stride predictor predicts if the selected entry's state is in "predict".

The DDISC is also compared against another branch predictor called *srcVP*. The *srcVP* predictor also uses predicted data values for branch prediction, but it obtains the value prediction from the closest prediction-eligible producer instructions in the DFG irrespective of their predictability; whereas, the DDISC predictor obtains the value prediction from the nearest predictable producer instructions in the DFG. The *srcVP* predictor uses the same value predictor used by the DDISC.

3.2 Experimental Results

Figure 4 shows the overall branch prediction statistics for a 20-bit gshare predictor, a 20-bit gshare-bimodal hybrid, a 20-bit gshare-srcVP hybrid, and a 20-bit gshare-DDISC hybrid. The Y-axis plots the percentage of branches that are mispredicted. Each benchmark has 4 histogram bars, corresponding to the 4 branch predictors simulated. It is important to note that the three hybrid predictors use roughly 3 times the hardware used by the gshare predictor.

From the results presented in Figure 4, it can be seen that all three hybrids perform better than gshare. The gshare-DDISC hybrid performs better than the gshare-bimodal hybrid for all benchmarks except `bzip2`. The gshare-DDISC hybrid performs better than the gshare-srcVP hybrid for all benchmarks except `bzip2, gzip and ijpeg`. These results are very encouraging.

Figure 5 shows the prediction statistics for the data value predictor used in the DDISC branch predictor. For each benchmark, there are 2 histogram

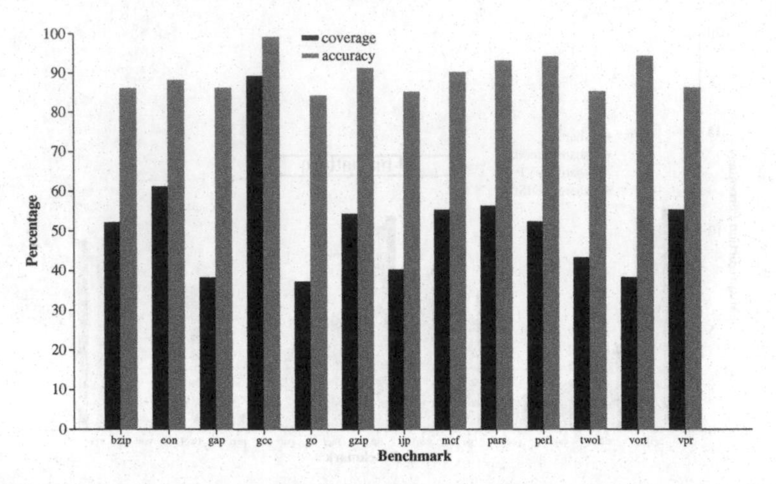

Fig. 5. Statistics of Data Value Predictor used in DDISC Branch Predictor

bars. The first bar indicates *coverage*, which represents the percentage of value-prediction eligible instructions that were actually predicted (because of high confidence). The second bar indicates *prediction accuracy*, which represents the percentage of predicted instructions that were correct. From these results, we can see that the coverage needs to be improved, possibly with better value predictors, such as context-based predictor.

4 Conclusions

Traditional branch predictors typically form the context of a branch-under-prediction from the outcomes of the last several instances of the same static branch or preceding global dynamic branches. In this paper we proposed to use contexts derived from instructions in the backward slice from the branch-under-prediction in the dataflow graph, contoured by the value predictable instructions, along with their predicted values. We proposed and investigated a hardware scheme to derive an improved context from the predicted values of previous instructions. Based on this idea, we proposed a novel predictor called *dynamic dataflow-inherited speculative context (DDISC) based branch predictor*. In order to verify the potential of the DDISC concept, we conducted a set of simulations with a gshare-DDISC hybrid predictor. These experimental results verify that the use of contexts based on dataflow yields significant reduction in branch mispredictions compared to a gshare predictor, ranging up to 40% (for go and gcc). This translates to an overall branch prediction accuracy of 89% to 99.5%. The DDISC predictor provides complementary predictability behavior compared to conventional branch predictors.

Acknowledgements

This work was supported by the U.S. National Science Foundation (NSF) through a CAREER grant (MIP 9702569) and a regular grant (CCR 0073582), and by the Intel Corporation. We are thankful to the reviewers for their helpful and insightful comments.

References

[1] B. Calder and G. Reinman, "A Comparative Survey of Load Speculation Architectures," *Journal of Instruction-Level Parallelism 1*, 2000. 591

[2] J. González and A. González, "Control-Flow Speculation through Value Prediction for Superscalar Processors," *Proc. PACT*, pp. 57-65, 1999. 588

[3] T.H. Heil, Z. Smith, and J.E. Smith, "Improving Branch Predictors by Correlating on Data Values," *Proc. Micro*, pp. 28-37, 1999. 588, 592

[4] M.H. Lipasti and J.P. Shen, "Exceeding the Dataflow Limit via Value Prediction," *Proc. 29th International Symposium on Microarchitecture (MICRO-29)*, pp. 226-237, 1996.

[5] S. McFarling, "Combining Branch Predictors," Technical Report TN-36, Digital Western Research Laboratory, June 1993.

[6] A. Moshovos and G.S. Sohi, "Streamlining Inter-operation Memory Communication via data Dependence Prediction," *Proc. 30th International Symposium on Microarchitecture (MICRO-30)*, 1997. 591

[7] R. Thomas and M. Franklin, "Using Dataflow Based Context for Accurate Prediction," *Proc. International Conference on Parallel Architectures and Compilation Techniques (PACT)*, 2001. 587

[8] G.S. Tyson and T.M. Austin, "Improving the Accuracy and Performance of Memory Communication Through Renaming," *Proc. 30th International Symposium on Microarchitecture (MICRO-30)*, 1997. 591

[9] K. Wang and M. Franklin, "Highly Accurate Data Value Prediction using Hybrid Predictors," *Proc. 30th International Symposium on Microarchitecture (MICRO-30)*, pp. 281-290, 1997.

[10] T.-Y. Yeh and Y.N. Patt, "Alternative Implementations of Two-Level Adaptive Branch Prediction," *Proc. 19th Annual International Symposium on Computer Architecture*, pp. 124-134, 1992.

[11] C.B. Zilles and G.S. Sohi, "Understanding the Backward Slices of Performance Degrading Instructions," *Proc. 27th International Symposium on Computer Architecture*, 2000.

Rehashable BTB: An Adaptive Branch Target Buffer to Improve the Target Predictability of Java Code

Tao Li, Ravi Bhargava and Lizy Kurian John

Laboratory for Computer Architecture
Department of Electrical and Computer Engineering
The University of Texas at Austin, Austin, TX 78712, USA
{tli3,ravib,ljohn}@ece.utexas.edu

Abstract. Java programs are increasing in popularity and prevalence on numerous platforms, including high-performance general-purpose processors. The dynamic characteristics of the Java runtime system present unique performance challenges for several aspects of microarchitecture design. In this work, we focus on the effects of indirect branches on branch target address prediction performance. Runtime bytecode translation, just-in-time compilation, frequent calls to the native interface libraries, and dependence on virtual methods increase the frequency of polymorphic indirect branches. Therefore, accurate target address prediction for indirect branches is very important for Java code. This paper characterizes the indirect branch behavior in Java processing and proposes an adaptive branch target buffer (BTB) design to enhance the predictability of the targets. Our characterization shows that a traditional BTB will frequently mispredict polymorphic indirect branches, significantly deteriorating predictor accuracy in Java processing. Therefore, we propose a Rehashable branch target buffer (R-BTB), which dynamically identifies polymorphic indirect branches and adapts branch target storage to accommodate multiple targets for a branch. The R-BTB improves the target predictability of indirect branches without sacrificing overall target prediction accuracy. Simulations show that the R-BTB eliminates 61% of the indirect branch mispredictions suffered with a traditional BTB for Java programs running in interpreter mode (46% in JIT mode), which leads to a 57% decrease in overall target address misprediction rate (29% in JIT mode). With an equivalent number of entries, the R-BTB also outperforms the previously proposed target cache scheme for a majority of Java programs by adapting to a greater variety of indirect branch behaviors.

1. Introduction

With the "write-once, run-anywhere" philosophy, Java applications are now prevalent on numerous platforms. This popularity has led to an increase in Java processing on general-purpose processors as well. However, the Java runtime system has unique execution characteristics that pose new challenges for high-performance design. One such area is branch target prediction. While many branches are easily predicted, indirect branches that jump to multiple targets are among the most difficult branches to predict with conventional branch target prediction hardware.

The current generation of microprocessors ubiquitously supports speculative execution by predicting the outcomes of the control flow transfer in programs. The trend toward wide issue and deeply pipelined designs increases the penalty for mispredicting control transfer. Therefore, accurate control flow prediction is a critical performance issue on current and future microprocessors. Current processors predict branch targets with a branch target buffer (BTB), which caches the most recently resolved target [5]. Most indirect branches are unconditional

S. Sahni et al. (Eds.): HiPC 2002, LNCS 2552, pp. 597-608, 2002.

jumps and predicting their branch direction is trivial. Therefore, indirect branch prediction performance is largely dependent on the target address prediction accuracy.

The execution of Java programs and the Java Runtime Environment results in more frequent indirect branching compared to other commonly studied applications. Runtime interpretation and just-in-time (JIT) compilation of bytecodes performed by the Java Virtual Machine (JVM) are subject to high indirect branch frequency. Common sources are switch statements and the numerous indirect function calls [3]. Moreover, to facilitate the modularity, flexibility and portability paradigms, many Java native interface routines are coded as dynamically shared libraries. Calls to these routines are implemented as indirect function calls. Finally, as an object oriented programming language, Java implements virtual methods to promote a clean, modular code design style. Virtual subroutines execute indirect branches using virtual method tables, which create additional indirect branches for most Java compilers.

Previous studies have concentrated mainly on the analysis and optimization of indirect branch prediction for SPEC integer and C++ programs [1][2][3]. Table 1 compares the indirect branch frequency found in Java processing with that found in the SPEC CINT95 C benchmarks. The indirect branch frequencies for Java are uniformly high, while only C programs that perform code compilation or interpretation (*gcc*, *li* and *perl*) show high indirect branch frequency. On average, 20% of branches in Java are indirect branches while only 8% are indirect branches in the SPEC CINT95 C benchmarks. In addition, compared with C++ programs [3], the Java workloads studied here execute indirect branches more frequently.

Table 1. Indirect Branch Frequency in Java and C Programs[1]

Benchmarks		% of Indirect Branches in Instruction Stream	
		Interpretation	Just-in-Time Compilation
Java (SPEC JVM98)	*db*	3.0	2.5
	jess	3.3	2.4
	javac	2.6	1.9
	jack	2.5	2.1
	mtrt	2.7	2.0
	compress	4.3	1.6
C (SPEC CINT95)	*go*	0.7	
	compress	0.4	
	m88ksim	0.8	
	gcc	1.1	
	ijpeg	0.2	
	li	2.0	
	perl	2.2	
	vortex	0.9	

The frequency is the percentage of all instructions that are indirect branches, which includes all control transfer instructions. The indirect branch instruction mix ratios in Java programs are presented for runs in interpreter-only mode and JIT mode.

Employing a complete system simulation framework, we further characterize the indirect branches in Java and study their impact on the underlying branch prediction hardware. Our characterization shows that a few critical polymorphic indirect branches can significantly deteriorate the BTB performance during Java execution. For example, the 10 most critical indirect branches are responsible for 75% of indirect branch mispredictions (on average for the studied SPEC JVM98 benchmarks). Therefore, a solution that can effectively handle target prediction for a small number of polymorphic branch sites could improve the BTB performance.

---

1 Sun JDK and SPECInt95 are compiled with MIPSpro C Compiler v7.3 with -O3 option. Simulations are performed on SimOS with IRIX5.3 OS. Initial instructions skipped: 1,000M; Instructions simulated: 200M. Input data set: S100 for SPECjvm98, ref for SPECInt95.

We propose a Rehashable BTB (R-BTB) scheme, which identifies critical polymorphic indirect branches and remembers them in a small separate structure called the Critical Indirect Branch Instruction Buffer (CIBIB). Targets for polymorphic branches promoted to the CIBIB are found by rehashing into the R-BTB target storage. This novel rehashing algorithm allows polymorphic branch targets to use the same resources as monomorphic branches without reducing overall branch target prediction accuracy. Simulations using SPEC JVM98 reveal that the R-BTB eliminates a significant portion of the indirect branch mispredictions versus a traditional BTB while reducing the overall branch target misprediction rate in both interpreter and JIT modes. In addition, the R-BTB outperforms an indirect branch target cache and BTB combination (target cache [1]) with comparable resources for five of the six Java benchmarks studied.

The rest of this paper is organized as follows. Section 2 describes the simulation-based experimental setup and the Java benchmarks. Section 3 provides insight into the indirect branch characteristics of Java execution. Section 4 presents the Rehashable BTB design. Section 5 evaluates the performance of the R-BTB by comparing its misprediction rate with that of a traditional BTB scheme and a combined BTB/target cache scheme. Section 6 discusses the related work. Finally, Section 7 summarizes the conclusions of this paper.

2. Experimental Methodology and Benchmarks

This section describes the simulation-based experimental setup and Java benchmarks used to evaluate the proposed Rehashable BTB scheme. To analyze the entire execution of the JVM and Java workloads, we use the SimOS full-system simulation framework [10] to study Java indirect branch characteristics. The simulation environment uses the IRIX 5.3 operating system. The Sun Java Development Kit ported by Silicon Graphics Inc. provides the Java runtime environment. The SPEC JVM98 [11] suite described in Table 2 is used for this research[2].

Table 2. SPEC JVM98 Benchmarks and their Indirect Branch Statistics

Benchmarks	Description		Indirect Branch Statistics	
			Static Sites	Dynamic Instances
db	Performs multiple database functions on a memory resident database	jit	5,786	2,514,766
		intr	4,116	2,815,831
jess	Java expert shell system based on NASA's CLIPS expert system	jit	7,249	7,496,669
		intr	4,205	11,132,086
javac	The JDK 1.0.2 Java compiler compiling 225,000 lines of code	jit	7,219	5,305,566
		intr	4,266	5,176,207
jack	Parser generator with lexical analysis, early version of what is now JavaCC	jit	7,480	31,348,141
		intr	3,998	31,037,413
mtrt	Dual-threaded raytracer	jit	7,015	24,523,313
		intr	4,097	54,533,384
compress	Modified Lempel-Ziv method (LZW) to compress and decompress large file	jit	5,726	36,698,515
		intr	3,964	40,948,296

We collect the system traces from a heavily instrumented SimOS MXS simulator and then feed them to our back-end simulators and profiling tool sets, which have been used for several of our research studies [6][7]. We simulate each benchmark on the SimOS MXS model until completion, except for the benchmark *compress* running in interpreter-only mode. In this case, we use the first 2,000M instructions. Table 2 reports the number of static and dynamic indirect

[2] We exclude the benchmark *mpegaudio* from our experiments because it failed to execute on the detailed model of SimOS.

branch call sites collected from our complete system simulation. Call returns are excluded because they can be predicted accurately with a return address stack. The execution of these benchmarks in both the JIT compiler (*jit*) and interpreter-only (*intr*) modes is analyzed. Choosing between the JIT and interpreter modes requires complex space and performance tradeoffs. Interpretation is still commonly used in state-of-the-art Java technologies and on resource-constrained platforms, so we present analysis for both scenarios.

3. Characterization of Indirect Branches in Java

In this section, we present our characterization of indirect branches in Java. The following analysis is performed with the JVM running in both interpreter mode and JIT mode.

3.1. Polymorphic vs. Monomorphic Indirect Branches

Indirect branches can be categorized as branches that only jump to one target during the course of execution (monomorphic branches) and those that jump to multiple targets (polymorphic branches). Polymorphic branches are the ones that make indirect branch target prediction difficult. Figure 1 reports the percentage of dynamic indirect branches that are monomorphic (target=1) and polymorphic (targets>=2). The remaining bars illustrate the degree of polymorphism. In the interpreter mode, over 50% of the executed indirect branches are polymorphic, on average. This is primarily due to a switch statement in the bytecode translation routine of the interpreter. In JIT mode, JVM spends no time interpreting bytecodes, but 25% of dynamic indirect branches are still polymorphic.

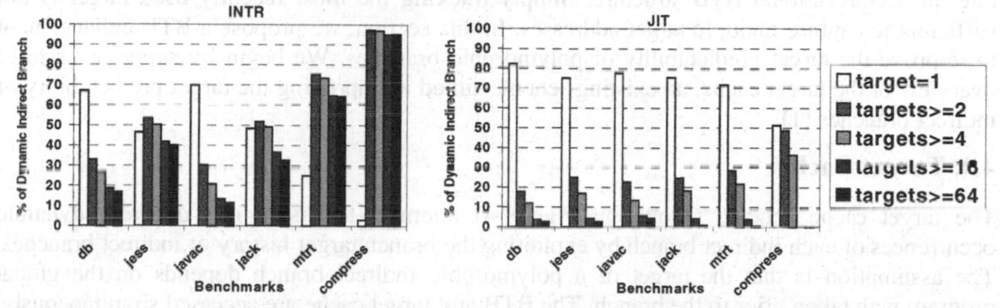

Fig. 1. Dynamic Indirect Branch Target Distribution

Although a large percentage of indirect branches are polymorphic, Figure 2 shows that a much smaller percentage of the static branches are polymorphic, less than 5% of all indirect branch sites. Therefore, a small buffer can capture many of the polymorphic indirect branches. This observation is exploited later in Section 4 when designing the R-BTB.

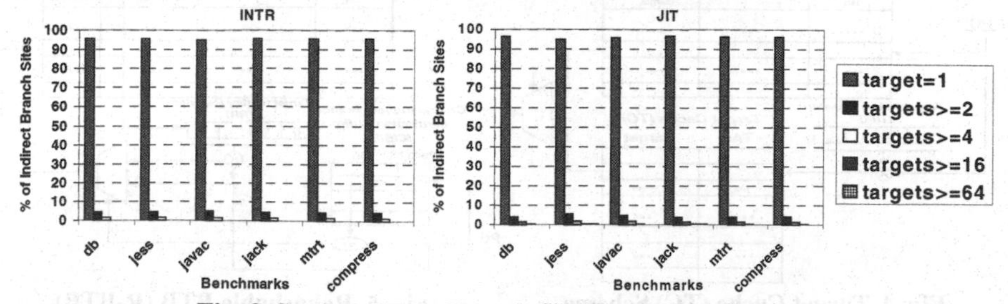

Fig. 2. Static Indirect Branch Site Target Distribution

3.2. The Impact of Polymorphic Indirect Branches

A few critical polymorphic indirect branches can deteriorate target prediction performance significantly. Figure 3 shows the misprediction rate for indirect branches using a traditional

BTB. The top portion of each bar represents the fraction of mispredictions due to the 10 most critical indirect branches. A study of the source code indicates that these polymorphic indirect branches come from code performing bytecode interpretation, calls to the dynamically shared native interface libraries and other JVM management routines. These critical polymorphic indirect branches show highly interleaved target transfer patterns, which cannot be predicted accurately with a conventional BTB structure [7].

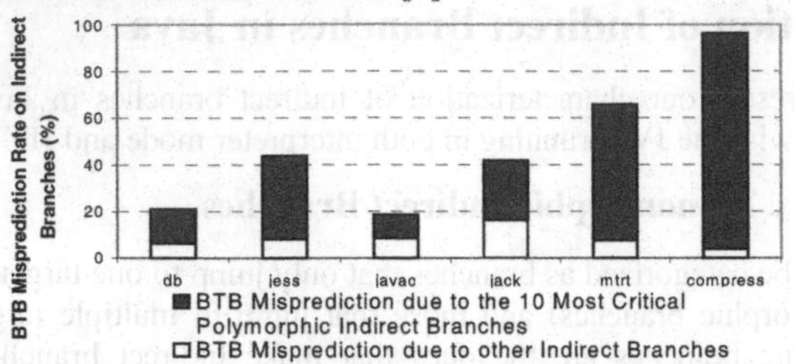

Fig. 3. Impact of Critical Polymorphic Indirect Branches

Benchmarks run in interpreter mode. BTB is 4-way with 2k entries

4. The Rehashable BTB

The previous section reveals that polymorphic indirect branches lead to a high misprediction rate on a conventional BTB structure. Simply tracking the most recently used target is not sufficient to capture multiple target addresses. In this section, we propose a BTB enhancement to improve the target predictability of polymorphic branches. We begin by supplying a brief overview of the target cache, an existing scheme aimed at improving the target predictability of indirect branches [1].

4.1. Target Cache

The target cache scheme (shown in Figure 4) attempts to distinguish different dynamic occurrences of each indirect branch by exploiting the branch target history of indirect branches. The assumption is that the target of a polymorphic indirect branch depends on the global program path taken prior to the branch. The BTB and target cache are accessed simultaneously. If an indirect branch is identified, the target address is taken from the target cache. Otherwise, the BTB produces the branch target.

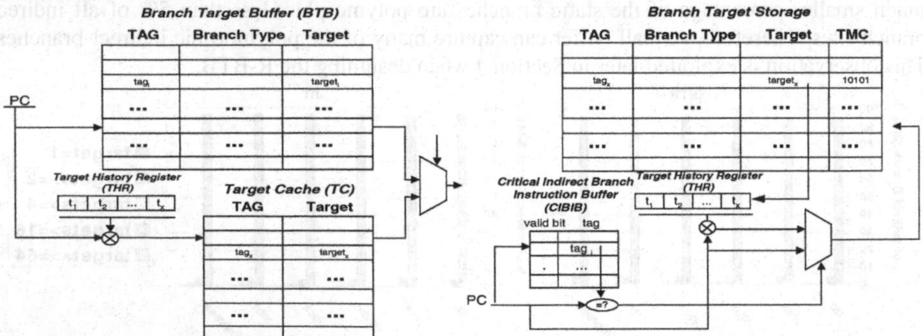

Fig. 4. Target Cache (TC) Scheme **Fig. 5. Rehashable BTB (R-BTB)**

In the target cache scheme, the number of entries allocated to the BTB and to the target cache is determined at design time. Because the indirect branch frequency changes between different programs, it is possible that the target cache resources are not always utilized efficiently. Our characterization of indirect branches in Java suggests that while the number of dynamic polymorphic targets varies widely between programs, the static number of polymorphic indirect branch sites is consistently low.

4.2. Rehashable BTB Design

We propose a Rehashable BTB (shown in Figure 5), which employs a small structure, the Critical Indirect Branch Instruction Buffer (CIBIB), to identify the performance-critical polymorphic indirect branches. Once these critical branches are identified, their targets are rehashed into multiple, separate entries in the R-BTB. Like the target cache, the R-BTB uses a target history register (THR) to collect path history. The path history in the THR is hashed with the critical branch PC to identify an entry in the R-BTB. The primary difference between the R-BTB and the target cache mechanism is that instead of using separate structures for storing the targets of indirect branches and the targets of direct branches, the Rehashable BTB uses the same structure. Therefore, the resources allocated to target prediction can be shared dynamically based on the frequency of polymorphic indirect branches instead of split statically based on a predetermined configuration.

As depicted by Figure 5, a CIBIB entry consists of a tag field for identifying critical branches. The branch target storage is similar to a traditional BTB augmented with a target miss counter (TMC). The TMC is incremented if a branch that hits in the BTB receives an incorrect target prediction. Once the TMC reaches a certain threshold, the branch is promoted to the CIBIB and its entry in the target storage is reclaimed.

Branches that reside in the CIBIB are critical polymorphic indirect branches. The R-BTB is still used to store their targets, but not in the traditional manner. Instead, the THR value is XORed with bits from the branch PC to choose a R-BTB entry. For example, for a 2048-entry, four-way R-BTB, a 11-bit index is generated. The most significant 9 bits are used to choose among the 512 sets and the lower two bits choose among the four ways.

The target history register stores a concatenation of partial target addresses. The THR can be maintained globally or locally. In a global configuration (as illustrated in Figure 5), the THR is updated with the targets of branches contained in the CIBIB, and all critical polymorphic branch sites share the same THR. In a local configuration, separate target patterns are maintained for each polymorphic branch site residing in the CIBIB. In this work, a global THR is used.

4.3. Target Prediction with the R-BTB

This section provides a detailed example of target prediction using the Rehashable BTB. Figure 6 is a corresponding illustration of this process. When a branch target is being predicted, the PC of the branch is sent to the CIBIB. If it hits in the CIBIB, the path history pattern collected in the THR along with the PC is used to generate a R-BTB entry index. If the branch PC misses in the CIBIB, it is used to index the R-BTB (Figure 6.a).

At runtime, the PCs of critical indirect branches with a high target misprediction rate are dynamically identified, removed from the branch target storage, and sent to the CIBIB (Figure 6.b). When a branch PC hits in the CIBIB, the target address is found in the rehashed entry of the R-BTB. This entry is located by XORing the branch PC with the THR value. By using the target history pattern as a hashing input, the multiple targets of critical polymorphic indirect branches are stored in different entries of the R-BTB (Figure 6.c). In this manner, the R-BTB houses targets of both indirect and direct branches.

5. Performance Evaluation of the R-BTB

In this section, we present the performance of a traditional BTB, a target cache scheme, and the R-BTB. The indirect branch misprediction rate and the overall branch prediction rate are compared for the different target prediction mechanisms. To illustrate the benefits of dynamic target storage allocation, several static configurations of the target cache scheme are also analyzed.

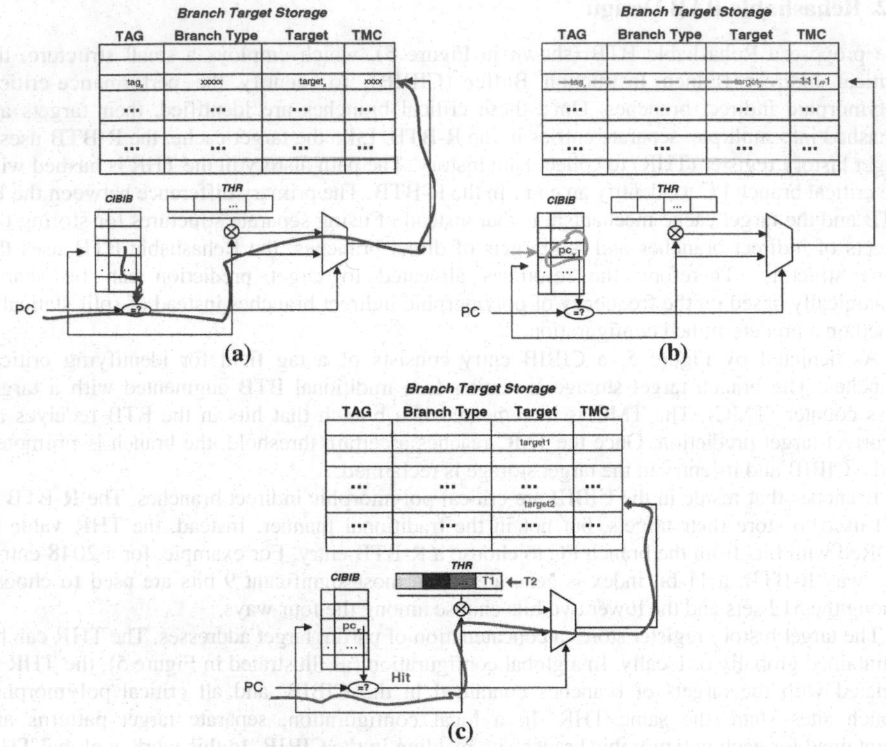

Fig. 6. Target Prediction with R-BTB

5.1 Evaluated Target Predictors

We examined the impact of several R-BTB factors such as THR entry configuration, TMC threshold, CIBIB size, and CIBIB associativity. Based on our experiments, we use a global THR, a TMC with a threshold of 512, and a 16-entry, direct-mapped CIBIB for our performance evaluation. The least significant three bits of the target address (bits 2-4 since bits 0-1 are always zero) are recorded and concatenated in the THR. The simple and small CIBIB configuration is chosen to reduce access latency. The other design parameters are optimized for performance.

All branch target prediction structures are allocated about 2048 entries [1][2] and are four-way set associative, unless specified. The target cache scheme shares resources evenly between a BTB and the target cache. We found that this is the best performing combination for the Java benchmarks, as discussed further in Section 5.4.

The target predictors in this section are also used to predict branch targets for branch types other than indirect branches. In addition to indirect branches, taken conditional branches access the target predictors in our evaluation. Although it depends on the architecture, target prediction is not always necessary for predicting fall-through paths of not-taken branches or for predicting direct branch targets.

5.2 Branch Target Prediction Performance

Tables 3 and 4 present the misprediction rates of indirect branch targets for the evaluated schemes in interpreter and JIT modes. The proposed R-BTB technique improves the misprediction rate for all of the benchmarks compared to a traditional BTB. On average, it reduces the misprediction rate versus a traditional BTB from 47.8% to 18.4% in interpreter mode and from 11.3% to 6.1% in JIT mode. The most drastic improvements are seen for the benchmarks *mtrt* and *compress*.

The R-BTB also improves the performance of indirect branches versus the target cache for five out of the six benchmarks. Only the program compress results in better target prediction for a target cache scheme. While the average performance is the same in interpreter mode, the R-BTB improves the misprediction rate in JIT mode from 11.4% to 6.1%. In fact, in JIT mode the target cache does not always perform better than a traditional BTB.

Table 3. Indirect Branch Target Misprediction Rates (Interpreter Mode)

Structures	Size (# of entries)	Indirect Misprediction Rate (%)					
		db	*jess*	*javac*	*jack*	*mtrt*	*compress*
BTB	2048	21.1	43.8	18.5	41.8	65.7	96.3
BTB + Tagged TC	1024+1024	11.0	20.4	15.0	28.7	21.7	13.2
R-BTB	2048 + 16-entry, DM CIBIB	7.9	20.1	8.6	23.2	21.0	29.7

Table 4. Indirect Branch Target Misprediction Rates (JIT Compilation Mode)

Structures	Size (# of entries)	Indirect Misprediction Rate (%)					
		db	*jess*	*javac*	*jack*	*mtrt*	*compress*
BTB	2048	7.9	11.7	12.3	14.3	13.1	8.8
BTB + Tagged TC	1024+1024	10.8	13.7	12.5	13.8	6.7	1.1
R-BTB	2048 + 16-entry, DM CIBIB	4.6	6.7	6.1	10	3.8	5.6

Table 5. Overall Branch Target Misprediction Rates (Interpreter Mode)

Structures	Size (# of entries)	Overall Branches Misprediction Rate (%)					
		db	*jess*	*javac*	*jack*	*mtrt*	*compress*
BTB	2048	3.2	9.8	3.4	10.6	14.8	35.2
BTB + Tagged TC	1024+1024	2.5	5.5	3.3	8.0	5.0	4.9
R-BTB	2048 + 16-entry, DM CIBIB	1.7	5.3	2.1	7.0	5.5	11.7

Table 6. Overall Branch Target Misprediction Rates (JIT Compilation Mode)

Structures	Size (# of entries)	Overall Branches Misprediction Rate (%)					
		db	*jess*	*javac*	*jack*	*mtrt*	*compress*
BTB	2048	1.5	2.8	2.4	3.1	2.4	2.8
BTB + Tagged TC	1024+1024	2.3	3.8	3.1	3.6	1.5	0.5
R-BTB	2048 + 16-entry, DM CIBIB	1.2	2.2	1.7	2.9	0.9	1.8

Improving indirect branch target prediction performance can only benefit the processor if the overall branch target prediction performance is also improved. Tables 5 and 6 present the overall branch target misprediction rates for interpreter and JIT modes. Versus a traditional BTB, in both interpreter and JIT modes the R-BTB improves overall branch performance for all of the benchmarks, and on average the reduces the overall branch target misprediction rate from 12.8% to 5.6% in interpreter mode and from 2.5% to 1.8% in JIT mode.

The R-BTB also outperforms the target cache scheme. In interpreter mode, the R-BTB produces a better misprediction rate for four out of the six benchmarks. The target cache does much better for the benchmark compress, which leads to an average improvement over the R-BTB, a 4.8% misprediction rate for the target cache versus 5.5% for the R-BTB. However, the R-BTB outperforms the target cache scheme for five out of six benchmarks in JIT mode, and reduces the average overall branch target misprediction rate from 2.4% to 1.8%. Once again, it is interesting to note that the target cache does worse than a traditional BTB for four of the six benchmarks.

5.3 Discussion of Performance Results

The performance results are different depending on the JVM mode of execution. In interpreter mode, 19.5% of all dynamic branches are indirect branches and 11.8% of all branches are polymorphic indirect branches. In this scenario, the target cache predicts indirect branch targets much better than a traditional BTB because it has dedicated half of its resources to handle indirect branches. Despite the reduction in resources for direct branches, the target cache scheme still easily outperforms the BTB overall.

In JIT mode, 10.5% of dynamic branches are indirect branches and only 3.2% are polymorphic branches. In this case, the traditional BTB and target cache have about the same average performance, and the BTB actually does overall branch target prediction better for four of the six benchmarks. There are many fewer indirect branches than in interpreter mode, so the ability to predict direct branches is important. Therefore, the 2048 shared entries of the traditional BTB provide more benefit than the 1024 dedicated entries of the target cache.

The advantage of the R-BTB is that it adapts to both cases. It allocates 2048 entries of target storage for all types of branches like the traditional BTB. However, using the CIBIB, it is able to identify critical polymorphic branches and rehash the multiple targets in the common target storage. Therefore, when the number of indirect branches is low, then the R-BTB behaves like a traditional BTB. When the number of polymorphic branches is high, then the R-BTB behaves in a similar manner to the target cache. On average, this adaptive behavior results in better overall target prediction accuracy, as shown.

The benchmark *compress* is the exception. The target cache does the best job of predicting branch targets for compress. This program has one of the largest percentages of indirect branches, 3.8% of all dynamic branches in JIT mode and 26.25% in interpreter mode. The more important characteristic is that *compress* has the highest degree of polymorphism. While the target cache and R-BTB have similar hashing schemes (based on THR and PC), the R-BTB is sharing the indirect target storage with other branches and this increases the chance for entry pollution or corruption. This is the scenario where smaller, dedicated indirect branch target storage proves beneficial. However, previous work [7] indicates that due to high target locality and a low number of polymorphic branch sites, the BTB corruption caused by rehashing and reuse of an already allocated BTB entry should be low.

5.4 Dynamic Target Storage Allocation versus Static Target Allocation

Previous work with the target cache [1] splits resources differently than in this paper. In previous work, the results are presented using a 2k-entry BTB with an additional target cache of 256, 512, and 1024 entries. However, these sizes are chosen based on the performance of C programs. As shown earlier, the SPEC CINT95 programs have a much lower percentage of indirect branches. In addition, the number of polymorphic branches and the degree of polymorphism are less for C programs versus Java programs. For example, the largest percentage of polymorphic branches (out of all branches) for a SPEC CINT95 program is 3.2% for *perl*, while the average for the JIT and interpreter modes of Java are 3.2% and 11.8% respectively. The R-BTB is better equipped to handle this variation in indirect branch behavior from workload to workload.

Figure 7 further states the case for a dynamic and adaptive scheme. Four different resource partitions are presented for the combined BTB and target cache scheme: 1024+1024 (as in Section 5.2), 2048+512 (as suggested in [1]), 2048+1024, and 2048+2048. In addition to the four-way configuration used earlier, a 16-way associative target cache is presented. A 4096-entry R-BTB is also presented for comparison in addition to the 2048-entry R-BTB from the previous sections.

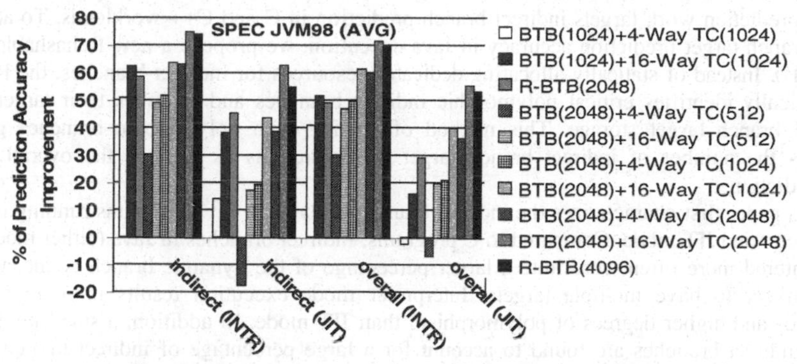

Fig. 7. Impact of Target Cache Resource Partitioning

Performance is with respect to a 4-way, 2048-entry BTB. TC stands for target cache

There are several important points to observe in this figure. The R-BTB does better than four-way target cache configurations with the same amount of target storage. In some cases, the R-BTB predicts branch targets more accurately than a target cache with more target storage entries and/or more associativity. The target cache configuration that is reported to do well on C programs (2048+512) does not do well for Java applications. This highlights the advantage of an R-BTB versus strategies that statically allocate target storage resources.

6. Related Work

Lee and Smith [5] did an early indirect branch prediction study, exclusively focusing on C code. As discussed earlier in this work, Chang et al. proposed several target cache schemes for indirect branches and their performance is evaluated using selected SPEC CINT95 programs. Hsieh et al. [4] studied the performance of Java code running in interpreter mode and observed that microarchitectural mechanisms, such as BTB, are not well utilized. However, their work does not provide an in-depth characterization on Java indirect branches. A related study [12] examined the effectiveness of using path history to predict target addresses of indirect branches to counter the effects of virtual method invocations in Java. The results are presented for small Java programs (e.g., richards and deltablue) and do not apply directly to all JVM execution modes. Recently, Li et al. [7] characterized control flow transfer in Java processing using full-system simulation and SPEC JVM98 benchmarks. However, no hardware optimization was proposed.

Driesen and Hölzle [2] investigated the performance of two-level and hybrid predictors dedicated exclusively to predicting indirect branch targets. Their work optimized for select SPEC CINT95 and C++ applications. However, like the target cache, this requires a static partitioning of target prediction resources. Driesen and Hölzle also proposed a cascaded predictor [3], which dynamically classifies and filters polymorphic indirect branches from a simple first-stage BTB into a second-stage history-based buffer. The primary differences between the cascaded predictor and the R-BTB are: 1) the R-BTB has a more strict filtering criteria for determining important polymorphic branches (512 misses versus one), and 2) the R-BTB stores polymorphic branch targets in the same structure as the monomorphic branch targets. While Driesen and Hölzle suggest using both of these mechanisms for indirect branches only, they could also be used for any type of target prediction.

7. Conclusion

Java execution results in more frequent execution of polymorphic indirect branches due to the nature of the language and the underlying runtime system. A traditional branch target buffer (BTB) is not equipped to predict multiple targets for one static branch, while previous indirect target prediction work targets indirect branch prediction in C and C++ workloads. To achieve high branch target prediction accuracy in Java execution, we propose a new Rehashable BTB (R-BTB). Instead of statically allocating dedicated resources for indirect branches, the R-BTB dynamically identifies critical polymorphic indirect branches and rehashes their targets into unified branch target storage. This method of dealing with polymorphic branches greatly reduces the number of indirect branch target mispredictions as well as the overall target misprediction.

This paper first characterizes the indirect branch behavior in Java programs running in both interpreter and JIT mode. Compared to C programs, indirect branches in Java (either mode) are encountered more often, constitute a larger percentage of the dynamic branch count, and are more likely to have multiple targets. Interpreter mode execution results in more indirect branches and higher degrees of polymorphism than JIT mode. In addition, a small number of static indirect branches are found to account for a large percentage of indirect branch target mispredictions. For example, the 10 most critical polymorphic branches cause about three-fourths of the indirect branch target mispredictions during Java execution.

The R-BTB copes with this behavior by identifying polymorphic branches that cause frequent mispredictions and rehashing their multiple targets into unified target storage. This is accomplished by augmenting a traditional target storage structure with a target history register for hashing, target miss counters to identify critical branches, and a small Critical Indirect Branch Instruction Buffer to store the critical polymorphic. The novelty of this scheme versus other indirect branch target prediction schemes is that it does not split target storage resources between indirect branches and direct branches. Instead, it utilizes one large storage table and rehashes the targets of polymorphic into this table, allowing the resource allocation to be determined dynamically by usage.

The R-BTB eliminates 61% of indirect branch target mispredictions caused by a traditional BTB for Java programs running in interpreter mode and eliminates 46% in JIT mode. Despite the possibility of introducing resource conflicts by rehashing, the overall branch target misprediction rate is improved as well. Compared to a target cache with comparable resources, the R-BTB predicts indirect branch targets more accurately for five out of six benchmarks. The R-BTB improves the overall branch prediction rate for four out of six benchmarks in interpreter mode and five out of six in JIT mode.

Acknowledgement

Ravi Bhargava is currently supported by an Intel Foundation Graduate Fellowship Award. This research is supported in part by the National Science Foundation under grant numbers 0113105 and 9807112, by a State of Texas Advanced Technology Program grant, and by Tivoli, Motorola, Intel, IBM and Microsoft Corporations.

References

[1] P. Y. Chang, E. H. and Y. N. Patt, Target Prediction for Indirect Jumps, In *Proceedings of the 24th International Symposium on Computer Architecture*, pages 274-283, 1997

[2] K. Driesen and U. Hölzle, Accurate Indirect Branch Prediction, In *Proceedings of the 25th Annual International Symposium on Computer Architecture*, pages 167-178, 1998

[3] K. Driesen, and U. Hölzle, The Cascaded Predictor: Economical and Adaptive Branch Target Prediction, In *Proceedings of the 31st Annual ACM/IEEE International Symposium on Microarchitecture*, pages 249-258, 1998

[4] C. H. A. Hsieh, M. T. Conte, T. L. Johnson, J. C. Gyllenhaal and W. W. Hwu, A Study of the Cache and Branch Performance Issues with Running Java on Current Hardware Platforms, In *Proceedings of COMPCON*, pages 211-216, 1997

[5] J. Lee and A. Smith, Branch Prediction Strategies and Branch Target Buffer Design, *IEEE Computer 17(1)*, 1984

[6] T. Li, L. K. John, N.Vijaykrishnan, A. Sivasubramaniam, J. Sabarinathan and A.Murthy, Using Complete System Simulation to Characterize SPECjvm98 Benchmarks, *In Proceedings of ACM International Conference on Supercomputing,* pages 22-33, 2000

[7] T. Li and L. K. John, Understanding Control Flow Transfer and its Predictability in Java Processing, In *Proceedings of the 2001 IEEE International Symposium on Performance Analysis of Systems and Software*, pages. 65-76, 2001

[8] T. Lindholm and F. Yellin, The Java Virtual Machine Specification, Second Edition, Addison Wesley, 1999

[9] R. Radhakrishnan, N. Vijaykrishnan, L. K. John and A. Sivasubramaniam, Architectural Issue in Java Runtime Systems, In *Proceedings of the 6th International Conference on High Performance Computer Architecture*, pages 387-398, 2000

[10] M. Rosenblum, S. A. Herrod, E. Witchel, and A. Gupta, Complete Computer System Simulation: the SimOS Approach, *IEEE Parallel and Distributed Technology: Systems and Applications*, vol.3, no.4, pages 34-43, Winter 1995

[11] SPEC JVM98 Benchmarks, http://www.spec.org/osg/jvm98/

[12] N. Vijaykrishnan and N. Ranganathan, Tuning Branch Predictors to Support Virtual Method Invocation in Java, In *Proceedings of the 5th USENIX Conference of Object-Oriented Technologies and Systems*, pages. 217-228, 1999

Return-Address Prediction in Speculative Multithreaded Environments

Mohamed Zahran[1] and Manoj Franklin[2]

[1] ECE Department, University of Maryland
College Park, MD 20742
mzahran@eng.umd.edu
[2] ECE Department and UMIACS, University of Maryland
College Park, MD 20742
manoj@eng.umd.edu

Abstract. There is a growing interest in the use of speculative multithreading to speed up the execution of sequential programs. In this execution model, threads are extracted from sequential code and are speculatively executed in parallel. This makes it possible to use parallel processing to speed up ordinary applications, which are typically written as sequential programs. This paper has two objectives. The first is to highlight the problems involved in performing accurate return address predictions in speculative multithreaded processors, where many of the subroutine call instructions and return instructions are fetched out of program order. A straightforward application of a return address stack popular scheme for predicting return addresses in single-threaded environments does not work well in such a situation. With out-of-order fetching of call instructions as well as return instructions, pushing and popping of return addresses onto and from the return address stack happen in a somewhat random fashion. This phenomena corrupts the return address stack, resulting in poor prediction accuracy for return addresses. The second objective of the paper is to propose a fixup technique for using the return address stack in speculative multithreaded processors. Our technique involves the use of a distributed return address stack, with facilities for repair when out-of-order pushes and pops happen. Detailed simulation results of the proposed schemes show significant improvements in the predictability of return addresses in a speculative multithreaded environment.

1 Introduction

There has been a growing interest in the use of speculative multithreading (SpMT) to speed up the execution of a single program [1] [4] [5] [9] [10] [12]. The compiler or the hardware extracts threads from a sequential program, and the hardware executes multiple threads in parallel, most likely with the help of multiple processing elements (PEs). Whereas a single-threaded processor can only extract parallelism from a group of adjacent instructions that fit in a dynamic scheduler, a speculative multithreaded processor can extract parallelism from

S. Sahni et al. (Eds.) HiPC 2002, LNCS 2552, pp. 609–619, 2002.

multiple, non-adjacent, regions of a dynamic program. For many non-numeric programs, SpMT may be the only option for exploiting parallelism [13].

Non-numeric programs typically tend to have a noticeable percentage of subroutine calls and returns. In order to obtain good performance for such programs, it is important to perform return address prediction [7]. That is, when a fetch unit encounters a return instruction, it must predict the target of that return, and perform speculative execution along the predicted path, instead of waiting for the return instruction to be executed. It is important to obtain high return address prediction accuracies in speculative multithreaded processors. Otherwise, many of the speculative execution happening in the processor will be along incorrect paths, and the advantages gained by speculative multithreading are lost. The traditional scheme for predicting return addresses is a *return address stack (RAS)* [7][11]. The RAS works based on the last-in first-out (LIFO) nature of subroutine calls and returns. When the fetch unit encounters a subroutine call, it pushes the return address to the top of the RAS; when the fetch unit encounters a return instruction, it pops the topmost entry from the RAS, and uses it as the predicted target of the return instruction. The fundamental assumption in the working of the RAS is that *instructions in the dynamic instruction stream are encountered by the fetch engine in program order.*

The above fundamental premise is violated in the case of speculative multithreaded processors, where multiple threads from a single sequential program are executed in parallel, causing call instructions (as well as return instructions) to be fetched in an order different from the dynamic program order. If all of the active threads share a common RAS, then pushes and pops to the RAS may happen out-of-order, thereby affecting the accuracy of return address predictions. The RAS mechanism, which works very well when accesses are done in correct order (especially with adequate fixup to handle branch mispredictions [6][11]), performs poorly when accesses are done in an incorrect order[1].

On the other hand, if each active thread has a private RAS that maintains only the return addresses of the calls encountered in that thread, then also the prediction accuracy is likely to be poor. This is because, in an SpMT environment, it is quite possible for a subroutine's call and return instructions to be present in different threads. This is likely to result in lower prediction accuracy, unless there is communication between the private RASes and proper repair mechanisms are included.

It is worthwhile to note that some repair mechanisms are beneficial even for using RAS in a single-threaded environment [6][11]. These repair mechanisms are useful for dynamically scheduled single-threaded processors. When several threads simultaneously update the RAS in a random order, these repair techniques will not yield good performance, as we show in this paper.

[1] Notice that this problem can be easily solved for multi-program multithreaded environments such as SMT [15] and Tera [2], by providing a separate RAS for each active thread. The same approach can be used in traditional multiprocessing environments where the parallelly executed threads do not have a sequential order.

This paper investigates return address prediction in SpMT processors. It analyzes different scenarios that lead to incorrect predictions, and investigates a technique to perform accurate return address prediction in SpMT processors.

The rest of this paper is organized as follows. Section 2 illustrates how speculative multithreading impacts a return address stack, and renders it unsuitable for accurate prediction of return addresses. Section 3 presents solutions to handle this problem, both for environments using constrained threads and for environments using unconstrained threads. Section 4 presents an experimental evaluation of the proposed schemes, and Section 5 presents the conclusions.

2 Inadequacy of a Shared Return Address Stack

The conventional RAS [7] was designed with a single-threaded processor in mind. We present two frequently occurring scenarios that corrupt the RAS in an SpMT processor in which all threads access a single RAS. These scenarios illustrate the need to use a distributed RAS, where each thread has its own RAS.

2.1 Constrained Threads

In speculative multithreaded architectures with a single RAS, the PEs can update the RAS in an incorrect order. For simplicity, let us first consider an SpMT environment in which each thread can have at most one call or return instruction. We call such threads *constrained threads*. SpMT processors that use constrained threads of this nature are the multiscalar [12], trace processor [10] and superthreaded processor [14]. Figure 1 shows two constrained dynamic threads[2] T0 and T1 that have been assigned to PE0 and PE1 respectively. Each PE (processing element) is nothing more than a small scale superscalar processor. PEs are connected together in a uni-directional ring [4]. Both threads have one call instruction each. If these two threads are executed in parallel, there is a possibility that the call instruction in T1 is fetched prior to fetching the call instruction in T0. If this happens, the return address of the call in T1 (i.e., B) is pushed to the RAS prior to pushing the return address of the call in T0 (i.e., A). Figure 1 shows this scenario where T0 is executed in PE0 and T1 in PE1.

2.2 Unconstrained Threads

If a single thread is permitted to have multiple call and/or return instructions as in [1][3][16][8], then the above problem gets exacerbated, as indicated in Figure 2. The figure shows two unconstrained dynamic threads, T0 and T1, that have been assigned to PE0 and PE1 respectively. PE 0 encounters its first call instruction and pushes return address `A` onto the RAS. PE 1 subsequently encounters its

[2] By "dynamic thread" we mean the path taken through the thread at run-time. A dynamic thread's instructions may not necessarily be in contiguous locations in the static program. In this paper, by "threads" we mean "dynamic threads" unless otherwise stated

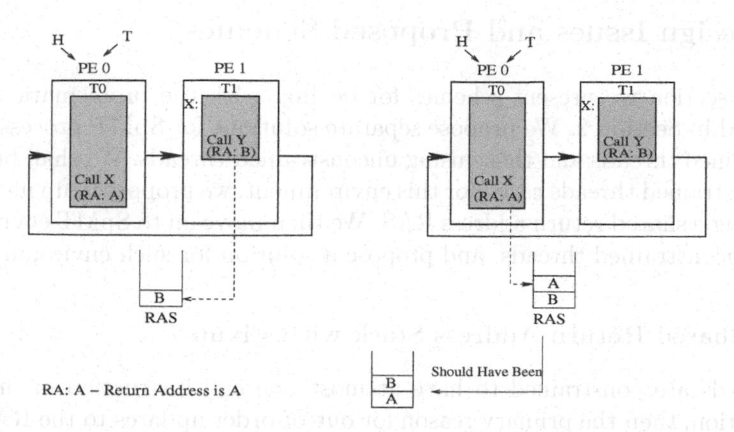

Fig. 1. Problem with Single RAS when using Constrained Threads

call instruction, and pushes address C onto the RAS. Finally, PE 0 encounters the second call instruction in its dynamic thread, and pushes address B. Now the RAS is incorrect, because the two addresses pushed by PE 0 must have been pushed *before* PE 1 pushed address C. The main problem here is that each thread is accessing the RAS, assuming that it is the only running thread. This is the main consequence of the RAS being designed with the single-threaded model in mind. In an SpMT processor, many threads will be concurrently accessing the RAS, frequently causing out-of-order updates to the RAS. If the manner in which the RAS is updated is not controlled, it gets corrupted very quickly, leading to poor performance.

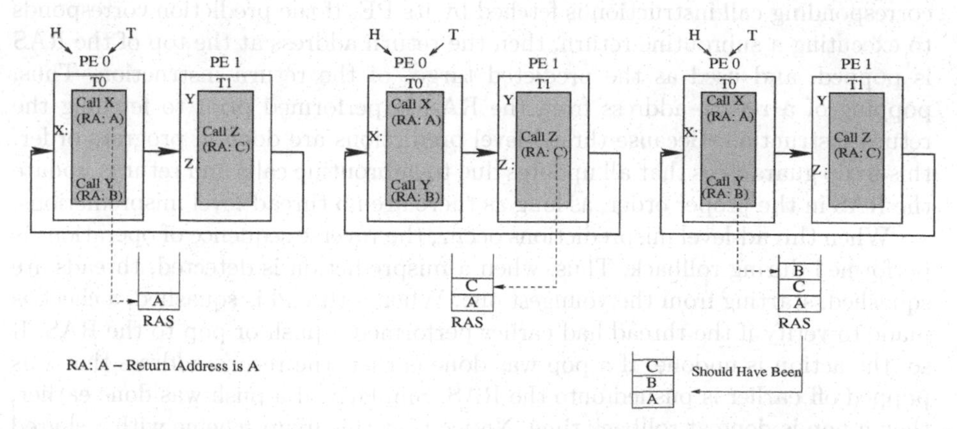

Fig. 2. Problems of Single RAS with Unconstrained Threads

3 Design Issues and Proposed Schemes

In this section we present schemes for dealing with the problematic scenarios depicted in Section 2. We propose separate solutions for SpMT processors using constrained threads and those using unconstrained threads. We shall begin with the constrained threads case. For this environment, we propose fixup mechanisms for a single shared return address RAS. We then move on to SpMT environments with unconstrained threads, and propose a solution for such environments.

3.1 Shared Return Address Stack with Fixup

If threads are constrained to have at most one call instruction or one return instruction, then the primary reason for out-of-order updates to the RAS is that some threads may have their call/return instruction fetched earlier than that of their predecessors. This can happen either because the thread is much shorter than its predecessors or because its call/return instruction is embedded early on in one of the control flow paths in the thread.

The fixup we propose to handle this situation is as follows: if a thread contains a subroutine call, the return address of the call (which is a constant) is noted along with other information pertaining to the thread. This is done at compile time, when threads are formed. Each thread is allowed to have a maximum number of targets. A detailed description on how threads are formed can be found in [4]. At run time, whenever a thread-level prediction is done to determine the successor thread (i.e., picking one of the possible targets of the current thread), a check is done to verify if that prediction amounts to executing a subroutine call or return. If the prediction amounts to executing a call, then the return address associated with that call is pushed onto the RAS, even before the corresponding call instruction is fetched by its PE. If the prediction corresponds to executing a subroutine return, then the return address at the top of the RAS is popped, and used as the predicted target of the return instruction. Thus, popping of a return address from the RAS is performed prior to fetching the return instruction. Because thread-level predictions are done in program order, this fixup guarantees that all updates due to subroutine calls and returns update the RAS in the proper order, as long as there are no thread-level mispredictions.

When thread-level mispredictions occur, the reverse sequence of operations is performed during rollback. Thus, when a misprediction is detected, threads are squashed, starting from the youngest one. When a thread is squashed, a check is made to verify if the thread had earlier performed a push or pop to the RAS. If so, the action is undone. If a pop was done earlier, the return address that was popped off earlier is pushed onto the RAS. Similarly, if a push was done earlier, then a pop is done at rollback time. Notice that this fixup scheme with a shared RAS works only if threads are constrained to have at most a single call/return instruction.

3.2 Distributed Return Address Stack

When a thread is allowed to have multiple call/return instructions, the shared RAS scheme with fixup can perform poorly, as illustrated in Figure 2. To handle this case, we propose a distributed RAS, where each PE has its own individual RAS, and each RAS communicates with its neighbors. The working of the distributed RAS can be summarized as follows:

- When a call instruction is fetched in a PE, its return address is pushed to the corresponding PE's RAS.
- When a return instruction is fetched in a PE, the return address is predicted by popping from the PE's RAS. If its RAS is empty, forwards the request to its predecessor PE.
- When a thread commits, its PE's RAS contents (if any) migrate to the bottom of the successor PE's RAS. It is to be noted that during our experiments, we have never experienced stack overflow. For all the benchmarks a stack of size 40 is more than enough.
- When a thread is squashed, to undo a push, its PE pops the top address from its RAS if it is not empty. If the RAS is empty, the forward the request to the *successor* RAS.
- When a thread is squashed, to undo a pop, a PE pushes onto its *own* RAS the return address that it had previously popped.

To adhere to the concept of the order imposed on the threads assigned to the PEs, the RASes must be connected in the same manner as the PEs. To be able to fulfill the other points, this connection must be bidirectional. In order to undo a pop, each PE must store the address that it had previously popped. In case a PE is executing a thread having multiple return instructions, then the PE stores all of the popped addresses in a *queue* called predicted PC. Figure 3 shows the entire scheme incorporated in an SpMT processor. By distributing the RAS in this manner, we can reduce the number of disruptions to the return address stack.

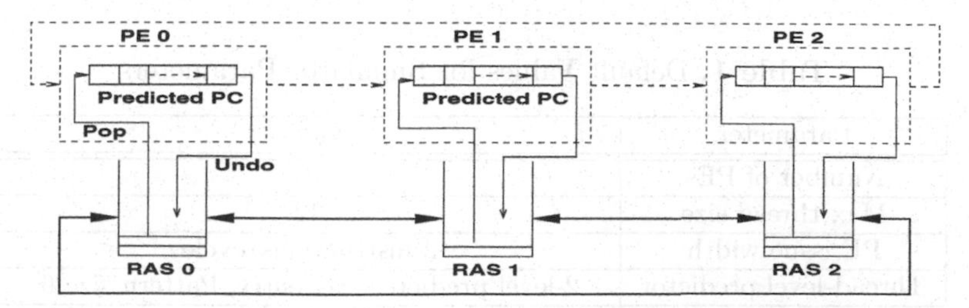

Fig. 3. Distributed Return Address Stack

4 Experimental Evaluation

In this section we present a detailed quantitative evaluation of the presented techniques. Such an evaluation is important to study the performance gain that can be obtained from these techniques and to see how efficient they are with standard benchmarks.

4.1 Experimental Methodology and Setup

Our experimental setup consists of a *detailed cycle-accurate execution-driven* simulator based on the MIPS-I ISA. The simulator accepts executable images of programs, and does cycle-by-cycle simulation; it is not trace driven. The simulator faithfully models all aspects of a speculative multithreaded microarchitecture.

Two different thread types were used: constrained and unconstrained. Table 1 shows the default hardware parameters used in the experiments. The benchmarks used are from SpecInt95 and SpecInt2000. Because of detailed cycle-accurate simulation, the experiments take a long time to run. Each data point is obtained by simulating the benchmark for 100 million instructions. To know the effect of return address prediction mechanisms on the performance of processors in general, we counted the number of return instructions in the first 100 million dynamic instructions. We found that `m88ksim`, `li` and `compress95` have the highest count; `bzip2` and `mcf` have the lowest number of return instructions, thus the impact of the return prediction may not be high for these two benchmarks.

4.2 Results for Shared RAS with Fixup

The first set of experiments simulate a speculative multithreaded architecture running constrained threads. These experiments use a single RAS. Each PE has only a single predict PC register, since this is the highest number of call/return allowed. Figure 4 shows the percentage of return mispredictions without and with fixup. As seen in the figure, the repairing mechanism leads to substantially fewer mispredictions in all the benchmarks. This highlights the impact of fixup

Table 1. Default Values for Simulator Parameters

Parameter	Value
Number of PEs	4
Max thread size	32
PE issue width	2 instructions/cycle
Thread-level predictor	2-level predictor, 1K entry, Pattern size 6
L1 - Icache	16KB, 4 -Way set assoc., 1 cycle access latency
L1 - Dcache	128KB 4-way set assoc., 2 cycle access latency
Functional unit latencies	Int/Branch: 1 cycle; Mul/Div: 10 cycles
RAS size	40 entries
RHT size	1K entries

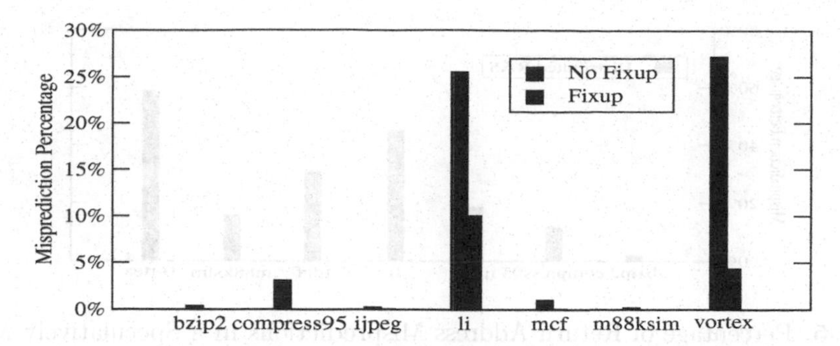

Fig. 4. Percentage of Return Address Mispredictions in Speculative Multithreaded Architecture Using a Shared RAS and Constrained Threads

operations in the RAS of an SpMT processor. Table 2 shows the speedup obtained using repairing mechanism over non-repair. On average, there is a speedup of 10.3%. We note that a small increase in the accuracy of `m88ksim` leads to speedup of 10.5%, and this is due to the fact that `m88ksim` has the largest number of return instructions. The same argument can be said about `compress95` which is on of highest benchmarks in terms of return instructions in the dynamic instructions stream. Furthermore, no speedup is obtained from `mcf` and this is due to the fact that it has the lowest number of return instructions in the 100M dynamic instructions stream among all the benchmarks. It is to be noted that a single return address misprediction can result in a big loss of performance. Because this leads all subsequent PEs to execute threads in a wrong path. This is quite different from branches, because a branch mispredictions may be enclosed in a thread and does not affect other threads.

4.3 Results for Distributed RAS

The next set of experiments deal with unconstrained threads and distributed RAS. In these experiments, we allow each thread to have a total of up to

Table 2. Speedup Obtained by RAS Fixup (Over that Without Fixup) in a Speculative Multithreaded Architecture Using a Shared RAS and Constrained Threads

Benchmark	Speedup
bzip2	3.7%
compress95	24.6%
ijpeg	4.7%
li	11.0%
mcf	0.0%
m88ksim	10.5%
vortex	13.8%

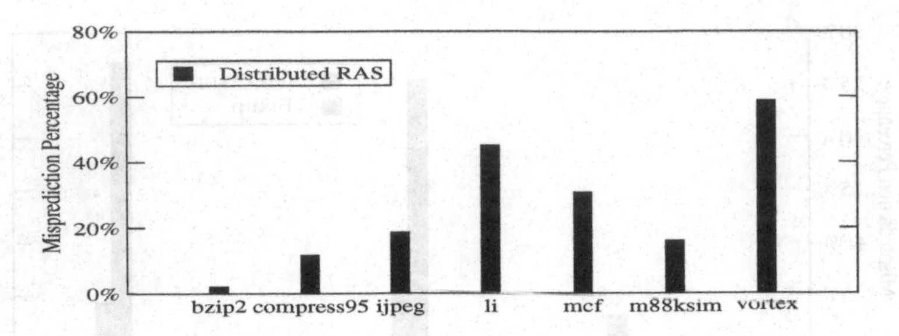

Fig. 5. Percentage of Return Address Mispredictions in a Speculatively Multithreaded Architecture Using a Distributed RAS and Unconstrained Threads

4 calls and returns. It is to be noted that the only constraint on the number of call/return instructions per thread is hardware availability, i.e., the number of entries in the predicted PC queue. Thus, we use a distributed RAS as explained in Section 3, with each PE having a predicted PC queue of length 4. Figure 5 shows the results we obtained. These results are a mixed bag. The mispredictions are low for `bzip2`, `compress95` and `m88ksim`, and high for `li` and `vortex`. As expected the number of mispredictions is higher than that obtained with constrained threads. This results in part from the complex scenarios that arise due to the larger threads. For example, if a PE has a return instruction and its stack is empty, then the PE will try to pop a return address from the predecessor RAS. If the predecessor has not yet executed all of its returns and calls, then the PE will certainly pop a wrong value, because at that point of time the stack is not yet complete and the correct address must be the top of the stack after all the calls.

5 Conclusions

Several researchers and a few vendors are working on speculative multithreaded (SpMT) processors, to harness additional levels of parallelism. When multiple threads are fetched and executed in parallel, return instructions belonging to these threads are very likely to be fetched out of order. This impacts the return address history recorded in a return address stack (RAS), and affects its capability to accurately predict return addresses. This paper demonstrated the problems associated with using a shared RAS.

We also investigated a technique for reducing the number of return address mispredictions in SpMT processors. The technique is for SpMT processors that work with simple threads, where each thread can have at most one call/return instruction. This scheme monitors thread-level predictions to determine if excecution of a thread involves taking a subroutine call or return, and updates the RAS in correct program order. The second technique we proposed involves the use of a distributed RAS mechanism, and is applicable for environments where a thread can have multiple calls and returns embedded in it. We described how the distributed RASes communicate with each other to collectively provide a RAS with fewer disruptions.

The results of our experiments show that for the constrained threads case, our technique provides very high return address prediction accuracies. For the unconstrained threads case, our distributed RAS scheme works well for some benchmarks. Further work is needed to improve the accuracy for the remaining benchmarks.

Acknowledgements

This work was supported by the U.S. National Science Foundation (NSF) through a CAREER grant (MIP 9702569) and a regular grant (CCR 0073582).

References

[1] H. Akkary and M. A. Driscoll. A dynamic multithreading processor. In *Proc. 31st Int'l Symposium on Microarchitecture*, 1998. 609, 611

[2] R. Alverson, D. Callahan, D. Cummings, B. Koblenz, A. Porterfield, and J. B. Smith. The tera computer system. In *Proc. International Conference on Supercomputing*, pages 1–6, 1990. 610

[3] A. Bhowmik and M. Franklin. A general compiler framework for speculative multithreading. In *Proc. ACM Symposium on Parallel Algorithms and Architectures (SPAA)*, August 2002. 611

[4] M. Franklin. *Multiscalar Processors.* Kluwer Academic Publishers, 2002. 609, 611, 613

[5] L. Hammond, B. Nayfeh, and K. Olukotun. A single-chip multiprocessor. *IEEE Computer*, 1997. 609

[6] S. Jourdan, J. Stark T-H. Hsing, and Y. N. Patt. The effects of mispredicted-path execution on branch prediction structures. In *Proc. Int'l Conference on Parallel Architectures and Compilation Techniques (PACT)*, October 1996. 610

[7] D. R. Kaeli and P. G. Emma. Branch history table prediction of moving target branches due to subroutine returns. In *Proc. 18th Int'l Symposium on Computer Architecture*, 1991. 610, 611

[8] V. Krishnan and J. Torrellas. Executing sequential binaries on a clustered multithreaded architecture with speculation support. In *Proc. Int'l Symposium on High Performance Computer Architecture (HPCA)*, 1998. 611

[9] P. Marcuello and A. Gonzalez. Clustered speculative multithreaded processors. In *Proc. Int'l Conference on Supercomputing*, pages 20–25, 1999. 609

[10] E. Rotenberg, Q. Jacobson, Y. Sazeides, and J. E. Smith. Trace processors. In *Proc. 30th Annual Symposium on Microarchitecture (Micro-30)*, pages 24–34, 1997. 609, 611

[11] K. Skadron, P. S. Ahuja, M. Martonosi, and D. W. Clark. Improving prediction for procedure returns with return-address-stack repair mechanisms. In *Proc. 31st Int'l Symposium on Microarchitecture*, pages 259–271, 1998. 610

[12] G. S. Sohi, S. E. Breach, and T. N. Vijaykumar. Multiscalar processors. In *Proc. 22nd Int'l Symposium on Computer Architecture (ISCA22)*, pages 414–425, 1995. 609, 611

[13] J. G. Steffan and T. C. Mowry. The potential for using thread-level data speculation to facilitate automatic parallelization. In *Proc. Int'l Symposium on High Performance Computer Architecture*, pages 2–13, 1998. 610

[14] J-Y. Tsai, J. Huang, C. Amlo, D. J. Lilja, and P-C. Yew. The superthreaded processor architecture. *IEEE Transactions on Computers*, 48(9):881–902, 1999. 611

[15] D. M. Tullsen, S. Eggers, and H. M. Levy. Simultaneous multithreading: Maximizing on-chip parallelism. In *Proc. 22th Int'l Symposium on Computer Architecture*, 1995. 610

[16] M. Zahran and M. Franklin. Hierarchical multi-threading for exploiting parallelism at multiple granularities. In *Proc. 5th Workshop on Multithreaded Execution, Architecture and Compilation (MTEAC-5)*, 2001. 611

HLSpower: Hybrid Statistical Modeling of the Superscalar Power-Performance Design Space

Ravishankar Rao[1], Mark H. Oskin[2], and Frederic T. Chong[1]

[1] University of California at Davis
[2] University of Washington

Abstract. As power densities increase and mobile applications become pervasive, power-aware microprocessor design has become a critical issue. We present HLSpower, a unique tool for power-aware design space exploration of superscalar processors. HLSpower is based upon HLS [OCF00], a tool which used a novel blend of statistical modeling and symbolic execution to accelerate performance modeling more than 100-1000X over conventional cycle-based simulators.
In this paper, we extend the HLS methodology to model energy efficiency of superscalars. We validate our results against the Wattch [BTM00] cycle-based power simulator. While minor second order power effects continue to require detailed cycle-by-cycle simulation, HLSpower is useful for large-scale exploration of the significant power-performance design space. For example, we can show that the instruction cache hit rate and pipeline depth interact with power efficiency in a non-trivial way as they are varied over significant ranges. In particular, we note that, while the IPC of a superscalar increases monotonically with both optimizations, the energy efficiency does not. We highlight the design capabilities by focusing on these non-monotonic contour graphs to demonstrate how HLSpower can help build intuition in power-aware design.

1 Introduction

The dramatic gains in microprocessor performance over the past decade have come at the cost of commensurate increases in power density. This increasing power density threatens to end our winning streak with Moore's Law. Furthermore, power consumption is already a critical issue for mobile platforms, cost-conscious consumer products, and high-end server farms. Battery life is the driving factor in mobile platforms. Consumer products, such as set-top boxes, would like to avoid the costly and noisy addition of a cooling fan. High-end servers farms would like to avoid the construction of special building spaces to support the weight and electrical needs of high-powered cooling units.

We clearly entered the era of power-aware design. Several circuit-level design techniques have been employed to reduce power, including clock-gating, voltage scaling, dynamic clock regulation and asynchronous logic [CYVA96] [MDG93]. However, researchers have only recently begun to study the application and

S. Sahni et al. (Eds.) HiPC 2002, LNCS 2552, pp. 620–629, 2002.

architectural impact on power consumption. While these studies have been productive, they have relied upon the ingenuity of their inventors and brute-force technique of cycle-level simulation [MKG98] [PLL+00] [pac].

In the authors previous work [OCF00] a new approach to performance simulation was introduced. This approach relied upon the combination of statistical models and traditional cycle-by-cycle simulation to achieve rapid but accurate performance estimates. In this paper we extend this simulation technology to model power as well as performance.

To demonstrate the power of this new simulation technology we have undertaken a number of studies of architectural trade-offs. While most reveal only the intuitively obvious result that power-efficiency closely tracks instruction execution efficiency there are some unique exceptions. This paper will present four of these trade-off studies that do not follow the general power-performance trend.

In the next section we present an overview of current architectural power research and simulation technology, followed in Section 3 by a description of HLS, a statistical superscalar performance simulator. Next in Section 4 we describe the extensions to HLS to create HLSpower, a statistical power and performance modeling tool. In Section 5 we validate HLSpower agaisnt Wattch, and demonstrate a couple of non-obvious design trade-offs and their affect on power. Finally we conclude in Section 6.

2 Background

As the importance of power-aware microprocessor design has grown, so has the research in this area. Several prior work has focussed on specific parts or issues of the processor. For instance [BKI98] and [KG97] have focussed on caches, while power benifits of value-based clock gating in integer ALUs has been studied here [BM99].

Tools such as PowerMill [H+95] have been developed in the CAD community to perform circuit-level power estimation.

Early design Stage Power and performance simulator(ESP) [SNT94] does architecture-level estimation specifically for CMOS RISC processors. Given an object code, ESP executes it, and at each clock cycle, determines the active components, and sums up the current for them. The value of these currents is precomputed for different hardware structures.

SimplePower [YVKI00] is a cycle-accurate RT level energy estimation tool. It uses transition sensitive energy models for energy estimation, coupled with analytical energy models for modelling memory system and on-chip buses. However, SimplePower currently models an in-order five stage pipelined datapath, assuming a perfect cache.

Cai-Lim [CL99] and the Wattch [BTM00] are two popular cycle-level simlator. These simulators provide power estimation through cycle-level simulation of a microprocessor. These simulators provide a significant improvement in speed over circuit-level simulation, at the expense of accuracy. However, being cycle-accurate, exploration of large design spaces is impractical, for benchmark applications. In the next section, we describe our statistical modeling techniques which allow performance estimation at significant speedups over cycle-level simulation. Our goal is to apply these techniques to power estimation, enabling the same design space explorations for power-aware design as we have previously achieved for performance-oriented design.

3 HLS

HLS is a hybrid simulator which uses statistical profiles of applications to model instruction and data streams. HLS takes as input a statistical profile of an application, dynamically generates a code base from the profile, and symbolically executes this statistical code on a superscalar microprocessor core. The use of statistical profiles greatly enhances flexibility and speed of simulation. For example, we can smoothly vary dynamic instruction distance or value predictability. This flexibility is only possible with a synthetic, rather than actual, code stream. Furthermore, HLS executes a statistical sample of instructions rather than an entire program, which dramatically decreases simulation time and enables a broader design space exploration which is not practical with conventional simulators. In this section, we provide a brief overview of the HLS approach. Details are available in [OCF00].

The key to the HLS approach lies in its mixture of statistical models and structural simulation. This mixture can be seen in Figure 1, where components of the simulator which use statistical models are shaded in gray. HLS does not

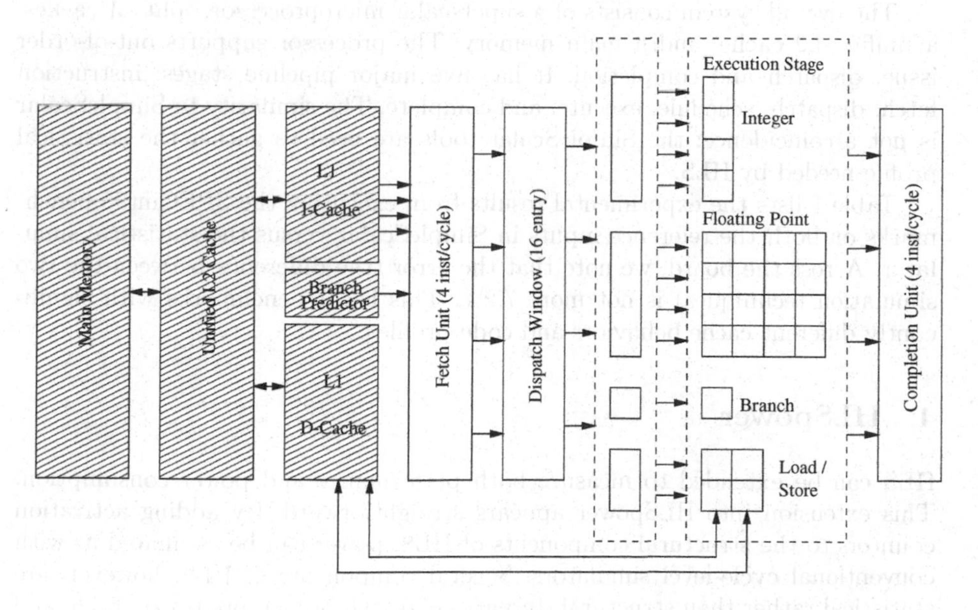

Fig. 1. Simulated Architecture

Table 1. Correlation between SimpleScalar and HLS Simulators for SPECint95 (ref input)

Benchmark	Simple-scale IPC	HLS IPC	HLS IPC σ	Error
perl	1.27	1.32	0.05	4.2%
compress	1.18	1.25	0.06	5.5%
gcc	0.92	0.96	0.03	3.9%
go	0.94	1.01	0.04	6.8%
ijpeg	1.67	1.73	0.06	3.9%
li	1.62	1.50	0.06	7.2%
m88ksim	1.16	1.14	0.03	1.5%
vortex	0.87	0.83	0.03	5.1%

simulate the precise order of instructions or memory accesses in a particular program. Rather, it uses a statistical profile of an application to generate a synthetic instruction stream. Cache behavior is also modeled with a statistical distribution.

Once the instruction stream is generated, HLS symbolically issues and executes instructions much as a conventional simulator does. The structural model of the processor closely follows that of the SimpleScalar tool set [BA97], a widely used processor simulator. This structure, however, is general and configurable enough to allow us to model and validate against a MIPS R10K processor [OCF00].

The overall system consists of a superscalar microprocessor, split L1 caches, a unified L2 cache, and a main memory. The processor supports out-of-order issue, dispatch and completion. It has five major pipeline stages: instruction fetch, dispatch, schedule, execute, and complete. The similarity to SimpleScalar is not a coincidence: the SimpleScalar tools are used to gather the statistical profile needed by HLS.

Table 1 lists the experimental results from executing the SPECint95 benchmarks on both the reference inputs in SimpleScalar versus the statistical simulator. Across the board, we note that the error (the difference between the two simulation techniques) is not more 7.2%. This is for benchmarks with significantly different cache behaviors and code profiles.

4 HLSpower

HLS can be extended to measure both performance and power consumption. This extension into HLSpower appears straightforward. By adding activation counters to the structural components of HLS, power can be estimated as with conventional cycle-level simulators. Several components of HLS, however, are statistical rather than structural. In particular, the branch prediction table and the caches are modeled by their statistical hit rate and accuracy, not their structure. This modeling gave us both speed and flexibility in HLS, but is problematic in power modeling.

Fortunately, we note that the accuracy and hit rate of the branch table and the caches are independent of most other parameters in HLS. We can use the SimpleScalar cycle-level simulator to produce a mapping of statistical parameter to actual structure. Once this mapping is created, it will apply to a wide range of design spaces, preserving the advantages of statistical simulation inherent in HLS.

For instance, we first obtained a plot of the cache hit rate and its size. In order to vary the cache hit rate in our statistical model, we change the cache configuration in our power models. Similarly, in order to change the branch prediction accuracy, a mapping of its size and the prediction rate was obtained. From this plot, to vary the prediction accuracy in our statistical model, we change the size in the power model.

4.1 Validation Results

We apply HLSpower towards modeling the Wattch simulator, a previously published cycle-by-cycle power simulator based upon Simplescalar. HLSpower is fully programmable in terms of queue sizes and inter-pipeline stage bandwidth; however, the baseline architecture was choosen to match the baseline SimpleScalar architecture. The various configuration parameters are summarized in Table 2. Table 3 compares HLSpower to Wattch. We can see that HLSpower can model the processor core within 16% of Wattch and that total power (including caches and branch tables) is within 22%. While these tolerances are not useful for detailed microarchitectural evaluation, we shall see in the next section that HLSpower predicts design space trends which correlate well with Wattch, yet are orders-of-magnitude faster to obtain.

As another level of comparison, we present a comparison of our models with published results of a high-end processor. We compare our model with Wattch and Alpha 21264[GBJ98]. Table 4, gives the results of relative power breakdowns of different hardware structures. As seen from the table, our results track Alpha and Wattch closely. The results presented are the average of different benchmarks.

5 Results

In this section, we give two examples of contour plots that can be generated using HLSpower to facilitate power-aware design space exploration. In these contour plots, we vary two design parameters, one on each axis, and plot average power efficiency. We compare plots generated with HLSpower and ones generated with Wattch for Perl, to provide additional validation of our model.

Power efficiency, in all our graphs has been normalized to the baseline architecture. Thus an efficiency greater than one, only indicates that it is more than that of the baseline architecture.

Table 2. Simulated Architecture configuration

Parameter	Value
Instruction fetch bandwidth	4 inst.
Instruction dispatch bandwidth	4 inst.
Dispatch window size	16 inst.
Integer functional units	4
Floating point functional units	4
Load/Store functional units	2
Branch units	1
Pipeline stages (integer)	1
Pipeline stages (floating point)	4
Pipeline stages (load/store)	2
Pipeline stages (branch)	1
L1 I-cache access time (hit)	1 cycle
L1 D-cache access time (hit)	1 cycle
L2 cache access time (hit)	6 cycles
Main memory access time (latency+transfer)	34 cycles
Fetch unit stall penalty for branch mis-predict	3 cycles
Fetch unit stall penalty for value mis-predict	3 cycles

Table 3. Correlation between Wattch and HLS power for SPECint95 (normalized to Perl)

Benchmark	Wattch		HLS-Power		Error	
	total	core only	total	core only	total	core only
perl	1.0	0.71	0.90	0.67	9.76%	4.4%
compress	1.32	0.91	1.12	0.87	15.15%	4.93%
gcc	0.92	0.66	0.76	0.64	17.39%	3.05%
ijpeg	1.29	0.94	1.10	0.83	14.5%	11.21%
li	1.17	0.81	1.12	0.82	4.27%	-1.61%
m88ksim	1.13	0.82	1.02	0.77	9.74%	5.84%
vortex	0.96	0.67	0.75	0.56	22.2%	16.49%
go	0.85	0.62	0.77	0.59	9.41%	5.61%

Although exact measurements of time were not performed, a simple analysis indiactes the considerable speedup HLSpower achieves over Wattch. The performance simulator HLS achieves about 100-1000x speedup over Simplescalar. Wattch has an overhead of power calculations over Simplescalar, and HLSpower has about the same overhead over HLS. Thus it is fairly straightforward to see that HLSpower is about 100-1000x faster than Wattch.

5.1 I-Cache vs. Branch Prediction

Our first set of graphs compare I-cache hit rate with branch prediction accuracy. A contour plot of the power efficiency, as the two parameters are varied, is gener-

Table 4. Comparison between HLSpower, Wattch and Reported breakdonws for Alpha 21264

Hardware Structure	Alpha 21264	Wattch	HLSpower
Caches	16.1%	15.3%	15.7%
Out-of-Order Issue Logic	19.3%	20.6%	23.3%
Memory Management Unit	8.6%	11.7%	9.2%
Integer Exec. Unit	10.8%	11.0%	6.5%
Floating Point Exec. Unit	10.8%	11.0%	6.7%
Total Clock Power	34.4%	30.4%	38.7%

ated. We plot the core power efficiency and the total power efficiency separately. The cahce hit rate, and the branch prediction accuracy were varied by changing their sizes in the power models, as obtained from the mapping (explained earlier). From Figure 2, as we might expect, power efficiency in the processor core increases as either the hit rate or the branch prediction accuracy increases. We can further see that the trends predicted by HLSpower are qualitatively similar to those found with Wattch.

We can see in Figure 3, however, that total processor efficiency decreases as I-cache hit rate increases, especially if branch prediction accuracy is not high. This non-monotonic behavior may seem initially surprising, but is actually a fairly intuitive tradeoff. If the branch predictor is not performing well, then the power used in the I-cache to fetch unneeded instructions actually decreases power effi-

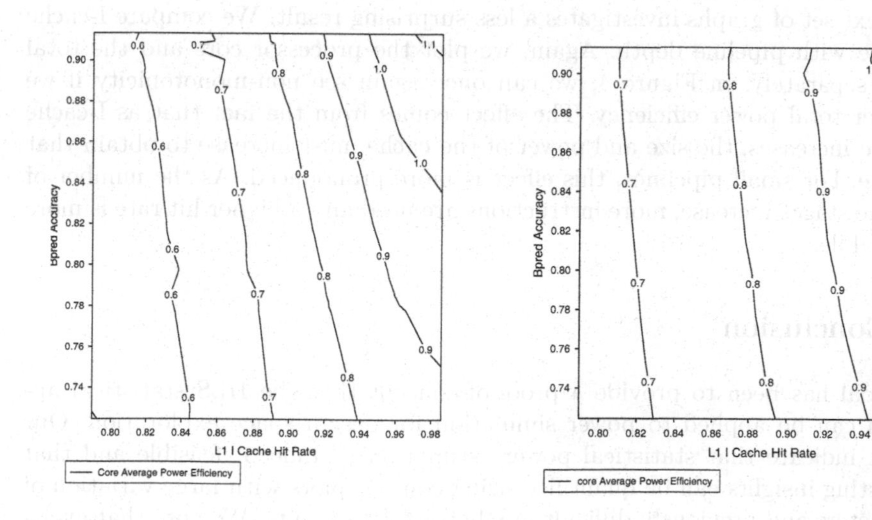

Fig. 2. HLS-Power (left) versus Wattch (right) (processor-core only)

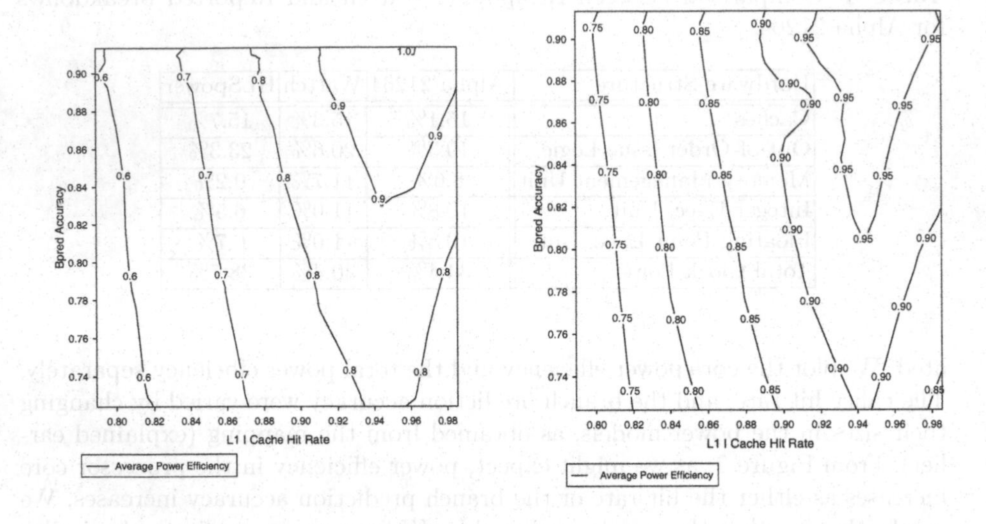

Fig. 3. HLS-Power(left) versus Wattch(right) (caches included)

ciency. Once again, we can see that the Wattch results agree qualitatively with HLSpower.

5.2 I-Cache vs. Pipelining

Our next set of graphs investigates a less surprising result. We compare I-cache hit rate with pipeline depth. Again, we plot the processor core and the total power separately. In Figure 4, we can once again see non-monotonicity if we consider total power efficiency. The effect comes from the fact that as I-cache hit rate increases, the size and power of the cache must increase to obtain that hit rate. For small pipelines, this effect is more pronounced. As the number of pipeline stages increase, more instructions are need and a higher hit rate is more worthwhile.

6 Conclusion

Our goal has been to provide a proof-of-concept that the HLS statistical approach can be applied to power simulation for design space exploration. Our results indicate that statistical power estimation is, indeed, feasible and that interesting insights can be quantified using contour plots with large variation of parameters and previously difficult numbers of data points. We note that cycle-level power simulation continues to improve [DLCD01], and these models can be incorporated into HLSpower in the future.

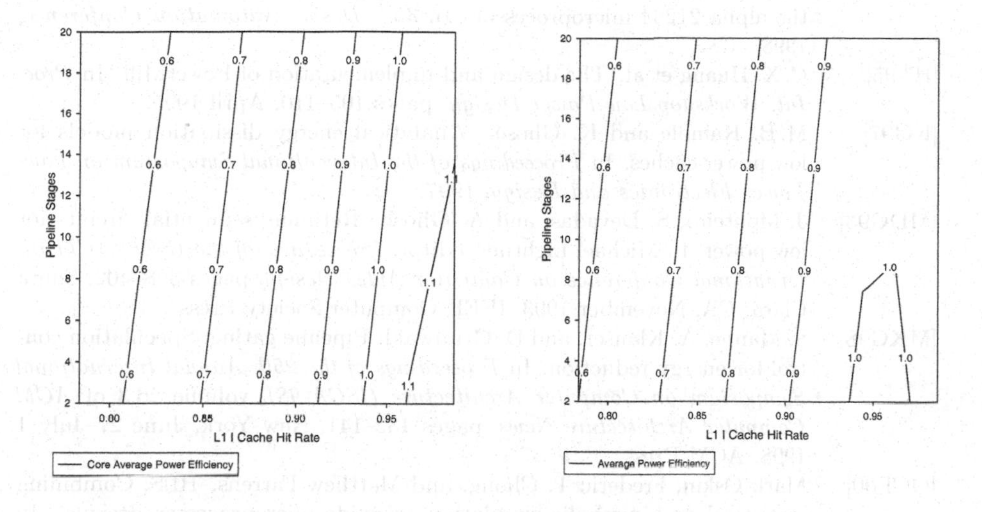

Fig. 4. HLSpower pipeline stages vs. I-cache hit rate – processor core (left) and total power (right)

References

[BA97] D. Burger and T. Austin. The SimpleScalar tool set, v2.0. *Comp Arch News*, 25(3), June 1997. 623

[BKI98] R. I. Bahar, T. Kelliher, and M. Irwin. Power and performance tradeoffs using various caching strategies. In *Proceedings of the International Symposium on Low-Power Electronics and Design*, 1998. 621

[BM99] D. Brooks and M. Martonosi. Dynamically exploiting narrow width operands to improve processor power and performance. In *Proceedings of the 5th International Symposium on High-Pwerformance Computer Architecture*, January 1999. 621

[BTM00] David Brooks, Vivek Tiwari, and Margaret Martonosi. Wattch: A framework for architectural-level power analysis and optimizations. In *Proceedings of the 27th Annual International Symposium on Computer Architecture*, pages 83–94, Vancouver, British Columbia, June 12–14, 2000. IEEE Computer Society and ACM SIGARCH. 620, 621

[CL99] G. Cai and C. H. Lim. Architectural level power/performance optimization and dynamic power estimation. Cool Chips Tutorial colocated with MICRO32, November 1999. 621

[CYVA96] Anantha Chandrakasan, Isabel Yang, Carlin Vieri, and Dimitri Antoniadis. Design considerations and tools for low-voltage digital system design. In *33rd Design Automation Conference (DAC'96)*, pages 113–118, New York, June 1996. Association for Computing Machinery. 620

[DLCD01] Ashutosh Dhodapkar, Chee How Lim, George Cai, and W. Robert Daasch. TEM^{2}P^{2}EST: A thermal enabled multi-model power/performance ESTimator. *Lecture Notes in Computer Science*, 2008:112–125, 2001. 627

[GBJ98] M. Gowan, L. Brio, and D. Jackson. Power considerations in the design of the alpha 21264 microprocessor. In 35th *Design Automation Conference*, 1998. 624

[H+95] C. X. Huang et al. The design and implementation of PowerMill. In *Proc. Int. Workshop Low-Power Design*, pages 105–110, April 1995. 621

[KG97] M. B. Kamble and K. Ghose. Analytical energy dissipation models for low power caches. In *Proceedings of the International Symposium on Low-Power Electronics and Design*, 1997. 621

[MDG93] J. Monteiro, S. Devadas, and A. Ghosh. Retiming sequential circuits for low power. In Michael Lightner, editor, *Proceedings of the IEEE/ACM International Conference on Computer-Aided Design*, pages 398–402, Santa Clara, CA, November 1993. IEEE Computer Society Press. 620

[MKG98] S. Manne, A. Klauser, and D. Grunwald. Pipeline gating: Speculation control for energy reduction. In *Proceedings of the 25th Annual International Symposium on Computer Architecture (ISCA-98)*, volume 26,3 of *ACM Computer Architecture News*, pages 132–141, New York, June 27–July 1 1998. ACM Press. 621

[OCF00] Mark Oskin, Frederic T. Chong, and Matthew Farrens. HLS: Combining statistical and symbolic simulation to guide microprocessor designs. In 27th *Annual International Symposium on Computer Architecture*, pages 71–82, 2000. 620, 621, 622, 623

[pac] proceedings of the second workshop on power-aware computer systems. To appear. 621

[PLL+00] Gi-Ho Park, Kil-Whan Lee, Jang-Soo Lee, Jung-Hoon Lee, Tack-Don Han, Shin-Dug Kim, Moon-Key Lee, Yong-Chun Kim, Seh-Woong Jeong, Hyung-Lae Roh, and Kwang-Yup Lee. Cooperative cache system: A low-power cache structure for embedded processor. In Anonymous, editor, *Cool Chips III: An International Symposium on Low-Power and High-Speed Chips, Kikai-Shinko-Kaikan, Tokyo, Japan April 24–25, 2000*, 2000. 621

[SNT94] T. Sato, M. Nagamatsu, and H. Tago. Power and performance simulator: Esp and its application for 100 mips/w class risc design. In *Low Power Electronics, 1994.Digest of Technical Papers*, pages 46–47, 1994. 621

[YVKI00] W. Ye, N. Vijaykrishnan, M. Kandemir, and M. J. Irwin. The design and use of simplepower: a cycle-accurate energy estimation tool. In *Design Automation Conference 2000*, pages 340–345, 2000. 621

Efficient Decomposition Techniques for FPGAs

Seok-Bum Ko[1] and Jien-Chung Lo[2]

[1] Department of Electrical Engineering, University of Saskatchewan
Saskatoon, SK S7N 5A9, Canada
[2] Department of Electrical and Computer Engineering, University of Rhode Island
Kingston, RI 02881, USA

Abstract. In this paper, we propose AND/XOR-based decomposition methods to implement parity prediction circuits efficiently in field programmable gate arrays (FPGAs). Due to the fixed size of the programmable blocks in an FPGA, decomposing a circuit into sub-circuits with appropriate number of inputs can achieve excellent implementation efficiency. The typical EDA tools deal mainly with AND/OR expressions and therefore are quite inefficient for the parity prediction functions since parity prediction function is inherently based on AND/XOR in nature. The Davio expansion theorem is applied here to the technology mapping method for FPGA. We design three different approaches: (1) Direct Approach, (2) AND/XOR Direct, and (3) Proposed Davio Approach and conduct experiments using MCNC benchmark circuits to demonstrate the effectiveness of Proposed Davio Approach. We formulate the parity prediction circuits for the benchmark circuits. The Proposed Davio Approach is superior to the typical methods for parity prediction circuits in terms of the number of CLBs. The proposed Davio expansion approach, which is based on AND/XOR expressions, is superior to the other common techniques in achieving realization efficiency. The proposed Davio approach only needs 21 CLBs for eight benchmark circuits. It takes only on average 2.75 CLBs or 20 % of the original area.

1 Introduction

Field programmable gate array (FPGA) has been widely used in many applications. Modern FPGAs are of extremely high density and are therefore vulnerable to defects and faults. We consider here the research problem of implementing reliable logic circuits in FPGA. These circuits, with a general structure shown in Figure 1, have the concurrent error detection (CED) capability such that hardware fault can be detected during the normal operations of the multiple output logic circuit. As the problems of parity checker design and the manipulation of the multiple output logic circuits are quite well known, we dedicate our effort in the efficient implementation of the parity prediction circuit.

RAM-based or re-programmable FPGAs have been used for reconfigurable computing and for design prototyping. Due to its unique characteristics, this type of FPGAs has been subject of fault-tolerant studies [1, 2]. Techniques have been proposed [3] to identify faults and defects in such FPGAs. Also, on-line

S. Sahni et al. (Eds.) HiPC 2002, LNCS 2552, pp. 630–639, 2002.

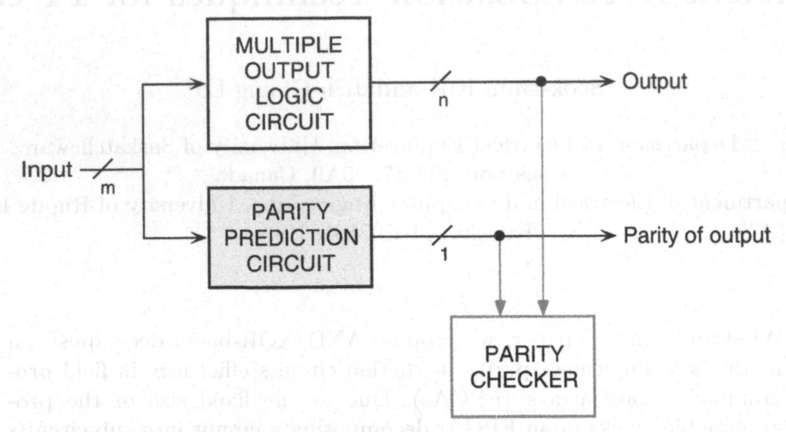

Fig. 1. Parity Prediction Circuits

testing technique [4] has been proposed to exercise the FPGA just before each reconfiguration occurs. Defective blocks are then avoided in the next configuration. Note that this technique differs from the CED technique discussed here.

An FPGA is made from a fixed number of fix-sized programmable or configurable blocks. In most cases, these configurable blocks consist of look-up tables or programmable universal logic gates. Therefore, when implementing a logic function, the realization complexity (the number of configurable blocks being used) has a stronger correlation to the number of inputs to the function than to the Boolean expression complexity. For instance, a configurable logic block (CLB) in a Xilinx's XC4010 [5] has two first stage 4-input look-up tables (LUTs) and a 3-input LUT at the second stage. Therefore, when implementing in XC4010, a 4-input Boolean function with one product term has the same realization complexity as a 4-input Boolean function with 10 product terms since both must fit one 4-input LUT. We remark here that such observation has been incorporated in most FPGA design automation tools. However, our problem is slightly different since we are dealing with the exclusive-OR (XOR) of multiple Boolean functions, i.e. the parity prediction function, as shown in Figure 1.

Previous studies, *e.g.* [6, 7], showed that parity code is very effective in protecting logic circuits. We will concentrate on the techniques for an efficient implementation of parity prediction circuit in FPGA. The key issue here is the effective decomposition of the parity prediction circuit so that an efficient realization in FPGA is possible. Specifically, we investigate the potential benefit of manipulating the parity prediction function in the form of AND/XOR expression. The superiority of such approach is quite intuitive since parity prediction is inherently XOR intensive. However, the existing EDA tools do not handle the AND/XOR expressions well. In fact, a typical EDA tool will first translate an AND/XOR expression into an AND/OR one before further processing.

Therefore, we derive here a process that will handle the AND/XOR expressions directly and independent of the existing EDA tools.

The Davio's (positive and negative) expansion theorems are applied here for the decomposition of the parity prediction function in its AND/XOR form. After examining their applications to the MCNC benchmark circuits, we discover that Davio expansion technique always yields the most efficient implementation results. This is quite reasonable as Davio expansion itself was based on AND/XOR expressions.

This paper is organized into five chapters. We provide an introduction in Chapter 1. The background is introduced in Chapter 2. In the following Chapter 3, we deal with technology mapping for AND/XOR expressions. Chapter 4 presents some benchmark results and discussions. Finally, Chapter 5 draws conclusions.

2 Background

2.1 Observations

The design of parity prediction circuit for any random logic circuit has been explained in several previous works [6, 7, 8]. In this paper, we will consider the following approach. First, we assume that the original circuit is described in VHDL. The VHDL description of the parity prediction circuit is then formed by simply changing the original outputs to the internal signals and then exclusive-ORing these internal signals to produce the parity bit. After logic synthesis process, these internal signals will ideally be minimized and completely removed from the circuit. Such approach ties the parity prediction function directly to the original output functions. Any discrepancy that may arise due to the don't-care conditions is avoided.

The synthesis result is very encouraging when working on several benchmark circuits. For instance, a parity prediction circuit for ***rd53*** needs only one CLB while the original ***rd53*** (5 inputs, 3 outputs, and 32 product terms) circuit will take three CLBs. An even more dramatic example is the benchmark circuit ***bw***. The circuit ***bw*** has 5 inputs, 28 outputs and 87 product terms. The original circuit will need 32 CLBs, while the parity prediction circuit will need only 1 CLB. This is roughly a 3% overhead in terms of the number of CLBs. These results suggested that the internal signals, 28 of them in the case of circuit ***bw***, have indeed been completely removed after the logic synthesis process.

However, when we examined the next benchmark circuit ***5xp1***, such efficiency no longer exists. The benchmark circuit ***5xp1*** has seven inputs and ten outputs. The original circuit will take 18 CLBs. The parity prediction circuit using the above direct method will occupy 33 CLBs if optimized for speed and 28 CLBs if optimized for area. The results are much worse than simple duplication. We believe the main reason of such bad results was due to the lack of AND/XOR expression handling capacity. Therefore, we need to utilize algorithms that handle AND/XOR directly.

The purpose of investigating parity prediction circuits in this paper is mainly to show the effectiveness of the proposed method on XOR intensive functions. While parity is an important code for CED, it is also an important basis for all linear error detecting and correcting codes. In addition, several important circuits in information technology are XOR intensive. As parity prediction circuits are the most representative case of XOR intensive circuits, this study hopes to provide efficient FPGA implementations for more the just parity prediction circuits.

2.2 AND/XOR Expressions and Minimization

Many researchers defined various classes of AND/XOR expressions. Arbitrary product terms combined by XORs are called an Exclusive-or-Sum-of-Products Expression (ESOP). The ESOP is the most general AND/XOR expression. No efficient minimization method is known, and iterative improvement methods are regularly used to obtain near minimal solutions [9]. This is yet another reason that typical EDA tools do not incorporate ESOP type considerations.

Parity prediction circuits can be realized with many fewer gates if XOR gates are available. Such circuits can be derived from AND/XOR circuits. So the minimization of ESOP, which corresponds to the minimization of AND/XORs, is very important. ESOP requires fewer products than SOPs to realize randomly generated functions and symmetric functions [10].

The ESOP represents a Boolean function in an AND/XOR relation rather than the typical AND/OR format. In other words, instead of sum-of-products, the Boolean functions are expressed in exclusive-OR-sum-of-products. Given an n-input Boolean function $f(x_1, x_2, \ldots, x_n)$, a function can be expressed in ESOP as follows:

$$f_l(x_1, x_2, \ldots, x_n) = p_{l,1} \oplus p_{l,2} \oplus \ldots \oplus p_{l,q} = \sum_{t=1}^{q} p_{l,t} \quad (1)$$

where $p_{l,t}$ is the t-th product terms of $f_l(x_1, x_2, \ldots, x_n)$. Then, the parity prediction function of the logic circuit can be written as:

$$P(x_1, x_2, \ldots, x_n) = \sum_{l=1}^{m} f_l(x_1, x_2, \ldots, x_n) = \sum_{l=1}^{m}\sum_{t=1}^{q} p_{l,t}. \quad (2)$$

The most remarkable feature of Equation (2) is that the boundaries between the original output, f_l, have been removed. The parity function is now a function of all the product terms. In [11], Even, Kohavi and Paz proposed an algorithm that minimized single output completely specified equations by the repeated application of four rules that linked product terms. The four rules are Merging:$ab \oplus \bar{a}b = b$, Exclusion:$ab \oplus b = \bar{a}b$, Increase of Order:$ab \oplus a = 1 \oplus a\bar{b}$, and Bridging:$ab \oplus \bar{a}\bar{b} = \bar{a} \oplus b$. Their strategy is to apply the rule which give the greatest benefit first: if any merging is possible then merging is applied, if not then exclusion, if neither of these then increase of order, and if none of these then bridging. We implemented this algorithm and extended it to handle multiple outputs.

There are many reasons that ESOP algorithms are not as popular in commercial EDA tools. As mentioned earlier, the lack of efficient minimization is a big issue. The minimizer algorithm described in Section 2.2 will give a solution in parity prediction circuit but not a guaranteed minimum solution. We note that XOR is not a primitive Boolean operator, such as AND, OR and NOT. Consequently, the hardware realization of XOR gate is not straightforward. In the CMOS realization, 2-input XOR gate needs about twice the area that that of a 2-input AND or an NOR gate. However, such disadvantage in hardware efficiency does not exist in FPGA implementation.

2.3 Overview of Xilinx XC4000

In this section, we will briefly describe the organization of the Xilinx XC 4000 FPGA family. Although our experimental results will base on this FPGA family, extension to other FPGA families, even from different vendors, may be easily obtained.

Xilinx XC4000 [12] consists of an array of CLBs embedded in a configurable interconncet structure and surrounded by configurable I/O blocks. The Xilinx XC4000 family consists of ten members. The family members differ in the number of CLBs, (ranging from 8×8 to 24×24), and I/O blocks, (ranging from 64-192). The typical capacity varies from 2000 to 13000 equivalent gates.

Xilinx XC4000 CLBs mainly consist of two 4-input LUTs, which are called F-LUT and G-LUT respectively, and one 3-input LUT, which is called H-LUT. A K-input LUT is a memory that can implement any Boolean function of K variables. The K inputs are used to address a $2K \times 1$-bit memory that stores the truth table of the Boolean function. Note that one XC4000 CLB, although it can accept 9 distinct inputs, is not equivalent to a 9-input LUT. A 9-input LUT can implement any functions of 9 distinct variables while one XC4000's CLB can implement any functions of up to 5 distinct inputs and some functions of 6 to 9 variables.

3 Technology Mapping for AND/XOR Expressions

Technology mapping is the logic synthesis task that is directly concerned with selecting the circuit elements used to implement the minimized circuit. Previous approaches to technology mapping have focused on using circuit elements from a limited set of simple gates. However, such approaches are inappropriate for complex logic blocks where each logic block can implement a large number of functions [13]. A K-input LUT can implement 2^{2^K} different functions. For values of K greater than 3, the number of different functions becomes too large for conventional technology mapping. Therefore, different approaches to technology mapping are required for LUT-based FPGAs.

An arbitrary logic function $f(x_1, x_2, \ldots, x_n)$ can be expanded as

$$f = \bar{x}_i f_i^0 \oplus x_i f_i^1 \quad (3)$$

$$f = f_i^0 \oplus x_i f_i^2 \quad (4)$$

$$f = f_i^1 \oplus \bar{x}_i f_i^2 \quad (5)$$

where $f_i^0 = f(x_1, x_2, \ldots, x_i = 0, \ldots, x_n)$, $f_i^1 = f(x_1, x_2, \ldots, x_i = 1, \ldots, x_n)$, and $f_i^2 = f_i^0 \oplus f_i^1$. Equations (3), (4) and (5) are called the Shannon expansion, the positive Davio expansion, and the negative Davio expansion, respectively.

As shown in equations (4) and (5), the Davio expansions can be seen as manipulations of the Shannon's expansion utilizing the XOR property. Based on our observation, $f_i^2 = f_i^0 \oplus f_i^1$ is usually no more complex than f_i^0 or f_i^1 itself. Therefore, the Davio expansions constantly lead to solutions with less literal counts. The resulting functions from the Davio expansions are, of course, AND/XOR expressions.

Based on the Davio expansions, we derived the following greedy-type algorithm. This algorithm attempts to decompose the given logic circuit into blocks or sub-functions, each with five or less inputs. This number is due to the Xilinx XC 4000 FPGA organization. For other FPGA families, different number may be used. Figure 2 shows the Davio-Decompose algorithm.

Step 1. If the number of inputs of the given logic circuit, f, is less than or equal to five, then exit the algorithm.
Step 2. If the number of inputs is greater than five, then collect the product terms that will require only the first five inputs on the input list.
Step 3. If such group cannot be found, try a different combination of five inputs from the input list.
Step 4. Repeat Step 3 until either a group of such product terms can be identified or no more possible input combinations available.
Step 5. If a group can be identified that contains product terms,$p_1, p_2, \ldots, p_t$, where $t \geq 1$, then we define $f_t = \sum_{i=1}^{t} = p_i$, where $\sum$ represents XOR-sum operation. If no such group can be found, then $f_t = 0$.
Step 6. Let $f = f_t \oplus f_u$, such that the logic circuit f is decomposed into f_t and f_u. We note that f_t can now be realized in one CLB since it has only five inputs (except, of course, when $f_t = 0$).
Step 7. Done if $f_u = 0$.
Step 8. Perform the Davio expansion on $f_u = 0$ by randomly picking an input variable still in the input list of $f_u = 0$, say x_a. Thus, $f_u = f_{u0} \oplus x_a f_{u2}$. Minimize f_u using minimizer (p) algorithm.
Step 9. Update the input list of f_u after the Davio expansion (since some variables may be cancelled out), and replace f with the new f_u.
Step 10. Repeat from Step 1.

Fig. 2. Davio-Decompose Algorithm

4 Some Benchmark Results and Discussions

The experiments are conducted on the MCNC benchmark circuits. The target is the Xilinx's XC4010 which has 400 CLBs with 7-20K equivalent gates. The software is the Xilinx's Foundation 2.1i with Synopsys' FPGA Express. Figure 3 shows the three different design paths of the experiments. Table 1 lists the results on applying the direct approach, the AND/XOR direct approach, and the Davio extension approach in terms of the number of CLBs. Here we list the number of CLBs since these are the direct evidence of the hardware efficiencies of these techniques.

By direct approach, we convert the VHDL description of the original circuit, PLA format into the VHDL description for the parity prediction circuit as shown in Figure 3. In some cases, the results are quite good, e.g., ***bw***, ***misex1***, ***rd53*** and ***rd73***. The XC4010 CLB can realize any 5-input functions, but also some 6-, 7-, 8-, and even 9-input functions [12, 14]. The complexity of the function is usually the key to realization efficiency. Hence, it is not surprised to see that such straightforward approach can achieve efficient implementation sometimes.

Next, we applied the AND/XOR direct method to the benchmark circuits. The AND/XOR expression can break the boundary of each internal signals which were the original outputs. After breaking the boundary, it can be minimized by different techniques [11, 15]. We implemented here the minimization algorithm described in Section 2.2 in C language. The minimized AND/XOR expression in VHDL is then processed by Synopsys's FPGA Express and Xilinx Foundation 2.1i. We observe that the number of CLBs is still related to the complexity of the original logic function. Compared to the direct approach, the AND/XOR direct approach shows better results except one case, circuit ***con1***.

Finally, we proposed the Davio approach to get the best result after getting the AND/XOR expressions. In all eight benchmark cases, the proposed Davio approach shows the best result. The Davio approach only needs 21 CLBs for eight benchmark circuits. It takes only on average 2.75 CLBs or 20 % of the original area. The positive and negative Davio expansion approaches yield the same results for these benchmark circuits. Hence, we list only Davio as the title of that column.

We remark here that the complexity of the parity prediction function is proportional to its original function. The complexity of the eight benchmark circuits can be seen from the number of inputs, outputs and product terms, and the number of CLBs columns in Table 1. Concludingly, the Davio approach can indeed provide the most efficient realization of parity function in all cases here. This is mainly accomplished by bypassing the AND/OR type mapping and with dedicated AND/XOR Davio expansion for technology mapping.

5 Conclusions

We have presented here XOR-based minimization and technology mapping techniques to the efficient realization of parity prediction functions in FPGA. We

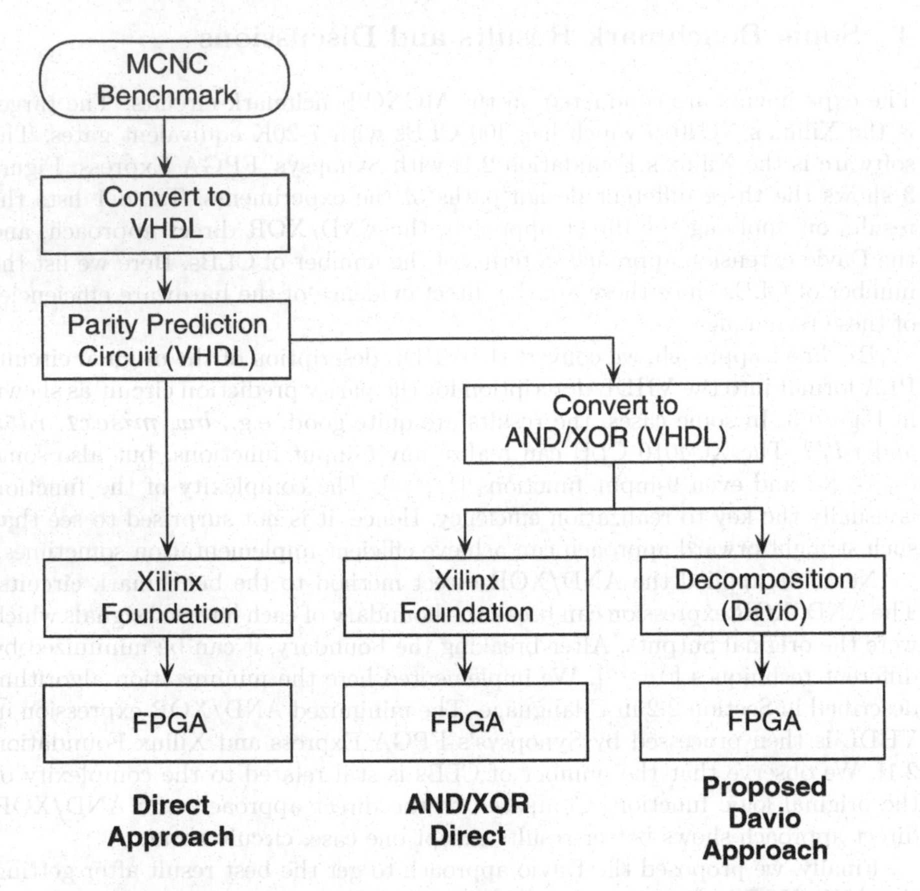

Fig. 3. Design flow for parity prediction circuit of combinational circuit

examine the applications of AND/XOR direct approach and the positive and negative Davio approaches to the MCNC benchmark circuits. These techniques all yield better results than the simple direct approach with the existing EDA tool. Among these techniques, the proposed Davio expansion approach gives the best results in all the examined cases. The proposed Davio expansion-based technology mapping method is an AND/XOR-based method for FPGA technology mapping. This technique can efficiently map XOR-intensive circuits, such as error detecting/correcting, data encryption/decryption, and arithmetic circuits, in FPGAs.

We emphasize that the proposed approaches are generic in nature. They can be used for classical concurrent error detection designs as well as any parity-based coding applications for reliability enhancement. The direct derivation of the parity function in VHDL also has interesting implications. We hope to eventually fully integrate these methods in the EDA tools so that reliable circuits can be systematically implemented.

Table 1. MCNC benchmark results in terms of number of CLBs

circuits	# in	# out	# product terms	Original		Direct Approach		AND/XOR Direct		**Proposed Davio Approach**
				S	A	S	A	S	A	
misex1	8	7	32	11	10	7	5	3	5	**3**
sqrt8	8	4	40	13	10	7	6	5	6	**3**
5xp1	7	10	75	18	15	33	28	17	17	**5**
inc	7	9	34	22	20	15	8	10	8	**5**
con1	7	2	9	3	2	3	2	3	4	**3**
bw	5	28	87	32	31	1	1	1	1	**1**
rd53	5	3	32	3	3	1	1	1	1	**1**
squar5	5	8	32	6	6	1	1	1	1	**1**
Total	52	71	341	108	97	68	52	41	43	**22**
Average	6.5	8.9	42.6	13.5	12.1	8.5	6.5	5.1	5.4	**2.75**

S: # of CLBs when optimized for *speed* in Xilinx Foundation 2.1i
A: # of CLBs when optimized for *area* in Xilinx Foundation 2.1i

References

[1] R. Cuddapah and M. Corba, Reconfigurable Logic for Fault Tolerance, Springer-Verlag, 1995. 630

[2] F. Hanchek and S. Dutt, "Methodologies for Tolerating Logic and Interconnect Faults in FPGAs," IEEE Trans. on Computers, Vol. 47, No. 1, pp. 15-33, Jan. 1998. 630

[3] W.K. Huang, F.J. Meyer, X. Chen, and F. Lombardi, "Testing Configurable LUT-Based FPGAs," IEEE Trans. on VLSI Systems, Vol. 47, No. 6, pp. 276-283, June 1998. 630

[4] M. Abramovici, C. Stroud, S. Wijesuriya, C. Hamilton, and V. Verma, "Using Roving STARs for On-Line and Diagnosis of FPGAs in Fault-Tolerant Applications," Proc. ITC, pp. 973-982, Oct. 1999. 631

[5] Xilinx Inc., `http://www.xilinx.com`. 631

[6] N.A. Touba, and E.J. McCluskey, "Logic Synthesis of Multilevel Circuits with Concurrent Error Detection," IEEE Transactions on Computer-Aided Design, Vol. 16, No. 7, pp. 783-789, Jul. 1997. 631, 632

[7] C. Bolchini, F. Salice and D. Sciuto, "A Novel methodology for Designing TSC Networks based on the Parity Bit Code," Proc. European Design and Test Conf., pp. 440-444, March 1997. 631, 632

[8] J.C. Lo, M. Kitakami and E. Fujiwara, "Reliable Logic Circuits using Byte Error Control Codes," Proc. Int'l Symposium on Defect and Fault Tolerance in VLSI Systems, pp. 286-294, October 1996. 632

[9] T. Sasao, "Logic Synthesis and Optimization," Kluwer Academic Publishers, 1998 633

[10] T. Sasao and P. Besslich, "On the complexity of MOD-2 Sum PLAs," IEEE Transactions on Computers, Vol. 32, No. 2, pp. 262-266, Feb. 1990. 633

[11] S. Even, I. Kohavi and A. Paz, "On minimal modulo-2 sums of products for switching functions," IEEE Transactions on Electronic Computers, EC-16:671-674, Oct. 1967. 633, 636

[12] Xinlinx Inc., Xilinx Data Book: XC4000E and XC4000X Series, May 1999. 634, 636

[13] J. Cong and Y. Ding, "Combinational Logic Synthesis for LUT Based Field Programmable Gate Arrays," ACM Transactions on Design Automation of Electronic Systems, Vol. 1, No. 2, pp. 145-204, April 1996. 634

[14] J. Cong and Y. -Y. Hwang, "Boolean Matching for Complex PLBs in LUT-based FPGAs with Application to Architecture Evaluation," Proc. ACM 6th Int'l Symposium on FPGA, pp. 27-34, Feb. 1998. 636

[15] M. Helliwell, and M. Perkowski, "A Fast Algorithm to Minimize Multi-Output Mixed-Polarity Generalized Reed-Muller Forms," Proc. ACM/IEEE Design Automation Conf., pp. 427-432, 1988. 636

Protocols for Bandwidth Management in Third Generation Optical Networks

Imrich Chlamtac

University of Texas at Dallas

Abstract

During the height of telecommunications boom all optical networking was seen as the next horizon, that would soon revolutionize the data communications world. Networking technology and service companies have been heavily investing in startups that promised to bring all optical, on demand and any bandwidth granularity solutions to the market in eighteen month or less. By the early 2001, there were several hundreds of companies focused on optical equipment development. including optical switches, optical cross-connects and similar, all coming with accompanying management software for the provisioning and control that would bring the final, all optical solution to all. A number of other service companies were focused on the delivery of next generation optically driven services, as exemplified by the emerging IXCs and CLECs in North America. The declared effect was to be a revolutionized networking, resulting in a complete shift to data based networking infrastructures.

The rationale for all of the optical enthusiasm was the promise of significantly more data carrying bandwidth, new optically based service provider revenue generating service opportunities, and simplified network infrastructures that greatly reduced service provider Capex and Opex costs. New optical equipment should reduce or eliminate the costly electronic regeneration of optical signals, and new optical switching and mux equipment promised to ease the burden of path provisioning, path monitoring, and path restoration of optical data carrying paths.

While certain optical technologies have, indeed, gained customer traction, by having proven themselves both economically and technologically, including 2.5Gb/s and 10Gb/s DWDM transport systems, together with certain associated EDFAs, optical switching, including fiber switching, wavelength switching, wavelength bundle switching, etc. have not. In some cases, technological barriers were more difficult to overcome than developers had originally thought, in others more time was needed to refine these technologies, and more time was needed to refine the distinguishable different needs in equipment cost, switching, speed, and amplifier needs, etc., of the access, metro core, regional core, and core (including long haul) networking infrastructures.

Definitely, more time was needed to develop a uniform (standards based) approach to the management of optical network infrastructures, for the wealth of product developments had also allowed a number of product specific, and incompatible network management systems to be created.

S. Sahni et al. (Eds.) HiPC 2002, LNCS 2552, pp. 643–644, 2002.

During 2002, we have come to believe that, with some exceptions, the optical revolution, as it was envisioned during the years 2000 and 2001, will not occur. The revolution called for the rapid replacement of electronically oriented networks by all optical networks, with the retention of electronics only at points of user ingress and egress. Instead some form of optical evolution will occur. This evolution to optical networks will occur at a slower, more rational, pace than was expected for the revolution.

We argue that carrier networks are currently changing through an evolutionary, and not a revolutionary, migration process. As a result, end-to-end networks with SONET, optical, and packet sub-networks will exist for some time to come. However, the gradual penetration of optical WDM systems, and the rapid growth of the Internet as a medium for global multimedia communications has resulted in a need for scalable network infrastructures which are capable of providing large amounts of bandwidth, on demand, with quality guarantees, at varying granularities, to support a diverse range of emerging applications. This means the DWDM network bandwidth in the emerging wavelength-routed systems, needs to be, managed, and provisioned effectively, and intelligently, and support the emerging services in heterogeneous networking environments, with no prevalent standard, hence presenting one of the bigger challenges for optical networking in the years to come.

Memory Architectures for Embedded Systems-On-Chip

Preeti Ranjan Panda[1] and Nikil D. Dutt[2]

[1] Dept. of Computer Science and Engineering, Indian Institute of Technology, Delhi
Hauz Khas, New Delhi - 110 016, India
panda@cse.iitd.ernet.in

[2] Center for Embedded Computer Systems, University of California, Irvine
Irvine, CA 92697-3425, USA
dutt@uci.edu

Abstract. Embedded systems are typically designed for one or a few target applications, allowing for customization of the system architecture for the desired system goals such as performance, power and cost. The memory subsystem will continue to present significant bottlenecks in the design of future embedded systems-on-chip. Using advance knowledge of the application's instruction and data behavior, it is possible to customize the memory architecture to meet varying system goals. On one hand, different applications exhibit varying memory behavior. On the other hand, a large variety of memory modules allow design implementations with a wide range of cost, performance and power profiles. The embedded system architect can thus explore and select custom memory architectures to fit the constraints of target applications and design goals. In this paper we present an overview of recent research in the area of memory architecture customization for embedded systems.

1 Introduction

The application-specific nature of embedded systems has caused a fresh look at architectural issues in recent times. Embedded systems implement a fixed application or set of related applications; consequently, the system architecture can be customized to suit the needs of the given application. This results in an architectural optimization strategy that is fundamentally different from that employed for general purpose processors. In the case of general purpose computer systems, the actual use to which the system will be put is not known, so the processors are designed for good average performance over a set of typical benchmark programs which cover a wide range of application with different behaviors. However, in the case of embedded systems, the features of the given application can be used to determine the architectural parameters. This becomes very important in modern embedded systems where power consumption is a crucial factor. For example, if an application does not use floating point arithmetic, then the floating point unit can be removed from the processor, thereby saving area and power in the implementation.

S. Sahni et al. (Eds.) HiPC 2002, LNCS 2552, pp. 647–662, 2002.

The memory subsystem is an important and interesting component of system designs that can benefit from customization. Unlike a general purpose processor where a standard cache hierarchy is employed, the memory hierarchy of embedded systems can be tailored in various ways. The memory can be selectively cached; the cache line size can be determined by the application; the designer can opt to discard the cache completely and choose specialized memory configurations such as FIFOs and stream buffers; and so on. The exploration space of different possible memory architectures is vast, and there have been attempts to automate or semi-automate this exploration process [1].

Memory issues have been separately addressed by disparate research groups: computer architects, compiler writers, and the CAD/embedded systems community.

Memory architectures have been studied extensively by computer architects. Memory hierarchy, implemented with cache structures, has received considerable attention from researchers. Cache parameters such as line size, associativity, and write policy, and their impact on typical applications have been studied in detail [2]. Recent studies have also quantified the impact of dynamic memory (DRAM) architectures [3].

Since architectures are closely associated with compilation issues, compiler researchers have addressed the problem of generating efficient code for a given memory architecture by appropriately transforming the program and data. Compiler transformations such as blocking/tiling are examples of such optimizations [4, 5].

Finally, researchers in the area of CAD/embedded systems have typically employed memory structures such as register files, static memory (SRAMs), and DRAMs in generating application specific designs.

While the optimizations identified by the architecture and compiler community are still applicable in embedded system design, the architectural flexibility available in the new context adds a new exploration dimension. To be really effective, these optimizations need to be integrated into the embedded system design process as well as enhanced with new optimization and estimation techniques.

We first present an overview of different memory architectures used in embedded systems, and then survey some of the ways in which these architectures have been customized.

2 Embedded Memory Architectures

In this section we outline some common architectures used in embedded memories, with particular emphasis on the more unconventional ones.

2.1 Caches

As application-specific systems became large enough to use a processor core as a building block, the natural extension in terms of memory architecture was the addition of instruction and data caches. Since the organization of typical caches is well known [2] we will omit the explanation. Caches have many parameters which can be customized for a given application. Some of these customizations are described in Section 3.

2.2 Scratch-Pad Memory

An embedded system designer is not restricted to using only a traditional cached memory architecture. Since the design needs to execute only a single application, we can use unconventional architectural variations that suit the specific application under consideration. One such design alternative is *Scratch Pad memory* [6, 1].

Scratch-Pad memory refers to data memory residing on-chip, that is mapped into an address space disjoint from the off-chip memory, but connected to the same address and data buses. Both the cache and Scratch-Pad memory (usually SRAM) allow fast access to their residing data, whereas an access to the off-chip memory requires relatively longer access times. The main difference between the Scratch-Pad SRAM and data cache is that the SRAM guarantees a single-cycle access time, whereas an access to the cache is subject to cache misses. The concept of Scratch Pad memory is an important architectural consideration in modern embedded systems, where advances in embedded DRAM technology have made it possible to combine DRAM and logic on the same chip. Since data stored in embedded DRAM can be accessed much faster and in a more power-efficient manner than that in off-chip DRAM, a related optimization problem that arises in this context is how to identify critical data in an application, for storage in on-chip memory.

Figure 1 shows the architectural block diagram of an application employing a typical embedded core processor [1], where the parts enclosed in the dotted rectangle are implemented in one chip, interfacing with an off-chip memory, usually realized with DRAM. The address and data buses from the *CPU core* connect to the *Data Cache, Scratch-Pad memory*, and the *External Memory Interface* (EMI) blocks. On a memory access request from the CPU, the data cache indicates a cache hit to the EMI block through the **C_HIT** signal. Similarly, if the SRAM interface circuitry in the Scratch-Pad memory determines that the referenced memory address maps into the on-chip SRAM, it assumes control of the data bus and indicates this status to the EMI through signal **S_HIT**. If both the cache and SRAM report misses, the EMI transfers a block of data of the appropriate size (equal to the *cache line size*) between the cache and the DRAM.

One possible data address space mapping for this memory configuration is shown in Figure 2, for a sample addressable memory of size N data words. Memory addresses $0 \ldots (P-1)$ map into the on-chip scratch pad memory, and have a single processor cycle access time. Memory addresses $P \ldots (N-1)$ map into the off-chip DRAM, and are accessed by the CPU through the data cache. A cache hit for an address in the range $P \ldots N-1$ results in a single-cycle delay, whereas a cache miss, which leads to a block transfer between off-chip and cache

[1] For example, the LSI Logic CW33000 RISC Microprocessor core [7].

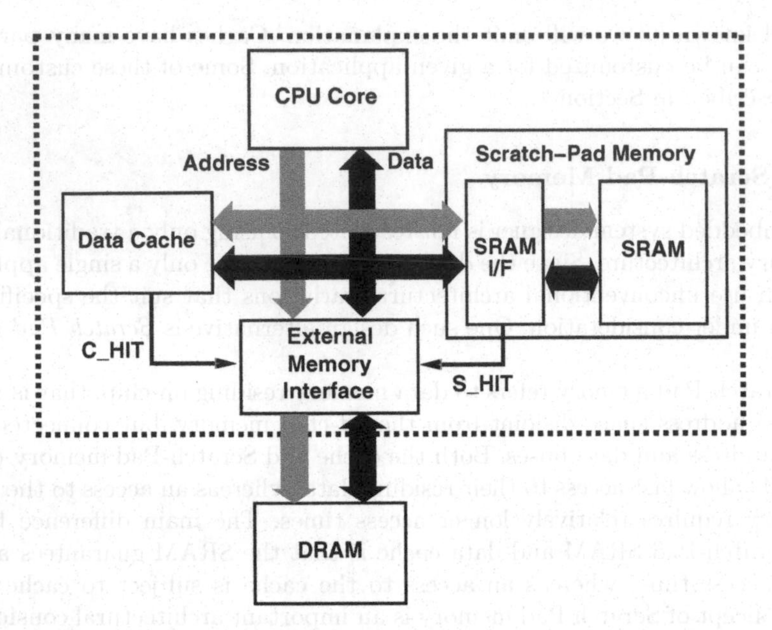

Fig. 1. Block Diagram of Typical Embedded Processor with Scratch-Pad Memory

memory, may result in a delay of say 10-20 processor cycles. We illustrate the use of this Scratch-Pad memory with the following example.

Example 1. A small (4×4) matrix of coefficients, *mask*, slides over the input image, *source*, covering a different 4×4 region in each iteration of y, as shown in Figure 3. In each iteration, the coefficients of the mask are combined with the region of the image currently covered, to obtain a weighted average, and the result, *acc*, is assigned to the pixel of the output array, *dest*, in the center of the covered region. If the two arrays *source* and *mask* were to be accessed through the data cache, the performance would be affected by cache conflicts. This problem can be solved by storing the small *mask* array in the Scratch-pad memory. This assignment eliminates all conflicts in the data cache – the data cache is now used for memory accesses to *source*, which are very regular. Storing *mask* on-chip, ensures that frequently accessed data is never ejected off-chip, thereby significantly improving the memory performance and energy dissipation.

The memory assignment described in [6] exploits this architecture by first determining a *Total Conflict Factor* (TCF) for each array based on the access frequency and possibility of conflict with other arrays, and then considering the arrays for assignment to scratch pad memory in the order of TCF/(array size), giving priority to high-conflict/small-size arrays.

Dynamic Data Transfers In the above formulation, the data stored in the Scratch-Pad Memory was statically determined. This idea can be extended to the

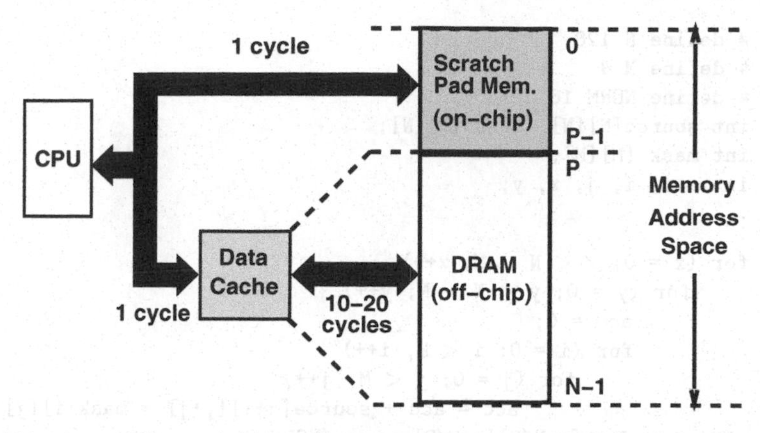

Fig. 2. Division of Data Address Space between Scratch Pad memory and off-chip memory

case of dynamic data storage. However, since there is no automatic hardware-controlled mechanism to transfer data between the scratch pad and the main memory, such transfers have to be explicitly managed by the compiler. In the technique proposed in [8], the tiling compiler optimization is modified by first moving the data tiles into scratch pad memory and moving it back to main memory after the computation is complete.

Storing Instructions in Scratch-Pad Memory A scratch-pad memory storing a small amount of frequently accessed data on-chip, has an equivalent in the instruction cache. The idea of using a small buffer to store blocks of frequently used instructions was first introduced in [9]. Recent extensions of this strategy are the Decoded Instruction Buffer [10] and the L-cache [11].

Researchers have also examined the possibility of storing both instructions and data in the scratch pad memory. In the formulation proposed in [12], the frequency of access of both data and program blocks are analyzed and the most frequently occurring among them are assigned to the scratch pad memory.

2.3 DRAM

DRAMs have been used in a processor-based environment for quite some time, but the context of its use in embedded systems – both from a hardware synthesis viewpoint, as well as from an embedded compiler viewpoint – have been investigated relatively recently. DRAMs offer better memory performance through the use of specialized access modes that exploit the internal structure and steering/buffering/banking of data within these memories. Explicit modeling of these specialized access modes allows the incorporation of such high-performance access modes into synthesis and compilation frameworks. New synthesis and compilation techniques have been developed that employ detailed knowledge of the

```
# define N 128
# define M 4
# define NORM 16
int source[N][N], dest [N][N];
int mask [M][M];
int acc, i, j, x, y;
:
for (x = 0; x < N - M; x++)
    for (y = 0; y < N - M; y++) {
        acc = 0;
        for (i = 0; i < M; i++)
            for (j = 0; j < M; j++)
                acc = acc + source[x+i][y+j] * mask[i][j];
        dest[x+M/2][y+M/2] = acc/NORM;
    }
```

(a)

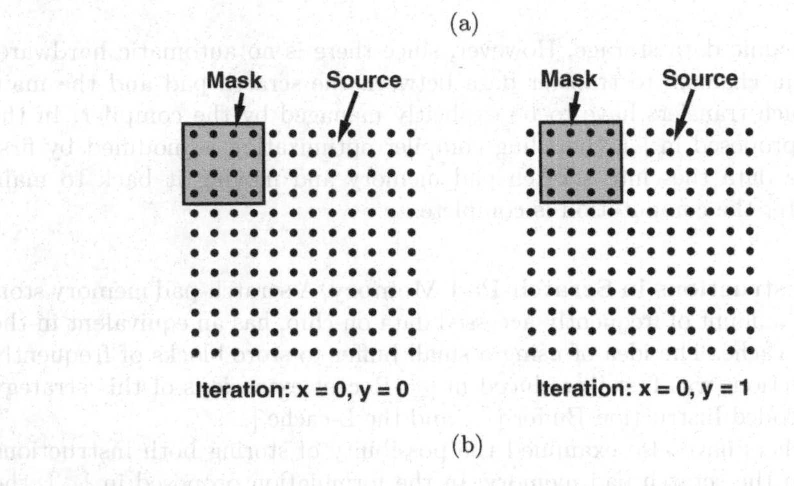

(b)

Fig. 3. (a)Procedure CONV (b) Memory access pattern in CONV

DRAM access modes and exploit advance knowledge of an embedded system's application to better improve system performance and power.

A typical DRAM memory address is internally split into a *row address* consisting of the most significant bits and a *column address* consisting of the least significant bits. The row address selects a page from the core storage and the column address selects an offset within the page to arrive at the desired word. When an address is presented to the memory during a READ operation, the entire page addressed by the row address is read into the page buffer, in anticipation of spatial locality. If future accesses are to the same page, then there is no need to access the main storage area since it can just be read off the page buffer, which acts like a cache. Thus, subsequent accesses to the same page are very fast.

A scheme for modeling the various memory access modes and using them to perform useful optimizations in the context of a Behavioral Synthesis environment is described in [13]. The main observation is that the input behavior's memory access patterns can potentially exploit the page mode (or other specialized access mode) features of the DRAM. The key idea is the representation of these specialized access modes as graph primitives that model individual DRAM access modes such as row decode, column decode, precharge, etc.; each DRAM family's specialized access modes are then represented using a composition of these graph primitives to fit the desired access mode protocol. These composite graphs can then be scheduled together with the rest of the application behavior, both in the context of synthesis, as well as for code compilation. For instance, some additional DRAM-specific optimizations discussed in [13] are:

Read-Modify-Write (R-M-W) Optimization that takes advantage of the R-M-W mode in modern DRAMs which provides support for a more efficient realization of the common case where a specific address is read, the data is involved in some computation, and then the output is written back to the same location.

Hoisting where the row-decode node is scheduled ahead of a conditional node if the first memory access in both branches are on the same page.

Unrolling optimization in the context of supporting the page mode accesses.

A good overview of the performance implications of the architectural features of modern DRAMs is found in [3].

Synchronous DRAM As DRAM architectures evolve, new challenges are presented to the automatic synthesis of embedded systems based on these memories. *Synchronous DRAM* represents an architectural advance that presents another optimization opportunity: multiple memory banks. The core memory storage is divided into multiple banks, each with its own independent page buffer, so that two separate memory pages can be simultaneously active in the multiple page buffers.

The problem of modeling the access modes of synchronous DRAMs is addressed in [14]. The modes include:

Burst mode read/write – fast successive accesses to data in the same page.

Interleaved row read/write modes – alternating burst accesses between banks.

Interleaved Column Access – alternating burst accesses between two chosen rows in different banks.

Memory bank assignment is performed by creating an interference graph between arrays and partitioning it into subgraphs so that data in each part is assigned to a different memory bank. The bank assignment algorithm is related to techniques such as [15] that address memory assignment for DSP processors such as the Motorola 56000 which has a dual-bank internal memory/register file [16, 17]. The bank assignment problem in [15] is targeted at scalar variables, and is solved in conjunction with register allocation by building a constraint graph that models the data transfer possibilities between registers and memories followed by a simulated annealing step.

[18] approached the SDRAM bank assignment problem by first constructing an *array distance table*. This table stores the *distance* in the DFG between each pair of arrays in the specification. A short distance indicates a strong correlation, possibly indicating that they might be, for instance, two inputs of the same operation, and hence, would benefit from being assigned to separate banks. The bank assignment is finally performed by considering array pairs in increasing order of their array distance information.

Whereas the previous discussion has focused primarily in the context of hardware synthesis, similar ideas have been employed to aggressively exploit the memory access protocols for compilers [19, 20]. In the traditional approach of compiler/architecture codesign, the memory subsystem was separated from the microarchitecture; the compiler typically dealt with memory operations using the abstractions of memory loads and stores, with the architecture (e.g., the memory controller) providing the interface to the (typically yet-unknown) family of DRAMs and other memory devices that would deliver the desired data. However in an embedded system, the system architect has advance knowledge of the specific memories (e.g., DRAMs) used; thus we can employ *memory-aware* compilation techniques [19] that exploit the specific access modes in the DRAM protocol to perform better code scheduling. In a similar manner, it is possible for the code scheduler to employ global scheduling techniques to hide potential memory latencies using knowledge of the memory access protocols, and in effect, improve the ability of the memory controller to boost system performance [20].

2.4 Special Purpose Memories

In addition to the general memories such as caches, and memories specific to embedded systems, such as scratch-pad, there exist various other types of custom memories that implement specific access protocols. Such memories include:

- LIFO (memory implementing Last-In-First-Out protocol).
- FIFO (memory implementing queue or First-In-First-Out protocol).
- CAM (content addressable memory).

3 Customization of Memory Architectures

We now survey some recent research efforts that address the exploration space involving on-chip memories. A number of distinct memory architectures could be devised to exploit different application specific memory access patterns efficiently. Even if we restrict the scope of the architecture to those involving on-chip memory only, the exploration space of different possible configurations is too large, making it infeasible to exhaustively simulate the performance and energy characteristics of the application for each configuration. Thus, exploration tools are necessary for rapidly evaluating the impact of several candidate architectures. Such tools can be of great utility to a system designer by giving fast initial feedback on a wide range of memory architectures [1].

3.1 Caches

Two of the most important aspects of data caches that can be customized for an application are: (1) the cache line size; and (2) the cache size. The customization of cache line size for an application is performed in [21] using an estimation technique for predicting the memory access performance – that is, the total number of processor cycles required for all the memory accesses in the application.

There is a tradeoff in sizing the cache line. If the memory accesses are very regular and consecutive, i.e, exhibit spatial locality, a longer cache line is desirable, since it minimizes the number of off-chip accesses and exploits the locality by pre-fetching elements that will be needed in the immediate future. On the other hand, if the memory accesses are irregular, or have large strides, a shorter cache line is desirable, as this reduces off-chip memory traffic by not bringing unnecessary data into the cache. The maximum size of a cache line is the DRAM page size.

The estimation technique uses data reuse analysis to predict the total number of cache hits and misses inside loop nests so that spatial locality is incorporated into the estimation. An estimate of the impact of conflict misses is also incorporated. The estimation is carried out for the different candidate line sizes and the best line size is selected for the cache.

The customization of the total cache size is integrated into the scratch pad memory customization described in the next section.

3.2 Scratch-Pad Memory

MemExplore [21], an exploration framework for optimizing the on-chip data memory organization addresses the following problem: given a certain amount of on-chip memory space, partition this into data cache and scratch pad memory so that the total access time and energy dissipation is minimized, i.e., the number of accesses to off-chip memory is minimized. In this formulation, an on-chip memory architecture is defined as a combination of the total size of on-chip memory used for data storage; the partitioning of this on-chip memory into: scratch memory, characterized by its size; and data cache, characterized by the cache size; and the cache line size. For each candidate on-chip memory size T, the technique considers different divisions of T into cache (size C) and scratch pad memory (size $S = T - C$), selecting only powers of 2 for C. The procedure described in Section 2.2 is used to identify the right data for storage in scratch pad memory. Among the data assigned to be stored in off-chip memory (and hence accessed through the cache), an estimation of the memory access performance is performed by combining an analysis of the array access patterns in

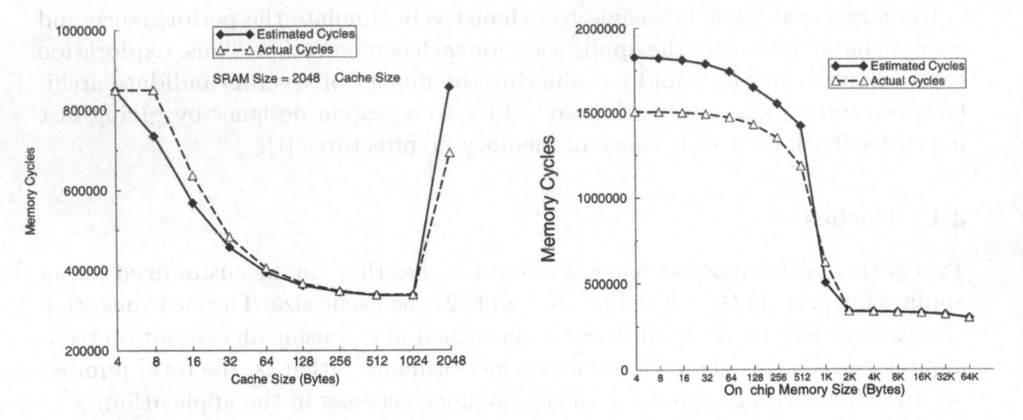

Fig. 4. *Histogram* Example (a) Variation of memory performance with different mixes of cache and Scratch-pad memory, for total on-chip memory of 2 KB (b) Variation of memory performance with total on-chip memory space

the application and an approximate model of the cache behavior. The result of the estimation is the expected number of processor cycles required for all the memory accesses in the application. For each T, the (C, L) pair that is estimated to maximize performance is selected.

Example 2. Typical exploration curves of the MemExplore algorithm are shown in Figure 4. Figure 4(a) shows that the ideal division of a 2K on-chip space is 1K scratch pad memory and 1K data cache. Figure 4(b) shows that very little performance improvement is observed beyond a total on-chip memory size of 2KB.

The exploration curves of Figure 4 are generated from fast analytical estimates, which are three orders of magnitude faster than actual simulations, and are independent of data size. This estimation capability is important in the initial stages of system design, where the number of possible architectures is large, and a simulation of each architecture is prohibitively expensive.

3.3 DRAM

The presence of embedded DRAMs adds several new dimensions to traditional architecture exploration. One interesting aspect of DRAM architecture that can be customized for an application is the banking structure.

Figure 5(a) illustrates a common problem with the single-bank DRAM architecture. If we have a loop that accesses in succession data from three large arrays A, B, and C, each of which is much larger than a page, then each memory access leads to a fresh page being read from the storage, effectively cancelling the benefits of the page buffer. This page buffer interference problem cannot be avoided if a fixed architecture DRAM is used. However, an elegant solution

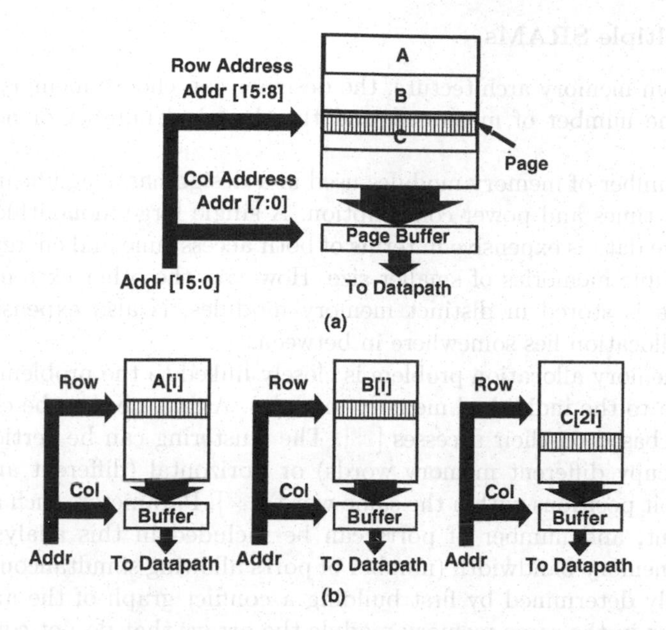

Fig. 5. (a) Arrays mapped to a single-bank memory (b) 3-bank memory architecture

to the problem is available if the banking configuration of the DRAM can be customized for the application [22]. Thus, in the example of Figure 5, the arrays can be assigned to separate banks as shown in Figure 5(b). Since each bank has its own private page buffer, there is no interference between the arrays, and the memory accesses do not represent a bottleneck.

In order to customize the banking structure for an application, we need to solve the memory bank assignment problem – determine an optimal banking structure (number of banks) and determine the assignment of each array variable into the banks such that the number of page misses is minimized. This objective optimizes both the performance as well as the energy dissipation of the memory subsystem. The memory bank customization problem is solved in [22] by modeling the assignment as a partitioning problem – partition a given set of nodes into a given number of groups such that a given criterion (bank misses in this case) is optimized. The partitioning proceeds by associating a cost of assigning two arrays into the same bank, determined by the number of accesses to the arrays and the loop count. If the arrays are accessed in the same loop, then the cost is high, thereby discouraging the partitioning algorithm from assigning them to the same bank. On the other hand, if two arrays are never accessed in the same loop, then they are candidates for assignment into the same bank. This pairing is associated with a low cost, guiding the partitioner to assign the arrays together.

3.4 Multiple SRAMs

In a custom memory architecture, the designer can choose memory parameters such as the number of memories, and the size, and number of ports on each memory.

The number of memory modules used in a design has a significant impact on the access times and power consumption. A single large monolithic memory to hold all the data is expensive in terms of both access time and energy dissipation than multiple memories of smaller size. However, the other extreme, where all array data is stored in distinct memory modules, is also expensive, and the optimal allocation lies somewhere in between.

The memory allocation problem is closely linked to the problem of assigning array data to the individual memory modules. Arrays need to be clustered into memories based on their accesses [23]. The clustering can be vertical (different arrays occupy different memory words) or horizontal (different arrays occupy different bit positions within the same word) [24]. Parameters such as bit-width, word count, and number of ports can be included in this analysis [25]. The required memory bandwidth (number of ports allowing simultaneous access) can be formally determined by first building a conflict graph of the array accesses and storing in the same memory module the arrays that do not conflict [26].

3.5 Special Purpose Memories

Special purpose memories such as stacks (LIFO), queues (FIFO), frame buffers, streaming buffers, etc. can be utilized when customizing the memory architecture for an application. Indeed, analysis of many large applications shows that a significant number of the memory references in data-intensive applications are made by a surprisingly small number of lines of code. Thus it is possible to customize the memory subsystem by tuning the memories for these segments of code, with the goal of improving performance, and also for reducing the power dissipation. In the approach described in [27], the application is first analyzed and different access patterns identified. Data for the most *critical* access patterns are assigned to memory modules that best fit the access pattern profiles. The system designer can then evaluate different cost/performance/power profiles for different realizations of the memory subsystem.

3.6 Processor-Memory Coexploration

Datapath Width and Memory Size The CPU's bit-width is an additional parameter that can be tuned during architectural exploration of customizable processors. [28] studied the relationship between the width of the processor data path and the memory subsystem. This relationship is important when different data types with different sizes are used in the application. The key observation made is that as datapath width is decreased, the data memory size decreases because of less wasted space. For example, storing 3-bit data in a 4-bit word instead of 8-bit word), but the instruction memory might increase. For example, storing 7-bit data in an 8-bit word requires only one instruction to access it, but requires two instructions if a 4-bit datapath is used. The authors use a RAM and ROM cost model to evaluate the cost of candidate bit-widths in a combined CPU-memory exploration.

Architectural Description Language (ADL) Driven Coexploration Processor Architecture Description Languages (ADLs) have been developed to allow for a language-driven exploration and software toolkit generation approach [29, 30]. Currently most ADLs assume an implicit/default memory organization, or are limited to specifying the characteristics of a traditional memory hierarchy. Since embedded systems may contain non-traditional memory organizations, there is a great need to model explicitly the memory subsystem for an ADL-driven exploration approach. A recent approach [31] describes the use of the EXPRESSION ADL [32] to drive Memory Architecture Exploration. The EXPRESSION ADL description of the processor-memory architecture is used to explicitly capture the memory architecture, including the characteristics of the memory modules (such as caches, DRAMs, SRAMs DMAs), the parallelism and pipelining present in the memory architecture (e.g., resources used, timings, access modes). Each such explicit memory architecture description is then used to automatically generate the information needed by the compiler [19, 20] to efficiently utilize the features in the memory architecture, and to generate a memory simulator, allowing feedback to the designer on the match between the application, the compiler and the memory architecture.

3.7 Split Spatial and Temporal Caches

Various specialized memory structures proposed over the years could be candidates for embedded systems. One such concept is split spatial/temporal caches.

Variables in real life applications present a wide variety of access patterns and locality types (for instance scalars, such as indexes, present usually high temporal and moderate spatial locality, while vectors with small stride present high spatial locality, and vectors with large stride present low spatial locality, and may or may not have temporal locality). Several approaches including [33] have proposed splitting a cache into a spatial cache and a temporal cache that store data structures with high temporal and high spatial locality respectively. These approaches rely on a dynamic prediction mechanism to route the data to either the spatial or the temporal caches, based on a history buffer. In an embedded system context, the approach of [34] uses a similar split-cache architecture, but allocates the variables statically to the different local memory modules, avoiding the power and area overhead of the dynamic prediction mechanism. Thus by targeting the specific locality types of the different variables, better utilization of the main memory bandwidth is achieved. The useless fetches due to locality mismatch are thus avoided. For instance, if a variable with low spatial locality is serviced by a cache with a large line size, a large number of the values read from the main memory will never be used. The approach in [34] shows that the memory bandwidth and memory power consumption can be reduced significantly.

4 Conclusion

The significant interest in embedded systems in recent times has caused researchers to study architectural optimizations from a new angle. With the full advance knowledge of the applications being implemented by the system, many design parameters can be customized. This is especially true of the memory subsystem where a vast array of different organizations can be employed for application specific systems and the designer is not restricted to the traditional cache hierarchy. The optimal memory architecture for an application specific system can be significantly different from the typical cache hierarchy of processors. We outlined different memory architectures relevant to embedded systems and strategies to customize them for a given application. While some of the analytical techniques are automated, a lot of work still remains to be performed in the coming years before a completely push-button methodology evolves for application-specific customization of the memory organization in embedded systems.

References

[1] Panda, P. R., Dutt, N. D., Nicolau, A.: Memory Issues in Embedded Systems-On-Chip: Optimizations and Exploration. Kluwer Academic Publishers, Norwell, MA (1999) 648, 649, 655

[2] Hennessy, J. L., Patterson, D. A.: Computer Architecture – A Quantitative Approach. Morgan Kaufman, San Francisco, CA (1994) 648, 649

[3] Cuppu, V., Jacob, B. L., Davis, B., Mudge, T. N.: A performance comparison of contemporary dram architectures. In: International Symposium on Computer Architecture, Atlanta, GA (1999) 222–233 648, 653

[4] Lam, M., Rothberg, E., Wolf, M. E.: The cache performance and optimizations of blocked algorithms. In: Proceedings of the Fourth International Conference on Architectural Support for Programming Languages and Operating Systems. (1991) 63–74 648

[5] Panda, P. R., Nakamura, H., Dutt, N. D., Nicolau, A.: Augmenting loop tiling with data alignment for improved cache performance. IEEE Transactions on Computers **48** (1999) 142–149 648

[6] Panda, P. R., Dutt, N. D., Nicolau, A.: On-chip vs. off-chip memory: The data partitioning problem in embedded processor-based systems. ACM Transactions on Design Automation of Electronic Systems **5** (2000) 682–704 649, 650

[7] LSI Logic Corporation Milpitas, CA: CW33000 MIPS Embedded Processor User's Manual. (1992) 649

[8] Kandemir, M. T., Ramanujam, J., Irwin, M. J., Vijaykrishnan, N., Kadayif, I., Parikh, A.: Dynamic management of scratch-pad memory space. In: Design Automation Conference, Las Vegas, NV (2001) 690–695 651

[9] Jouppi, N. P.: Improving direct-mapped cache performance by the addition of a small fully-associative cache and prefetch buffers. In: International Symposium on Computer Architecture, Seattle, WA (1990) 364–373 651

[10] Bajwa, R. S., Hiraki, M., Kojima, H., Gorny, D. J., Nitta, K., Shridhar, A., Seki, K., Sasaki, K.: Instruction buffering to reduce power in processors for signal processing. IEEE Transactions on VLSI Systems **5** (1997) 417–424 651
[11] Bellas, N., Hajj, I. N., Polychronopoulos, C. D., Stamoulis, G.: Architectural and compiler techniques for energy reduction in high-performance microprocessors. IEEE Transactions on VLSI Systems **8** (2000) 317–326 651
[12] Steinke, S., Wehmeyer, L., Lee, B. S., Marwedel, P.: Assigning program and data objects to scratchpad for energy reduction. In: Design, Automation & Test in Europe, Paris, France (2002) 651
[13] Panda, P. R., Dutt, N. D., Nicolau, A.: Incorporating DRAM access modes into high-level synthesis. IEEE Transactions on Computer Aided Design **17** (1998) 96–109 653
[14] Khare, A., Panda, P. R., Dutt, N. D., Nicolau, A.: High-level synthesis with SDRAMs and RAMBUS DRAMs. IEICE Transactions on fundamentals of electronics, communications and computer sciences **E82-A** (1999) 2347–2355 653
[15] Sudarsanam, A., Malik, S.: Simultaneous reference allocation in code generation for dual data memory bank asips. ACM Transactions on Design Automation of Electronic Systems **5** (2000) 242–264 653, 654
[16] Saghir, M. A. R., Chow, P., Lee, C. G.: Exploiting dual data-memory banks in digital signal processors. In: International conference on Architectural Support for Programming Languages and Operating Systems, Cambridge, MA (1996) 234–243 654
[17] Cruz, J. L., Gonzalez, A., Valero, M., Topham, N.: Multiple-banked register file architectures. In: International Symposium on Computer Architecture, Vancouver, Canada (2000) 315–325 654
[18] Chang, H. K., Lin, Y. L.: Array allocation taking into account SDRAM characteristics. In: Asia and South Pacific Design Automation Conference, Yokohama (2000) 497–502 654
[19] Grun, P., Dutt, N., Nicolau, A.: Memory aware compilation through accurate timing extraction. In: Design Automation Conference, Los Angeles, CA (2000) 316–321 654, 659
[20] Grun, P., Dutt, N., Nicolau, A.: MIST: An algorithm for memory miss traffic management. In: IEEE/ACM International Conference on Computer Aided Design, San Jose, CA (2000) 431–437 654, 659
[21] Panda, P. R., Dutt, N. D., Nicolau, A.: Local memory exploration and optimization in embedded systems. IEEE Transactions on Computer Aided Design **18** (1999) 3–13 655
[22] Panda, P. R.: Memory bank customization and assignment in behavioral synthesis. In: Proceedings of the IEEE/ACM International Conference on Computer Aided Design, San Jose, CA (1999) 477–481 657
[23] Ramachandran, L., Gajski, D., Chaiyakul, V.: An algorithm for array variable clustering. In: European Design and Test Conference, Paris (1994) 262–266 658
[24] Schmit, H., Thomas, D. E.: Synthesis of application-specific memory designs. IEEE Transactions on VLSI Systems **5** (1997) 101–111 658
[25] Jha, P. K., Dutt, N.: Library mapping for memories. In: European Design and Test Conference, Paris, France (1997) 288–292 658
[26] Wuytack, S., Catthoor, F., Jong, G. D., Man, H. D.: Minimizing the required memory bandwidth in vlsi system realizations. IEEE Transactions on VLSI Systems **7** (1999) 433–441 658

[27] Grun, P., Dutt, N., Nicolau, A.: Apex: Access patter based memory architecture customization. In: Proceedings International Symposium on System Synthesis, Montreal, Canada (2001) 25–32 658
[28] Shackleford, B., Yasuda, M., Okushi, E., Koizumi, H., Tomiyama, H., Yasuura, H.: Memory-cpu size optimization for embedded system designs. In: Design Automation Conference. (1997) 658
[29] Tomiyama, H., Halambi, A., Grun, P., Dutt, N., Nicolau, A.: Architecture description languages for systems-on-chip design. In: Proceedings 6th Asia Pacific Conference on Chip Design Languages, Fukuoka (1999) 109–116 659
[30] Halambi, A., Grun, P., Tomiyama, H., Dutt, N., Nicolau, A.: Automatic software toolkit generation for embedded systems-on-chip. In: Proceedings ICVC'99, Korea (1999) 659
[31] Mishra, P., Grun, P., Dutt, N., Nicolau, A.: Processor-memory co-exploration driven by a memory-aware architecture description language. In: VLSIDesign, Bangalore (2001) 659
[32] Halambi, A., Grun, P., Ganesh, V., Khare, A., Dutt, N., Nicolau, A.: Expression: A language for architecture exploration through compiler/simulator retargetability. In: Proceedings DATE'99, Munich, Germany (1999) 659
[33] Gonzales, A., Aliagas, C., Valero, M.: A data cache with multiple caching strategies tuned to different types of locality. In: International Conference on Supercomputing, Barcelona, Spain (1995) 338–347 659
[34] Grun, P., Dutt, N., Nicolau, A.: Access pattern based local memory customization for low power embedded systems. In: Design, Automation, and Test in Europe, Munich (2001) 659

Structured Component Composition Frameworks for Embedded System Design*

Sandeep K. Shukla[1], Frederic Doucet[2], and Rajesh K. Gupta[2]

[1] Electrical and Computer Engineering Department
Virginia Tech, Blacksburg, VA 24061
[2] Information and Computer Science Department, University of California
Irvine, CA 92697

Abstract. The increasing integration of system-chips is leading to a widening gap in the size and complexity of the chip-level design and the design capabilities. A number of advances in high-level modeling and validation have been proposed over the past decade in an attempt to bridge the gap in design productivity. Prominent among these are advances in *Abstraction* and *Reuse* and structured design methods such as Component-Based Design and Platform-Based Design. In this paper, we present an overview of the recent advances in reuse, abstraction, and component frameworks. We describe a compositional approach to high-level modeling as implemented in the BALBOA project.

1 Introduction

Due to advances in microelectronic processing and devices, while fabricating millions of transistors on chip has become easier, the functionalities implemented using these devices have steadily been growing, outstripping manual design capabilities and the capacity of design automation tools. A number of strategies are being explored by the microelectronic designers in an attempt to improve the design productivity and the quality of designs through advances in modeling and validation techniques. Raising the abstraction level at which designs are entered and validated has a direct impact on the design quality and design time. Consequently, much of the recent effort in the area has been focused on the specification methodologies and languages for modeling designs at the system level. The focus of this paper is on component composition frameworks, which provide support for correct composition of existing components and automatic or semi-automatic means for validation checks. We show the techniques that are used to improve component reuse and design productivity using an example framework currently under development.

2 Components and Virtual Components for SOCs

A system-on-chip or SOC refers to a complete system from an end application point of view. In other words, a SOC represents implementation of a complete

* This reasearch was partially funded by SRC, FCAR, NSF, CAL-MICRO, CORE

S. Sahni et al. (Eds.) HiPC 2002, LNCS 2552, pp. 663–678, 2002.

application on a single chip consisting of a range of building blocks from processors, memory, to communication and networking elements. There may be a bottom-up or a top-down approach to building an application into an SOC. In a bottom up approach, various functionalities of the system are mapped to existing components, and then the components are connected together in such a way that the resulting system has the required functionalities and performance characteristics. In a top-down approach, a functional specification of the application is refined to obtain architectural specification, and synthesized into implementation either by automated synthesis tools (probably in the future), or by manual refinements to synthesizable descriptions.

In the bottom-up approach the application functionality can often be structured into various hardware and software components which can provide parts of the functionality, and as a whole meet the functional and performance goals. Top-down approaches could also yield to refinements that are then mapped to various hardware/software components. So, from that perspective, a component may then be a piece of functionality implemented in software or as a dedicated piece of silicon hardware or a combination of the two. A component may be virtual in that it represents a well-defined functionality without an associated hardware/hardware implementation. The phrase 'virtual component' is used to describe reusable IP components which are composed with other components (as if plugging them onto virtual sockets), similar to real hardware components are plugged into real sockets on a board [22]. A virtual component may be turned into a concrete implementation by instantiating and appropriately combining with other components. To quote from the VSIA's statement on the purpose of standardization effort for virtual socket interfaces [22]

> *One solution for this dilemma (productivity gap) is to design with pre-designed blocks, much as is now done using off-the-shelf IC's on printed circuit boards. The pre-designed blocks are a form of Intellectual Property (IP), which is variously referred to as IP, IP blocks, cores, system-level blocks (SLB), macros, system level macros (SLMs), or Virtual Components (VCs - the VSIA term for these elements).*

A virtual component is then defined to be a reusable piece of functionality, that is, its functions may be reused in other applications. However, not every component can be reused across all applications. Reusability requires not only matching of functionality, but an ability to compose the component functionalities in a way that correctly implements the end application. This requires specific component capabilities that we shall discuss later in the context of composition frameworks (tools and methods) that enable SOC application development.

2.1 Component Composition Frameworks

A composition framework provides reasoning capabilities and tools that enable a system designer to compose components into a specific application. These capabilities include selection of the correct components, automated creation of correct interfaces, simulation of the composed design, testing and validation for behavioral correctness and equivalence checks. A limited form of Component composition is common in purely software systems where environments often known as Integrated Development Environments or IDEs are used to facilitate component selection and composition. Compared to software IDEs such as Microsoft Visual studio, hardware component composition frameworks are more difficult to build. Part of the complexity is due to the various ways in which the integrated circuit blocks are represented, designed and composed. Due to a common model of the execution machine and commonly accepted compiler conventions, software reusable components can be composed fairly easily either statically during compilation or even during runtime. Even with this ease in compilation and runtime library linking, correct application functionality remains a challenge. In the absence of such compiler and middleware conventions, ensuring component composition for silicon hardware is a challenge despite well-understood physics and logic of interface circuits. At higher abstraction levels, often a connection between components is created through limited set of ports and signals in, what is often known as, *structural design for SOCs.* Such a composition presupposes a structural representation for the components. Even if a component is not structural, but behavioral, it can often be composed using special components (e.g., protocol modules) interconnecting the components. Further, one often is faced with composing components described at different levels of abstraction, one component at behavioral level (e.g., as an algorithm) and another in a register transfer description. Composability of models can be defined along a number of modeling dimensions [7]:

1. *Temporal detail*: which expresses the degree of precision of the ordering of the modeled events. This includes partial-ordered event accurate models, token-cycle accurate models, instruction-cycle accurate models, clock-cycle accurate models, clock-phase accurate models and so on.
2. *Data value detail*: which expresses the representation or format of data values specified in a model. Data values could be enumerated values, word-level values, bit-true representations, etc.
3. *Functional detail*: which expresses the level of detail in the functionality of a model, ranging from mathematical formulae to detailed intermediate operations (gate-level or instruction level).
4. *Structural detail*: which expresses the level of detail in the structure of a model, ranging from single-block code to multiple levels.

To ensure systematic composability of models, it is important to address how the composition is resolved along each of these dimensions. This is often achieved by creating *wrappers* around the existing library components to enable communication of data values between different modules and co-ordination between them. Since we are dealing with component descriptions in programming languages, the wrappers here refer to pieces of code that enable reuse of existing component models. Using programming languages, there are several ways in which such wrappers can be built. A common strategy is by using inheritance

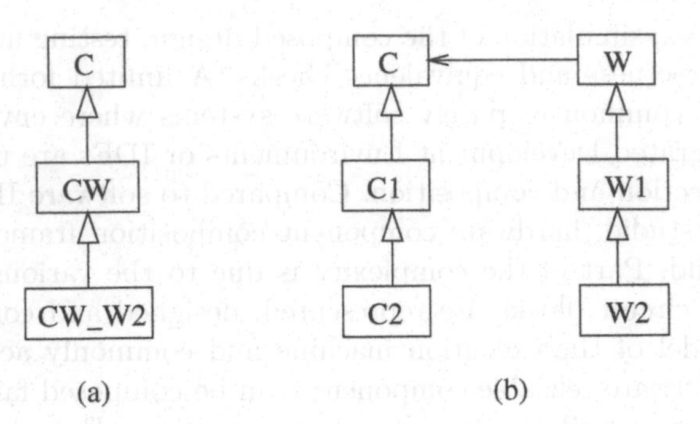

Fig. 1. Wrapper implementation strategies: (a) by inheritance (b) by composition

available in most object-oriented programming languages. In this approach, the wrapper is programmed by manually inserting code to align various design axes inside the inherited class. The component and the wrapper have a *common self* in this implementation; i.e. the wrapper and the component are the same object. As a result, the interoperability issues related to typing are resolved at compile time, and the wrapper and component have strong dependencies.

An alternative is to use a wrapper that, if needed, *delegates* to the design component. In this case, the component is not modified, and the wrapper and the component are two distinct objects. Modules from different libraries can be imported as is, and dynamically placed in wrappers at runtime.

While inheritance can quickly achieve interoperability in some cases, it is not a recommended approach for many reasons. For one, it may actually hinder reusability in the long term. Figure 1(a) illustrates the UML class diagram of the typical problem of inheritance-based composition (the *common-self* problem [21]). If a designer wants to reuse a component of class `C`, the class can be specialized by inheritance to a subclass `CW` to implement the wrapper functionality. Let us suppose that the behavior of the original class `C` is modified in the class `CW` by adding more functionality or by re-implementing a virtual function. If the class `CW` is to be reused in a different context, then it can be also inherited into a class `CW_W2` that implements more wrapper code to interoperate in the new integration context. The problem in this scenario is that all the three classes have a common self, and the original component has to be modified in every reuse context, via inheritance. Also, substituting an object of class `C` with a `CW` (which is legal in C++) may introduce subtle side effects. For instance, the wrapper code may modify the state of the component in ways that may not be obvious to the designer. Further, when reusing an inherited component, the class hierarchy must also be copied. By contrast, if in a reuse environment, the original component being reused is unaltered, then it keeps its identity distinct from the identity of the wrapper that contains the code for the interoperability. Figure 1(b) shows the UML diagram of how a wrapper hierarchy can be built for composition (the open arrow indicates an association). In this case, the wrappers are separate from the component object hierarchy, and the interoperability interface remains separated in the wrappers, and any call to functionality of the original component is delegated from the wrapper to the component. Researchers have suggested various ways of generating wrappers, or interface protocols. One such method which is often used in making various tools interoperate is by scripting. Examples of such scripting-based interoperability can be found in [19].

Among the prominent component composition frameworks Ptolemy [17] takes a very different approach to composition. Ptolemy views the interoperability between components as the problem of interoperating between their models of computations. As per our understanding, though not explicitly, but implicitly Ptolemy merges the four dimensions into the notion of models of computation. Ptolemy defines 'model of computation (MOC)' domains [12], and any component that can be described with Ptolemy belongs to one of the MOC domains. Ptolemy composes and simulates models by virtue of a hierarchy of *domain directors* that controls the simulation of a component encapsulated in its own domain [24, 13]. The global director can resolve the exchange of data values between the domains. Ptolemy's approach is clean and elegant, but it requires that the components be designed in the style of actors that conform to one of the Ptolemy domains. Also, it is not clear, how to use this framework in composing existing IP components, designed in multiple different languages, styles, and levels of abstractions, unless their MOCs are identified correctly.

2.2 Component Composition Frameworks: Desirable Features

A good Component Composition Framework (CCF) provides a composition language, and capabilities for dynamic composition, simulation and verification. The Composition language, either visual or textual, should be able to ask for components from the component library and should not have to worry about implementation types. The choice of types may be different between simulation and synthesis tasks. Automated type inference and type matching is useful not only along datavalue dimension but along other dimensions of interoperability. Hence, the framework must have automated support for selecting the correct type that makes the composition possible, with limited user intervention. However, just datatype matching is not enough, because interfaces may match in datatypes along the ports, but behaviors displayed at those ports may be drastically different due to differeing implementation, different models of computation etc.

The composition should be dynamic, in the sense, that one does not have to go through a recompile-test cycle when new components are added or replaced. Usually such a framework can be implemented in scripting, but that might sacrifice the efficiency of simulation, testing, coverage etc. So one must be able to simulate the composition of composed objects in compiled domain, without having to recompile the whole design for every change in the composition.

Given the importance of formal verification, a framework should at least be able to partially verify or enable constructively correct interface composition, or allow synthesis of the intervening protocol between the two interfaces.

In order to be able to compose components at different levels of abstraction, and/or models of computation, such meta information should be available about the components at a meta-level such that either the framework may use such meta-data for automated matching of components, or at least, can allow users to understand implications of composing two arbitrary components. We now examine the requirements imposed by these features.

Composition Languages: Unlike programming languages used for behavioral specification of components, the role of a composition language is to instantiate and connect the components. The component model describes the connections by dictating how and when things can be composed. A connection could also be thought of as a "relation" among components. Allen and Garlan [1] have proposed to consider connector separately from components in order to capture and isolate component interactions. There have been many papers about methodologies and benefits for interface-based design [18], the separation of communication from computation. In component frameworks, this has to be pushed further as component interaction, typing and modeling dimensions have to be separated from computation [7].

Partial Typing and Typing Abstraction at the Architectural Level: This is the ability to be typeless at the composition level, where connections and relations should be loosely defined. This helps the conceptual design of SOCs by saving the effort of manually specifying every detail during the architectural exploration. For instance, when changing a bit width of a control word it should be abstracted in a "virtual connection" with a "virtual type". Connections can be abstracted by loose typing. This can even be pushed further as components can be abstracted, where multiple versions of the same thing could be hidden behind a general facade, with varying types of assumptions on the environments. In order to convert the virtual architecture into something that can be simulated, or into an implementation, depending on the way the framework is used. This includes data type matching and behavioral type matching, and can include the automation of the verification of the validity of a composition and interface verifications. Many co-simulation environments have been implemented, bus-functional models are often used for interfacing. Guerin et. al. [11] provide a good perspective for co-simulation in a SystemC backplane environment. They use a mixed-level interface as a transducer between protocols for communication on different levels of abstraction. SystemC and SpecC do programming level integration, while tools like System Studio [20] use graphical integration by linking ports. In software engineering research, architecture description languages (ADL) [14] have addressed parts of the problems of orthogonolizing component definition from component assembly and to solve typing mismatches [10]. Colif [6] provides an architecture description language (ADL), for describing topologies.

2.3 Platform-Based Design and Component-Based Design

Platform-based design (PBD) [25] is often defined as the creation of a stable core-based architecture that can be rapidly extended, customized for a range of applications, and delivered to the customer for rapid deployment. Platform-based design requires a "standard" architecture, to which components are interfaced, and to which wrapper can be generated and PBD often uses "put()/get()" interfaces. Generally, a PBD provides structure to pure component-based design by providing architectural constraints on SOC implementations.

Coral [3] and Colif [5] use standard architectures and effectively implement platform-based design in the architectural domain using the interface unit as the CoreConnect bus (or some other industry standard bus like an AMBA bus). These approaches use channels as connector, and use channel refinement to attach properties to the connector in order to pick an interface implementation from a library, that is compatible with the standard bus. On a more general note, bus interfacing, or wrapper generation are more complicated in general component frameworks than in platform frameworks due to the absence of interface standards and connectors.

3 Component Composition in BALBOA

The BALBOA [2] [9] component composition environment is a layered environment that provides a component model with introspection and *partial* typing capabilities. Components are composed dynamically using wrappers. These automatically generated wrappers use 'split-level' interfaces to implement the composition rules, dynamic type determination and type inference algorithms. Split-level programming refers to system model generation and component programming in two different levels that are strongly connected by a matching class hierarchy and methods [16] . Split-level programming relieves the system engineers of programming artifacts and software engineering concerns and lets them focus on system architecture. The BALBOA component composition framework is used to build system models with an architectural perspective. The BALBOA framework is used for the following two different tasks:

1. **Architectural Design**: the system architect builds the overall system architecture by instantiating, connecting, configuring components, and establishing relationships among components;
2. **Component Design**: the library designer implements components or virtial components to populate the IP library using a programming language, such as C++. The implementation is restricted as much as possible on modeling a behavior or a structure.

The design of a library component has to be done by a designer who understands the language (e.g., C++/SystemC) The design of an architecture is done by a system architect, where the focus is on module instantiation and interconnection by using the architectural support in the component integration

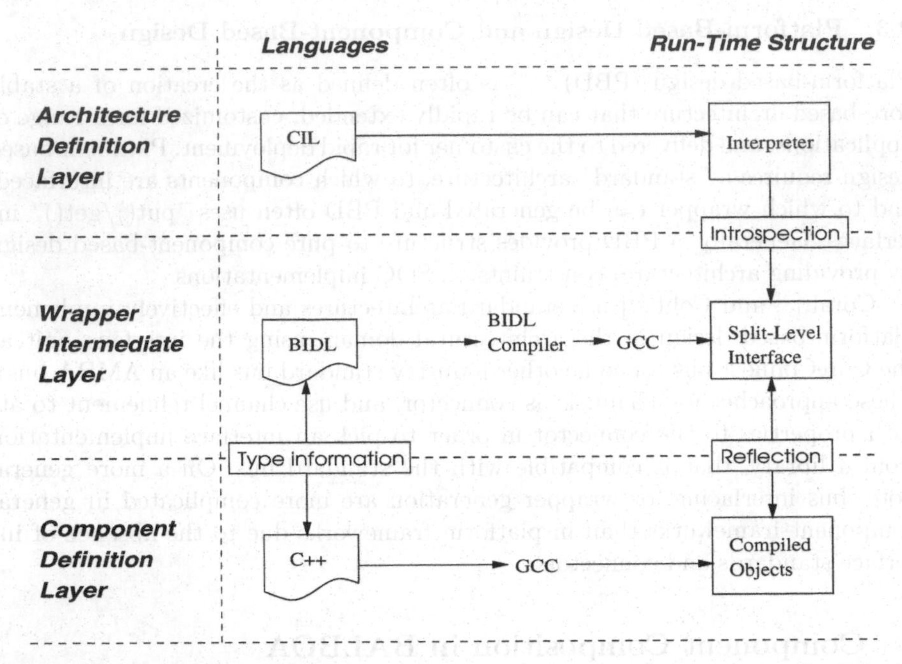

Fig. 2. Layering in the BALBOA environment: the languages are on the left side and the run-time structures on the right side of the figure

language (CIL). Figure 2 shows the layers in BALBOA. Languages are on the left side and the run-time structure on the right side. The description of the layers of Figure 2 is as follows.

The Architecture Definition Layer is where architectural structure is assembled from components using the Component Integration Language (CIL). The CIL is very close to an ADL but it implements a component model for component compositions and connections, an object model for object compositions, aggregations and associations. The CIL is built with Tcl, OTcl and TclCL extensions.

Tcl is used for the procedural and variable scripting basics. OTcl [23] provides object-oriented extensions to Tcl, to specify classes and instantiate objects. OTcl structures can be introspected by the "info" commands to query a class for its list of instances, an object for its type, list of attributes, and list of methods. TclCl is the link between OTcl and the C++ classes for combined manipulation. The CIL also uses a type system to abstract component types. Because the CIL is interpreted, we also refer to this layer as the *interpreted layer*.

The Component Definition Layer consists of a set of IP components stored in libraries. Any C++ class or object can be placed in this layer without needing to derive from a specific C++ class. Ideally, this layer can accomodate

C++ IP models in a range of libraries without affecting the implementation of the two other upper layers. This layer is also called the compiled layer.

The Intermediate Wrapper Layer is the link between the interpreted and the compiled layer. Each C++ objects instantiated in the environment is contained and manipulated by a *split-level interfaces* (SLI). This wrapper provides the mechanism for manipulation of the compiled object by the scripting layer. The split level interface implements the *reflection* and the *introspection* capabilities [4] of the environment. The reflection is the capability of the split-level interface to read or write the attributes, and to invoke the methods of the compiled object. Introspection is the capability of the CIL language to query the reflected information of a component, and to understand its own structure.

The information that is being reflected and introspected is generated by the BALBOA Interface Definition Language (BIDL) compiler. The BIDL compiler translates and expands the description of the type of the component to a format that the interpreter can understand. The BIDL has a role similar to the CPP preprocessor- however it does not do macro expansions but a customization of the split-level interface framework specific to every component.

One of the novelties of the BALBOA environment is the separation between component definition and architecture elaboration through split programming to take advantage of weaker typing dependencies for typing abstractions at the architectural level. Typing abstraction means that it is possible to reduce the type dependencies of the strongly typed compiled C++ layer, through careful type management at the wrapper level. The split-level interfaces can implement the type inference to keep the CIL description focused on component instantiations, compositions and connections.

BALBOA Typing Our motivation in providing the hardware designer to design in a "loose" typing environment comes from the observation that strict typing in C++ often results in excessive programming effort on the part of the system designer. This effort is particularly notable in case of using predefined IP libraries, where a component port type may be restricted to a subset of possible interface types. Our goal is to be to able to provide an environment that makes it easy to instantiate components and connect ports without specifying the C++ types completely. This is what we call *partial/loose* typing capability at the interpretive layer of our design environment. However, actual instantiation and running of a simulation cannot work without instantiating the correct concrete C++ types. We address this by type inferencing. In [8] we show that this problem is in general NP-Complete and we provide an efficient heuristic for solving the problem.

3.1 Using BALBOA

Component Design Figure 3 illustrates the tool flow and design process for the component designer. The lower part of the figure illustrates the flow for

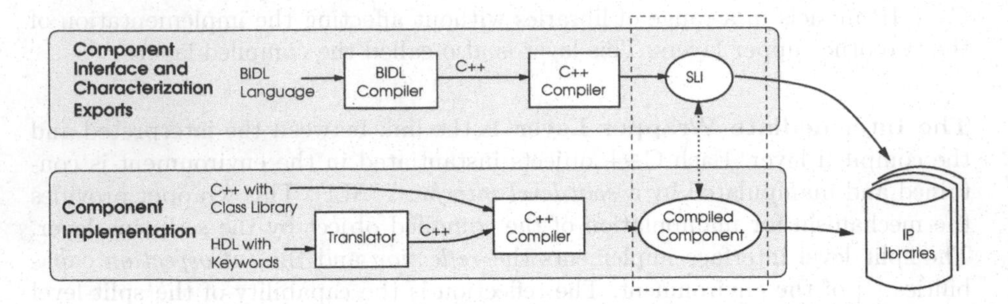

Fig. 3. Component design tool flow: write the component in C++, use the BIDL to characterize/export it and generate a SLI

component implementations in C++. The upper part of the figure shows the flow for the component characterization and the exportation of the interface of the components to the interpreted domain. The BIDL compiler generates C++ code to create and configure the split-level interface of a component and to generate the type system information and the specific code for the delayed instantiation and delayed typing. The delayed type instantiation happens after the type inference problem is resolved as alluded to in the previoius subsection. The BIDL compiler also generates the object model configuration specific to the component. The principal steps for using BIDL to export a C++ class to the interpreter are the following: the designer uses the header of the class into the BIDL description and removes the part to be hidden from the interpreted domain. Keywords are also added to configure the generation, such as component families, versioning and template classes handling and specification of available types. From the point of view of system architect, the component and the split-level interface can be the same entity, as shown in Figure 3 by the vertical dashed rectangle.

Figure 4 illustrates the internal architecture of a BALBOA component consisting of four blocks: the internal object (for example, such as a SystemC object or any other C++ object or component), the type system information (with an object model), the interpreted attributes/methods (that can be a reflection of the compiled attributes/methods), and the split-level interface routines. As shown in Figure 4, the split-level interfaces are the links between the interpreted domain and the compiled domain. Composition requests from the CIL script language are only interpreted in the SLIs. However, the simulation commands are delegated to the compiled components. Usually, the simulation control flow is kept only in the compiled layer because interpreted command execution in the SLI can be slow. However, the SLI layer can also interact with the simulation. For example, in our libraries we have a number of stimuli generators and monitors that use the interpreter control flow to compute stimuli and check assertion during the simulation.

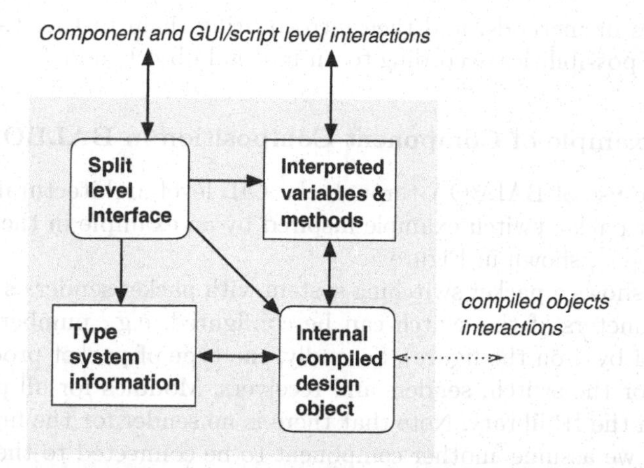

Fig. 4. Internal architecture for a BALBOA component

Component Integration Language (CIL) The CIL is somewhere in between a module interconnection language and an architecture description language. This is because the CIL is used to build connections, and to build new components or compose attributes or behaviors to existing ones. The basic composition unit in CIL is an entity. For example, a component called `c1` is instantiated with the command:

```
Entity c1
```

This component can be composed of a subcomponent `c2` by the command:

```
Entity c1.c2
```

The result of this command is the instantiation of an entity named `c2` inside `c1`. The syntax for the composition is the dot "." operator, which is also used to navigate hierarchies. The CIL implements introspection [4], which is the capacity of an object to query itself to know its structure, attribute and methods. It is similar to self-inspection. For instance, Tcl provides introspection capabilities with the "info" procedure, and Java also provides introspection through the reflection packages. The BALBOA environment implements and extends these models to add introspection using a `query` method to the split-level interfaces. The following characteristics of a component can be queried: name, SLI type, C++ type, kind, attributes and methods. For example, the following query:

```
c1 query attributes
=> c2
```

returns the list of attributes for component `c1`. In this case, there is only the `c2` attribute that is returned as result of the command. This attribute is visible in the interpreted domain, but other attributes might be present in the compiled domain, but not visible if they were not exported. Complex commands can be built to query each subcomponent for information. For instance: The environment's use model for design assembly is built through introspection, looking for attributes or methods, and then introspecting them further to find out the composition possibilities according to an internal object model.

3.2 An Example of Component Composition in BALBOA

We illustrate use of BALBOA through the CIL level architectural composition of a compact packet switch example inspired by an example in the SystemC-2.0 distribution [20], shown in Figure 5.

Figure 5 shows a packet switching system with packet senders `s` and receivers `r`. The parameters of the switch can be configured, e.g., number of ports, up to n by m (4 by 4 on the figure). Secondly, the type of packet processed can be configured for the switch, senders and receivers. Modules for all possible types are stored in the IP library. Note that there is no sender for the first port of the switch since we assume another component to be connected to the first port.

Figure 6 shows the CIL listing for a switch topology composition. Line 1 sets a variable to "4" for the number of ports. Line 3 instantiates the switch component, parametrized for 4 ports. Line 2 sets a variable to `Pkt` for the type of packets processed, and line 4 instantiates a signal with that sub-type. Lines 5-7 instantiate clocks for the senders, the switch and the receivers and lines 8, 13 and 19 connect the clocks to those components. The `for` loop on line 9 is parametrized to iterate for every port to instantiate a sender and receiver and connect them to the input and output ports of the switch. Lines 11 and 15 instantiate a sender and a receiver, lines 12 and 16 instantiate the signal connectors and lines 17, 18, 20 and 21 establish the connections between the components.

When the `pkt_in0` signal is connected to the switch, the split-level interface of the switch will pick the packet switch type with four ports that processes the `Pkt` packet type, among all possible switch implementation types in the library. The types for the signals, senders and receivers will be inferred to transmit and process the `Pkt` types. It is required that these types be defined in the libraries for

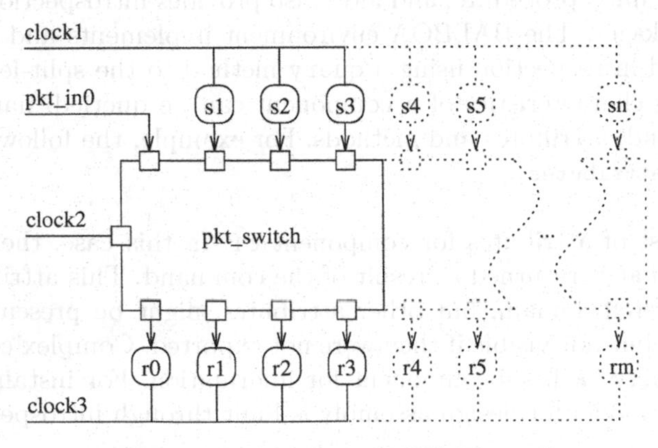

Fig. 5. Packet Switch CIL Example

```
1  set NUMBER_OF_PORTS 4X4
2  set PACKET_TYPE     Pkt
3  Pkt_Switch pkt_switch -number_of_ports $NUMBER_OF_PORTS
4  Signal     pkt_in0 -subtypes {$PACKET_TYPE}
5  Clock      clock1 -period 75 -duty_cycle 0.5 -start_time  0.0
6  Clock      clock2 -period 30 -duty_cycle 0.5 -start_time 10.0
7  Clock      clock3 -period 15 -duty_cycle 0.5 -start_time  0.0
8  connect pkt_switch.CLK to clock2
9  for {set i 0} {$i<$NUMBER_OF_PORTS} {incr i} {
10   if {$i>0} {
11     Sender   s$i -id $i
12     Signal   pkt_in$i
13     connect  s$i.CLK          to clock1
14   }
15   Receiver r$i -id $i
16   Signal   pkt_out$i
17   connect  s$i.pkt_out      to pkt_in$i
18   connect  r$i.pkt_in       to pkt_out$i
19   connect  r$i.CLK          to clock3
20   connect  pkt_switch.in$i  to pkt_in$i
21   connect  pkt_switch.out$i to pkt_out$i
22 }
```

Fig. 6. CIL listing for a 4 ports packet switch composition

the split level interface to instantiate them. Of course, the same topology can be built using only C++. By comparison the description in C++ can be quite large (above 100 lines). One can find the C++ listing of a nonparameterized version of this example in the SystemC downloadable distribution (in the examples directory) [20]. Because of the regular structure of the packet switch structure with respect to the number of ports and types, the usage of the CIL leverages the following advantages for flexibility and abstraction:

1. *Static parametrization for regular structures*:
   In this example, the design structure generation is parametrized with respect to the number of ports and use the static for loop of the CIL in the parametrization. The for loop instantiates and connects the surrounding signals, sender and receiver components for every port of the switch.
2. *Name expansion for regular structures*:
   Names are expanded by the interpreter with interpreted variable values. Component names for the signals, senders and receivers are expanded by the interpreter: `pkt_in$i` is expanded with the value of the iteration counter `pkt_in0`, `pkt_in1`, etc. for the design structure of the example.
3. *Type inference*: The components and connections are introspected by the environment and the split-level interfaces. In this example, the components will be picked by the enviroment to process the `Pkt` data type for the switch, the signals, and the senders and receivers.

This example illustrates how BALBOA composition is done in a typeless fashion and how it enables a separation of concerns of type compatibility from concerns of composition and the architectural structure. It also has the advantage of avoiding recompilation cycles when changing parameter values. BALBOA has been used in moderately complex designs that have been taken from high level specification all the way to harware level description [9].

4 Closing Remarks

Component Composition Frameworks (CCFs) represent an exciting development in the area of high-level modeling of complex SOC functionalities. A successful adoption of CCF is likely to have a direct impact on the succesful management of complexity of the new generations of SOC designs. However, there are several technical challenges that must be overcome. The chief among them are: ensuring inherent composability and reuse of SOC components. The problem extends beyond large scale program constructions in software engineering where several advances in architectural modeling and design environments have occurred. The challenge is due to the diversity of the computation models, levels of abstractions used and the notion of correctness applicable to SOC components. Thanks to advances in understanding of models of computation and their cosimulations (as exemplified by Ptolemy) an important aspect of the problem seems to have been addressed well. Challenges remain, however, in aspects related to encapsulation and reusability of components. The BALBOA framework addresses this aspect of the problem by essentially deconstructing the task of component creation from component composition. The underlying programming and automatic wrapper generation capabilities are built upon known advances in software engineering, namely, reflection and introspection of the components and composition by delegation. The focus of our ongoing effort is to understand and develop techniques that can raise the level of abstraction used in interface composition, and exploit the system-level verification opportunities present in such an approach. We may view our approach as a bottom-up approach of SOC construction using reusable IP. On the other hand an enhancement in Ptolemy framework whereby Ptolemy actors can be mapped to existing IP components can be seen as a top-down approach. We are currently working in combining the two approaches to obtain a component composition framework that works both ways. We plan to enhance the top-down approach by raising abstraction a level further by means of aspect-oriented specification techniques[15].

References

[1] R. Allen and D. Garlan. A Formal Basis for Architectural Connection. *ACM Transactions on Software Engineering and Methodology*, July 1997. 668

[2] Balboa Project. Component Composition Environment. `http://www.ics.uci.edu/ ~balboa`. 669

[3] R. A. Bergamaschi, S. Bhattacharya, R. Wagner, C. Fellenz, M. Muhlada, F. White, W. R. Lee, and J.-M. Daveau. Automating the Design of SOCs Using Cores. *IEEE Design and Test of Computers*, 18(5):32–44, September-October 2001. 669

[4] Frank Buschmann, Regine Meunier, Hans Rohnert, Peter Sommerlad, and Michael Stal. *Pattern Oriented Software Architecture: A System of Patterns.* John Wiley and Sons, 1996. 671, 673

[5] W. Cesario, A. Baghdadi, L. Gauthier, D. Lyonnard, G. Nicolescu, Y. Paviot, S. Yoo, A. A. Jerraya, and M. Diaz-Nava. Component-based design approach for multicore socs. In *Proc. IEEE/ACM Design Automation Conf.*, 2002. 669

[6] W. Cesario, G. Nicolescu, L. Gauthier, D. Lyonnard, and A. Jerraya. Colif: a Multilevel Design Representation for Application-Specific Multiprocessor System-on-Chip Design. In *Proc. Int. Workshop on Rapid System Prototyping*, 2001. systemc. 668

[7] Frederic Doucet, Rajesh Gupta, Masato Otsuka, Patrick Schaumont, and Sandeep Shukla. Interoperability as a Design Issue in C++ Based Modeling Environments. In *Proc. Int. Symposium on System Synthesis*, 2001. 665, 668

[8] Frederic Doucet, Sandeep Shukla, and Rajesh Gupta. BALBOA: A Component-Based Design Environment for Composition and Simulation of System Level Models. *Submitted for Publication*, 2002. 671

[9] Frederic Doucet, Sandeep Shukla, Rajesh Gupta, and Masato Otsuka. An Environment for Dynamic Component Composition for Efficient Co-Design. In *Proc. Design Automation and Test in Europe Conf.*, 2002. 669, 676

[10] David Garlan, Robert Allen, and John Ockerbloom. Architectural Mismatch: Why Reuse Is So Hard. *IEEE Software*, November 1995. 668

[11] Patrice Gerin, Sungjoo Yoo, Gabriela Nicolescu, and Ahmed A. Jerraya. Scalable and Flexible Cosimulation of SoC Designs with Heterogeneous Multi-Processor Target. In *Proc. Asia-South Pacific Design Automation Conf.*, 2001. 668

[12] Edward A. Lee and Alberto Sangiovanni-Vincentelli. A Framework for Comparing Models of Computation. *IEEE Trans. on Computer-Aided Design of Integrated Circuits and Systems*, December 1998. 667

[13] Edward A. Lee and Yuhong Xiong. System-Level Types for Component-Based Design. Technical Report ERL M00/8, UCB, Febuary 2000. 667

[14] Nenad Medvidovic and Richard N. Taylor. A Classification and Comparison Framework for Software Architecture Description Languages. *IEEE Trans. on Software Engineering*, January 2000. 668

[15] M. Mousavi, M.R.V. Chaudron, G. Russello, M. Reniers, T. Basten, A. Corsaro, S. Shukla, R. Gupta, and D. Schmidt". Using Aspect-GAMMA in Design and Verification of Embedded Systems. In *Proc. High Level Design Validation and Test Workshop*, 2002. 676

[16] John K. Ousterhout. Scripting: Higher-Level Programming for the 21st Century. *IEEE Computer*, March 1998. 669

[17] The Ptolemy 2 Project, UC Berkeley, `http://ptolemy.eecs.berkeley.edu/`. 667

[18] James A. Rowson and Alberto Sangiovanni-Vincentelli. Interface-Based Design. In *Proc. IEEE/ACM Design Automation Conf.*, 1997. 668

[19] Simplified wrapper and interface generator (SWIG) `http://www.swig.org`. 667

[20] SystemC. OSCI. `http://www.systemc.org`. 668, 674, 675

[21] Clemens Szyperski. *Component Software: Beyond Object Oriented Programming.* Addison-Wesley, 1998. 666

[22] Virtual Socket Interface Alliance, `http://www.vsi.org` . 664

[23] David Wetherall and Christopher J. Lindblad. Extending Tcl for Dynamic Object-Oriented Programming. In *Tcl/Tk Workshop*, 1995. 670
[24] Y. Xiong and E. A. Lee. An Extensible Type System For Component Based Design. In *Sixth International Conference on Tools and Algorithms for the Construction and Analysis of Systems*, 2000. 667
[25] CoWare N2C `http://www.coware.com`. 669

Low Power Distributed Embedded Systems: Dynamic Voltage Scaling and Synthesis*

Jiong Luo and Niraj K. Jha

Department of Electrical Engineering, Princeton Univ., Princeton
NJ, USA 08544
{jiongluo,jha}@ee.princeton.edu

Abstract. In this paper, we survey multi-objective system synthesis algorithms for low power real-time systems-on-a-chip (SOCs), distributed and wireless client-server embedded systems, distributed embedded systems with reconfigurable field-programmable gate arrays (FPGAs), as well as distributed systems of SOCs. Many of these synthesis algorithms target simultaneous optimization of different cost objectives, including system price, area and power consumption. Dynamic voltage scaling has proved to be a powerful technique for reducing power consumption. We also survey several dynamic voltage scaling techniques for distributed embedded systems containing voltage-scalable processors. The dynamic voltage scaling algorithms can be embedded in the inner-loop of a system synthesis framework and provide feedback for system-level design space exploration. Besides voltage-scalable processors, dynamically voltage-scalable links have also been proposed for implementing high performance and low power interconnection networks for distributed systems. We survey relevant techniques in this area as well.

1 Introduction

Reducing power consumption is one of the crucial design considerations for modern electronic systems, in order to reduce chip packaging and cooling costs, to increase system reliability, as well as to extend battery lifetime of portable systems. Power optimization can be targeted at different levels of the design hierarchy during the design process. We concentrate our attention on two specific areas at the system level: system synthesis and scheduling.

System synthesis is the problem of automatic generation of an embedded system architecture from a given embedded system specification. A system synthesis algorithm selects hardware and software processing elements (PEs) upon which tasks execute. It also generates a communication architecture, composed of communication links, which connect the different PEs in the system architecture. In addition, the system assigns each task (communication event) to a PE (communication link). Finally, schedules are produced for tasks and communication events, such that real-time constraints are met.

* Acknowledgments: This work was supported by DARPA under contracts DAAB07-00-C-L516 and DAAB07-02-C-P302.

S. Sahni et al. (Eds.) HiPC 2002, LNCS 2552, pp. 679–693, 2002.

Reducing the price of a system (as well as area for SOCs) has been the traditional concern of system synthesis. The growing importance of power consumption, however, requires the system synthesis algorithm to explore the design space in a multi-objective way, with optimization of system price, area and power consumption addressed simultaneously. In Section 2, we introduce a series of embedded system synthesis approaches which consider power consumption as one of the design objectives. These approaches address unique concerns for different types of systems, including real-time SOCs, distributed and wireless client-server embedded systems, distributed systems of SOCs, as well as distributed embedded systems with reconfigurable FPGAs.

Dynamic voltage scaling refers to dynamic adjustment of the supply voltage to the minimum level required for a PE to work at a desired clock frequency. There are many commercially available voltage-scalable processors, including Intel's Xscale [1], Transmeta's Crusoe [2], and AMD's mobile processors with AMD PowerNow! technology support [3]. Voltage scaling has been widely acknowledged as a powerful technique for trading off power consumption and delay. In Section 3, we introduce several variable voltage scheduling approaches targeting real-time distributed embedded systems. Dynamically voltage-scalable links have also been proposed recently for implementing high performance and low power interconnection networks for distributed systems. We survey relevant techniques in this area in Section 4. Section 5 presents conclusions and future work.

2 System Synthesis Algorithms

In this section, we survey system synthesis algorithms for distributed embedded systems. The input specification of real-time distributed systems is frequently given in terms of a set of task graphs. A task graph is a directed acyclic graph in which a node is associated with a task and an edge is associated with the amount of data transferred between tasks. A resource library consisting of different PE and communication link types is given. The worst-case execution time and average power consumption of each task, for each PE from the resource library on which the task can run, are usually provided. A PE can be a general-purpose processor, FPGA or application-specific integrated circuit (ASIC). The period associated with a task graph indicates the time interval between its successive executions. A hard (or soft) deadline, by which time the task must (or had better) complete execution, is given for each sink node and some intermediate nodes. Hard deadlines have to be met, while soft deadlines can be violated. A multi-rate system consists of multiple task graphs with different periods.

Three tasks must be performed by any system synthesis algorithm:

- Allocation: Determine the quantity of each type of PE and determine the communication architecture.
- Assignment: Assign each task (communication event) to a PE (communication link).

– Scheduling: Generate a valid schedule for all the tasks and communication events, which guarantees all real-time constraints.

Optimal system synthesis is an intractable problem. Allocation/assignment and scheduling have each been proven to be NP-complete for distributed systems. Earlier work involved mixed integer linear programming [4], which provides optimal solutions. However, its application is limited to very small task graphs. Iterative improvement algorithms are also available [5, 6], which start from a sub-optimal initial solution and iteratively make local changes to the initial solution. The work in [7] employs simulated annealing and tabu search for hardware/software partitioning. The work in [8] targets assignment and scheduling for a distributed embedded system of heterogeneous multiprocessors, with a hierarchical memory model incorporated. However, the above works do not incorporate power dissipation as a design objective. Next, we survey some system synthesis approaches, which target low power distributed embedded systems, spanning a wide range of system types, including distributed embedded systems, SOCs, wireless client-server systems, as well as distributed embedded systems composed of SOCs.

2.1 Low Power Distributed Embedded System Synthesis

We first survey system synthesis algorithms for low power distributed embedded systems.

COSYN: A constructive algorithm based system synthesis technique [9], called COSYN, was one of the first works to take power dissipation into consideration during system synthesis. Constructive algorithms build a system by incrementally adding components. COSYN starts with multi-rate periodic task graphs with real-time constraints as well as a resource library, and produces a low cost heterogeneous distributed embedded system meeting the constraints. It models various forms of inter-PE communication links, such as point-to-point, bus, local area network (LAN), etc. It employs a task clustering technique to reduce the complexity of assignment. COSYN can optimize system cost while guaranteeing real-time constraints. An extension of COSYN, termed COSYN-LP, targets synthesis of low power embedded systems. COSYN provides a very fast system synthesis approach based on heuristics. However, in general, constructive algorithms may get trapped in local minima, especially when large systems are constructed. COSYN also suffers from an inability to explore the design space in a true multi-objective way.

MOGAC: In [10], a system synthesis algorithm, called MOGAC, for low power distributed heterogeneous embedded systems, is presented. It partitions and schedules embedded system specifications consisting of multi-rate periodic task graphs. It uses a general model for bus and point-to-point communication links. It uses an adaptive multi-objective genetic algorithm that can escape local minima. Price and power consumption are simultaneously optimized while hard real-time constraints are guaranteed. The multi-objective optimization strategy allows a single co-synthesis run to produce multiple solutions which trade off different architectural features, in order to explore the *Pareto-optimal* set of the design space. The *Pareto-optimal* set refers to those solutions which are better than others in at least one design feature.

MOGAC maintains a pool of solutions which evolve in parallel over time. Solutions evolve from one generation to the next, through reproduction, random mutation as well as information crossover. Solutions are encoded in various strings. This is done in such a way that the locality of solutions is maintained (i.e., closely connected features of the solution are kept as neighbors in the string). Architectures with equivalent PE and communication resource allocations are clustered together. However, the link connectivity strings of solutions, and the task and communication assignments in the same cluster may differ. Crossover of assignment and link connectivity strings occurs between solutions in the same cluster. Mutation of these strings is performed on individual solutions within a cluster. This helps prevent assignment crossover from leading to invalid architectures. Mutation of the allocation string can happen to a single cluster. Crossover of allocation strings occurs between entire clusters, destroying the solutions within the clusters.

Optimization in MOGAC is multi-objective. A solution's cost is evaluated using a multi-dimensional vector, which includes parameters such as number of unscheduled tasks, number of unscheduled communication events, price of the architecture, and power consumption. Instead of collapsing all the system costs into one variable, it uses *Pareto-rank* to rank different solutions. The *Pareto-rank* of a solution is the number of other solutions which do not dominate it. A solution dominates another if it is superior in every dimension of the cost vector. Multiple solutions which are good in different cost dimensions are kept in the solution pool and evolve in parallel. It uses a Boltzmann trial to compute the probability of choosing solution B over solution A based on a global temperature T as follows

$$(1 + e^{(rank_A - rank_B)/T})^{-1} \tag{1}$$

Initially, T is high, which allows lower-ranked solutions to also have a chance to remain in the solution pool, in order to escape local minima. As T decreases, higher-ranked solutions have a much better chance to remain in the solution pool than lower-ranked solutions. This allows the algorithm to converge to good solutions in the neighborhood.

Synthesis of Distributed Embedded Systems with Dynamically Reconfigurable FPGAs: The work presented in [11] describes a multi-objective system synthesis algorithm which uses dynamically reconfigurable FPGAs. In such FPGAs, parts of the FPGA can be reconfigured in a few clock cycles, without disturbing the execution of the remaining logic. Thus, they offer a parallel and flexible hardware platform. In [11], a two-dimensional, multi-rate cyclic scheduling algorithm is presented for embedded systems containing such FPGAs. It is two-dimensional because it needs to solve the problem in both the space and time domains - where to place the different tasks in the FPGA and when to execute them.

A dynamically reconfigurable FPGA is modeled as a frame-based parallel hardware platform. In this model, a frame is the atomic unit that can be dynamically reconfigured. The reconfiguration of one frame and the execution of other frames can be done in parallel. However, the reconfiguration of multiple frames can only be done sequentially. Each task can be implemented on several contiguous frames. For each frame, the task has a specific configuration pattern. A task's configuration patterns have to be loaded into the FPGA reconfiguration memory before its execution. The scheduling algorithm uses a list scheduling approach, with task priorities based on real-time constraints and reconfiguration overhead. It has a location assignment policy to take care of reconfiguration prefetch, configuration pattern reutilization, configuration pattern eviction, and avoidance of FPGA reconfiguration memory fragmentation. The scheduling algorithm is carefully designed, to not only minimize schedule length, but to also reduce reconfiguration power significantly. Reconfiguration power can be a very substantial percentage of overall FPGA power. The synthesis system is based on an evolutionary algorithm, and can simultaneously optimize both system price and power consumption, with the power overhead of FPGA reconfiguration incorporated.

2.2 Low Power SOC Synthesis

Initial work on low power SOCs dealt with SOC architectures containing a single CPU and a single ASIC. These are the works surveyed first.

One-CPU-One-ASIC SOC Architectures: The work in [12] considers a fixed SOC architecture composed of a processor, ASIC, instruction cache, data cache and main memory. It investigates hardware/software partitioning of the input specification between the processor and ASIC. It also investigates the effect of different memory hierarchy configurations, such as the data cache size, the instruction cache size, the cache associativities and line sizes, and main memory size, on power dissipation of the SOC. An MPEG-2 encoder is chosen as a case study. The extension of this method in [13] assumes the same SOC architecture, with its hardware part (ASIC) fixed beforehand. It investigates the design space of software transformations and memory hierarchy configurations. The software transformations include procedure inling and loop unrolling. It investigates the effect of interactions between software transformations and cache and memory configurations on performance and overall system energy. It explores the design space in a multi-objective way through a genetic algorithm. Another extension of this approach is given in [14].

MOCSYN: A comprehensive SOC synthesis algorithm [15], called MOCSYN, targets synthesis of single-chip systems based on multiple intellectual property

(IP) cores. The IP cores can be protocol processors, general-purpose processors, micro-controllers, digital signal processors, memory, etc. MOCSYN uses an adaptive multi-objective genetic algorithm as well. Integrated circuit price, power consumption, and area are optimized under hard real-time constraints. Although its optimization framework is similar to that of MOGAC, MOCSYN addresses several problems unique to single-chip systems. It solves the problem of clock selection for different cores (i.e., selecting the best clock frequencies for cores), subject to a trade-off between execution time and power consumption. It provides a bus formation algorithm, subject to a trade-off between ease of layout and reduction of bus contention. In addition, it derives a block placement for the cores within its inner loop, allowing accurate estimation of global wire delays, wire power consumption as well as silicon area at the system level. It follows the locally synchronous globally asynchronous paradigm.

Functional Partitioning for Distributed Embedded Systems of SOCs: As the functionality being implemented by embedded systems is rapidly increasing, one may be required to implement a distributed system of SOCs (in which system nodes are themselves SOCs). The challenging issues involved in this problem are partitioning of a system specification into different SOCs, synthesis of each partition, as well as synthesis of the communication architecture connecting different SOCs. The work presented in [16] merges functional partitioning and system synthesis (allocation, assignment, and scheduling) into a unified framework.

This work uses a genetic algorithm and is integrated into the SOC synthesis tool, MOCSYN. The solutions are encoded in strings and classified into clusters. Each cluster has the same allocation and partition, and is represented by several integer arrays. The number of arrays is determined by the number of SOCs, the length of each array is determined by the number of core types, and an entry in the array indicates the number of instances of the corresponding core type allocated. Within a cluster, different solutions have different assignments. The task assignment is encoded as a three-tuple index, consisting of the core type, core instance, and the SOC to which the task is assigned. Hence, partitioning is combined with both allocation and assignment. The mutation for allocation operates on arrays within a cluster. The crossover for allocation operates on two randomly selected arrays from two clusters. The mutation and crossover for task assignment operate on solutions within one cluster. Under a genetic optimization framework, this work can produce multiple distributed SOC-based embedded system architectures that trade off the overall distributed system price and power consumption, under real-time constraints and an area constraint for each SOC.

2.3 COWLS: Synthesis of Distributed Wireless, Low-Power Client-Server Systems

In [17], a system synthesis algorithm, called COWLS, is presented, which targets embedded systems composed of servers and low-power clients. Clients and servers can communicate with each other through wireless communication links. COWLS allows both hard and soft real-time constraints. Since clients are often portable mobile systems, reducing their power consumption is crucial. COWLS simultaneously optimizes the price of the client-server system, the power consumption of the clients, bandwidth requirements, and the response times of soft tasks which have only soft deadlines, without violating any hard real-time constraints. The optimization engine in COWLS, again, is a multi-objective evolutionary algorithm.

3 Variable Voltage Scheduling

One of the tasks in system synthesis is to generate a valid schedule which can meet the real-time constraints and guarantee the task precedence relationships in the input specification. When voltage scaling is incorporated, the scheduling algorithm should also be able to intelligently assign frequencies and voltages to voltage-scalable processors running different tasks in the distributed embedded system, in order to maximize energy savings, without violating any real-time constraints. Variable voltage scheduling is a powerful technique for trading off delay and power consumption, as illustrated through the following equations. For today's deep sub-micron CMOS technology, the processor clock frequency, f, can be expressed in terms of the supply voltage, V_{dd} , and threshold voltage, V_t, as follows (k is a constant) [18]

$$f = k(V_{dd} - V_t)^{\alpha}/V_{dd} \tag{2}$$

where $1 < \alpha \leq 2$.

The power consumption, P, can be expressed in terms of the clock frequency, f, switching activity, N, capacitance, C, and the supply voltage, V_{dd}, as:

$$P = 1/2fNCV_{dd}^2 \tag{3}$$

which can be proven to be a convex function of f [19].

There is a rich literature addressing variable voltage scheduling for a set of independent tasks on a single processor [20]. The work in [19] exploits the characteristics of convex functions for optimization and gives an off-line algorithm, which generates a minimum-energy preemptive schedule for a set of independent tasks. The minimum-energy preemptive schedule, which is based on an earliest deadline first scheme, tries to achieve a uniform scaling of voltage levels of different tasks. The work in [21] points out that variations in power consumption among tasks invalidate the conclusion in [19] that a uniform scaling is optimal. It proposes an iterative slack allocation algorithm based on the Lagrange Multiplier method. The work in [22] proposes a compiler-directed variable voltage scheduling algorithm for a single task based on program regions. It tries to find memory-bound program regions where a CPU slowdown may be hidden by the memory hierarchy access latency. Other works addressing variable voltage scheduling of independent tasks on a single processor include those in [23, 24, 25, 26].

There is also work addressing variable voltage scaling for tasks with precedence relationships in distributed systems. The work in [27] is based on a hybrid global/local search optimization. It uses a genetic algorithm with simulated heating for global search, and hill climbing and Monte Carlo techniques for local search. The work in [28] formulates the problem as an integer programming problem. Other works are mainly based on list scheduling. List scheduling has been widely accepted for scheduling of distributed systems and can provide close to optimal results in many cases [29]. To adapt a list scheduling algorithm to a variable voltage scheme, there are two aspects that need to be addressed. The first one is to derive an efficient way to assign priorities to different scheduling events. The second one is to allocate slack efficiently to maximize energy savings. In a heterogeneous distributed embedded system, slack allocation should take into account variations in switched capacitances for running different tasks on different PEs. In the following sections, we introduce several list-scheduling based approaches.

3.1 LEneS: Task-Scheduling for Low-Energy Systems Using Variable Voltage Processors

LEneS is a list scheduling algorithm that uses a special priority function, which can trade off energy reduction and delay [30]. The schedule is constructed step by step. At each step, a ready task is selected based on its assigned priority and is scheduled in a time step at which the partial schedule can achieve a maximum probabilistic energy reduction. This work is novel in the aspect that probabilistic evaluation of energy reduction is exploited when the schedule decisions are only partially determined. However, the complexity of this approach can be high due to the number of discrete time steps that need to be evaluated. Moreover, probabilistic evaluation of energy reduction of a partial schedule may not yield the best decision for the final schedule.

3.2 Work by Schmitz et al.

The work in [31] uses a genetic algorithm to optimize task assignment, a genetic list scheduling algorithm to optimize the task priority, and an iterative slack allocation scheme, which allocates a small time unit to the task which leads to the most energy reduction in each step. The performance and complexity of this approach is dependent on the size of the time unit. This is determined empirically. The usage of a small time unit for task extension may lead to large computational complexity.

3.3 Critical-Path Based Analysis

Motivated by the work in [19], the work in [32] and its extended version [33] proposes a variable voltage scheduling scheme based on a static priority list scheduling algorithm and effective slack allocation via critical path analysis.

The slack allocation algorithm evaluates all the paths in a graph constructed by taking into consideration all the precedence relationships of the tasks and resource constraints of the distributed system. It locates the most critical path that minimizes the ratio of the total slack on that path to the total worst-case execution time on that path. The critical path is extended such that the available slack is uniformly distributed. Then the nodes on the critical path are deleted and the timing constraints of other nodes are restricted by the extended critical path. The algorithm continues until no path can be extended anymore. The critical path analysis algorithm determines a slack allocation for all the tasks, and the voltages and clock frequencies of tasks can be determined based on the slack allocation. This algorithm can be applied to single processor systems as well and is optimal for the case of non-preemptive fixed priority scheduling on a single processor. Although critical path based analysis, in conjunction with scheduling event execution order optimization, can achieve promising results, as shown in [32], it cannot target the scenario in which various tasks have different switching activities, or target heterogeneous systems in which different voltage-scalable PEs have different voltage scaling characteristics.

3.4 Energy-Gradient Driven Iterative Slack Allocation for Heterogeneous Distributed Embedded Systems

The variable voltage scheduling approach presented in [34] addresses variations in power consumption of different tasks and characteristics of different voltage-scalable PEs in an effective and efficient manner. It uses simulated annealing to optimize task priorities and a fast slack allocation scheme based on energy gradients.

Given the number of clock cycles, η_i, for executing task i with switching activity, N_i, its energy consumption, E_i, under supply voltage, V_i, and clock frequency, f_i, is given by

$$E_i = t_i * (1/2 f_i N_i C V_i^2) \tag{4}$$

where

$$t_i = \eta_i / f_i \tag{5}$$

is the task's corresponding execution time under clock frequency f_i, and C is the capacitance.

When the execution time of task i is extended by dt time unit, the clock period can be extended by dt/η_i. Correspondingly, the supply voltage can be scaled down based on Equation (2), and the energy consumption of task i is reduced as well. This work defines the energy gradient, G_i, as the negative of the derivative of energy consumption with respect to the execution time of task i under supply voltage V_i:

$$G_i(V_i, N_i) = -\frac{\partial E_i(V_i, N_i)}{\partial t} = 1/k * N_i C * \frac{V_i(V_i - V_t)^{(1+\alpha)}}{V_i(\alpha - 1) + V_t} \tag{6}$$

Suppose the total slack in the system is *total_slack*. If the overall slack is represented as an interval $[0, total\_slack]$, then the energy reduction resulting from voltage scaling, after the overall execution times of all the tasks are extended by *total_slack*, can be represented as:

$$\delta E = \int_0^{total\_slack} G(t)dt \tag{7}$$

where $G(t)dt$ refers to the energy reduction achieved by allocating dt unit of slack $[t, t+dt]$ from overall slack to execution of some task. An optimal energy reduction is achieved by maximizing the integral of energy gradients over the overall slack.

One can observe that the energy gradient function at a given switching activity is a monotonically decreasing function with respect to the supply voltage. For a single processor, an optimal slack allocation can be achieved by always allocating slack to the set of extensible tasks with the highest energy gradient. A task is defined as non-extensible if extending its execution time would lead to violations of its own deadline or that of other tasks. The allocation of slack to a set of tasks at the highest energy gradient level should be done in a balanced way such that their energy gradients can be kept at the same level. For distributed systems, the situation is complicated by precedence relationships among tasks and resource contentions on PEs and communication links. However, the slack allocation strategy described above is still very effective as a heuristic.

The above slack allocation scheme iteratively decreases supply voltages of tasks. The earliest start time of a task is defined as the earliest time by which it can begin its execution without violating its arrival time constraint and precedence relationships, and the latest finish time of a task is defined as the latest possible time by which an event must complete its execution without violating its deadline as well as that of any other events and without violating any precedence relationships. It exploits the earliest start times and latest finish times of tasks, to evaluate the validity of the schedule when multiple tasks need to update their execution times (due to voltage scaling), as well as to annul any invalid updates, in linear time. It shows significant improvement in complexity, compared to previous work [31], without sacrificing performance.

3.5 Dynamic Power-Conscious Joint Scheduling of Periodic Task Graphs and Aperiodic Tasks

Besides periodic task graphs, the embedded system may also contain soft aperiodic tasks. A soft aperiodic task is invoked for execution at any time and only has a soft deadline. One only needs to minimize the response times of such tasks. Except for the work in [32, 33], all the above works only target static scheduling of periodic tasks, whose characteristics are based on an off-line worst-case analysis. They do not consider soft aperiodic tasks in the system, and also do not take care of the actual execution times (which may be smaller than the specified worst-case execution times) observed on-line. The work in [32, 33] proposes dynamic power-conscious joint scheduling of periodic task graphs and soft aperiodic tasks. It performs resource reclaiming when the actual execution time of a task is smaller than its worst-case execution time, as well as slack stealing by using slack in periodic tasks to facilitate the service of soft aperiodic tasks. The voltages and frequencies of periodic tasks are adjusted adaptively to the on-line workload, by considering dynamic execution time variations and service of aperiodic tasks.

4 Dynamic Voltage Scaling with Voltage-Scalable Links

In this section, we survey various techniques which exploit dynamically voltage-scalable communication links. Reducing latency and increasing throughput were conventionally the primary goals for interconnection network design. However, as the demand for system bandwidth, and hence correspondingly the power dissipation, is increasing, interconnection networks are becoming power/energy limited as well. Some previous work has addressed power modeling of interconnection networks [35, 36]. The channels of interconnection networks are composed of high-speed links. In chip-to-chip networks, link circuitry power has been shown to be a significant contributor to interconnection network power. For example, designers of Alpha 21364 [37] estimate that the router and links consume 25W out of the total chip power of 125W, with the router core consuming 7.6W, link circuitry consuming 13.3W, clocks consuming 2W, and other circuitry consuming the remaining (from [36]). To reduce the link power consumption, variable-frequency links have been proposed for both parallel links [38] and serial links [39], which can adaptively regulate the supply voltage to a desired link frequency, in order to exploit the variations in bandwidth requirement. A high-speed link is composed of a transmitter, which converts digital binary signals into electrical signals; a receiver, which converts electrical signals back to digital data; a signaling channel and a timing recovery block which compensates for delay through the signaling channel. In addition, a variable-frequency link needs an adaptive power-supply regulator which can provide the minimum supply voltage for a desired clock frequency, and replicate the supply voltage to the I/O subsystem. It also requires the I/O transceiver to be able to operate in this dynamically scaled supply voltage environment. In [40], a digital sliding controller design has been presented for the adaptive power-supply regulator.

The work in [41] investigates dynamic voltage scaling policies for links in high-performance interconnection networks. It exploits a history-based algorithm, where each router predicts future communication traffic based on past link utilization. The link frequency is dynamically adjusted based on this prediction and a voltage transition overhead. It uses an exponential weighted average to combine both short-term and long-term utilization history. The algorithm only relies on local traffic information. Therefore, the hardware in each router to implement the dynamic voltage scaling algorithm can be kept quite simple. A flit-level interconnection network simulator is used to evaluate the proposed algorithm. The simulator supports pipelined virtual-channel (VC) routers with credit-based flow control, and both deterministic and adaptive routing algorithms. This approach can realize up to 4.3X power savings with an average 27.4% latency increase and 2.5% throughput reduction. This work is applicable to both general-purpose multi-processor systems as well as distributed embedded systems.

5 Conclusions and Future Work

In this paper, we surveyed techniques for synthesis of distributed embedded systems and dynamic voltage scaling for processors and communication links. Next, we point towards two new directions which deserve further investigation for low power distributed embedded systems.

- As the technology is advancing, the supply voltage for processors is decreasing. This diminishes the impact of dynamic voltage scaling on processor power consumption. Leakage power is increasing and becoming comparable to dynamic power consumption as technologies continue to scale. Therefore, as pointed out in [42], a combined technique which can address both dynamic power and leakage power is required. Adaptive body biasing is a promising technique which can reduce leakage current exponentially. Simultaneous dynamic voltage scaling and adaptive reverse body biasing has been proposed in [42] to control both dynamic power and leakage power. This achieves better results than using dynamic voltage scaling alone. This technique needs to be extended to distributed embedded systems, in order to effectively reduce both leakage power and dynamic power at the system level.
- The work in [41] only addresses non-real-time traffic when performing dynamic voltage scaling for links. However, interconnection networks may need to support both real-time and non-real-time traffic. Real-time traffic requires quality of service (QoS) guarantees, such as satisfaction of delay, delay jitter and throughput constraints. This requires that dynamic voltage scaling policies should not only be able to trade off delay and power consumption, as illustrated in [41], but also be able to meet QoS constraints for real-time traffic. The history-based approach alone is not sufficient for this purpose. A method needs to be developed which is able to handle the mixed traffic type.

References

[1] **http://www.intel.com/design/intelxscale/** 680
[2] **http://www.transmeta.com** 680
[3] **http://www.amd.com** 680
[4] S. Prakash and S. Parker "SOS: Synthesis of Application-Specific Heterogeneous Multiprocessor Systems," *J. Parallel & Distributed Computing*, vol. 16, pp. 338-351, Dec. 1992. 681
[5] T.-Y Yen and W. Wolf, "Communication synthesis for distributed embedded systems," in *Proc. Int. Conf. Computer-Aided Design*, pp. 288-294, Nov. 1995. 681
[6] J. Hou and W. Wolf, "Partitioning methods for hardware-software co-design," in *Proc. Int. Workshop hardware/Software Codesign*, pp. 70-76, Mar. 1996. 681

[7] P. Eles, Z. Peng, and K. Kuchcinski, "System level hardware/software partitioning based on simulated annealing and tabu search," *Kluwer J. Design Automation for Embedded Systems*, vol. 2, no. 1, pp. 5-32, Jan. 1997. 681

[8] Y. Li and W. Wolf, "A task-level hierarchical memory model for system synthesis of multiprocessors," in *Proc. Design Automation Conf.*, pp. 153-156, June 1997. 681

[9] B. P. Dave, G. Lakshminarayana, and N. K. Jha, "COSYN: Hardware-software co-synthesis of embedded systems," in *Proc. Design Automation Conf.*, pp. 703-708, June 1997. 681

[10] R.P. Dick and N.K. Jha, "MOGAC: A multiobjective genetic algorithm for hardware-software co-synthesis of hierarchical heterogeneous distributed embedded systems," *IEEE Trans. Computer-Aided Design*, vol. 17, no. 10, pp. 920-935, Oct. 1998. 681

[11] L. Shang and N.K. Jha, "Hardware-software co-synthesis of low power real-time distributed embedded systems with dynamically reconfigurable FPGAs," in *Proc. Int. Conf. VLSI design*, pp. 345-352, Jan. 2002. 682

[12] J. Henkel and Y. Li, "Energy-conscious HW/SW-partitioning of embedded systems: A case study of an MPEG-2 encoder," in *Proc. Int. Wkshp. HW/SW Co-design*, pp. 23-27, Mar. 1998. 683

[13] Y. Li and J. Henkel, "A framework for estimating and minimizing energy dissipation of embedded HW/SW systems," in *Proc. Design Automation Conf.*, pp. 188-193, June 1998. 683

[14] J. Henkel, "A low power hardware/software partitioning approach for core-based embedded systems," in *Proc. Design Automation Conf.*, pp. 122-127, June 1999. 683

[15] R.P. Dick and N.K. Jha, "MOCSYN: Multiobjective core-based single-chip system synthesis," in *Proc. Design, Automation & Test in Europe Conf.*, pp. 263-270, Mar. 1999. 683

[16] Y. Fei and N.K. Jha, "Functional partitioning for low-power distributed systems of systems-on-a-chip," in *Proc. Int. Conf. VLSI Design*, pp. 274-279, Jan. 2002. 684

[17] R.P. Dick and N.K. Jha, "COWLS: Hardware-software co-synthesis of distributed wireless low-power embedded client-server systems," in *Proc. Int. Conf. VLSI Design*, pp. 114-120, Jan. 2000. 684

[18] K.A. Bowman, B.L. Austin, J.C. Eble, X. Tang, and J.D. Meindl, "A physical alpha-power law MOSFET model," *IEEE J. Solid-State Circuits*, vol. 34, pp. 1410-1414, Oct. 1999. 685

[19] F. Yao, A. Demers, and S. Shenker, "A scheduling model for reduced CPU energy," in *Proc. Symp. Foundations of Computer Science*, pp. 374-382, Oct. 1995. 685, 686

[20] N.K. Jha, "Low power system scheduling and synthesis," in *Proc. Int. Conf. Computer-Aided Design*, pp. 259-263, Nov. 2001. 685

[21] A. Manzak and C. Chakrabarti, "Variable voltage task scheduling algorithms for minimizing energy," in *Proc. Int. Symp. Low Power Electronics & Design*, pp. 279-282, Aug. 2001. 685

[22] C. Hsu, U. Kremer, and M. Hsiao, "Compiler-directed dynamic voltage/frequency scheduling for energy reduction in microprocessors," in *Proc. Int. Symp. Low Power Electronics & Design*, pp. 275-280, Aug. 2001. 685

[23] T. Pering, T. Burd, and R. Brodersen, "The simulation and evaluation of dynamic voltage scaling algorithms," in *Proc. Int. Symp. Low Power Electronics & Design*, pp. 76-81, Aug. 1998. 685

[24] I. Hong, D. Kirovski, G. Qu, M. Potkonjak, and M. B. Srivastava, "Power optimization of variable-voltage core-based systems," *IEEE Trans. Computer-Aided Design*, vol. 18, no. 12, pp. 1702-1714, Dec. 1999. 685
[25] G. Quan and X. Hu, "Energy efficient fixed-priority scheduling for real-time systems on variable voltage processors," in *Proc. Design Automation Conf.*, pp. 828-833, June 2001. 685
[26] J. Pouwelse, K. Langendoen, and H. Sips, "Energy priority scheduling for variable voltage processors," in *Proc. Int. Symp. Low Power Electronics & Design*, pp. 28-35, Aug. 2001. 685
[27] N. Bambha, S. S. Bhattacharyya, J. Teich, and E. Zitzler, "Hybrid search strategies for dynamic voltage scaling in embedded multiprocessors," in *Proc. Int. Workshop Hardware/Software Co-Design*, pp. 243-248, Apr. 2001. 686
[28] Y. Zhang, X. Hu, and D. Chen, "Task scheduling and voltage selection for energy minimization," in *Proc. Design Automation Conf.*, pp. 183-188, June 2002. 686
[29] Y. Kwok and I. Ahmad, "Dynamic critical-path scheduling: An effective technique for allocating task graphs to multiprocessors," *IEEE Trans. Parallel & Distributed Systems*, vol. 7, no. 5, pp. 506-521, May 1996. 686
[30] F. Gruian and K. Kuchcinski, "LEneS: Task-scheduling for low-energy systems using variable voltage processors," in *Proc. Asian South Pacific Design Automation Conf.*, pp. 449-455, Jan. 2001. 686
[31] M. T. Schmitz and B. M. Al-Hashimi, "Considering power variations of DVS processing elements for energy minimisation in distributed systems," in *Proc. Int. Symp. System Synthesis*, pp. 250-255, Oct. 2001. 686, 688
[32] J. Luo and N. K. Jha, "Static and dynamic variable voltage scheduling algorithms for real-time heterogeneous distributed embedded systems," in *Proc. Asian and South Pacific Design Automation Conf.*, pp. 719-724, Apr. 2002. 686, 687, 688
[33] J. Luo and N. K. Jha, "Variable voltage joint scheduling of periodic task graphs and aperiodic tasks in real-time distributed embedded systems," Tech. Rep. CE-J02-002, Dept. of Electrical Engineering, Princeton Univ., Apr. 2002. 686, 688
[34] J. Luo and N. K. Jha, "Power-profile driven variable voltage scaling for heterogeneous distributed real-time embedded systems," Tech. Rep. CE-J02-001, Dept. of Electrical Engineering, Princeton Univ., Jan. 2002. 687
[35] C. Partel, S. Chai, S. Yalamanchili, and D. Schimmel, "Power-constrained design of multiprocessor interconnection networks," in *Proc. Int. Conf. Computer Design*, pp. 408-416, Oct. 1997. 689
[36] H. Wang, L.-S. Peh, and S. Malik, "A power model for routers: Modeling Alpha 21364 and InfiniBand routers," in *Proc. Hot Interconnects 10*, Aug. 2002. 689
[37] S. Mukherjee, P. Bannon, S. Lang, A. Spink, and D. Webb, "The ALPHA 21364 network architecture," in *Proc. Hot Interconnects 9*, Aug. 2001. 689
[38] G. Wei, J. Kim, D. Liu, S. Sidiropoulos, and M. Horowitz, "A variable-frequency parallel I/O interface with adaptive power-supply regulation," *J. Solid-State Circuits*, vol. 35, no. 11, pp. 1600-1610, Nov. 2000. 689
[39] J. Kim and M. Horowitz "Adaptive supply serial links with sub-1V operation and per-pin clock recovery," in *Proc. Int. Solid-State Circuits Conf.*, pp. 216-217, Feb. 2002. 689
[40] J. Kim and M. Horowitz, "An efficient digital sliding controller for adaptive power supply regulation," in *Proc. Int. Symp. VLSI Circuits*, pp. 133-136, June 2001. 689
[41] L. Shang, L.-S. Peh and N. K. Jha, "Power-efficient interconnection networks: dynamic voltage scaling with links," *Computer Architecture Letters*, vol. 1, no. 2, pp. 2-5, July 2002. 689, 690

[42] S. M. Martin, K. Flautner, T. Mudge, and D. Blaauw, "Combined dynamic voltage scaling and adaptive body biasing for low power microprocessors under dynamic workloads," in *Proc. Int. Conf. Computer-Aided Design*, Nov. 2002. 690

The Customization Landscape for Embedded Systems

Sudhakar Yalamanchili[1,2]

[1] Proceler Inc.
3350 Riverwood Parkway, Atlanta, Ga. 30339
[2] Center for Experimental Research in Computer Systems
School of Electrical and Computer Engineering, Georgia Institute of Technology
Atlanta, Georgia 30332
sudha@ece.gatech.edu

Abstract

The explosive growth of embedded systems in existing and emerging application domains is accompanied by unique constraints and performance requirements along multiple dimensions such as speed, power, and real-time behavior. Typically, these requirements are encapsulated in a few important components, or *kernels*, for example, for audio and video encoding/decoding, data compression, and encryption. Software implementations of these kernels executing on embedded microprocessors are inadequate to meet the requirements. Customization plays a central role in meeting these unique and specialized needs and has historically been realized by the development of custom hardware solutions such as fully custom application specific integrated circuits (ASICs).

However, each new generation of semiconductor technology is increasing the non-recurring engineering (NRE) costs associated with ASIC development and thereby making it feasible for only high volume applications and those with product lifetimes amenable to a long time to market design cycle. The increasing pervasiveness of embedded systems encompasses applications with smaller product lifetimes, shorter development cycles, and increasing cost containment pressures. Sustaining the continued growth demanded by the market will require technology to deliver performance characteristics of custom solutions while overcoming the twin hurdles of high NRE costs and long time to market of current ASIC solutions. Further, the embedded market is comprised of application segments covering a range of costs, volumes, and time to market needs. For example, components of 3G phones, set top boxes, color laser printers, color copiers, and network attached storage devices span a range of prices from on the order of a few hundred to a few thousand dollars and a range of volumes ranging from thousands to tens of millions of units. Consequently the approaches to customization are diverse.

As new products and services are envisioned, these evolving workload requirements will place more pressure on the need to innovate in overcoming the high NRE costs and time to market considerations of custom hardware solutions. The challenge for future embedded systems is to reconcile the demands of hardware customization, cost, and time to market constraints. This talk presents the emerging landscape of solutions, identifies key trends, and describes some representative approaches.

S. Sahni et al. (Eds.) HiPC 2002, LNCS 2552, p. 693, 2002.

Parallel Computations of Electron-Molecule Collisions in Processing Plasmas

B. Vincent McKoy and Carl Winstead

California Institute of Technology
Pasadena, CA 91125

Abstract. In the plasmas used in semiconductor fabrication, collisions between electrons and polyatomic molecules produce reactive fragments that drive etching and other processes at the wafer surface. Extensive and reliable data on electron-molecule collisions are therefore essential to simulations of plasma reactors. For the low electron energies and polyatomic gases of interest, both measurements and calculations are difficult, and many needed cross sections are lacking. However, rapid advances in computer speeds now make such calculations feasible.Because the fastest computers are highly parallel,both a formulation that accounts well for the physics of low-energy collisions and an implementation that is efficient on parallel architectures are required. We will give an overview of our formulation of the electron-molecule collision problem and of its implementation and performance on parallel machines, and of some results of its application.

S. Sahni et al. (Eds.) HiPC 2002, LNCS 2552, p. 697, 2002.

Computing Challenges and Systems Biology

Srikanta P. Kumar[1], Jordan C. Feidler[2], and Henrietta Kulaga[3]

[1] The Defense Advanced Research Projects Agency
3701 North Fairfax Drive, Arlington, VA 22203-1714, USA
[2] The MITRE Corporation
7515 Colshire Drive McLean, VA 22102-7508, USA
[3] GEOMET Technologies, Inc.
20251 Century Boulevard Germantown, MD 20874-1192, USA

Abstract. Over the past two decades the manner in which science is conducted in biology and related disciplines has been changing dramatically. Automation has led to a vast amount of data being generated by new experimental techniques, such as genomic sequencing and DNA microarrays. To extract the scientific insights buried in this high volume of data, life science researchers are employing increasingly sophisticated information technology approaches to create data analysis and simulation tools that run on high performance computing (HPC) platforms. The domain is one rich in compute-intensive applications: determining gene sequence, predicting macromolecular structure, understanding the temporal dynamics of protein folding, modeling molecular interactions, simulating the behavior of cell-signaling cascades and genetic regulatory networks, and the real-time manipulation of 3D rendered volumes of organs derived from structural magnetic resonance imaging scans. Future trends in the life sciences are expected to expand upon this set to include computation performed using hybrid bio-nano devices, personalized medicine, mining of heterogeneous data sets stored at disparate sites, and knowledge discovery in patient medical records. The talk will review high performance computing needs in emerging systems biology applications, the potential impact of bio-nano research on future high performance computing designs, and work on related topics at DARPA.

S. Sahni et al. (Eds.) HiPC 2002, LNCS 2552, p. 701, 2002.

Visual Programming for Modeling and Simulation of Biomolecular Regulatory Networks*

Rajeev Alur[1], Calin Belta[1], Franjo Ivančić[1], Vijay Kumar[1], Harvey Rubin[1], Jonathan Schug[1], Oleg Sokolsky[1], and Jonathan Webb[2]

[1] Hybrid Systems Group, University of Pennsylvania
Philadelphia, PA 19104, USA
[2] BBN Technologies
Cambridge, MA 02138, USA

Abstract. In this paper we introduce our new tool BioSketchPad that allows visual progamming and modeling of biological regulatory networks. The tool allows biologists to create dynamic models of networks using a menu of icons, arrows, and pop-up menus, and translates the input model into Charon, a modeling language for modular design of interacting multi-agent hybrid systems. Hybrid systems are systems that are characterized by continuous as well as discrete dynamics. Once a Charon model of the underlying system is generated, we are able to exploit the various analysis capabilities of the Charon toolkit, including simulation and reachability analysis. We illustrate the advantages of this approach using a case study concerning the regulation of bioluminescence in a marine bacterium.

1 Introduction

We now know that approximately 30,000 to 40,000 genes control and regulate the human body. The recent completion of a rough draft of the human genome and the complete sequence of *Drosophila melanogaster*, *Caenorhabditis elegans*, *Mycobacterium tuberculosis*, and numerous other sequencing projects provide a vast amount of genomic data for further refinement and analysis [17]. These landmarks in human scientific achievement promise remarkable advances in our understanding of fundamental biological processes. To achieve this goal, we must develop the ability to model, analyze, and predict the effect of the products of specific genes and genetic networks on cell and tissue function [12].

Traditional models and simulations of metabolic and cellular control pathways are based on either continuous or discrete dynamics [6, 15]. However, many important systems in biology are *hybrid* – they involve both discrete and continuous dynamics. At the molecular level, the fundamental process of inhibitor proteins turning off the transcription of genes by RNA polymerase reflects a switch

* This research was supported in part by NSF grants CDS-97-03220 and EIA-01-30797, and DARPA grant F30602-01-2-0563.

S. Sahni et al. (Eds.) HiPC 2002, LNCS 2552, pp. 702–712, 2002.

between two continuous processes. This is perhaps most clearly manifested in the classic genetic switch observed in the λ-phage [16], where we see two distinct behaviors, *lysis* and *lysogeny*, each with different mathematical models. At the cellular level, we can best describe cell growth and division in a eukaryotic cell as a sequence of four processes, each being a continuous process being triggered by a set of conditions or events [14]. At the inter-cellular level, we can even view cell differentiation as a hybrid system [9].

In all of these examples, a hybrid approach that combines elements of discrete and continuous dynamics is necessary to model, analyze, and simulate the system. This paper addresses the modeling of networks of biochemical reactions, and the control of molecular and cellular functions with genetic regulatory apparatus, and a set of theories, algorithms, and methodologies that allows biologists to analyze and characterize such systems.

2 Background

Our approach to modeling the different elements (biomolecules, cells, proteins) and their interactions is based on modern concepts in software engineering and control theory derived from the literature on hybrid systems [1].

A hybrid system consists of a set of *hybrid automata*. A hybrid automaton is characterized by a *continuous state* $x \in \mathbb{R}^n$ and a collection of *discrete modes*. Each mode consists of a set of *ordinary differential equations* (ODEs) that govern the evolution of the continuous state x and a set of *invariants* that describe the conditions under which the ODEs are valid. We can write the ODEs of a particular mode as $\dot{x} = f(x, z)$, where $z \in \mathbb{R}^p$ is the information from other hybrid automata running concurrently as part of the overall hybrid system. The definition of a hybrid automaton includes *transitions* among its modes. A transition specifies source and destination modes, an enabling condition called *guard*, and the associated discrete *update* of variables.

As an illustrative example, consider the description of a simple hybrid system in Figure 1. It consists of one hybrid automaton that contains two discrete modes q_1 and q_2, and the continuous variable x which evolves under the differential equation $\dot{x} = f_1(x)$ in discrete mode q_1, and $\dot{x} = f_2(x)$ in mode q_2. The invariant sets associated with the locations q_1 and q_2 are $g_1(x) \leq 0$ and $g_2(x) \leq 0$, respectively. The hybrid system continuously evolves in discrete mode q_1 according to the differential equation $\dot{x} = f_1(x)$, as long as x remains inside the invariant set $g_1(x) \leq 0$. If during the continuous flow, it happens that x belongs in the

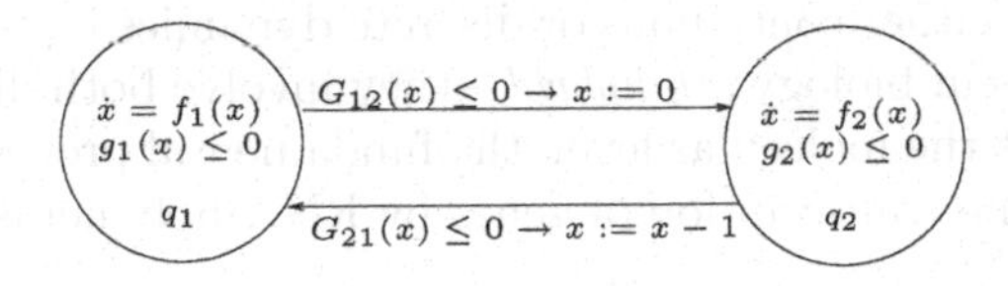

Fig. 1. A simple hybrid system

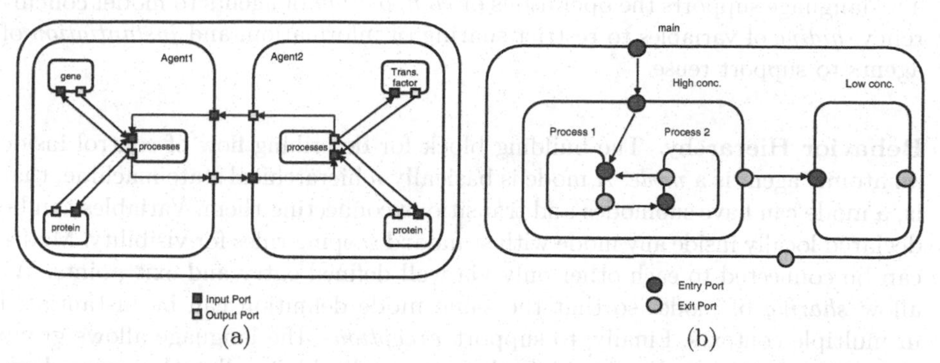

Fig. 2. The architectural hierarchy in CHARON is based on agents. New agents can be formed by parallel composition of two or more agents (left). The behavioral hierarchy is described by modes. Sequential composition of modes models the switching between behaviors (right)

guard set $G_{12}(x) \leq 0$, then the transition from q_1 to q_2 is enabled. If the transition is enabled, the system can switch from the mode q_1 to q_2. The variable x is reset to zero and then evolves according to the differential equation $f_2(x)$.

Hybrid system models are increasingly being used for modeling embedded systems, in particular for automotive control systems, avionics and robotics. Modern object-oriented design paradigms such as *Unified Modeling Language* (UML) allow specifications of the architecture and control at high levels of abstraction in a modular fashion [5]. Emerging tools such as RationalRose (see `www.rational.com`) support modeling, simulation, and code generation. Tools such as MATLAB and SIMULINK (see `www.mathworks.com`) allow the modeling and simulation of systems with models of continuous dynamics combined with state-machine-based models of discrete states.

All these paradigms and tools are directly applicable to modeling and analysis of biological regulatory networks. In the next two sections, we describe CHARON, a programming language for modeling such systems, and BIOSKETCHPAD, an easy-to-use, interactive tool that allows users to create CHARON models of regulatory networks.

3 Modeling Hybrid Systems in CHARON

We have developed the programming language CHARON [1], for modeling and analyzing hybrid systems (see Figure 2). The language incorporates ideas from concurrency theory (languages such as CSP [13]), object-oriented software design notations (such as Statecharts [11] and UML [5]), and formal models for hybrid systems. The key features of CHARON are:

Architectural Hierarchy. The building block for describing the system architecture is an *agent* that communicates with its environment via shared variables.

The language supports the operations of *composition* of agents to model concurrency, *hiding* of variables to restrict sharing of information, and *instantiation* of agents to support reuse.

Behavior Hierarchy. The building block for describing flow of control inside an atomic agent is a *mode*. A mode is basically a hierarchical state machine, that is, a mode can have submodes and transitions connecting them. Variables can be declared locally inside any mode with standard scoping rules for visibility. Modes can be connected to each other only via well-defined entry and exit points. We allow *sharing* of modes so that the same mode definition can be instantiated in multiple contexts. Finally, to support *exceptions*, the language allows group transitions from default exit points that are applicable to all enclosing modes.

Discrete Updates. Discrete updates are specified by *guarded actions* labeling transitions connecting the modes. Actions can have calls to externally defined Java functions which can be used to write complex data manipulations. It also allows us to mimic stochastic aspects through randomization.

Continuous Updates. Some of the variables in CHARON can be declared *analog*, and they flow continuously during continuous updates that model passage of time. The evolution of analog variables can be constrained in three ways: *differential* constraints (e.g. by equations such as $\dot{x} = f(x, u)$), *algebraic* constraints (e.g. by equations such as $y = g(x, u)$), and *invariants* (e.g. $|x - y| \leq \varepsilon$) which limit the allowed durations of flows. Such constraints can be declared at different levels of the mode hierarchy.

4 BIOSKETCHPAD

The BIOSKETCHPAD (BSP) is an interactive tool for modeling and designing biomolecular and cellular networks with a simple, easy to use, graphical front end leveraging powerful tools from control theory, hybrid systems, and software engineering. Models developed using BSP can be automatically converted to CHARON input files.

4.1 Nodes in BIOSKETCHPAD

The elements of a BSP model diagram are nodes and connecting edges. The nodes are either species or processes. Processes include chemical reactions, regulation, and cell growth. The edges describe relations between species and processes.

Species Nodes. A species node is identified by a tuple of parameters which are listed below. The different parameters support the kinds of common biological distinctions between similar species. Thus Ca and $Ca^{+}2$ will have the same name value, but have different values for charge. A particular tuple of parameters can be represented by one or more graphical instance of a species node.

The *name* of the species, i.e., "Ca", "alcohol dehydrogenase", or "notch" is one parameter of a species node. Other parameters are the *type* of the species such as "gene" or "protein", the *physical location* of the species, which may be "cell membrane" or "cell nucleus" reflecting different physical areas of interest where concentrations of a particular compound may be different. In addition, the *n-mer* polymerization of the species, the *state*, and the *electrical charge* of a species are parameters. Finally, when appropriate, the user can specify an *initial concentration* for a species which will be used in simulation runs. The initial concentration can be specified in a consistent unit system for the model as a whole or the particular species.

Reaction Nodes. A reaction node represents some kind of interaction which may alter the concentration of one or more species. Each reaction node must have at least one node attached to it. Further constraints are placed on attached nodes for specific types of reactions. Species nodes attached to a reaction are considered either input or output nodes.

The *type* of the reaction represented by the node may, for example, be "transformation" or "transcription." Some types of reactions may be restricted in the types of species they may operate on, specific species properties may be required, or there may be restrictions on reaction geometry. Sources and sinks are special types of reaction nodes.

Regulation Nodes. A regulation node is used to modulate the rate of a reaction by the concentration of one or more species. It must have at least one output node which must be a reaction and at least one input species node.

There are currently three functional forms for regulation nodes, "weighted sum," "product," and "tabular" and they represent the mathematical operation performed for the node. Species inputs reference the concentration of the species. Regulation node inputs reference the computed value of the regulation function.

4.2 Models for Transformation, Transcription and Translation

Rate laws for transformation, transport, transcription and translation reaction types are the basic reaction formulations in the BSP. Rate laws can be reversible or irreversible, and can be regulated by specified species.

The BSP allows the specification of a variety of rate laws. For the sake of simplicity, we will consider the case of a *reversible mass action* (see Figure 3). For a reaction of the form

$$aA + bB \leftrightarrow cC + dD \tag{1}$$

the reaction rate is given by

$$\nu = k_f * [A]^\alpha * [B]^\beta - k_r * [C]^\gamma * [D]^\delta, \tag{2}$$

where k_f and k_r stand for the forward and reverse rate constant, and $[A]$ and $[B]$ are the concentrations of species A and B. The constants a, b, c, and d are the stoichiometric coefficients, whereas α, β, γ and δ are called order of reaction and are determined experimentally. Terms in the ODE for the concentrations of reaction participants are constructed as follows:

$$\begin{aligned} d[A]/dt &= -a * \nu, \\ d[B]/dt &= -b * \nu, \\ d[C]/dt &= c * \nu, \text{ and} \\ d[D]/dt &= d * \nu. \end{aligned} \tag{3}$$

The BSP also allows the definition of *tabulated functions.* Tabulated functions allow the specification of different behaviors in various parts of the state space. These functions introduce *hybrid* behavior into the underlying CHARON models. The general form of a tabulated functional form mass action is shown below for a reversible reaction of the form shown previously.

$$\nu = k_f * f_1([A],[B],[C],[D]) * [A]^\alpha * [B]^\beta - k_r * f_2([A],[B],[C],[D]) * [C]^\gamma * [D]^\delta. \tag{4}$$

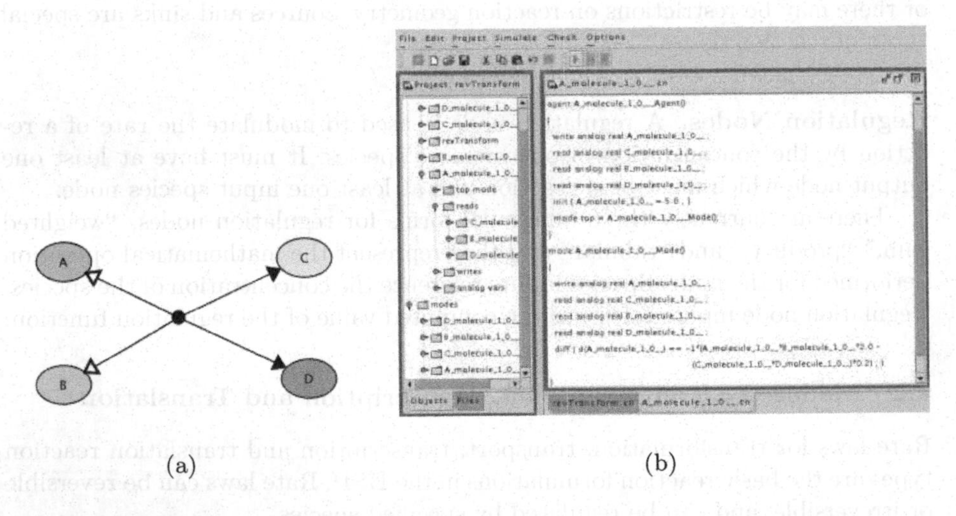

Fig. 3. A simple reversible mass action model with four species. (a) The model as a snapshot of BSP. (b) The automatically generated CHARON model representing the BSP input model in (a). The CHARON toolkit editor on the right shows the model of species A. The left panel of the CHARON toolkit is the project window showing the internal tree representation of the CHARON model

The tabulated function may also include the concentration of any regulating species as an independent variable.

In addition to the reactions described above, BSP also models *irreversible mass action*, *regulated mass action*, *Michelis-Menton reactions*, and *cell growth*. The BSP translates such an input model automatically into an equivalent CHARON model, which is shown in Figure 3 (b).

We next introduce a case study in which we show how a fairly complex biological phenomenom can be modeled using BSP and CHARON.

5 Case Study: Luminescence Regulation in V. Fischeri

Vibrio fischeri is a marine bacterium which can be found both as free-living organism and as a symbiont of some marine fish and squid. As a free-living organism, *V. fisheri* exists at low densities and appears to be non-luminescent. As a symbiont, the bacteria live at high densities and are, usually, luminescent.

5.1 Description of the Regulatory System

The luminescence in *V. fischeri* is controlled by the transcriptional activation of the *lux* genes. The *lux* regulon is organized in two transcriptional units (see Figure 4). The leftward operon contains the *luxR* gene encoding the protein[1] LuxR, a transcriptional regulator of the system. The rightward operon contains seven genes *luxICDABEG*. The transcription of the *luxI* gene results in the production of protein LuxI. This protein is required for endogenous production of *autoinducer*, Ai, a small membrane-permeant signal molecule (acyl-homoserine lactone). The genes *luxA* and *luxB* code for the luciferase subunits, which in turn are responsible for luminescence. *luxC*, *luxD*, and *luxE* code for polypeptides of the fatty-acid reductase, which generates aldehyde substrate for luciferase. Along with LuxR and LuxI, cAMP receptor protein (CRP) plays an important role in controlling luminescence. The network of biochemical reactions in the cell is as follows: The autoinducer *Ai* binds to protein LuxR to form a complex *Co* which binds to the *lux box*. The *lux box* is in the middle of a regulatory region between the two transcriptional units (operons). This region also contains a binding site for CRP. The transcription from the *luxR* promoter is activated by the binding of CRP to its binding site, and the transcription of the *luxICDABEG* by the binding of *Co* to the *lux box*. However, growth in the levels of *Co* and cAMP/CRP inhibit *luxR* and *luxICDABEG* transcription, respectively.

5.2 BIOSKETCHPAD Model and Simulation in CHARON

The regulatory network described above can be easily drawn in the BSP as shown in Figure 4. Notice the figure shows different icons for the gene, the

[1] We use italics (*e.g.*, *luxR*) to indicate the genes and plain font to denote the protein expressed by the gene (*e.g.*, LuxR).

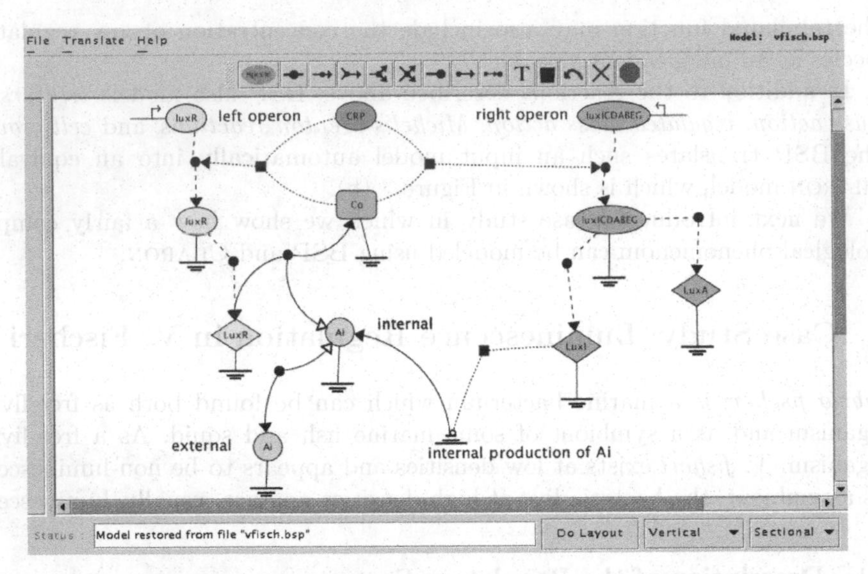

Fig. 4. Snapshot of the BIOSKETCHPAD model of the luminescence regulatory system in *V. fischeri*. The top line represents buttons used to perform standard operations, such as "create a new species" or "add a two-to-one edge". The bottom line holds a status message and buttons to perform automatic layout operation

transcribed mRNA, and the translated protein, all of which play an important role in the network. The transcriptional regulation of the two operons by CRP and *Co* is modeled using the tabular function approach, as described before.

The tool then translates the graphical model into equivalent CHARON code. Each species (mRNA, protein, complex) is modeled by an agent. The complete model is the parallel composition of all the agents. Each agent has only one mode (biological behavior) except for mRNAs luxR and luxICDABEG, which are regulated by signals given in tabular form. The differential equations describing the dynamics of the system are computed according to the edges in the model and the parameters specified by the user and translated into CHARON.

The above model was simulated starting from zero initial conditions for all the species except for external autoinducer Ai_e, which was set at 100nM. As expected, the non-zero initial value of external autoinducer determines an increase in the concentration of internal autoinducer (see Figure 5). The positive feedback loop LuxR - Ai_i turns on the luminescence gene. For the sake of brevity, we omit the parameters used during this simulation run and instead refer the reader to details on the website (`www.cis.upenn.edu/biocomp`).

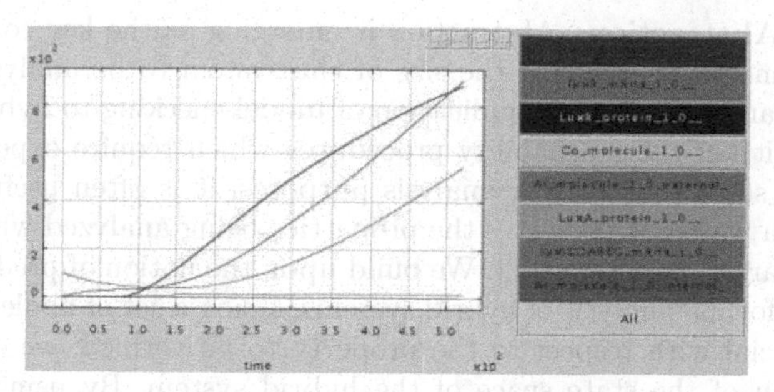

Fig. 5. Snapshot of CHARON showing the time-evolution of the concentrations in nano-molars. More details are available in online papers at `www.cis.upenn.edu/biocomp`

6 Discussion

In this paper we introduced a new tool BIOSKETCHPAD that allows visual programming and modeling of biological regulatory networks through an easy-to-use interactive, icon-based interface. BIOSKETCHPAD allows the automatic generation of CHARON models. The underlying formal semantics of CHARON models allow symbolic in addition to more traditional numerical analysis. We will give a brief overview of our ongoing research directed toward the formal analysis of hybrid systems.

Simulation with Accurate Event Detection. Simulation of hybrid systems is challenging because it is difficult to develop an efficient integration algorithm that also guarantees the detection of transitions between modes. The commonly used technique of just checking the value of the guards at integration points can cause the simulator to miss critical events. It has been shown that such inaccuracies can lead to grossly inaccurate simulations due to the discontinuous nature of hybrid systems. We have developed a method [8] which is guaranteed to detect enabling of all transitions, and is at least as efficient as conventional, adaptive integration algorithms.

Multirate Simulation. Many systems naturally evolve on different time scales. Traditional numerical integration methods force all coupled differential equations to be integrated using the same step size. The idea behind multi-rate integration method is to use larger step sizes for the slow changing sets of differential equations and smaller step sizes for the differential equations evolving on the fast time scale. To implement such a scheme we need to show how to accommodate coupling between the sets of fast and slow equations when they are integrated asynchronously and how to schedule the order of integration. In [7] we introduce a multi-rate algorithm suited to hybrid system simulation.

Predicate Abstraction. Abstraction is emerging as the key to formal verification as a means of reducing the size of the system to be analyzed.The main obstacle towards an effective application of model checking to hybrid systems is the complexity of the reachability procedures which require expensive computations over sets of states. For analysis purposes, it is often useful to abstract a system in a way that preserves the properties being analyzed while hiding the details that are of no interest [3]. We build upon the notion of predicate abstraction [10] for formal analysis of hybrid systems. Using a set of boolean predicates, that are crucial with respect to the property to be verified, we construct a finite partition of the state space of the hybrid system. By using conservative reachability approximations we guarantee that if the property holds in the abstracted system, then it also holds in the concrete system represented by the hybrid system [2].

Reachability Analysis. While reachability analysis for hybrid systems is, in general, intractable, we are working on methods that efficiently partition the state space for *multi-affine*, hybrid systems into discrete states. For example, equation (2) is multi-affine when the orders of reaction are unity. In other words, they are affine in each variable. Our preliminary work in this direction is described in a forthcoming paper [4].

We envision reporting on a more complete toolbox with the software tools described above, tightly integrated with BioSketchPad in forthcoming publications. We believe such a toolbox can have a significant impact on post-genomics biology research.

References

[1] R. Alur, T. Dang, J. Esposito, R. Fierro, Y. Hur, F. Ivančić, V. Kumar, I. Lee, P. Mishra, G. Pappas, and O. Sokolsky. Hierarchical hybrid modeling of embedded systems. In *Embedded Software, First Intern. Workshop*, LNCS 2211. 2001. 703, 704

[2] R. Alur, T. Dang, and F. Ivančić. Reachability analysis of hybrid systems via predicate abstraction. In *Hybrid Systems: Computation and Control, Fifth International Workshop*, LNCS 2289, pages 35–48. Springer-Verlag, March 2002. 711

[3] R. Alur, T. Henzinger, G. Lafferriere, and G. Pappas. Discrete abstractions of hybrid systems. *Proceedings of the IEEE*, 88(7):971–984, July 2000. 711

[4] C. Belta, L. Habets, and V. Kumar. Control of multi-affine systems on rectangles with applications to hybrid biomolecular networks. *CDC 2002*, Dec. 2002. 711

[5] G. Booch, I. Jacobson, and J. Rumbaugh. *Unified Modeling Language User Guide*. Addison Wesley, 1997. 704

[6] M. Elowitz and S. Leibler. Asynthetic oscillatory network of transciptional regulators. *Nature*, 403:335–338, January 2000. 702

[7] J. Esposito and V. Kumar. Efficient dynamic simulation of robotic systems with hierarchy. In *Intl. Conf. on Robotics and Automation*, pages 2818–2823, 2001. 710

[8] J. Esposito, V. Kumar, and G. Pappas. Accurate event detection for simulating hybrid systems. In *Hybrid Systems : Computation and Control*, LNCS 2034, 2001. 710

[9] R. Ghosh and C. J. Tomlin. Lateral inhibition through delta-notch signaling: A piecewise affine hybrid model. In *HSCC*, Rome, Italy, Mar 28-30 2001. 703

[10] S. Graf and H. Saidi. Construction of abstract state graphs with PVS. In *Proc. 9th Intl. Conf. on Computer Aided Verification*, LNCS 1254, 1997. 711

[11] D. Harel. Statecharts: A visual formalism for complex systems. *Science of Computer Programming*, 8:231–274, 1987. 704

[12] L. H. Hartwell, J. J. Hopfield, S. Leibler, and A. W. Murray. From molecular to modular cell biology. *Nature*, 402((6761 Suppl)):C47–52, December 1999. 702

[13] C. A. R. Hoare. *Communicating Sequential Processes.* Prentice-Hall, 1985. 704

[14] B. Lewin. *Genes VII.* Oxford University Press, 2000. 703

[15] P. Mendes and D. B. Kell. Non-linear optimization of biochemical pathways: applications to metabolic engineering and parameter estimation. *Bioinformatics*, 10:869–883, 1998. 702

[16] M. Ptashne. *A Genetic Switch: Phage λ and Higher Organisms.* Cell Press and Blackwell Science, 1992. 703

[17] J. C. Venter et al. The sequence of the human genome. *Science*, 291(5507):1304–51., 2001. 702

Framework for Open Source Software Development for Organ Simulation in the Digital Human

M. Cenk Cavusoglu[1], Tolga Goktekin[1], Frank Tendick[2], and S. Shankar Sastry[1]

[1] Dept. of Electrical Eng. and Computer Sci., University of California, Berkeley
[2] Dept. of Surgery, University of California, San Francisco
{mcenk,goktekin,tendick,sastry}@eecs.berkeley.edu

The current state of the field of medical simulation is characterized by scattered research projects using a variety of models that are neither interoperable nor independently verifiable models. Individual simulators are frequently built from scratch by individual research groups without input and validation from a larger community. The challenge of developing useful medical simulations is often too great for any individual group since expertise is required from different fields, from molecular and cell biology, from anatomy and physiology, and from computer science and systems theory.

Open source, open architecture software development model provides an attractive framework as it addresses the needs for interfacing models from multiple research groups and facilitates the ability to critically examine and validate quantitative biological simulations.

The focus of our research is to explore the feasibility of developing open source, open architecture models of different levels of granularity and spatio-temporal scale for a project that has been labeled the Digital Human project, an initiative which aims to build a complete functioning library of interactive views and simulations of human anatomy, physiology, pathology, histology and genomics. While the emphasis of our research is on how the simulations that we develop will allow for the interconnections between individual organ simulations, and between different types of physical processes within a given organ, we will develop our tools on a specific test bed application: the construction of a heart model for simulation of heart surgery.

We are developing a draft API for organ models in surgical simulation. The focus of the initial draft is on the interface for the mechanical models of organs. Several heart models have been selected from the literature to determine the requirements on the open source simulator framework to be able to include these models in the simulator. These models are actually very low level and detailed models of heart, and are beyond the proposed scope of a surgical simulator. In particular, the Peskin-McQueen heart model, which is a model of the muscle mechanics coupled with hemodynamics, is being studied in order to determine the proper paradigm to handle coupled models within the simulator, even though a comparatively higher-level model will actually be used in this project.

Parallel to the effort on the API development, we are developing a surgical simulator framework, which is based on our earlier VESTA surgical training simulator test-bed, to accommodate open source release, and to follow the open architecture specifications of the API mentioned above. The simulation environment

S. Sahni et al. (Eds.) HiPC 2002, LNCS 2552, pp. 713-714, 2002.

is also being designed to support parallelization of parts of the computation, for example the finite element computations for mechanical deformations of organs.

As a test-bed application for the API and surgical simulation environment development, we are also developing a basic model and simulation of heart. The model being developed will include basic electromechanical, circulatory, and physiological behavior of the heart.

Reachability Analysis of Delta-Notch Lateral Inhibition Using Predicate Abstraction

Inseok Hwang, Hamsa Balakrishnan, Ronojoy Ghosh, and Claire Tomlin

Hybrid Systems Laboratory, Department of Aeronautics and Astronautics
Stanford University, Stanford, CA 94305
{ishwang,hamsa,ronojoy,tomlin}@stanford.edu

Abstract. This paper examines the feasibility of predicate abstraction as a method for the reachability analysis of hybrid systems. A hybrid system can be abstracted into a purely discrete system by mapping the continuous state space into an equivalent finite discrete state space using a set of Boolean predicates and a decision procedure in the theory of real closed fields. It is then possible to find the feasible transitions between these states. In this paper, we propose new conditions for predicate abstraction which greatly reduce the number of transitions in the abstract discrete system. We also develop a computational technique for reachability analysis and apply it to a biological system of interest (the Delta-Notch lateral inhibition problem).

1 Introduction

A hybrid system has both discrete and continuous transitions, and so has an infinite number of state transitions over any continuous time interval. Although there exist methods to calculate the reachable sets of hybrid systems by solving the system equations, the infinite nature of the state space makes these methods computationally very expensive. It is appealing, therefore, to find a method by which we could extract equivalent finite state models of these systems and use them to find approximate reachable sets of the original systems.

Predicate abstraction has emerged as a powerful technique to extract such models from complex, infinite state models (Das et al[3], Graf et al[6]). Discrete abstraction and reachability analysis of hybrid systems based on predicate abstraction have been proposed in earlier work (Tiwari et al[12], Alur et al[1], Sokolsky et al[11], Lafferriere et al[8]). The extracted finite state model, called the Abstract Discrete System (ADS), is said to be an over-approximation of the original system if the original system satisfies any property of interest that is satisfied by the ADS.

In this paper, inspired by Tiwari et al[12][13], we return to our hybrid model of Delta-Notch lateral inhibition (Ghosh et al[5]), and attempt to use the above methods to perform reachability analysis. We propose new conditions for predicate abstraction, beyond those of Tiwari et al[12], that greatly reduce the number of transitions in the abstract discrete system given a set of polynomials. We then

S. Sahni et al. (Eds.) HiPC 2002, LNCS 2552, pp. 715–724, 2002.

develop a computational technique for the reachability analysis of hybrid systems. Finally, we analyze the Delta-Notch lateral inhibition problem using the technique thus developed.

This paper is organized as follows: the problem description, modeling, and analytical results of the Delta-Notch lateral inhibition problem are presented in Sections 2-3; the predicate abstraction algorithm and its implementation are described in Section 4 and Section 5 respectively. Conclusions are presented in Section 6.

2 Delta-Notch Signaling

The biological process under study, intercellular signaling, results in cellular differentiation in embryonic tissue, which is a complex control process regulated by a set of developmental genes, most of which are conserved in form and function across a wide spectrum of organisms. Found in almost all multicellular organisms from an early embryo stage, intercellular signaling is a feedback network which interrelates the fate of a single cell and its neighbors in a population of homogeneous cells. Among the various signaling channels, the Delta-Notch protein pathway in particular has gained wide acceptance as the arbiter of cell fate for an incredibly varied range of organisms (Artavanis-Tsakonas et al[2]).

Delta and Notch are both transmembrane proteins that are active only when cells are in direct contact, in a densely packed epidermal layer for example (Lewis[9]). Delta is a ligand that binds and activates its receptor Notch in neighboring cells. The activation of Notch in a cell affects the production of Notch ligands (i.e. Delta) both in itself and its neighbors, thus forming a feedback control loop. In the case of lateral inhibition, high Notch levels suppress ligand production in the cell and thus a cell producing more ligands forces its neighboring cells to produce less. The Delta-Notch signaling mechanism has been found to cause pattern formation in many different biological systems, like the South African claw-toed frog (*Xenopus laevis*) embryonic skin (Marnellos et al[10]) and the eye R3/R4 photoreceptor differentiation and planar polarity in the fruit fly *Drosophila melanogaster* (Fanto et al[4]). An example of the distinctive "salt-and-pepper" pattern formed due to lateral inhibition is the *Xenopus* epidermal layer where a regular set of ciliated cells form within a matrix of smooth epidermal cells as seen in Figure 1(a).

3 Model and Analytical Results

To model the regulation of intracellular Delta and Notch protein concentrations through the feedback network, experimentally observed rules governing the biological phenomenon have to be implemented. Firstly, since Delta and Notch are transmembrane proteins, cells have to be in direct contact for Delta-Notch signaling to occur. This implies that a cell is directly affected by, and directly affects in turn, only immediate neighbors. Secondly, Notch production is turned on by high Delta levels in the immediate neighborhood of the cell and Delta production

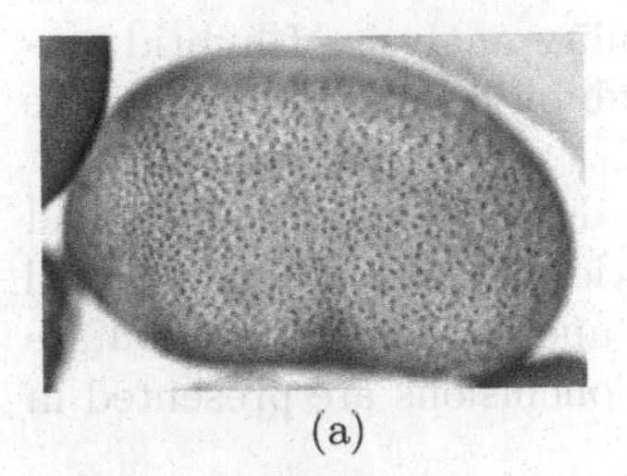

(a)

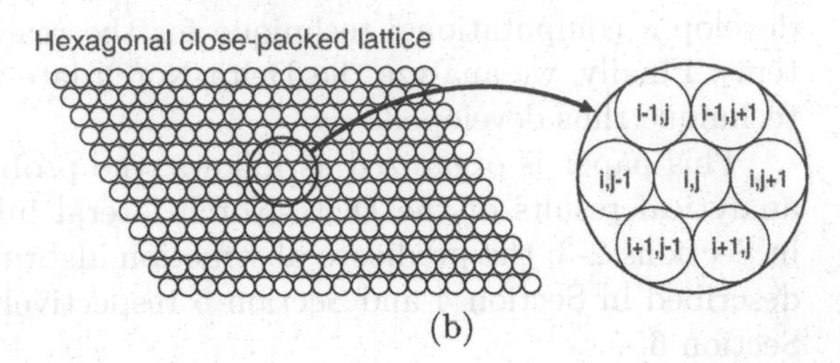

(b)

Fig. 1. (a) *Xenopus* embryo labeled by a marker for ciliated cell precursors seen as black dots. Photograph courtesy of P. D. Vize (*The Xenopus Molecular Marker Resource*, `http://vize222.zo.utexas.edu`). (b) Hexagonal close-packed layout scheme for cells in two dimensional arrays

is switched on by low Notch concentrations in the same cell. Thirdly, at steady state, a cell with high Delta levels must have low Notch level and vice versa. This is essential for differentiation to occur, for cells with high Delta levels in steady state become ciliated, and cells with high Notch in steady state remain smooth. Finally, both Delta and Notch proteins decay exponentially through normal proteolysis. In the model, the cells are assumed to be hexagonal close packed, i. e. each cell has six neighbors in contact with it (Figure 1(b)). We will denote as "biologically consistent" our computational results which comply with these biological observations.

Each biological cell is modeled as a four state piecewise affine hybrid automaton. The four states capture the property that Notch and Delta protein production can be individually switched on or off at any given time. It is assumed that there is no command-actuation delay in the mode switching. The formal definition of the hybrid automaton is given by:

$$H_1 = (Q_1, X_1, \Sigma_1, V_1, Init_1, f_1, Inv_1, R_1)$$
$$Q_1 = \{q_1, q_2, q_3, q_4\}$$
$$X_1 = (v_D, v_N)^T \in \Re^2$$
$$\Sigma_1 = \left\{ u_D, u_N : u_D = -v_N, u_N = \sum_{i=1}^{6} v_D^i \right\}$$
$$V_1 = \emptyset$$
$$Init_1 = Q_1 \times \{X_1 \in \Re^2 : v_D, v_N > 0\}$$
$$f_1(q,x) = \begin{cases} [-\lambda_D v_D; -\lambda_N v_N]^T & \text{if } q = q_1 \\ [R_D - \lambda_D v_D; -\lambda_N v_N]^T & \text{if } q = q_2 \\ [-\lambda_D v_D; R_N - \lambda_N v_N]^T & \text{if } q = q_3 \\ [R_D - \lambda_D v_D; R_N - \lambda_N v_N]^T & \text{if } q = q_4 \end{cases}$$
$$Inv_1 = \{q_1, \{u_D < h_D, u_N < h_N\}\} \cup \{q_2, \{u_D \geq h_D, u_N < h_N\}\}$$
$$\cup \{q_3, \{u_D < h_D, u_N \geq h_N\}\} \cup \{q_4, \{u_D \geq h_D, u_N \geq h_N\}\}$$

$$R_1 : \begin{bmatrix} R_1\left(q_1, \{u_D \geq h_D \wedge u_N < h_N\}\right) \in q_2 \times \Re^2 \\ R_1\left(q_1, \{u_D < h_D \wedge u_N \geq h_N\}\right) \in q_3 \times \Re^2 \\ \vdots \\ R_1\left(q_4, \{u_D < h_D \wedge u_N \geq h_N\}\right) \in q_3 \times \Re^2 \end{bmatrix}$$

where, v_D and v_N: Delta and Notch protein concentrations, respectively, in a cell; v_D^i: Delta protein concentration in i^{th} neighboring cell; λ_D and λ_N: Delta and Notch protein decay constants respectively; R_D and R_N: constant Delta and Notch protein production rates, respectively; h_D and h_N: switching thresholds for Delta and Notch protein production, respectively. R_D, R_N, λ_D and λ_N are experimentally-determined constants. The switching thresholds h_D and h_N are unknown and possible ranges for them are derived in Ghosh et al[5], which are biologically consistent. In the single cell, $v_D^i = 0, \forall i \in \{1, \ldots 6\}$. The inputs u_D and u_N are the physical realization of the protein regulatory properties in the model outlined before.

The two cell hybrid automaton H_2 is the composition of two single cell automata, to form a model with four continuous states and 16 discrete modes. Here, $v_D^1 \neq 0$ for each of the two cells, and thus the Delta level of each cell is communicated to its neighbor to control Notch production. Modeling the full two dimensional layer of cells involves composing $N \times N$ single cell hybrid automata.

Both the single and two cell hybrid automata were analyzed in Ghosh et al. [5], to obtain constraints on the range of the protein kinetic parameters and switching thresholds for biologically feasible equilibria to exist:

$$h_D, h_N : -\frac{R_N}{\lambda_N} < h_N \leq 0 \wedge 0 < h_N \leq \frac{R_D}{\lambda_D}$$

The two cell automaton was also shown to have a Zeno state with a particular Zeno execution that is an invariant: $v_{D_1} - v_{D_2} \wedge v_{N_1} - v_{N_2}$.

4 Predicate Abstraction

In this section, we review the predicate abstraction techniques proposed in Tiwari et al[12] and propose new conditions to refine transitions in the ADS. We then develop a method of finding approximate backward reachable sets of the equilibria. A flow-chart describing this procedure is shown in Figure 2.

Any trajectory of a hybrid system can be resolved into discrete and continuous transitions. We first consider the predicate abstraction of the continuous state space. A continuous dynamical system can be represented by a tuple $(X, InitX, f, Inv)$, where $X \in \Re^n$ is the set of continuous states, $InitX$ is a set of initial states, $f : X \to TX$ is the continuous dynamics, and Inv is the invariant set. We assume $InitX$, f, and Inv are polynomials of continuous states.

4.1 Construction of the Abstract Discrete System

Given a continuous system, following the methods of Tiwari et al[12], we construct the ADS, $(Q, InitQ, \delta)$ where Q is a finite set of discrete states, $InitQ$ is

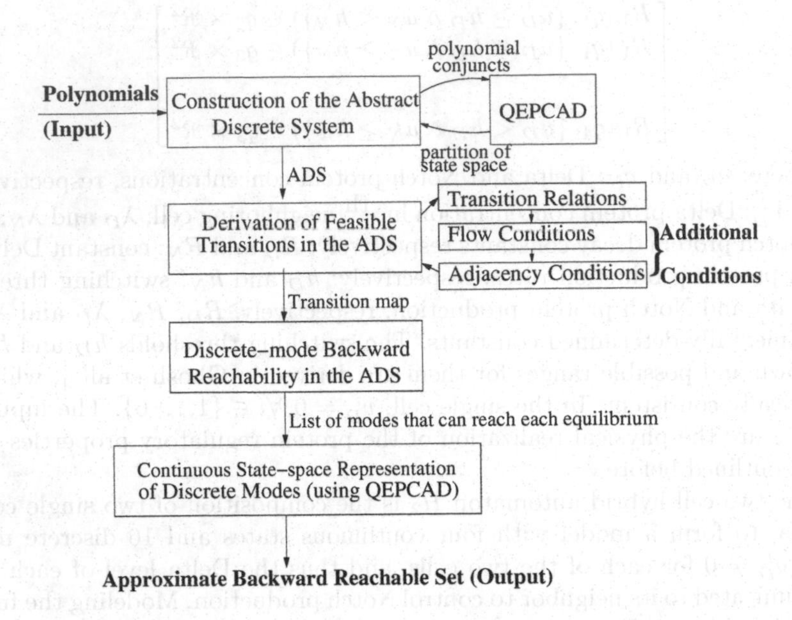

Fig. 2. Flow-chart describing the process of calculating approximate backward reachable sets using predicate abstraction techniques

the set of discrete initial states, and δ is the set of transitions. We first construct an initial set P_0 which contains polynomials of interest (for example: state variables, governing equations, invariants, guards, or any properties we may wish to verify). We then construct a finite set P of polynomials following the inference rule: if a polynomial $p \in P$, then we add the derivative of p with respect to time, $\dot{p}$, to the set P, if $\dot{p}$ is not a constant multiple of any existing polynomial in P. This process may be terminated either when the set P saturates (ie. taking derivatives of the elements in P adds no new polynomials) or at a convenient time. Given this finite set P of polynomials, the abstract discrete states (ψ_i) are described by conjunctions of all these polynomials evaluated over the domain $\{pos, neg, zero\}$ *i.e.* conjunctions of these polynomial inequalities. An abstract discrete state is therefore a truth-invariant region in $\Re^n$ for all polynomial inequalities from the set P. The larger the number of polynomials in P, the larger the number of discrete states, and the finer the abstraction.

It is clear from the above that if we begin with n polynomials in our original system, we could have a total of 3^n discrete modes in our abstracted system. However, many (in fact, most) of these modes are physically infeasible regions, *i.e.* there is no solution to the conjuncts of these polynomial inequalities in real space. For example, consider the set $P = \{x, x-1\}$. Although this would theoretically allow 2^3 discrete regions, regions represented by $[x < 0 \bigwedge (x-1) > 0]$ and $[x = 0 \bigwedge (x-1) > 0]$ can clearly not exist in real space. We use a decision making procedure in the theory of closed fields, QEPCAD (Hong[7]), to eliminate these modes. Since QEPCAD is restricted to polynomial inputs, we can apply this procedure only to systems where the dynamics, invariants and initial states are polynomial functions of the state variables. We are similarly constrained to verify polynomial properties of the system.

4.2 Derivation of Feasible Transitions

The transition relations (transition map) in the ADS can be obtained using the three-step process outlined below. The first has been proposed in Tiwari et al[12]; we propose the other two.

1. **Transition relations:** *We add an abstract transition $(\psi_1, \psi_2) \in \delta$ if the signs of the derivatives are consistent with the transition.*
   For example, for any $p \in P$, If $p < 0$ is a conjunct in ψ_1,
   (a) If $\dot{p} < 0$, then $p < 0$ is a conjunct in ψ_2.
   (b) If $\dot{p} = 0$, then $p < 0$ is a conjunct in ψ_2.
   (c) If $\dot{p} > 0$, then either $p < 0$ or $p = 0$ is a conjunct in ψ_2.
   (d) If the valuation of $\dot{p} < 0$ cannot be determined from ψ_1, then either $p > 0$ or $p = 0$ is a conjunct in ψ_2.
   Similar conditions exist for the cases when $p > 0$ and $p = 0$ are conjuncts in ψ_1 (Tiwari et al.[12]). The state transitions in the ADS are thus determined by the signs of the derivatives of polynomials with respect to time *i.e. flow directions.* The abstract discrete transitions obtained by the above rules can be refined by eliminating abstract states and transitions which do not satisfy invariant conditions.

2. **Flow conditions:** *We include the signs of the derivatives of **all** the polynomials in the abstraction, and thus eliminate condition (d) in the above transition relations.*
   In the abstraction procedure in Tiwari et al[12], we have no information about the flow direction of some polynomials in P unless it is saturated. As a result, we allow transitions which arise from the ambiguity in the sign of these derivatives. On the other hand, if we were to use the sign of these derivatives in the finding the abstract transitions, but not in forming the ADS, we could eliminate a large number of infeasible transitions without increasing computational time substantially. This problem was not specifically encountered in the examples used in Tiwari et al[12] because the system of polynomials saturated. For a more general system, however, this condition greatly refines the transition map.

3. **Adjacency conditions:** *We only allow abstract transitions between adjacent regions.*
   The abstraction procedure maps the boundaries of different regions in the original system onto separate abstract states. We allow an abstraction of a continuous transition from a state only if it is either transitioning to itself or to the abstraction of a boundary adjoining the original state. Even

for a one-dimensional system, physically infeasible discrete transitions (discontinuous jumps) may occur in the ADS if the adjacency condition is not enforced.

With the two additional conditions, physically infeasible transitions in the ADS are eliminated and thus the complexity of the ADS may be dramatically reduced.

The predicate abstraction procedure for continuous systems can be easily extended to hybrid systems by considering discrete states in the original hybrid system as new abstract discrete states in the ADS as well. Then, discrete states in the ADS are pairs of the original discrete states and the abstract discrete states of the original continuous states. The transition relation in the ADS can be obtained by combining the discrete transitions in the original hybrid systems and the abstract transitions for continuous states. We find that an ADS is an over-approximation of the original system, *i.e.* any transitions that are possible in the original system are definitely manifested in the ADS.

4.3 Reachability

The reachability analysis is carried out in two phases.

1. **Backward reachability in the ADS:** Given the transition map, we compute the approximate backward reachable set of each equilibrium in the discrete state space, *i.e.* the set of modes that can reach a given equilibrium mode in the ADS.

2. **Continuous state-space representation of backward reachable set:** We combine the regions corresponding to the modes of the ADS that can reach a particular equilibrium point (mode) and find a continuous state-space representation of the backward reachable set of that equilibrium point. There are several ways to do this; we use QEPCAD for this in this paper.

5 Implementation

In this section we perform reachability analysis of the Delta-Notch model using the methods outlined above. We consider a model in which the protein kinetic parameters (λ's and R's) are set to unity. We assume that the switching thresholds are $h_D = -0.5$ and $h_N = 0.2$ which were found in Ghosh et al[5] to be in the range which produces sensible biological results. We compute reachable sets using predicate abstraction with the new conditions. In this reachability analysis, self-loops in the transition map are eliminated in all modes except those which correspond to the equilibria of the original system.

5.1 Single-Cell Hybrid Automaton

The single-cell Delta-Notch lateral inhibition problem has two state variables (v_D and v_N) and 5 polynomials (the state variables, their derivatives, and a switching line). We impose the additional condition that v_D and v_N (Delta and Notch concentrations) are positive. For the case where $u_N \geq h_N$, we find that the invariant boundaries correspond to sections of the switching lines themselves, and are polynomials in our original set, and is hence discrete modes of the ADS. The backward reachable set of the equilibrium ($v_D = 0$, $v_N = 1$) is simply the positive quadrant, as expected from Ghosh et al[5].

5.2 Two-Cell Network

In the two-cell Delta-Notch lateral inhibition problem, we begin with 12 polynomials (state variables: $v_{D_1}, v_{N_1}, v_{D_2}, v_{N_2}$, their derivatives: $\dot{v}_{D_1}, \dot{v}_{N_1}, \dot{v}_{D_2}, \dot{v}_{N_2}$, and the switching conditions: $v_{N_1}+u_D$, $v_{D_1}-u_N$, $v_{N_2}+u_D$,$v_{D_2}-u_N$) in 4 dimensions. These polynomials are those that are derived from the dynamic equations and the switching conditions alone. We assume that the input threshold values satisfy the conditions proposed by Ghosh et al[5], i.e.,

$$h_D, h_N : -\frac{R_N}{\lambda_N} < h_N \leq 0 \wedge 0 < h_N \leq \frac{R_D}{\lambda_D}$$

As in Ghosh et al[5], we find that $u_D = -0.5$ and $u_N = 0.5$ satisfy these conditions. However, the abstraction in this form alone is not enough for us to find the invariant regions, since the boundaries of these regions do not correspond to any of the polynomials in the original set. From phase portrait analysis, we include projections of the invariant boundaries ($v_{D_1} - v_{D_2}$ and $v_{N_1} - v_{N_2}$) and their derivatives as predicates in our polynomial set. We now have 18 polynomials in 4 variables. The system has 2 equilibrium points, ($v_{D_1} = 1, v_{N_1} = 0, v_{D_2} = 0, v_{N_2} = 1$) and ($v_{D_1} = 0, v_{N_1} = 1, v_{D_2} = 1, v_{N_2} = 0$). Performing reachability analysis on this transition set, we find that we can divide the infinite state space into four sets: Set 1 is the backward reachable set of the first equilibrium, Set 2 is the backward reachable set of the other equilibrium, Set 3 is the invariant set and Set 4 is a region of ambiguity which is backwards reachable from either equilibrium. We also have an explicit way of mathematically describing the different sets in higher dimensions.

Set	Backward Reachable Set of	Equivalent region
1	$v_{D_1} = 1, v_{N_1} = 0, v_{D_2} = 0, v_{N_2} = 1$	$(v_{D_2} - v_{D_1} \leq 0) \wedge (v_{N_2} - v_{N_1} \geq 0) \wedge \{[(v_{D_1} \neq 1 \vee v_{N_1} > 0 \vee v_{D_2} > 0 \vee v_{N_2} - v_{N_1} \neq 1) \wedge v_{N_2} \neq v_{N_1}] \vee [v_{D_2} - v_{D_1} < 0 \wedge v_{N_2} = v_{N_1}]\}$
2	$v_{D_1} = 0, v_{N_1} = 1, v_{D_2} = 1, v_{N_2} = 0$	$(v_{D_2} - v_{D_1} \geq 0) \wedge (v_{N_2} - v_{N_1} \leq 0) \wedge \{[(v_{D_1} \neq 0 \vee v_{N_1} < 1 \vee v_{D_2} \neq 1 \vee v_{N_2} \neq 0) \wedge (\wedge v_{D_2} \neq v_{D_1})] \vee [(v_{N_1} > 1 \wedge (v_{N_1} > 0 \wedge v_{D_2} = v_{D_1})) \wedge v_{N_2} \neq v_{N_1}]\}$
3	Neither equilibrium (Invariant region)	$v_{D_2} - v_{D_1} = 0 \wedge v_{N_2} - v_{N_1} = 0$
4	Either equilibrium (Ambiguous region)	$[(v_{D_2} - v_{D_1} \geq 0) \wedge (v_{N_2} - v_{N_1} \geq 0)] \vee [(v_{D_2} - v_{D_1} \leq 0) \wedge (v_{N_2} - v_{N_1} \leq 0)]$

However, we find that when we take the projections on the $v_{N_1} = v_{N_2}$ or $v_{D_1} = v_{D_2}$ planes, this region of ambiguity (Set 4) disappears. Since it is difficult to visualize regions in 4 dimensions, we study projections along planes of interest. A comparison to the phase portraits analyzed in Ghosh et al[5] is shown in Figure 3. Since the system is deterministic, we know for certain that

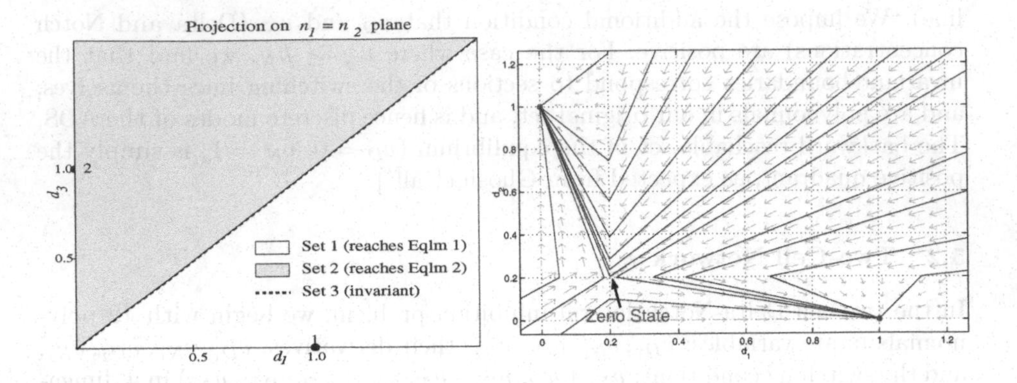

Fig. 3. Comparison of projections on the $v_{N_1} = v_{N_2}$ plane, from predicate abstraction (left) and simulations from [5] (right)

Set 4 must be an empty set. However, with the chosen predicates, we can only reduce the region of ambiguity to one-eighth of the infinite continuous state space. We believe that adding more polynomials relevant to the dynamics of the system (obtained from further analysis and simulation) will help us further prune this ambiguous region. We are currently working on addressing this issue.

We must comment, at this point, on the restrictions imposed on the complexity of the problem by the use of QEPCAD for predicate abstraction. The time-complexity of the quantifier elimination procedure employed here is doubly exponential in the number of variables, and polynomial in the number of polynomials in our set. This restricts us to problems in a small number of dimensions. Also, since we are dealing with a large number of polynomials in each step of our elimination procedure, we employ a hierarchical application of QEPCAD while eliminating physically infeasible modes.

6 Conclusions

Predicate abstraction has been proposed as a means of finding approximate backward reachable sets of the equilibria of hybrid systems in this paper. Since the Abstract Discrete System (ADS) is an over-approximation of the original system, this reachability analysis is conservative. We have proposed new conditions in checking for feasible transitions in the ADS, which greatly reduces the number of transitions in the discrete transition map. We have also implemented this as a computational technique for the reachability analysis of the Delta-Notch lateral inhibition system. Approximate backward reachable sets for this system are computed efficiently and elegantly using this technique. However, we find the fact that the accuracy of reachability analysis using predicate abstraction greatly depends on the choice of polynomials for abstraction makes it important to have information about a given system *a priori* (from analysis and simulations) to get good results in the reachability analysis.

References

[1] R. Alur, T. Dang, and F. Ivančić. Reachability analysis of hybrid sytems via predicate abstraction. In C. Tomlin and M. Greenstreet, editors, *5th International Workshop, Hybrid Systems: Computation and Control*, volume LNCS 2289, pages 35–48, Stanford, CA, USA, 2002. Springer. 715

[2] Spyros Artavanis-Tsakonas, Matthew D. Rand, and Robert J. Lake. Notch signaling: cell fate control and signal integration in development. *Science*, 284:770–776, April 1999. 716

[3] S. Das, D. Dill, and S. Park. Experience with predicate abstraction. In *Computer Aided Verification, 11th international Conference*, volume LNCS 1633, 1999. 715

[4] Manolis Fanto and Marek Mlodzik. Asymmetric notch activation specifies photoreceptors R3 and R4 and planar polarity in the *Drosophila* eye. *Nature*, 397:523–526, February 1999. 716

[5] R. Ghosh and C. Tomlin. Lateral inhibition through delta-notch signaling: A piecewise affine hybrid model. In *4th International Workshop, Hybrid Systems: Computation and Control*, volume LNCS 2034, pages 232–246, Rome, Italy, 2001. 715, 718, 721, 722, 723

[6] S. Graf and H. Saidi. Construction of abstract state graphs with PVS. In *Computer Aided Verification, 9th International Conference*, volume LNCS 1254, 1997. 715

[7] H. Hong. An improvement of the projection operator in cylindrical algebraic decomposition. In *Proceedings of ISAAC 90*, pages 261–264, 1990. 719

[8] G. Lafferriere, G.J. Pappas, and S. Yovine. Symbolic reachability computation for families of linear vector fields. *Journal of Symbolic Computation*, 32(3):231–253, September 2001. 715

[9] Julian Lewis. Notch signalling and the control of cell fate choices in vertebrates. *Seminars in Cell & Developmental Biology*, 9:583–589, 1998. 716

[10] G. Marnellos, G. A. Deblandre, E. Mjolsness, and C. Kintner. Delta-notch lateral inhibitory patterning in the emergence of ciliated cells in *Xenopus*: experimental observations and a gene network model. In *Pacific Symposium on Biocomputing*, pages 5:326–337, 2000. 716

[11] O. Sokolsky and H. Hong. Qualitative modeling of hybrid systems. In *Proceedings of the Monterey Workshop*, June 2001. 715

[12] A. Tiwari and G. Khanna. Series of abstractions for hybrid automata. In C. Tomlin and M. Greenstreet, editors, *5th International Workshop, Hybrid Systems: Computation and Control*, volume LNCS 2289, pages 465–478, Stanford, CA, USA, 2002. 715, 718, 720

[13] A. Tiwari and P. Lincoln. Automated techniques for stability analysis of delta notch lateral inhibition mechanism. 2002. 715

A Symbolic Approach to Modeling Cellular Behavior

Bhubaneswar Mishra

[1] Courant Institute, New York University
251 Mercer Street, New York, NY 10012, USA
mishra@nyu.edu
http://www.cs.nyu.edu/cs/faculty/mishra/
[2] Watson School of Biological Sciences, Cold Spring Harbor Laboratory
Demerec Building, 1 Bungtown Road, Cold Spring Harbor, NY 11724, USA

Abstract. The author examines the connection between classical differential algebra of Ritt and Kolchin and differential algebraic models of biochemical systems–in particular, the models generated by S-system of Savageau. Several open problems of both biological and mathematical significance are proposed.

1 Introduction

Unprecedented advances in genomics have made it possible for the first time for a biologist to access enormous amounts of information at the genomic level for a number of organisms, including human, mouse, arabidopsis, fruit fly, yeast and *E. coli*. These developments are at the heart of the many renewed ambitious attempts by the biologists to understand the functional roles of a group of genes using powerful computational algorithms and high-throughput microbiological protocols. The freshly emerging field of systems biology and its sister field of bioinformatics focuses on creating a finely detailed picture of biology at the cellular level by combining the *part-lists* (e.g., genes, regulatory sequences, and other objects from an annotated genome), with the observations of transcriptional states of a cell (using Microarrays) and translational states of the cell (using new proteomic tools). In the process it has become self-evident that the mathematical foundations of these systems need to be explored exhaustively and accurately. In this paper, we describe the basic structure of the underlying differential-algebraic system and the mathematical and computational problem they naturally lead to.

1.1 Outline

1. **S-systems:** Section two gives a short biological introduction and then describes Savageau-Voit approach to model bio-chemical reactions based on S-systems. (See [11, 10].)

S. Sahni et al. (Eds.) HiPC 2002, LNCS 2552, pp. 725–732, 2002.

2. **Canonical Forms:** Section three provides a canonical description of an S-system in terms of a system of differential binomial equations and set of linear equality constraints. This formulation suggests that a biological system can always be described as a differential system evolving on a linear subspace of a high-dimensional embedding space.
3. **Differential Algebra:** Section four describes the elimination theoretic approaches from Ritt-Kolchin differential algebra that can be used in this context to understand the input-output behavior of a bio-chemical system.
4. **Open Questions:** The paper concludes with a short description of the open questions.

2 S-systems

The genome of an organism is a collection of its genes, encoded by four chemical *bases* in its DNA (Deoxyribo Nucleic Acid), and forms the genetic core of a cell. The genes ultimately encode for the protein (a chain of amino acids) and in turn, the genes are regulated by transcription factors and other operons, many of which are proteins. The sequence of amino acids, specified by the DNA through transcription and translation processes, determine the three dimensional structure and biochemical properties of the proteins as well as the nature of their interactions. Furthermore, mRNA stability, protein degradation, post-translational modifications and many other bio-chemical processes tightly regulate the time-constants involved in the resulting bio-chemical machinery. Proteins also associate in complexes to form *dimers* (pair of proteins), *trimers* (triplets) and *multimers*. An *isoform* of a protein is a slightly different protein with closely related sequence, and often share similar functional properties, e.g., enzymatic reactions, but are regulated differently.

An *enzyme*, E, is a protein which can enhance the activity of a chemical reaction by attaching to a *substrate*, A, and making the formation of the *product*, P, energetically easier.

$$E + A \leftrightarrows EA \to E + P.$$

In general, equations of these kind take the form

$$A + B \rightleftharpoons^{K_+}_{K_-} C + D,$$

and the rate of change of A's concentration is given by the difference of the "synthesis rate" ($K_-[C][D]$) and the "degradation rate" ($K_+[A][B]$).

$$\frac{d[A]}{dt} = K_-[C][D] - K_+[A][B]. \tag{1}$$

Using a system of first order differential equations (in explicit form), one can construct a general model of a rather complex biochemical reaction involving many genes and proteins. One such model is Savageau-Voit S-system, whose ingredients are n dependent variables, denoted $X_1, \ldots, X_n$ and m independent variables $X_{n+1}, \ldots, X_m$ with $D_1, \ldots, D_{n+m}$ being the domains where these $n +$

m variables take value. In addition the differential equations may need to be constrained by algebraic equations corresponding to stoichiometric constraints, or conserved rates for concentrations.

The basic differential equations of the system are of the form:

$$\dot{X}_i(t) = V_i^+(X_1(t), \ldots, X_{n+m}(t)) - V_i^-(X_1(t), \ldots, X_{n+m}(t)), \tag{2}$$

for each dependent variable X_i (see [11]). The functions V^+ and V^- are arbitrary rational functions over $\mathbb{R}$. The set of algebraic constraints take the form

$$\{C_j(X_1(t), \ldots, X_{n+m}(t)) = 0\} \tag{3}$$

3 Canonical Forms

However, one can rewrite (recast) the system of equations as the one shown above in a much more simpler manner. We show that every such system admits a canonical form involving first order ordinary differential equations with binomial terms and linear constraints.

Theorem 1 *Every bio-chemical system arising from an S-system model can be expressed in a canonical form involving $r > n + m$ variables Z_1, Z_2, ..., Z_r:*

$$\begin{bmatrix} \dot{Z}_1 \\ \dot{Z}_2 \\ \vdots \\ \dot{Z}_r \end{bmatrix} = \begin{bmatrix} m_1^+(\mathbf{Z}) - m_1^-(\mathbf{Z}) \\ m_2^+(\mathbf{Z}) - m_2^-(\mathbf{Z}) \\ \vdots \\ m_r^+(\mathbf{Z}) - m_r^-(\mathbf{Z}) \end{bmatrix}, \tag{4}$$

$$\begin{bmatrix} a_{11}\, a_{12} \cdots a_{1r} \\ a_{21}\, a_{22} \cdots a_{2r} \\ \vdots \quad \vdots \quad \ddots \quad \vdots \\ a_{s1}\, a_{s2} \cdots a_{sr} \end{bmatrix} \begin{bmatrix} Z_1 \\ Z_2 \\ \vdots \\ Z_r \end{bmatrix} = \begin{bmatrix} 0 \\ 0 \\ \vdots \\ 0 \end{bmatrix}, \tag{5}$$

where m_i^+'s and m_i^-'s are ratios of monomials and a_{ij}'s are constants in $\mathbb{R}[Z_1$, ..., $Z_r]$ with positive coefficients.

PROOF:

Starting from the original description, one can derive a description in the canonical form by repeated applications of the following rules:

1. Assume that an equation is given as

$$\dot{X}(t) = \frac{p(X(t))}{q(X(t))},$$

where the right hand side of the explicit form is a rational function.

$$p = \alpha_1 m_1^+ + \cdots + \alpha_k m_k^+ - \beta_1 m_1^- + \cdots + \beta_l m_l^-$$
$$q = \alpha_1' m_1'^+ + \cdots + \alpha_k' m_{k'}'^+ - \beta_1' m_1'^- + \cdots + \beta_l' m_{l'}'^-,$$

where m^+'s and m^-'s are power-products with arbitrary powers and positive valued coefficients α's and β's.
Replace the above equation by the following system:

$$\begin{aligned} \dot{X} &= p(X(t))y(t)^{-1} \\ \dot{c_1} &= q(X(t)) - y(t)^{-1} \\ c_1 &= 0. \end{aligned}$$

2. An algebraic constraint of the form

$$r(X(t)) = \gamma_1 m_1 + \cdots + \gamma_k m_k = 0,$$

is replaced by

$$\begin{aligned} \dot{c_2} &= r(X(t)) \\ c_2 &= 0. \end{aligned}$$

3. Finally, an equation of the form

$$\begin{aligned} \dot{X}(t) &= \alpha_1 m_1^+ + \cdots + \alpha_k m_k^+ - \beta_1 m_1^- - \cdots - \beta_l m_l^+ \\ &= [\alpha_1 m_1^+ - (1/k)W(t)] + \cdots + [\alpha_k m_k^+ - (1/k)W(t)] \\ &\quad - [\beta_1 m_1^- - (1/l)W(t)] - \cdots - [\beta_l m_l^+ - (1/l)W(t)] \end{aligned}$$

is replaced by

$$\begin{aligned} \dot{\Gamma}_i(t) &= \alpha_i m_i^+ - (1/k)W(t), \qquad 1 \le i \le k, \\ \dot{\Gamma}_i(t) &= \beta_i m_i^- - (1/l)W(t), \qquad k+1 \le i \le k+l, \\ X(t) - \Gamma_1(t) - \cdots - \Gamma_k(t) &+ \Gamma_{k+1}(t) + \cdots + \Gamma_{k+l}(t) = 0. \end{aligned}$$

Repeated applications of these three rules to any S-system of equations not in the canonical form terminates after finitely many steps and results in the desired final canonical form. □

4 Differential Algebra

The semantics for a bio-chemical reaction then can be given by the evolution equations in the explicit form, or more geometrically, by the *trajectory* semantics where all possible evolution paths of the system are explicitly represented. A more compact geometric picture can be given in terms of the *distributions*, e.g., the *classical phase portraits* represented as a vector field. For instance, a simple model of a circadian clock can be represented in terms of the mRNA level of *per*, M, and corresponding protein levels of PER at various degrees of phosphorylation, P_0, P_1 and P_2, and in terms of its location inside the nucleus P_N or cytoplasm:

$$\dot{M} = v_s \frac{K_1^n}{(K_1^n + P_N^n)} - v_m \frac{M}{(K_m + M)}$$

$$\begin{aligned}
\dot{P}_0 &= k_s M - V_1 \frac{P_0}{(K_1 + P_0)} + V_2 \frac{P_1}{(K_2 + P_1)} \\
\dot{P}_1 &= V_1 \frac{P_0}{(K_1 + P_0)} - V_2 \frac{P_1}{(K_2 + P_1)} - V_3 \frac{P_1}{(K_3 + P_1)} + V_4 \frac{P_2}{(K_4 + P_2)} \\
\dot{P}_2 &= V_3 \frac{P_1}{(K_3 + P_1)} - V_4 \frac{P_2}{(K_4 + P_2)} - k_1 P_2 + k_2 P_N - v_d \frac{P_2}{(k_d + P_2)} \\
\dot{P}_N &= k_1 P_2 - k_2 P_N \\
P_t &= P_0 + P_1 + P_2 + P_N
\end{aligned}$$

Its phase portrait then can be analyzed to determine if the system has a stable and robust limit-cycle (e.g., by applying "Bendixon criteria," etc.).

Another approach is to describe the system in terms of an automaton, whose state can be represented as a finite-dimensional vector $S(t)$ and its transition from $S(t)$ and $S(t + \Delta t)$ can be determined by following the trajectory starting at state $S(t)$:

$$\int_t^{t+\Delta t} F(S(\tau))\, d\tau,$$

subject to the constraints on the system. Wherever an appropriate numerical integrator is available, such an automata can be numerically described by the "traces" of the numerical integrator. In order to keep the complexity of such an automaton simple one can obtain "approximate versions" of the automaton by discretization and collapse operations that hide all or some of the "internal states."

An ultimate example of collapsing involves hiding all the internal state variables and just describing the evolution of outputs in terms of its input. Here, one describes the system in terms of its *input-output relation* that describes only the relation between the control inputs and the output variables starting from a redundant state-space description. From an algebraic point of view, this is exactly the problem of *variable elimination* and comes under the subject of *elimination theory*. Thus all the theories related to *standard bases*, *characteristic sets* and *differential-algebraic resultants* play important roles.

Assume that the system (SISO) is described as shown below:

$$\begin{aligned}
\dot{x}_1 &= p_1(X, u, \dot{u}, \ldots, u^{(k)}) \\
&\vdots \\
\dot{x}_r &= p_r(X, u, \dot{u}, \ldots, u^{(k)}) \\
0 &= q_1(X, u) \\
&\vdots \\
0 &= q_s(X, u) \\
y &= h(X, u)
\end{aligned}$$

Consider the following differential ideal I in the differential ring $\mathbb{R}\{X, u, y\}$:

$$I = [\dot{x}_1 - p_1, \ldots, \dot{x}_r - p_r, q_1, \ldots, q_s, y - h].$$

The input-output relation is then obtained by finding the contraction I^c of the ideal I to the ring $\mathbb{R}\{u, y\}$. The generators of $I^c = I \cap \mathbb{R}\{u, y\}$ give the differential polynomials involving u and y. However, the underlying algorithmic questions for differential algebraic elimination remain largely unsolved.

Example Consider the following system (adapted from Forsman [4]):

$$A \to B,$$

with the following kinetic equations:

$$[\dot{B}] = [A]^{0.5} - [B]^{0.5}.$$

The input u controls the concentration $[A]$ as follows:

$$[\dot{A}] = u[A]^{-2} - [A]^{-1.5},$$

and the output y is simply $[B]$:

$$y = [B].$$

We can simplify the above system to a polynomial system by following transformations:

$$x_1^2 = [A] \quad \text{and} \quad x_2^2 = [B].$$

Thus,

$$I = [2x_1^5\dot{x}_1 + x_1 - u, 2x_2\dot{x}_2 + x_2 - x_1, x_2^2 - y].$$

After eliminating x_1 and x_2, we obtain the following input-output relation:

$$\begin{aligned}
&(20\dot{y}^8y^2 - 4\dot{y}^{10}y - 40\dot{y}^6y^3 + 40\dot{y}^4y^4 - 20\dot{y}^2y^5 + 4y^6)\ddot{y}^2 \\
&+ (4u\dot{y}^5y - 4\dot{y}^6y - 20\dot{y}^4y^2 + 40u\dot{y}^3y^2 + 20\dot{y}^2y^3 + 20u\dot{y}y^3 + 4y^4)\ddot{y} \\
&- \dot{y}^2y^5 + 5\dot{y}^4y^4 - 10\dot{y}^6y^3 + 20u\dot{y}^3y^2 + 10\dot{y}^8y^2 + y^2 - 8\dot{y}^6y + 10u\dot{y}^5y \\
&- u^2y + 2u\dot{y}y - \dot{y}^2y - 5\dot{y}^{10}y + \dot{y}^{12} + 8\dot{y}^2y^3 + 2u\dot{y}y^3 = 0.
\end{aligned}$$

□

5 Open Questions

Several interesting questions remain to be further explored.

1. **Reactions Models:** We have primarily focused on a simple ODE model (Differential Algebraic Equations, DAE) and narrowed this even further to a model based on S-systems. Does this imply that there is a significant deviation from reality? How can a stochastic model representing small number of molecules interacting pair-wise and randomly be incorporated?
2. **Hybrid Systems:** Certain interactions are purely discrete and after each such interaction, the system dynamics may change. Such a hybrid model implies that the underlying automaton must be modified for each such mode. How do these enhancements modify the basic symbolic model?

3. **Spatial Models:** The cellular interactions are highly specific to their spatial locations within the cell. How can these be modeled with symbolic cellular-automata? How can we account for dynamics due to changes to the cell volume? The time constants associated with the diffusion may vary from location to location; how can that be modeled?
4. **State Space (Product Space):** A number of interacting cells can be modeled by product automata. In addition to the classical "state-explosion problem" we also need to pay attention to the variable structure due to i) Cell division, ii) Apoptosis and iii) Differentiation.
5. **Communication:** How do we model the communication among the cells mediated by the interactions among the extra-cellular factor and external receptor pairs?
6. **Hierarchical Models:** Finally, as we go to more and more complex cellular processes, a clear understanding can only be obtained through modularized hierarchical models. What are the ideal hierarchical models? How do we model a population of cells with related statistics?
7. **Simulation:** If a biologist wishes to obtain a visualization based on numerical simulation, how can we take advantage of the underlying symbolic description?
8. **Symbolic Verification:** If a biologist wishes to reason about the system with logical queries in an appropriate query language (e.g., temporal logic), what are the best query languages? What are the best algorithms that take advantage of the symbolic structures? What are the correct way to solve problems associated with i) Model Equivalence, ii) Experimental Analysis, and iii) Reachability Analysis?

References

[1] Brockett, R.: Nonlinear Systems and Differential Geometry. Proceedings of the IEEE, **64** (1976): 61–72.

[2] Carrá Ferro, G.: Gröbner Bases and Differential Ideals. Proceedings of AAECC-5, Lecture Notes in Computer Science, Springer-Verlag, (1987): 129–140.

[3] Diop, S.: Elimination in Control Theory. Math. Control Signals Systems, **4** (1991): 17–32.

[4] Forsman, K.: *Constructive Commutative Algebra in Nonlinear Control Theory*, Linköping Studies in Science and Technology, Dissertation, No. 261, Department of Electrical Engineering, Linköping University, Linköping, Sweden, 1992. 730

[5] Gallo, G., Mishra, B., and Ollivier, F.: Some Constructions in Rings of Differential Polynomials. Proceedings of AAECC-9, Lecture Notes in Computer Science, Springer-Verlag, **539** (1991): 171–182.

[6] Kolchin, E. R.: On the Basis Theorem for Differential Systems. Transactions of the AMS, **52** (1942): 115–127.

[7] Mishra, B.: Computational Differential Algebra. Geometrical Foundations of Robotics, (Ed. Jon Selig), World-Scientific, Singapore, **Lecture 8** (2000): 111-145.

[8] Ritt, J.F.: *Differential Equations from the Algebraic Standpoint*, AMS Coloq. Publ. 14, New York 1932.

[9] Seidenberg, A.: An Elimination Theory for Differential Algebra. University of California, Berkeley, Publications in Mathematics, **3** (1956): 31–65.
[10] Savageau, M. A.: *Biochemical System Analysis: A Study of Function and Design in Molecular Biology* . Addison-Wesley, 1976. 725
[11] Voit, E. O.: *Computational Analysis of Biochemical Systems*. Cambridge, 2000. 725, 727

Author Index